$P(A|B)$ — Probability that event A will occur given event B has occurred

q — $1 - p$, probability of failure for the binomial experiments

$\hat{q}$ — $1 - \hat{p}$ where $\hat{p}$ is the sample proportion.

$\overline{q}$ — $1 - \overline{p}$ where $\overline{p}$ is the pooled sample proportion for two samples

Q_1, Q_2, Q_3 — First, second, and third quartiles, respectively

R — Number of rows in a contingency table

r — Linear correlation coefficient for sample data

r^2 — Coefficient of determination

r_{12} — Simple linear correlation between x_1 and x_2

r_{y1} — Simple linear correlation between y and x_1

r_{y2} — Simple linear correlation between y and x_2

R^2 — Coefficient of multiple determination

$\overline{R}^2$ — Adjusted coefficient of multiple determination

ρ — (Greek letter *rho*) Linear correlation coefficient for population data

S — Sample space

s — Sample standard deviation

s^2 — Sample variance

s_b — Estimator of σ_b

s_d — Standard deviation of the paired differences for a sample

$s_{\overline{d}}$ — Estimator of $\sigma_{\overline{d}}$

s_e — Standard deviation of errors for the sample regression model

s_p — Pooled standard deviation

$s_{\hat{p}}$ — Estimator of $\sigma_{\hat{p}}$

$s_{\hat{p}_1 - \hat{p}_2}$ — Estimator of $\sigma_{\hat{p}_1 - \hat{p}_2}$

$s_{\overline{x}}$ — Estimator of $\sigma_{\overline{x}}$

$s_{\overline{x}_1 - \overline{x}_2}$ — Estimator of $\sigma_{\overline{x}_1 - \overline{x}_2}$

$s_{\hat{y}_m}$ — Estimator of $\sigma_{\hat{y}_m}$

$s_{\hat{y}_p}$ — Estimator of $\sigma_{\hat{y}_p}$

Σ — (Greek letter capital *sigma*) summation notation

σ — (Greek letter lower-case *sigma*) Population standard deviation

σ^2 — Population variance

σ_b — Standard deviation of the sampling distribution of b

σ_d — Standard deviation of the paired differences for the population

$\sigma_{\overline{d}}$ — Standard deviation of the sampling distribution of $\overline{d}$

σ_ϵ — Standard deviation of errors for the population regression model

$\sigma_{\hat{p}}$ — Standard deviation of the sampling distribution of $\hat{p}$

$\sigma_{\hat{p}_1 - \hat{p}_2}$ — Standard deviation of the sampling distribution of $\hat{p}_1 - \hat{p}_2$

$\sigma_{\overline{x}}$ — Standard deviation of the sampling distribution of $\overline{x}$

$\sigma_{\overline{x}_1 - \overline{x}_2}$ — Standard deviation of the sampling distribution of $\overline{x}_1 - \overline{x}_2$

$\sigma_{\hat{y}_m}$ — Standard deviation of $\hat{y}$ when estimating $\mu_{y|x}$

$\sigma_{\hat{y}_p}$ — Standard deviation of $\hat{y}$ when predicting y_p

t — The t distribution

T — Trend component of a time series

T_i — Sum of the values included in sample i in one-way ANOVA

x — (1) Variable; (2) random variable; (3) independent variable in a regression model

$\overline{x}$ — Sample mean

y — (1) Variable; (2) dependent variable in a regression model

Y — Value of the time series at a given time

$\hat{y}$ — Estimated or predicted value of y using a regression model

$\hat{Y}$ — Estimated trend value of a time series

z — Units of the standard normal distribution

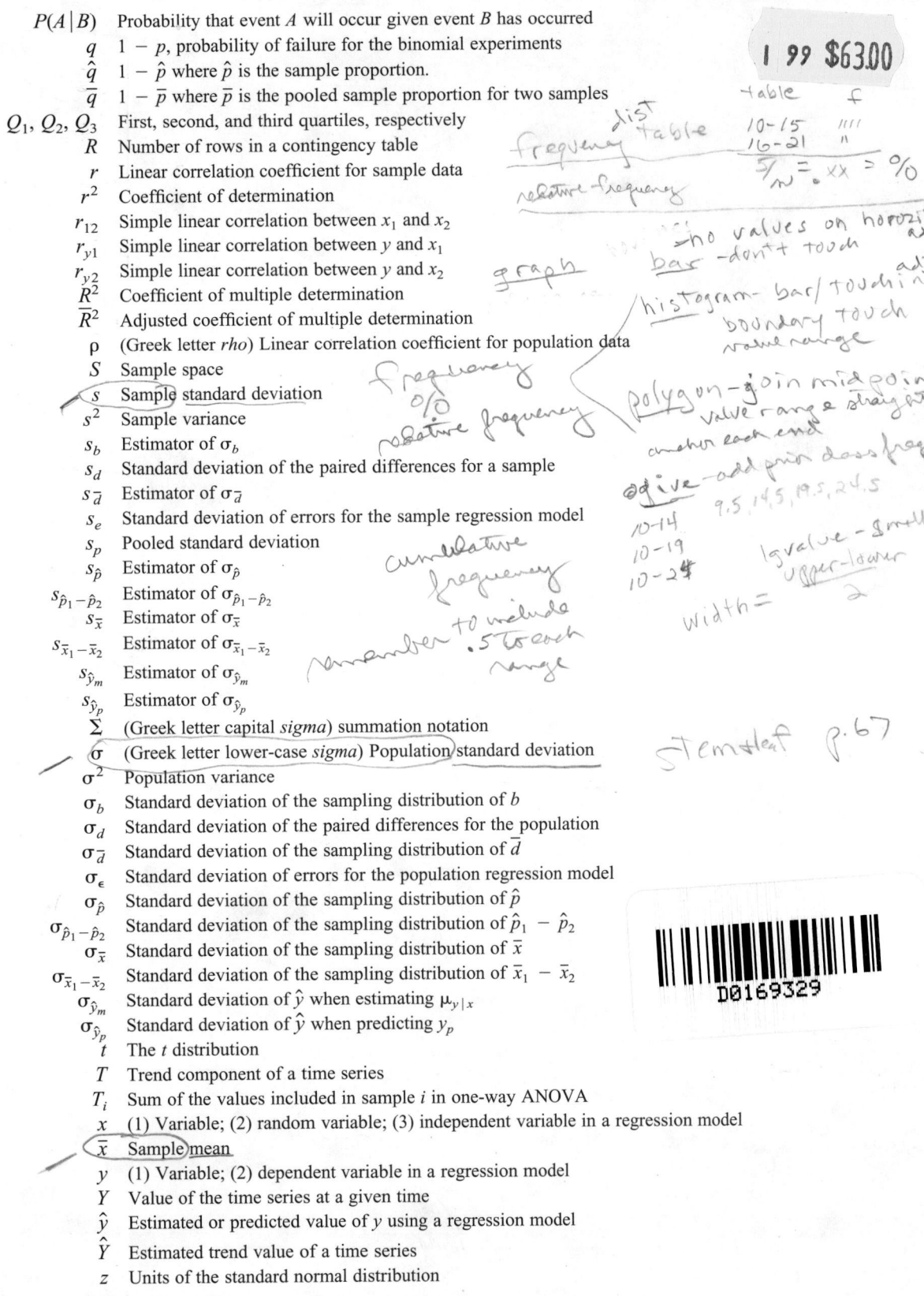

(Handwritten margin notes:) frequency dist table 10-15 //// , 16-21 // , f/n = .xx = % ; relative frequency ; graph ; bar — no values on horizontal axis, don't touch ; histogram — bar/touching, boundary touch, value range, adjacent ; polygon — join midpoints on value range straight line, anchor each end ; ogive — add prior class frequency, 9.5, 14.5, 19.5, 24.5, 10-14, 10-19, 10-24 ; lg value — small value, upper-lower, width = /2 ; frequency %, relative frequency, cumulative frequency, remember to include .5 to each range ; stem+leaf p.67

M = midpoint

ungrouped X, $a-b$

grouped $\dfrac{M}{\dfrac{a-b}{2}}$

2.09 Range = g value − small value

ungrouped mean − p93

$\dfrac{\sum x}{n} = \dfrac{\text{sum of values}}{\text{\# of values}}$

median 6.98 − rank all values ascending order in data set
middle value (turn) in position

mean $\bar{x} = \dfrac{\sum x}{n}$

ungrouped median $= \dfrac{n+1}{2}$

p119 grouped mean $\mu = \dfrac{\sum mf}{M}$

$0 =$

p125

$\Delta \dfrac{x}{2}$ midpoint

class X group	f	m	mf	m²f
24−39	III	31.5		
40−55	IIII	47.5		
	+ ___ M=		+ ___ Σ mf	+ ___ Σ m²f

CHAPTER 5 · DISCRETE RANDOM VARIABLES AND THEIR PROBABILITY DISTRIBUTIONS

- Mean of a discrete random variable x: $\mu = \Sigma x P(x)$
- Standard deviation of a discrete random variable x:

$$\sigma = \sqrt{\Sigma x^2 P(x) - \mu^2}$$

- n factorial: $n! = n(n-1)(n-2)\dots3\cdot2\cdot1$
- Number of combinations of n items selected x at a time:

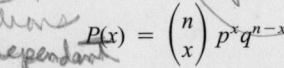

$$\binom{n}{x} = \frac{n!}{x!\,(n-x)!}$$

- Binomial probability formula:

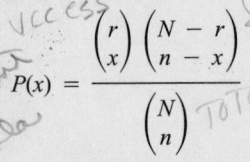

$$P(x) = \binom{n}{x} p^x q^{n-x}$$

- Mean and standard deviation of the binomial distribution:

$$\mu = np \quad \text{and} \quad \sigma = \sqrt{npq}$$

- Hypergeometric probability formula:

$$P(x) = \frac{\binom{r}{x}\binom{N-r}{n-x}}{\binom{N}{n}}$$

where
- N = total number of elements in the population
- r = number of successes in the population
- $N - r$ = number of failures in the population
- n = number of trials (sample size)
- x = number of successes in n trials
- $n - x$ = number of failures in n trials

- Poisson probability formula:

$$P(x) = \frac{\lambda^x e^{-\lambda}}{x!}$$

- Mean, variance, and standard deviation of the Poisson probability distribution:

$$\mu = \lambda, \quad \sigma^2 = \lambda, \quad \text{and} \quad \sigma = \sqrt{\lambda}$$

CHAPTER 6 · CONTINUOUS RANDOM VARIABLES AND THEIR PROBABILITY DISTRIBUTIONS

- z value for an x value:

$$z = \frac{x - \mu}{\sigma}$$

- Value of x when μ, σ, and z are known: $x = \mu \pm z\sigma$
- Uniform probability distribution:
Let x be a continuous random variable that is uniformly distributed in an interval a to b. The probability that x assumes a value in the interval c to d is

$$P(c \le x \le d) = \frac{d - c}{b - a}$$

- Mean and standard deviation of the uniform probability distribution:

$$\mu = \frac{a+b}{2} \quad \text{and} \quad \sigma = \sqrt{\frac{(b-a)^2}{12}}$$

- Exponential probability distribution:
If λ is the mean number of occurrences per unit of time and x is the lapse time between two successive occurrences, then,

$$P(x \ge a) = e^{-\lambda a}, \quad P(x \le a) = 1 - e^{-\lambda a},$$
$$\text{and} \quad P(a \le x \le b) = e^{-\lambda a} - e^{-\lambda b}$$

where $\lambda > 0$, $a > 0$, and $b > a$.

CHAPTER 7 · SAMPLING DISTRIBUTIONS

- Mean of $\bar{x}$: $\mu_{\bar{x}} = \mu$
- Standard deviation of $\bar{x}$ when $n/N \le .05$: $\sigma_{\bar{x}} = \sigma/\sqrt{n}$
- z value for $\bar{x}$:

$$z = \frac{\bar{x} - \mu}{\sigma_{\bar{x}}}$$

- Population proportion: $p = x/N$
- Sample proportion: $\hat{p} = x/n$
- Mean of $\hat{p}$: $\mu_{\hat{p}} = p$
- Standard deviation of $\hat{p}$ when $n/N \le .05$: $\sigma_{\hat{p}} = \sqrt{pq/n}$
- z value for $\hat{p}$:

$$z = \frac{\hat{p} - p}{\sigma_{\hat{p}}}$$

CHAPTER 8 · ESTIMATION OF THE MEAN AND PROPORTION

- Margin of error for the point estimation of μ:

$$\pm 1.96\,\sigma_{\bar{x}} \quad \text{or} \quad \pm 1.96\,s_{\bar{x}}$$

where $\sigma_{\bar{x}} = \sigma/\sqrt{n}$ and $s_{\bar{x}} = s/\sqrt{n}$

- Confidence interval for μ for a large sample:

$$\bar{x} \pm z\,\sigma_{\bar{x}} \quad \text{if } \sigma \text{ is known}$$
$$\bar{x} \pm z\,s_{\bar{x}} \quad \text{if } \sigma \text{ is not known}$$

where $\sigma_{\bar{x}} = \sigma/\sqrt{n}$ and $s_{\bar{x}} = s/\sqrt{n}$

- Confidence interval for μ for a small sample:

$$\bar{x} \pm t\,s_{\bar{x}} \quad \text{where} \quad s_{\bar{x}} = s/\sqrt{n}$$

- Margin of error for the point estimation of p:

$$\pm 1.96\,s_{\hat{p}} \quad \text{where} \quad s_{\hat{p}} = \sqrt{\hat{p}\hat{q}/n}$$

- Confidence interval for p for a large sample:

$$\hat{p} \pm z\,s_{\hat{p}} \quad \text{where} \quad s_{\hat{p}} = \sqrt{\hat{p}\hat{q}/n}$$

- Maximum error of the estimate for μ:

$$E = z\,\sigma_{\bar{x}} \quad \text{or} \quad z\,s_{\bar{x}}$$

- Determining sample size for estimating μ: $n = z^2\sigma^2/E^2$
- Maximum error of the estimate for p:

$$E = z\,s_{\hat{p}} \quad \text{where} \quad s_{\hat{p}} = \sqrt{\hat{p}\,\hat{q}/n}$$

- Determining sample size for estimating p: $n = z^2 p\,q/E^2$

CHAPTER 9 · HYPOTHESIS TESTS ABOUT THE MEAN AND PROPORTION

- Test statistic z for a test of hypothesis about μ for a large sample:

$$z = \frac{\bar{x} - \mu}{\sigma_{\bar{x}}} \quad \text{if } \sigma \text{ is known, where } \sigma_{\bar{x}} = \frac{\sigma}{\sqrt{n}}$$

or $\quad z = \dfrac{\bar{x} - \mu}{s_{\bar{x}}} \quad$ if σ is not known, where $s_{\bar{x}} = \dfrac{s}{\sqrt{n}}$

- Test statistic for a test of hypothesis about μ for a small sample:

$$t = \frac{\bar{x} - \mu}{s_{\bar{x}}} \quad \text{where} \quad s_{\bar{x}} = \frac{s}{\sqrt{n}}$$

- Test statistic for a test of hypothesis about p for a large sample:

$$z = \frac{\hat{p} - p}{\sigma_{\hat{p}}} \quad \text{where } \sigma_{\hat{p}} = \sqrt{\frac{p\,q}{n}}$$

CHAPTER 10 · ESTIMATION AND HYPOTHESIS TESTING: TWO POPULATIONS

- Mean of the sampling distribution of $\bar{x}_1 - \bar{x}_2$:

$$\mu_{\bar{x}_1 - \bar{x}_2} = \mu_1 - \mu_2$$

- Confidence interval for $\mu_1 - \mu_2$ for two large and independent samples:

$(\bar{x}_1 - \bar{x}_2) \pm z\, \sigma_{\bar{x}_1 - \bar{x}_2} \quad$ if σ_1 and σ_2 are known

or $\quad (\bar{x}_1 - \bar{x}_2) \pm z\, s_{\bar{x}_1 - \bar{x}_2} \quad$ if σ_1 and σ_2 are not known

where $\quad \sigma_{\bar{x}_1 - \bar{x}_2} = \sqrt{\dfrac{\sigma_1^2}{n_1} + \dfrac{\sigma_2^2}{n_2}} \quad$ and $\quad s_{\bar{x}_1 - \bar{x}_2} = \sqrt{\dfrac{s_1^2}{n_1} + \dfrac{s_2^2}{n_2}}$

- Test statistic for a test of hypothesis about $\mu_1 - \mu_2$ for two large and independent samples:

$$z = \frac{(\bar{x}_1 - \bar{x}_2) - (\mu_1 - \mu_2)}{\sigma_{\bar{x}_1 - \bar{x}_2}}$$

If σ_1 and σ_2 are not known, then replace $\sigma_{\bar{x}_1 - \bar{x}_2}$ by its point estimator $s_{\bar{x}_1 - \bar{x}_2}$.

- For two small and independent samples taken from two populations with equal standard deviations:

Pooled standard deviation:

$$s_p = \sqrt{\frac{(n_1 - 1)\,s_1^2 + (n_2 - 1)\,s_2^2}{n_1 + n_2 - 2}}$$

Estimate of the standard deviation of $\bar{x}_1 - \bar{x}_2$:

$$s_{\bar{x}_1 - \bar{x}_2} = s_p\,\sqrt{\frac{1}{n_1} + \frac{1}{n_2}}$$

Confidence interval for $\mu_1 - \mu_2$:

$$(\bar{x}_1 - \bar{x}_2) \pm t\, s_{\bar{x}_1 - \bar{x}_2}$$

Test statistic: $\quad t = \dfrac{(\bar{x}_1 - \bar{x}_2) - (\mu_1 - \mu_2)}{s_{\bar{x}_1 - \bar{x}_2}}$

- For two paired or matched samples:

Sample mean for paired differences: $\quad \bar{d} = \dfrac{\Sigma d}{n}$

Sample standard deviation for paired differences:

$$s_d = \sqrt{\frac{\Sigma d^2 - \dfrac{(\Sigma d)^2}{n}}{n - 1}}$$

Mean and standard deviation of the sampling distribution of $\bar{d}$:

$$\mu_{\bar{d}} = \mu_d \quad \text{and} \quad s_{\bar{d}} = \frac{s_{\bar{d}}}{\sqrt{n}}$$

Confidence interval for μ_d:

$$\bar{d} \pm t\, s_{\bar{d}} \quad \text{where} \quad s_{\bar{d}} = \frac{s_d}{\sqrt{n}}$$

Test statistic for a test of hypothesis about μ_d:

$$t = \frac{\bar{d} - \mu_d}{s_{\bar{d}}}$$

- For two large and independent samples, confidence interval for $p_1 - p_2$:

$$(\hat{p}_1 - \hat{p}_2) \pm z\, s_{\hat{p}_1 - \hat{p}_2}$$

where $\quad s_{\hat{p}_1 - \hat{p}_2} = \sqrt{\dfrac{\hat{p}_1\,\hat{q}_1}{n_1} + \dfrac{\hat{p}_2\,\hat{q}_2}{n_2}}$

- For two large and independent samples, for a test of hypothesis about $p_1 - p_2$ with $H_0: p_1 - p_2 = 0$:

Pooled sample proportion:

$$\bar{p} = \frac{x_1 + x_2}{n_1 + n_2} \quad \text{or} \quad \frac{n_1\,\hat{p}_1 + n_2\,\hat{p}_2}{n_1 + n_2}$$

Estimate of the standard deviation of $\hat{p}_1 - \hat{p}_2$:

$$s_{\hat{p}_1 - \hat{p}_2} = \sqrt{\bar{p}\,\bar{q}\left(\frac{1}{n_1} + \frac{1}{n_2}\right)}$$

Test statistic: $\quad z = \dfrac{(\hat{p}_1 - \hat{p}_2) - (p_1 - p_2)}{s_{\hat{p}_1 - \hat{p}_2}}$

CHAPTER 11 · CHI-SQUARE TESTS

- Expected frequency for a category for a goodness-of-fit test:
$$E = np$$

- Degrees of freedom for a goodness-of-fit test:

$df = k - 1 \quad$ where k is the number of categories

- Expected frequency for a cell for an independence or homogeneity test:

$$E = \frac{(\text{Row total})\,(\text{Column total})}{n}$$

- Degrees of freedom for a test of independence or homogeneity:

$$df = (R - 1)\,(C - 1)$$

where R and C are the total number of rows and columns, respectively, in the contingency table

- Test statistic for a goodness-of-fit test and a test of independence or homogeneity:

$$\chi^2 = \Sigma\,\frac{(O - E)^2}{E}$$

- Confidence interval for the population variance σ^2:

$$\frac{(n - 1)\,s^2}{\chi_{\alpha/2}^2} \quad \text{to} \quad \frac{(n - 1)\,s^2}{\chi_{1 - \alpha/2}^2}$$

- Test statistic for a test of hypothesis about σ^2:

$$\chi^2 = \frac{(n - 1)\,s^2}{\sigma^2}$$

CHAPTER 12 · ANALYSIS OF VARIANCE

Let:

k = the number of different samples (or treatments)
n_i = the size of sample i
T_i = the sum of the values in sample i
n = the number of values in all samples
$\quad = n_1 + n_2 + n_3 + \cdots$
Σx = the sum of the values in all samples
$\quad = T_1 + T_2 + T_3 + \cdots$
Σx^2 = the sum of the squares of values in all samples

- For the F distribution:

Degrees of freedom for the numerator = $k - 1$
Degrees of freedom for the denominator = $n - k$

- Between-samples sum of squares:

$$SSB = \left(\frac{T_1^2}{n_1} + \frac{T_2^2}{n_2} + \frac{T_3^2}{n_3} + \cdots\right) - \frac{(\Sigma x)^2}{n}$$

- Within-samples sum of squares:

$$SSW = \Sigma x^2 - \left(\frac{T_1^2}{n_1} + \frac{T_2^2}{n_2} + \frac{T_3^2}{n_3} + \cdots\right)$$

- Total sum of squares:
SST = SSB + SSW =

$$\Sigma x^2 - \frac{(\Sigma x)^2}{n}$$

- Variance between samples: $\quad MSB = \dfrac{SSB}{k - 1}$

- Variance within samples: $\quad MSW = \dfrac{SSW}{n - k}$

- Test statistic for a one-way ANOVA test: $\quad F = \dfrac{MSB}{MSW}$

CHAPTER 13 · SIMPLE LINEAR REGRESSION

- Simple linear regression model:
$$y = A + Bx + \epsilon$$

- Estimated simple linear regression model:
$$\hat{y} = a + bx$$

- Sum of squares of xy, xx, and yy:

$$SS_{xy} = \Sigma xy - \frac{(\Sigma x)\,(\Sigma y)}{n},$$

$$SS_{xx} = \Sigma x^2 - \frac{(\Sigma x)^2}{n}, \text{ and } \quad SS_{yy} = \Sigma y^2 - \frac{(\Sigma y)^2}{n}$$

- Least squares estimates of A and B:

$$b = \frac{SS_{xy}}{SS_{xx}} \text{ and } \quad a = \bar{y} - b\bar{x}$$

- Standard deviation of the sample errors:

$$s_e = \sqrt{\frac{SS_{yy} - b\,SS_{xy}}{n - 2}}$$

- Error sum of squares: $\quad SSE = \Sigma e^2 = \Sigma(y - \hat{y})^2$

- Total sum of squares: $\quad SST = \Sigma y^2 - \dfrac{(\Sigma y)^2}{n}$

- Regression sum of squares: $\quad SSR = SST - SSE$

- Coefficient of determination: $\quad r^2 = b\,\dfrac{SS_{xy}}{SS_{yy}}$

- Confidence interval for B:

$$b \pm t\,s_b \quad \text{where} \quad s_b = \frac{s_e}{\sqrt{SS_{xx}}}$$

- Test statistic for a test of hypothesis about B:

$$t = \frac{b - B}{s_b}$$

- Linear correlation coefficient:

$$r = \frac{SS_{xy}}{\sqrt{SS_{xx}\,SS_{yy}}}$$

- Confidence interval for $\mu_{y|x}$:

$$\hat{y} \pm t\,s_{\hat{y}_m} \quad \text{where} \quad s_{\hat{y}_m} = s_e\sqrt{\frac{1}{n} + \frac{(x_0 - \bar{x})^2}{SS_{xx}}}$$

- Prediction interval for y_p:

$$\hat{y} \pm t\,s_{\hat{y}_p} \quad \text{where} \quad s_{\hat{y}_p} = s_e\sqrt{1 + \frac{1}{n} + \frac{(x_0 - \bar{x})^2}{SS_{xx}}}$$

CHAPTER 14 · MULTIPLE REGRESSION

- Multiple (linear) regression model:
$$y = A + B_1x_1 + B_2x_2 + B_3x_3 + \cdots + B_kx_k + \epsilon$$

- Estimated multiple (linear) regression model:
$$\hat{y} = a + b_1x_1 + b_2x_2 + b_3x_3 + \cdots + b_kx_k$$

- The $(1 - \alpha)100\%$ confidence interval for B_i:
$$b_i \pm t\,s_{b_i}$$

- Test statistic t for a test of hypothesis about a single B_i:

$$t = \frac{b_i - B_i}{s_{b_i}}$$

- Test statistic F for a test of overall significance of the multiple regression model:

$$F = \frac{SSR/k}{SSE/(n - k - 1)} \quad \text{or} \quad \frac{MSR}{MSE}$$

CHAPTER 15 · TIME SERIES ANALYSIS

- Additive time series model:
$$Y = T + C + S + I$$

where T, C, S, and I are the trend, cyclical, seasonal, and irregular components, respectively, and Y is the value of the time series at a given time.

- Multiplicative time series model:
$$Y = T \cdot C \cdot S \cdot I$$

- Regression model for linear trend:
$$Y = A + Bt + \epsilon$$

- Estimated linear trend line:
$$\hat{Y} = a + bt$$

- Multiplicative time seris model for annual data (no seasonal variations):
$$Y = T \cdot C \cdot I$$

- Seasonal index for a period:

$$\frac{\text{Mean for that period}}{\text{Sum of means for all periods}} \times \left(\begin{array}{c}\text{Number} \\ \text{of periods}\end{array}\right)$$

- Deseasonalized value for a period:

$$\frac{\text{Actual value for that period}}{\text{Seasonal Index for that period}}$$

[handwritten top margin: mean = (upper boundary − lower boundary) / 2]

[handwritten top margin: with %, find K to find interval]

[handwritten top margin: with % Know K find lower + upper boundary interval]

KEY FORMULAS
Prem S. Mann · Statistics for Business and Economics

CHAPTER 2 · ORGANIZING DATA

- Relative frequency of a class = $f/\Sigma f$
- Percentage of a class = (Relative frequency) $\times$ 100
- Class midpoint or mark = (Upper limit + Lower limit)/2
- Class width = Upper boundary − Lower boundary
- Cumulative relative frequency

$$= \frac{\text{Cumulative frequency}}{\text{Total observations in the data set}}$$

- Cumulative percentage

$$= \text{(Cumulative relative frequency)} \times 100$$

CHAPTER 3 · NUMERICAL DESCRIPTIVE MEASURES

[handwritten: N = population] *[handwritten: n = sample or size]*

- Mean for ungrouped data: $\mu = \Sigma x/N$ and $\bar{x} = \Sigma x/n$
- Mean for grouped data: $\mu = \Sigma mf/N$ and $\bar{x} = \Sigma mf/n$
  where m is the midpoint and f is the frequency of a class
- Median for ungrouped data

$$= \text{Value of the } \left(\frac{n+1}{2}\right) \text{th term in a ranked data set}$$

- Range = Largest value − Smallest value
- Variance for ungrouped data:

$$\sigma^2 = \frac{\Sigma x^2 - \frac{(\Sigma x)^2}{N}}{N} \quad \text{and} \quad s^2 = \frac{\Sigma x^2 - \frac{(\Sigma x)^2}{n}}{n-1}$$

where σ^2 is the population variance and s^2 is the sample variance

- Standard deviation for ungrouped data: *[handwritten: population / sample]*

$$\sigma = \sqrt{\frac{\Sigma x^2 - \frac{(\Sigma x)^2}{N}}{N}} \quad \text{and} \quad s = \sqrt{\frac{\Sigma x^2 - \frac{(\Sigma x)^2}{n}}{n-1}} \quad \text{df}$$

where σ and s are the population and sample standard deviations, respectively

- Variance for grouped data:

$$\sigma^2 = \frac{\Sigma m^2 f - \frac{(\Sigma mf)^2}{N}}{N} \quad \text{and} \quad s^2 = \frac{\Sigma m^2 f - \frac{(\Sigma mf)^2}{n}}{n-1}$$

- Standard deviation for grouped data: *[handwritten: population / sample]*

$$\sigma = \sqrt{\frac{\Sigma m^2 f - \frac{(\Sigma mf)^2}{N}}{N}} \quad \text{and} \quad s = \sqrt{\frac{\Sigma m^2 f - \frac{(\Sigma mf)^2}{n}}{n-1}}$$

- Coefficient of variation:

$$\text{CV} = \frac{\sigma}{\mu} \times 100\% \quad \text{or} \quad \frac{s}{\bar{x}} \times 100\%$$

- Chebyshev's theorem:
  For any number k greater than 1, at least $(1 - 1/k^2)$ of the values for any distribution lie within k standard deviations of the mean. *[handwritten: = %]*

[handwritten: P.127]

[handwritten: find % of Total observation which lies within given interval about the mean %= 1 − 1/(K)²]

[handwritten right column top: with %, find K to find interval]

- Empirical rule:
  For a specific bell-shaped distribution, about 68% of the observations fall in the interval $(\mu - \sigma)$ to $(\mu + \sigma)$, about 95% fall in the interval $(\mu - 2\sigma)$ to $(\mu + 2\sigma)$, and about 99.7% fall in the interval $(\mu - 3\sigma)$ to $(\mu + 3\sigma)$. *[handwritten: (2)]*

- Interquartile range: $\text{IQR} = Q_3 - Q_1$ *[handwritten: $\mu \pm K\sigma$]*
  where Q_3 is the third quartile and Q_1 is the first quartile

- The kth percentile:

$$P_k = \text{Value of the } \left(\frac{kn}{100}\right) \text{th term in a ranked data set}$$

- Percentile rank of x_i

$$= \frac{\text{Number of values less than } x_i}{\text{Total number of values in the data set}} \times 100$$

[handwritten: $K = \dfrac{\text{upper boundary} - \mu \,(\text{mean})}{\sigma \,(\text{standard deviation})}$]

CHAPTER 4 · PROBABILITY *[handwritten: P.157]*

- Classical probability rule for a simple event:

$$P(E_i) = \frac{1}{\text{Total number of outcomes}}$$

- Classical probability rule for a compound event:

$$P(A) = \frac{\text{Number of outcomes in } A}{\text{Total number of outcomes}}$$

- Relative frequency as an approximation of probability:

$$P(A) = \frac{f}{n}$$

- Conditional probability of an event:

$$P(A|B) = \frac{P(A \text{ and } B)}{P(B)} \quad \text{and} \quad P(B|A) = \frac{P(A \text{ and } B)}{P(A)}$$

- Condition for independence of events:

$$P(A) = P(A|B) \quad \text{and/or} \quad P(B) = P(B|A)$$

- For complementary events: $P(A) + P(\bar{A}) = 1$
- Multiplication rule for dependent events:

$$P(A \text{ and } B) = P(A)\,P(B|A)$$

- Multiplication rule for independent events:

$$P(A \text{ and } B) = P(A)\,P(B)$$

- Joint probability of two mutually exclusive events:

$$P(A \text{ and } B) = 0$$

- Addition rule for mutually nonexclusive events:

$$P(A \text{ or } B) = P(A) + P(B) - P(A \text{ and } B)$$

- Addition rule for mutually exclusive events:

$$P(A \text{ or } B) = P(A) + P(B)$$

- Bayes' theorem:
  The revised probability of event A_1, after additional information on another event B is obtained, is calculated as follows.

$$P(A_1|B) = \frac{P(A_1)\,P(B|A_1)}{P(A_1)\,P(B|A_1) + P(A_2)\,P(B|A_2)}$$

[handwritten: $K = \dfrac{\text{upper boundary} - \mu \,(\text{mean})}{\sigma \,(\text{standard dev})}$]

STATISTICS FOR BUSINESS AND ECONOMICS

p.127

rank + group data

$Q_1 = 25\%$ $Q_2 = 50\%$ $Q_3 = 75\%$

p.131 quartile

$$\tilde{x} = L + \frac{j}{c} \cdot c$$

L = lower boundary

j = # needed from next to group to reach median

f = frequency of last class used

c = interval

value of percentile

p.134   1) rank data   ascending order

$K = \%$ given $n = \#$ in data bank

$$\frac{Kn}{100} = \text{position in rank}$$
(location)

add #'s on each side $\div 2 =$

% less than lower #
side

% higher than
higher #
side

STATISTICS FOR BUSINESS AND ECONOMICS

PREM S. MANN
EASTERN CONNECTICUT STATE UNIVERSITY

JOHN WILEY & SONS, INC.
NEW YORK · CHICHESTER · BRISBANE · TORONTO · SINGAPORE

ACQUISITIONS EDITOR	Brad Wiley II
DEVELOPMENTAL EDITOR	Joan Carrafiello
MARKETING MANAGER	Susan Elbe
SENIOR PRODUCTION EDITOR	Charlotte Hyland
DESIGNER	Laura Nicholls
MANUFACTURING MANAGER	Susan Stetzer
ILLUSTRATION COORDINATOR	Jaime Perea
COVER PHOTO	Peter Menzel, Stock Boston

This book was set in New Times Roman by CRWaldman and printed and bound by R.R. Donelley. The cover was printed by Phoenix Color Corp.

Library of Congress Cataloging in Publication Data:
Mann, Prem S.
    Statistics for Business and Economics / Prem S. Mann.
      p. cm.
    Includes index.
    ISBN 0–471–58969–1
    1. Social sciences—Statistical methods. 2. Commercial statistics. 3. Economics—Statistical methods. 4. Statistics—Problems, exercises, etc. I. Title.
    HA29.M2456   1995
    519.5—dc20   94-26247
                     CIP

Printed in the United States of America

10 9 8 7 6 5 4

To my wife Sarabjeet,
daughter Harpreet,
and sons Kulwinder and Sukhwinder

PREFACE

Today, companies of all sizes use statistical methods in making decisions. Consequently, the study of statistical methods has taken on a prominent role in the education of students majoring in business and economics. *Statistics for Business and Economics* is written for a first course in statistics for business and economics students. Although it is primarily intended for a one-semester course, it can be used for two semesters at some colleges and universities. A strong background in mathematics is not required of students using this text, with elementary algebra being the only prerequisite.

The goal of this text is to present statistics in a clear and interesting way. Three major characteristics of this text support this goal: the realistic content of its examples and exercises that draw on a comprehensive range of applications from business and economics, the clarity and brevity of its presentation, and the soundness of its pedagogical approach. These characteristics are exhibited through the interplay of a variety of significant text features. The following is a description of these features and their purpose.

MAIN FEATURES OF THIS TEXT

Style and Pedagogy   **Clear and Concise Exposition**   The explanation of statistical methods and concepts is clear and concise. Moreover, the style is more user-friendly and readable. In chapter introductions and in transition from section to section, new ideas are related to those discussed earlier.

Abundant Examples   **Examples**   The text contains a wealth of examples, a total of 211 in 15 chapters. The examples are usually given in a format showing a problem and its solution. They are well sequenced and thorough, displaying all facets of concepts. Furthermore, the examples capture students' interest because they cover a wide variety of relevant topics within the areas of business and economics. They are based on situations practicing statisticians encounter in business and economics environments. Finally, a large number of examples are based on real data that are taken from sources such as books, government and private data sources and reports, magazines, newspapers, and professional journals.

Realistic Settings   **Solutions**   A clear, concise solution follows each problem presented in an example. When the solution to an example involves many steps, it is presented in a step-by-step format. For instance, examples related to tests of hypotheses contain five steps that are consistently used to solve such examples in all chapters. Thus, procedures are presented in the concrete settings of applications rather than as isolated abstractions. Frequently, solutions contain highlighted remarks that recall and reinforce ideas critical to the solution of the problem. Such remarks add to the clarity of presentation.

Guideposts   **Margin Notes for Examples**   Appearing in the margin beside each example is a brief description of what is being done in that example. Students can use these margin notes to assist them as they read through sections and to quickly locate appropriate model problems as they work through exercises.

Frequent Use of Diagrams Concepts can often be made more understandable by describing them visually, with the help of diagrams. This text uses diagrams frequently to help students understand concepts and solve problems. For example, tree diagrams are used extensively in Chapters 4 and 5 to assist in explaining probability concepts and in computing probabilities. Similarly, solutions to all examples about tests of hypotheses contain diagrams showing rejection regions, nonrejection regions, and critical values.

Highlighting Definitions of important terms, formulas, and key concepts are enclosed in color boxes so students can easily locate them. A similar use of color is found in the *Using MINITAB* sections where a color tint highlights MINITAB commands, their explanations, and their usage along with MINITAB solutions. Important terms appear in the text either in boldface or italic type.

Cautions Certain items need special attention. These may deal with potential trouble spots that commonly cause errors. Or they may deal with ideas that students often overlook. Special emphasis is placed on such items through the headings: ''Remember,'' ''An Observation,'' or ''Warning.'' An icon is used to identify such items.

Realistic Applications **Case Studies** Case studies, which appear in many chapters, provide additional illustrations of the application of statistics in the areas of business and economics. Most of these case studies are based on articles published in journals, magazines, or newspapers. All case studies are based on real data.

Abundant Exercises **Exercises and Supplementary Exercises** The text contains an abundance of exercises, a total of 1336 in 15 chapters (excluding Computer Assignments). Moreover, a large number of these exercises contain several parts. Exercise sets appearing at the end of each section (or sometimes at the end of two or three sections) include problems on the topics of that section. These exercises are divided into two parts: **Concepts and Procedures** that emphasize key ideas and techniques, and **Applications** that use these ideas and techniques in concrete settings. Supplementary exercises appear at the end of each chapter and contain exercises on all sections and topics discussed in that chapter. A large number of these exercises are based on real data taken from varied data sources such as books, government and private data sources and reports, magazines, newspapers, and professional journals. Exercises given in the text provide not only a source of practice but also interesting information from real data and insight into business and economics areas. Finally, the exercise sets also contain many problems that demand critical thinking skills. The answers to selected odd-numbered exercises appear in the ''Answers Section'' at the back of the book. Optional exercises are indicated by an asterisk (*).

Quality Control Applications One of the most important applications of statistics is in the area of quality control. Hence, this text does not isolate work on quality control in one chapter. Instead, quality control examples and exercises are presented throughout several chapters so that students can see a broad range of statistical concepts and procedures applied to this important area. Such exercises are identified in the Teacher's Edition by the marginal annotation ''quality control application.''

More Challenging Exercises A set of especially challenging exercises appears at the end of each block of three chapters. These exercises, which are optional, use material presented in the three previous chapters, or any material already covered. These exercise blocks appear after Chapters 3, 6, 9, 12, and 15. Answers to these problems are not given in the "Answers Section" at the end of the book. They appear only in the *Instructor's Solutions Manual*. These exercises have been written by Professor Daniel S. Miller of Central Connecticut State University.

Summary and Review   **Key Formulas**   Each chapter contains a list of key formulas used in that chapter. This list appears toward the end of each chapter (before the "Supplementary Exercises").

Glossary   Each chapter has a glossary that lists the key terms introduced in that chapter, along with a brief explanation of each term. Almost all the terms that appear in boldface type in the text are in the glossary.

Self-Review Tests   Each chapter contains a "Self-Review Test," which appears immediately after the "Supplementary Exercises." These problems can help students to test their grasp of the concepts and skills presented in respective chapters and to monitor their understanding of statistical methods. The problems marked by an asterisk (*) in the self-review tests are **optional**. The answers to almost all problems of the self-review tests appear in the "Answers Section."

Formula Card   A formula card that contains key formulas from all chapters is included at the beginning of the book.

Technology   **Computer Usage**   Another feature of this text is the detailed instructions on the use of **MINITAB**.[1] A "Using MINITAB" section follows 13 of the 15 chapters. In addition, all of Chapter 14 is based on using and analyzing MINITAB solutions. Each of these "Using MINITAB" sections contains a detailed description of the MINITAB commands that are used to perform the statistical analysis presented in that chapter. In addition, each section contains several illustrations that demonstrate how MINITAB can be used to solve statistical problems. These MINITAB instructions and illustrations are so complete that students will not need to purchase any other MINITAB supplement. Computer assignments are also given at the end of each MINITAB section so that students can further practice MINITAB. A total of 58 computer assignments are contained in these sections.

Calculator Usage   The text contains many footnotes that explain how a calculator can be used to evaluate complex mathematical expressions.

Data Sets   Three data sets appear in Appendix A. These data sets, collected from different sources, contain information on many variables. They can be used to perform statistical analysis with statistical computer software such as MINITAB. **These data sets are available from the publisher on a diskette in MINITAB format and in ASCII format.**

Teacher's Annotated Edition   A *Teacher's Annotated Edition* of this text is also available to instructors. This edition contains a large number of annotations that provide teaching tips and suggestions that can be helpful in identifying items to be emphasized or areas of difficulty for students.

---

[1]MINITAB is a registered trademark of Minitab, Inc., 3081 Enterprise Drive, State College, PA 16801. Phone: 814-238-3280; fax: 814-238-4383; telex: 881612. The author would take this opportunity to thank Minitab, Inc., for their help.

OPTIONAL SECTIONS

Because each instructor has different preferences, the text does not indicate optional sections. This decision has been left to the instructor. Instructors may cover the sections or chapters that they think are important. However, the *Instructor's Solutions Manual* presents a few alternative one-semester syllabi.

Complete Learning System

SUPPLEMENTS

The following supplements are available to accompany this text:

Instructor's Solutions Manual This manual contains complete solutions to all exercises, the self-review test problems, and the ''More Challenging Exercises.''

Students' Solutions Manual This manual contains complete solutions to all of the odd-numbered exercises and to all the self-review test problems.

Study Guide and Review This guide contains review material about studying and learning patterns for a first course in statistics for business and economics. Special attention is given to the critical material of each chapter. Review of mathematical notation, formulas, and table reading is also included.

Printed Test Bank The printed copy of the test bank contains a large number of multiple choice questions, essay questions, and quantitative problems for each chapter.

Computerized Test Bank All questions that are in the printed ''Test Bank'' are available on a diskette. This diskette can be obtained from the publisher.

Data Diskette A diskette that contains all data sets given in Appendix A is available to the adopters of the text. This diskette is available in MINITAB format and in ASCII format.

ACKNOWLEDGMENTS

I thank the following reviewers whose comments and suggestions were invaluable in improving the manuscript.

K. S. Asal	Broward Community College, Coconut Creek, Florida
Louise Audette	Manchester Community College, Manchester, Connecticut
Joan Bookbinder	Johnson & Wales University, Providence, Rhode Island
Dean Burbank	Gulf Coast Community College, Panama City, Florida
Jayanta Chandra	University of Notre Dame, Notre Dame, Indiana
Fred H. Dorner	Trinity University, San Antonio, Texas
Frank Goulard	Portland Community College, Portland, Oregon
Robert Graham	Jacksonville State University, Jacksonville, Alabama
A. Eugene Hileman	Northeastern State University, Tahlequah, Oklahoma
Jean Johnson	Governors State University, University Park, Illinois
Linda Kohl	University of Michigan, Ann Arbor, Michigan
Carlos de la Lama	San Diego City College, San Diego, California
Richard McGowan	University of Scranton, Scranton, Pennsylvania
Daniel S. Miller	Central Connecticut State University, New Britain, Connecticut
Robert A. Nagy	University of Wisconsin, Green Bay, Wisconsin
Paul T. Nkansah	Florida Agricultural and Mechanical University, Tallahassee, Florida

Chester Piascik	Bryant College, Smithfield, Rhode Island
Phillis Schumacher	Bryant College, Smithfield, Rhode Island
Ronald Schwartz	Wilkes University, Wilkes-Barre, Pennsylvania
Larry Stephens	University of Nebraska, Omaha, Nebraska
Jean Weber	University of Arizona, Tucson, Arizona
K. Paul Yoon	Fairleigh Dickinson University, Madison, New Jersey

My special thanks to Professor Gerald Geissert, of Eastern Connecticut State University, for writing annotations for the Teacher's Edition, checking solutions to all examples for mathematical accuracy, reading proofs, and preparing the index. I thank Professor Daniel S. Miller for writing the ''More Challenging Exercises'' and Dr. Naitee Ting for preparing the answers to the odd-numbered exercises for the ''Answers Section'' that appears at the end of the book. In addition, I thank Eastern Connecticut State University for all the support I received, including a one-semester sabbatical leave. I thank many of my students, especially Thenmalar (Thane) Meyyappan, who were of immense help in the preparation of the manuscript. I also offer my thanks to my colleagues, friends, and family whose support was a source of encouragement during the period when I spent long hours on this project.

It is my pleasure to thank all the professionals at John Wiley with whom I enjoyed working. Among them are Wayne Anderson, Publisher in the College Division; Brad Wiley II, Editor; Joan C. Carrafiello, former Development Editor; Charlotte Hyland, Senior Production Editor; Lucille Buonocore, Production Manager; Susan Elbe, Marketing Manager; Laura Nicholls, Designer; Jaime Perea, Illustration Coordinator; and many others. My special thanks go to Joan C. Carrafiello, whose help was omnipresent. She supervised this manuscript through all stages and was very helpful during all times of need.

Any suggestions from readers for future editions would be greatly appreciated.

Prem S. Mann

CONTENTS

Chapter 3 **NUMERICAL DESCRIPTIVE MEASURES**   90

Chapter 5 | **DISCRETE RANDOM VARIABLES AND THEIR PROBABILITY DISTRIBUTIONS** | **216**

Chapter 8 **ESTIMATION OF THE MEAN AND PROPORTION**   **406**

CASE STUDIES

STATISTICS FOR BUSINESS AND ECONOMICS

1 INTRODUCTION

The study of statistics has become more popular during the past two decades. The increasing availability of computers and statistical software packages has enlarged the role of statistics as a tool for empirical research. As a result, statistics is used for research in almost all professions, from medicine to sports. Today, college students in almost all disciplines, including business and economics, are required to take at least one statistics course.

Every field of study has its own terminology. Statistics is no exception. This introductory chapter explains the basic terms of statistics. These terms will bridge our understanding of the concepts and techniques presented in subsequent chapters.

1.1   WHAT IS STATISTICS?

The word **statistics** has two meanings. In the more common usage, statistics refers to numerical facts. The numbers that represent the income of a family, the number of cars sold at a dealership during the past month, the number of employees of a company, and the starting salary of a typical college graduate are examples of statistics in this sense of the word. A 1988 article in the *U.S. News & World Report* declared "Statistics are an American obsession."[1] During the 1988 baseball World Series between the Los Angeles Dodgers and the Oakland A's, NBC commentator Joe Garagiola reported to the viewers numerical facts about the players' performances. In response, fellow commentator Vin Scully said, "I love it when you talk statistics." In these examples, the word *statistics* refers to numbers. The following examples present a few statistics about U.S. companies.

1. Johnson & Johnson sold goods worth $13,753 million in 1992.
2. The market value of AT&T was $75,700 million as of March 5, 1993.
3. General Motors incurred a loss of $2620.6 million in 1992.
4. Mr. Anthony O'Reilly, CEO of H.J. Heinz, received a total pay (salary, bonus, and long-term compensation) of $75,085,000 in 1991.
5. IBM spent $5083 million on R&D (research and development) in 1992.
6. The total assets of Citicorp were $213,701 million as of 1992.
7. United Technologies exported products worth $3451 million in 1992.

Figure 1.1 presents statistics on alcoholism and its monetary and nonmonetary costs in the United States.

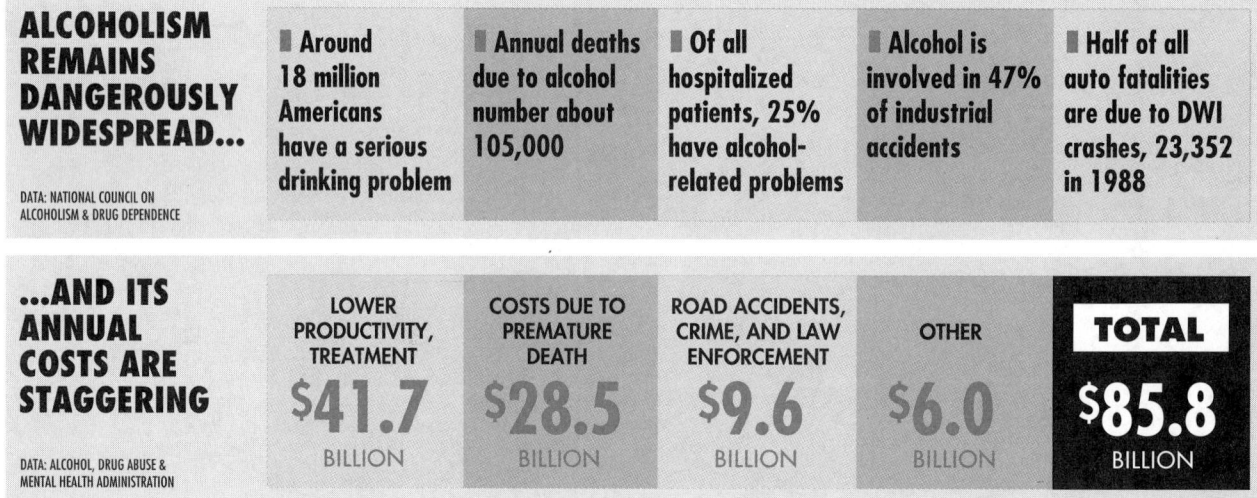

ALCOHOLISM REMAINS DANGEROUSLY WIDESPREAD...

DATA: NATIONAL COUNCIL ON ALCOHOLISM & DRUG DEPENDENCE

▮ Around 18 million Americans have a serious drinking problem

▮ Annual deaths due to alcohol number about 105,000

▮ Of all hospitalized patients, 25% have alcohol-related problems

▮ Alcohol is involved in 47% of industrial accidents

▮ Half of all auto fatalities are due to DWI crashes, 23,352 in 1988

...AND ITS ANNUAL COSTS ARE STAGGERING

DATA: ALCOHOL, DRUG ABUSE & MENTAL HEALTH ADMINISTRATION

LOWER PRODUCTIVITY, TREATMENT	COSTS DUE TO PREMATURE DEATH	ROAD ACCIDENTS, CRIME, AND LAW ENFORCEMENT	OTHER	TOTAL
$41.7 BILLION	**$28.5** BILLION	**$9.6** BILLION	**$6.0** BILLION	**$85.8** BILLION

Figure 1.1   Statistics on alcoholism and its monetary and nonmonetary costs. Reproduced from *Business Week*, March 25, 1991, with permission. Copyright © McGraw-Hill, Inc., 1991.

[1]"The Numbers Racket: How Polls and Statistics Lie," *U.S. News & World Report*, July 11, 1988, pp. 44–47.

The second meaning of statistics refers to the field or discipline of study. In this sense of the word, statistics is defined as follows.

STATISTICS

Statistics is a group of methods that are used to collect, analyze, present, and interpret data and to make decisions.

Every day we make decisions that may be personal, business related, or of some other kind. Usually these decisions are made under conditions of uncertainty. Many times, the situations or problems we face in the real world have no precise or definite solution. Statistical methods help us to make scientific and intelligent decisions in such situations. Decisions made by using statistical methods are called *educated guesses*. Decisions made without using statistical (or scientific) methods are *pure guesses* and, hence, may prove to be unreliable.

Like almost all fields of study, statistics has two aspects: theoretical and applied. *Theoretical* or *mathematical statistics* deals with the development, derivation, and proof of statistical theorems, formulas, rules, and laws. *Applied statistics* involves the applications of those theorems, formulas, rules, and laws to solve real-world problems. This text is concerned with applied statistics and not with theoretical statistics. By the time you finish studying this book, you will learn how to think statistically and how to make educated guesses.

1.2 TYPES OF STATISTICS

Broadly speaking, applied statistics can be divided into two areas: *descriptive statistics* and *inferential statistics*.

1.2.1 DESCRIPTIVE STATISTICS

Suppose we have information on the 1992 total sales of 100 companies. In statistical terminology, the whole set of numbers that represents the sales of companies is called a **data set**, the name of each company is called an **element**, and the sales of each company is called an **observation**. (These terms are defined in more detail in Section 1.4.)

A data set in its original form is usually very large. Consequently, such a data set is not very helpful in drawing conclusions or making decisions. It is easier to draw conclusions from summary tables and diagrams than from the original version of a data set. So, we reduce data to a manageable size by constructing tables, drawing graphs, or calculating summary measures such as averages. The portion of statistics that helps us to do this type of statistical analysis is called **descriptive statistics**.

DESCRIPTIVE STATISTICS

Descriptive statistics consists of methods for organizing, displaying, and describing data by using tables, graphs, and summary measures.

Case Studies 1–1 and 1–2 present examples of descriptive statistics. Case Study 1–1 shows a graphical display of data and Case Study 1–2 gives a table that is constructed to organize a data set.

CASE STUDY 1–1   FRESHMEN'S INTEREST IN BUSINESS

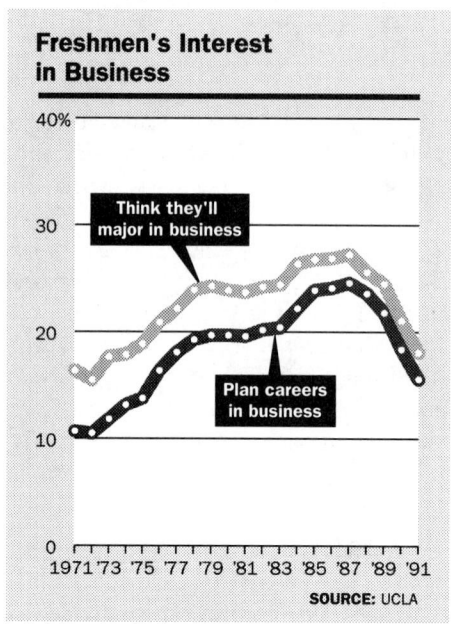

Reprinted with permission, The Chronicle of Higher Education

The Higher Education Research Institute at the University of California at Los Angeles conducts a study of freshmen entering college each fall. The above graph shows the percentage of freshmen entering college during 1971 to 1991 who intended to major in business and who planned careers in business. As we can observe, these percentages increased steadily from 1971 to 1987 and reached their peak level in 1987. However, these percentages have continuously declined since then.

Source: The American Freshman: Twenty-Five Year Trends, Dey, Astin, Korn, Higher Education Research Institute, UCLA. Copyright © UCLA, 1992. Chart reproduced with permission.

CASE STUDY 1-2 PERCENTAGE OF U.S. FAMILIES WHO BELONG TO DIFFERENT INCOME GROUPS

Income	Percent
Under $15,000	16.8
$15,000 to $24,999	16.4
$25,000 to $49,999	36.3
$50,000 to $99,999	25.1
$100,000 and over	5.4

Source: U.S. Bureau of the Census.

The above table lists the percentage of U.S. families who belong to different income groups. For example, according to the U.S. Bureau of the Census, 16.8% of families in the United States have a yearly income of less than $15,000, 16.4% have an income of $15,000 to $24,999, and so forth. At the upper end, 5.4% of families have a yearly income of $100,000 or more.

Both Chapters 2 and 3 cover descriptive statistical methods. In Chapter 2, we learn how to construct graphs and tables like the ones presented in Case Studies 1–1 and 1–2. In Chapter 3, we learn to calculate numerical summary measures such as averages.

1.2.2 INFERENTIAL STATISTICS

In statistics, the collection of all elements of interest is called a **population**. The selection of a few elements from this population is called a **sample**. (Population and sample are discussed in more detail in Section 1.3.)

A major portion of statistics deals with making decisions, inferences, predictions, and forecasts about populations based on results obtained from samples. For example, we may make some decisions about the views of all college and university students on the role of ethics in business based on the views of 1000 students selected from a few colleges and universities. As another example, suppose a company receives a shipment of parts from a manufacturer that are to be used in CD players manufactured by this company. To check the quality of the whole shipment, the company will select a few items from the shipment, inspect them, and make a decision. The area of statistics that deals with such decision-making procedures is referred to as **inferential statistics**. This branch of statistics is also called *inductive reasoning* or *inductive statistics*.

INFERENTIAL STATISTICS

Inferential statistics consists of methods that use sample results to help make decisions or predictions about a population.

Chapters 8 through 13 and parts of Chapter 7 deal with inferential statistics.

Probability, which gives a measurement of the likelihood that a certain outcome will occur, acts as a link between descriptive and inferential statistics. Probability is used to make statements about the occurrence or nonoccurrence of a certain event under uncertain conditions. Probability and probability distributions are discussed in Chapters 4 through 6 and parts of Chapter 7.

EXERCISES

Concepts and Procedures

1.1   Briefly describe the two meanings of the word *statistics*.

1.2   Briefly explain the types of statistics.

1.3   POPULATION VERSUS SAMPLE

We will encounter the terms *population* and *sample* on almost every page of this text.[2] Consequently, understanding the meaning of each of these two terms and the difference between the two is crucial.

Suppose a statistician is interested in knowing

1.   The percentage of all U.S. families who earn less than $20,000 a year
2.   The 1992 gross sales of all companies in Canada
3.   The prices of all statistics books published in the United States during the past five years

In these examples, the statistician is interested in *all* families, *all* firms, and *all* statistics books. Each of these groups is called the population for the respective example. In statistics, a population does not necessarily mean a collection of people. It can, in fact, be a collection of people or of any kind of items such as books, television sets, or cars. The population of interest is usually called the **target population**.

> **POPULATION OR TARGET POPULATION**
>
> A population consists of all elements—individuals, items, or objects—whose characteristics are being studied. The population that is being studied is also called the *target population*.

Most of the time, decisions are made based on portions of populations. For example, the various election polls conducted in the United States to estimate the percentage of voters favoring various candidates in any presidential election are based on only a few hundred or a few thousand voters selected from across the country. In this case, the population consists of all registered voters in the United States. The sample is made up of the few hundred or few thousand voters who are included in an opinion poll. Thus, the collection of a few elements selected from a population is called a **sample**.

---

[2]To learn more about sampling and sampling techniques, refer to Chapter 7.

> **SAMPLE**
>
> A portion of the population selected for study is referred to as a sample.

Figure 1.2 illustrates the selection of a sample from a population.

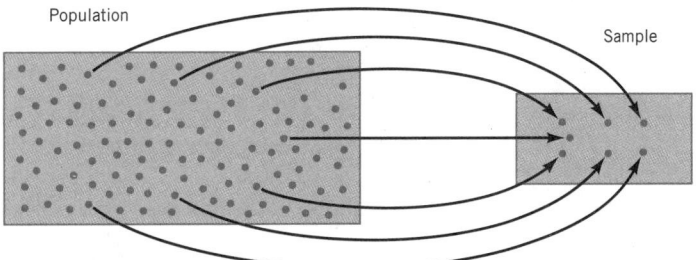

Figure 1.2 Population and sample.

The collection of information from the elements of a population or a sample is called a **survey**. A survey that includes every element of the target population is called a **census**. Often the size of the target population is large. Hence, in practice, a census is rarely taken because it is very expensive and time consuming. In many cases it is even impossible to identify each element of the target population. Usually, to conduct a survey, we select a sample and collect the required information from the elements included in that sample. We then make decisions based on this sample information. Such a survey that is conducted on a sample is called a **sample survey**. As an example, if we collect information on the 1993 gross sales of all businesses in Connecticut, it will be referred to as a census. On the other hand, if we collect information on the 1993 gross sales of 50 businesses from Connecticut, it will be called a sample survey.

> **CENSUS AND SAMPLE SURVEY**
>
> A survey that includes every member of the population is called a census. The technique of collecting information from a portion of the population is called a sample survey.

The purpose of conducting a sample survey is to make decisions about the corresponding population. It is important that the results obtained from a sample survey closely match the results that we would obtain by conducting a census. Otherwise, decisions derived from a sample survey will not apply to the corresponding population. As an example, to find the average income of families living in New York City by conducting a sample survey, the sample must contain families who belong to different income groups in almost the same proportion as they exist in the population. Such a sample is called a **representative sample**. Inferences derived from a representative sample will be more reliable.

> **REPRESENTATIVE SAMPLE**
>
> A sample that represents the characteristics of the population as closely as possible is called a representative sample.

Case Study 1–3, based on a sample of 22,000 male U.S. physicians, describes whether taking aspirin reduces the risk of heart attack.

CASE STUDY 1-3   ASPIRIN REDUCES THE RISK OF HEART ATTACK

The Steering Committee of the Physicians' Health Study Research Group conducted an experiment to investigate if taking one adult-size aspirin every other day reduces the risk of heart attack.

The researchers selected a group of 22,000 male U.S. physicians, aged 40 to 84, from 59,000 volunteers. The men were randomly assigned to receive either aspirin or placebo (a dummy pill). The experiment was conducted over a period of five years. According to the editorial by Arnold S. Relman in *The New England Journal of Medicine*,

> The total number of myocardial infarctions among the physicians taking aspirin had been reduced by nearly half. Strokes, on the other hand, were slightly, although not significantly, more numerous among the aspirin takers. Aspirin had no effect on the total number of vascular deaths or deaths from all causes. When the numbers of important vascular events were combined (nonfatal myocardial infarctions plus nonfatal strokes plus vascular deaths from all causes), those receiving aspirin still had a significant 23 percent reduction in risk.

For Class Discussion
If the results of this study are assumed to be true, what economic impact would it have on the companies that produce aspirin?

Source: Arnold S. Relman, "Aspirin for the Primary Prevention of Myocardial Infarctions." *The New England Journal of Medicine*, 318(4), January 28, 1988, pp. 245–246; The Steering Committee of the Physicians' Health Study Research Group, "Preliminary Report: Findings from the Aspirin Component of the Ongoing Physicians' Health Study," pp. 262–264.

From Case Study 1–3, can we deduce that aspirin always works for all people to reduce the risk of heart attack? In other words, was the sample selected for this study representative of the general population? Case Study 1–4 provides an answer to this question.

CASE STUDY 1-4   ASPIRIN: YES, NO, MAYBE?

. . . Why were only doctors chosen for the study? Doctors are an accessible, fairly uniform group of skilled medical observers who could be tracked accurately by questionnaire. But it is worth thinking about the selection process. Of the 59,000 doctors who originally volunteered, 26,000 were excluded because of disqualifying medical conditions, such as a previous heart attack or aspirin intolerance. Of the remaining 33,000, a third dropped out for various reasons. It's legitimate to wonder at what rate the dropouts had heart attacks, since the answer could have bearing on the heart attack rate among the men studied. Are these doctors typical of the general population? Clearly they are not. . . .

Does what applies to white males apply to everyone? The report raised this question and concluded that ''there seems little reason to suspect that the biologic effects of aspirin would be materially different in other populations with comparable or higher risks of cardiovascular disease.'' But though the biologic effect may be the same, its relative importance may well depend on what other risk factors are present.

For example, women are a special group when it comes to heart attacks. Presumably, also, the majority of the doctors were white. Yet we know that the risk factors for black men are different from those of white men. Blacks have a 33% greater chance of having hypertension—a significant risk factor for heart attack—than whites. So the answer to this question is tentative at best.

Source: Excerpted from ''Aspirin: Yes, No, Maybe?'' *University of California, Berkeley Wellness Letter,* 4(7), April 1988, p. 1. Reprinted by permission. Copyright © Health Letter Associates, 1988.

A sample may be random or nonrandom. In a **random sample**, each element of the population has some chance of being included in the sample. However, in a nonrandom sample this may not be the case.

RANDOM SAMPLE

A sample drawn in such a way that each element of the population has some chance of being selected is called a random sample.

One way to select a random sample is by lottery or draw. For example, if we are to select five employees from a total of 50, we write each of the 50 names on a separate piece of paper. Then we place all 50 slips in a box and mix them thoroughly. Finally, we randomly draw five slips from the box. The five names drawn will give a random sample. On the other hand, if we arrange all 50 names alphabetically and then select the first five names on the list, it would be a nonrandom sample because the employees listed sixth to fiftieth have no chance of being included in the sample.

A sample may be selected with or without replacement. In sampling **with replacement**, each time we select an element from the population, we put it back in the population before we select the next element. Thus, in sampling with replacement, the population will contain the same number of items each time a selection is made. As a result, we may select the same item more than once in such a sample. Consider a box that contains 25 balls of different colors. Suppose we draw a ball, record its color, and put it back in the box before drawing the next ball. Every time we draw a ball from this box, the box contains 25 balls. This is an example of sampling with replacement.

Sampling **without replacement** occurs when the selected element is not replaced in the population. In this case, each time we select an item, the size of the population is reduced by one element. Thus, we cannot select the same item more than once in this type of sampling. Most of the time, samples taken in statistics are without replacement. Consider an opinion poll based on a certain number of voters selected from the population of all eligible voters. In this case, the same voter would not be selected more than once. Therefore, this is an example of sampling without replacement.

EXERCISES

Concepts and Procedures

1.3 Briefly explain the terms population, sample, representative sample, random sample, sampling with replacement, and sampling without replacement.

1.4 Explain which of the following constitute a population and which constitute a sample.

    a. Weekly salaries of all employees of a company
    b. Yield of potatoes for 10 acres of land
    c. Cattle owned by 100 farmers in Iowa
    d. Number of days missed by all employees of a company during the past month

1.5 Explain which of the following constitute a population and which constitute a sample.

    a. Assets of five banks selected from a state
    b. Number of computers sold during the past week at all computer stores in Chicago
    c. Number of VCRs owned by all families
    d. Weights of 100 packages

1.6 Give one example each of sampling with and sampling without replacement.

1.7 Briefly explain the difference between a census and a sample survey. Why is conducting a sample survey preferable to conducting a census?

1.4 BASIC TERMS

It is very important to understand the meaning of some basic terms that will be used frequently in this text. This section explains the meaning of an element (or member), a variable, an observation, and a data set. An element and a data set were briefly defined earlier in Section 1.2. This section defines these terms formally and illustrates them with the help of an example.

Table 1.1 gives information on 1992 profits (in millions of U.S. dollars) of six international companies. We can call this group of companies a sample of six companies. Each company listed in this table is called an **element** or a **member** of the sample. Table 1.1 contains information on six elements. Note that elements are also called *observational units*.

Table 1.1 Profits of Six International Companies

Company	1992 Profits ←——Variable (millions of dollars)
Exxon, United States	4800
Toyota Motor, Japan	2222
Daimler-Benz, Germany	891
An element→ Guinness, Britain or a member	818 ←——An observation or measurement
Northern Telecom, Canada	548
Fiat Group, Italy	375

Source: Business Week, July 12, 1993.

> **ELEMENT OR MEMBER**
>
> An element or member of a sample or population is a specific subject or object (for example, a person, firm, item, state, or country) about which the information is collected.

The *1992 profits of companies* in our example is called a **variable**. The *1992 profits* is a characteristic of companies that we are investigating.

> **VARIABLE**
>
> A variable is a characteristic under study that assumes different values for different elements. In contrast to a variable, the value of a *constant* is fixed.

A few other examples of variables are the incomes of households, the number of houses built in a city per month during the past year, the makes of cars owned by people, the gross sales of companies, and the number of insurance policies sold by a salesperson per day during the past month.

In general, a variable assumes different values for different elements, as does 1992 profits of six companies in Table 1.1. For some elements, however, the value of the variable may be the same. For example, if we collect information on incomes of households, these households are expected to have different incomes, although some of them may have the same income.

A variable is often denoted by x, y, or z. For instance, in Table 1.1, the 1992 profits of companies may be denoted by any one of these letters. Starting with Section 1.7, we will begin to use these letters to denote variables.

Each of the values representing the 1992 profits of six companies in Table 1.1 is called an **observation** or **measurement**.

> **OBSERVATION OR MEASUREMENT**
>
> The value of a variable for an element is called an observation or measurement.

According to Table 1.1, the 1992 profits of Guinness were $818 million. The value $818 million is an observation or measurement. Table 1.1 contains six observations, one for each of the six companies.

The information given in Table 1.1 on 1992 profits of companies is called the **data** or a **data set**.

> **DATA SET**
>
> A data set is a collection of observations on one or more variables.

Another example of a data set would be a list of prices of 25 recently sold homes.

EXERCISES

Concepts and Procedures

1.8 Explain the meaning of an element, a variable, an observation, and a data set.

1.9 The following table lists the 1992 total pay (salary, bonus, and other compensation) of the CEOs of five companies.

Company	Pay (millions of dollars)
Merck	16.0
Philip Morris	11.5
Black & Decker	10.2
Walt Disney	7.5
Ford Motor	5.1
Kellogg	2.6

Source: Fortune, June 14, 1993. © 1993 Time Inc. All rights reserved.

Briefly explain the meaning of a member, a variable, a measurement, and a data set with reference to this table.

1.10 The following table lists the recent per capita gross national products of eight countries.

Country	Per Capita GNP (in U.S. dollars)
United States	$20,998
Canada	18,635
Austria	15,266
Singapore	15,108
Japan	14,311
Britain	13,732
Bahamas	11,293
New Zealand	11,550

Source: Fortune, July 27, 1992. © 1992 Time Inc. All rights reserved.

Briefly explain the meaning of a member, a variable, an observation, and a data set with reference to this table.

1.11 Refer to the data set given in Exercise 1.9.
    a. What is the variable for this data set?
    b. How many observations does this data set contain?
    c. How many elements does this data set contain?

1.12 Refer to the data set given in Exercise 1.10.
    a. What is the variable for this data set?
    b. How many observations does this data set contain?
    c. How many elements does this data set contain?

1.5 TYPES OF VARIABLES

In Section 1.4, we learned that a variable is a characteristic under investigation that assumes different values for different elements. The incomes of families, stock prices, gross sales of companies, prices of college textbooks, and makes of cars owned by families are a few examples of variables.

A variable may be classified as quantitative or qualitative. These two types of variables are explained next.

1.5.1 QUANTITATIVE VARIABLES

Some variables can be measured numerically whereas others cannot. A variable that can assume numerical values is called a **quantitative variable**.

QUANTITATIVE VARIABLE

A variable that can be measured numerically is called a quantitative variable. The data collected about a quantitative variable are called *quantitative data*.

Incomes, gross sales, prices of homes, number of cars owned, stock prices, and accidents are examples of quantitative variables since each of them can be expressed numerically. For instance, the income of a family may be $31,520.75 per year, the gross sales for a company may be $567 million for the past year, and so forth. Such quantitative variables can be classified as either *discrete variables* or *continuous variables*.

Discrete Variables

The values that a certain quantitative variable can assume may be countable or not. For example, we can count the number of cars owned by a family but we cannot count the income of a family. The variable whose values are countable is called a **discrete variable**. Note that there are no possible intermediate values between consecutive values of a discrete variable.

DISCRETE VARIABLE

A variable whose values are countable is called a discrete variable. In other words, a discrete variable can assume only certain values with no intermediate values.

For example, the number of cars sold on any day at a car dealership is a discrete variable because the number of cars sold must be 0, 1, 2, 3, ... The number of cars sold cannot be between 0 and 1, or between 1 and 2. A few other examples of discrete variables are the number of people visiting a bank on any day, the number of cars in a parking lot, the number of cattle owned by a farmer, and the number of employees of a company.

Continuous Variables

A **continuous variable** can assume any value over a certain range, and we cannot count these values.

CONTINUOUS VARIABLE

A variable that can assume any numerical value over a certain interval or intervals is called a continuous variable.

The time taken to serve a customer by a bank teller is an example of a continuous variable because it can assume any value, let us say, between 5 seconds and 20 minutes. The time taken by the teller can be 3.5 minutes, 3.48 minutes, or 3.482 minutes. (Theoretically, we can measure time as precisely as we want.) Similarly, the length of an iron rod manufactured by a company can be measured to the tenth of an inch or to the hundredth of an inch. However, neither time nor length can be counted in a discrete fashion. A few other examples of continuous variables are total assets of banks, weights of packages, amount of soda in 12-ounce cans, and the yield of potatoes per acre. Note that any variable that involves money is considered a continuous variable.

1.5.2   QUALITATIVE OR CATEGORICAL VARIABLES

Variables that cannot be measured numerically but can be divided into different categories are called **qualitative** or **categorical variables**.

> **QUALITATIVE OR CATEGORICAL VARIABLE**
>
> A variable that cannot assume a numerical value but can be classified into two or more nonnumeric categories is called a qualitative or categorical variable. The data collected about such a variable are called *qualitative data*.

For example, the status of an undergraduate college student is a qualitative variable because a student can fall into any one of four categories: freshman, sophomore, junior, or senior. A few other examples of qualitative variables are the gender of a person, hair color, and the make of a car. Figure 1.3 illustrates the types of variables.

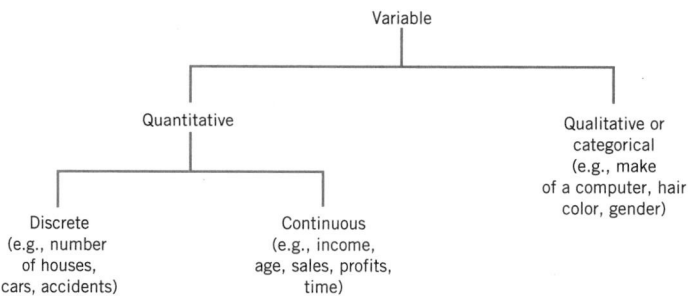

Figure 1.3   Types of variables.

1.6   SCALES OF MEASUREMENT

Data can also be classified based on scales of measurement, also called levels of measurement. There are four scales of measurement: nominal, ordinal, interval, and ratio data. The nominal scale data are of the lowest level, and ratio scale data are of the highest level. These levels are described next.

NOMINAL SCALE

Data that are divided into different categories only for identification purposes are said to have a **nominal scale**.

> **NOMINAL SCALE**
>
> The nominal scale applies to data that are divided into different categories, and these categories are used only for identification purposes.

An example of nominal scale would be different makes of cars. The names given to different makes of cars, such as Town car, Cadillac Sedan de Ville, Jaguar, and Mercedes, are for identification purposes only. We can neither rank these categories nor do any arithmetic operations (such as addition, subtraction, multiplication, or division) to such data. The categories for such a variable can be listed in any order. Other examples of nominal data are names of companies, gender of people, marital status, and models of computers. We can assign numeric codes to different categories such as Male = 1 and Female = 2. However, code 2 does not mean it is superior or inferior to code 1.

ORDINAL SCALE

Data that can be divided into different categories that can be ranked are said to have an **ordinal scale**.

> **ORDINAL SCALE**
>
> The ordinal scale applies to data that are divided into different categories that can be ranked.

For example, suppose in a survey people are asked to evaluate a product as excellent, good, or poor. These categories possess the characteristic that they can be ranked. We know that *excellent* has the highest rank and *poor* has the lowest rank. However, as in the case of a nominal scale, arithmetic operations are meaningless for ordinal scale data. The categories can be coded in this scale of measurement, and these codes refer to superior or inferior ranking. For example, if we code *excellent* = 1, *good* = 2, and *poor* = 3, then we know that 1 means the most superior and 3 indicates the most inferior.

INTERVAL SCALE

Data that can be ranked and for which we can find the difference between two values are said to have an **interval scale**.

> **INTERVAL SCALE**
>
> The interval scale applies to data that can be ranked and for which the difference between two values can be calculated and interpreted.

For example, temperature possesses an interval scale. Suppose on a day in December, it is 80° in Los Angeles and 30° in Boston. We can rank these data and state that it is warmer in Los Angeles than in Boston. We can also find the difference between the two values and say that it is 50° warmer in Los Angeles than in Boston. Another example of data that have an interval scale is scores in a test. A student who scores 90 is probably a better student than someone who scores 70. We can say the difference between the scores of these two students is 20.

Data with an interval scale contain a zero point, but it does not have any importance. For example, a zero temperature in itself has no meaning; that is, it does not mean complete absence of warmness. Similarly a zero score does not mean that the student has no intelligence. The zero point is chosen arbitrarily.

Another characteristic of data with an interval scale is that a ratio does not make sense for such data. For example, we cannot say that it is twice as warm if it is 100° than if it is 50°. Similarly we cannot say that a student who scores 90 in a test is twice as intelligent as a student who scores 45.

RATIO SCALE

Data that possess all the characteristics of interval scale data and for which a ratio of two values has a meaning is said to have a **ratio scale**.

> **RATIO SCALE**
>
> The ratio scale applies to data that can be ranked and for which all arithmetic operations (addition, subtraction, multiplication, and division) can be done.

Suppose the 1993 gross sales of Company A are $120 million and those of Company B are $60 million. We can say that 1993 gross sales of Company A are larger than those of B, that Company A sold goods worth $60 million more than Company B, and that Company A's gross sales are twice as much as those of Company B. A few other examples of such data are incomes of people, expenditures, and prices.

Note that a zero point has a meaning in ratio scale data. We know that a company that has zero sales did not sell any product. A person with zero income has no income.

EXERCISES

Concepts and Procedures

1.13 Explain the meaning of the following terms.

    a. Quantitative variable   b. Qualitative variable
    c. Discrete variable   d. Continuous variable
    e. Quantitative data   f. Qualitative data

1.14 Explain which of the following variables are quantitative and which are qualitative.

    a. Number of earners in a family   b. Color of cars
    c. Marital status of people   d. Square footage of houses

1.15 Explain which of the following variables are quantitative and which are qualitative.

    a. Number of homes owned   b. Rent paid by tenants
    c. Types of cars owned by families   d. Monthly phone bills

1.16 Classify the quantitative variables in Exercise 1.14 as discrete or continuous.

1.17 Classify the quantitative variables in Exercise 1.15 as discrete or continuous.

1.18 Indicate the scale of measurement for each of the variables mentioned in Exercise 1.14.

1.19 Indicate the scale of measurement for each of the variables mentioned in Exercise 1.15.

1.7 SUMMATION NOTATION

Sometimes mathematical notation helps to express a mathematical relationship concisely. This section describes the **summation notation** that is used to denote the sum of values.

Suppose a sample consists of five books and the prices of these five books are $25, $60, $37, $53, and $16. The variable *price of a book* can be denoted by x. The prices of five books can be written as follows.

$$\text{Price of the first book} = x_1 = \$25$$

$$\uparrow$$

Subscript of x denotes
the number of the book

Similarly,

$$\text{Price of the second book} = x_2 = \$60$$

$$\text{Price of the third book} = x_3 = \$37$$

$$\text{Price of the fourth book} = x_4 = \$53$$

$$\text{Price of the fifth book} = x_5 = \$16$$

In the above notation, x represents the price and the subscript denotes a particular book. Now, suppose we want to add the prices of all five books. Then,

$$x_1 + x_2 + x_3 + x_4 + x_5 = 25 + 60 + 37 + 53 + 16 = \$191$$

The uppercase Greek letter Σ (pronounced *sigma*) is used to denote the sum of all values. Using Σ notation, we can write the foregoing sum as follows.

$$\Sigma x = x_1 + x_2 + x_3 + x_4 + x_5 = \$191$$

The notation Σx in this expression represents the sum of all the values of x.

Using summation notation: one variable.

EXAMPLE 1–1 Suppose the ages of four managers are 35, 47, 28, and 60 years. Find (a) Σx (b) $\Sigma(x - 6)$ (c) $(\Sigma x)^2$ (d) Σx^2

Solution Let x_1, x_2, x_3, and x_4 be the ages (in years) of the first, second, third, and fourth manager, respectively. Then,

$$x_1 = 35, \quad x_2 = 47, \quad x_3 = 28, \quad \text{and} \quad x_4 = 60$$

(a) $\Sigma x = x_1 + x_2 + x_3 + x_4 = 35 + 47 + 28 + 60 = \mathbf{170}$

(b) To calculate $\Sigma(x - 6)$, first we subtract 6 from each value of x and then add the resulting values. Thus,

$$\Sigma(x - 6) = (x_1 - 6) + (x_2 - 6) + (x_3 - 6) + (x_4 - 6)$$

$$= (35 - 6) + (47 - 6) + (28 - 6) + (60 - 6)$$

$$= 29 + 41 + 22 + 54 = \mathbf{146}$$

(c)   Note that $(\Sigma x)^2$ is the square of the sum of all x values. Thus,

$$(\Sigma x)^2 = (170)^2 = \mathbf{28{,}900}$$

(d)   The expression Σx^2 is the sum of the squares of x values. To calculate Σx^2, we first square each of the x values and then add these squared values. Thus,

$$\Sigma x^2 = (35)^2 + (47)^2 + (28)^2 + (60)^2 = 1225 + 2209 + 784 + 3600 = \mathbf{7818}$$

Using summation notation: two variables.

EXAMPLE 1–2   The following table lists four pairs of m and f values:

m	12	15	20	30
f	5	9	10	16

Compute the following:

(a) Σm (b) Σf^2 (c) Σmf (d) $\Sigma m^2 f$ (e) $\Sigma(m - 5)^2 f$

Solution   We can write

$$m_1 = 12, \quad m_2 = 15, \quad m_3 = 20, \quad m_4 = 30,$$
$$f_1 = 5, \quad\;\; f_2 = 9, \quad\;\; f_3 = 10, \quad\;\; f_4 = 16$$

(a) $\Sigma m = 12 + 15 + 20 + 30 = \mathbf{77}$

(b) $\Sigma f^2 = (5)^2 + (9)^2 + (10)^2 + (16)^2 = 25 + 81 + 100 + 256 = \mathbf{462}$

(c)   To compute Σmf, we multiply the corresponding values of m and f and add the products as follows:

$$\Sigma mf = m_1 f_1 + m_2 f_2 + m_3 f_3 + m_4 f_4$$
$$= 12(5) + 15(9) + 20(10) + 30(16) = \mathbf{875}$$

(d)   To calculate $\Sigma m^2 f$, we square each m value, then multiply the corresponding m^2 and f values, and add the products.

$$\Sigma m^2 f = (m_1)^2 f_1 + (m_2)^2 f_2 + (m_3)^2 f_3 + (m_4)^2 f_4$$
$$= (12)^2(5) + (15)^2(9) + (20)^2(10) + (30)^2(16) = \mathbf{21{,}145}$$

The calculations done in parts (a) through (d) to find the values of Σm, Σf^2, Σmf, and $\Sigma m^2 f$ can be performed in tabular form, as shown in Table 1.2.

Table 1.2

m	f	f^2	mf	$m^2 f$
12	5	$5 \times 5 = 25$	$12 \times 5 = 60$	$12 \times 12 \times 5 = 720$
15	9	$9 \times 9 = 81$	$15 \times 9 = 135$	$15 \times 15 \times 9 = 2{,}025$
20	10	$10 \times 10 = 100$	$20 \times 10 = 200$	$20 \times 20 \times 10 = 4{,}000$
30	16	$16 \times 16 = 256$	$30 \times 16 = 480$	$30 \times 30 \times 16 = 14{,}400$
$\Sigma m = 77$	$\Sigma f = 40$	$\Sigma f^2 = 462$	$\Sigma mf = 875$	$\Sigma m^2 f = 21{,}145$

The columns of Table 1.2 can be explained as follows:

1.   The first column lists the values of m. The sum of these values gives $\Sigma m = 77$.

2. The second column lists the values of f. The sum of this column gives $\Sigma f = 40$.
3. The third column lists the squares of the f values. For example, the first value, 25, is the square of 5. The sum of the values in this column gives $\Sigma f^2 = 462$.
4. The fourth column records products of the corresponding m and f values. For example, the first value, 60, in this column is obtained by multiplying 12 by 5. The sum of the values in this column gives $\Sigma mf = 875$.
5. Next, the m values are squared and multiplied by the corresponding f values. The resulting products, denoted by m^2f, are recorded in the fifth column of the table. For example, the first value, 720, is obtained by squaring 12 and multiplying this result by 5. The sum of the values in this column gives $\Sigma m^2 f = 21{,}145$.

(e) The value of $\Sigma(m - 5)^2 f$ is computed as follows:

$$\Sigma(m - 5)^2 f = (m_1 - 5)^2 f_1 + (m_2 - 5)^2 f_2 + (m_3 - 5)^2 f_3 + (m_4 - 5)^2 f_4$$

$$= (12 - 5)^2(5) + (15 - 5)^2(9) + (20 - 5)^2(10) + (30 - 5)^2(16)$$

$$= 13{,}395$$

We can do these computations to calculate $\Sigma(m - 5)^2 f$ in a tabular form, as shown in Table 1.3.

Table 1.3

m	f	$(m - 5)$	$(m - 5)^2$	$(m - 5)^2 f$
12	5	$12 - 5 = 7$	49	$49 \times 5 = 245$
15	9	$15 - 5 = 10$	100	$100 \times 9 = 900$
20	10	$20 - 5 = 15$	225	$225 \times 10 = 2{,}250$
30	16	$30 - 5 = 25$	625	$625 \times 16 = 10{,}000$
$\Sigma m = 77$	$\Sigma f = 40$			$\Sigma(m - 5)^2 f = 13{,}395$

The third column of Table 1.3 lists the $(m - 5)$ values, which are obtained by subtracting 5 from the values listed in the column labeled m. The fourth column lists the squares of $(m - 5)$ values that are recorded in the third column. The fifth column contains the products of the corresponding values of $(m - 5)^2$ and f. The sum of the values listed in the fifth column gives the value of $\Sigma(m - 5)^2 f$. ■

EXERCISES

Applications

1.20 The scores of five students in a statistics class are 75, 80, 97, 91, and 63. Find
   a. Σx b. $\Sigma(x - 12)$ c. $(\Sigma x)^2$ d. Σx^2

1.21 The number of cars owned by six families are 3, 2, 1, 4, 1, and 2. Find
   a. Σx b. $\Sigma(x - 1)$ c. $(\Sigma x)^2$ d. Σx^2

1.22 The electric heating bills for February 1994 for four families were $122, 72, 96, and 110. Find
   a. Σx b. $\Sigma(x - 25)$ c. $(\Sigma x)^2$ d. Σx^2

1.23 The weights of seven packages mailed by a company are 7, 9, 6, 12, 10, 9, and 8 pounds. Find
   a. Σx b. $\Sigma(x - 4)$ c. $(\Sigma x)^2$ d. Σx^2

1.24 The following table lists five pairs of m and f values.

m	5	10	15	20	25
f	12	8	5	16	4

Compute the value of each of the following:

    a. Σm b. Σf^2 c. Σmf d. $\Sigma m^2 f$ e. $\Sigma(m-15)^2 f$

1.25 The following table lists six pairs of m and f values.

m	3	6	9	12	15	18
f	16	11	6	8	4	14

Calculate the value of each of the following:

    a. Σf b. Σm^2 c. Σmf d. $\Sigma m^2 f$ e. $\Sigma(m-10)^2 f$

1.26 The following table lists five pairs of x and y values.

x	15	20	11	8	5
y	10	7	14	9	18

Compute a. Σx b. Σy c. Σxy d. Σx^2 e. Σy^2

1.27 The following table lists six pairs of x and y values.

x	4	18	25	9	12	20
y	12	5	14	7	12	8

Compute a. Σx b. Σy c. Σxy d. Σx^2 e. Σy^2

GLOSSARY

Census A survey that includes all members of the population.

Continuous variable A (quantitative) variable that can assume any numerical value over a certain interval or intervals.

Data or **data set** Collection of observations or measurements on a variable.

Descriptive statistics Collection of methods that are used for organizing, displaying, and describing data using tables, graphs, and summary measures.

Discrete variable A (quantitative) variable whose values are countable.

Element or **member** A specific subject or object included in a sample or population.

Inferential statistics Collection of methods that help make decisions about a population based on sample results.

Interval scale Data that can be ranked and for which we can find the difference between two values are said to have an interval scale.

Nominal scale Data that are divided into different categories that are used for identification purposes only are said to have a nominal scale.

Observation or **measurement** The value of a variable for an element.

Ordinal scale Data that can be divided into different categories that can be ranked are said to have an ordinal scale.

Population or **target population** The collection of all elements whose characteristics are being studied.

Qualitative or **categorical data**   Data generated by a qualitative variable.

Qualitative or **categorical variable**   A variable that cannot assume numerical values but is classified into two or more categories.

Quantitative data   Data generated by a quantitative variable.

Quantitative variable   A variable that can be measured numerically.

Random sample   A sample drawn in such a way that each element of the population has some chance of being included in the sample.

Ratio scale   Data that can be ranked and for which all arithmetic operations can be performed are said to have a ratio scale.

Representative sample   A sample that contains the characteristics of the corresponding population.

Sample   A portion of the population of interest.

Sample survey   A survey that includes elements of a sample.

Statistics   Collection of methods that are used to collect, analyze, present, and interpret data and to make decisions.

Survey   Collecting data on the elements of a population or sample.

Variable   A characteristic under study or investigation that assumes different values for different elements.

SUPPLEMENTARY EXERCISES

1.28   The following table gives the average hourly earnings (in current dollars) of production or non-supervisory workers on private nonagricultural payrolls in the United States for 9 months, from June 1992 through February 1993.

Month and Year	Average Hourly Earnings
June 1992	$10.53
July 1992	10.53
August 1992	10.56
September 1992	10.66
October 1992	10.69
November 1992	10.73
December 1992	10.71
January 1993	10.78
February 1993	10.78

Source: U.S. Bureau of Labor Statistics.

Describe the meaning of a variable, a measurement, and a data set with reference to this table.

1.29   The following table gives the per capita health expenditures for 1990 for seven countries.

Country	Per Capita Health Expenditure (in U.S. dollars)
Britain	$ 974
Canada	1730
France	1543
Germany	1487
Italy	1234
Japan	1171
United States	2566

Source: Fortune, July 27, 1992. © 1992 Time Inc. All rights reserved.

Describe the meaning of an element, a variable, a measurement, and a data set with reference to this table.

1.30 Indicate which of the following examples refer to a population and which refer to a sample.

    a. A group of five employees selected from a company to represent it at a conference
    b. Total items produced on a machine during one week
    c. Yearly expenditure on clothes for 50 persons
    d. Number of houses sold by all 10 employees of a real estate agency

1.31 Indicate which of the following examples refer to a population and which refer to a sample.

    a. Ages of CEOs of all companies in New York City
    b. Prices of 100 houses selected from Chicago
    c. Number of subscribers to five magazines
    d. Salaries of all employees of a bank

1.32 State which of the following is an example of sampling with replacement and which is of sampling without replacement.

    a. Selecting 10 patients out of 100 to test a new drug
    b. Selecting one professor to be a member of the university senate and then selecting one professor from the same group to be a member of the curriculum committee

1.33 The number of ties owned by six managers are 10, 9, 14, 12, 7, and 4. Find

    a. Σx **b.** $\Sigma(x - 6)$ **c.** $(\Sigma x)^2$ **d.** Σx^2

1.34 The number of hours worked during the past week by five employees of a company are 43, 39, 44, 31, and 40. Find

    a. Σx **b.** $\Sigma(x - 8)$ **c.** $(\Sigma x)^2$ **d.** Σx^2

1.35 The following table lists five pairs of m and f values.

m	3	16	11	9	20
f	7	32	17	12	34

Compute the value of each of the following.

    a. Σm **b.** Σf^2 **c.** Σmf **d.** $\Sigma m^2 f$ **e.** $\Sigma(m - 5)^2 f$

1.36 The following table lists six pairs of x and y values.

x	7	11	8	4	14	18
y	5	15	7	10	9	12

Compute the value of each of the following.

    a. Σy **b.** Σx^2 **c.** Σxy **d.** $\Sigma x^2 y$ **e.** $\Sigma(x - 3)^2 y$

SELF-REVIEW TEST

1. A population in statistics means

    a. a collection of all men and women
    b. a collection of all subjects or objects of interest
    c. a collection of all people living in a country

2. A sample in statistics means

    a. a portion of the people selected from the population of a country

    b. a portion of the people selected from the population of an area

    c. a portion of the population of interest

3. Indicate which of the following is an example of a sample with replacement or a sample without replacement.

    a. Ten students are selected from a statistics class in such a way that as soon as a student is selected his or her name is deleted from the list before the next student is selected.

    b. A box contains 5 balls of different colors. A ball is drawn from this box, its color is recorded, and it is put back in the box before the next ball is drawn. This experiment is repeated 12 times.

4. Indicate which of the following variables are quantitative and which are qualitative. Classify the quantitative variables as discrete or continuous.

    a. Brand of coffee   **b.** Number of TV sets owned by families

    c. Weekly earnings of employees

5. The following table gives the starting salaries of five recent college graduates.

Name	Salary
Matt	$29,200
Lucia	22,450
Alison	27,920
Warren	32,350
Lori	28,100

Explain the meaning of a member, a variable, a measurement, and a data set with reference to this table.

6. The values (in thousands of dollars) of the cars owned by six persons are 13, 9, 3, 28, 7, and 16. Calculate

    a. Σx **b.** Σx^2 **c.** $(\Sigma x)^2$ **d.** $\Sigma(x - 4)$ **e.** $\Sigma(x - 3)^2$

7. The following table lists five pairs of m and f values.

m	3	6	9	12	15
f	15	25	40	20	12

Calculate **a.** Σm **b.** Σf **c.** Σm^2 **d.** Σmf **e.** $\Sigma m^2 f$

USING MINITAB[3]
AN INTRODUCTION

In recent years the use of computers has significantly increased in almost every aspect of life. Such usage of computers has reduced the computation time for quantitative analysis to a negligible amount.

In the real world, when doing a statistical analysis, we usually deal with hundreds or thousands of observations. For this reason, it is either very time consuming or almost impossible to make all the required calculations manually. The use of computers is of invaluable assistance in such situations. Consequently, learning to use a statistical software package has become an important part of learning statistics.

A large number of statistical software packages, both for mainframe computers and for microcomputers, have been developed in recent years. Most of these software packages are user-friendly.

Four of the major statistical software packages are BMDP, MINITAB, SAS, and SPSS.[4] All four of these packages are available for mainframe computers and for microcomputers. Besides these four packages, a large number of software packages have been developed for personal computers. The *Using MINITAB* sections, given at the end of most of the chapters of this text, provide brief instructions on how to use MINITAB to solve statistical problems. Four formal manuals that explain MINITAB commands in detail are

1. *MINITAB Reference Manual*, Release 8, MINITAB Inc., November 1991.
2. *MINITAB Reference Manual*, Release 9 for Windows, MINITAB Inc., July 1993.
3. *MINITAB Graphics Manual*, Release 9 for Windows, MINITAB Inc., July 1993.
4. Barbara F. Ryan, Brian L. Joiner, and Thomas A. Ryan, Jr., *MINITAB Handbook*, 2nd ed., PWS-KENT Publishing Company, Boston, 1985.

STARTING AND ENDING A MINITAB SESSION

MINITAB is available for both **mainframe computers** and **microcomputers**. The MINITAB commands are the same for both systems. However, procedures to start a system and to access MINITAB are different for the two systems. If you are using a mainframe computer system, the first step is to *log on* to the system using an account number and a password. Your instructor or an assistant in the computer center can explain how to use your school's main-

---

[3]The MINITAB commands illustrated in this textbook are for the PC version of MINITAB. Some of these commands may not work for other versions of MINITAB, especially the student version of MINITAB.
[4]BMDP is a registered trademark of BMDP Statistical Software, Inc. MINITAB is a registered trademark of Minitab, Inc. SAS is a registered trademark of SAS Institute, Inc. SPSS is a registered trademark of SPSS, Inc.

frame computer system. Also, remember that you must *log off* the system at the end of each session. You must also ask either your instructor or an assistant at the computer center how to obtain a *hard (printed) copy* of the MINITAB *output*. After you *log on* to the mainframe system, the next step is to enter the MINITAB *environment*. Again, your instructor or an assistant at the computer center can show you how to enter the MINITAB *environment*.

If you are working on a microcomputer, you need to know how to *load* MINITAB and how to enter the MINITAB *environment*. Your instructor or the computer lab assistant can explain this to you.

Once you have completed these formalities and entered the MINITAB *environment*, your computer terminal will display the message shown in Figure 1.4.

Figure 1.4

```
MTB >
```

The message 'MTB >' is called the MINITAB *prompt*. At this point, the computer is ready to receive MINITAB *commands*. The MINITAB commands are entered next to the MINITAB prompt, and the *enter/return* key is pressed after completing each command. The MINITAB commands can be entered in uppercase letters, lowercase letters, or in any combination of the two. To end a MINITAB session, type **STOP** next to the MINITAB prompt and hit the *enter/return* key.

PRINTING A FILE

Any information that you enter on the computer terminal is called *input*, and the solution provided by MINITAB in response to your commands is called *output*.

If you are working on a microcomputer (an IBM PC or compatible or Macintosh), the PAPER command shown in Figure 1.5 will send all your subsequent MINITAB commands and MINITAB output to the printer.

Figure 1.5

```
MTB > PAPER
```

If you want to print everything from the very beginning, type PAPER at the very first MINITAB prompt and hit the *enter/return* key. When you want to stop printing, type NOPAPER and hit the *enter/return* key. Note that the printer must be turned on before you type the PAPER command. The current screen can also be printed by pressing the key marked 'Print Screen'. The problem with the PAPER command is that it ties up the printer. This command should be used when there is only one person using the printer.

The OUTFILE command can be used to save MINITAB output to print at a later time. To do so, type OUTFILE followed by the name of the file enclosed within single quotes, as shown in Figure 1.6, at the very first MINITAB prompt.

Figure 1.6

```
MTB > OUTFILE 'FILENAME'
```

This command will save the file as FILENAME.LIS in your main directory. After you are finished with MINITAB, leave the MINITAB environment and print the FILENAME.LIS file. Note that you will substitute the name you give to your file for FILENAME in the above command.

ENTERING DATA INTO MINITAB

Usually you need to enter some data before you do any statistical analysis. Illustration M1−1 demonstrates how you can enter data using MINITAB commands.

Illustration M1−1   Following are the scores of seven students in a statistics test.

$$86 \quad 91 \quad 74 \quad 80 \quad 65 \quad 97 \quad 79$$

To enter data into MINITAB, you can use the SET or READ command. The SET command is used to enter data on one variable only. The READ command can be used to enter data on one or more variables. The data are always entered into a column that is denoted by C followed by a number (e.g., C1, C2, C3, . . .). The notation used to enter a constant is K followed by a number (e.g., K1, K2, K3, . . .). A MINITAB command that begins with **NOTE** is not processed by MINITAB. This command is only for the information of the user. Using the NOTE command, you can enter any comments for your information. The set of commands presented in Figure 1.7 shows how to enter data on scores of seven students into MINITAB.

Figure 1.7   Entering data into MINITAB.

```
MTB > OUTFILE 'SCORES'  ←── { This command will create a file named 'SCORES.LIS' in
                                 the main directory, which will contain the solution.
MTB > NOTE: DATA ON SCORES OF SEVEN STUDENTS
MTB > SET C1 ←── This command instructs MINITAB that you are to enter data in column C1.
DATA > 86 91 74 80 } When using the SET command, you can enter these data values in as
DATA > 65 97 79      many rows as you want.
DATA > END ←── This command indicates the end of data entry.
MTB > STOP ←── This command ends the MINITAB session.
```

In the MINITAB display of Figure 1.7, the "DATA >" prompt indicates that MINITAB is ready for data entry. When the SET command is used to enter data on one variable, as shown in the display in Figure 1.7, all values can be entered in one row or in any number of rows. However, if the READ command is used to enter data on one variable, only one value should be entered in each row. After entering all data values, type END at the next "DATA >" prompt to instruct MINITAB that all data values have been entered.   ▄

You can give a column a name using the **NAME** command. The MINITAB command given in Figure 1.8 will name the data entered in column C1 as 'SCORES'.

Figure 1.8

```
MTB > NAME C1 'SCORES'
```

The **PRINT C1** command will show the data on scores on the computer screen as shown in Figure 1.9.

Figure 1.9

```
MTB > PRINT C1
C1
     86    91    74    80    65    97    79
```

The PRINT 'SCORES' command (after naming C1 as 'SCORES') will also print these data values on the computer screen. Note that when you use the assigned name for a column, such as SCORES in our example, this name must always be enclosed within single quotes.

SAVING A MINITAB FILE

If you plan to use the data you entered into MINITAB again at some later time, you need to save it as a file before you end the current MINITAB session. To save a data file, you must give it a name. Suppose you call the data file of Illustration M1–1 'SCORES'. If you are using a microcomputer, any of the commands given in Figure 1.10 will save this file. Use only one of these commands, depending on whether you want to save your file on the hard disk, disk drive A, or disk drive B, respectively. Note that the file name is enclosed within single quotes.

Figure 1.10   Saving a MINITAB file.

```
MTB > SAVE 'SCORES' ←——   { This command will save the data file as SCORES.MTW on the
                          { hard disk.
MTB > SAVE 'A:SCORES' ←—— { This command will save the data file on the floppy disk in
                          { drive A.
MTB > SAVE 'B:SCORES' ←—— { This command will save the data file on the floppy disk in
                          { drive B.
```

MINITAB will attach the extension ''.MTW'' (which stands for MINITAB Worksheet) at the end of the file name when you save a file, unless you use another extension.

If you are using a mainframe computer, usually the MINITAB command given in Figure 1.11 will save a file. However, consult your instructor to find out the exact command.

Figure 1.11

```
MTB > SAVE 'SCORES'
```

RETRIEVING A MINITAB FILE

To work on the SCORES file at a later date, the RETRIEVE command will bring it back into the current worksheet.

Figure 1.12 Retrieving a MINITAB file.

```
MTB > RETRIEVE 'SCORES' ←—— This command is used if the saved file is on the hard disk.
MTB > RETRIEVE 'A:SCORES' ←—— ⎰ This command is used if the saved file is on a floppy
                                ⎱ disk in drive A.
MTB > RETRIEVE 'B:SCORES' ←—— ⎰ This command is used if the saved file is on a floppy
                                ⎱ disk in drive B.
```

MINITAB recognizes a command from the first four letters. Thus, if you use the command RETR 'SCORES' instead of RETRIEVE 'SCORES', the computer will respond with the same answer. This is true of all MINITAB commands. Remember that whenever you use the file name, either to save it or to retrieve it or to work on it, you must enclose the file name within single quotes.

The INFORMATION command, shown in Figure 1.13, helps you know what is in a retrieved file.

Figure 1.13

```
MTB > INFO
```

SELECTING A SAMPLE USING MINITAB

The MINITAB SAMPLE command can be used to select a sample from a population. Suppose the data on seven scores entered in column C1 in the MINITAB display of Figure 1.7 belong to a population. Using the procedure of Figure 1.14, a sample of three scores can be selected from this population.

Figure 1.14 Selecting a sample.

```
MTB > NOTE: SELECTING A SAMPLE OF 3 SCORES
MTB > SAMPLE 3 FROM C1 PUT IN C2 ←—— ⎧ This command instructs MINITAB to take
                                      ⎪ a sample of three observations (without
                                      ⎨ replacement) from the data of column C1
                                      ⎩ and put those in column C2.
MTB > PRINT C2 ←—— ⎰ This command will print the sample data of column C2 on the computer
                    ⎱ terminal screen.
C2
   91    65    97 ←—— These values give the required sample.
```

Remember that if you already have entered data on a variable in column C1 and you want to enter new data on another variable into MINITAB, the new data must be entered in a new column, say, C2. If you enter the new data in column C1, the data entered in C1 previously will be lost.

ENTERING DATA ON TWO VARIABLES

Illustration M1–2 explains how data on two variables can be entered into MINITAB using the READ command.

Illustration M1-2 Suppose you need to enter the following data on heights and weights of six persons into MINITAB.

Height (inches)	Weight (pounds)
69	178
67	135
65	121
71	210
68	149
66	142

To enter these data using the READ command, enter both the height and weight of each person in one row. The above table contains information on six persons. Therefore, enter these data in six rows, each row containing information on the height and weight of one person, as shown in Figure 1.15.

Figure 1.15 Entering data on two variables.

```
MTB  > NOTE: DATA ON HEIGHTS AND WEIGHTS
MTB  > READ C1 C2  ⟵  ⎰ This command instructs MINITAB that you are to enter data on two
                       ⎱ variables in two columns, C1 and C2.

DATA > 69  178
DATA > 67  135
DATA > 65  121
DATA > 71  210
DATA > 68  149
DATA > 66  142
DATA > END
```

ADDITIONAL MINITAB COMMANDS

Figure 1.16 gives some additional MINITAB commands and their explanations. (Note that these commands will not be used in the sequence they are presented here.)

Figure 1.16 Some additional MINITAB commands.

```
MTB > HELP HELP  ⟵  This command can be used to seek help about MINITAB commands.
MTB > HELP COMMANDS  ⟵  This command also provides help about MINITAB commands.
MTB > HELP OVERVIEW  ⟵  This command can be used for an overview of MINITAB.
MTB > COPY C1 TO C2  ⟵  ⎰ This command will copy all data values from column C1 to
                          ⎱ column C2.
MTB > ERASE C2  ⟵  This command will delete all data values entered in column C2.
MTB > DELETE ROW 2 C1  ⟵  This command will delete the second value entered in column C1.
MTB > DELETE ROW 2 C1-C2  ⟵  ⎰ This command will delete the data values entered in
                              ⎱ the second row of columns C1 and C2.
MTB > INSERT BETWEEN 2 AND 3 C1-C2  ⟵  ⎰ This command will insert a new row
                                         ⎨ between the second and third rows for
                                         ⎩ columns C1 and C2.
```

(Figure 1.16 *continued on next page*)

```
MTB > LET C1(4) = 10  ←——  This command will replace the fourth entry in column C1 with 10.
MTB > SORT C1 PUT IN C3  ←——  { This command will sort the data of column C1 in an
                                 increasing order and put those in column C3.
MTB > ADD C1 C2 PUT IN C4  ←——  { This command will add the corresponding values of
                                   columns C1 and C2 and put the new data in column C4.
MTB > SUBTRACT C2 FROM C1 PUT IN C5  ←——  { This command will subtract each
                                             value of column C2 from the
                                             corresponding value of column C1
                                             and put the new data in column C5.
MTB > MULTIPLY C1 BY C2 PUT IN C6  ←——  { This command will multiply the
                                           corresponding values of columns C1
                                           and C2 and put the new data in column
                                           C6.
MTB > DIVIDE C1 BY C2 PUT IN C7  ←——  { This command will divide each value of
                                         column C1 by the corresponding value of
                                         column C2 and put the new data in column
                                         C7.
MTB > LET C8 = C1*C2  ←——  { This command will multiply the corresponding values of
                             columns C1 and C2 and put the new data in column C8.
MTB > LET C9 = C1**2  ←——  { This command will square each value entered in column C1
                             and put the new data in column C9.
MTB > ADD 5 TO C1 PUT IN C10  ←——  { This command will add 5 to each value of
                                     column C1 and put the new data in column C10.
MTB > SUBTRACT 8 FROM C1 PUT IN C11  ←——  { This command will subtract 8 from
                                            each value of column C1 and put
                                            the new data in column C11.
MTB > MULTIPLY C1 BY 2 PUT IN C12  ←——  { This command will multiply each value
                                          of column C1 by 2 and put the new
                                          data in column C12.
MTB > DIVIDE C1 BY 3 PUT IN C13  ←——  { This command will divide each value of
                                        column C1 by 3 and put the new data in
                                        column C13.
```

The MINITAB HELP command followed by a specific command provides information about that command. For example, the command HELP SET can be used to find information about the SET command and its usage.

COMPUTER ASSIGNMENTS

M1.1 The following table gives the hours worked and the salary for the past week for five workers.

Hours Worked	Salary (dollars)
42	725
33	483
28	355
47	790
40	820

Try the following commands for this assignment and analyze the output for each command.

```
MTB  > OUTFILE 'ASSIGN1'
MTB  > NOTE: DATA ON HOURS AND SALARY
MTB  > READ C1 C2
DATA > 42 725
DATA > 33 483
DATA > 28 355
DATA > 47 790
DATA > 40 820
DATA > END
MTB  > PRINT C1 C2
MTB  > NAME C1 'HOURS' C2 'SALARY'
MTB  > PRINT 'HOURS' 'SALARY'
MTB  > LET K1 = SUM(C1)
MTB  > PRINT K1
MTB  > LET K2 = SUM(C2)
MTB  > PRINT K2
MTB  > PRINT K1-K2
MTB  > LET C3 = C1*C1
MTB  > PRINT C3
MTB  > LET K3 = SUM(C3)
MTB  > PRINT K3
MTB  > LET C4 = C2*C2
MTB  > LET K4 = SUM(C4)
MTB  > PRINT C1-C4
MTB  > PRINT K1-K4
MTB  > SAMPLE 3 FROM C1-C2 PUT IN C5-C6
MTB  > PRINT C5-C6
MTB  > SAVE 'ASSIGN1'
MTB  > DIR
MTB  > RETRIEVE 'ASSIGN1'
MTB  > INFO
MTB  > STOP
C:\> MINITAB> PRINT 'ASSIGN1.LIS'
```

2 ORGANIZING DATA

In addition to thousands of private organizations and individuals, a large number of U.S. government agencies such as the Bureau of the Census, the Bureau of Labor Statistics, the National Agricultural Statistics Service, the National Center for Education Statistics, the National Center for Health Statistics, and the Bureau of Justice Statistics conduct hundreds of surveys every year. The data collected from each of these surveys comprise hundreds of thousands of pages. In their original form, these data sets may be so large that they do not make sense to most of us. Descriptive statistics, however, supplies the techniques that help to condense large data sets by using tables, graphs, and summary measures. We see such tables, graphs, and summary measures in newspapers and magazines every day. At a glance, these tabular and graphical displays present information on every aspect of life. Consequently, descriptive statistics is of immense importance because it provides efficient and effective methods for summarizing and analyzing information.

This chapter explains how to organize and display data using tables and graphs. We will learn how to prepare frequency distribution tables for qualitative and quantitative data; how to construct bar graphs, pie charts, histograms, and polygons for such data; how to prepare stem-and-leaf displays; and how to display time-series data.

2.1   RAW DATA

When data are collected, the information obtained from each member of a population or sample is recorded in the sequence in which it becomes available. This sequence of data recording is random and unranked. Such data, before they are grouped or ranked, are called **raw data**.

RAW DATA

Data recorded in the sequence in which they are collected and before they are processed or ranked are called raw data.

Suppose we collect information on the ages (in years) of 50 students enrolled in the school of business at a university. The data values, in the order they are collected, are recorded in Table 2.1. For instance, the first student's age is 21, the second student's age is 19 (second number in the first row), and so forth. The data given in Table 2.1 are quantitative raw data.

Table 2.1   Ages of 50 Students

21	19	24	25	29	34	26	27	37	33	18	20	19
22	19	19	25	22	25	23	25	19	31	19	23	18
23	19	23	26	22	28	21	20	22	22	21	20	19
21	25	23	18	37	27	23	21	25	21	24		

Suppose we ask the same 50 students about their student status. The responses of the students are recorded in Table 2.2. In this table, F, SO, J, and SE are the abbreviations for freshman, sophomore, junior, and senior, respectively. This is an example of qualitative (or categorical) raw data.

Table 2.2   Status of 50 Students

J	F	SO	SE	J	J	SE	J	J	J	F	F	J
F	F	F	SE	SO	SE	J	J	F	SE	SO	SO	F
J	F	SE	SE	SO	SE	J	SO	SO	J	J	SO	F
SO	SE	SE	F	SE	J	SO	F	J	SO	SO		

The data presented in Tables 2.1 and 2.2 are also called *ungrouped data*. An ungrouped data set contains information on each member of a sample or population individually.

2.2   ORGANIZING AND GRAPHING QUALITATIVE DATA

This section discusses how to organize and display qualitative (or categorical) data. A data set is organized and displayed by constructing tables and by making graphs, respectively.

2.2.1   FREQUENCY DISTRIBUTIONS

A sample of 100 students enrolled in the school of business at a university were asked what they intended to do after graduation. Forty-four said they wanted to work for private companies/businesses; 16 said they wanted to work for the federal government; 23 wanted to work for state or local governments; and 17 intended to start their own businesses. Table 2.3 lists the type of employment and the number of students who intend to engage in each type of employment. In this table, the variable is the *type of employment*, which is a qualitative variable. The categories (representing the type of employment) listed in the first column of the table are mutually exclusive. In other words, each of the 100 students belongs to one and only one of these categories. The number of students who belong to a certain category is called the *frequency* of that category. A **frequency distribution** exhibits how the frequencies are distributed over various categories. Table 2.3 is called a *frequency distribution table* or simply a *frequency table*.

Table 2.3 Type of Employment Students Intend to Engage In

Variable ⟶ **Type of Employment**	**Number of Students** ⟵ Frequency column
Private companies/businesses	44
Category → Federal government	16 ⟵ Frequency
State/local government	23
Own business	17
	Sum = 100

FREQUENCY DISTRIBUTION FOR QUALITATIVE DATA

A frequency distribution for qualitative data lists all categories and the number of elements that belong to each of the categories.

Example 2–1 illustrates how a frequency distribution table is constructed for qualitative data.

Constructing frequency distribution table for qualitative data.

EXAMPLE 2-1   A sample was taken of 25 high school seniors who were planning to go to college. Each of the students was asked which of the following majors he or she intended to choose: business, economics, management information systems (MIS), behavioral sciences (BS), other. The responses of these students are listed below.

Economics	MIS	Economics	Business	Business
Business	Business	Other	Other	Other
BS	BS	MIS	Other	MIS
Other	Business	MIS	Business	Other
Economics	MIS	Other	Other	MIS

Construct a frequency distribution table for these data.

Solution   Note that the *major a student intends to choose* is the variable in this example. This variable is classified into five categories: business, economics, MIS, BS, and other. We record these categories in the first column of Table 2.4. Then we read each student's response from the given data and mark a *tally*, denoted by the symbol |, in the second column of Table 2.4 next to the corresponding category. For example, the first student intends to major in economics. We show this in the frequency distribution table by marking a tally in the second column next to the category *Economics*. Note that the tallies are marked in blocks of fives for counting convenience. Finally, we record the total of tallies for each category in the third column of the table. This column is called the column of frequencies and is usually denoted by f. The sum of the entries in the frequency column gives the sample size or total frequency. In Table 2.4, this total is 25, which is the sample size.

Table 2.4   Frequency Distribution of Majors

Major	Tally	Frequency (f)
Business	⊤⊦⊦ I	6
Economics	III	3
MIS	⊤⊦⊦ I	6
BS	II	2
Other	⊤⊦⊦ III	8
		Sum = 25

2.2.2   RELATIVE FREQUENCY AND PERCENTAGE DISTRIBUTIONS

The **relative frequency** of a category is obtained by dividing the frequency of that category by the sum of all frequencies. Thus, the relative frequency shows what fractional part or proportion of the total frequency belongs to the corresponding category. A *relative frequency distribution* lists the relative frequencies for all categories.

RELATIVE FREQUENCY OF A CATEGORY

$$\text{Relative frequency of a category} = \frac{\text{Frequency of that category}}{\text{Sum of all frequencies}}$$

The **percentage** for a category is obtained by multiplying the relative frequency of that category by 100. A *percentage distribution* lists the percentages for all categories.

PERCENTAGE

$$\text{Percentage} = (\text{Relative frequency}) \cdot 100$$

Constructing relative frequency and percentage distributions.

EXAMPLE 2-2   Construct the relative frequency and percentage distributions for Table 2.4.

Solution   The relative frequencies and percentages for Table 2.4 are calculated and listed in Table 2.5.

Based on Table 2.5, we can state that .24 or 24% of the students in the sample said that they intend to major in business. By adding the percentages for the first two categories, we can state that 36% of the students said they intend to major in business or economics. The other numbers in Table 2.5 can be interpreted the same way.

Notice that the sum of the relative frequencies is always 1.00 (or approximately 1.00 if the relative frequencies are rounded), and the sum of the percentages is always 100 (or approximately 100 if the percentages are rounded).

Table 2.5 Relative Frequency and Percentage Distributions Table

Major	Relative Frequency	Percentage
Business	6/25 = .24	.24 (100) = 24
Economics	3/25 = .12	.12 (100) = 12
MIS	6/25 = .24	.24 (100) = 24
BS	2/25 = .08	.08 (100) = 8
Others	8/25 = .32	.32 (100) = 32
	Sum = 1.00	Sum = 100

2.2.3 GRAPHICAL PRESENTATION OF QUALITATIVE DATA

All of us have heard the saying ''a picture is worth a thousand words.'' A graphic display can reveal at a glance the main characteristics of a data set. The bar graph and the pie chart are two types of graphs used to display qualitative data.

Bar Graphs

To construct a **bar graph** (also called a *bar chart*), we mark the various categories on the horizontal axis as in Figure 2.1. Note that all categories are represented by intervals of equal width. We mark the frequencies on the vertical axis. Then we draw one bar for each category such that the height of the bar represents the frequency of the corresponding category. We leave a small gap between adjacent bars. Figure 2.1 gives the bar graph for the frequency distribution of Table 2.4.

BAR GRAPH

A graph made of bars whose heights represent the frequencies of respective categories is called a bar graph.

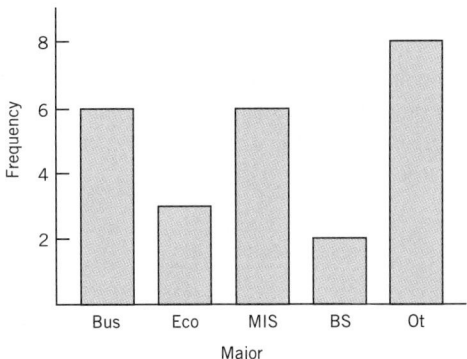

Figure 2.1 Bar graph for the frequency distribution of Table 2.4.

The bar graphs for relative frequency and percentage distributions can be drawn simply by marking the relative frequencies or percentages, instead of the class frequencies, on the vertical axis.

Sometimes, in a bar graph the categories are marked on the vertical axis and the frequencies on the horizontal axis.

CASE STUDY 2-1   RECORD RATE OF FAILURE

In 1991, a total of 87,266 businesses failed in the United States. As shown in the chart, 6,100 of them were wholesale businesses, 6,534 belonged to the manufacturing sector, 11,815 belonged to the construction industry, 17,008 were retail businesses, and 22,663 belonged to the services sector. The remaining 23,146 businesses belonged to other industries.

USA SNAPSHOTS ®

A look at statistics that shape your finances

Record rate of failures
A record 87,266 businesses failed in the USA last year, up 44% from 1990. Industries with the highest number of failures:

6,100   Wholesale
6,534   Manufacturing
11,815   Construction
17,008   Retail
22,663   Services

Source: Dun & Bradstreet   By Ron Coddington, USA TODAY

The following table, which is constructed using the above chart, lists the types of industries and the number of businesses that failed in each of those industries in 1991. The frequency column in this table represents the number of failed businesses that belonged to different industries.

Industry	f
Wholesale	6,100
Manufacturing	6,534
Construction	11,815
Retail	17,008
Services	22,663
Other	23,146
	$\Sigma f = 87,266$

Source: *USA TODAY*, March 10, 1992. Copyright © 1992, *USA TODAY*. Chart reprinted with permission.

Pie Charts

A **pie chart** is more commonly used to display percentages, although it can be used to display frequencies or relative frequencies. The whole pie (or circle) represents the total sample or population. The pie is divided into different portions that represent the percentages of the population or sample belonging to different categories.

> **PIE CHART**
>
> A circle divided into portions that represent the relative frequencies or percentages of a population or a sample belonging to different categories is called a pie chart.

As we know, a circle contains 360 degrees. To construct a pie chart, we multiply 360 by the relative frequency for each category to obtain the degree measure or size of the angle for the corresponding category. Table 2.6 shows the calculation of angle sizes for various categories of Table 2.5.

Table 2.6   Calculating Angle Sizes for the Pie Chart

Major	Relative Frequency	Angle Size
Business	.24	360 (.24) = 86.4
Economics	.12	360 (.12) = 43.2
MIS	.24	360 (.24) = 86.4
BS	.08	360 (.08) = 28.8
Others	.32	360 (.32) = 115.2
	Sum = 1.00	Sum = 360

Figure 2.2 shows the pie chart for the percentage distribution of Table 2.5, which uses the angle sizes calculated in Table 2.6.

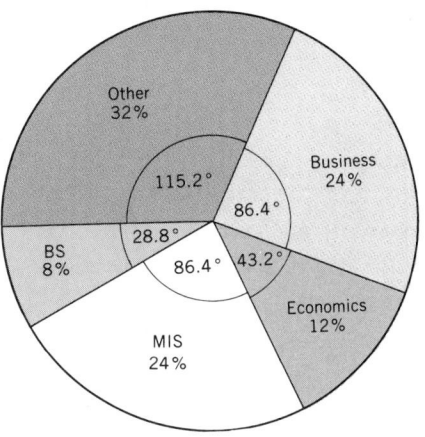

Figure 2.2   Pie chart for the percentage distribution of majors.

CASE STUDY 2-2   1991 MARKET SHARES OF SNEAKERS

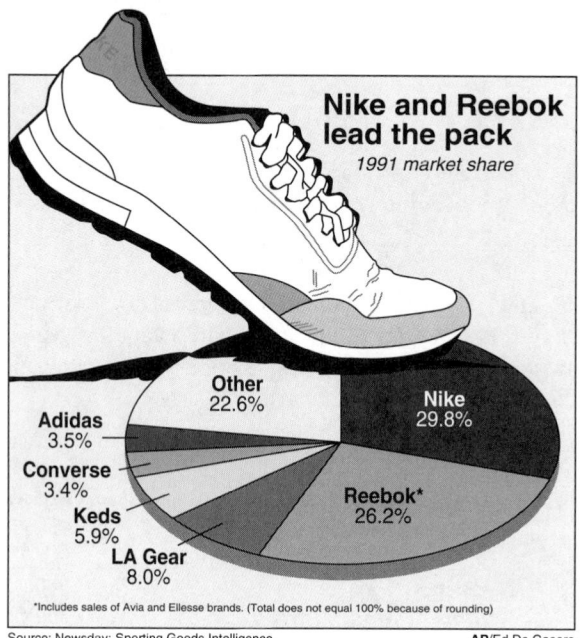

The above pie chart shows the 1991 U.S. sneaker market shares of various companies. Of the total sales of sneakers in the United States in 1991, 29.8% were sold by Nike, 26.2% by Reebok, 8.0% by LA Gear, 5.9% by Keds, 3.4% by Converse, and 3.5% by Adidas; and 22.6% were sold by other companies.

Source: *The Hartford Courant*, July 10, 1992. Copyright © 1992, Wide World Photos Inc. The chart reprinted with permission of Wide World Photos Inc.

EXERCISES

Concepts and Procedures

2.1 Why do we need to group data in the form of a frequency table? Explain briefly.

2.2 How are the relative frequencies and percentages of categories obtained from the frequencies of categories? Illustrate with the help of an example.

2.3 The following data give the results of a sample survey. The letters A, B, and C represent the three categories.

A	B	B	A	C	B	C	C	C	A	C	B	C	C
C	B	C	C	A	A	B	C	C	B	C	B	B	C	A

a. Prepare a frequency distribution table.
b. Calculate the relative frequencies and percentages for all categories.
c. What percentage of the elements in this sample belong to category B?
d. What percentage of the elements in this sample belong to categories A or C?
e. Draw a bar graph for the frequency distribution.

2.4 The following data give the results of a sample survey. The letters Y, N, and D represent the three categories.

D	N	N	Y	Y	Y	N	Y	D	N	Y	Y	Y	Y
N	Y	Y	N	N	Y	N	Y	Y	N	D	N	Y	N
Y	Y	Y	Y	N	N	Y	Y	N	N	D	Y		

a. Prepare a frequency distribution table.
b. Calculate the relative frequencies and percentages for all categories.
c. What percentage of the elements in this sample belong to category Y?
d. What percentage of the elements in this sample belong to categories N or D?
e. Draw a pie chart for the percentage distribution.

Applications

2.5 Data on the status of 50 students given in Table 2.2 of Section 2.1 are reproduced below.

J	F	SO	SE	J	J	SE	J	J	J
F	F	J	F	F	F	SE	SO	SE	J
J	F	SE	SO	SO	F	J	F	SE	SE
SO	SE	J	SO	SO	J	J	SO	F	SO
SE	SE	F	SE	J	SO	F	J	SO	SO

a. Prepare a frequency distribution table.
b. Calculate the relative frequencies and percentages for all categories.
c. What percentage of these students are juniors or seniors?
d. Draw a bar graph for the frequency distribution.

2.6 The following are the responses of 20 students of a business statistics class who were asked to evaluate their instructor. The students were asked to choose one of five answers: Excellent (E), Above average (AA), Average (A), Below average (B), and Poor (P).

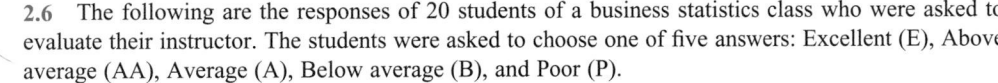

AA	B	A	A	E	AA	P
E	AA	B	E	AA	E	B
E	A	B	P	AA	E	

a. Construct a frequency distribution table.
b. Calculate the relative frequencies and percentages for all categories.
c. What percentage of these students ranked this instructor as excellent or above average?
d. Draw a bar graph for the relative frequency distribution.

2.7 Fifteen workers of a company were asked if they thought the salaries of CEOs of U.S. companies were too high. The responses of the workers are listed below. (H, N, and D indicate that a worker thinks the salaries of CEOs are too high, not too high, or that the worker has no opinion/does not know, respectively.)

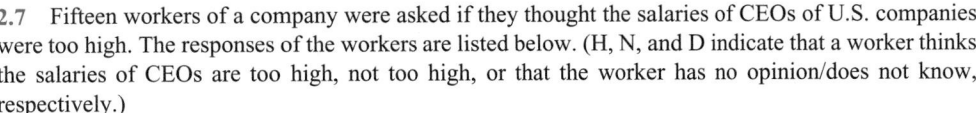

H	H	D	N	H	H	N	H	D	H	N	H	H	N	D

a. Prepare a frequency distribution table.
b. Calculate the relative frequencies and percentages for all categories.
c. What percentage of the workers in this sample think that the salaries of CEOs are too high?
d. Draw a pie chart for the percentage distribution.

2.8 Twelve persons were asked to taste two types of soft drinks, *A* and *B*, and indicate if the taste of *A* was superior (S), the same (M), or inferior (I) to that of *B*. Their responses are listed below.

S   I   I   M   M   S   M   M   S   I   S   S

    a. Construct a frequency distribution table.
    b. Calculate the relative frequencies and percentages for all categories.
    c. Draw a pie chart for the percentage distribution.

2.9 The following table, based on House Ways and Means Committee research, lists the reasons that people go on welfare and the percentage of people who go on welfare for each of these reasons.

Reason	Percentage
Divorce/Separation	45
Child born to unmarried woman	30
Earnings of mother fall	12
Earnings of others in family fall	3
Other reasons	10

Draw a pie chart for this percentage distribution.

2.10 According to Neilsen Marketing Research, the total sales of cosmetic products in the United States were $2120 million in 1991. The following table lists the total 1991 sales of major companies making cosmetic products.

Company	Total Sales (millions of dollars)
Procter & Gamble	721.9
Revlon	477.7
Maybelline	350.2
L'Oreal	184.7
Other	385.5
	Total = 2120

Calculate the 1991 market shares of all companies listed in the table and draw a pie chart for the market shares. (*Hint:* To calculate the market shares, divide the total sales of each company by the total sales of all companies.)

2.3 ORGANIZING AND GRAPHING QUANTITATIVE DATA

In the previous section we learned how to group and display qualitative data. This section explains how to group and display quantitative data.

2.3.1 FREQUENCY DISTRIBUTIONS

Table 2.7 gives the weekly earnings of 100 employees of a large company. The first column of this table lists the *classes*, which represent the (quantitative) variable *the weekly earnings*. For quantitative data, an interval that includes all the values that fall within two numbers, the lower and upper limits, is called a **class**. Note that the classes always represent a variable. As we can observe, the classes are nonoverlapping; that is, each value on earnings belongs to one and only one class. The second column in the table lists the number of employees who have earnings within each class. For example, nine employees of this company earn $301 to

$400 per week. The numbers listed in the second column of this table are called the **frequencies**, which give the number of values that belong to different classes. The frequencies are denoted by f.

Table 2.7 Weekly Earnings of 100 Employees of a Company

Weekly Earnings (dollars)	Number of Employees f
301 to 400	9
401 to 500	16
501 to 600	33
601 to 700	20
701 to 800	14
801 to 900	8

Variable → Weekly Earnings (dollars)   Frequency column ← Number of Employees f

Third class → 501 to 600   33 ← Frequency of the third class

801 to 900 — Lower limit of the sixth class, Upper limit of the sixth class

For quantitative data, the frequency of a class represents the number of values in the data set that fall in that class. Table 2.7 contains six classes. Each class has a *lower limit* and an *upper limit*. The values 301, 401, 501, 601, 701, and 801 give the lower limits and the values 400, 500, 600, 700, 800, and 900 give the upper limits of the six classes, respectively. The data presented in Table 2.7 are an illustration of a **frequency distribution table** for quantitative data. Data presented in the form of a frequency distribution table are called **grouped data**.

FREQUENCY DISTRIBUTION FOR QUANTITATIVE DATA

A frequency distribution for quantitative data lists all the classes and the number of values that belong to each class. Data presented in the form of a frequency distribution are called *grouped data*.

To find the midpoint of the upper limit of the first class and the lower limit of the second class in Table 2.7, we divide the sum of these two limits by 2. Thus, this midpoint is

$$\frac{400 + 401}{2} = 400.5$$

The value 400.5 is called the *upper boundary* of the first class and the *lower boundary* of the second class. By using this technique, we can convert the class limits of Table 2.7 to **class boundaries**, which are also called *real class limits*. The second column of Table 2.8 lists the boundaries for Table 2.7.

Table 2.8 Class Boundaries, Class Widths, and Class Midpoints for Table 2.7

Class Limits	Class Boundaries	Class Width	Class Midpoint
301 to 400	300.5 to less than 400.5	100	350.5
401 to 500	400.5 to less than 500.5	100	450.5
501 to 600	500.5 to less than 600.5	100	550.5
601 to 700	600.5 to less than 700.5	100	650.5
701 to 800	700.5 to less than 800.5	100	750.5
801 to 900	800.5 to less than 900.5	100	850.5

> **CLASS BOUNDARY**
>
> The class boundary is given by the midpoint of the upper limit of one class and the lower limit of the next class.

The difference between the two boundaries of a class gives the **class width**. The class width is also called the **class size**.

> **CLASS WIDTH**
>
> $$\text{Class width} = \text{Upper boundary} - \text{Lower boundary}$$

Thus, in Table 2.8,

$$\text{Width of the first class} = 400.5 - 300.5 = 100$$

The class widths for the frequency distribution of Table 2.7 are listed in the third column of Table 2.8. Each class in Table 2.8 (and Table 2.7) has the same width of 100.

The **class midpoint** or **mark** is obtained by dividing the sum of the two limits (or the two boundaries) of a class by 2.

> **CLASS MIDPOINT OR MARK**
>
> $$\text{Class midpoint or mark} = \frac{\text{Lower limit} + \text{Upper limit}}{2}$$

Thus, the midpoint of the first class in Table 2.7 or Table 2.8 is calculated as follows.

$$\text{Midpoint of the first class} = \frac{301 + 400}{2} = 350.5$$

The class midpoints for the frequency distribution of Table 2.7 are listed in the fourth column of Table 2.8.

Note that in Table 2.8, when we write classes using class boundaries, we write *to less than* in order to ensure that each value belongs to one and only one class. As we can see, the upper boundary of the preceding class and the lower boundary of the succeeding class are the same.

2.3.2 CONSTRUCTING FREQUENCY DISTRIBUTION TABLES

While constructing a frequency distribution table, the following three major decisions need to be made.

Number of Classes

Usually the number of classes for a frequency distribution table varies from 5 to 20, depending mainly on the number of observations in the data set.[1] It is preferable to have more classes

---

[1] One rule to help decide about the number of classes is the Sturge's formula, which is
$$c = 1 + 3.3 \log n$$
where c is the number of classes and n is the number of observations in the data set. The value of $\log n$ can be obtained by entering the value of n on the calculator and pressing the *log* key.

as the size of a data set increases. The decision about the number of classes is arbitrarily made by the data organizer.

Class Width

Although it is not uncommon to have classes of different sizes, most of the time it is preferable to have the same width for all classes. To determine the class width when all classes are of the same size, first find the difference between the largest and the smallest values in the data. Then, the approximate width of a class is obtained by dividing this difference by the number of desired classes.

CALCULATION OF CLASS WIDTH

$$\text{Approximate class width} = \frac{\text{Largest value} - \text{Smallest value}}{\text{Number of classes}}$$

Usually this approximate class width is rounded to a convenient number, which is then used as the class width. Note that rounding this number may slightly change the number of classes initially intended.

Lower Limit of the First Class or the Starting Point

Any convenient number, which is equal to or less than the smallest value in the data set, can be used as the lower limit of the first class.

Example 2–3 illustrates the procedure for constructing a frequency distribution table for quantitative data.

Constructing frequency distribution table for quantitative data.

EXAMPLE 2–3   The price–earnings ratio for a company is an indication of whether the stock of that company is undervalued or overvalued. If the price–earnings ratio is low, then the price of the stock is low compared to the earnings of that company and, consequently, the stock of that company is undervalued. The converse is true if the price–earnings ratio is high for a company. The following data, which appeared in the 1993 Special Issue of *Business Week*, give the price–earnings ratios for 25 companies.[2] These price–earnings ratios are based on 1992 earnings and March 5, 1993 stock prices.

31	13	12	22	27	33	17	26	
16	22	18	13	16	23	20	18	
22	15	26	12	20	21	23	27	30

Construct a frequency distribution table.

Solution   In these data, the minimum value is 12 and the maximum value is 33. Suppose we decide to group these data using five classes of equal width. Then,

$$\text{Approximate width of a class} = \frac{33 - 12}{5} = 4.2$$

Suppose we round this approximate width to a convenient number, say, 5. The lower limit of the first class can be taken as 12 or any number less than 12. Suppose we take 10 as the

---

[2]The data for the 25 companies are entered (by row) in the following order: American Express, Apple Computer, BankAmerica, Campbell Soups, Chrysler, Du Pont, Exxon, Gillette, General Electric, Hewlett-Packard, H.J. Heinz, J.C. Penney, Johnson & Johnson, Kellogg, McDonald's, Merck, Mobil, Nike, Pepsico, Philip Morris, Procter & Gamble, Quaker Oats, The Gap, Toys 'R' Us, and Walt Disney.

lower limit of the first class. Then, our classes will be

$$10-14, \qquad 15-19, \qquad 20-24, \qquad 25-29, \qquad \text{and} \qquad 30-34$$

We record these five classes in the first column of Table 2.9.

Table 2.9 Frequency Distribution of Price–Earnings Ratios

Price–Earnings Ratio	Tally	f
10–14	||||	4
15–19	⊞|	6
20–24	⊞ |||	8
25–29	||||	4
30–34	|||	3
		$\Sigma f = 25$

Now we read each value from the given data and make a tally mark in the second column of Table 2.9 next to the corresponding class. The first value in our original data is 31, which belongs to the 30–34 class. To record it, we make a tally mark in the second column of Table 2.9 next to the 30–34 class. We continue this process until all the data values have been read and entered in the tally column. Note that tallies are marked in blocks of fives for counting convenience. After the tally column is completed, we count the tally marks for each class and write those numbers in the third column. This is the *column of frequencies* and is denoted by f. These frequencies represent the number of companies that belong to each of the five different classes. For example, 4 of the 25 companies have a price–earnings ratio of 10 to 14, and so forth.

In Table 2.9, we can denote the frequencies of the five classes by $f_1, f_2, f_3, f_4,$ and f_5, respectively. Therefore,

$$f_1 = \text{frequency of the first class} = 4$$

Similarly,

$$f_2 = 6, \qquad f_3 = 8, \qquad f_4 = 4, \qquad \text{and} \qquad f_5 = 3$$

Using the Σ notation (see Section 1.7 of Chapter 1), we can denote the sum of the frequencies of all classes by Σf. Hence,

$$\Sigma f = f_1 + f_2 + f_3 + f_4 + f_5 = 4 + 6 + 8 + 4 + 3 = 25$$

The number of observations in a sample is usually denoted by n. Thus, for the sample data, Σf is equal to n. The number of observations in a population is denoted by N. Consequently, Σf is equal to N for population data. Because the data set on price–earnings ratios in Table 2.9 is for only 25 companies, it represents a sample. Therefore, in Table 2.9 we can denote the sum of frequencies by n instead of Σf. ▬

Note that when we present the data in the form of a frequency distribution table, as in Table 2.9, we lose the information on individual observations. We cannot know the exact price–earnings ratio of any particular company from Table 2.9. All we know is that for four companies the price–earnings ratios are in the interval 10 to 14, and so forth.

2.3.3 RELATIVE FREQUENCY AND PERCENTAGE DISTRIBUTIONS

Using Table 2.9, we can compute the *relative frequency* and *percentage* columns the same way we did for qualitative data in Section 2.2.2. The relative frequencies and percentages for a quantitative data set are obtained as follows.

RELATIVE FREQUENCY AND PERCENTAGE

$$\text{Relative frequency of a class} = \frac{\text{Frequency of that class}}{\text{Sum of all frequencies}} = \frac{f}{\Sigma f}$$

$$\text{Percentage} = (\text{Relative frequency}) \cdot 100$$

Example 2–4 illustrates how to construct relative frequency and percentage distributions.

Constructing relative frequency and percentage distributions.

EXAMPLE 2–4 Calculate the relative frequencies and percentages for Table 2.9.

Solution The relative frequencies and percentages for Table 2.9 are calculated and listed in the third and fourth columns, respectively, of Table 2.10. Note that the class boundaries are listed in the second column of Table 2.10.

Table 2.10 Relative Frequency and Percentage Distributions for Table 2.9

Price–Earnings Ratio	Class Boundaries	Relative Frequency	Percentage
10–14	9.5 to less than 14.5	4/25 = .16	16
15–19	14.5 to less than 19.5	6/25 = .24	24
20–24	19.5 to less than 24.5	8/25 = .32	32
25–29	24.5 to less than 29.5	4/25 = .16	16
30–34	29.5 to less than 34.5	3/25 = .12	12
		Sum = 1.00	Sum = 100%

From Table 2.10, we can make statements about the percentage of companies with price–earnings ratios within a certain interval. For example, 16% of the companies had price–earnings ratios of 10 to 14. By adding the percentages for the first two classes, we can state that 40% of the companies had price–earnings ratios of 10 to 19. Similarly, by adding the percentages of the last two classes, we can state that 28% of the companies had price–earnings ratios of 25 to 34. ▪

2.3.4 GRAPHING GROUPED DATA

Grouped (quantitative) data can be displayed by using a *histogram* or a *polygon*. This section describes how to construct such graphs. We can also draw a pie chart to display the percentage distribution for a quantitative data set. The procedure to construct a pie chart is similar to the one for qualitative data explained in Section 2.2.3; it will not be repeated in this section.

Histograms

A **histogram** is a certain kind of graph that can be drawn for a frequency distribution, a relative frequency distribution, or a percentage distribution. To draw a histogram, we first mark classes on the horizontal axis and frequencies (or relative frequencies or percentages) on the vertical axis. Notice that either class limits or class boundaries can be used to mark classes on the horizontal axis. Next, we draw a bar for each class so that its height represents the frequency of that class. The bars in a histogram are drawn adjacent to each other without leaving any gap between them. A histogram is called a **frequency histogram**, a **relative frequency histogram**, or a **percentage histogram** depending on whether the frequencies, relative frequencies, or percentages are marked on the vertical axis.

HISTOGRAM

A histogram is a graph in which classes are marked on the horizontal axis and either the frequencies, relative frequencies, or percentages are marked on the vertical axis. The frequencies, relative frequencies, or percentages are represented by the heights of the bars. In a histogram, the bars are drawn adjacent to each other.

Figures 2.3 and 2.4 show the frequency and the relative frequency histograms, respectively, for the data of Tables 2.9 and 2.10 of Sections 2.3.2 and 2.3.3. The two histograms look alike because they represent the same data. A percentage histogram can be drawn for the percentage distribution of Table 2.10 by marking the percentages on the vertical axis.

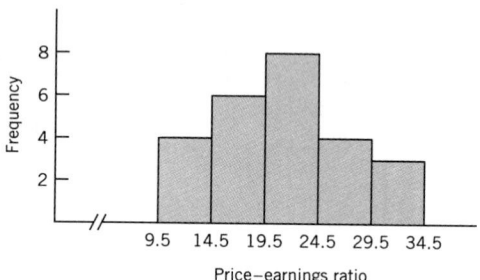

Figure 2.3   Frequency histogram for Table 2.9.

The symbol "–//–" used in the horizontal axis of both Figures 2.3 and 2.4 represents a break, called the **truncation**, in the horizontal axis. It indicates that the entire horizontal axis is not shown in these figures. As can be noticed, the zero to 9.5 portion of the horizontal axis has been omitted in each figure.

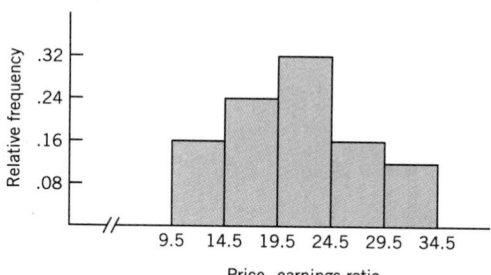

Figure 2.4   Relative frequency histogram for Table 2.10.

In Figures 2.3 and 2.4, we have used class boundaries to mark classes on the horizontal axis. However, we can show the intervals on the horizontal axis by using the class limits instead of the class boundaries.

CASE STUDY 2-3   FEW GET BIG BUCKS

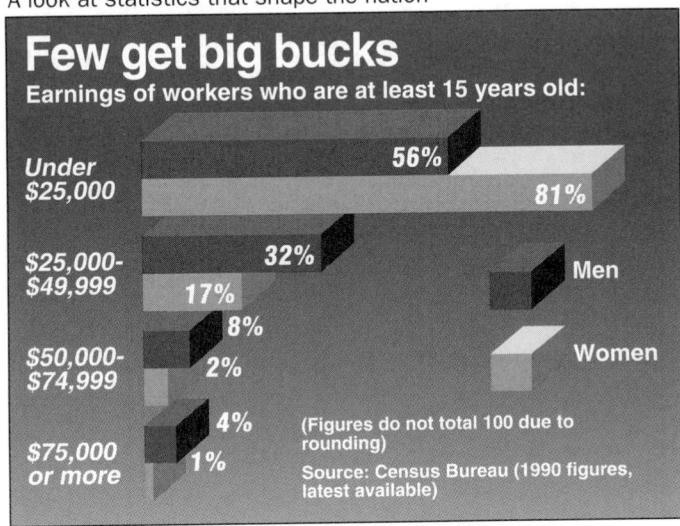

By Marty Baumann, USA TODAY

The above chart, reproduced from *USA TODAY*, shows the 1990 earnings of men and women workers who are 15 years of age and over. The following table is prepared from this chart.

1990 Earnings (dollars)	Percentage of Men 15 Years Old or Over	Percentage of Women 15 Years Old or Over
Under $25,000	56	81
$25,000 to $49,999	32	17
$50,000 to $74,999	8	2
$75,000 or more	4	1

Thus, 56% of men and 81% of women, aged 15 or over, earned less than $25,000 in 1990. On the other hand, only 4% of men and 1% of women, aged 15 or over, earned $75,000 or more in 1990.

Notice that there is no lower limit for the first class and no upper limit for the fourth class. Such classes are called *open-ended classes.*

Source: *USA TODAY*, May 26, 1992. Copyright © 1992, *USA TODAY*. Chart reprinted with permission.

Polygons

A polygon is another device that can be used to present quantitative data in graphic form. To draw a frequency polygon, we first mark a dot above the midpoint of each class at a height equal to the frequency of that class. This is the same as marking the midpoint at the top of each bar in a histogram. Next, we mark two more classes, one at each end, and mark their midpoints. Note that these two classes have zero frequencies. In the last step, we join the adjacent dots with straight lines. The resulting line graph is called a **frequency polygon** or simply a **polygon**.

A polygon with relative frequencies marked on the vertical axis is called a *relative frequency polygon*. Similarly, a polygon with percentages marked on the vertical axis is called a *percentage polygon*.

> **POLYGON**
>
> A graph formed by joining the midpoints of the tops of successive bars in a histogram with straight lines is called a polygon.

Figure 2.5 shows the frequency polygon for the frequency distribution of Table 2.9.

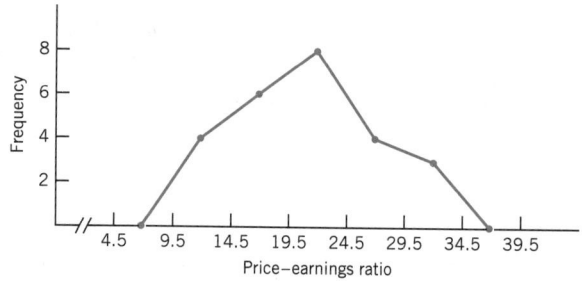

Figure 2.5   Frequency polygon for Table 2.9.

For a very large data set, as the number of classes is increased (and the width of classes is decreased), the frequency polygon eventually becomes a smooth curve. Such a curve is called a *frequency distribution curve* or simply a *frequency curve*. Figure 2.6 shows the frequency curve for a large data set with a large number of classes.

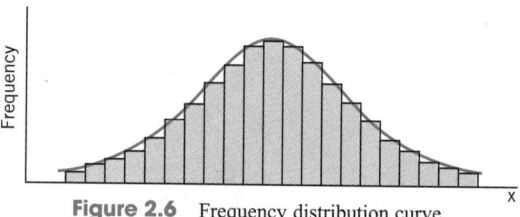

Figure 2.6   Frequency distribution curve.

2.3.5   MORE ON CLASSES AND FREQUENCY DISTRIBUTIONS

This section presents two alternative methods for writing classes to construct a frequency distribution for quantitative data.

Less Than Method for Writing Classes

The classes in frequency distribution given in Table 2.9 for the data on price–earnings ratios for 25 companies were written as 10–14, 15–19, etc. Alternatively, we can write the classes in a frequency distribution table using the *less than* method. The technique for writing classes shown in Table 2.9 is more commonly used for data sets that do not contain fractional values. The *less than* method is more appropriate when a data set contains fractional values. Example 2–5 illustrates the *less than* method.

Constructing frequency distribution using less than *method.*

EXAMPLE 2–5 The following data give the hourly wage rates for a sample of 30 workers selected from a large company.

12.50	9.45	13.85	7.25	8.70	14.60	6.25
5.70	15.80	12.30	10.40	13.50	9.50	7.40
6.90	14.50	11.40	10.80	9.25	13.25	16.25
8.45	10.30	11.50	9.75	12.45	7.50	10.80
11.75	9.90					

Construct a frequency distribution table. Calculate the relative frequencies and percentages for all classes.

Solution The minimum value in this data set is 5.70 and the maximum value is 16.25. Suppose we decide to group these data using six classes of equal width. Then,

$$\text{Approximate width of a class} = \frac{16.25 - 5.70}{6} = 1.758$$

We round this number to a more convenient number, say, 2. Then, we take 2 as the width of each class. If we start the first class at 5, the classes will be written as *5 to less than 7, 7 to less than 9*, and so on. The six classes, which cover all the data, are recorded in the first column of Table 2.11. The second column of that table shows the tallies. The third column lists the frequencies of classes. A value in the data set that is 5 or larger but less than 7 belongs to the first class, and a value that is 7 or larger but less than 9 falls in the second class, and so on. The relative frequencies and percentages for classes are recorded in the fourth and fifth columns, respectively, of Table 2.11. Note that due to rounding, the sum of the relative frequencies is 1.001 and the sum of percentages is 100.1.

Table 2.11 Hourly Wage Rates of 30 Workers

Hourly Wage Rate (dollars)	Tallies	f	Relative Frequency	Percentage
5 to less than 7	|||	3	.100	10.0
7 to less than 9	卌	5	.167	16.7
9 to less than 11	卌 ||||	9	.300	30.0
11 to less than 13	卌		6	.200	20.0
13 to less than 15	卌	5	.167	16.7
15 to less than 17	||	2	.067	6.7
		$\Sigma f = 30$	Sum = 1.001	Sum = 100.1

A histogram and a polygon for the data of Table 2.11 can be drawn in the same way as for the data of Tables 2.9 and 2.10.

Single-Valued Classes

If the observations in a data set assume only a few distinct values, it may be appropriate to prepare a frequency distribution table using *single-valued classes*, that is, classes that are made of single values and not of intervals. This technique is especially useful in cases of discrete data with only a few possible values. Example 2–6 exhibits such a situation.

Constructing frequency distribution using single-valued classes.

EXAMPLE 2–6   The U.S. Bureau of Labor Statistics conducts a survey called the Interview Survey every quarter of the year.[3] The following data give the number of vehicles owned by 40 randomly selected households from the 1990 Interview Survey public-use tape, which contains information on 20,517 households.

5	1	1	2	0	1	1	2
1	1	1	3	3	0	2	5
1	2	3	4	2	1	2	2
1	2	2	1	1	1	4	2
1	1	2	1	1	4	1	3

Construct a frequency distribution table for these data using single-valued classes.

Solution   The observations in this data set assume only six distinct values: 0, 1, 2, 3, 4, and 5. Each of these six values is used as a class in the frequency distribution Table 2.12, and these six classes are listed in the first column of that table. To obtain the frequencies of these classes, the observations in the data that belong to each class are counted and the results are recorded in the second column of Table 2.12. Thus, in these data, 2 households own no vehicle, 18 own one vehicle each, 11 own two vehicles each, and so on.

Table 2.12   Frequency Distribution of Vehicles Owned

Vehicles Owned	Number of Households (f)
0	2
1	18
2	11
3	4
4	3
5	2
	$\Sigma f = 40$

The data of Table 2.12 can also be displayed by drawing a bar graph, as shown in Figure 2.7. To construct a bar graph, we mark the classes, as intervals, on the horizontal axis with a little gap between consecutive intervals. The bars represent the frequencies of respective classes.

The frequencies of Table 2.12 can be converted to relative frequencies and percentages the same way as in Table 2.10. Then, a bar graph can be constructed to display the relative

[3]In this and subsequent chapters, the data that refer to Interview and Diary Surveys are taken from the public-use tapes containing data from the following sources: *Interview Survey:* Bureau of Labor Statistics 1990 Interview Survey based on 20,517 consumer units containing information on families, households, and individuals; *Diary Survey:* Bureau of Labor Statistics 1990 Diary Survey based on 11,735 consumer units containing information on families, households, and individuals. The latest year for which these tapes are available at this time is 1990.

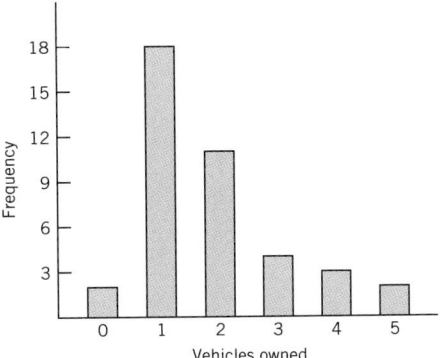

Figure 2.7 Bar graph for vehicles owned.

frequency or percentage distribution by marking the relative frequencies or percentages, respectively, on the vertical axis.

2.4 SHAPES OF HISTOGRAMS

A histogram can assume any one of a large number of shapes. The most common of these shapes are

1. Symmetric
2. Skewed
3. Uniform or rectangular

A **symmetric histogram** is identical on both sides of its central point. The histograms shown in Figure 2.8 are symmetric around the dashed lines that represent their central points.

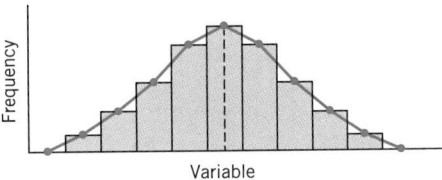 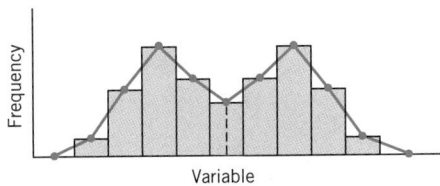

Figure 2.8 Symmetric histograms.

A **skewed histogram** is nonsymmetric. For a skewed histogram, the tail on one side is longer than the tail on the other side. A **skewed to the right histogram** has a longer tail on the right side (see Figure 2.9a). A **skewed to the left histogram** has a longer tail on the left side (see Figure 2.9b).

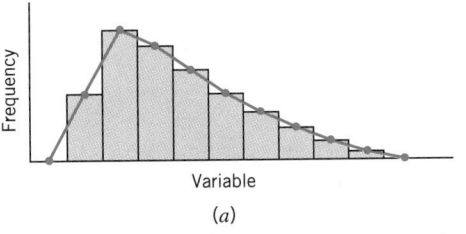 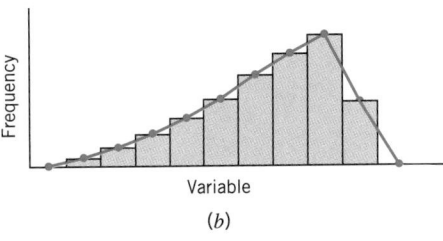

(a)   (b)

Figure 2.9 (a) A histogram skewed to the right. (b) A histogram skewed to the left.

A **uniform** or **rectangular histogram** has the same frequency for each class. Figure 2.10 is an illustration of such a case.

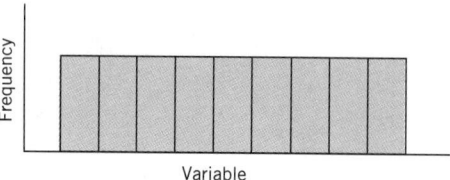

Figure 2.10 A histogram with uniform distribution.

Figures 2.11*a* and 2.11*b* display symmetric frequency curves. Figures 2.11*c* and 2.11*d* show frequency curves skewed to the right and to the left, respectively.

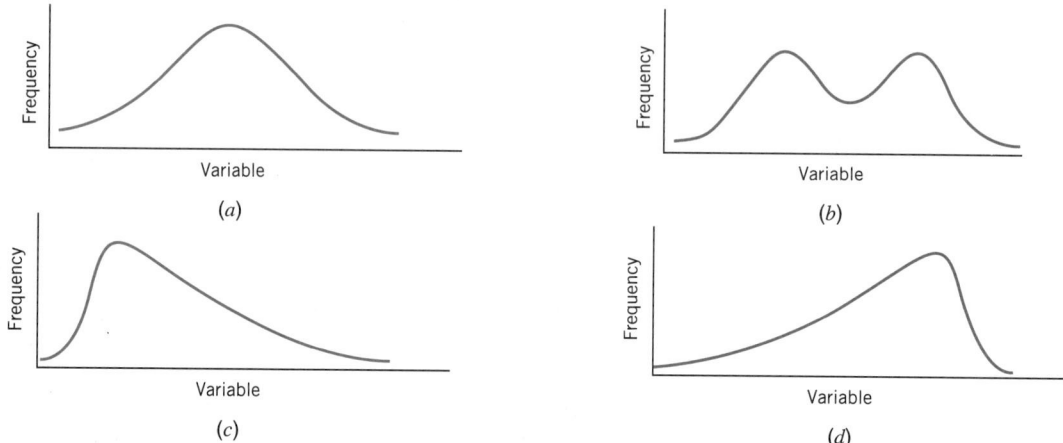

Figure 2.11 (*a*) and (*b*) Symmetric frequency curves. (*c*) Frequency curve skewed to the right. (*d*) Frequency curve skewed to the left.

☞ **WARNING**

Describing data using graphs helps to give us insights into the main characteristics of the data. But graphs, unfortunately, can also be used, intentionally or unintentionally, to distort the facts and to deceive the reader. Following are two ways to manipulate graphs to convey a particular opinion or impression.

1. *Changing the scale* either on one or on both axes, that is, shortening or stretching one or both of the axes

2. *Truncating the frequency axis*, that is, starting the frequency axis at a number greater than zero

When interpreting a graph, we should be very cautious. We should observe carefully whether the frequency axis has been truncated or whether any axis has been unnecessarily shortened or stretched.

The two charts given in Case Study 2–4 show two views of a one-day 40-point drop in the Dow Jones industrial average. Due to the different scales used on the vertical axis, the two charts give different impressions. If we draw a third chart that shows the complete vertical axis, the 40-point drop will be almost unnoticeable.

CASE STUDY 2-4   TWO VIEWS OF A ONE-DAY, 40-POINT DROP IN THE DOW

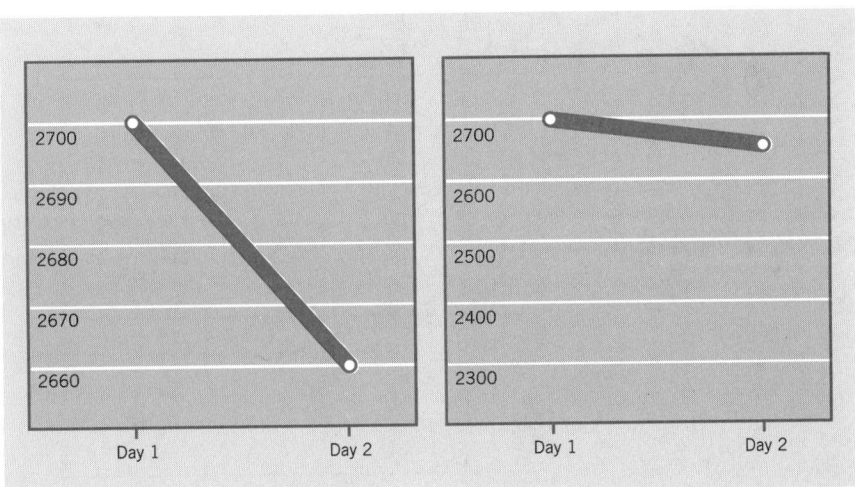

The charts . . . track a hypothetical one-day decline in the Dow Jones industrial average. The one on the left paints a picture of a disastrous day on Wall Street. The one on the right shows a ho-hum trading session.

In fact, both illustrate an identical 40-point drop in the Dow. The difference is the scale that's used. The plummeting line is on a graph with a narrow range of 40 points. The same 1.5% decline from 2700 to 2660 is barely perceptible when plotted on a 400-point scale. Always check the scale before being soothed or alarmed by what you think you see.

Source: Excerpted with permission from the August 1990 issue of *Changing Times Magazine.* Copyright © 1990 The Kiplinger Washington Editors, Inc.

EXERCISES

Concepts and Procedures

2.11   Briefly explain the three decisions that have to be made to group a data set in the form of a frequency distribution table.

2.12   How are the relative frequencies and percentages of classes obtained from the frequencies of classes? Illustrate with the help of an example.

2.13   Three methods—writing classes using limits, using the *less than* method, and grouping data using single-valued classes—were discussed to group quantitative data into classes. Explain these three methods and give one example of each.

Applications

2.14   The following table gives the frequency distribution of weekly earnings for a sample of 100 employees selected from a large company.

Weekly Earnings (dollars)	Number of Employees
200 to 349	12
350 to 499	23
500 to 649	31
650 to 799	19
800 to 949	15

a. Find the class boundaries and class midpoints.
b. Do all classes have the same width? If yes, what is that width?
c. Prepare the relative frequency and percentage distribution columns.
d. What percentage of these employees earn $650 or more per week?

2.15 The following table gives the frequency distribution of ages for all 50 employees of a company.

Age	Number of Employees
18 to 30	12
31 to 43	17
44 to 56	14
57 to 69	7

a. Find the class boundaries and class midpoints.
b. Do all classes have the same width? If yes, what is that width?
c. Prepare the relative frequency and percentage distribution columns.
d. What percentage of the employees of this company are 43 years old or younger?

2.16 A data set on the political contributions (rounded to the nearest dollar) made during the past one year by 200 households has a lowest value of $1 and a highest value of $683. Suppose we want to group these data into seven classes of equal widths.

a. Assuming we take the lower limit of the first class as $1 and the width of each class equal to $100, write the class limits for all seven classes.
b. Write the class boundaries and class midpoints.

2.17 A data set on weekly expenditures (rounded to the nearest dollar) on bakery products for a sample of 500 households has a minimum value of $1 and a maximum value of $18. Suppose we want to group these data into five classes of equal widths.

a. Assuming we take the lower limit of the first class as $1 and the upper limit of the fifth class as $20, write the class limits for all five classes.
b. Determine the class boundaries and class widths.
c. Find the class midpoints.

2.18 The following data give the 1992 gross sales (in billions of dollars) for a sample of 20 U.S. companies.

Company	1992 Sales (billions of dollars)	Company	1992 Sales (billions of dollars)
Philip Morris	50	GTE	20
General Electric	62	Abbott Laboratories	8
Wal-Mart Stores	55	Pepsico	22
Merck	10	Pfizer	7
Coca-Cola	13	American Home Products	8

(*Continued on next page*)

Company	1992 Sales (billions of dollars)	Company	1992 Sales (billions of dollars)
AT&T	65	Mobil	57
Bristol-Myers Squibb	11	Amoco	28
Procter & Gamble	30	Hewlett-Packard	17
Johnson & Johnson	14	Bell Atlantic	13
Du Pont	38	Boeing	30

a. Construct a frequency distribution table with the classes 1–17, 18–34, 35–51, and 52–68.
b. Prepare the relative frequency and percentage columns for the table of part a.
c. Based on the frequency distribution, can you say whether the data are symmetric or skewed?
d. What percentage of companies had 1992 sales of $35 billion or higher?

2.19 The following data give the amount in checking accounts at the time of the survey for 30 households randomly selected from the 1990 Interview Survey.

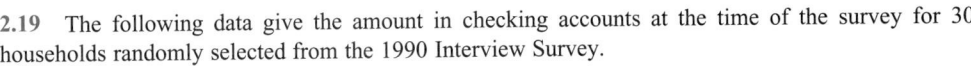

500	100	650	1917	2200	500	180	3000
1500	1300	319	1500	1102	405	124	1000
134	2000	150	800	200	750	300	2300
40	1200	500	900	20	160		

a. Construct a frequency distribution table. Take the classes as 1–750, 751–1500, 1501–2250, 2251–3000.
b. Calculate the relative frequencies and percentages for all classes.
c. Based on the frequency distribution, can you say whether the data are symmetric or skewed?
d. What percentage of the households in this sample had more than $1500 in their checking accounts?

2.20 Nixon Corporation manufactures computer terminals. The following data give the number of computer terminals produced at the company for a sample of 30 days.

24	32	27	23	33	33	29	25	23	28
21	26	31	22	27	33	27	23	28	29
31	35	34	22	26	28	23	35	31	27

a. Construct a frequency distribution table using the classes 21–23, 24–26, 27–29, 30–32, and 33–35.
b. Calculate the relative frequencies and percentages for all classes.
c. Construct a histogram and a polygon for the percentage distribution.
d. For what percentage of the days is the number of computer terminals produced in the interval 27 to 29?

2.21 The following data give the number of computer keyboards assembled at the Twentieth Century Electronics Company for a sample of 25 days.

45	52	48	41	56	46	44	42	48	53	51	53	51
48	46	43	52	50	54	47	44	47	50	49	52	

a. Construct a frequency distribution table using the classes 41–44, 45–48, 49–52, and 53–56.
b. Calculate the relative frequencies and percentages for all classes.
c. Construct a histogram and a polygon for the percentage distribution.
d. For what percentage of the days is the number of computer keyboards assembled in the interval 53 to 56?

Exercises 2.22 through 2.26 are based on the following data.

The following table, based on the American Chamber of Commerce Researchers Association Survey for the first quarter of 1993, gives the prices of five items in 25 urban areas across the United States. (See Data Set II of Appendix A.)

City	Apartment Rent	Price of House	Phone Bill	Cost of a Hospital Room	Price of Wine
Huntsville (AL)	441	105,111	23.69	276.67	5.95
Anchorage (AK)	720	155,913	15.91	532.50	6.13
Phoenix (AZ)	531	100,748	17.79	358.14	4.40
Little Rock (AR)	412	92,750	23.63	221.80	4.95
San Diego (CA)	900	223,600	12.65	527.88	5.41
Denver (CO)	604	125,200	19.72	433.00	4.39
Orlando (FL)	556	104,112	20.41	333.56	5.13
Bloomington (IN)	518	105,613	21.14	433.00	4.67
Ames (IA)	490	110,325	18.03	295.00	5.36
New Orleans (LA)	508	88,000	22.90	318.80	4.52
Minneapolis (MN)	503	128,112	20.74	476.60	5.18
Lincoln (NE)	432	92,925	16.46	275.00	5.18
Manchester (NH)	667	125,750	21.04	379.50	4.24
Albuquerque (NM)	547	130,550	20.86	327.00	4.93
Reno-Sparks (NV)	606	142,300	13.50	445.72	3.97
Albany (NY)	622	116,363	36.47	287.40	5.91
Charlotte (NC)	412	115,800	18.12	293.67	4.69
Cincinnati (OH)	570	121,329	20.30	351.25	5.49
Salem (OR)	490	108,656	17.91	325.00	4.51
Sioux Falls (SD)	541	99,890	23.11	310.00	5.07
Memphis (TN)	425	91,297	20.04	246.20	5.91
Houston (TX)	579	96,900	18.35	326.36	5.64
Salt Lake City (UT)	505	92,281	17.95	375.00	4.95
Charleston (WV)	551	126,000	29.93	289.60	6.05
Green Bay (WI)	519	114,500	16.54	283.33	3.67

Explanation of variables

Apartment rent   Monthly rent of an unfurnished two-bedroom apartment (excluding all utilities except water), 1½ or 2 baths, approximately 950 square feet

Price of house   Purchase price of a new house with 1800 square feet of living area, on an 8000-square-foot lot in an urban area with all utilities

Phone bill   Monthly telephone charges for a private residential line (customer owns instruments)

Cost of a hospital room   Average cost per day of a semiprivate room in a hospital

Price of wine   Price of Paul Masson Chablis, 1.5-liter bottle

2.22   a.   Prepare a frequency distribution table with five classes for apartment rents.
       b.   Construct the relative frequency and percentage distribution columns.
       c.   Write the class midpoints.

2.23   a.   Prepare a frequency distribution table with five classes for the purchase prices of houses.
       b.   Calculate the relative frequencies and percentages for all classes.
       c.   Write the class midpoints.

2.24   a.   Prepare a frequency distribution table for phone bills.
       b.   Construct the relative frequency and percentage distribution columns.
       c.   Draw a histogram and a polygon for the relative frequency distribution.

2.25 a. Prepare a frequency distribution table for the average cost per day of rooms in hospitals.
  b. Calculate the relative frequencies and percentages for all classes.
  c. Draw a histogram and a polygon for percentage distribution.

2.26 a. Prepare a frequency distribution table using four classes for the price of wine.
  b. Construct the relative frequency and percentage distribution columns.

2.27 The following data give the number of new cars sold at a dealership during a 20-day period.

8	5	12	3	9	10	6	3	8	8
4	6	10	11	7	7	3	5	9	11

  a. Prepare a frequency distribution table with five classes.
  b. Calculate the relative frequencies and percentages for all classes.
  c. Construct a histogram and a polygon for the percentage distribution.

2.28 The following data give the market value (in billions of dollars) of 30 companies as of March 5, 1993.

Company	Market Value (billions of dollars)	Company	Market Value (billions of dollars)
American Express	12.2	IBM	31.5
Ameritech	20.0	J.C. Penney	9.7
BankAmerica	17.9	Johnson & Johnson	26.5
Boeing	11.7	McDonald's	18.7
Caterpillar	5.9	Microsoft	23.1
CBS	2.8	Motorola	16.2
Chase Manhattan	5.1	Nike	5.3
Chrysler	13.6	Pepsico	33.1
Cigna	4.3	Pfizer	19.1
Citicorp	9.6	Procter & Gamble	36.5
Dow Jones	3.1	Sears, Roebuck	18.3
Ford Motor	24.1	Travelers	3.9
General Motors	27.6	Walt Disney	24.3
Gillette	13.1	Wells Fargo	5.6
H.J. Heinz	11.2	Xerox	7.8

Source: Business Week, 1993 Special Issue.

  a. Construct a frequency distribution table using the *less than* method to write classes. Take $2 billion as the lower boundary of the first class and 7 as the width of each class.
  b. Calculate the relative frequencies and percentages for all classes.
  c. What percentage of the companies in this sample had a market value of $30 billion or higher on March 5, 1993?
  d. From the frequency distribution of part a, can you tell whether the data are symmetric or skewed? If skewed, are they skewed to the left or right?

2.29 The following data give the annual income (in thousands of dollars) before taxes for 36 families randomly selected from the 1990 Interview Survey. (These families were selected from those who had positive income before taxes.)

21.6	13.0	25.6	27.9	50.0	18.1
10.1	21.5	10.0	72.8	58.2	15.4
7.2	27.0	32.2	45.0	95.0	27.8
92.8	8.4	45.3	76.0	28.6	9.3
30.6	19.0	25.5	27.5	9.7	15.1
6.3	44.5	24.0	13.0	61.7	16.0

a. Construct a frequency distribution table using the *less than* method to write classes. Take $0 thousand as the lower boundary of the first class and $20 thousand as the width of each class.
b. Calculate the relative frequencies and percentages for all classes.
c. What percentage of the families in this sample had an income of $80 thousand or higher in 1990?
d. From the frequency distribution of part a, can you tell whether the data are symmetric or skewed? If skewed, are they skewed to the left or right?

2.30 The following data give the number of bedrooms in homes owned or rented by 30 families randomly selected from the 1990 Interview Survey.

3	5	2	3	2	3	1	2	1	3
4	1	4	3	1	3	3	2	2	3
3	4	3	1	2	4	2	2	5	3

a. Prepare a frequency distribution table for these data using single-valued classes.
b. Calculate the relative frequencies and percentages for all classes.
c. How many homes in this sample have 2 or 3 bedrooms?
d. Draw a bar graph for the frequency distribution.

2.31 The following data give the number of earners per household for 40 households randomly selected from the 1990 Diary Survey.

2	1	1	4	0	1	1	0	2	2
1	2	1	1	3	2	2	2	0	1
2	2	0	3	2	2	2	2	2	1
1	1	2	2	1	0	4	2	2	2

a. Prepare a frequency distribution table for these data using single-valued classes.
b. Calculate the relative frequencies and percentages for all classes.
c. What percentage of the households in this sample have one or two earners?
d. Draw a bar graph for the relative frequency distribution.

2.32 The following table lists the frequency distribution for a data set.

Classes	Frequency
6–10	64
11–15	70
16–20	86
21–25	80
26–30	72
31–35	58

Draw two histograms for these data, one without truncating the frequency axis and the second by truncating the frequency axis. In the second case, mark the frequencies on the vertical axis starting with 55. Briefly comment on the two histograms.

2.33 The following table lists the frequency distribution for a data set.

Classes	Frequency
21–30	120
31–40	140
41–50	115
51–60	105
61–70	95

Draw two histograms for these data, one without truncating the frequency axis and the second by truncating the frequency axis. In the second case, mark the frequencies on the vertical axis starting with 90. Briefly comment on the two histograms.

2.5 CUMULATIVE FREQUENCY DISTRIBUTIONS

Consider again Example 2–3 of Section 2.3.2 about the price–earnings ratios of 25 companies. Suppose we want to know how many companies have price–earnings ratios of less than 20. Such a question can be answered using a **cumulative frequency distribution**. Each class in a cumulative frequency distribution table gives the total number of values that fall below a certain value. A cumulative frequency distribution is constructed for quantitative data only.

> **CUMULATIVE FREQUENCY DISTRIBUTION**
>
> A cumulative frequency distribution gives the total number of values that fall below the upper boundary of each class.

In a *less than* cumulative frequency distribution table, each class has the same lower limit but a different upper limit. Example 2–7 illustrates the procedure to prepare a cumulative frequency distribution.

Constructing cumulative frequency distribution table.

EXAMPLE 2–7 Using the frequency distribution of Table 2.9, reproduced below, prepare a cumulative frequency distribution for the price–earnings ratios of 25 companies.

Price–Earnings Ratio	f
10–14	4
15–19	6
20–24	8
25–29	4
30–34	3

Solution Table 2.13 gives the cumulative frequency distribution for the price–earnings ratios of 25 companies. As we can observe, 10 (which is the lower limit of the first class in Table 2.9) is taken as the lower limit of each class in Table 2.13. The upper limits of all classes in Table 2.13 are the same as those in Table 2.9. To obtain the cumulative frequency of a class, we have added the frequency of that class in Table 2.9 to the frequencies of all preceding classes. The cumulative frequencies are recorded in the third column of Table 2.13. The second column of this table lists the class boundaries.

Table 2.13 Cumulative Frequency Distribution of Price–Earnings Ratios of 25 Companies

Class Limits	Class Boundaries	Cumulative Frequency
10–14	9.5 to less than 14.5	4
10–19	9.5 to less than 19.5	4 + 6 = 10
10–24	9.5 to less than 24.5	4 + 6 + 8 = 18
10–29	9.5 to less than 29.5	4 + 6 + 8 + 4 = 22
10–34	9.5 to less than 34.5	4 + 6 + 8 + 4 + 3 = 25

From Table 2.13, we can determine the number of observations that fall below the upper boundary of a class. For example, from Table 2.13, 22 companies had price–earnings ratios of 29 or lower.   ▬

The **cumulative relative frequencies** are obtained by dividing the cumulative frequencies by the total number of observations in the data. The **cumulative percentages** are obtained by multiplying the cumulative relative frequencies by 100.

CUMULATIVE RELATIVE FREQUENCY AND CUMULATIVE PERCENTAGE

$$\text{Cumulative relative frequency} = \frac{\text{Cumulative frequency}}{\text{Total observations in the data set}}$$

$$\text{Cumulative percentage} = (\text{Cumulative relative frequency}) \cdot 100$$

Table 2.14 contains both the cumulative relative frequencies and the cumulative percentages for Table 2.13. We can observe from this table, for example, that 72% of the companies had price–earnings ratios of 24 or lower.

Table 2.14   Cumulative Relative Frequency and Cumulative Percentage Distributions of Price–Earnings Ratios

Class Limits	Cumulative Relative Frequency	Cumulative Percentage
10–14	4/25 = .16	16
10–19	10/25 = .40	40
10–24	18/25 = .72	72
10–29	22/25 = .88	88
10–34	25/25 = 1.00	100

OGIVES

When plotted on a diagram, the cumulative frequencies give a curve that is called an **ogive** (pronounced o-jive). Figure 2.12 gives an ogive for the cumulative frequency distribution of Table 2.13. To draw the ogive in Figure 2.12, the variable, price–earnings ratio, is marked

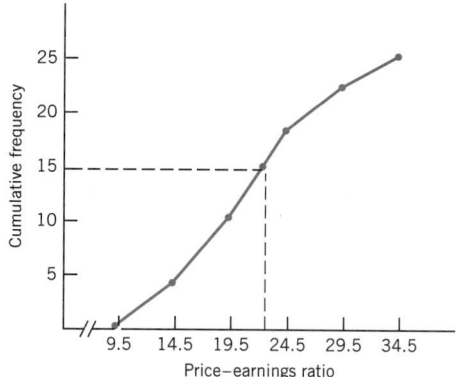

Figure 2.12   Ogive for Table 2.13.

on the horizontal axis and the cumulative frequencies on the vertical axis. Then, the dots are marked above the upper boundaries of various classes at the heights equal to the corresponding cumulative frequencies. The ogive is obtained by joining consecutive points with straight lines. Note that the ogive starts at the lower boundary of the first class and ends at the upper boundary of the last class.

> **OGIVE**
>
> An ogive is a curve drawn for the cumulative frequency distribution by joining with straight lines the dots marked above the upper boundaries of classes at heights equal to the cumulative frequencies of respective classes.

One advantage of an ogive is that it can be used to approximate the cumulative frequency for any interval. For example, to find the number of companies with 22 or lower price–earnings ratios, first draw a vertical line from 22 on the horizontal axis up to the ogive. Then draw a horizontal line from the point where this line intersects the ogive to the vertical axis. This point gives the cumulative frequency of the class 10 to 22. In Figure 2.12, this cumulative frequency is approximately 15. Therefore, approximately 15 companies had a price–earnings ratio of 22 or less.

We can draw an ogive for cumulative relative frequency and cumulative percentage distributions in the same way as we did for the cumulative frequency distribution.

EXERCISES

Concepts and Procedures

2.34 Briefly explain the concept of cumulative frequency distribution. How are the cumulative relative frequencies and cumulative percentages calculated?

2.35 Explain for what kind of frequency distribution an ogive is drawn. Can you think of any use of an ogive? Explain.

Applications

2.36 The following table, reproduced from Exercise 2.14, gives the frequency distribution of weekly earnings for a sample of 100 employees selected from a company.

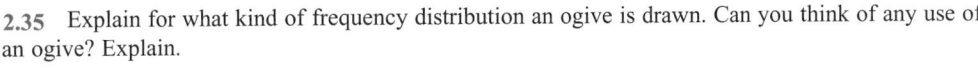

Weekly Earnings (dollars)	Number of Employees
200 to 349	12
350 to 499	23
500 to 649	31
650 to 799	19
800 to 949	15

a. Prepare a cumulative frequency distribution table.
b. Calculate the cumulative relative frequencies and cumulative percentages for all classes.
c. What percentage of the workers earned $649 or less per week?
d. Draw an ogive for the cumulative percentage distribution.
e. Using the ogive, find the percentage of workers who earned $600 or less per week.

2.37 The following table, reproduced from Exercise 2.15, gives the frequency distribution of ages for all 50 employees of a company.

Age	Number of Employees
18 to 30	12
31 to 43	17
44 to 56	14
57 to 69	7

    a. Prepare a cumulative frequency distribution table.
    b. Calculate the cumulative relative frequencies and cumulative percentages for all classes.
    c. What percentage of the employees of this company are 44 years of age or older?
    d. Draw an ogive for the cumulative percentage distribution.
    e. Using the ogive, find the percentage of employees who are 40 years old or younger.

2.38 Using the frequency distribution table constructed in Exercise 2.18, prepare the cumulative frequency, cumulative relative frequency, and cumulative percentage distributions.

2.39 Using the frequency distribution table constructed in Exercise 2.19, prepare the cumulative frequency, cumulative relative frequency, and cumulative percentage distributions.

2.40 Using the frequency distribution table constructed in Exercise 2.20, prepare the cumulative frequency, cumulative relative frequency, and cumulative percentage distributions.

2.41 Prepare the cumulative frequency, cumulative relative frequency, and cumulative percentage distributions using the frequency distribution table of Exercise 2.21.

2.42 Refer to Exercise 2.22. Construct the cumulative frequency, cumulative relative frequency, and cumulative percentage distributions by using the frequency distribution table of that exercise.

2.43 Prepare the cumulative frequency, cumulative relative frequency, and cumulative percentage distributions using the frequency distribution constructed in Exercise 2.23.

2.44 Using the frequency distribution table constructed for the data of Exercise 2.28, prepare the cumulative frequency, cumulative relative frequency, and cumulative percentage distributions. Draw an ogive for the cumulative frequency distribution. Using the ogive, find the (approximate) number of companies in this sample that had a market value of less than $20 billion on March 5, 1993.

2.45 Refer to the frequency distribution table constructed in Exercise 2.29. Prepare the cumulative frequency, cumulative relative frequency, and cumulative percentage distributions by using that table. Draw an ogive for the cumulative percentage distribution. Using the ogive, find the (approximate) percentage of families in this sample who had an annual income of less than $50 thousand in 1990.

2.6 STEM-AND-LEAF DISPLAYS

Another technique used to present quantitative data in condensed form is the **stem-and-leaf display**. An advantage of a stem-and-leaf display over a frequency distribution is that by preparing a stem-and-leaf display we do not lose information on individual observations. A stem-and-leaf display is constructed only for quantitative data.

> **STEM-AND-LEAF DISPLAYS**
>
> In a stem-and-leaf display of quantitative data, each value is divided into two portions— a stem and a leaf. Then the leaves for each stem are shown separately in a display.

Example 2–8 describes the procedure for constructing a stem-and-leaf display.

Constructing stem-and-leaf display for two-digit numbers.

EXAMPLE 2–8 Refer to Example 2–3. The data on price–earnings ratios of 25 companies are reproduced below from that exercise.

31	13	12	22	27	33	17	26
16	22	18	13	16	23	20	18
22	15	26	12	20	21	23	27	30

Construct a stem-and-leaf display.

Solution To construct a stem-and-leaf display for these price–earnings ratios, we split each price–earnings ratio into two parts. The first part contains the first digit, which is called the *stem*. The second part contains the second digit, which is called the *leaf*. Thus, for the price–earnings ratio of the first company, which is 31, 3 is the stem and 1 is the leaf. For the price–earnings ratio of the second company, which is 13, the stem is 1 and the leaf is 3. We observe from these data that stems for all price–earnings ratios are 1, 2, and 3 because all the price–earnings ratios lie in the range 12 to 33. To obtain a stem-and-leaf display, we draw a vertical line and write the stems on the left side of it arranged in increasing order, as shown in Figure 2.13.

After we have listed the stems, we read the leaves for each of the price–earnings ratios and record them next to the corresponding stems on the right side of the vertical line. For example, for the first price–earnings ratio we write the leaf 1 next to the stem 3; for the second price–earnings ratio we write the leaf 3 next to the stem 1. The stem-and-leaf display of these two price–earnings ratios is shown in Figure 2.13.

```
Stems
  │
  ↓
  1 │ 3 ⟵── Leaf for 13
  2 │
  3 │ 1 ⟵── Leaf for 31    Figure 2.13  Stem-and-leaf display.
```

Now we read all the price–earnings ratios and write the leaves on the right side of the vertical line in the rows of corresponding stems. The complete stem-and-leaf display for the price–earnings ratios is shown in Figure 2.14.

```
1 │ 3 2 7 6 8 3 6 8 5 2
2 │ 2 7 6 2 3 0 2 6 0 1 3 7      Figure 2.14  Stem-and-leaf display of
3 │ 1 3 0                         price–earnings ratios.
```

By looking at the stem-and-leaf display of Figure 2.14, we can observe how the data values are distributed. For example, stem 2 has the highest frequency, followed by stems 1 and 3.

The leaves of the stem-and-leaf display of Figure 2.14 are *ranked* (in increasing order) and presented in Figure 2.15.

```
1 │ 2 2 3 3 5 6 6 7 8 8
2 │ 0 0 1 2 2 2 3 3 6 6 7 7      Figure  2.15  Ranked  stem-and-leaf
3 │ 0 1 3                         display of price–earnings ratios.
```

From Figure 2.15, we observe that stems of 1 and 2 contain too many leaves, as compared to stem 3. In such cases, we can write the leaves for each of these stems in two rows, the first row containing the leaves 0 through 4 and the second row from 5 through 9. Figure 2.16 gives such a stem-and-leaf display for the data on price–earnings ratios.

```
1 | 2 2 3 3
1 | 5 6 6 7 8 8
2 | 0 0 1 2 2 2 3 3
2 | 6 6 7 7
3 | 0 1 3
```

Figure 2.16 Stem-and-leaf display of price–earnings ratios.

As mentioned earlier, one advantage of a stem-and-leaf display is that we do not lose information on individual observations. We can rewrite the individual price–earnings ratios of the 25 companies from the stem-and-leaf displays of Figures 2.14, 2.15, or 2.16. By contrast, the information on individual observations is lost when data are grouped into a frequency distribution table.

Constructing stem-and-leaf display for three- and four-digit numbers.

EXAMPLE 2–9   The following data give the monthly rents paid by a sample of 30 households selected from a city.

429	585	732	675	550	989	1020	620	750	660
540	578	956	1030	1070	930	871	765	880	975
650	1020	950	840	780	870	900	800	750	820

Construct a stem-and-leaf display for these data.

Solution   Each of the values in the given data set contains either three or four digits. We will take the first digit for three-digit numbers and the first two digits for four-digit numbers as stems. Then, we will use the last two digits of each number as a leaf. Thus, for the first value, which is 429, the stem is 4 and the leaf is 29. The stems for the entire data are 4, 5, 6, 7, 8, 9, and 10. They are recorded on the left side of the vertical line in Figure 2.17. The leaves for various numbers are recorded on the right side.

```
 4 | 29
 5 | 85 50 40 78
 6 | 75 20 60 50
 7 | 32 50 65 80 50
 8 | 71 80 40 70 00 20
 9 | 89 56 30 75 50 00
10 | 20 30 70 20
```

Figure 2.17 Stem-and-leaf display of rents.

Sometimes a data set may contain too many stems, with each stem containing only a few leaves. In such cases, we may want to condense the stem-and-leaf display by *grouping the stems*. The following example describes this procedure.

Preparing a grouped stem-and-leaf display.

EXAMPLE 2-10   The following is the stem-and-leaf display prepared for the stock prices of 21 companies.

```
1 | 3 5
2 | 2 5 6
3 | 0 1
4 | 2 3 6
5 | 0
6 | 5 6
7 | 0 3 9
8 | 1 5 7
9 | 2 6
```

Prepare a new stem-and-leaf display by grouping the stems.

Solution   To condense the given stem-and-leaf display, we can combine the first three rows, the middle three rows, and the last three rows, thus getting the stems 1–3, 4–6, and 7–9. The leaves for each stem of a group are separated by an asterisk (*), as shown in Figure 2.18. Thus, the leaves 3 and 5 in the first row of Figure 2.18 correspond to stem 1; the leaves 2, 5, and 6 correspond to stem 2; and leaves 0 and 1 belong to stem 3.

```
1–3 | 3 5 * 2 5 6 * 0 1
4–6 | 2 3 6 * 0 * 5 6
7–9 | 0 3 9 * 1 5 7 * 2 6
```

Figure 2.18   Grouped stem-and-leaf display.

If a stem does not contain a leaf, this can be indicated in the grouped stem-and-leaf display by two consecutive asterisks. For example, in the following stem-and-leaf display there is no leaf for 3, that is, there is no number in the 30s. The numbers in this display are 21, 25, 43, 48, and 50.

```
2–5 | 1 5 * * 3 8 * 0
```

EXERCISES

Concepts and Procedures

2.46   Briefly explain how a stem-and-leaf display for a data set is prepared. You may use an example to illustrate.

2.47   What advantages does preparing a stem-and-leaf display have over grouping a data set using a frequency distribution? Give one example.

2.48   Prepare a ranked stem-and-leaf display for the following data.

35	23	29	13	18	34	39	17	26	31
33	36	29	21	25	38	33	34	37	24

2.49   Prepare a ranked stem-and-leaf display for the following data. (*Hint:* To prepare a stem-and-leaf display, each number in this data set can be written as a three-digit number. For example, 32 can be written as 032 for which the stem is 0 and the leaf is 32.)

111	233	217	279	123	186	32	363	216	325
239	344	229	67	159	282	162	126	291	316

2.50 Consider the following stem-and-leaf display.

```
4 | 3 6
5 | 0 1 4 5 9
6 | 3 4 6 7 7 7 8 9
7 | 2 2 3 5 6 6
8 | 0 7 8 9
```

Write the data set that is represented by this stem-and-leaf display.

2.51 Consider the following stem-and-leaf display.

```
2–3 | 18 45 56 * 29 67 83 97
4–5 | 04 27 33 71 * 23 37 51 63 81 92
6–8 | 22 36 47 55 78 89 * * 10 41
```

Write the data set that is represented by this stem-and-leaf display.

Applications

2.52 The following data give the time (in minutes) that each of 20 workers took to complete an assembly job.

55	49	53	59	38	56	39	58	47	53
58	42	37	43	47	44	55	51	46	45

Construct a stem-and-leaf display for these data. Arrange the leaves for each stem in increasing order.

2.53 Following are the GMAT scores of 12 students who took this test recently.

585	490	410	569	381	623	595	497	473	480	549	632

Prepare a stem-and-leaf display. Arrange the leaves for each stem in increasing order.

2.54 Reconsider the data on 1992 sales (in billions of dollars) for a sample of 20 U.S. companies given in Exercise 2.18. Prepare a stem-and-leaf display for those data. Arrange the leaves for each stem in increasing order.

2.55 Reconsider the data on the number of computer terminals produced at the Nixon Corporation for a sample of 30 days given in Exercise 2.20. Prepare a stem-and-leaf display for those data. Arrange the leaves for each stem in increasing order.

2.56 Reconsider the data on the number of computer keyboards assembled at the Twentieth Century Electronics Company given in Exercise 2.21. Prepare a stem-and-leaf display for those data. Arrange the leaves for each stem in increasing order.

2.57 The following data give the time (in minutes) taken to commute from home to work for 20 workers.

10	50	65	33	48	5	11	23	37	26
26	32	17	7	13	19	29	43	21	22

Construct a stem-and-leaf display for these data. Arrange the leaves for each stem in increasing order.

2.58 The following data give the monthly grocery expenditures (in dollars) for 35 households selected from the 1990 Diary Survey.

435	391	161	130	109	261	109
65	174	152	326	217	130	87
130	304	283	283	304	391	348
261	161	652	348	326	283	187
152	174	174	217	174	78	196

a. Prepare a stem-and-leaf display for these data using the last two digits as leaves.
b. Condense the stem-and-leaf display by grouping the stems as 0–1, 2–3, and 4–6.

2.59 The following data give the annual earnings (rounded to thousands of dollars) of 35 households selected from the 1990 Diary Survey.

30	5	29	7	20	31	28	10
13	24	25	7	22	15	15	22
78	79	99	30	80	80	75	63
24	35	25	90	35	33	70	63
35	17	9					

a. Prepare a stem-and-leaf display for these data.
b. Condense the stem-and-leaf display by grouping the stems as 0–2, 3–5, and 6–9.

2.7 GRAPHING TIME-SERIES DATA

In this section we discuss the difference between cross-section and time-series data and how to plot time-series data.

2.7.1 CROSS-SECTION VERSUS TIME-SERIES DATA

Data can be classified as cross-section or time-series data. First we explain these two types of data and then describe how to graph time-series data.

Cross-Section Data

Cross-section data contain information on different elements of a population or sample for the *same* period of time. The information on incomes of 100 families for the year 1992 is an example of cross-section data. All examples of data taken so far in this chapter have been those of cross-section data.

> **CROSS-SECTION DATA**
>
> Cross-section data are data collected on different elements at the same point in time or for the same period of time.

Table 2.15 gives the total assets (in millions of dollars) of five banks as of 1992. This is an example of cross-section data.

Table 2.15 Assets of Five Banks

Bank	Total Assets (millions of dollars)
Citicorp	213,701
BankAmerica	180,646
Chase Manhattan	95,862
Wells Fargo	52,537
First Chicago	49,281

Source: Business Week, 1993 Special Issue.

Time-Series Data

Time-series data contain information on the same element for *different* periods of time. Information on U.S. exports for the years 1975 to 1990 is an example of time-series data.

> **TIME-SERIES DATA**
>
> Time-series data are data collected on the same element for the same variable at different points in time or for different periods of time.

Data given in Table 2.16 are an example of time-series data. This table gives the total value of U.S. exports (in billions of dollars) for the years 1986 through 1991. The figures for 1991 are predicted, not actual.

Table 2.16   U.S. Exports for the Years 1986 to 1991

Year	Exports (billions of dollars)
1986	319.2
1987	364.0
1988	444.2
1989	504.9
1990	550.4
1991	593.3

Source: Economic Report of the President, February 1992.

2.7.2   GRAPHING TIME-SERIES DATA

Example 2–11 exhibits the graphing of time-series data. All the frequency distributions discussed so far in this chapter contained information on one variable only. However, a time-series data set contains information on two variables. For instance, Example 2–11 contains information on two variables, time (year) and unemployment rate.

Graphing time-series data.

EXAMPLE 2–11   The following table gives the U.S. unemployment rate (measured as a percentage of the civilian labor force who were unemployed) for the period 1980 through 1991.

Year	Unemployment Rate	Year	Unemployment Rate
1980	7.1	1986	7.0
1981	7.6	1987	6.2
1982	9.7	1988	5.5
1983	9.6	1989	5.3
1984	7.5	1990	5.5
1985	7.2	1991	6.7

Source: Economic Report of the President, 1992.

Display these data in graphic form.

Solution   The given time-series data can be displayed by drawing a *line graph* or a *bar graph*. A line graph is usually used for time-series data only. To draw a line graph, we mark the years on the horizontal axis and the unemployment rate on the vertical axis, as in Figure 2.19. Then we mark a dot above each year at a height equal to the unemployment rate for that year. Finally, we join consecutive points with straight lines.

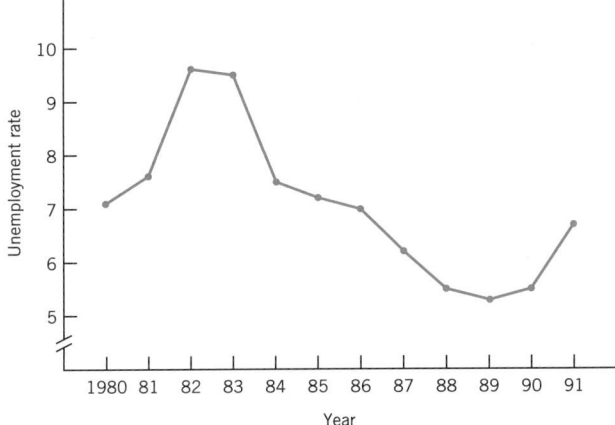

Figure 2.19   Line graph for unemployment rates.

From the graph in Figure 2.19 we can observe how the unemployment rate changed or fluctuated during the period 1980 through 1991. In general, a time-series line graph reveals the trend of a variable over a period of time. Note that in Figure 2.19 the vertical axis is truncated and, hence, the variation in unemployment rate over time is magnified.

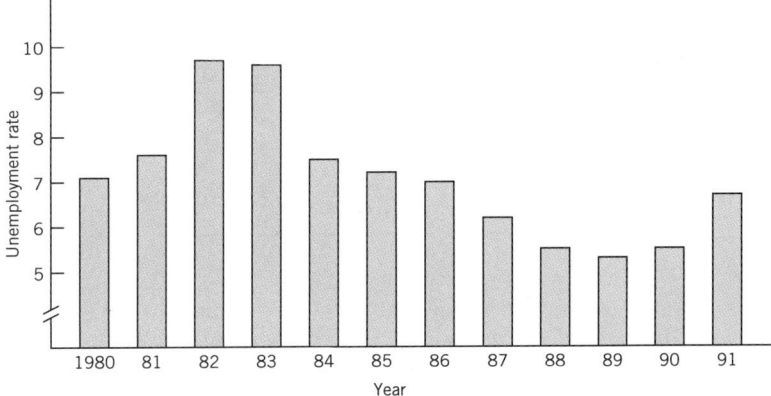

Figure 2.20   Bar graph for unemployment rates.

Another device used to display time-series data is the bar graph. Figure 2.20 shows the bar graph for the data on unemployment rates. In this bar graph, each year is represented by an interval on the horizontal axis and the unemployment rate is marked on the vertical axis. The unemployment rate for a specific year is given by the height of the bar for that year.

☞   **WARNING**

Time-series data can be mistreated by manipulation in an attempt to prove different opinions. Case Study 2–5 shows how time-series graphs can be used to give erroneous impressions.

CASE STUDY 2-5   WHAT IS WRONG WITH THIS PICTURE?

. . . Congress has been suspicious that this Administration, in particular, was more interested in cooking the facts to fit its ideology than in doing objective research, especially on the subject of education spending.

Proof that such suspicions are justified comes from a graph incorrectly attributed . . . to the U.S. Department of Education's Center for Education Statistics (CES) that was distributed at a conference by a high Department official and at a House Budget Committee hearing at which the Secretary of Education testified. The handout tries to convince us that from 1963 to 1988 there was a direct connection between the *rise* in elementary and secondary school spending and the *decline* in SAT scores. The graph makes a great visual impact. . . .

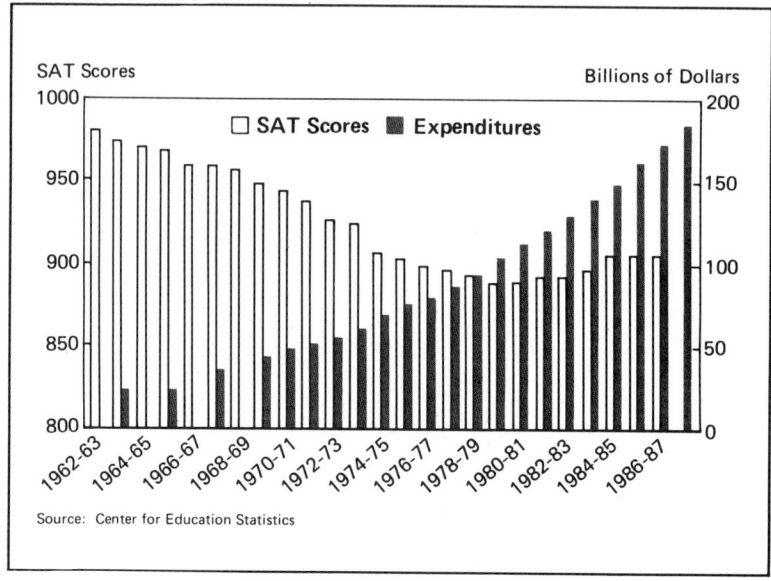

Source: Center for Education Statistics

What's wrong with this picture? In the July 1988 issue of *ETS Policy Notes*, Dr. Joan Baratz-Snowden, then Director of the Policy Information Center of the Educational Testing Service, provides the answer: 'The graph is as inaccurate as it is dramatic.'. . . Baratz-Snowden points out, the Administration used current dollars rather than constant dollars adjusted for inflation. A more accurate graph done by ETS shows that with inflation figured in, 'expenditures have barely risen during the last 20 years.' Although spending has been relatively constant, our public schools have added a host of special programs for disadvantaged, handicapped, non–English-speaking, and other students who were previously neglected and, sometimes, shut out.

Second, although SAT scores have gone down, the Education Department graph vastly exaggerated the decline. Baratz-Snowden shows us how. SAT scores go from 400–1600, but the graph uses an 800–1000 scale. . . . The Department also doesn't bother to tell us that the students who take the exam don't represent a random sample of our student population. Forexample, it used to be that only the brightest students were encouraged to take the SAT. Now the test-takers are more numerous and diverse, and this affects the average scores.

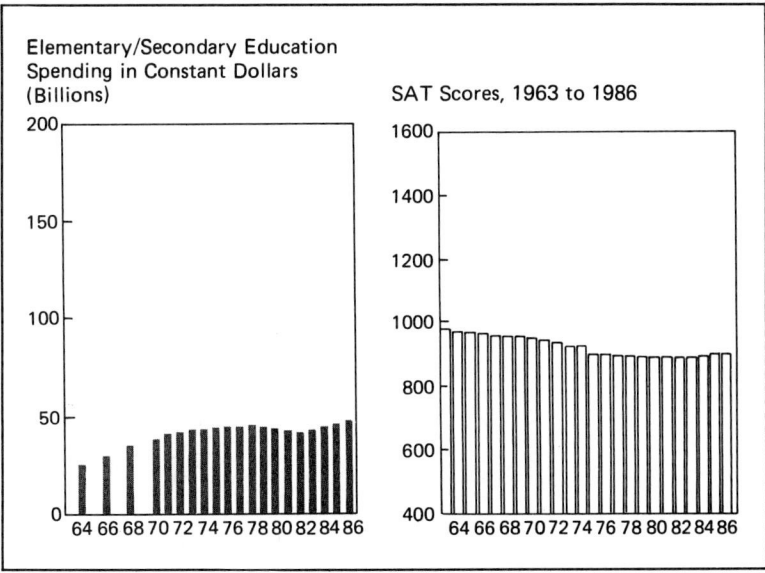

Elementary/Secondary Education
Spending in Constant Dollars
(Billions)

SAT Scores, 1963 to 1986

Source: Albert Shanker: "What Is Wrong With This Picture? Lies, Damned Lies, and Statistics." *Chance*, 2(1), 1989, pp. 34–35. Copyright © 1989 by Springer-Verlag New York, Inc. Reprinted with permission.

EXERCISES

Concepts and Procedures

2.60 Explain the difference between cross-section and time-series data. Give an example of each of these two types of data.

2.61 Classify the following as cross-section or time-series data.
- a. Gross sales of IBM for the period 1970 to 1993
- b. Weights of 20 chickens
- c. Poverty rates in the United States for 1975 to 1993
- d. Auto insurance premiums paid by 100 drivers

2.62 Classify the following as cross-section or time-series data.
- a. Average prices of houses in 100 cities
- b. Salaries of 50 employees
- c. Gross national product of the United States for the years 1980 to 1993
- d. Number of workers employed by a company each year from 1985 to 1993

Applications

2.63 The following table gives the average monthly electric bill per farm for the period 1977 through 1985.

Year	Average Monthly Bill per Farm (dollars)
1977	48.10
1978	53.20
1979	55.10
1980	61.90
1981	69.80
1982	83.90
1983	92.50
1984	98.00
1985	104.00

Source: National Agricultural Statistics Service.

Draw a line graph for these data.

2.64   The following table gives the female (civilian) labor force participation rate for selected years.

Year	Female Labor Force Participation Rate
1960	37.7
1965	39.3
1970	43.3
1975	46.3
1980	51.5
1985	54.5
1990	57.5

Source: Economic Report of the President, 1992.

Draw a line graph for these data.

2.65   The following table gives the total earnings (in millions of dollars) of Apple Computer, Inc., for the years 1988 to 1992.

Year	Earnings (millions of dollars)
1988	400
1989	454
1990	475
1991	310
1992	530

Draw a bar graph for these data.

2.66   The following table gives the U.S. imports of goods and services (in billions of dollars) for the period 1985 through 1991.

Year	Imports of Goods and Services (billions of dollars)
1985	417.6
1986	451.7
1987	507.1
1988	552.2
1989	587.8
1990	624.8
1991	620.4

Draw a bar graph for these data.

GLOSSARY

Bar graph   A graph made of bars whose heights represent the frequencies of respective categories.

Class   An interval that includes all the values in a (quantitative) data set that fall within two numbers, the lower and upper limits of the class.

Class boundary   The midpoint of the upper limit of one class and the lower limit of the next class.

Class frequency   The number of values in a data set that belong to a certain class.

Class midpoint or **mark**   Obtained by dividing the sum of the lower and upper limits (or boundaries) of a class by 2.

Class width or **size**   The difference between the two boundaries of a class.

Cross-section data   Data collected on different elements at the same point in time or for the same period of time.

Cumulative frequency   The frequency of a class that includes all values in a data set that fall below the upper boundary of that class.

Cumulative frequency distribution   A table that lists the total number of values that fall below the upper boundary of each class.

Cumulative relative frequency   The cumulative frequency of a class divided by the total number of observations.

Cumulative percentage   The cumulative relative frequency multiplied by 100.

Frequency distribution   A table that lists all the categories or classes and the number of values that belong to each of these categories or classes.

Grouped data   A data set presented in the form of a frequency distribution.

Histogram   A graph in which classes are marked on the horizontal axis and either frequencies, relative frequencies, or percentages are marked on the vertical axis. The frequencies, relative frequencies, or percentages of various classes are represented by bars that are drawn adjacent to each other.

Ogive   A curve drawn for cumulative frequency distribution.

Percentage   The percentage for a class or category is obtained by multiplying the relative frequency of that class or category by 100.

Pie chart   A circle divided into portions that represent the relative frequencies or percentages of different categories or classes.

Polygon   A graph formed by joining the midpoints of the tops of successive bars in a histogram by straight lines.

Raw data   Data recorded in the sequence in which they are collected and before they are processed.

Relative frequency   The frequency of a class or category divided by the sum of all frequencies.

Skewed to the left histogram   A histogram with a longer tail on the left side.

Skewed to the right histogram   A histogram with a longer tail on the right side.

Stem-and-leaf display   A display of data in which each value is divided into two portions, a stem and a leaf.

Symmetric histogram   A histogram that is identical on both sides of its central point.

Time-series data   Data that give the values for the same variable for the same element at different points in time or for different periods of time.

Uniform or **rectangular histogram**   A histogram with the same frequency for all classes.

KEY FORMULAS

1. **Relative frequency of a class**

$$\text{Relative frequency of a class} = \frac{\text{Frequency of that class}}{\text{Sum of all frequencies}} = \frac{f}{\Sigma f}$$

2. **Percentage of a class**

$$\text{Percentage} = (\text{Relative frequency}) \cdot 100$$

3. **Class midpoint or mark**

$$\text{Class midpoint} = \frac{\text{Upper limit} + \text{Lower limit}}{2}$$

4. **Class width or size**

$$\text{Class width} = \text{Upper boundary} - \text{Lower boundary}$$

5. **Cumulative relative frequency**

$$\text{Cumulative relative frequency} = \frac{\text{Cumulative frequency}}{\text{Total observations in the data set}}$$

6. **Cumulative percentage**

$$\text{Cumulative percentage} = (\text{Cumulative relative frequency}) \cdot 100$$

SUPPLEMENTARY EXERCISES

2.67 The following data give the educational attainment of 40 persons selected from the 1990 Interview Survey. In the data, E stands for an elementary school education, H for high school but less than high school graduate, HG for a high school graduate, S for some college but less than college graduate, C for a college graduate, and M for more than 4 years of college.

S	M	H	S	E	HG	C	S
H	HG	HG	M	M	HG	S	S
S	S	HG	H	C	E	E	H
S	H	S	S	M	M	S	HG
M	S	E	HG	S	H	C	HG

a. Prepare a frequency distribution table for these data.
b. Calculate the relative frequency and percentage distributions.
c. Draw a bar graph for the relative frequency distribution and a pie chart for the percentage distribution.
d. What percentage of the people in this sample have more than 4 years of college?

2.68 The following data list the reasons why 30 (nonworking) persons, selected from the 1990 Interview Survey, were not working at the time of the survey. In the data, I stands for ill, disabled, or unable to work; T for taking care of home/family; G for going to school; C for could not find work; and R for retired.

R	R	T	R	I	I	T	R	T	R
C	T	R	R	T	I	R	R	T	R
I	R	G	C	T	R	R	R	I	R

a. Prepare a frequency distribution table for these data.

    b. Calculate the relative frequency and percentage distributions.

    c. Draw a bar graph for the frequency distribution and a pie chart for the percentage distribution.

    d. What percentage of the people in this sample are retired?

2.69 The following data give the number of television sets owned by 40 randomly selected households.

1	1	2	3	2	4	1	3	2	1
3	0	2	1	2	3	2	3	2	2
1	2	1	1	1	3	1	1	1	2
2	4	2	3	1	3	1	2	2	4

    a. Prepare a frequency distribution table for these data using single-valued classes.

    b. Compute the relative frequency and percentage distributions for all categories.

    c. Draw a bar graph for the frequency distribution.

    d. What percentage of the households own 2 or more television sets?

2.70 The following data give the number of computer courses taken by 30 business majors who recently graduated from a university.

2	3	2	3	1	4	2	2	3	4
2	3	4	1	2	3	2	1	4	2
1	2	3	1	1	3	2	2	4	1

    a. Prepare a frequency distribution table for these data using single-valued classes.

    b. Compute the relative frequency and percentage distributions for all categories.

    c. Draw a bar graph for the relative frequency distribution.

    d. What percentage of these graduates took 2 or 3 computer courses?

2.71 The following data give the amount of public assistance or welfare (in hundreds of dollars) received by 25 households randomly selected from the 1990 Interview Survey. (These households were selected from those who received some public assistance or welfare.)

85	4	20	73	19	29	54	44	
41	33	10	45	10	35	55	28	
23	83	48	49	99	18	58	53	78

    a. Construct a frequency distribution table with the classes 1–20, 21–40, 41–60, 61–80, and 81–100.

    b. Prepare the relative frequency and percentage columns for this table.

    c. Draw a histogram and a polygon for the frequency distribution.

    d. Based on the frequency histogram, can you say if the data are symmetric or skewed?

2.72 The following data give the amount (in hundreds of dollars) paid for child support during the 12-month period by 25 persons randomly selected from the 1990 Interview Survey. (These persons were selected from those who paid such support.)

48	11	21	94	30	74	36	16	
60	13	88	12	24	18	53	48	
18	15	12	60	32	60	30	15	48

    a. Construct a frequency distribution table with the classes 1–20, 21–40, 41–60, 61–80, and 81–100.

    b. Compute the relative frequencies and percentages for all classes.

    c. What percentage of the persons in this sample paid $6100 or more for child support?

    d. What are the class boundaries and the width of the fifth class?

2.73 The following data give the charitable contributions (in dollars) made during the 12-month period before the survey by 30 households randomly selected from the 1990 Interview Survey. (These households were selected from those who made such contributions.)

505	25	10	100	100	50	25	390	120
50	116	750	364	100	173	60	200	90
736	100	50	120	90	300	90	150	25
60	75	24						

a. Construct a frequency distribution table. Take $1 as the lower limit of the first class and 200 as the width of each class.
b. Calculate the relative frequencies and percentages for all classes.
c. What percentage of the households in this sample made charitable contributions of more than $400?

2.74 The following data give the number of orders received for a sample of 30 hours at the Timesaver Mail Order Company.

34	44	31	52	41	47	38	35	32	39
28	24	46	41	49	53	57	33	27	37
30	27	45	38	34	46	36	30	47	50

a. Construct a frequency distribution table. Take 23 as the lower limit of the first class and 7 as the width of each class.
b. Calculate the relative frequencies and percentages for all classes.
c. For what percentage of the hours in this sample was the number of orders more than 36?

2.75 The following data give the annual incomes (in thousands of dollars) for 40 production managers randomly selected from large companies.

57.6	63.3	47.3	72.5	41.2	66.1	59.6	68.5
73.3	39.4	44.5	84.9	53.7	37.7	63.3	77.4
60.2	55.9	43.1	35.6	49.3	67.4	79.2	71.9
48.8	73.2	76.0	64.3	51.8	73.5	48.8	63.5
81.5	72.7	69.4	51.5	77.5	67.9	46.1	65.1

a. Construct a frequency distribution table with the classes as $30 to less than $40 thousand, $40 to less than $50 thousand, . . . and $80 to less than $90 thousand.
b. Calculate the relative frequencies and percentages for all classes.
c. What is the width of each class?

2.76 The following data give the weekly expenditures (in dollars) on bakery products for 30 households randomly selected from the 1990 Diary Survey. (These households were selected from those who incurred such expenses.)

4.23	3.86	5.79	4.27	1.40	2.49	7.60
4.91	8.16	14.60	8.61	1.64	5.15	7.12
9.61	18.70	9.05	10.66	4.72	9.92	3.77
1.99	1.49	12.99	18.93	10.15	6.83	11.02
7.12	16.11					

a. Construct a frequency distribution table using the *less than* method to write classes. Take $0 as the lower boundary of the first class and 4 as the width of each class.
b. Calculate the relative frequencies and percentages for all classes.

2.77 Refer to Exercise 2.71. Prepare the cumulative frequency, cumulative relative frequency, and cumulative percentage distributions by using the frequency distribution table of that exercise.

2.78 Refer to Exercise 2.72. Prepare the cumulative frequency, cumulative relative frequency, and cumulative percentage distributions using the frequency distribution table constructed for the data of that exercise.

2.79 Prepare the cumulative frequency, cumulative relative frequency, and cumulative percentage distributions by using the frequency distribution table constructed in Exercise 2.75. Draw an ogive for the cumulative percentage distribution.

2.80 Using the frequency distribution constructed for the data of Exercise 2.76, construct the cumulative frequency, cumulative relative frequency, and cumulative percentage distributions. Draw an ogive for the cumulative frequency distribution.

2.81 Refer to Exercise 2.71. Prepare a stem-and-leaf display for the data of that exercise.

2.82 Construct a stem-and-leaf display for the data given in Exercise 2.72.

2.83 Refer to Exercise 2.73. Prepare a stem-and-leaf display for the data given in that exercise.

2.84 Prepare a stem-and-leaf display for the data given in Exercise 2.74.

2.85 The following table gives the gross sales for a company for the past five years.

Year	Gross Sales (millions of dollars)
1988	24
1989	28
1990	36
1991	42
1992	53

Draw two bar graphs for these data, one without truncating the axis with gross sales and the second by truncating the axis with gross sales. In the second case, mark the gross sales on the vertical axis starting with 20. Briefly comment on the two bar graphs.

2.86 The following table gives the price of the stock of a company at the beginning of each of the past six years.

Year	Price of the Stock
1988	$34
1989	47
1990	42
1991	53
1992	58
1993	60

Draw two bar graphs for these data, one without truncating the axis with stock price and the second by truncating the axis with stock price. In the second case, mark the price of the stock on the vertical axis starting with 30. Briefly comment on the two bar graphs.

SELF-REVIEW TEST

1. Briefly explain the difference between ungrouped and grouped data. Give one example of each type of such data.

2.   The following table gives the frequency distribution of the duration (in minutes) of 100 long-distance calls made by persons using TVI long-distance service.

Duration (minutes)	Frequency
0 to 4	8
5 to 9	22
10 to 14	35
15 to 19	20
20 to 24	15

Circle the correct answer for each of the following questions, which are based on this table.

a. The number of classes in the table is   5, 100, 80
b. The class width is   4, 5, 10
c. The midpoint of the third class is   11.5, 12, 12.5
d. The lower boundary of the second class is   4.5, 5, 5.5
e. The upper limit of the second class is   8.5, 9, 9.5
f. The sample size is   5, 100, 50
g. The relative frequency of the first class is   .04, .08, .16

3.   Briefly explain and illustrate with the help of graphs a symmetric histogram, a histogram skewed to the right, and a histogram skewed to the left.

4.   Classify the following as cross-section or time-series data.

a. Money spent on food during the past month by 25 families   cross
b. The value of goods imported by the United States each year from 1980 to 1993   Time
c. 1993 property tax rates for 10 cities   cross

5.   Twenty elementary school children were asked if they live with both parents (B), father only (F), mother only (M), or someone else (S). The responses of the children are as follows.

M   B   B   M   F   S   B   M   F   B
B   F   B   M   M   B   B   F   B   M

a. Construct a frequency distribution table.
b. Write the relative frequencies and percentages for all categories.
c. What percentage of the children in this sample live with mother only?
d. Draw a bar graph for the frequency distribution and a pie chart for the percentages.

6.   The following data set gives the number of years for which 24 workers have been with their current employers.

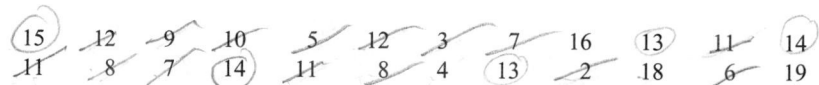

15   12   9   10   5   12   3   7   16   13   11   14
11   8   7   14   11   8   4   13   2   18   6   19

a. Construct a frequency distribution table. Take 1 as the lower limit of the first class and 4 as the width of each class.
b. Calculate the relative frequencies and percentages for all classes.
c. What percentage of the employees have been with their current employers for 8 or fewer years?
d. Draw the frequency histogram and polygon.

7.   Refer to the frequency distribution prepared in Problem 6. Prepare the cumulative percentage distribution using that table. Draw an ogive for the cumulative percentage distribution. Using the ogive, find the percentage of employees in the sample who have been with their current employers for 11 or fewer years.

8. Construct a stem-and-leaf display for the following data, which give the time (in minutes) taken by an accountant to prepare 20 income tax returns.

34	21	67	53	18	38	45	56	62	48
75	58	43	69	56	38	71	50	42	36

9. Consider the following stem-and-leaf display.

```
3 | 0 3 7
4 | 2 4 6 7 9
5 | 1 3 3 6
6 | 0 7 7
7 | 1 9
```

Write the data set that was used to construct this stem-and-leaf display.

10. The following table gives the number of new houses built in a small town each year from 1988 to 1993.

Year	Number of Houses Built
1988	105
1989	140
1990	128
1991	92
1992	74
1993	40

Draw a bar graph and a line graph for these data.

USING MINITAB

The MINITAB commands HISTOGRAM and STEM-AND-LEAF are used to construct a histogram and a stem-and-leaf display, respectively, for any quantitative data. The use of these commands is explained below.

HISTOGRAM AND FREQUENCY DISTRIBUTION

To construct a histogram for a data set, first we enter the data in column C1 using the SET command. Then we instruct MINITAB to construct a histogram for the data of column C1 using the MINITAB command given in Figure 2.21.

Figure 2.21   MINITAB command for constructing a histogram.

```
MTB > HISTOGRAM C1 ←——— {This command instructs MINITAB to construct a histogram for the
                          data of column C1.
```

The HISTOGRAM command will also list the midpoints and frequencies of various classes. Illustration M2−1 describes the procedure to construct a histogram using MINITAB.

Illustration M2−1   Refer to the data on price−earnings ratios of 25 companies given in Example 2−3. A histogram for those data is constructed as in Figure 2.22.

Figure 2.22   Histogram for Illustration M2−1.

```
MTB  > NOTE: HISTOGRAM FOR DATA ON PRICE-EARNINGS RATIOS
MTB  > SET C1
DATA > 31   13   12   22   27   33   17   26   16   22   18   13
DATA > 16   23   20   18   22   15   26   12   20   21   23   27   30
DATA > END
MTB  > HISTOGRAM C1

Histogram of C1    N = 25
```

(Continued on next page)

```
Midpoint     Count
      12         2     ** ←─────┐     ┌─ These asterisks represent the bar of the histogram for the
                                      └  first class.
      14         2     **
      16         3     ***
      18         3     ***
      20         2     **
      22         4     ****
      24         2     **
      26         2     **
      28         2     **
      30         1     *
      32         1     *
      34         1     *
```

These numbers give the frequencies of classes.

These numbers represent the midpoints of classes.

In the MINITAB output of Figure 2.22, the first column gives the midpoints of various classes. Using these midpoints, we can write the classes as 11 to less than 13, 13 to less than 15, . . . , and 33 to less than 35. The second column of the MINITAB output gives the frequencies of various classes. The asterisks in the third column represent the bars of the histogram. Note that the bars are drawn horizontally and not vertically. ▄

In Illustration M2-1, MINITAB itself decided on the number of classes and their widths. In Example 2–3 of Section 2.3.2 (see Table 2.9), we used the classes as 10–14, 15–19, . . . , 30–34 to group these data. The midpoints of these classes are 12, 17, . . . , 32, and the width of each class is 5. Suppose we decide to use the same class limits and width for our MINITAB histogram. MINITAB commands of Figure 2.23 will give us a histogram with these classes for the data entered in column C1.

Figure 2.23   MINITAB commands to construct a histogram.

```
MTB  > HISTOGRAM C1;   ←──── Note the semicolon at the end of this command.
SUBC > INCREMENT = 5;  ←──── This subcommand indicates class width.
SUBC > START = 12.     ←── ┌This subcommand instructs MINITAB that the midpoint of the first
                           └class is 12.
```

Observe these MINITAB commands carefully for semicolons and periods. Note that when we put a semicolon at the end of a MINITAB command, it instructs MINITAB that a subcommand (SUBC) is to follow with some additional information. A semicolon at the end of a subcommand indicates that another subcommand is to follow with some more information. A period at the end of a subcommand instructs MINITAB that all MINITAB commands and subcommands have been entered. In the foregoing set of MINITAB commands, we instructed MINITAB to construct a histogram for the data of column C1 with each class having a width of 5 and the first class having a midpoint of 12. Note that the class width in MINITAB is indicated using the INCREMENT command and the midpoint of the first class by using the START command. The MINITAB output of Figure 2.24 is obtained by using these commands.

Figure 2.24   Histogram for Illustration M2–1.

```
MTB  > NOTE: HISTOGRAM FOR DATA ON PRICE-EARNINGS RATIOS
MTB  > HISTOGRAM C1;
SUBC > INCREMENT = 5;
SUBC > START = 12.

Histogram of C1    N = 25

Midpoint    Count
    12.00      4    ****
    17.00      6    ******
    22.00      8    ********
    27.00      4    ****
    32.00      3    ***
```

The MINITAB DOTPLOT command plots all the data values. The dotplot in Figure 2.25 is for the data entered in column C1 in Illustration M2–1.

Figure 2.25   Dotplot for the data of Illustration M2–1.

```
MTB > DOTPLOT C1


                                            .
      :  :     .  : .  :     :  . :  :      :  :       .  .      .
   -+---------+---------+---------+---------+---------+-----C1
   12.0      16.0      20.0      24.0      28.0      32.0
```

STEM-AND-LEAF DISPLAY

To prepare a stem-and-leaf display, first the data are entered in column C1 using the SET command. Then the commands of Figure 2.26 are used to construct a stem-and-leaf display for those data.

Figure 2.26   MINITAB commands to prepare a stem-and-leaf display.

```
MTB  > STEM-AND-LEAF C1; ←——— {This command instructs MINITAB to make a
                                stem-and-leaf display for the data of column C1.
SUBC > INCREMENT = 10. ←——— {This subcommand tells MINITAB that the distance between
                              any two consecutive stems is 10 units.
```

Illustration M2–2 describes the procedure to construct a stem-and-leaf display by using MINITAB.

Illustration M2-2   Refer to data on price–earnings ratios for 25 companies given in Example 2–8 of Section 2.6. A stem-and-leaf display for those data is prepared as in Figure 2.27.

Figure 2.27   Stem-and-leaf display for Illustration M2–2.

```
MTB > NOTE: STEM-AND-LEAF DISPLAY FOR PRICE-EARNINGS RATIOS
MTB > SET C1
DATA > 31  13  12  22  27  33  17  26  16  22  18  13
DATA > 16  23  20  18  22  15  26  12  20  21  23  27  30
DATA > END
MTB > STEM-AND-LEAF C1;
SUBC > INCREMENT = 10.

  Stem-and-leaf of C1            N = 25
  Leaf unit = 1.0
                                  └──── Number of observations in the data.

     10      1  2233566788
    (12)     2  001222336677
     3       3  013

           └── These numbers are the leaves.
          └── These numbers are the stems.
     └── These numbers, called depths, are explained in text.
```

In the MINITAB display of Figure 2.27, the STEM-AND-LEAF C1 command instructs MINITAB to plot a stem-and-leaf display for the data entered in column C1. The subcommand INCREMENT = 10 indicates the distance between any two consecutive stems. As a consequence of this command, the stems will be 1 (for the numbers in teens), 2 (for the numbers in 20s), and so forth.

In the MINITAB printout of Figure 2.27, $N = 25$ is the number of observations in the data. LEAF UNIT = 1.0 means that the decimal point is after one leaf digit in the printout. Thus, the first number is 12, the second is 12, the third number is 13, and so on. If the leaf unit is .10, the numbers in the above stem-and-leaf display will be 1.2, 1.2, 1.3, and so on. (See Computer Assignment M2.8 as an example of this case.) On the other hand, if the leaf unit is 10, then the numbers will be 120, 120, 130, and so on.

In the stem-and-leaf display of Figure 2.27, if we change the increment to 5, the MINITAB will list the leaves 0 through 4 for each stem in the first row and the leaves 5 through 9 for the same stem in the second row. The output in Figure 2.28 is obtained by changing the increment to 5 in the above set of commands.

Figure 2.28   Stem-and-leaf display for Illustration M2–2.

```
MTB > NOTE: STEM-AND-LEAF DISPLAY FOR PRICE-EARNING RATIOS
MTB > SET C1
DATA > 31  13  12  22  27  33  17  26  16  22  18  13
DATA > 16  23  20  18  22  15  26  12  20  21  23  27  30
DATA > END
MTB > STEM-AND-LEAF C1;
SUBC > INCREMENT = 5.
```

(Continued on next page)

```
Stem-and-leaf of C1              N = 25
Leaf unit = 1.0

     4     1   2233
    10     1   566788
    (8)    2   00122233
     7     2   6677
     3     3   013
```

The numbers in the first column of the MINITAB display in Figure 2.28 are called *depths* and they give the cumulative frequencies from above and below. The depth, which appears in parentheses (8 in this MINITAB output), gives the number of leaves in the row that contains the median value. The depths before this row give the total number of leaves in the corresponding row and before it. Thus, the first depth, which is 4, gives the number of leaves that belong to stem 1 of the first row. The second number, which is 10, gives the cumulative number of leaves that belong to the first two rows, both of which actually belong to stem 1. After the stem that contains the median, the depths are cumulative from the bottom of the stem-and-leaf display. For example, the depth of 7 for the second row from the bottom indicates the total number of leaves that belong to the last two rows of the stem-and-leaf display. The last depth, which is 3, indicates the number of leaves for stem 3. ▬

COMPUTER ASSIGNMENTS

M2.1 Refer to Data Set I of Appendix A. Using MINITAB, construct a histogram for the data on recent share prices of companies given in column C6.

M2.2 Refer to Data Set I of Appendix A. Using MINITAB, construct a histogram for the data on price–earnings ratios of companies given in column C7.

M2.3 Refer to data on telephone charges given in column C7 of Data Set II of Appendix A. From that data set, select the 6th value and then select every 10th value after that (i.e., select the 6th, 16th, 26th, 36th, . . . values). This subsample will give you 20 measurements. (Such a sample taken from a population is called a *systematic random sample*.) Using MINITAB, construct a histogram for the phone charges for these 20 observations.

M2.4 Refer to Data Set III of Appendix A. Using MINITAB, construct a histogram for the data on income before taxes given in column C3. Use the classes $0 to less than $20,000, $20,000 to less than $40,000, $40,000 to less than $60,000, and so on.

M2.5 Refer to Data Set I of Appendix A. Using MINITAB, prepare a stem-and-leaf display for the data on recent share prices of companies given in column C6.

M2.6 Refer to Data Set I of Appendix A. Using MINITAB, prepare a stem-and-leaf display for the data on price–earnings ratios of companies given in column C7.

M2.7 Using MINITAB, prepare a stem-and-leaf display for the data given in Exercise 2.58.

M2.8 The following data give the weights (in pounds) of 20 of the parcels mailed from a post office during the past one week.

1.8	7.5	8.2	3.4	5.1	9.3	1.9	2.5	7.3	5.8
6.2	8.6	2.0	6.3	8.5	0.7	3.8	7.3	5.2	3.7

Using MINITAB, prepare a stem-and-leaf display for these data. In the subcommand, use INCREMENT = 1. Observe the value of the leaf unit in the output.

3

NUMERICAL DESCRIPTIVE MEASURES

In Chapter 2 we discussed how to organize and display large data sets. The techniques presented in that chapter, however, are not helpful when we need to describe verbally the main characteristics of a data set. The numerical summary measures, such as the ones that give the center and spread of a distribution, provide us with the main features of a data set. For example, the techniques learned in Chapter 2 can help us to graph the data on family incomes. However, we may want to know the income of a "typical" family (given by the center of the distribution), the spread of the distribution of incomes, or the location of a family with a specific income. Figure 3.1 shows these three concepts. Such questions can

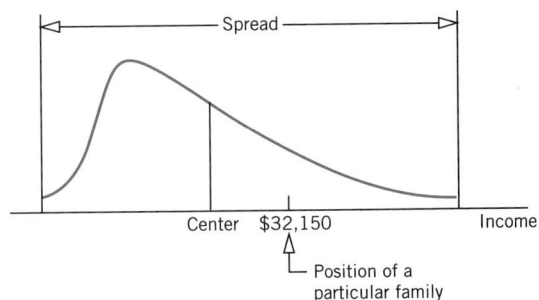

Figure 3.1

be answered using the numerical summary measures discussed in this chapter. Included among these are (1) measures of central tendency, (2) measures of dispersion, and (3) measures of position.

3.1   MEASURES OF CENTRAL TENDENCY FOR UNGROUPED DATA

We often represent a data set by numerical summary measures, usually called the *typical values*. For example, Table 3.1 gives the summary measures for data representing characteristics of female and male business executives (vice-president or higher rank). These summary measures are based on a 1992 survey of female executives and a 1989 survey of male executives, both conducted by UCLA and Korn/Ferry International.

Table 3.1   A Profile of the Average Business Executive

Characteristic	Women	Men
Age	44	52
Salary	$187,000	$289,000
Years at company	12	17
Percentage married	69%	91%
Number of children	1	3
Hours worked per week	56	56
Have employed spouse	81%	55%

Source: USA TODAY, June 30, 1993. Copyright © 1993
USA TODAY. Table reproduced with permission.

As we can observe, these summary measures (averages and percentages) are based on data sets that contained hundreds of business executives. Data obtained on each variable for women and men, respectively, are represented by one number. The summary measures that give the averages are called the **measures of central tendency**. A measure of central tendency gives the center of a histogram or a frequency distribution curve. This section discusses three different measures of central tendency: the mean, the median, and the mode. We will learn how to calculate each of these measures for ungrouped data. Recall from Chapter 2 that data containing information on each member of the population or sample individually are called *ungrouped data*, whereas *grouped data* refer to the data presented in the form of a frequency distribution table.

3.1.1   MEAN

The **mean**, also called the *arithmetic mean*, is the most frequently used measure of central tendency. This book will use the words *mean* and *average* synonymously. For ungrouped data, the mean is obtained by dividing the sum of all values by the number of values in the data set.

$$\text{Mean} = \frac{\text{Sum of all values}}{\text{Number of values}}$$

The mean calculated for sample data is denoted by $\bar{x}$ (read as ''*x* bar''), and the mean calculated for population data is denoted by μ (Greek letter *mu*). We know from the discussion in Chapter 2 that the number of values in a data set is denoted by *n* for a sample and by *N* for a population. In Chapter 1 we learned that a variable is denoted by *x* and the sum of all values of *x* is denoted by Σx. Using these notations, we can write the following formulas for the mean.

> **MEAN FOR UNGROUPED DATA**
>
> The mean for ungrouped data is obtained by dividing the sum of all values by the number of values in the data set. Thus,
>
> $$\text{Mean for population data:} \quad \mu = \frac{\Sigma x}{N}$$
>
> $$\text{Mean for sample data:} \quad \bar{x} = \frac{\Sigma x}{n}$$
>
> where Σx is the sum of all values, N is the population size, n is the sample size, μ is the population mean, and $\bar{x}$ is the sample mean.

Calculating sample mean for ungrouped data.

EXAMPLE 3-1 The following data give the 1992 profits (in millions of dollars) of a sample of five companies.

Company	1992 Profits (millions of dollars)
General Electric	4725
Coca-Cola	1884
AT&T	3807
Philip Morris	4939
Johnson & Johnson	1625

Source: Business Week, 1993 Special Issue.

Find the mean of the 1992 profits for these companies.

Solution The variable in this example is the *1992 profits of companies*. Let us denote it by x. Then the five values of x are

$$x_1 = 4725, \quad x_2 = 1884, \quad x_3 = 3807, \quad x_4 = 4939, \quad \text{and} \quad x_5 = 1625$$

where x_1 represents the 1992 profits of the first company, x_2 denotes the 1992 profits of the second company, and so on. The sum of the profits of these five companies is

$$\Sigma x = x_1 + x_2 + x_3 + x_4 + x_5$$
$$= 4725 + 1884 + 3807 + 4939 + 1625 = 16{,}980$$

Note that the given information is on only five companies. Hence, it represents a sample. Because the data set contains five values, $n = 5$. Substituting the values of Σx and n in the sample formula, the mean of the 1992 profits of these five companies is

$$\bar{x} = \frac{\Sigma x}{n} = \frac{16{,}980}{5} = \textbf{\$3396.00 million}$$

Thus, these five companies earned an average profit of $3396.00 million in 1992. ▪

Physically, the mean is the point that balances a histogram. If we consider the mean as a fulcrum, the histogram will balance on the fulcrum. This is shown in Figure 3.2 for data of Example 3-1.

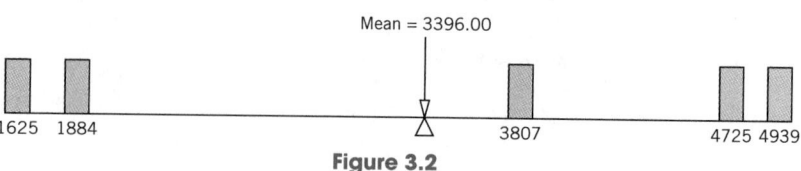

Figure 3.2

Calculating population mean for ungrouped data.

EXAMPLE 3-2   The following are the weekly earnings of all eight employees of a small company.

$450   530   825   370   615   480   910   560

Find the mean weekly earnings of these employees.

Solution   Since the given data set includes all eight employees of the company, it represents the population. Hence, $N = 8$.

$$\Sigma x = 450 + 530 + 825 + 370 + 615 + 480 + 910 + 560 = 4740$$

The population mean is

$$\mu = \frac{\Sigma x}{N} = \frac{4740}{8} = \textbf{\$592.50}$$

Thus, the mean weekly earnings of the employees of this company are $592.50.   ■

Reconsider Example 3–2. Now if we take a sample of three employees from this company and calculate the mean weekly earnings of those three employees, this mean will be denoted by $\bar{x}$. Suppose the three values included in the sample are $530, $480, and $910. Then, the mean weekly earnings for this sample are

$$\bar{x} = (530 + 480 + 910)/3 = \textbf{\$640}$$

If we take a second sample of three employees of this company, the value of $\bar{x}$ will (most likely) be different. Suppose the second sample includes the values $450, $825, and $480. Then, the mean weekly earnings for this sample are

$$\bar{x} = (450 + 825 + 480)/3 = \textbf{\$585}$$

Consequently, we can state that the value of the population mean μ is constant. However, the value of the sample mean $\bar{x}$ varies from sample to sample. The value of $\bar{x}$ for a particular sample will depend on what values of the population are included in that sample.

Sometimes a data set may contain a few very small or a few very large values. Such values are called **outliers** or **extreme values**.

OUTLIERS OR EXTREME VALUES

Values that are very small or very large relative to the majority of the values in a data set are called outliers or extreme values.

A major shortcoming of the mean as a measure of central tendency is that it is very sensitive to outliers. Example 3–3 illustrates this point.

Illustrating the effect of an outlier on the mean.

EXAMPLE 3–3   The following table lists the 1991 total pay (salary and bonus plus long-term compensation) of the chief executive officers (CEOs) of five U.S. companies.

Chief Executive Officer	Company	Total Pay of the CEO (thousands of dollars)
Louis V. Gerstner Jr.	RJR Nabisco	6,776
Lawrence G. Rawl	Exxon	9,266
William E. Lamothe	Kellogg	9,383
H. Brewster Atwater Jr.	General Mills	6,469
Anthony O'Reilly	H.J. Heinz	75,085 ←——— An outlier

Source: Business Week, May 4, 1992.

Notice that the total pay of the CEO of H.J. Heinz is very large compared to the total pay of each of the other four CEOs. Hence, it is an outlier. Show how the inclusion of this outlier affects the value of the mean.

Solution   If we do not include the CEO of H.J. Heinz (the outlier), the mean pay of the remaining four CEOs is

$$\text{Mean} = (6776 + 9266 + 9383 + 6469)/4 = \textbf{\$7973.50 thousand}$$

Now, to see the impact of the outlier on the value of the mean, we include the pay of the CEO of H.J. Heinz and find the mean pay of the five CEOs. This mean is

$$\text{Mean} = (6,776 + 9,266 + 9,383 + 6,469 + 75,085)/5$$
$$= \textbf{\$21,395.80 thousand}$$

Thus, including the pay of the CEO of H.J. Heinz causes nearly a threefold increase in the mean pay as it changes from $7,973.50 thousand to $21,395.80 thousand. ■

The above example should encourage us to be cautious. We should remember that the mean is not always the best measure of central tendency because it is heavily influenced by outliers. Sometimes other measures of central tendency give a more accurate impression of a data set.

Case Study 3–1 gives a comparative look at the mean annual compensations of CEOs in the United States, Japan, and Germany. It also gives the mean annual incomes of families in the United States with different characteristics.

CASE STUDY 3-1   A LOOK AT THE AVERAGES

The CEO pay gap flap

Average U.S. CEO of large company is paid 85 to 160 times as much as the average worker.

Average Japanese CEO of large company is paid 10 to 20 times as much as the average worker.

A look at the averages

Average annual CEO compensation, including pay and benefits, for companies with revenue of $1 billion or more:

USA	**$3 million**
Japan	**$600,000**
Germany	**$1.1 million**

Average annual CEO compensation, including pay and benefits, for companies with revenue of about $250 million:

USA	**$748,000**
Japan	**$370,000**
Germany	**$364,500**

Source: Towers Perrin, USA TODAY research

By Marty Baumann, USA TODAY

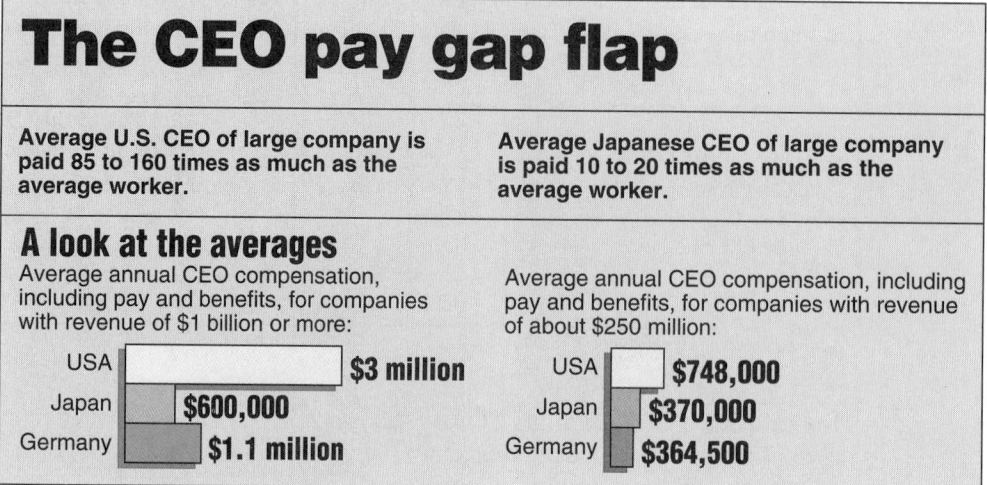

The two bar graphs show the mean annual compensation of CEOs of companies with revenue of $1 billion or more and for companies with revenue of about $250 million for the United States, Japan, and Germany. According to *USA TODAY*, the average U.S. CEO of a large company is paid 85 to 160 times as much as the average worker. In contrast to this, the average Japanese CEO of a large company is paid 10 to 20 times as much as the average worker.

Table 3.2 lists the average incomes for 1991 of U.S. households with certain characteristics. According to the U.S. Bureau of the Census, "A household consists of all persons who occupy a housing unit. . . . A household includes the related family members and all the unrelated persons, if any, . . . who share the housing unit. A person living alone in a housing unit, or a group of unrelated persons sharing a housing unit as partners, is also counted as a household.''

Table 3.2   1991 Average Incomes of U.S. Households

Characteristics of Households	1991 Mean Income (in dollars)
1. All households	37,922
2. Nonfarm households	37,935
3. Farm households	37,150
4. Family households:	43,704
Married-couple families	48,589
Male householder, no wife present	36,936
Female householder, no husband present	23,535
5. Nonfamily households:	24,292
Male householder	29,746
Female householder	20,074
6. Number of earners:	
No earners	16,389
One earner	31,352
Two earners or more	53,130
7. Educational attainment:	
Less than 9th grade	18,142
9th to 12th grade (no diploma)	22,925
High school graduate or equivalent	33,275
Some college, no degree	40,167
Associate degree	44,041
Bachelor's degree	56,134
Master's degree	63,375
Professional degree	91,765
Doctorate degree	79,902

Source: The bar graphs are reproduced with permission from *USA TODAY*, January 9, 1992. Copyright © *USA TODAY* 1992. The cartoon is reproduced with permission of Gary Huck. Copyright © Huck/UE/Huck-Konopacki Cartoons. The table is based on the *Money Income of Households, Families, and Persons in the United States: 1991*, Current Population Reports, Series P-60, No. 180, U.S. Bureau of the Census, August 1992.

3.1.2   MEDIAN

Another important measure of central tendency is the **median**. It is defined as follows.

MEDIAN

The median is the value of the middle term in a data set that has been ranked in increasing order.

As is obvious from the definition of the median, it divides a ranked data set into two equal parts. The calculation of the median consists of the following two steps.

1. Rank the given data set in increasing order.
2. Find the middle term. The value of this term is the median.[1]

The position of the middle term in a data set with n values is obtained as follows.

$$\text{Position of the middle term} = \frac{n + 1}{2}$$

Thus, we can redefine the median as follows.

MEDIAN FOR UNGROUPED DATA

$$\text{Median} = \text{the value of the } \left(\frac{n + 1}{2}\right)\text{th term in a ranked data set}$$

If the given data set represents a population, replace n by N.

If the number of observations in a data set is *odd*, then the median is given by the value of the middle term in the ranked data. If the number of observations is *even*, then the median is given by the average of the values of the two middle terms.

Calculating median for ungrouped data: odd number of data values.

EXAMPLE 3-4   The following table lists the market values (share price on March 5, 1993, multiplied by the number of common shares outstanding) of five U.S. discount and fashion retailing companies.

Company	Market Value (millions of dollars)
Kmart	10,007
Toys 'R' Us	11,548
J.C. Penney	9,693
Wal-Mart Stores	75,275
Sears, Roebuck	18,296

Source: Business Week 1993 Special Issue.

Find the median of the market values of these companies.

Solution   First, we rank the data in increasing order as follows.

$$9{,}693 \qquad 10{,}007 \qquad 11{,}548 \qquad 18{,}296 \qquad 75{,}275$$

There are five observations in the data set. Consequently, $n = 5$ and

$$\frac{n + 1}{2} = \frac{5 + 1}{2} = 3$$

---

[1]The value of the middle term in a data set ranked in decreasing order will also give the value of the median.

Therefore, the median is the value of the third term in the ranked data.

$$9{,}693 \qquad 10{,}007 \qquad \boxed{11{,}548} \qquad 18{,}296 \qquad 75{,}275$$

$$\uparrow$$

Median

Thus, the median market value for this sample of five companies is $11,548 million, which is the market value of Toys 'R' Us. ■

Calculating median for ungrouped data: even number of data values.

EXAMPLE 3–5 The following table lists the number of patents granted by the U.S. government to 12 automobile companies in 1991.

Company	Number of Patents Granted
General Motors	506
Ford Motor	234
Chrysler	97
Nissan Motor	385
Porsche	50
Honda Motor	249
Mazda Motor	210
Volvo	23
Daimler-Benz	275
Toyota Motor	257
Mitsubishi Motors	36
BMW	13

Source: *Business Week*, August 3, 1992.

Find the median for the number of patents granted to these companies.

Solution First, we rank the data in increasing order as follows.

$$13 \quad 23 \quad 36 \quad 50 \quad 97 \quad 210 \quad 234 \quad 249 \quad 257 \quad 275 \quad 385 \quad 506$$

There are 12 values in the data set. Hence, $n = 12$ and

$$\frac{n + 1}{2} = \frac{12 + 1}{2} = 6.5$$

Therefore, the median is given by the mean of the sixth and seventh values in the ranked data.

$$13 \quad 23 \quad 36 \quad 50 \quad 97 \quad \boxed{210} \quad \boxed{234} \quad 249 \quad 257 \quad 275 \quad 385 \quad 506$$

$$\uparrow$$

$$\text{Median} = \frac{210 + 234}{2} = \mathbf{222}$$

Thus, the median number of patents granted by the U.S. government to these automobile companies in 1991 was 222. ■

The median gives the center of a histogram, with half of the data values to the left of the median and half to the right of the median. The advantage of using the median as a measure of central tendency is that it is not influenced by outliers. Consequently, the median is preferred over the mean as a measure of central tendency for data sets that contain outliers. Case Study 3–2 is an example of such a case.

CASE STUDY 3–2   MEDIAN SALE PRICES OF EXISTING SINGLE-FAMILY HOMES FOR METROPOLITAN AREAS

The Economics and Research Division of the National Association of Realtors obtains data on existing single-family home sales from more than 600 Boards/Associations of Realtors and multiple-listing systems across the country. Based on these data, the National Association of Realtors publishes descriptive statistics and graphs representing the characteristics of the existing single-family homes for the United States, for different regions of the country, for individual states, and for selected metropolitan areas. These data appear in the Association's monthly journal *Home Sales*, which is now called *Real Estate Outlook*. Table 3.3 gives the median sale prices of existing single-family homes for the year 1992 for a few selected metropolitan areas from a recent issue of *Home Sales*.

Table 3.3   Median Sale Prices of Existing Single-family Homes for 1992 for Selected Metropolitan Areas

Metropolitan Area	Median Price (in dollars)
Baltimore, MD	113,400
Boston, MA	171,100
Chattanooga, TN	73,000
Chicago, IL	136,800
Denver, CO	96,200
Houston, TX	80,200
Los Angeles area, CA	213,200
New York City area, NY/NJ/CT	172,700
Philadelphia, PA	117,000
San Francisco Bay area, CA	254,800
Washington, DC	157,800

Table 3.4 gives the mean and median prices of existing single-family homes for the United States and its four regions for the year 1992.

Table 3.4   Mean and Median Sale Prices of Existing Single-family Homes

	All Regions	Northeast	Midwest	South	West
Mean price	$130,900	$165,100	$95,200	$116,500	$175,100
Median price	$103,700	$140,000	$81,700	$92,100	$143,800

As the distribution of sale prices of existing single-family homes is expected to be skewed to the right, the mean price is affected by extreme values at the upper end. Hence, the mean price is higher than the median price of single-family homes for the United States and for each of the four regions. In such a case, the median is a better measure of central tendency.

Source: Home Sales, 7(6), June 1993. Copyright © 1993 NATIONAL ASSOCIATION OF REALTORS®, Washington, D.C. Data reproduced with permission. *Home Sales* is now called *REAL ESTATE OUTLOOK: Market Trends and Insights*. These data cannot be reproduced without the written permission of the NATIONAL ASSOCIATION OF REALTORS®.

3.1.3 MODE

Mode is a French word that means *fashion*—an item that is most popular or common. In statistics, the mode represents the most common value in a data set.

> **MODE**
>
> The mode is the value that occurs with the highest frequency in a data set.

Calculating mode for ungrouped data.

EXAMPLE 3–6 The following data give the number of years eight employees have been with their current employers.

$$11 \quad 9 \quad 13 \quad 6 \quad 8 \quad 9 \quad 20 \quad 3$$

Find the mode.

Solution In this data set, 9 occurs twice and each of the remaining values occurs only once. Because 9 occurs with the greatest frequency, it is the mode. Therefore,

$$\text{Mode} = \textbf{9 years}$$ ■

A major shortcoming of the mode is that a data set may have no or more than one mode, whereas it will have only one mean and only one median. For instance, a data set with each value occurring only once has no mode. A data set with only one value occurring with highest frequency has only one mode. The data set in this case is called **unimodal**. A data set with two values occurring with the same (highest) frequency has two modes. The distribution, in this case, is said to be **bimodal**. If more than two values in a data set occur with the same (highest) frequency, then the data set contains more than two modes and it is said to be **multimodal**.

Data set with no mode.

EXAMPLE 3–7 Last year's incomes of five randomly selected families were $26,150, 65,750, 34,985, 47,490, and 13,740. Find the mode.

Solution As each value in this data set occurs only once, this data set contains no mode. ■

Data set with two modes.   **EXAMPLE 3-8**   The prices of the same brand of television set at eight stores are found to be $495, 486, 503, 495, 470, 505, 470, and 499. Find the mode.

Solution   In this data set, each of the two values $495 and $470 occurs twice and each of the remaining values occurs only once. Therefore, this data set has two modes: **$495** and **$470**.   ▬

Data set with three modes.   **EXAMPLE 3-9**   The ages of 10 randomly selected students from a class are 21, 19, 27, 22, 29, 19, 25, 21, 22, and 30. Find the mode.

Solution   This data set has three modes: **19**, **21**, and **22**. Each of these three values occurs with a (highest) frequency of 2.   ▬

One advantage of the mode is that it can be calculated for both kinds of data, quantitative and qualitative, whereas the mean and median can be calculated only for quantitative data.

Finding mode for qualitative data.   **EXAMPLE 3-10**   The status of five students, who are members of the student senate at a college, are senior, sophomore, senior, junior, senior. Find the mode.

Solution   As *senior* occurs more frequently than the other categories, it is the mode for this data set. However, we cannot calculate the mean and median for this data set.   ▬

To sum up, we cannot conclude which of the three measures of central tendency is a better measure overall. Each of them may be better under different situations. Probably the mean is the most used measure of central tendency followed by the median. The mean has the advantage that its calculation includes each value of the data set. The median is a better measure when a data set includes outliers. The mode is simple to locate, but it is not of much use in practical applications.

3.1.4   RELATIONSHIP BETWEEN THE MEAN, MEDIAN, AND MODE

As discussed in Chapter 2, two of the many shapes that a histogram or a frequency distribution curve can assume are symmetric and skewed. This section describes the relationship between the mean, median, and mode for three such histograms and frequency curves. Knowing the values of the mean, median, and mode can give us some idea about the shape of a frequency curve.

1.   For a symmetric histogram and frequency curve with one peak (see Figure 3.3), the values of the mean, median, and mode are identical and they lie at the center of the distribution.

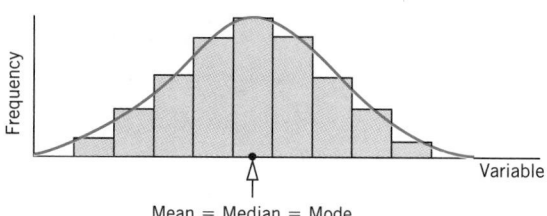

Figure 3.3   Mean, median, and mode for a symmetric histogram and a frequency curve.

2. For a histogram and a frequency curve skewed to the right (see Figure 3.4), the value of the mean is the largest, that of the mode is the smallest, and the value of the median lies between these two. (Notice that the mode always occurs at the peak point.) The value of the mean is the largest in this case because it is sensitive to outliers that occur in the right tail. These outliers pull the mean to the right.

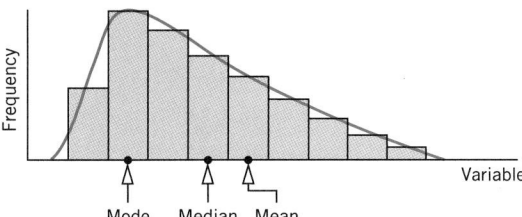

Figure 3.4 Mean, median, and mode for a histogram and a frequency curve skewed to the right.

3. If a histogram and a distribution curve are skewed to the left (see Figure 3.5), the value of the mean is the smallest and that of the mode is the largest, with the value of the median lying between these two. In this case, the outliers in the left tail pull the mean to the left.

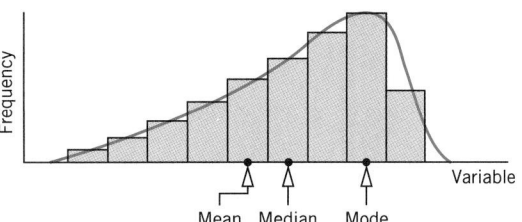

Figure 3.5 Mean, median, and mode for a histogram and a frequency curve skewed to the left.

EXERCISES

Concepts and Procedures

3.1 Explain how the value of the median is determined for a data set that contains an odd number of observations and for a data set that contains an even number of observations.

3.2 Briefly explain the meaning of an outlier. Is the mean or the median a better measure of central tendency for a data set that contains an outlier? Illustrate with the help of an example.

3.3 Using an example, show how an outlier can affect the value of the mean.

3.4 Which of the three measures of central tendency (the mean, the median, and the mode) can be calculated for quantitative data only, and which one can be calculated for both quantitative and qualitative data? Illustrate with examples.

3.5 Which of the three measures of central tendency (the mean, the median, and the mode) can assume more than one value for a data set? Give an example of a data set for which this summary measure assumes more than one value.

3.6 Is it possible for a (quantitative) data set to have no mean, no median, or no mode? Give an example of a data set for which this summary measure does not exist.

3.7   Explain the relationship between the mean, median, and mode for symmetric and skewed histograms. Illustrate these relationships with graphs.

3.8   Prices of cars have a distribution that is skewed to the right with outliers in the right tail. Which of the measures of central tendency is the best to summarize this data set? Explain.

3.9   The following data set belongs to a population.

$$5 \quad -7 \quad 2 \quad 0 \quad -9 \quad 12 \quad 10 \quad 7$$

Calculate the mean, median, and mode.

3.10   The following data set belongs to a sample.

$$14 \quad 11 \quad -10 \quad 8 \quad 8 \quad -16$$

Calculate the mean, median, and mode.

Applications

Exercises 3.11 through 3.15 are based on the following data.

The following table, based on the American Chamber of Commerce Researchers Association Survey, gives the prices of five items in 12 urban areas across the United States. (See Data Set II of Appendix A.)

City	Apartment Rent	Price of House	Phone Bill	Cost of a Hospital Room	Price of Wine
Huntsville (AL)	441	105,111	23.69	276.67	5.95
Phoenix (AZ)	531	100,748	17.79	358.14	4.40
San Diego (CA)	900	223,600	12.65	527.88	5.41
Denver (CO)	604	125,200	19.72	433.00	4.39
Bloomington (IN)	518	105,613	21.14	433.00	4.67
New Orleans (LA)	508	88,000	22.90	318.80	4.52
Manchester (NH)	667	125,750	21.04	379.50	4.24
Albuquerque (NM)	547	130,550	20.86	327.00	4.93
Albany (NY)	622	116,363	36.47	287.40	5.91
Charlotte (NC)	412	115,800	18.12	293.67	4.69
Salem (OR)	490	108,656	17.91	325.00	4.51
Charleston (WV)	551	126,000	29.93	289.60	6.05

Explanation of variables

Apartment rent   Monthly rent of an unfurnished two-bedroom apartment (excluding all utilities except water), 1½ or 2 baths, approximately 950 square feet

Price of house   Purchase price of a new house with 1800 square feet of living area and on an 8000-square-foot lot in an urban area with all utilities

Phone bill   Monthly telephone charges for a private residential line (customer owns instruments)

Cost of a hospital room   Average cost per day of a semiprivate room in a hospital

Price of wine   Price of Paul Masson Chablis, 1.5-liter bottle

3.11   Calculate the mean and median for data on apartment rents.

3.12   Find the mean and median for data on the prices of houses.

3.13   Calculate the mean and median for data on phone bills. Do these data have a mode?

3.14   Calculate the mean and median for data on the cost of hospital rooms. Do these data have a mode?

3.15   Find the mean and median for data on the price of wine.

3.16 The following data give the number of car thefts that occurred in a city during the past 12 days.

6   3   7   11   5   3   8   7   2   6   9   13

Find the mean, median, and mode.

3.17 The following data give the farm value (in millions of dollars) of wheat produced in 10 states in 1990. (*Source:* National Agricultural Statistics Service.) The data, entered by row, is for the states of Idaho, Illinois, Kansas, Montana, Minnesota, North Dakota, Oklahoma, South Dakota, Texas, and Washington.

100   91   472   146   139   385   202   128   130   150

Calculate the mean and median. Do these data have a mode? Why or why not?

3.18 The following table lists the prices of seven different brands and models of air conditioners.

Brand and Model of Air Conditioner	Price (dollars)
Panasonic CW-601JU	405
General Electric AME06LA	400
Quasar HQ5061DW	435
Sharp AF-602M6	375
Sears Kenmore 76069	370
Whirlpool ACQ062XW	410
Signature KMJ-5816	325

Source: Consumer Reports, July 1992, p. 422. Copyright © 1992 by Consumer Union of United States, Inc., Yonkers, NY. Adapted and reproduced with permission. Results and conclusions not endorsed by Consumer Union.

Find the mean and median. Do these data have a mode? Why or why not?

3.19 The following table lists the 1992 total assets (in millions of dollars) of seven U.S. companies.

Company	Total Assets (millions of dollars)
General Electric	192,876
GTE	42,100
American Home Products	7,010
Chevron	33,970
Walt Disney	11,986
Ford Motor	180,545
Sears, Roebuck	111,664

Source: Business Week 1993 Special Issue.

Calculate the mean and median. Do these data have a mode?

3.20 Nixon Corporation manufactures computer terminals. The following data give the number of computer terminals produced at the company for a sample of 10 days.

24   32   27   23   35   33   29   21   23   28

Calculate the mean, median, and mode for these data.

3.21 The following data give the number of computer keyboards assembled at the Twentieth Century Electronics Company for a sample of 12 days.

| 45 | 52 | 48 | 41 | 56 | 46 | 44 | 42 | 48 | 53 | 43 | 50 |

Calculate the mean, median, and mode for these data.

3.22 The following data give the amount (in dollars) of personal taxes paid during the 12-month period before the survey by seven households selected from the 1990 Diary Survey mentioned in Chapter 2. (These households were selected from those who paid taxes.)

| 1,038 | 2,604 | 895 | 17,411 | 1,400 | 3,428 | 5,447 |

   a. Find the mean and median for these data.
   b. Do these data contain an outlier? If yes, drop this value and recalculate the mean and median. Which of the two summary measures changes by a larger amount when you drop the outlier?
   c. Is the mean or the median a better summary measure for these data?

3.23 The following data give the savings accounts balances (in dollars) at the time of the survey for nine households selected from the 1990 Interview Survey mentioned in Chapter 2. (These households were selected from those who had savings accounts.)

| 2,500 | 245 | 70,000 | 1,200 | 150 | 2,200 | 700 | 2,000 | 3,200 |

   a. Calculate the mean and median for these data.
   b. Do these data contain an outlier? If yes, drop this value and recalculate the mean and median. Which of the two summary measures changes by a larger amount when you drop the outlier?
   c. Is the mean or the median a better summary measure for these data?

3.24 The following data give the estimated market value (in dollars) of all stocks, bonds, mutual funds, and other such securities held at the time of the survey by eight households selected from the 1990 Interview Survey. (These households were selected from those who held such securities.)

| 3,250 | 11,028 | 8,900 | 90,000 | 2,250 | 18,000 | 6,000 | 12,000 |

   a. Calculate the mean and median for these data. Do these data have a mode?
   b. Do these data contain an outlier? If yes, drop this value and recalculate the mean and median. Which of the two summary measures changes by a larger amount when you drop the outlier?
   c. Is the mean or the median a better summary measure for these data?

3.25 At Smith's Foundry, the machine that makes 3-inch-long nails is adjusted from time to time to make sure that the length of these nails is within certain limits. (Note that all nails produced by this machine will not be exactly 3 inches long.) From time to time, the quality control officer takes a sample of 13 nails from the production line and calculates the mean length of these nails. If this sample mean is either smaller than 2.95 inches or larger than 3.05 inches, the machine is stopped and adjusted.

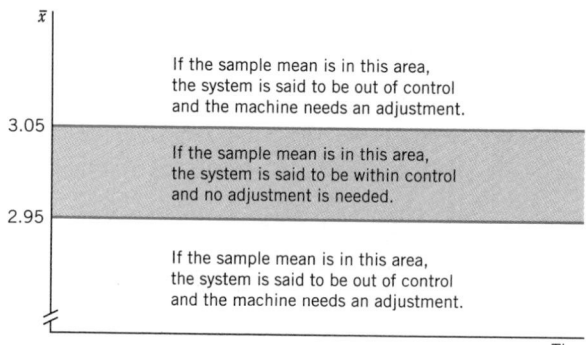

One such recent sample of 13 nails produced the following data on their lengths.

3.05	3.15	3.30	2.95	3.35	3.25	3.20
2.90	3.25	3.15	3.05	3.10	3.25	

Based on this sample data, do you think the machine needs an adjustment?

3.26 One of the items produced by Soho International is detergents. A machine that fills 64-ounce liquid detergent jugs is adjusted from time to time to make sure that the mean amount of liquid detergent in all such jugs is not less than 64 ounces. (Note that all 64-ounce jugs filled by this machine will not contain exactly 64 ounces of detergent.) Quite often, the quality control officer takes a sample of 10 jugs from the production line and measures the amount of detergent in each of these jugs and calculates the sample mean. The quality control officer has established the rule that if this sample mean is less than 64.5 ounces, the machine will be stopped and adjusted. A sample of 10 such jugs taken on a certain day produced the following data on the amount of detergent in these jugs.

64.24	63.75	64.10	63.85	64.10
64.45	63.90	64.29	64.20	64.35

Based on this sample data, do you think the machine needs an adjustment?

*3.27 One property of the mean is that if we know the means and sample sizes of two (or more) data sets, we can calculate the *combined mean* of both (or all) data sets. The combined mean for two data sets is calculated by using the formula

$$\text{Combined mean} = \bar{x} = \frac{n_1\bar{x}_1 + n_2\bar{x}_2}{n_1 + n_2}$$

where n_1 and n_2 are the sample sizes of the two data sets and $\bar{x}_1$ and $\bar{x}_2$ are the means of the two data sets, respectively. Suppose a sample of 10 statistics books gave a mean price of $41 and a sample of 8 mathematics books gave a mean price of $43. Find the combined mean. (*Hint:* For this example $n_1 = 10$, $n_2 = 8$, $\bar{x}_1 = \$41$, $\bar{x}_2 = \$43$.)

*3.28 For any data, the sum of all values is equal to the product of the sample size and mean, that is, $\Sigma x = n\bar{x}$. Suppose the average amount of money spent on shopping by 10 persons during a given week is $85.50. Find the total amount of money spent on shopping by these 10 persons.

*3.29 The mean age of six persons is 46. The ages of five of these six persons are 57, 39, 44, 51, and 37. Find the age of the sixth person.

*3.30 Consider the following two data sets.

Data Set I:	12	25	37	8	41
Data Set II:	19	32	44	15	48

Notice that each value of the second data set is obtained by adding 7 to the corresponding value of the first data set. Calculate the mean for each of these two data sets. Comment on the relationship between the two means.

*3.31 Consider the following two data sets.

Data Set I:	4	8	15	9	11
Data Set II:	8	16	30	18	22

Notice that each value of the second data set is obtained by multiplying the corresponding value of the first data set by 2. Calculate the mean for each of these two data sets. Comment on the relationship between the two means.

*3.32   The *trimmed mean* is calculated by dropping a certain percentage of values from each end of a ranked data set. The trimmed mean is especially useful as a measure of central tendency when a data set contains a few outliers at each end. Suppose the following data give the ages of 10 employees of a company.

$$47 \quad 53 \quad 38 \quad 26 \quad 39 \quad 49 \quad 19 \quad 67 \quad 31 \quad 23$$

To calculate the 10% trimmed mean, first rank these data values in increasing order, then drop 10% of the smallest values and 10% of the largest values. The mean of the remaining 80% of the values will give the 10% trimmed mean. As this data set contains 10 values, 10% of 10 is 1. Hence, drop the smallest value and the largest value from this data set. The mean of the remaining 8 values will be the 10% trimmed mean. Calculate the 10% trimmed mean for this data set.

3.2   MEASURES OF DISPERSION FOR UNGROUPED DATA

The measures of central tendency, such as the mean, median, and mode, do not reveal the whole picture of the distribution of a data set. Two data sets with the same mean may have completely different spreads. The variation among values of observations for one data set may be much larger or smaller than for the other data set. (Note that the words *dispersion*, *spread*, and *variation* have the same meaning.) Consider the following two data sets on the ages of all workers for each of two small companies.

$$\begin{array}{lccccccc}
\text{Company 1:} & 47 & 38 & 35 & 40 & 36 & 45 & 39 \\
\text{Company 2:} & & 70 & 33 & 18 & 52 & 27 &
\end{array}$$

The mean age of workers of each of these two companies is the same, 40 years. If we do not know the ages of individual workers for these two companies and are told only that the mean age of the workers for both companies is the same, we may deduce that the workers of these two companies have a similar age distribution. But, as we can observe, the variation in the workers' ages in each of these two companies is very different. As illustrated in the diagram, the ages of the workers of the second company have a much larger variation than the ages of the workers of the first company.

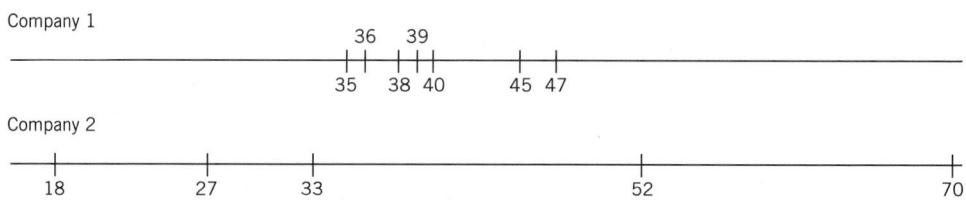

Thus, the mean, median, or mode is usually not by itself a sufficient measure to reveal the shape of the distribution of a data set. We also need a measure that can provide some information about the variation among data values. The measures that help us to know about the spread of a data set are called the **measures of dispersion**. The measures of central tendency and dispersion taken together give a better picture of a data set than the measures of central tendency alone. This section discusses four measures of dispersion: range, variance, standard deviation, and coefficient of variation.

3.2.1 RANGE

The **range** is the simplest measure of dispersion to calculate. It is obtained by taking the difference between the largest and the smallest values in a data set.

RANGE FOR UNGROUPED DATA

$$\text{Range} = \text{Largest value} - \text{Smallest value}$$

Calculating range for ungrouped data.

EXAMPLE 3–11 The following data give the farm value (in millions of dollars) of the production of corn for 10 states for the year 1989.

State	Farm Value of Corn (millions of dollars)	State	Farm Value of Corn (millions of dollars)
Iowa	3108	Ohio	804
Kansas	349	Pennsylvania	272
Kentucky	335	South Dakota	391
Michigan	490	Texas	393
Missouri	495	Washington	684

Find the range for this data set.

Solution The highest farm value of corn for a state in this data set is $3108 million and the lowest farm value of corn for a state is $272 million. Therefore,

$$\text{Range} = \text{Largest value} - \text{Smallest value} = 3108 - 272 = \textbf{\$2836 million}$$

Thus, the farm values of the production of corn for these 10 states are spread over a range of $2836 million.

The range, like the mean, has the disadvantage of being influenced by outliers. In Example 3–11, if the state of Iowa with a value of $3108 million is dropped, the range decreases from $2836 million to $532 million. Consequently, the range is not a good measure of dispersion to use for a data set that contains outliers.

Another disadvantage of using the range as a measure of dispersion is that its calculation is based on two values only: the largest and the smallest. All other values in a data set are ignored while calculating the range. Thus, the range is not a very satisfactory measure of dispersion.

3.2.2 VARIANCE AND STANDARD DEVIATION

The **standard deviation** is the most used measure of dispersion. The value of the standard deviation tells how closely the values of a data set are clustered around the mean. In general, a lower value of the standard deviation for a data set indicates that the values of that data set are spread over a relatively smaller range around the mean. On the other hand, a large value of the standard deviation for a data set indicates that the values of that data set are spread over a relatively larger range around the mean.

The standard deviation is obtained by taking the positive square root of the **variance**. The variance calculated for population data is denoted by σ^2 (read as sigma squared),[2] and

---

[2]Note that Σ is uppercase sigma and σ is lowercase sigma of the Greek alphabet.

the variance calculated for sample data is denoted by s^2. Consequently, the standard deviation calculated for population data is denoted by σ, and the standard deviation calculated for sample data is denoted by s. Following are the *basic formulas* that are used to calculate the variance.[3]

$$\sigma^2 = \frac{\Sigma(x - \mu)^2}{N} \quad \text{and} \quad s^2 = \frac{\Sigma(x - \bar{x})^2}{n - 1}$$

where σ^2 is the population variance and s^2 is the sample variance.

The quantity $x - \mu$ or $x - \bar{x}$ in the above formulas is called the *deviation* of x value from the mean.

The sum of the deviations of x values from the mean is always zero. That is, $\Sigma(x - \mu) = 0$ and $\Sigma(x - \bar{x}) = 0$. For example, suppose the midterm scores of a sample of four students are 82, 95, 67, and 92. Then, the mean score for these four students is

$$\bar{x} = (82 + 95 + 67 + 92)/4 = 84$$

The deviations of the four scores from the mean are calculated in Table 3.5.

Table 3.5

x	$x - \bar{x}$
82	$82 - 84 = -2$
95	$95 - 84 = +11$
67	$67 - 84 = -17$
92	$92 - 84 = +8$
	$\Sigma(x - \bar{x}) = 0$

As we can observe from Table 3.5, the sum of the deviations of x values from the mean is zero, that is, $\Sigma(x - \bar{x}) = 0$. For this reason we square the deviations to calculate the variance and standard deviation.

From the computational point of view, it is easier and more efficient to use *short-cut formulas* to calculate the variance and standard deviation. By using the short-cut formula, we reduce the computation time and round off errors. The short-cut formulas for calculating the variance and standard deviation are as follows.

SHORT-CUT FORMULAS FOR THE VARIANCE AND STANDARD DEVIATION FOR UNGROUPED DATA

$$\sigma^2 = \frac{\Sigma x^2 - \dfrac{(\Sigma x)^2}{N}}{N} \quad \text{and} \quad s^2 = \frac{\Sigma x^2 - \dfrac{(\Sigma x)^2}{n}}{n - 1}$$

where σ^2 is the population variance and s^2 is the sample variance.

The standard deviation is obtained by taking the positive square root of the variance.

The population standard deviation: $\qquad \sigma = \sqrt{\sigma^2}$

The sample standard deviation: $\qquad s = \sqrt{s^2}$

---

[3]From the formula for σ^2, it can be stated that the population variance is the mean of the squared deviations of x values from the mean. However, this is not true for the variance calculated for a sample data set.

Note that the denominator in the formula for population variance is N but that in the formula for sample variance it is $n - 1$.[4]

Calculating variance and standard deviation for ungrouped data.

EXAMPLE 3–12   The following table lists the annual energy cost (in dollars) of five basic top-freezer models of refrigerators.

Refrigerator Model	Annual Energy Cost (dollars)
General Electric TBH18JP	71
Hotpoint CTX18EP	81
Sears Kenmore 60821	79
Montgomery Ward 19514	93
Whirlpool ET18zkxx	76

Source: Consumer Reports, July 1992, p. 462. Copyright © 1992 by Consumer Union of United States, Inc., Yonkers, NY. Adapted and reproduced with permission. Results and conclusions not endorsed by Consumers Union.

Find the variance and standard deviation for data on annual energy costs.

Let x denote the annual energy cost of a basic top-freezer model of refrigerator. The values of Σx and Σx^2 are calculated in Table 3.6.

Table 3.6

x	x^2
71	5041
81	6561
79	6241
93	8649
76	5776
$\Sigma x = 400$	$\Sigma x^2 = 32{,}268$

The calculation of variance involves the following steps.

Step 1.   *Calculate Σx*

The sum of the entries in the first column of Table 3.6 gives the value of Σx, which is 400.

Step 2.   *Find Σx^2*

The value of Σx^2 is obtained by squaring each value of x and then adding the squared values. The results of this step are shown in the second column of Table 3.6. Notice that $\Sigma x^2 = 32{,}268$.

Step 3.   *Determine the variance*

Substitute all the values in the variance formula and simplify. Because the given data belong to a sample of five refrigerators, we use the formula for the sample variance.

$$s^2 = \frac{\Sigma x^2 - \dfrac{(\Sigma x)^2}{n}}{n - 1} = \frac{32{,}268 - \dfrac{(400)^2}{5}}{5 - 1} = \frac{32{,}268 - 32{,}000}{4} = \mathbf{67}$$

---

[4]The reason that the denominator in the sample formula is $n - 1$ and not n is the following. The sample variance underestimates the population variance when the denominator in the sample formula for variance is n. However, the sample variance does not underestimate the population variance if the denominator in the sample formula for variance is $n - 1$. In Chapter 8 we will learn that $n - 1$ is called the degrees of freedom.

Step 4.   *Obtain the standard deviation*

The standard deviation is obtained by taking the positive square root of the variance.

$$s = \sqrt{67} = \textbf{\$8.19}$$

Thus, the standard deviation of the annual energy costs of these five refrigerators is $8.19.   ▃▃

☞   **TWO OBSERVATIONS**

1. **The values of variance and standard deviation are never negative.** That is, the numerator in the formula for variance should never produce a negative value. Usually the values of variance and standard deviation are positive, but if a data set has no variation then the variance and standard deviation are both zero. For example, if four persons in a group are of the same age, say 35 years, then the four values in the data set are

$$35 \quad 35 \quad 35 \quad 35$$

   If we calculate the variance and standard deviation for these data, their values will be zero. This will be so because there is no variation in the values of this data set.

2. **The measurement units of variance are always the square of the measurement units of the original data.** This is so because the original values are squared to calculate the variance. In Example 3–12, the measurement units of the original data are dollars. However, the measurement units of the variance are squared dollars, which, of course, does not make any sense. Thus, the variance of the annual energy costs of five refrigerators in Example 3–12 is 67 squared dollars. But the measurement units of the standard deviation are the same as the measurement units of the original data because the standard deviation is obtained by taking the square root of the variance.

Calculating variance and standard deviation for ungrouped data.

EXAMPLE 3–13   Following are the 1993 earnings (in thousands of dollars) before taxes for all six employees of a small company.

$$29.50 \quad 16.20 \quad 35.45 \quad 21.35 \quad 49.70 \quad 24.60$$

Calculate the variance and standard deviation for these data.

Solution   Let x denote the 1993 earnings before taxes of employees of this company. The values of Σx and Σx^2 are calculated in Table 3.7.

Table 3.7

x	x^2
29.50	870.2500
16.20	262.4400
35.45	1256.7025
21.35	455.8225
49.70	2470.0900
24.60	605.1600
$\Sigma x = 176.80$	$\Sigma x^2 = 5920.4650$

Since the data on earnings are for *all* employees of this company, we will use the population formula to compute the variance. Thus, the variance is

$$\sigma^2 = \frac{\Sigma x^2 - \dfrac{(\Sigma x)^2}{N}}{N} = \frac{5920.4650 - \dfrac{(176.80)^2}{6}}{6} = \mathbf{118.4597}$$

The standard deviation is obtained by taking the (positive) square root of the variance.

$$\sigma = \sqrt{118.4597} = \textbf{\$10.884 thousand} = \textbf{\$10,884}$$

Thus, the standard deviation of the 1993 earnings of all six employees of this company is $10,884. ▬

☞ **WARNING**

Note that Σx^2 is not the same as $(\Sigma x)^2$. The value of Σx^2 is obtained by squaring the x values and adding them. The value of $(\Sigma x)^2$ is obtained by squaring the value of Σx.

The uses of the standard deviation are discussed in Section 3.4. Later chapters will explain how the mean and the standard deviation taken together can help in making inferences about the population.

3.2.3 COEFFICIENT OF VARIATION

One disadvantage of the standard deviation as a measure of dispersion is that it is a measure of absolute variability and not of relative variability. Sometimes we may need to compare the variability for two different data sets that have different units of measurement. In such cases, a measure of relative variability is preferable. One such measure is the **coefficient of variation**.

COEFFICIENT OF VARIATION

The coefficient of variation, denoted by CV, expresses standard deviation as a percentage of the mean and is computed as follows.

$$\text{For population data:} \qquad CV = \frac{\sigma}{\mu} \times 100\%$$

$$\text{For sample data:} \qquad CV = \frac{s}{\bar{x}} \times 100\%$$

Note that the coefficient of variation does not have any units of measurement as it is always expressed as a percent.

Calculating coefficient of variation.

EXAMPLE 3–14 The yearly salaries of all employees working for a company have a mean of $42,350 and a standard deviation of $3,820. The years of schooling for the same employees have a mean of 15 years and a standard deviation of 2 years. Is the relative variation in the salaries higher or lower than that in years of schooling for these employees?

Solutions Because the two variables (salary and years of schooling) have different units of measurement (dollars and years, respectively), we cannot compare the two standard deviations. Hence, we calculate the coefficient of variation for each data set.

$$\text{CV for salaries} = \frac{\sigma}{\mu} \times 100\% = \frac{3,820}{42,350} \times 100\% = \mathbf{9.02\%}$$

$$\text{CV for years of schooling} = \frac{\sigma}{\mu} \times 100\% = \frac{2}{15} \times 100\% = \mathbf{13.33\%}$$

Thus, the standard deviation for salaries is 9.02% of its mean and that for years of schooling is 13.33% of its mean. Since the coefficient of variation for salaries has a lower value than the coefficient of variation for years of schooling, the salaries have a lower relative spread than the years of schooling. ▬

3.2.4 POPULATION PARAMETER AND A SAMPLE STATISTIC

A numerical measure such as the mean, median, mode, range, variance, or standard deviation calculated for a population data set is called the *population parameter*, or simply a **parameter**. A summary measure calculated for a sample data set is called the *sample statistic*, or simply a **statistic**. Thus, μ and σ are population parameters and $\bar{x}$ and s are sample statistics. As an illustration, $\bar{x} = \$3396.00$ million in Example 3–1 is a sample statistic and $\mu = \$592.50$ in Example 3–2 is a population parameter. Similarly, $s = \$8.19$ in Example 3–12 is a sample statistic whereas $\sigma = \$10,884$ in Example 3–13 is a population parameter.

EXERCISES

Concepts and Procedures

3.33   The range, as a measure of spread, has the disadvantage of being influenced by outliers. Illustrate this with an example.

3.34   Can the standard deviation have a negative value? Explain.

3.35   When is the value of the standard deviation for a data set zero? Give one example. Calculate the standard deviation for this example and show that its value is zero.

3.36   Briefly explain the difference between a population parameter and a sample statistic. Give one example of each of these.

3.37   The following data set belongs to a population.

$$5 \quad -7 \quad 2 \quad 0 \quad -9 \quad 12 \quad 10 \quad 7$$

Calculate the range, variance, standard deviation, and coefficient of variation.

3.38   The following data set belongs to a sample.

$$14 \quad 11 \quad -10 \quad 8 \quad 8 \quad -16$$

Calculate the range, variance, standard deviation, and coefficient of variation.

3.39 The following are the mean and standard deviation for two data sets.

$$\text{Data Set I:} \quad \mu = 60 \quad \text{and} \quad \sigma = 12$$
$$\text{Data Set II:} \quad \mu = 400 \quad \text{and} \quad \sigma = 50$$

Is the relative variation in Data Set I higher or lower than that in Data Set II?

3.40 The following are the mean and standard deviation for two data sets.

$$\text{Data Set I:} \quad \mu = 250 \quad \text{and} \quad \sigma = 35$$
$$\text{Data Set II:} \quad \mu = 110 \quad \text{and} \quad \sigma = 22$$

Is the relative variation in Data Set I higher or lower than that in Data Set II?

3.41 The following data give the weekly food expenditures for a sample of five families.

$$\$82 \quad 116 \quad 65 \quad 75 \quad 92$$

Find the mean for these data. Calculate the deviations of the data values from the mean. Is the sum of these deviations zero?

3.42 A sample of 7 business statistics books produced the following data on their prices.

$$\$56 \quad 47 \quad 68 \quad 55 \quad 71 \quad 52 \quad 62$$

Calculate the range, variance, standard deviation, and coefficient of variation.

3.43 The following data, reproduced from Exercise 3.16, give the number of car thefts that occurred in a city during the past 12 days.

$$6 \quad 3 \quad 7 \quad 11 \quad 5 \quad 3 \quad 8 \quad 7 \quad 2 \quad 6 \quad 9 \quad 13$$

Calculate the range, variance, standard deviation, and coefficient of variation for these data.

3.44 The following table, reproduced from Exercise 3.18, lists the prices of seven different brands and models of air conditioners.

Brand and Model of Air Conditioner	Price (dollars)
Panasonic CW-601JU	405
General Electric AME06LA	400
Quasar HQ5061DW	435
Sharp AF-602M6	375
Sears Kenmore 76069	370
Whirlpool ACQ062XW	410
Signature KMJ-5816	325

Source: Consumer Reports, July 1992, p. 422. Copyright © 1992 by Consumer Union of United States, Inc., Yonkers, NY. Adapted and reproduced with permission. Results and conclusions not endorsed by Consumer Union.

Calculate the range, variance, standard deviation, and coefficient of variation for the data on prices of these seven air conditioners.

3.45 The following table lists the earnings per share (in dollars) for six U.S. companies for the year 1992.

Company	Earnings per Share (dollars)
General Electric	5.51
Wal-Mart Stores	.87
IBM	− 12.03
Pepsico	1.61
Amoco	1.71
Eastman Kodak	3.06

Source: Business Week 1993 Special Issue.

Find the range, variance, standard deviation, and coefficient of variation for the data on earnings per share for these six companies.

3.46 The following table lists the 1991 total compensation (which includes salary, bonus, stock gains, and other payments) of the chief executive officers of seven U.S. companies.

Company	1991 Compensation of the CEO (millions of dollars)
General Dynamics	4.5
Morgan Stanley	7.5
Toys 'R' Us	16.4
United HealthCare	8.0
American Express	1.8
BankAmerica	4.0
US Surgical	23.3

Source: Forbes, May 25, 1992.

Calculate the range, variance, standard deviation, and coefficient of variation for the data on compensations of these CEOs.

3.47 The following table gives the 1992 gross sales (rounded to billions of dollars) for a sample of eight U.S. companies.

Company	1992 Gross Sales (billions of dollars)
Philip Morris	50
General Electric	62
Pfizer	7
Merck	10
Coca-Cola	13
AT&T	65
Hewlett-Packard	17
Johnson & Johnson	14

Source: Business Week 1993 Special Issue.

Find the range, variance, standard deviation, and coefficient of variation for the data on 1992 gross sales of these companies.

3.48 The following data give the number of cars that stopped at a service station during each of the 10 hours observed.

29   35   42   31   24   18   16   27   39   34

Find the range, variance, standard deviation, and coefficient of variation.

3.49 The following data give the number of new cars sold at a dealership during a 12-day period.

13   5   9   6   8   11   9   15   4   11   7   5

Find the range, variance, standard deviation, and coefficient of variation.

3.50 The following data give the weekly expenditures (in dollars) on bakery products for 10 households randomly selected from the 1990 Diary Survey. (These households were selected from those who incurred such expenses.)

$5.79   4.27   3.86   2.49   6.83   8.16   14.60   7.12   9.92   3.77

Find the range, variance, standard deviation, and coefficient of variation for these data.

3.51 The following data give the charitable contributions (in dollars) made during the 12-month period before the survey by 12 households randomly selected from the 1990 Interview Survey. (These households were selected from those who made such contributions.)

$25   100   50   390   120   116   750   173   90   300   150   60

Find the range, variance, standard deviation, and coefficient of variation for these data.

3.52 The yearly salaries of all employees working for a company have a mean of $38,575 and a standard deviation of $4,345. The ages of the same employees have a mean of 43 years and a standard deviation of 7 years. Is the relative variation in salaries higher or lower than that in ages?

3.53 The assets of 500 U.S. companies as of the last day of 1993 had a mean of $11,270 million and a standard deviation of $2,780 million. The price–earnings ratios of the same companies as of the last day of 1993 had a mean of 31 and a standard deviation of 8. Was the relative variation in assets higher or lower than that in price–earnings ratios on the last day of 1993?

3.54 The following data give the hourly wage rate of eight employees of a company.

$12   12   12   12   12   12   12   12

Calculate the standard deviation. Is its value zero? If yes, why?

3.55 Refer to Exercise 3.25 about Smith's Foundry. It is necessary not only to keep the mean length of these nails within a certain limit but also to keep the standard deviation of their lengths within predetermined limits so that the lengths of these nails do not vary a great deal. From time to time when the quality control officer takes a sample of 13 nails from the production line to check for the mean length of nails, he also checks for the standard deviation. If the standard deviation for this sample is greater than .10 inches, the machine is stopped and adjusted. (The value of .10 inches is called the upper limit of the standard deviation.)

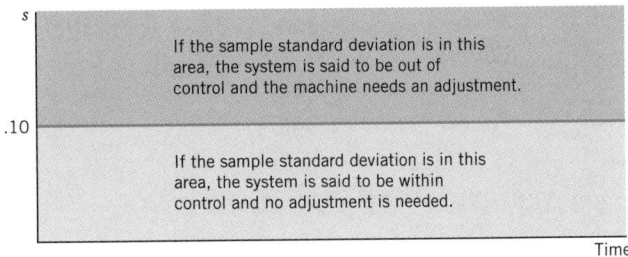

One such recently taken sample of 13 nails produced the following data on their lengths.

3.05   3.15   3.30   2.95   3.35   3.25   3.20
2.90   3.25   3.15   3.05   3.10   3.25

Based on the standard deviation of these sample data, do you think the machine needs to be adjusted?

3.56   Refer to Exercise 3.26 about Soho International. It is necessary not only to make sure that the mean amount of detergent in 64-ounce jugs is never less than 64 ounces but also to make sure that the standard deviation of the amounts of detergent in these jugs does not exceed a certain limit. From time to time when the quality control officer takes a sample of 10 jugs from the production line and measures the amount of detergent in each of these jugs to check for the mean, she also checks for the standard deviation. If the standard deviation of this sample is greater than .15 ounces, the machine is stopped and adjusted. A sample of 10 such jugs taken on a certain day produced the following data on the amount of detergent in these jugs.

64.24	63.75	64.10	63.85	64.10
64.45	63.90	64.29	64.20	64.35

Based on the standard deviation of these sample data, do you think the machine needs an adjustment?

*3.57   Consider the following two data sets.

Data Set I:	12	25	37	8	41
Data Set II:	19	32	44	15	48

Note that each value of the second data set is obtained by adding 7 to the corresponding value of the first data set. Calculate the standard deviation for each of these two data sets using the formula for sample data. Comment on the relationship between the two standard deviations.

*3.58   Consider the following two data sets.

Data Set I:	4	8	15	9	11
Data Set II:	8	16	30	18	22

Note that each value of the second data set is obtained by multiplying the corresponding value of the first data set by 2. Calculate the standard deviation for each of these two data sets using the formula for population data. Comment on the relationship between the two standard deviations.

3.3   MEAN, VARIANCE, AND STANDARD DEVIATION FOR GROUPED DATA

In Sections 3.1.1 and 3.2.2, we learned how to calculate the mean, variance, and standard deviation for ungrouped data. In this section we will learn how to calculate the mean, variance, and standard deviation for grouped data.

3.3.1   MEAN FOR GROUPED DATA

We learned in Section 3.1.1 that the mean is obtained by dividing the sum of all values by the number of values in a data set. However, if the data are given in the form of a frequency table, we will no longer know the values of individual observations. Consequently, in such cases, we cannot obtain the sum of individual values. We find an approximation for the sum of these values using the procedure explained in the next paragraph and example. The formulas used to calculate the mean for grouped data are as follows.

MEAN FOR GROUPED DATA

$$\text{Mean for population data:} \quad \mu = \frac{\Sigma mf}{N}$$

$$\text{Mean for sample data:} \quad \bar{x} = \frac{\Sigma mf}{n}$$

where m is the midpoint and f is the frequency of a class.

To calculate the mean for grouped data, first find the midpoint of each class and then multiply the midpoints by the frequencies of the corresponding classes. The sum of these products, denoted by Σmf, gives an approximation for the sum of all values. To find the value of the mean, divide this sum by the total number of observations in the data.

Calculating population mean for grouped data.

EXAMPLE 3–15 The following table gives the frequency distribution of the daily commuting time (in minutes) from home to work for all 25 employees of a company.

Daily Commuting Time (minutes)	Number of Employees
0 to less than 10	4
10 to less than 20	9
20 to less than 30	6
30 to less than 40	4
40 to less than 50	2

Calculate the mean of the daily commuting times.

Solution Note that because the data set includes all 25 employees of the company, it represents the population. Table 3.8 shows the calculation of Σmf. Note that in Table 3.8, m denotes the midpoint of the classes.

Table 3.8

Daily Commuting Time (minutes)	f	m	mf
0 to less than 10	4	5	20
10 to less than 20	9	15	135
20 to less than 30	6	25	150
30 to less than 40	4	35	140
40 to less than 50	2	45	90
	$N = 25$		$\Sigma mf = 535$

To calculate the mean, we first find the midpoint of each class. The class midpoints are recorded in the third column of Table 3.8. The products of the midpoints and the corresponding frequencies are listed in the fourth column of that table. The sum of the fourth column, denoted by Σmf, gives the approximate total daily commuting time (in minutes) for all 25 employees. The mean is obtained by dividing this sum by the total frequency. Therefore,

$$\mu = \frac{\Sigma mf}{N} = \frac{535}{25} = \textbf{21.40 minutes}$$

Thus, the employees of this company spend an average of 21.40 minutes a day commuting from home to work.

What do the numbers 20, 135, 150, 140, and 90 in the column labeled mf in Table 3.8 represent? We know from this table that 4 employees spend 0 to less than 10 minutes commuting per day. Assuming that the time spent commuting by these 4 employees is evenly spread in the interval 0 to less than 10, the midpoint of this class (which is 5) gives the mean time spent commuting by these 4 employees. Hence, $4(5) = 20$ is the approximate total time (in minutes) spent commuting per day by these 4 employees. Similarly, 9 employees spend 10 to less than 20 minutes commuting per day and the total time spent commuting by these 9 employees is approximately 135 minutes a day. The other numbers in this column can be interpreted in the same way. Note that these numbers give the approximate commuting times for these employees based on the assumption of an even spread within classes. The total commuting time for all 25 employees is approximately 535 minutes. Consequently, 21.40 minutes is an approximate and not the exact value of the mean. We can find the exact value of the mean only if we know the exact commuting time for each of the 25 employees of the company.

Calculating sample mean for grouped data.

EXAMPLE 3–16   The following table gives the frequency distribution of the number of orders received each day during the past 50 days at the office of a mail-order company.

Number of Orders	Number of Days
10–12	4
13–15	12
16–18	20
19–21	14

Calculate the mean.

Solution   Because the data set includes only 50 days, it represents a sample. The value of Σmf is calculated in Table 3.9.

Table 3.9

Number of Orders	f	m	mf
10–12	4	11	44
13–15	12	14	168
16–18	20	17	340
19–21	14	20	280
	$n = 50$		$\Sigma mf = 832$

The value of the sample mean is

$$\bar{x} = \frac{\Sigma mf}{n} = \frac{832}{50} = \textbf{16.64 orders}$$

Thus, this mail-order company receives an average of 16.64 orders per day.

3.3.2 VARIANCE AND STANDARD DEVIATION FOR GROUPED DATA

Following are the *basic formulas* used to calculate the population and sample variances for grouped data.

$$\sigma^2 = \frac{\Sigma f(m - \mu)^2}{N} \qquad \text{and} \qquad s^2 = \frac{\Sigma f(m - \bar{x})^2}{n - 1}$$

where σ^2 is the population variance, s^2 is the sample variance, and m is the midpoint of a class.

In either case, the standard deviation is obtained by taking the positive square root of the variance.

Again, the *short-cut formulas* are more efficient for calculating the variance and standard deviation.

SHORT-CUT FORMULAS FOR THE VARIANCE AND STANDARD DEVIATION FOR GROUPED DATA

$$\sigma^2 = \frac{\Sigma m^2 f - \dfrac{(\Sigma mf)^2}{N}}{N} \qquad \text{and} \qquad s^2 = \frac{\Sigma m^2 f - \dfrac{(\Sigma mf)^2}{n}}{n - 1}$$

where σ^2 is the population variance, s^2 is the sample variance, and m is the midpoint of a class.

The standard deviation is obtained by taking the positive square root of the variance.

The population standard deviation: $\sigma = \sqrt{\sigma^2}$

The sample standard deviation: $s = \sqrt{s^2}$

Examples 3–17 and 3–18 illustrate the use of these formulas to calculate the variance and standard deviation.

Calculating population variance and standard deviation for grouped data.

EXAMPLE 3–17   The following table, reproduced from Example 3–15, gives the frequency distribution of the daily commuting time (in minutes) from home to work for all 25 employees of a company.

Daily Commuting Time (minutes)	Number of Employees
0 to less than 10	4
10 to less than 20	9
20 to less than 30	6
30 to less than 40	4
40 to less than 50	2

Calculate the variance and standard deviation.

Solution   All the steps needed to calculate the variance and standard deviation for grouped data are shown below.

Table 3.10

Daily Commuting Time (minutes)	f	m	mf	m^2f
0 to less than 10	4	5	20	100
10 to less than 20	9	15	135	2,025
20 to less than 30	6	25	150	3,750
30 to less than 40	4	35	140	4,900
40 to less than 50	2	45	90	4,050
	$N = 25$		$\Sigma mf = 535$	$\Sigma m^2f = 14{,}825$

Step 1.   *Calculate the value of Σmf*

To calculate the value of Σmf, first find the midpoint m of each class (see the third column in Table 3.10) and then multiply the corresponding class midpoints and class frequencies (see the fourth column in Table 3.10). The value of Σmf is obtained by adding these products. Thus,

$$\Sigma mf = 535$$

Step 2.   *Find the value of Σm^2f*

To find the value of Σm^2f, square each m value and multiply this squared value of m by the corresponding frequency (see the fifth column in Table 3.10). The sum of these products (that is, the sum of the fifth column in Table 3.10) gives Σm^2f. Hence,

$$\Sigma m^2f = 14{,}825$$

Step 3.   *Calculate the variance*

Because the data set includes all 25 employees of the company, it represents the population. Therefore, we will use the formula for the population variance. Thus,

$$\sigma^2 = \frac{\Sigma m^2f - \dfrac{(\Sigma mf)^2}{N}}{N} = \frac{14{,}825 - \dfrac{(535)^2}{25}}{25} = \frac{3376}{25} = \mathbf{135.04}$$

Step 4.   *Calculate the standard deviation*

To obtain the standard deviation, take the (positive) square root of the variance.

$$\sigma = \sqrt{\sigma^2} = \sqrt{135.04} = \mathbf{11.62\ minutes}$$

Thus, the standard deviation of the daily commuting times for these employees is 11.62 minutes.   ■

Note that the values of the variance and standard deviation calculated in Example 3–17 for the grouped data are approximations. The exact values of the variance and standard deviation can be obtained only by using the ungrouped data on the daily commuting times of the 25 employees.

EXAMPLE 3-18 The following table, reproduced from Example 3–16, gives the frequency distribution of the number of orders received each day during the past 50 days at the office of a mail-order company.

Number of Orders	f
10–12	4
13–15	12
16–18	20
19–21	14

Calculate the variance and standard deviation.

Solution All the information required for the calculation of the variance and standard deviation appears in Table 3.11.

Table 3.11

Number of Orders	f	m	mf	m^2f
10–12	4	11	44	484
13–15	12	14	168	2,352
16–18	20	17	340	5,780
19–21	14	20	280	5,600
	$n = 50$		$\Sigma mf = 832$	$\Sigma m^2f = 14{,}216$

Because the data set includes only 50 days, it represents a sample. Hence, we will use the sample formulas to calculate the variance and standard deviation. By substituting the values in the formula for the sample variance, we obtain

$$s^2 = \frac{\Sigma m^2f - \dfrac{(\Sigma mf)^2}{n}}{n - 1} = \frac{14{,}216 - \dfrac{(832)^2}{50}}{50 - 1} = \textbf{7.5820}$$

Hence, the standard deviation is

$$s = \sqrt{s^2} = \sqrt{7.5820} = \textbf{2.75 orders}$$

Thus, the standard deviation of the number of orders received at the office of this mail-order company during the past 50 days is 2.75. ◼

EXERCISES

Concepts and Procedures

3.59 Are the values of the mean and standard deviation that are calculated using grouped data exact or approximate values of the mean and standard deviation, respectively? Explain.

3.60 Using the population formulas, calculate the mean, variance, and standard deviation for the following grouped data.

x	2–4	5–7	8–10	11–13	14–16
f	5	9	14	7	5

3.61 Using the sample formulas, find the mean, variance, and standard deviation for the grouped data displayed in the following table.

x	f
0 to less than 4	17
4 to less than 8	23
8 to less than 12	15
12 to less than 16	11
16 to less than 20	8
20 to less than 24	6

Applications

3.62 The following table gives the frequency distribution of the amount of the telephone bills for the month of January 1993 for a sample of 50 families.

Amount of Telephone Bill (dollars)	Number of Families
20 to less than 40	8
40 to less than 60	13
60 to less than 80	17
80 to less than 100	9
100 to less than 120	3

Calculate the mean, variance, and standard deviation.

3.63 The following table gives the frequency distribution of entertainment expenditures (in dollars) incurred by 50 families during the past week.

Entertainment Expenditure (dollars)	Number of Families
0 to less than 10	5
10 to less than 20	10
20 to less than 30	15
30 to less than 40	12
40 to less than 50	5
50 to less than 60	3

Find the mean, variance, and standard deviation.

3.64 The time taken to serve each of a sample of 100 customers at a bank was observed. The following table gives the frequency distribution of service time for these 100 customers.

Service Time (minutes)	Number of Customers
0 to less than 2	18
2 to less than 4	30
4 to less than 6	24
6 to less than 8	16
8 to less than 10	8
10 to less than 12	4

Calculate the mean, variance, and standard deviation.

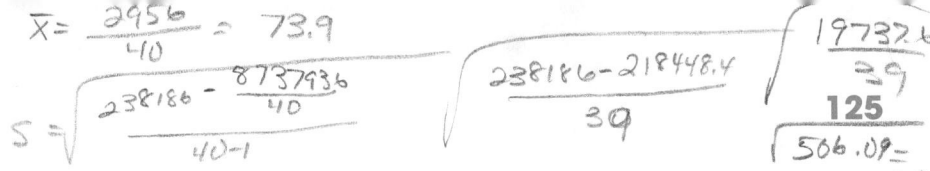

(handwritten, top of page) $\bar{X} = \dfrac{2956}{40} = 73.9$

$S = \sqrt{\dfrac{238186 - \dfrac{8737936}{40}}{40-1}}$

$\dfrac{238186 - 218448.4}{39}$

$\dfrac{197376}{39}$

125

$\dfrac{506.09 =}{}$

3.65 The following table gives the frequency distribution of total hours spent studying statistics during the semester for a sample of 40 university students enrolled in an introductory business statistics course during Spring 1993.

(handwritten left margin)

mean $\bar{X} = \dfrac{\Sigma mf}{N}$

$S = \sqrt{\dfrac{\Sigma m^2 f - \dfrac{(\Sigma mf)^2}{N}}{N-1}}$

standard deviation

$\bar{X} = 73.9$

$S = 22.50$ ✓

$L = \dfrac{55+56}{2} = 55.5$

(handwritten heading: grouped class X)

Hours of Study	Number of Students
24 to less than 40 39	3
40 to less than 56 55	5
56 to less than 72 71	10
72 to less than 88 87	12
88 to less than 104 103	5
104 to less than 120 119	5

(handwritten annotations to the right)

M	mf	m²f
31.5	94.5	
47.5	237.5	
63.5	635	
79.5	954	
95.5	477.5	
111.5	557.5	

$N = 40 \qquad \Sigma mf = 2956 \qquad \Sigma m^2 f = 238186$

Find the mean, variance, and standard deviation.

3.66 The following table gives the frequency distribution of the number of personal computers sold during the past month at 40 computer stores located in New York City.

(handwritten: C = interval = 16)

Computers Sold	Number of Stores
4 to 12	6
13 to 21	9
22 to 30	14
31 to 39	7
40 to 48	4

Calculate the mean, variance, and standard deviation. Give a brief interpretation of the values in the column labeled mf in your table of calculations. What does Σmf represent?

3.67 The following table gives the frequency distribution of the mortgage payment per month for a sample of 40 home owners.

Monthly Mortgage Payment (hundreds of dollars)	Number of Home Owners
4 to less than 8	2
8 to less than 12	9
12 to less than 16	16
16 to less than 20	8
20 to less than 24	5

Calculate the mean, variance, and standard deviation. Give a brief interpretation of the values in the column labeled mf in your table of calculations. What does Σmf represent?

3.68 The following table gives information on the amount (in dollars) of the electric bills for March 1993 for a sample of 50 families.

Amount of Electric Bill (dollars)	Number of Families
0 to less than 20	5
20 to less than 40	16
40 to less than 60	11
60 to less than 80	10
80 to less than 100	8

Find the mean, variance, and standard deviation. Give a brief interpretation of the values in the column labeled mf in your table of calculations. What does Σmf represent?

3.69 For 50 airplanes that arrived late at an airport during a week, the time by which they were late was observed. In the following table, x denotes the time (in minutes) by which an airplane was late and f denotes the number of airplanes.

x	f
0 to less than 20	14
20 to less than 40	18
40 to less than 60	9
60 to less than 80	5
80 to less than 100	4

Find the mean, variance, and standard deviation. Give a brief interpretation of the values in the column labeled mf in your table of calculations. What does Σmf represent?

3.70 The following table gives the frequency distribution of the number of cars owned by 100 households.

Number of Cars Owned	Number of Households
0	12
1	40
2	30
3	15
4	3

Calculate the mean, variance, and standard deviation. (*Hint:* The classes in this example are single-valued. These values of classes [the number of cars owned] will be used as values of m in the formula for the mean, variance, and standard deviation.)

3.71 The following table gives the number of television sets owned by 80 households.

Number of Television Sets Owned	Number of Households
0	4
1	33
2	28
3	10
4	5

Find the mean, variance, and standard deviation. (*Hint:* The classes in this example are single-valued. These values of classes [the number of television sets owned] will be used as values of m in the formula for the mean, variance, and standard deviation.)

3.4 USE OF STANDARD DEVIATION

By using the mean and standard deviation, we can find the proportion or percentage[5] of the total observations that fall within a given interval about the mean. This section briefly discusses Chebyshev's theorem and the empirical rule, both of which demonstrate this use of the standard deviation.

---

[5]Proportion refers to the relative frequency of a category or class and is expressed as a decimal. Percentage is the relative frequency multiplied by 100.

3.4.1 CHEBYSHEV'S THEOREM

Chebyshev's theorem gives a lower bound for the area under a curve between two points that are on opposite sides of the mean and at the same distance from the mean.

CHEBYSHEV'S THEOREM

For any number k greater than 1, at least $(1 - 1/k^2)$ of the data values lie within k standard deviations of the mean.

Figure 3.6 illustrates Chebyshev's theorem.

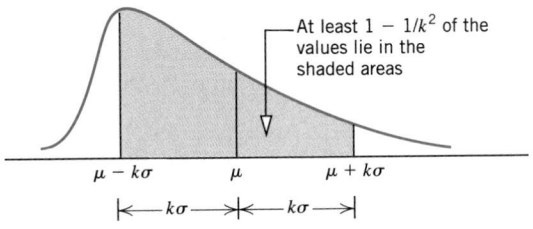

Figure 3.6   Chebyshev's theorem.

Thus, if $k = 2$, then,

$$1 - \frac{1}{k^2} = 1 - \frac{1}{(2)^2} = 1 - \frac{1}{4} = 1 - .25 = .75 \text{ or } 75\%$$

Therefore, according to Chebyshev's theorem, at least .75 or 75% of the values of a data set lie within two standard deviations of the mean. This is shown in Figure 3.7.

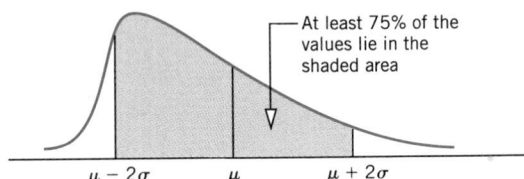

Figure 3.7   Percentage of values within two standard deviations of the mean for Chebyshev's theorem.

If $k = 3$, then,

$$1 - \frac{1}{k^2} = 1 - \frac{1}{(3)^2} = 1 - \frac{1}{9} = 1 - .11 = .89 \text{ or } 89\% \text{ approximately}$$

Therefore, according to Chebyshev's theorem, at least .89 or 89% of the values fall within three standard deviations of the mean. This is shown in Figure 3.8.

Although in Figures 3.6 through 3.8 we have used the population notation for the mean and standard deviation, the theorem applies to both sample and population data. Note that Chebyshev's theorem is applicable to a distribution of any shape. However, Chebyshev's

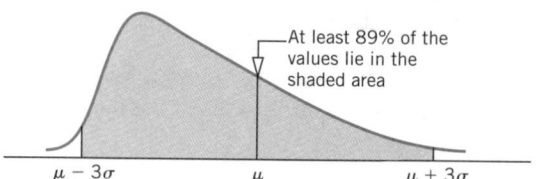

Figure 3.8 Percentage of values within three standard deviations of the mean for Chebyshev's theorem.

theorem can be used only for $k > 1$. This is so because when $k = 1$, the value of $1 - 1/k^2$ is zero, and when $k < 1$ the value of $1 - 1/k^2$ is negative.

Applying Chebyshev's theorem.

EXAMPLE 3-19 According to the U.S. Bureau of the Census, the mean income of female householders with no husband present was $23,380 in 1990. Assume that the mean and standard deviation of 1990 incomes of all such householders were $23,380 and $4,200, respectively. Using Chebyshev's theorem, find at least what percentage of all female householders with no husband present had 1990 incomes between $14,980 and $31,780.

Solution Let μ and σ be the mean and the standard deviation of the 1990 incomes of all female householders with no husband present. We are given that

$$\mu = \$23,380 \quad \text{and} \quad \sigma = \$4,200$$

To find the percentage of all female householders with no husband present whose 1990 incomes were between $14,980 and $31,780, the first step is to determine k. As shown below, each of the two points, $14,980 and $31,780, is $8,400 away from the mean.

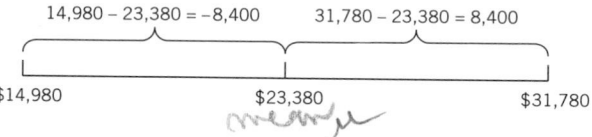

The value of k is obtained by dividing the distance between the mean and each point by the standard deviation. Thus,

$$k = 8400/4200 = 2$$

$$1 - \frac{1}{k^2} = 1 - \frac{1}{(2)^2} = 1 - \frac{1}{4} = 1 - .25 = .75 \text{ or } \mathbf{75\%}$$

Hence, according to Chebyshev's theorem, at least 75% of all female householders with no husband present had 1990 incomes between $14,980 and $31,780. Figure 3.9 shows the area that corresponds to this interval.

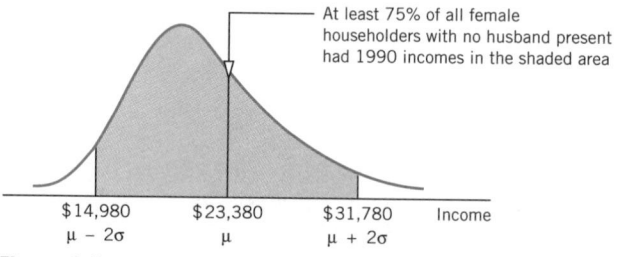

Figure 3.9 Percentage of female householders with no husband present with 1990 incomes between $14,980 and $31,780.

3.4.2 EMPIRICAL RULE

Whereas Chebyshev's theorem is applicable to any kind of distribution, the **empirical rule** applies only to a specific type of distribution called a *bell-shaped distribution*, as shown in Figure 3.10. More will be said about such a distribution in Chapter 6, where it will be called a *normal curve*. In this section, only the following three rules for such a curve are given.

EMPIRICAL RULE

For a bell-shaped distribution, approximately

1. 68% of the observations lie within one standard deviation of the mean
2. 95% of the observations lie within two standard deviations of the mean
3. 99.7% of the observations lie within three standard deviations of the mean

---

Figure 3.10 illustrates the empirical rule. Again, the empirical rule applies to population data as well as sample data.

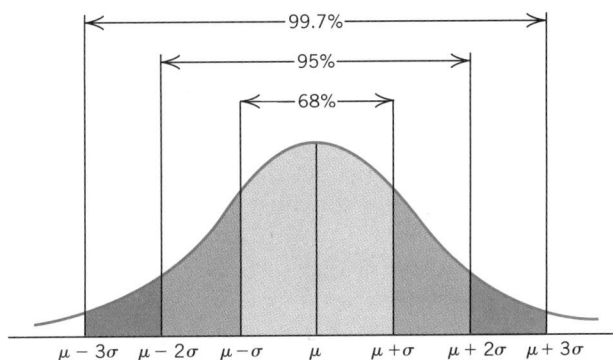

Figure 3.10 Illustration of the empirical rule.

Applying the empirical rule.

EXAMPLE 3-20   Refer to Example 3–19. Now assume that the 1990 incomes of all female householders with no husband present have a bell-shaped distribution with a mean of $23,380 and a standard deviation of $4,200. Determine the approximate percentage of all female householders with no husband present whose 1990 incomes were between $14,980 and $31,780.

Solution   We will use the empirical rule to find the required percentage because the distribution of incomes has a bell-shaped curve. For this distribution,

$$\mu = \$23,380 \quad \text{and} \quad \sigma = \$4,200$$

As shown in Example 3–19, each of the two points, $14,980 and $31,780, is $8,400 away from the mean. Dividing $8,400 by $4,200, we convert the distance between each of the two points and the mean in terms of standard deviation. Thus, the distance between $14,980 and $23,380 and between $31,780 and $23,380 is each 2σ. Consequently, as shown in Figure 3.11, the area from $14,980 to $31,780 is equal to the area from $\mu - 2\sigma$ to $\mu + 2\sigma$.

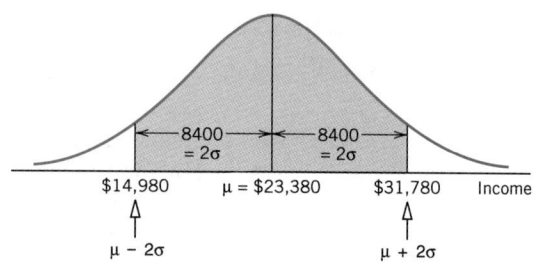

Figure 3.11   Percentage of female householders with no husband present with 1990 incomes between $14,980 and $31,780.

As the area within two standard deviations of the mean is approximately 95% for a bell-shaped curve, approximately 95% of all female householders with no husband present had incomes between $14,980 and $31,780 in 1990.   ■■

EXERCISES

Concepts and Procedures

3.72   Briefly explain Chebyshev's theorem and how it is applied.

3.73   Briefly explain the empirical rule. To what kind of distribution is it applied?

3.74   A sample of 2000 observations has a mean of 74 and a standard deviation of 12. Using Chebyshev's theorem find at least what percentage of the observations fall in the intervals $\bar{x} \pm 2s$, $\bar{x} \pm 2.5s$, and $\bar{x} \pm 3s$, respectively.

3.75   A large population has a mean of 230 and a standard deviation of 41. Using Chebyshev's theorem find at least what percentage of the observations fall in the intervals $\mu \pm 2\sigma$, $\mu \pm 2.5\sigma$, and $\mu \pm 3\sigma$, respectively.

3.76   A large population has a mean of 310 and a standard deviation of 37. Using the empirical rule, find what percentage of the observations fall in the intervals $\mu \pm 1\sigma$, $\mu \pm 2\sigma$, and $\mu \pm 3\sigma$, respectively.

3.77   A sample of 3000 observations has a mean of 82 and a standard deviation of 16. Using the empirical rule, find what percentage of the observations fall in the intervals $\bar{x} \pm 1s$, $\bar{x} \pm 2s$, and $\bar{x} \pm 3s$, respectively.

Applications

3.78   According to *American Banker*, the mean amount owed (including mortgages) by households in the United States is $71,500 (*USA TODAY*, June 10, 1992). Assume that the debts of all households in the United States have a mean of $71,500 and a standard deviation of $20,700. Using Chebyshev's theorem, find at least what percentage of all households have a debt of

    a. $30,100 to $112,900   **b.** $19,750 to $123,250   **c.** $9,400 to $133,600

3.79   The 1992 gross sales of all firms in a large city have a mean of $2.3 million and a standard deviation of $.6 million. Using Chebyshev's theorem, find at least what percentage of firms in this city had 1992 gross sales of

    a. $1.1 to $3.5 million   **b.** $.8 to $3.8 million   **c.** $.5 to $4.1 million

3.80   According to the U.S. Bureau of the Census, the mean income of households was $37,403 in 1990. Assume that the current incomes of all households have a mean of $37,403 and a standard deviation of $8,450.

    a.   Using Chebyshev's theorem, find at least what percentage of all households have an income of

        i. $16,278 to $58,528   **ii.** $12,053 to $62,753

  *b.   Using Chebyshev's theorem, find the interval that contains the incomes of at least 75% of all households.

3.81 The mean monthly mortgage paid by all home owners in a city is $1365 with a standard deviation of $240.

    a. Using Chebyshev's theorem, find at least what percentage of all home owners in this city pay a monthly mortgage of

        i. $885 to $1845   ii. $645 to $2085

    *b. Using Chebyshev's theorem, find the interval that contains monthly mortgage payments of at least 84% of all home owners.

3.82 The mean life of a certain brand of auto batteries is 44 months with a standard deviation of 3 months. Assume that the lives of all auto batteries of this brand have a bell-shaped distribution. Using the empirical rule, find the percentage of auto batteries of this brand that have a life of

    a. 41 to 47 months   **b.** 38 to 50 months   **c.** 35 to 53 months

3.83 According to the National Education Association, the mean salary of public school teachers was $34,413 in 1992. Assume that the salaries of all public school teachers have a bell-shaped distribution with a mean of $34,413 and a standard deviation of $3,500. Using the empirical rule, find the percentage of public school teachers whose 1992 salaries were between

    a. $27,413 and $41,413   **b.** $30,913 and $37,913   **c.** $23,913 and $44,913

3.84 According to Metropolitan Life Insurance Company's claims data for 1990, the average hospital's and physician's charges for a coronary artery bypass graft for patients aged 35 to 64 were $43,370 (*Statistical Bulletin*, 73(3), July–September 1992). Assume that the hospital's and physician's charges for all coronary artery bypass grafts for 1990 have a bell-shaped distribution with a mean of $43,370 and a standard deviation of $4,100.

    a. Using the empirical rule, find the percentage of 1990 coronary artery bypass grafts for which such charges were between

        i. $35,170 and $51,570   ii. $39,270 and $47,470

    *b. Using the empirical rule, find the interval that contains the hospital's and physician's charges for 99.7% of all coronary artery bypass grafts for 1990.

3.85 The ages of cars owned by all employees of a large company have a bell-shaped distribution with a mean of 7 years and a standard deviation of 2 years.

    a. Using the empirical rule, find the percentage of cars owned by these employees that are

        i. 5 to 9 years old   ii. 1 to 13 years old

    *b. Using the empirical rule, find the interval that contains the ages of the cars owned by 95% of all employees of this company.

3.5 MEASURES OF POSITION

A **measure of position** determines the position of a single value in relation to other values in a sample or a population data set. There are many measures of position. However, only quartiles, percentiles, and percentile rank are discussed in this section.

3.5.1 QUARTILES AND INTERQUARTILE RANGE

Quartiles are the summary measures that divide a ranked data set into four equal parts. Three measures will divide any data set into four equal parts. These three measures are the **first quartile** (denoted by Q_1), the **second quartile** (denoted by Q_2), and the **third quartile** (denoted by Q_3). The data should be ranked in increasing order before the quartiles are determined. The quartiles are defined as follows.

QUARTILES

Quartiles are three summary measures that divide ranked data into four equal parts. The second quartile is the same as the median of a data set. The first quartile is the value of the middle term among the observations that are less than the median, and the third quartile is the value of the middle term among the observations that are greater than the median.

Figure 3.12 describes the positions of the three quartiles.

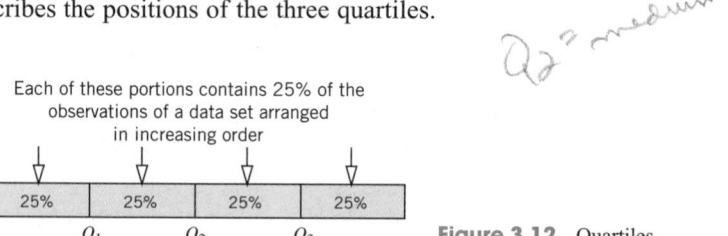

$Q_2 = $ medium

Each of these portions contains 25% of the
observations of a data set arranged
in increasing order

25%	25%	25%	25%
Q_1	Q_2	Q_3	

25%   50%   75%

Figure 3.12   Quartiles.

Approximately 25% of the values in a ranked data set are less than Q_1 and about 75% are greater than Q_1. The second quartile, Q_2, divides a ranked data set into two equal parts; hence, the second quartile and the median are the same. Approximately 75% of the data values are less than Q_3 and about 25% are greater than Q_3.

The difference between the third quartile and the first quartile for a data set is called the **interquartile range (IQR)**.

INTERQUARTILE RANGE

The difference between the third and the first quartiles gives the interquartile range. That is,

$$IQR = \text{Interquartile range} = Q_3 - Q_1$$

Examples 3–21 and 3–22 show the calculation of the quartiles and the interquartile range.

*Finding quartiles and
interquartile range.*

EXAMPLE 3–21   The following data give the price–earnings ratios of 12 U.S. companies.[6] (*Source: Business Week* 1993 Special Issue.)

16   38   18   20   20   18   22   34   7   58   31   19

(a)   Find the values of the three quartiles. Where does the price–earnings ratio of 34 lie in relation to these quartiles?

(b)   Find the interquartile range.

---

[6]The data for the 12 companies are entered (by row) in the following order: General Electric, Wal-Mart Stores, Merck, AT&T, Procter & Gamble, Pfizer, Mobil, Microsoft, Boeing, Home Depot, American Express, and Bristol-Myers Squibb.

Solution

Finding quartiles for even number of data values.

(a) First, we rank the given data in increasing order. Then we calculate the three quartiles as follows.

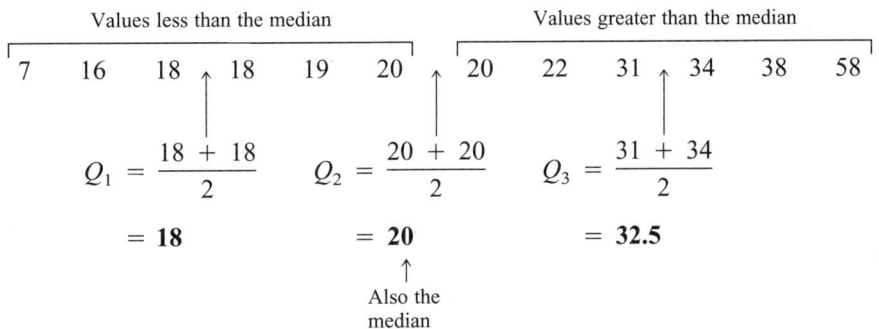

The value of Q_2, which is also the median, is given by the value of the middle term in the ranked data. In the data of this example, this value is given by the average of the sixth and seventh terms. Consequently, Q_2 is 20. The value of Q_1 is given by the value of the middle term of six values that fall below the median (or Q_2). Thus, it is obtained by taking the average of the third and fourth terms. So, Q_1 is 18. The value of Q_3 is given by the value of the middle term of the six values that fall above the median. In the data of this example, Q_3 is obtained by taking the average of the ninth and tenth terms, and it is 32.5.

The value of $Q_1 = 18$ indicates that the price–earnings ratios of (approximately) 25% of the companies in this sample are less than 18 and those of (approximately) 75% of the companies are greater than this value. Similarly, we can state that the price–earnings ratios of (about) half the companies are less than 20 (which is Q_2) and those of the other half are greater than this value. The value of $Q_3 = 32.5$ indicates that the price–earnings ratios of (approximately) 75% of the companies in this sample are less than 32.5 and those of (approximately) 25% of the companies are greater than this value.

Thus, we can state that the price–earnings ratio of 34 lies in the top 25% of these values.

Finding interquartile range.

(b) The interquartile range is given by the difference between the values of the third and the first quartiles. Thus,

$$\text{IQR} = \text{Interquartile range} = Q_3 - Q_1 = 32.5 - 18 = \textbf{14.5} \quad \blacksquare$$

Finding quartiles and interquartile range.

EXAMPLE 3–22 The following are the ages of nine employees of an insurance company.

$$47 \quad 28 \quad 39 \quad 51 \quad 33 \quad 37 \quad 59 \quad 24 \quad 33$$

(a) Find the values of the three quartiles. Where does the age of 28 fall in relation to the ages of these employees?

(b) Find the interquartile range.

Solution

Finding quartiles for odd number of data values.

(a) First we rank the given data in increasing order. Then we calculate the three quartiles as follows.

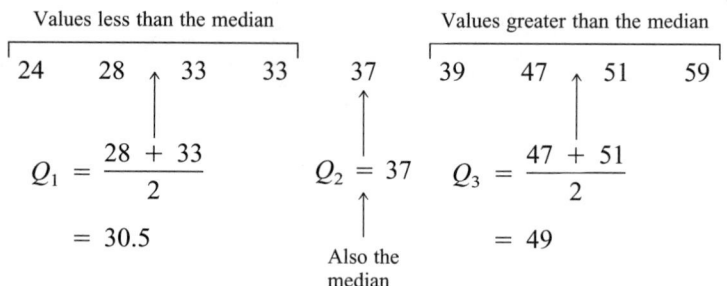

Thus, the values of the three quartiles are

$$Q_1 = \textbf{30.5}, \qquad Q_2 = \textbf{37}, \qquad \text{and} \qquad Q_3 = \textbf{49}$$

The age of 28 falls in the lowest 25% of the ages.

Finding interquartile
range.

(b)   The interquartile range is

$$\text{IQR} = \text{Interquartile range} = Q_3 - Q_1 = 49 - 30.5 = \textbf{18.5} \qquad \blacksquare$$

3.5.2   PERCENTILES AND PERCENTILE RANK

Percentiles are the summary measures that divide a ranked data set into 100 equal parts. Each (ranked) data set has 99 percentiles that divide it into 100 equal parts. The data should be ranked in increasing order to compute percentiles. The kth percentile is denoted by P_k, where k is an integer in the range 1 to 99. For instance, the 25th percentile is denoted by P_{25}. Figure 3.13 shows the positions of the 99 percentiles.

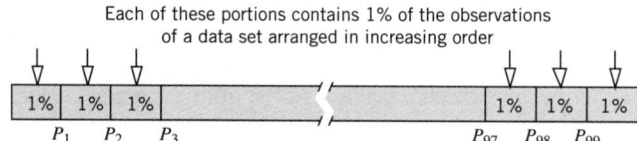

Figure 3.13   Percentiles.

Thus, the kth percentile, P_k, can be defined as a value in a data set such that about $k\%$ of the measurements are smaller than the value of P_k and about $(100 - k)\%$ of the measurements are greater than the value of P_k.

The approximate value of the kth percentile is determined by the following formula.

PERCENTILES

The value of the kth percentile, denoted by P_k, is

$$P_k = \text{the value of the } \left(\frac{kn}{100}\right)\text{th term in a ranked data set}$$

where k denotes the number of the percentile and n represents the sample size.

Example 3–23 describes the procedure to calculate the percentiles.

Finding percentile for a data set.

EXAMPLE 3-23   The following data, reproduced from Example 3–21, give the price–earnings ratios of 12 U.S. companies.

16	38	18	20	20	18	22	34	7	58	31	19

Find the value of the 62nd percentile. Give a brief interpretation of the 62nd percentile.

Solution   First, we rank the given data in increasing order.

$$
7 \quad 16 \quad 18 \quad 18 \quad 19 \quad 20 \quad 20 \quad 22 \quad 31 \quad 34 \quad 38 \quad 58
$$

The position of the 62nd percentile is

$$
\frac{kn}{100} = \frac{62(12)}{100} = 7.44\text{th term}
$$

The value of 7.44th term can be approximated by the average of the seventh and eighth terms in the ranked data. Therefore,

$$
P_{62} = 62\text{nd percentile} = \frac{20 + 22}{2} = \mathbf{21}
$$

Thus, we can state that the price–earnings ratios of approximately 62% of the companies in this sample are less than 21 and those of the remaining 38% are greater than 21.

Note that if a data set contains only a few observations, then the number of values less than the 62nd percentile may not be exactly 62% and the number of values greater than the 62nd percentile may not be exactly 38%. For example, in our data on the price–earnings ratios of 12 companies, 7 (which is approximately 58% of 12) are less than 21 and 5 (which is approximately 42% of 12) are greater than 21. However, these percentages will be more accurate in a larger data set.   ■

We can also calculate the **percentile rank** for a particular value x_i of a data set by using the following formula. The percentile rank of x_i gives the percentage of values in the data set that are smaller than x_i.

PERCENTILE RANK OF A VALUE

$$
\text{Percentile rank of } x_i = \frac{\text{Number of values less than } x_i}{\text{Total number of values in the data set}} \cdot 100
$$

Example 3–24 shows how the percentile rank is calculated for a data value.

Finding percentile rank for a data value.

EXAMPLE 3-24   Refer to the price–earnings ratios of 12 U.S. companies given in Example 3–23, which are reproduced below.

16	38	18	20	20	18	22	34	7	58	31	19

Find the percentile rank for the price–earnings ratio of 31. Give a brief interpretation of this percentile rank.

Solution   First, we rank the given data in increasing order.

$$7 \quad 16 \quad 18 \quad 18 \quad 19 \quad 20 \quad 20 \quad 22 \quad 31 \quad 34 \quad 38 \quad 58$$

In this data set, 8 of the 12 price–earnings ratios are less than 31. Hence,

$$\text{Percentile rank of 31} = \frac{8}{12} \cdot 100 = \textbf{66.67\%}$$

Rounding this answer to the nearest integral value, we can state that about 67% of the companies in this sample have price–earnings ratios lower than 31. Consequently, 33% of these companies have price–earnings ratios of 31 or higher.   ▬

EXERCISES

Concepts and Procedures

3.86   Briefly describe how the three quartiles are calculated for a data set. Illustrate by calculating the three quartiles for two examples, one with an odd number of observations and the second with an even number of observations.

3.87   Explain how the interquartile range is calculated. Give one example.

3.88   Briefly describe how the percentiles are calculated for a data set.

3.89   Explain the concept of the percentile rank for an observation of a data set.

Applications

3.90   The following data, reproduced from Example 2–3 of Chapter 2, give the recent price–earnings ratios for 25 U.S. companies.

31	13	12	22	27	33	17	26	
16	22	18	13	16	23	20	18	
22	15	26	12	20	21	23	27	30

a.   Find the values of the three quartiles.
b.   Calculate the (approximate) value of the 40th percentile.
c.   Find the percentile rank of 23.

3.91   The following data give the number of computer keyboards assembled at the Twentieth Century Electronics Company for a sample of 25 days.

45	52	48	41	56	46	44	42	48	53
51	53	51	48	46	43	52	50	54	47
44	47	50	49	52					

a.   Calculate the values of the three quartiles.
b.   Determine the (approximate) value of the 53rd percentile.
c.   Find the percentile rank of 50.

3.92   The following data give the hours worked last week by 30 employees of a company.

42	45	40	38	35	47	40	27	39	43
40	53	23	51	42	48	40	36	51	40
48	34	21	40	31	34	16	39	41	36

a.   Calculate the values of the three quartiles.
b.   Find the (approximate) value of the 79th percentile.
c.   Calculate the percentile rank of 39.

3.93   The following data, reproduced from Exercise 3.16, give the number of car thefts that occurred in a city during the past 12 days.

6   3   7   11   5   3   8   7   2   6   9   13

    a.   Determine the values of the three quartiles.
    b.   Calculate the (approximate) value of the 55th percentile.
    c.   Find the percentile rank of 8.

3.94   The following data, reproduced from Exercise 2.18 of Chapter 2, give the 1992 gross sales (in billions of dollars) for a sample of 20 U.S. companies.

50	62	55	10	13	65	11	30	14	38
20	8	22	7	8	57	28	17	13	30

    a.   Determine the values of the three quartiles.
    b.   Find the (approximate) value of the 80th percentile. Give a brief interpretation of this percentile.
    c.   What percentage of these companies had gross sales of $38 billion or higher in 1992? Answer by finding the percentile rank of 38.

3.95   Nixon Corporation manufactures computer terminals. The following data give the number of computer terminals produced at the company for a sample of 30 days.

24	32	27	23	33	33	29	25	23	28
21	26	31	20	27	33	27	23	28	29
31	35	34	22	26	28	23	35	31	27

    a.   Calculate the values of the three quartiles.
    b.   Find the (approximate) value of the 65th percentile. Give a brief interpretation of this percentile.
    c.   For what percentage of the days was the number of computer terminals produced 32 or higher? Answer by finding the percentile rank of 32.

3.96   The following data give the number of new cars sold at a dealership during a 20-day period.

8	5	12	3	9	10	6	3	8	8
4	6	10	11	7	7	3	5	9	11

    a.   Calculate the values of the three quartiles.
    b.   Find the (approximate) value of the 25th percentile. Give a brief interpretation of this percentile.
    c.   Find the percentile rank of 10. Give a brief interpretation of this percentile rank.

3.97   The following data, reproduced from Exercise 2.28 of Chapter 2, give the market values (in billions of dollars) of 30 U.S. companies as of March 5, 1993.

12.2	20.0	17.9	11.7	5.9	2.8	5.1	13.6	4.3	9.6
3.1	24.1	27.6	13.1	11.2	31.5	9.7	26.5	18.7	23.1
16.2	5.3	33.1	19.1	36.5	18.3	3.9	24.3	5.6	7.8

    a.   Determine the values of the three quartiles.
    b.   Find the (approximate) value of the 40th percentile. Give a brief interpretation of this percentile.
    c.   What percentage of these companies had a market value of $11.7 billion or higher on March 5, 1993? Answer by finding the percentile rank of $11.7 billion.

3.6 BOX-AND-WHISKER PLOT

A **box-and-whisker plot** gives a graphic presentation of data using five measures: the median, the first quartile, the third quartile, and the smallest and the largest values in the data set between the lower and the upper inner fences. A box-and-whisker plot can help us visualize the center, the spread, and the skewness of a data set. It also helps detect outliers. We can compare the different distributions by making box-and-whisker plots for each of them.

> **BOX-AND-WHISKER PLOT**
>
> A plot that shows the center, spread, and skewness of a data set. It is constructed by drawing a box and two whiskers that use the median, the first quartile, the third quartile, and the smallest and the largest values in the data set between the lower and the upper inner fences.

Example 3–25 explains all the steps needed to make a box-and-whisker plot.

Constructing a box-and-whisker plot.

EXAMPLE 3–25   The following are the incomes (in thousands of dollars) for a sample of 12 households.

$$23 \quad 17 \quad 32 \quad 60 \quad 22 \quad 52 \quad 29 \quad 38 \quad 42 \quad 92 \quad 27 \quad 46$$

Construct a box-and-whisker plot for these data.

Solution   The following five steps are performed to construct a box-and-whisker plot.

Step 1.   First, rank the data in increasing order and calculate the values of the median, the first quartile, the third quartile, and the interquartile range. The ranked data are

$$17 \quad 22 \quad 23 \quad 27 \quad 29 \quad 32 \quad 38 \quad 42 \quad 46 \quad 52 \quad 60 \quad 92$$

For these data,

$$\text{Median} = (32 + 38)/2 = 35$$
$$Q_1 = (23 + 27)/2 = 25$$
$$Q_3 = (46 + 52)/2 = 49$$
$$\text{IQR} = Q_3 - Q_1 = 49 - 25 = 24$$

Step 2.   Find the points that are $1.5 \times \text{IQR}$ below Q_1 and $1.5 \times \text{IQR}$ above Q_3. These two points are called the lower and upper inner fences, respectively.

$$1.5 \times \text{IQR} = 1.5 \times 24 = 36$$
$$\text{Lower inner fence} = Q_1 - 36 = 25 - 36 = -11$$
$$\text{Upper inner fence} = Q_3 + 36 = 49 + 36 = 85$$

Step 3.   Determine the smallest and the largest values in the given data set within the two inner fences. These two values for our example are as follows.

$$\text{Smallest value within the two inner fences} = 17$$
$$\text{Largest value within the two inner fences} = 60$$

Step 4. Draw a horizontal line and mark the income levels on it such that all the values in the given data set are covered. Above the horizontal line, draw a box with its left side at the position of the first quartile and the right side at the position of the third quartile. Inside the box, draw a vertical line at the position of the median. The result of this step is shown in Figure 3.14.

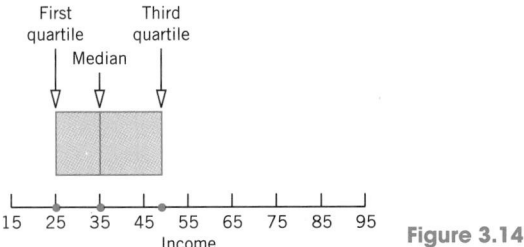

Figure 3.14

Step 5. By drawing two lines, join the points of the smallest and the largest values within the two inner fences to the box. These values are 17 and 60 in this example as listed in Step 3. The two lines that join the box to these two values are called *whiskers*. A value that falls outside the two inner fences is shown by marking an asterisk and is called an outlier. This completes the box-and-whisker plot, as shown in Figure 3.15.

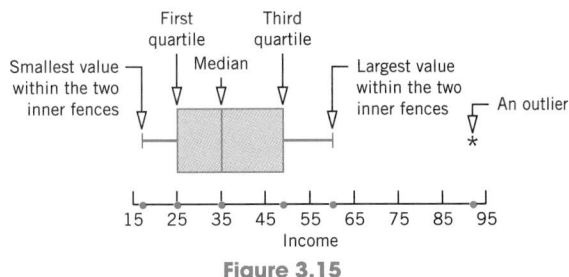

Figure 3.15

In Figure 3.15, about 50% of the data values fall within the box, about 25% of the values fall on the left side of the box, and about 25% fall on the right side of the box. Also, 50% of the values fall on the left side of the median and 50% lie on the right side of the median. The data of this example are skewed to the right because the lower 50% of the values are spread over a smaller range than the upper 50% of the values.

The observations that fall outside the two inner fences are called outliers. These outliers can be classified into two kinds of outliers—mild and extreme outliers. To do so, we define two outer fences—a lower outer fence at $3.0 \times$ IQR below the first quartile and an upper outer fence at $3.0 \times$ IQR above the third quartile. If an observation is outside either of the two inner fences but within either of the two outer fences, it is called a mild outlier. However, an observation that is outside either of the two outer fences is called an extreme outlier. For this example, the outer fences occur at -47 and 121. Since 92 is outside the upper inner fence but inside the upper outer fence, it is a mild outlier. ■

For a symmetric data set, the line representing the median will be in the middle of the box and the spread of the values will be almost over the same range on both sides of the box.

EXERCISES

Concepts and Procedures

3.98   Briefly explain what summary measures are used to construct a box-and-whisker plot.

3.99   Prepare a box-and-whisker plot for the following data.

36	43	28	52	41	59	47	61
24	55	63	73	32	25	35	49
31	22	61	42	58	65	98	34

Does this data set contain any outliers?

3.100   Prepare a box-and-whisker plot for the following data.

11	8	26	31	62	19	7	3	14	75
33	30	42	15	18	23	29	13	16	6

Does this data set contain any outliers?

Applications

3.101   The following data, reproduced from Example 2–3 of Chapter 2, give the recent price–earnings ratios for 25 U.S. companies.

31	13	12	22	27	33	17	26	
16	22	18	13	16	23	20	18	
22	15	26	12	20	21	23	27	30

Prepare a box-and-whisker plot. Comment on the skewness of these data.

3.102   The following data give the number of computer keyboards assembled at the Twentieth Century Electronics Company for a sample of 25 days.

45	52	48	41	56	46	44	42	48	53
51	53	51	48	46	43	52	50	54	47
44	47	50	49	52					

Prepare a box-and-whisker plot. Comment on the skewness of these data.

3.103   The following data, reproduced from Exercise 3.16, give the number of car thefts that occurred in a city during the past 12 days.

6	3	7	11	5	3	8	7	2	6	9	13

Make a box-and-whisker plot. Comment on the skewness of these data.

3.104   The following data, reproduced from Exercise 2.18 of Chapter 2, give the 1992 gross sales (in billions of dollars) for a sample of 20 U.S. companies.

50	62	55	10	13	65	11	30	14	38
20	8	22	7	8	57	28	17	13	30

Prepare a box-and-whisker plot. Comment on the skewness of these data.

3.105 Nixon Corporation manufactures computer terminals. The following data give the number of computer terminals produced at the company for a sample of 30 days.

24	32	27	23	33	33	29	25	23	28
21	26	31	20	27	33	27	23	28	29
31	35	34	22	26	28	23	35	31	27

Prepare a box-and-whisker plot. Comment on the skewness of these data.

3.106 The following data, reproduced from Exercise 2.28 of Chapter 2, give the market values (in billions of dollars) of 30 U.S. companies as of March 5, 1993.

12.2	20.0	17.9	11.7	5.9	2.8	5.1	13.6	4.3	9.6
3.1	24.1	27.6	13.1	11.2	31.5	9.7	26.5	18.7	23.1
16.2	5.3	33.1	19.1	36.5	18.3	3.9	24.3	5.6	7.8

Make a box-and-whisker plot. Comment on the skewness of these data.

GLOSSARY

Bimodal distribution   A distribution that has two modes.

Box-and-whisker plot   A plot that shows the center, spread, and skewness of a data set by drawing a box and two whiskers using the median, the first quartile, the third quartile, and the smallest and the largest values in the data set between the lower and the upper inner fences.

Chebyshev's theorem   For any number k greater than 1, at least $(1 - 1/k^2)$ of the values for any distribution lie within k standard deviations of the mean.

Coefficient of variation   A measure of relative variability that expresses standard deviation as a percentage of the mean.

Empirical rule   For a specific bell-shaped distribution, about 68% of the observations fall in the interval $(\mu - \sigma)$ to $(\mu + \sigma)$, about 95% fall in the interval $(\mu - 2\sigma)$ to $(\mu + 2\sigma)$, and about 99.7% fall in the interval $(\mu - 3\sigma)$ to $(\mu + 3\sigma)$.

First quartile   The value in a ranked data set such that about 25% of the measurements are smaller than this value and about 75% are larger. It is the median of the values that are smaller than the median of the whole data set.

Interquartile range   The difference between the third and the first quartiles.

Mean   A measure of central tendency calculated by dividing the sum of all values by the number of values in the data set.

Measures of central tendency   Measures that describe the center of a distribution. The mean, median, and mode are three of the measures of central tendency.

Measures of dispersion   Measures that give the spread of a distribution. The range, variance, standard deviation, and coefficient of variation are four such measures.

Measures of position   Measures that determine the position of a single value in relation to other values in a data set. Quartiles, percentiles, and percentile rank are examples of measures of position.

Median   The value of the middle term in a ranked data set. The median divides a ranked data set into two equal parts.

Mode   A value (or values) that occurs with highest frequency in a data set.

Multimodal distribution   A distribution that has more than two modes. Bimodal is a special case of a multimodal distribution with two modes.

Outliers or **extreme values**   Values that are very small or very large relative to the majority of the values in a data set.

Parameter   A summary measure calculated for population data.

Percentile rank   The percentile rank of a value gives the percentage of values in the data set that are smaller than this value.

Percentiles   Ninety-nine values that divide a ranked data set into 100 equal parts.

Quartiles   Three summary measures that divide a ranked data set into four equal parts.

Range   A measure of spread obtained by taking the difference between the largest and the smallest values in a data set.

Second quartile   Middle or second of the three quartiles that divide a ranked data set into four equal parts. About 50% of the values in the data set are smaller and about 50% are larger than the second quartile. The second quartile is the same as the median.

Standard deviation   A measure of spread that is given by the positive square root of the variance.

Statistic   A summary measure calculated for sample data.

Third quartile   Third of the three quartiles that divide a ranked data set into four equal parts. About 75% of the values in a data set are smaller than the value of the third quartile and about 25% are larger. It is the median of the values that are greater than the median of the whole data set.

Unimodal distribution   A distribution that has only one mode.

Variance   A measure of spread.

KEY FORMULAS

1. **Mean for ungrouped data**

$$\text{For population data:} \quad \mu = \frac{\Sigma x}{N}$$

$$\text{For sample data:} \quad \bar{x} = \frac{\Sigma x}{n}$$

2. **Mean for grouped data**

$$\text{For population data:} \quad \mu = \frac{\Sigma mf}{N}$$

$$\text{For sample data:} \quad \bar{x} = \frac{\Sigma mf}{n}$$

where m is the midpoint and f is the frequency of a class.

3. **Median for ungrouped data**

$$\text{Median} = \text{Value of the } \left(\frac{n+1}{2}\right)\text{th term in a ranked data set}$$

4. **Range**

$$\text{Range} = \text{Largest value} - \text{Smallest value}$$

5. **Variance for ungrouped data**

$$\text{For population data:} \quad \sigma^2 = \frac{\Sigma x^2 - \frac{(\Sigma x)^2}{N}}{N}$$

$$\text{For sample data:} \quad s^2 = \frac{\Sigma x^2 - \frac{(\Sigma x)^2}{n}}{n-1}$$

6. **Variance for grouped data**

$$\text{For population data:} \quad \sigma^2 = \frac{\Sigma m^2 f - \dfrac{(\Sigma m f)^2}{N}}{N}$$

$$\text{For sample data:} \quad s^2 = \frac{\Sigma m^2 f - \dfrac{(\Sigma m f)^2}{n}}{n - 1}$$

where m is the midpoint and f is the frequency of a class.

7. **Standard deviation**

$$\text{For population data:} \quad \sigma = \sqrt{\sigma^2}$$

$$\text{For sample data:} \quad s = \sqrt{s^2}$$

8. **Coefficient of variation**

$$\text{For population data:} \quad CV = \frac{\sigma}{\mu} \times 100\%$$

$$\text{For sample data:} \quad CV = \frac{s}{x} \times 100\%$$

9. **Interquartile range**

$$IQR = \text{Interquartile range} = Q_3 - Q_1$$

10. **Percentiles**

The kth percentile is given by:

$$P_k = \text{Value of the } \left(\frac{kn}{100}\right)\text{th term in a ranked data set}$$

11. **Percentile rank of a value**

The percentile rank for a particular value x_i of a data set is calculated as follows.

$$\text{Percentile rank of } x_i = \frac{\text{Number of values less than } x_i}{\text{Total number of values in the data set}} \cdot 100$$

SUPPLEMENTARY EXERCISES

3.107 The following data give the income (in dollars) from interest for 10 households selected from the 1990 Diary Survey. (These households were selected from those who had positive income from interest.)

4,000	645	520	34,899	6,868
1,500	278	85	4,000	2,043

a. Find the mean and median. Do these data have a mode?
b. Does this data set contain any outlier? If yes, drop this value and recalculate the mean and median. Which of the two summary measures changes by a larger amount when you drop the outlier?
c. Is the mean or the median a better summary measure for these data?

3.108 The following data give the amount (in dollars) of alimony paid during the 12-month period by five persons selected from the 1990 Interview Survey. (These persons were selected from those who made such payments.)

30,000	13,600	1,200	10,800	16,800

    a. Calculate the mean and median. Do these data have a mode?

    b. Does this data set contain any outlier(s)? If yes, drop the outlier(s) and recalculate the mean and median. Which of these measures changes by a larger amount when you drop the outlier(s)?

3.109 The following data give the 1989 production of corn (in millions of bushels) for seven states. (*Source:* National Agricultural Statistics Service.) The data, entered in that order, are for the states of California, Indiana, Minnesota, Ohio, South Dakota, Texas, and Wisconsin.

$$27 \quad 692 \quad 700 \quad 342 \quad 191 \quad 148 \quad 311$$

    a. Calculate the mean and median. Do these data have a mode?

    b. Find the range, variance, standard deviation, and coefficient of variation.

3.110 The following data give the occupational expenses (in hundreds of dollars) incurred during the 12-month period before the survey by 10 households selected from the 1990 Diary Survey. The occupational expenses include such costs as union dues, tools, uniforms, business or professional association dues, licenses, or permits. (These households were selected from those who incurred such expenses.)

$$1.5 \quad .8 \quad 5.6 \quad 10.0 \quad 5.0 \quad 3.4 \quad 5.3 \quad 14.1 \quad 16.0 \quad 2.6$$

    a. Calculate the mean and median. Do these data have a mode?

    b. Find the range, variance, standard deviation, and coefficient of variation.

3.111 The following data give the 1987 production of strawberries (in hundred thousands of pounds) for eight states. (*Source:* National Agricultural Statistics Service.) The data, entered in that order, are for the following states: Arkansas, Louisiana, Michigan, New Jersey, North Carolina, Oregon, Washington, and Wisconsin.

$$11 \quad 43 \quad 92 \quad 42 \quad 46 \quad 56 \quad 25 \quad 44$$

    a. Calculate the mean and median. Do these data have a mode?

    b. Find the range, variance, standard deviation, and coefficient of variation.

3.112 The following data give the annual earnings (in thousands of dollars) before taxes for 11 households selected from the 1990 Interview Survey. (These households were selected from those who had positive earnings.)

$$21 \quad 13 \quad 28 \quad 50 \quad 68 \quad 56 \quad 21 \quad 32 \quad 45 \quad 43 \quad 15$$

    a. Calculate the mean and median for these data. Do these data contain a mode?

    b. Find the range, variance, standard deviation, and coefficient of variation.

3.113 The following table gives the frequency distribution of hourly wages (in dollars) for all 120 employees of a company.

Hourly Wage	Number of Employees
4 to less than 6	16
6 to less than 8	34
8 to less than 10	26
10 to less than 12	22
12 to less than 14	17
14 to less than 16	5

Calculate the mean, variance, and standard deviation. Are the values of these summary measures the population parameters or the sample statistics?

3.114 The following table gives the distribution of the amounts (in dollars) that 100 randomly selected persons said they will spend on a wedding gift for their closest friend.

Amount	Number of Persons
0 to less than 50	8
50 to less than 100	27
100 to less than 150	31
150 to less than 200	23
200 to less than 250	11

Find the mean, variance, and standard deviation. Are the values of these summary measures population parameters or sample statistics?

3.115 The following table gives the distribution of the number of hours that 80 persons spent learning to drive before passing the driving test.

Number of Hours	Number of Persons
5 to 7	4
8 to 10	9
11 to 13	16
14 to 16	23
17 to 19	17
20 to 22	11

Calculate the mean, variance, and standard deviation. Are the values of these summary measures population parameters or sample statistics?

3.116 The following table gives the distribution of the number of days for which all 40 employees of a company were absent during the last year.

Number of Days Absent	Number of Employees
0 to 2	13
3 to 5	14
6 to 8	6
9 to 11	4
12 to 14	3

Calculate the mean, variance, and standard deviation. Are the values of these summary measures population parameters or sample statistics?

3.117 The mean time taken to learn the basics of a word processor by all students is 200 minutes with a standard deviation of 20 minutes.

- **a.** Using Chebyshev's theorem, find at least what percentage of students will learn the basics of this word processor in
  - i.   160 to 240 minutes   ii.   140 to 260 minutes
- **b.** Using Chebyshev's theorem, find the interval that contains the time taken by at least 75% of all students to learn this word processor.

3.118 According to the American Association of University Professors, the mean salary of university and college professors is $53,540. Assume that the standard deviation of these salaries is $2,950.

- **a.** Using Chebyshev's theorem, find at least what percentage of university and college professors have their salaries between
  - i. $47,640 and $59,440   ii. $44,690 and $62,390
- **b.** Using Chebyshev's theorem, find the interval that contains the salaries of at least 84% of all university and college professors.

3.119 Refer to Exercise 3.117. Suppose the time taken to learn the basics of this word processor by all students has a bell-shaped distribution with a mean of 200 minutes and a standard deviation of 20 minutes.

    a. Using the empirical rule, find the percentage of students who will learn the basics of this word processor in

        i. 180 to 220 minutes   ii. 160 to 240 minutes

    b. Using the empirical rule, find the interval that contains the time taken by 99.7% of all students to learn this word processor.

3.120 Refer to Exercise 3.118. Assume that the salaries of university and college professors have a bell-shaped distribution with a mean of $53,540 and a standard deviation of $2,950.

    a. Using the empirical rule, find the percentage of university and college professors whose salaries are between

        i. $47,640 and $59,440   ii. $44,690 and $62,390

    b. Using the empirical rule, find the interval that contains the salaries of 68% of all university and college professors.

3.121 Refer to the data of Exercise 3.110 on the occupational expenses (in hundreds of dollars) incurred during the 12-month period before the survey by 10 households selected from the 1990 Diary Survey.

    a. Determine the values of the three quartiles.

    b. Calculate the (approximate) value of the 25th percentile.

    c. Find the percentile rank of 5.6.

3.122 Refer to the data of Exercise 3.112 on the annual earnings (in thousands of dollars) before taxes for 11 households selected from the 1990 Interview Survey.

    a. Determine the values of the three quartiles.

    b. Calculate the (approximate) value of the 70th percentile.

    c. Find the percentile rank of 45.

3.123 The following data give the ages of 15 employees of a company.

36	47	23	55	42	31	27	19
38	65	52	47	39	25	44	

Prepare a box-and-whisker plot. Is this data set skewed in any direction? If yes, is it skewed to the right or to the left?

3.124 The following data give the prices (in thousands of dollars) of 16 recently sold houses in an area.

141	163	127	104	197	203	113	179
256	228	183	119	133	199	871	191

Make a box-and-whisker plot. Comment on the skewness of this data set.

SELF-REVIEW TEST

1. The value of the middle term in a ranked data set is called the

    a. mean   **b.** median   **c.** mode

2. Which of the following summary measures is/are influenced by extreme values?

    a. mean   **b.** median   **c.** mode   **d.** range

3. Which of the following summary measures can be calculated for qualitative data?

    a. mean   **b.** median   **c.** mode

4. Which of the following can have more than one value?

    a. mean   **b.** median   **c.** mode

5. Which of the following is given by the difference between the largest and the smallest values of a data set?

    a. variance   **b.** range   **c.** mean

6. Which of the following is the mean of the squared deviations of x values from the mean?

    a. standard deviation   **b.** population variance   **c.** sample variance

7. The values of the variance and standard deviation are

    a. never negative   **b.** always positive   **c.** never zero

8. A summary measure calculated for the population data is called

    a. a population parameter   **b.** a sample statistic   **c.** an outlier

9. A summary measure calculated for the sample data is called

    a. a population parameter   **b.** a sample statistic   **c.** a boxplot

10. Chebyshev's theorem can be applied to

    a. any distribution   **b.** bell-shaped distributions only
    c. skewed distributions only

11. The empirical rule can be applied to

    a. any distribution   **b.** bell-shaped distributions only
    c. skewed distributions only

12. The first quartile is a value in a ranked data set such that

    a. about 75% of the values are smaller and about 25% are larger than this value
    b. about 50% of the values are smaller and about 50% are larger than this value
    c. about 25% of the values are smaller and about 75% are larger than this value

13. The third quartile is a value in a ranked data set such that

    a. about 75% of the values are smaller and about 25% are larger than this value
    b. about 50% of the values are smaller and about 50% are larger than this value
    c. about 25% of the values are smaller and about 75% are larger than this value

14. The 75th percentile is a value in a ranked data set such that

    a. about 75% of the values are smaller and about 25% are larger than this value
    b. about 25% of the values are smaller and about 75% are larger than this value

15. The following data give the number of times 10 persons used their credit cards during the past three months.

$$9 \quad 6 \quad 22 \quad 14 \quad 2 \quad 18 \quad 7 \quad 3 \quad 11 \quad 6$$

Calculate the mean, median, mode, range, variance, standard deviation, and coefficient of variation.

16. The mean, as a measure of central tendency, has the disadvantage of being influenced by extreme values. Illustrate this point with an example.

17. The range, as a measure of spread, has the disadvantage of being influenced by extreme values. Illustrate this point with an example.

18. When is the value of the standard deviation for a data set zero? Give one example of such a data set. Calculate the standard deviation for that data set to show that it is zero.

19. The following table gives the frequency distribution of the number of computers sold during the past 25 weeks at a computer store.

Computers Sold	Frequency
4 to 9	2
10 to 15	5
16 to 21	10
22 to 27	5
28 to 33	3

    a. What does the frequency column in the table represent?
    b. Calculate the mean, variance, and standard deviation.

20. The cars owned by all people living in a city are on average 7.3 years old with a standard deviation of 2.2 years.

    a. Using Chebyshev's theorem, find at least what percentage of the cars in this city are
       i. 1.8 to 12.8 years old   ii. .7 to 13.9 years old
    b. Using Chebyshev's theorem, find the interval that contains the ages of cars owned by at least 75% of all people in this city.

21. The ages of cars owned by all people living in a city have a bell-shaped distribution with a mean of 7.3 years and a standard deviation of 2.2 years.

    a. Using the empirical rule, find the percentage of cars in this city that are
       i. 5.1 to 9.5 years old   ii. .7 to 13.9 years old
    b. Using the empirical rule, find the interval that contains the ages of cars owned by 95% of all people in this city.

22. The following data give the number of books purchased by 16 adults during the past one year.

8	12	20	16	0	11	18	4
10	6	17	24	15	9	2	6

    a. Calculate the three quartiles.
    b. Find the (approximate) value of the 68th percentile. Give a brief interpretation of this value.
    c. Calculate the percentile rank of 16. Give a brief interpretation of this value.

23. Make a box-and-whisker plot for the data on the number of books purchased by 16 adults during the past one year given in Problem 22. Comment on the skewness of this data set.

∗24. The mean weekly wages of a sample of 15 employees of a company are $435. The mean weekly wages of a sample of 20 employees of another company are $490. Find the combined mean for these 35 employees.

∗25. The mean price of five cars is $16,563. The prices of four of these five cars are $8,090, $27,835, $13,299, and $15,455. Find the price of the fifth car.

∗26. Following are the prices (in thousands of dollars) of 10 houses sold recently in a city.

179	166	58	207	287	149	193	2534	163	238

Calculate the 10% trimmed mean for this data set. Do you think the 10% trimmed mean is a better summary measure for these data than the simple mean (i.e., the mean of all 10 values)? Briefly explain why, or why not.

∗27. Consider the following two data sets.

Data Set I:	8	16	20	35
Data Set II:	5	13	17	32

Note that each value of the second data set is obtained by subtracting 3 from the corresponding value of the first data set.

   a. Calculate the mean for each of these two data sets. Comment on the relationship between the two means.

   b. Calculate the standard deviation for each of these two data sets. Comment on the relationship between the two standard deviations.

USING MINITAB

The MINITAB commands given in Figure 3.16 can be used to do the statistical analysis learned in this chapter. We have assumed that the data are entered in column C1.

Figure 3.16 MINITAB commands to find summary measures.

```
MTB > MEAN C1 ←—— This command will compute the mean.
MTB > MEDIAN C1 ←—— This command will compute the median.
MTB > RANGE C1 ←—— This command will compute the range.
MTB > STDEV C1 ←—— This command will compute the standard deviation.
MTB > DESCRIBE C1 ←—— This command will compute the various summary measures.
MTB > BOXPLOT C1 ←—— This command will prepare the box-and-whisker plot.
```

Using each of these commands displays the corresponding MINITAB output on the screen as soon as the *enter* key is hit.

Illustration M3-1   The following data, reproduced from Example 3–5, give the number of patents granted by the U.S. government to 12 automobile companies in 1991.

| 506 | 234 | 97 | 385 | 50 | 249 | 210 | 23 | 275 | 257 | 36 | 13 |

Suppose we want to find the mean, median, range, and standard deviation and prepare a box-and-whisker plot for these data.

The use of the MINITAB commands mentioned in Figure 3.16 to calculate the mean, median, range, and standard deviation is illustrated in Figure 3.17 for these data.

Figure 3.17 Finding summary measures.

```
MTB  > NOTE: CALCULATING DESCRIPTIVE MEASURES
MTB  > SET C1
DATA > 506 234 97 385 50 249 210 23 275 257 36 13
DATA > END

MTB  > MEAN C1
    MEAN = 194.58
```

(Continued on next page)

```
MTB  > MEDIAN C1
     MEDIAN = 222.00

MTB  > RANGE C1
     RANGE = 493.00

MTB  > STDEV C1
     ST.DEV. = 155.61
```

The DESCRIBE command can also be used to find some of the summary measures. The MINITAB output shown in Figure 3.18 is obtained in response to this command.

Figure 3.18   Using MINITAB DESCRIBE command.

```
MTB > DESCRIBE C1

              N      MEAN    MEDIAN    TRMEAN    STDEV    SEMEAN
C1           12     194.6     222.0     181.6    155.6      44.9


            MIN       MAX        Q1        Q3
C1         13.0     506.0      39.5     270.5
```

Most of the entries in the MINITAB output of Figure 3.18 are self-explanatory. The fourth entry, labeled TRMEAN (called the trimmed mean and briefly explained in Exercise 3.32) is the mean of the data when approximately 5% of the values at each end of the ranked data are dropped. In this example, it is calculated by dropping the smallest and the largest values of the data set. The sixth entry, SEMEAN, gives the standard error (also called the standard deviation) of the mean, which will be discussed in Chapter 7. The last two entries, Q1 and Q3, give the first and the third quartiles, respectively. Again, note that the SET command to enter the data is not repeated.

The BOXPLOT command gives the box-and-whisker plot. In Figure 3.19, for the box-and-whisker plot for the data of column C1, the '' + '' sign inside the box indicates the position of the median. Note that we have used the data that have already been entered in column C1.

Figure 3.19   Preparing a box-and-whisker plot.

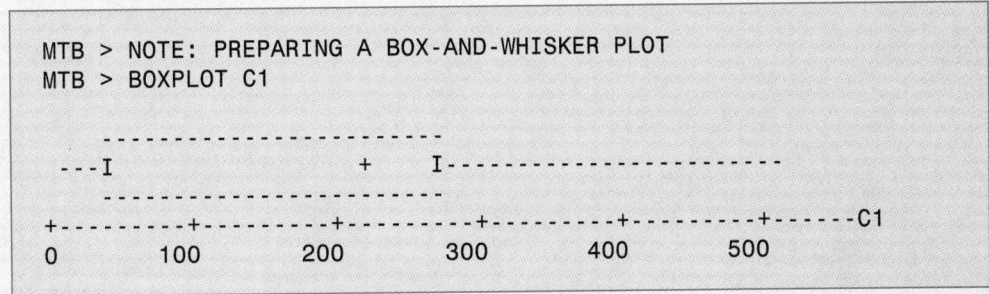

COMPUTER ASSIGNMENTS

M3.1   Refer to Data Set I of Appendix A. Using MINITAB, find the mean, median, range, and standard deviation for the data on recent share prices of companies given in column C6.

M3.2   Refer to Data Set I of Appendix A. Using MINITAB, find the mean, median, range, and standard deviation for the data on price–earnings ratios of companies given in column C7.

M3.3   Refer to data on phone charges given in column C7 of Data Set II of Appendix A. From that data set, select the 6th value and then select every 10th value after that (i.e., select the 6th, 16th, 26th, 36th, . . . values). This subsample will give you 20 measurements. (Such a sample taken from a population is called a *systematic random sample*.) Using the MINITAB DESCRIBE command, find the mean, median, standard deviation, first quartile, and third quartile for the phone charges for these 20 observations.

M3.4   Refer to Data Set III of Appendix A. Using the MINITAB DESCRIBE command, find the mean, median, standard deviation, first quartile, and third quartile for the data on income before taxes given in column C3.

M3.5   Refer to Data Set I of Appendix A. Using MINITAB, prepare a box-and-whisker plot for the data on recent share prices of companies given in column C6.

M3.6   Refer to Data Set I of Appendix A. Using MINITAB, prepare a box-and-whisker plot for the data on price–earnings ratios of companies given in column C7.

MORE CHALLENGING EXERCISES (Optional)
CHAPTERS 1 TO 3

1. An automaker would like to design a new model of a specific car that would be affordable to middle income families. Consider the annual incomes before taxes of a sample of 200 families given in Column 3 of Data Set III in Appendix A.

   a. Using both graphical and numerical methods, organize and interpret these data. Write a report summarizing your results.

   b. What measure of central tendency would you suggest the automaker use in determining a typical middle income family? Explain.

2. An executive of a record company suggests that the characteristics of recordings that provide optimal profit differ depending on the music category. A sample of 12 randomly selected single rock music CDs and a sample of 14 randomly selected single country music CDs have the following total lengths (in minutes).

Rock Music	Country Music
43.0	45.3
44.3	40.2
63.8	42.8
32.8	33.0
54.2	33.5
51.3	37.7
64.8	36.8
36.1	34.6
33.9	33.4
51.7	36.5
36.5	43.3
59.7	31.7
	44.0
	42.7

   a. Make a box plot of each of the data sets and use them to discuss the similarities and differences between the lengths of rock and country music CDs.

   b. Compute the various descriptive statistics you have learned for each sample. How do they compare?

3. To generate revenue, the senior class at a large university plans to sell specially designed jackets that can be ordered from the manufacturer in small, medium, and large sizes. The senior class advisor suggests ordering an equal number of each of the three sizes. Clothes sizes for students of this university are not available, however their weights are. The following data give the weights (in pounds) of a random sample of 44 students. (In these data, F and M indicate whether the student is a female or a male.)

123 F	195 M	138 M	115 F	179 M	119 F	148 F
147 F	180 M	146 F	179 M	189 M	175 M	108 F
193 M	114 F	179 M	147 M	108 F	128 F	164 F
174 M	128 F	159 M	193 M	204 M	125 F	133 F
115 F	168 M	123 F	183 M	116 F	182 M	174 M
102 F	123 F	99 F	161 M	162 M	155 F	202 M
110 F	132 M					

a. Construct a stem and leaf display of these weights.
b. Can you explain why these data appear the way they do?
c. Compute the three measures of central tendency.
d. Which measure of central tendency is most informative?
e. Based on these data, do you think it would be prudent to order an equal number of the three sizes of jackets? Explain.

4. Actuaries for an insurance company must determine a premium for a new type of insurance. A random sample of 40 potential purchasers of this type of insurance were found to have suffered the following losses during the past year. These losses would have been covered by this insurance if it were available.

$100	32	0	0	470	50	0	14,589	212	93
0	0	1127	421	0	87	135	420	0	250
12	0	309	0	177	295	501	0	143	0
167	398	54	0	141	0	3709	122	0	0

a. Use the graphical and numerical methods you have learned to organize and interpret these data. Write a report summarizing the characteristics of losses for the insurance company.
b. Which measure of central tendency and variation would you suggest the actuaries use when determining the premium for this insurance? Explain.

5. Many individuals and companies are choosing to relocate to areas of the country where they realize a lower cost of living, favorable employment situations, and lower tax burdens. Consider the prices per pound of T-bone steaks given in Column 2 of Data Set II in Appendix A.

a. To compare the prices of T-bone steaks in the southern states of Alabama, Georgia, Louisiana, Mississippi, and Tennessee with the desert states of Arizona, Nevada, and New Mexico,
   i. determine the two corresponding data sets
   ii. make a box plot of the data in each group
   iii. compute the descriptive measures you feel will allow you to compare the price per pound of T-bone steaks for these two groups of cities
b. Write a report summarizing the results of your comparison of the data sets in part a.

6. The SAT scores of the applicants to a community college have a bell-shaped distribution with a mean of 900 and a standard deviation of 60. The admissions office, which bases acceptance solely on SAT scores, advertises an 84% acceptance rate. Cathy Henry is suing the college because she feels she should have been accepted with her SAT score of 780. Should Cathy's attorney hire a statistician to substantiate her case? That is, would a statistician find that she has been unjustly denied acceptance? If not, what would the acceptance rate have to be for Cathy to be accepted?

7. A portfolio of 1000 television commercials has a mean air time of 30 seconds and a standard deviation of 10 seconds, with no commercial over one minute. At a certain television station, the net revenue generated by commercials that are between 20 and 40 seconds is $100 each, between 10 and 20 seconds or between 40 and 50 seconds is $80 each, and between 0 and 10 seconds or between 50 and 60 seconds is $50 each.

a. If the air times of the commercials in this portfolio have a bell-shaped distribution, approximately how much revenue will the portfolio generate?
b. If nothing is known about the distribution of the air times of these commercials, at least how much revenue will the portfolio generate?

8. An investor must decide on one of four types of investments, all of which require the same capital. The investor collects the following sample data on the gain (in thousands of dollars) for each type of investment.

	Investment Type		
I	**II**	**III**	**IV**
14	6	34	13
31	10	26	37
1	35	20	15
20	35	8	12
6	2	35	6
35	30	1	13
	1	34	18
	24		

Rank the four types of investments in order of preference for a conservative investor. Use descriptive summary measures and graphs to answer this question.

9. The final examination scores of students in an elementary business statistics course are summarized in the following frequency distribution table.

Scores	Frequency
81–90	6
91–100	24
101–110	33
111–120	30
121–130	30
131–140	9
141–150	9

Construct an ogive for cumulative percentage and use it to approximate

    a. the 80th percentile
    b. the percentile rank of someone who scored 95 on the examination
    c. the interquartile range
    d. the median

10. Find the missing entries in the following frequency distribution table.

Class Limits	Frequency	Relative Frequency	Cumulative Frequency	Cumulative Percentage
8 to —	—	—	—	25
— to —	—	.05	—	—
— to —	—	—	9	—
— to —	—	.30	15	—
— to 32	—	—	—	—
	—	—		

4 PROBABILITY

We often make statements about probability. The following are a few such examples.

1. There is a 30% chance that a newly started business will fail within five years.
2. The chance that the auto sales will be higher next year is 70%.
3. There is a 75% chance that the price of AT&T stock will be higher next year.
4. The chance that the new car I am buying is not a lemon is 95%.

Probability, which measures the likelihood that an event will occur, is an important part of statistics. It is the basis of inferential statistics, which will be introduced in later chapters. In inferential statistics, we make decisions under conditions of uncertainty. Probability theory is used to evaluate the uncertainty involved in those decisions. For example, estimating next year's sales for a company is based on many assumptions, some of which may happen to be true and others may not. Probability theory will help us make decisions under such conditions of imperfect information and uncertainty. Combining probability and probability distributions (which are discussed in Chapters 5 through 7) with descriptive statistics will help us make decisions about populations based on information obtained from samples. This chapter presents the basic concepts of probability and the rules for computing probability.

4.1 EXPERIMENT, OUTCOMES, AND SAMPLE SPACE

Quality control inspector Jack Cook of Tennis Products Company picks up a tennis ball from the production line to check whether it is good or defective. Jack Cook's act of inspecting a tennis ball is an example of a statistical **experiment**. The result of his inspection will be that the ball is either "good" or "defective." Each of these two observations is called an **outcome** (also called a *basic* or *final outcome*) of the experiment, and these outcomes taken together constitute the **sample space** for this experiment.

EXPERIMENT, OUTCOMES, AND SAMPLE SPACE

An experiment is a process that, when performed, results in one and only one of many observations. These observations are called the outcomes of the experiment. The collection of all outcomes for an experiment is called a sample space.

A sample space is denoted by **S**. The sample space for the example of inspecting a tennis ball is written as

$$S = \{\text{good, defective}\}$$

The elements of a sample space are called **sample points**.

Table 4.1 lists a few examples of experiments, their outcomes, and their sample spaces.

Table 4.1 Examples of Experiments, Outcomes, and Sample Spaces

Experiment	Outcomes	Sample Space
Toss a coin once	Head, Tail	$S = \{\text{Head, Tail}\}$
Roll a die once	1,2,3,4,5,6	$S = \{1,2,3,4,5,6,\}$
Toss a coin twice	*HH, HT, TH, TT*	$S = \{HH, HT, TH, TT\}$
Birth of a baby	Boy, Girl	$S = \{\text{Boy, Girl}\}$
Take a test	Pass, Fail	$S = \{\text{Pass, Fail}\}$
Select a student	Male, Female	$S = \{\text{Male, Female}\}$

The sample space for an experiment can also be described by drawing either a Venn diagram or a tree diagram. A **Venn diagram** is a picture (a closed geometric shape such as a rectangle, a square, or a circle) that depicts all the possible outcomes for an experiment. In a **tree diagram**, each outcome is represented by a branch of the tree. Venn and tree diagrams help us understand probability concepts by presenting them visually. Examples 4–1 through 4–3 describe how to draw these diagrams for statistical experiments.

Venn and tree diagrams: one toss of a coin.

EXAMPLE 4–1 Draw the Venn and tree diagrams for the experiment of tossing a coin once.

Solution This experiment has two possible outcomes: head and tail. Consequently, the sample space is given by

$$S = \{H, T\} \qquad \text{where } H = \text{Head}, \qquad T = \text{Tail}$$

To draw a Venn diagram for this example, we draw a rectangle and mark two points inside this rectangle that represent the two outcomes, head and tail. The rectangle is labeled

S because it represents the sample space (see Figure 4.1a). To draw a tree diagram, we draw two branches starting at the same point, one representing the head and the second representing the tail. The two final outcomes are listed at the end of the branches (see Figure 4.1b).

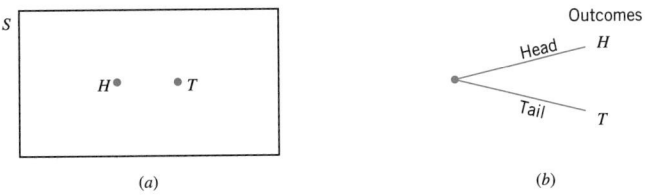

(a)   (b)

Figure 4.1   (a) Venn diagram and (b) tree diagram for one toss of a coin.

Venn and tree diagrams: two tosses of a coin.

EXAMPLE 4–2   Draw the Venn and tree diagrams for the experiment of tossing a coin twice.

Solution   This experiment can be split into two parts: the first toss and the second toss. Suppose the first time the coin is tossed we obtain a head. Then, on the second toss, we can still obtain a head or a tail. This gives us the two outcomes: *HH* (head on both tosses) and *HT* (head on the first toss and tail on the second toss). Now suppose we observe a tail on the first toss. Again, either a head or a tail can occur on the second toss, giving the remaining two outcomes: *TH* (tail on the first toss and head on the second toss) and *TT* (tail on both tosses). Thus, the sample space for two tosses of the coin is

$$S = \{HH, HT, TH, TT\}$$

The Venn and tree diagrams are given in Figure 4.2. Both these diagrams show the sample space for this experiment.

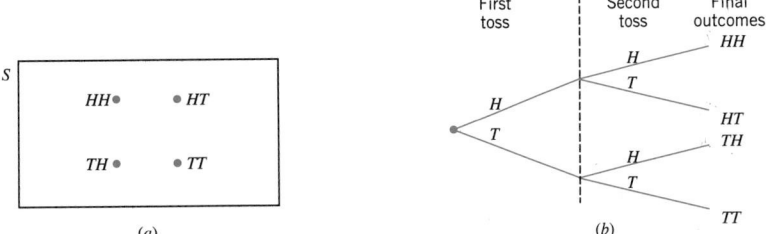

(a)   (b)

Figure 4.2   (a) Venn diagram and (b) tree diagram for two tosses of a coin.

Venn and tree diagrams: two selections.

EXAMPLE 4–3   Suppose we randomly select two employees from a company and observe whether the employee selected each time is a man or a woman. Write all the outcomes for this experiment. Draw the Venn and tree diagrams for this experiment.

Solution   Let us denote the selection of a man by M and that of a woman by W. We can compare the selection of two employees to two tosses of a coin. Just as each toss of a coin can result in one of two outcomes, head or tail, each selection from the employees of this company can result in one of two outcomes, man or woman. As we can see from the Venn and tree diagrams of Figure 4.3, there are four final outcomes: *MM, MW, WM, WW*. Hence, the sample space is written as

$$S = \{MM, MW, WM, WW\}$$

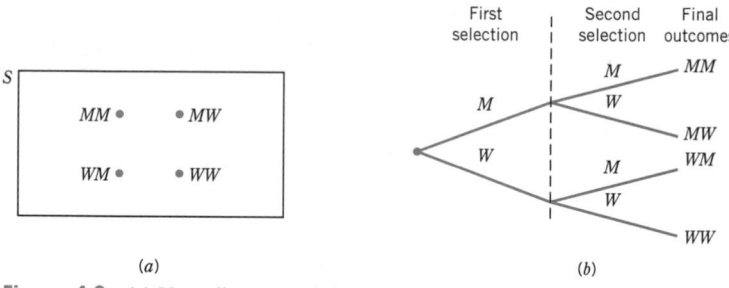

Figure 4.3 (*a*) Venn diagram and (*b*) tree diagram for selecting two employees.

4.1.1 SIMPLE AND COMPOUND EVENTS

An **event** consists of one or more of the outcomes of an experiment.

> **EVENT**
>
> An event is a collection of one or more of the outcomes of an experiment.

An event may be a *simple event* or a *compound event*. A simple event is also called an *elementary event*, and a compound event is also called a *composite event*.

Simple Event

Each of the final outcomes for an experiment is called a **simple event**. In other words, a simple event includes one and only one outcome. Usually, a simple event is denoted by E_1, E_2, E_3, and so forth. However, we can denote it by any of the other letters too, that is, by A, B, C, and so forth.

> **SIMPLE EVENT**
>
> An event that includes one and only one of the (final) outcomes for an experiment is called a simple event and is usually denoted by E_i.

Example 4–4 describes simple events.

Illustrating simple events.

EXAMPLE 4–4 Reconsider Example 4–3 about selecting two employees from a company and observing whether the employee selected each time is a man or a woman. Each of the final four outcomes (*MM*, *MW*, *WM*, and *WW*) for this experiment is a simple event. These four events can be denoted by E_1, E_2, E_3, and E_4, respectively. Thus,

$$E_1 = (MM) \quad E_2 = (MW) \quad E_3 = (WM) \quad \text{and} \quad E_4 = (WW)$$

Compound Event

A **compound event** consists of more than one outcome.

> **COMPOUND EVENT**
>
> A compound event is a collection of more than one outcome for an experiment.

Compound events are denoted by $A, B, C, D, \ldots$, or by $A_1, A_2, A_3, \ldots, B_1, B_2, B_3, \ldots$, and so forth. The following examples describe compound events.

Illustrating compound event: two selections.

EXAMPLE 4–5 Reconsider Example 4–3 about selecting two employees from a company and observing whether the employee selected each time is a man or a woman. Let A be the event that at most one man is selected. Event A will occur if either no man or one man is selected. Hence, the event A is given by

$$A = \{MW, WM, WW\}$$

Since event A contains more than one outcome, it is a compound event. The Venn diagram of Figure 4.4 gives a graphic presentation of compound event A.

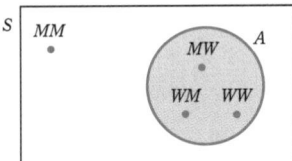

Figure 4.4 Venn diagram for event A.

Illustrating simple and compound events: two selections.

EXAMPLE 4–6 In a group of people, some are in favor of free trade and others are against it. Two persons are selected at random from this group and asked whether they are in favor of or against free trade. How many distinct outcomes are possible? Draw a Venn diagram and a tree diagram for this experiment. List all the outcomes included in each of the following events and mention whether they are simple or compound events.

(a) Both persons are in favor of free trade.
(b) At most one person is against free trade.
(c) Exactly one person is in favor of free trade.

Solution Let

$$F = \text{a person is in favor of free trade}$$

$$A = \text{a person is against free trade}$$

This experiment has the following four outcomes.

$$FF = \text{both persons are in favor of free trade}$$

$$FA = \text{the first person is in favor and the second is against}$$

$$AF = \text{the first person is against and the second is in favor}$$

$$AA = \text{both persons are against free trade}$$

The Venn and tree diagrams in Figure 4.5 show these four outcomes.

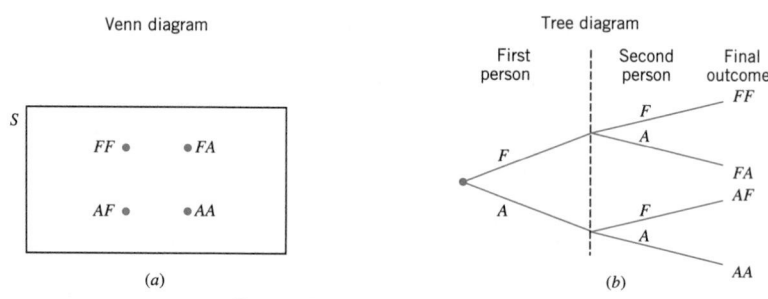

Figure 4.5 Venn and tree diagrams.

(a) The event "both persons are in favor of free trade" will occur if *FF* is obtained. Thus,

Both persons are in favor of free trade = {*FF*}

Because this event includes only one of the final four outcomes, it is a simple event.

(b) The event "at most one person is against free trade" will occur if either none or one of the persons selected is against free trade. Consequently,

At most one person is against free trade = {*FF, FA, AF*}

Because this event includes more than one outcome, it is a compound event.

(c) The event "exactly one person is in favor of free trade" will occur if one of the two persons selected is in favor and the other is against free trade. Hence, it includes the following two outcomes.

Exactly one person is in favor of free trade = {*FA, AF*}

Because this event includes more than one outcome, it is a compound event. ■

EXERCISES

Concepts and Procedures

4.1 Define the following terms: experiment, outcome, sample point, simple event, compound event.

4.2 List the simple events for the following statistical experiments in a sample space *S*.
  a. One roll of a die.
  b. Three tosses of a coin.
  c. One toss of a coin and one roll of a die.

4.3 A box contains three items that are labeled A, B, and C. Two items are selected at random (without replacement) from this box. List all the possible outcomes for this experiment. Write the sample space *S*.

Applications

4.4 Two students are randomly selected from a business statistics class, and it is observed whether or not they suffer from math anxiety. How many total outcomes are possible? Draw a tree diagram for this experiment. Draw a Venn diagram.

4.5 A hat contains a few red and a few green marbles. If two marbles are randomly drawn and the colors of these marbles are observed, how many total outcomes are possible? Draw a tree diagram for this experiment. Show all the outcomes in a Venn diagram.

4.6 A test contains two multiple-choice questions. If a student makes a random guess to answer each question, how many outcomes are possible? Depict all these outcomes in a Venn diagram. Also draw a tree diagram for this experiment. (*Hint:* Consider two outcomes for each question—either the answer is correct or it is wrong.)

4.7 A box contains a certain number of computer parts, a few of which are defective. Two parts are selected at random from this box and inspected to determine if they are good or defective. How many total outcomes are possible? Draw a tree diagram for this experiment.

4.8 In a group of people, some are in favor of a tax increase on rich people to reduce the federal deficit and others are against it. Three persons are selected at random from this group and their opinions in favor or against raising such taxes are noted. How many total outcomes are possible? Write these outcomes in a sample space S. Draw a tree diagram for this experiment.

4.9 Draw a tree diagram for three tosses of a coin. List all outcomes for this experiment in a sample space S.

4.10 Refer to Exercise 4.4. List all the outcomes included in each of the following events. Indicate which are simple and which are compound events.

    a. Both students suffer from math anxiety.
    b. Exactly one student suffers from math anxiety.
    c. The first student does not suffer and the second suffers from math anxiety.
    d. None of the students suffers from math anxiety.

4.11 Refer to Exercise 4.5. List all the outcomes included in each of the following events. Indicate which are simple and which are compound events.

    a. Both marbles are of different colors.
    b. At least one marble is red.
    c. Not more than one marble is green.
    d. The first marble is green and the second is red.

4.12 Refer to Exercise 4.6. List all the outcomes included in each of the following events and mention which are simple and which are compound events.

    a. Both answers are correct.
    b. At most one answer is wrong.
    c. The first answer is correct and the second is wrong.
    d. Exactly one answer is wrong.

4.13 Refer to Exercise 4.7. List all the outcomes included in each of the following events. Indicate which are simple and which are compound events.

    a. At least one part is good.
    b. Exactly one part is defective.
    c. The first part is good and the second is defective.
    d. At most one part is good.

4.14 Refer to Exercise 4.8. List all the outcomes included in each of the following events and mention which are simple and which are compound events.

    a. At most one person is against a tax increase on rich people.
    b. Exactly two persons are in favor of a tax increase on rich people.
    c. At least one person is against a tax increase on rich people.
    d. More than one person is against a tax increase on rich people.

4.2 CALCULATING PROBABILITY

Probability, which gives the likelihood of occurrence of an event, is denoted by P. The probability that a simple event E_i will occur is denoted by $P(E_i)$, and the probability that a compound event A will occur is denoted by $P(A)$.

> **PROBABILITY**
>
> Probability is a numerical measure of the likelihood that a specific event will occur.

☞ TWO PROPERTIES OF PROBABILITY

The following are two important properties of probability.

1. The probability of an event always lies in the range zero to 1.

Whether it is a simple or a compound event, the probability of an event is never less than zero or greater than 1. Using mathematical notation, we can write this property as follows.

$$0 \leq P(E_i) \leq 1$$

$$0 \leq P(A) \leq 1$$

An event that cannot occur has zero probability; such an event is called an **impossible event**. An event that is certain to occur has a probability equal to 1 and is called a **sure event**. That is,

For an impossible event M: $P(M) = 0$
For a sure event C: $P(C) = 1$

2. The sum of the probabilities of all simple events (or final outcomes) for an experiment, denoted by $\Sigma P(E_i)$, is always 1.

Thus, for an experiment

$$\Sigma P(E_i) = P(E_1) + P(E_2) + P(E_3) + \ldots = 1$$

From this property, for the experiment of one toss of a coin

$$P(H) + P(T) = 1$$

For the experiment of two tosses of a coin

$$P(HH) + P(HT) + P(TH) + P(TT) = 1$$

For one game of football by a National Football League team

$$P(\text{Win}) + P(\text{Loss}) + P(\text{Tie}) = 1$$

4.2.1 THREE CONCEPTUAL APPROACHES TO PROBABILITY

There are three conceptual approaches to probability: (1) classical probability, (2) the relative frequency concept of probability, and (3) the subjective probability concept. These three concepts of probability are explained next.

Classical Probability

Outcomes that have the same probability of occurrence are called **equally likely outcomes**. The classical probability rule is applied to compute the probabilities of events for an experiment all of whose outcomes are equally likely.

> **EQUALLY LIKELY OUTCOMES**
>
> Two or more outcomes (or events) that have the same probability of occurrence are said to be equally likely outcomes (or events).

According to the **classical probability rule**, the probability of a simple event is equal to 1 divided by the total number of outcomes for the experiment. This is obvious, as the sum of the probabilities of all final outcomes for an experiment is 1, and all the final outcomes are equally likely. On the other hand, the probability of a compound event A is equal to the number of outcomes favorable to event A divided by the total number of outcomes for the experiment.

> **CLASSICAL PROBABILITY RULE**
>
> $$P(E_i) = \frac{1}{\text{Total number of outcomes for the experiment}}$$
>
> $$P(A) = \frac{\text{Number of outcomes favorable to } A}{\text{Total number of outcomes for the experiment}}$$

Examples 4–7 through 4–9 illustrate how probabilities of events are calculated using the classical probability rule.

Calculating probability of a simple event.

EXAMPLE 4–7 Find the probability of obtaining a head and the probability of obtaining a tail for one toss of a coin.

Solution The two outcomes, head and tail, are equally likely outcomes. Therefore,[1]

$$P(\text{head}) = \frac{1}{\text{Total number of outcomes}} = \frac{1}{2} = .50$$

Similarly,
$$P(\text{tail}) = \frac{1}{2} = .50$$

Calculating probability of a compound event: one roll of a die.

EXAMPLE 4–8 Find the probability of obtaining an even number in one roll of a die.

Solution This experiment has six outcomes: 1, 2, 3, 4, 5, and 6. All these outcomes are equally likely. Let A be an event that an even number is observed on the die. Event A includes three outcomes: 2, 4, and 6. If any one of these three numbers is obtained, event A is said to occur. Hence,

$$P(A) = \frac{\text{Number of outcomes included in } A}{\text{Total number of outcomes}} = \frac{3}{6} = .50$$

---

[1]If the final answer for the probability of an event does not terminate within three decimal places, usually it will be rounded to three decimal places.

EXAMPLE 4-9   Tozier & Sons employs 100 workers, of whom 60 are men and 40 are women. Suppose one of the employees is randomly selected to represent them on the management consultation committee. What is the probability that this member is a woman?

Solution   Because the selection is to be made randomly, each of the 100 employees of the company has the same probability of being selected. Consequently, this experiment has a total of 100 equally likely outcomes. Forty of these 100 outcomes are included in the event that "a woman is selected." Hence,

$$P(\text{a woman is selected}) = \frac{40}{100} = .40$$

Relative Frequency Concept of Probability

Suppose we want to calculate the following probabilities.

1. The probability that the next car that comes out of an auto factory is a "lemon"
2. The probability that a randomly selected family has an annual income of more than $50,000
3. The probability that a randomly selected person owns stocks and bonds
4. The probability that a randomly selected company incurred losses last year
5. The probability that we will observe a 1-spot if we roll a loaded die

These probabilities cannot be computed using the classical probability rule because the various outcomes for the corresponding experiments are not equally likely. For example, the next car manufactured at an auto factory may or may not be a lemon. The two outcomes, "it is a lemon" and "it is not a lemon," are not equally likely. If they were, then (approximately) half the cars manufactured by this company would be lemons, and this might prove disastrous to the survival of the firm.

Although the various outcomes for each of these experiments are not equally likely, each of these experiments can be performed again and again to generate data. In such cases, to calculate probabilities we either use past data or generate new data by performing the experiment a large number of times. The relative frequency of an event is used as an approximation for the probability of that event. This method of assigning a probability to an event is called the **relative frequency concept of probability**. Because relative frequencies are determined by performing an experiment, the probabilities calculated using relative frequencies may change almost each time an experiment is repeated. For example, every time a new sample of 500 cars is selected from the production line of an auto factory, the number of lemons in those 500 cars is expected to be different. However, the variation in the percentage of lemons will be small if the sample size is large.

RELATIVE FREQUENCY AS AN APPROXIMATION OF PROBABILITY

If an experiment is repeated n times and an event A is observed f times, then, according to the relative frequency concept of probability:

$$P(A) = \frac{f}{n}$$

Examples 4-10 and 4-11 illustrate how the probabilities of events are approximated using the relative frequencies.

Approximating probability by relative frequency: sample data.

EXAMPLE 4-10 Ten of the 500 randomly selected cars manufactured at a certain auto factory are found to be lemons. Assuming that the lemons are manufactured randomly, what is the probability that the next car manufactured at this auto factory is a lemon?

Solution Let n denote the total number of cars in the sample and f the number of lemons in n. Then,

$$n = 500 \quad \text{and} \quad f = 10$$

Using the relative frequency concept of probability, we obtain:

$$P(\text{next car is a lemon}) = \frac{f}{n} = \frac{10}{500} = .02$$

This probability is actually the relative frequency of lemons in 500 cars. Table 4.2 lists the frequency and relative frequency distributions for this example.

Table 4.2 Frequency and Relative Frequency Distributions for the Sample of Cars

Car	f	Relative Frequency
Good	490	$490/500 = .98$
Lemon	10	$10/500 = .02$
	$n = 500$	Sum $= 1.0$

The column of relative frequencies in Table 4.2 is used as the column of approximate probabilities. Thus, from the relative frequency column:

$$P(\text{next car is a lemon}) = .02$$

and
$$P(\text{next car is a good car}) = .98$$

Note that relative frequencies are not probabilities but approximate probabilities. However, if the experiment is repeated again and again, this approximate probability of an outcome obtained from the relative frequency will approach the actual probability of that outcome. This is called the **Law of Large Numbers**.

LAW OF LARGE NUMBERS

If an experiment is repeated again and again, the probability of an event obtained from the relative frequency approaches the actual or theoretical probability.

Approximating probability by relative frequency.

EXAMPLE 4-11 Allison wants to determine the probability that a randomly selected family from New York State owns a home. How would she determine this probability?

Solution There are two outcomes for a randomly selected family from New York State: "This family owns a home" or "this family does not own a home." These two events are not equally likely. (Note that these two outcomes will be equally likely if exactly half of the families in New York State own homes and exactly half do not own homes.) Hence, the classical probability rule cannot be applied. However, we can repeat this experiment again and again. In other words, we can select a sample of families from New York State and observe whether or not each of them owns a home. Hence, we will use the relative frequency approach to probability.

Suppose Allison selects a random sample of 1000 families from New York State and observes that 670 of them own homes and 330 do not own homes. Then,

$$n = \text{sample size} = 1000$$

and $\qquad f = \text{number of families who own homes} = 670$

Consequently,

$$P(\text{a randomly selected family owns a home}) = \frac{f}{n} = \frac{670}{1000} = \mathbf{.670}$$

Again, note that .670 is just an approximation of the probability that a randomly selected family from New York State owns a home. Every time Allison repeats this experiment she may obtain a different probability for this event. However, because the sample size (1000) in this example is large, the variation is expected to be very small. ▬

Subjective Probability

Many times we face experiments that neither have equally likely outcomes nor can be repeated to generate data. In such cases, we cannot compute the probabilities of events using the classical probability rule or the relative frequency concept. For example, consider the following probabilities of events.

1. The probability that Carol, who is taking business statistics this semester, will earn an A in this course
2. The probability that the Dow Jones industrial average will be higher at the end of the next trading day
3. The probability that the Los Angeles Raiders will win the Super Bowl next season
4. The probability that Joe will lose the lawsuit that he has filed against his landlord

Neither the classical probability rule nor the relative frequency concept of probability can be applied to calculate probabilities for these examples. All these examples belong to experiments that have neither equally likely outcomes nor the potential of being repeated. For example, Carol, who is taking business statistics this semester, will take the test (or tests) only once and based on that she will either earn an A or not. The two events ''she will earn an A'' and ''she will not earn an A'' are not equally likely. The probability assigned to an event in such cases is called **subjective probability**. It is based on the individual's own judgment, experience, information, and belief. Carol may assign a high probability to the event that she will earn an A in business statistics, whereas her instructor may assign a low probability to the same event.

SUBJECTIVE PROBABILITY

Subjective probability is the probability assigned to an event based on subjective judgment, experience, information, and belief.

Subjective probability is assigned arbitrarily. It is usually influenced by the biases, preferences, and experience of the person assigning the probability.

According to the U.S. Nuclear Regulatory Commission, the probability that a core melt-down like the one at the Chernobyl nuclear power station in the Ukraine (the former USSR) will occur in the United States within the next 20 years is .50. This is an example of subjective probability because it is based on the research of the commission members and not on the performance of any experiment.

4.2.2 ODDS

Another concept related to probability is that of odds. The odds are obtained by finding the ratio of the probability that an event will occur to the probability that this event will not occur. Case Study 4−1 is an example of the application of odds.

CASE STUDY 4-1 PROBABILITY AND ODDS

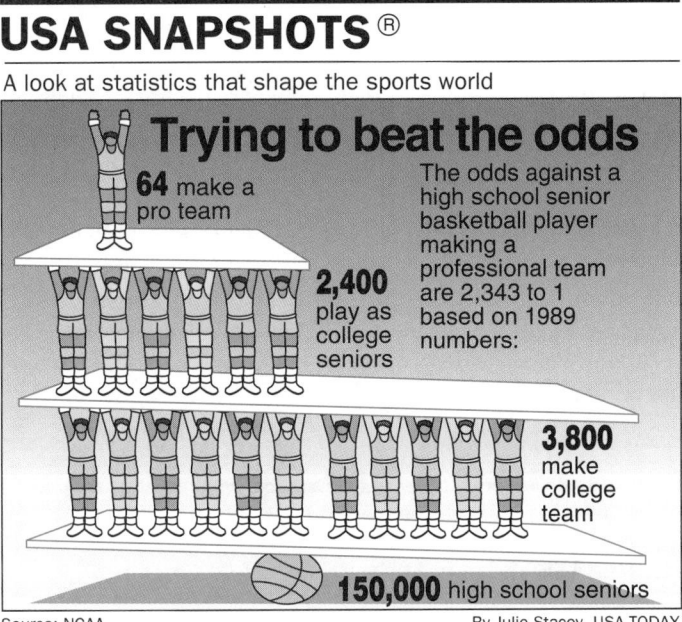

USA SNAPSHOTS ®

A look at statistics that shape the sports world

Trying to beat the odds

64 make a pro team

The odds against a high school senior basketball player making a professional team are 2,343 to 1 based on 1989 numbers:

2,400 play as college seniors

3,800 make college team

150,000 high school seniors

Source: NCAA

By Julie Stacey, USA TODAY

According to the information given in the chart, the odds against a high school senior basketball player making a professional team, based on 1989 data, are 2343 to 1. In other words, out of every 2344 (which is the sum of the two odds, 2343 and 1) high school senior basketball players, 2343 will not be able to make a professional team and one will make it. The probabilities of the two events, that a high school senior basketball player will not make a professional team and that he will make it, are calculated by dividing the respective odds by the sum of the two odds. Let

N = a high school senior basketball player will not make a pro team

M = a high school senior basketball player will make a pro team

Then,

$$P(N) = \frac{2343}{2343 + 1} = .9995733$$

$$P(M) = \frac{1}{2343 + 1} = .0004266$$

If we take the ratio of $P(N)$ and $P(M)$, we obtain the odds.

How are these odds calculated from the data given in the chart? As we can observe, out of 150,000 high school senior basketball players, 64 made a pro team and 149,936 (which is 150,000 minus 64) did not. Hence, the odds against a high school senior basketball player making a pro team are 149,936 to 64. Dividing these two numbers by 64, the odds become approximately 2343 to 1. Note that we can also state that the odds in favor of a high school senior basketball player making a pro team are 1 to 2343.

Quizzes

1. According to 1989 data, out of 265,000 high school senior football players, only 215 made a professional football team (*USA TODAY*, March 22, 1990). Calculate the odds against a high school senior football player making a professional football team. What are the odds in favor of a high school senior football player making a professional football team?

2. In the United States, one out of every 8 workers is a manager or an administrator (*Harper's Magazine*, September 1992). In other words, the odds in favor of a U.S. worker being a manager or an administrator are 1 to 7. Calculate the probability that a randomly selected worker from the United States is a manager or an administrator.

3. In Japan, one out of every 27 workers is a manager or an administrator (*Harper's Magazine*, September 1992). Calculate the odds in favor of a Japanese worker being a manager or an administrator. Calculate the probability that a randomly selected worker from Japan is a manager or an administrator.

Source: Chart reproduced with permission from *USA TODAY*, March 21, 1990. Copyright © 1990, *USA TODAY*.

EXERCISES

Concepts and Procedures

4.15 Briefly explain the two properties of probability.

4.16 Briefly describe an impossible and a sure event. What is the probability for the occurrence of each of these two events?

4.17 Briefly explain the three approaches to probability. Give one example of each of these three approaches.

4.18 Briefly explain for what kind of experiments we use the classical approach to calculate probabilities of events and for what kind of experiments we use the relative frequency approach.

4.19 Which of the following values cannot be probabilities of events and why?

$$1/5, \quad .97, \quad -.35, \quad 1.56, \quad 5/3, \quad 0.0, \quad -2/7, \quad 1.0$$

4.20 Which of the following values cannot be probabilities of events and why?

$$.46, \quad 2/3, \quad -.09, \quad 1.42, \quad .56, \quad 9/4, \quad -1/4, \quad .02$$

4.21 A statistical experiment has eight equally likely outcomes. What is the probability of each of these outcomes?

4.22 A statistical experiment that has five outcomes was repeated 200 times. The five outcomes occurred 33, 56, 47, 28, and 36 times, respectively. Find the probabilities of each of the five outcomes.

Applications

4.23 Suppose a family is selected at random from New York City. Consider the following two outcomes: This family's yearly income is less than $50,000; this family's yearly income is $50,000 or higher. Are these two outcomes equally likely? Explain why or why not. If you are to find the probabilities of these two outcomes, what approach or procedure will you use?

4.24 A class has a total of 35 students. Of them, 15 are business majors and 20 are nonbusiness majors. Suppose one student is selected at random from this class. Consider the following two events: This student is a business major; this student is a nonbusiness major. If you are to find the probabilities of these two events, will you use the classical approach or the relative frequency approach? Explain why.

4.25 The president of a company has a hunch that there is a .80 probability that the company will be successful if it introduces a new brand of ice cream. Is this a case of classical, relative frequency, or subjective probability? Explain why.

4.26 In a group of 50 persons, 15 have made investments in stocks and bonds. Find the probability that a randomly selected person from this group has made investments in stocks and bonds.

4.27 In a group of 50 executives, 27 have a type A personality. If one executive is selected at random from this group, what is the probability that this executive has a type A personality?

4.28 A company that plans to hire one new employee has prepared a final list of six candidates, all of whom are equally qualified. Two of these six candidates are women. If the company decides to select at random one person out of these six candidates, what is the probability that this person will be a woman? What is the probability that this person will be a man? Do these two probabilities add up to 1.0? If yes, why?

4.29 A large company has 840 employees and 520 of them are union members. If one employee of this company is selected at random, what is the probability that this employee is not a union member? What is the probability that this employee is a union member? Do these two probabilities add up to 1.0? If yes, why?

4.30 According to a recent estimate of the U.S. Internal Revenue Service, there were 1650 thousand business partnerships in the United States. Of them, 890 thousand earned profits and 760 thousand incurred losses. If one partnership is selected at random from these 1650 thousand partnerships, what is the probability that this partnership

    a. earned profits   **b.** incurred losses?

4.31 According to the Administrative Office of the U.S. Courts, in 1988 a total of 594,567 bankruptcy petitions were filed in the United States. Of these, 68,501 were business petitions. If one of these 594,567 petitions is randomly selected, what is the probability that it is

    a. a business bankruptcy petition

    b. a nonbusiness bankruptcy petition?

4.32 A sample of 500 large companies showed that 80 of them offer free psychiatric help to their employees who suffer from psychological problems. If one company is selected at random from this sample, what is the probability that this company offers free psychiatric help to its employees who suffer from psychological problems? What is the probability that this company does not offer free psychiatric help to its employees who suffer from psychological problems? Do these two probabilities add up to 1.0? If yes, why?

4.33 A sample of 400 large companies showed that 120 of them offer free health fitness centers to their employees within the company premises. If one company is selected at random from this sample, what is the probability that this company offers a free health fitness center to its employees within the company premises? What is the probability that this company does not offer a free health fitness center to its employees within the company premises? Do these two probabilities add up to 1.0? If yes, why?

4.34 A sample of 1000 families showed that 34 of them own no cars, 208 own one car each, 376 own two cars each, 265 own three cars each, and 117 own four or more cars each. Write the frequency distribution table for this problem. Calculate the relative frequencies for all categories. Suppose one family is randomly selected from these 1000 families. Find the probability that this family owns

    a.   two cars   b.   four or more cars

4.35 In a sample of 500 families, 95 have a yearly income of less than $20,000, 272 have a yearly income of $20,000 to $50,000, and the remaining families have a yearly income of more than $50,000. Write the frequency distribution table for this problem. Calculate the relative frequencies for all classes. Suppose one family is randomly selected from these 500 families. Find the probability that this family has a yearly income of

    a.   less than $20,000   b.   more than $50,000

4.36 Suppose you want to find the (approximate) probability that a randomly selected family from Los Angeles earns more than $75,000 a year. How will you find this probability? What procedure will you use? Explain briefly.

4.37 Suppose you have a loaded die and you want to find the (approximate) probabilities of different outcomes for this die. How will you find these probabilities? What procedure will you use? Explain briefly.

4.3 COUNTING RULE

The experiments dealt with so far in this chapter have had only a few outcomes, which were easy to list. However, for experiments with a large number of outcomes, it may not be easy to list all outcomes. In such cases, we may use the **counting rule** to find the total number of outcomes.

COUNTING RULE

If an experiment consists of three steps and if the first step can result in m outcomes, the second step in n outcomes, and the third step in k outcomes, then,

$$\text{Total outcomes for the experiment} = m \cdot n \cdot k$$

The counting rule can easily be extended to apply to an experiment with less or more than three steps.

Applying the counting rule: 3 steps.

EXAMPLE 4–12 Suppose we toss a coin three times. This experiment has three steps: the first toss, the second toss, and the third toss. Each step has two outcomes: a head and a tail. Thus,

$$\text{Total outcomes for three tosses of a coin} = 2 \times 2 \times 2 = \mathbf{8}$$

The eight outcomes for this experiment are *HHH, HHT, HTH, HTT, THH, THT, TTH,* and *TTT.* ▬

Applying the counting rule: 2 steps.

EXAMPLE 4–13 A prospective car buyer can choose between a fixed or a variable interest rate and can also choose a payment period of 36 months, 48 months, or 60 months. How many total outcomes are possible?

Solution This experiment is made up of two steps: choosing an interest rate and selecting a loan payment period. There are two outcomes (a fixed or a variable interest rate) for the first step and three outcomes (a payment period of 36 months, 48 months, or 60 months) for the second step. Hence,

$$\text{Total outcomes} = 2 \times 3 = 6 \qquad ▬$$

Applying the counting rule: 16 steps.

EXAMPLE 4–14 A National Football League team will play 16 games during a regular season. Each game can result in one of three outcomes: a win, a loss, or a tie. The total possible outcomes for 16 games are calculated as follows.

$$\text{Total outcomes} = 3 \cdot 3 \cdot 3 \cdot 3 \cdot 3 \cdot 3 \cdot 3 \cdot 3 \cdot 3 \cdot 3 \cdot 3 \cdot 3 \cdot 3 \cdot 3 \cdot 3 \cdot 3$$

$$= 3^{16} = \textbf{43,046,721}$$

One of the 43,046,721 possible outcomes is all 16 wins.[2] ▬

4.4 MARGINAL AND CONDITIONAL PROBABILITIES

Suppose all 100 employees of a company were asked whether they are in favor of or against paying high salaries to chief executive officers (CEOs) of U.S. companies. Table 4.3 gives a two-way classification of the responses of these 100 employees.

Table 4.3 Two-way Classification of the Responses of Employees

	In Favor	Against
Male	15	45
Female	4	36

Table 4.3 gives the distribution of 100 employees based on two variables or characteristics: sex (male or female) and opinion (in favor or against). Such a table is called a *contingency table*. In Table 4.3, each box that contains a number is called a *cell*. Notice that there are four cells in Table 4.3. Each cell gives the frequency for two characteristics. For example, 15 employees in this group possess two characteristics: They are male and in favor of paying high salaries to CEOs. We can interpret the numbers in other cells the same way.

By adding the row of totals and the column of totals to Table 4.3, we write Table 4.4.

Suppose one employee is selected at random from these 100 employees. This employee may be classified either on the basis of sex alone or on the basis of opinion. If only one characteristic is considered at a time, the employee selected can be a male, a female, in favor,

---

[2]Using a calculator to evaluate 3^{16}: If your calculator contains a y^x or an x^y key, you can use that key to simplify 3^{16} as follows: First enter 3 on the calculator, then press the y^x key, next enter 16, and finally press the "=" key. The screen of the calculator will display 43046721 as the answer.

Table 4.4   Two-way Classification of Employees

	In Favor	Against	Total
Male	15	45	60
Female	4	36	40
Total	19	81	100

or against. The probability of each of these four characteristics or events is called **marginal probability** or *simple probability*. These probabilities are called marginal probabilities because they are calculated by dividing the corresponding row margins (totals for the rows) or column margins (totals for the columns) by the grand total.

MARGINAL PROBABILITY

Marginal probability is the probability of a single event without consideration of any other event. Marginal probability is also called *simple probability*.

For Table 4.4, the four marginal probabilities are calculated as follows.

$$P(\text{male}) = \frac{\text{Number of males}}{\text{Total number of employees}} = \frac{60}{100} = .60$$

As we can observe, the probability that a male will be selected is obtained by dividing the total of the row labeled ''Male'' (60) by the grand total (100).

Similarly, $P(\text{female}) = 40/100 = .40$

$$P(\text{in favor}) = 19/100 = .19$$

and $P(\text{against}) = 81/100 = .81$

These four marginal probabilities are shown along the right side and along the bottom of Table 4.5.

Table 4.5   Listing the Marginal Probabilities

	In Favor (A)	Against (B)	Total	
Male (M)	15	45	60	$P(M) = 60/100 = .60$
Female (F)	4	36	40	$P(F) = 40/100 = .40$
Total	19	81	100	
	$P(A) = 19/100$ $= .19$	$P(B) = 81/100$ $= .81$		

Now suppose that one employee is selected at random from these 100 employees. Furthermore, assume that it is known that this (selected) employee is a male. In other words, the event that the employee selected is a male has already occurred. What is the probability that the employee selected is in favor of paying high salaries to CEOs? This probability is written as follows.

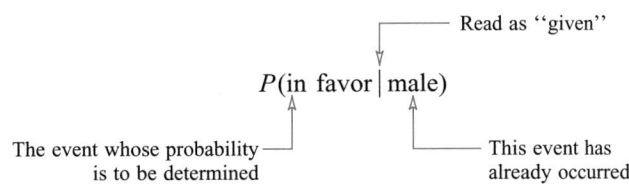

This probability, $P(\text{in favor} \mid \text{male})$, is called the **conditional probability** of "in favor" and it is read as "the probability that the employee selected is in favor given that this employee is a male."

CONDITIONAL PROBABILITY

Conditional probability is the probability that an event will occur given that another event has already occurred. if A and B are two events, then the conditional probability of A is written as

given

$$P(A \mid B) \qquad p \cdot 190$$

and read as "the probability of A given that B has already occurred."

Calculating conditional probability: two-way table.

EXAMPLE 4–15 Compute the conditional probability $P(\text{in favor} \mid \text{male})$ for the data on 100 employees given in Table 4.4.

Solution The probability $P(\text{in favor} \mid \text{male})$ is the conditional probability that a randomly selected employee is in favor given that this employee is a male. It is known that the event "male" has already occurred. Based on the information that the employee selected is a male, we can infer that the employee selected must be one of the 60 males and, hence, must belong to the first row of Table 4.4. Therefore, we are concerned only with the first row of that table.

	In Favor	**Against**	**Total**
Male	15	45	60

Males who are in favor

Total number of males

The required conditional probability is calculated as follows.

$$P(\text{in favor} \mid \text{male}) = \frac{\text{Number of males who are in favor}}{\text{Total number of males}} = \frac{15}{60} = .25$$

As we can observe from this computation of conditional probability, the total number of males (the event that has already occurred) is written in the denominator and the number of males who are in favor (the event whose probability we are to find) is written in the numerator.

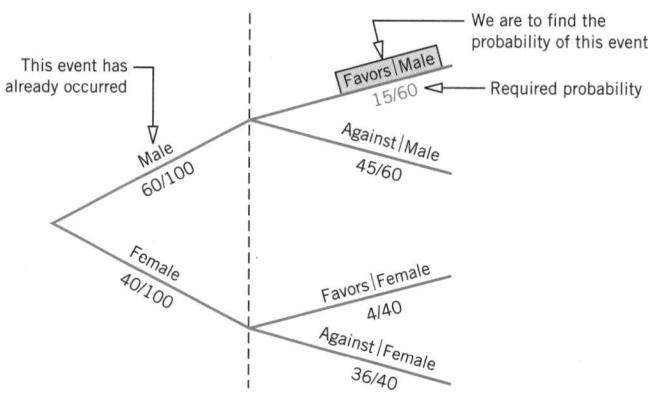

Figure 4.6   Tree diagram.

Note that we are considering the row of the event that has already occurred. The tree diagram in Figure 4.6 illustrates Example 4–15.   ■

Calculating conditional probability: two-way table.

EXAMPLE 4–16   For the data of Table 4.4, calculate the conditional probability that a randomly selected employee is a female given that this employee is in favor of paying high salaries to CEOs.

Solution   We are to compute the probability

$$P(\text{female} \mid \text{in favor})$$

Since it is known that the employee selected is in favor of paying high salaries to CEOs, this employee must belong to the first column (the column labeled ''in favor'') and must be one of the 19 employees who are in favor.

In Favor
15
4 ←——Females who are in favor
19 ←——Total number of employees who are in favor

Hence, the required probability is

$$P(\text{female} \mid \text{in favor}) = \frac{\text{Number of females who are in favor}}{\text{Total number of employees who are in favor}}$$

$$= \frac{4}{19} = \textbf{.211}$$

The tree diagram in Figure 4.7 illustrates this example.

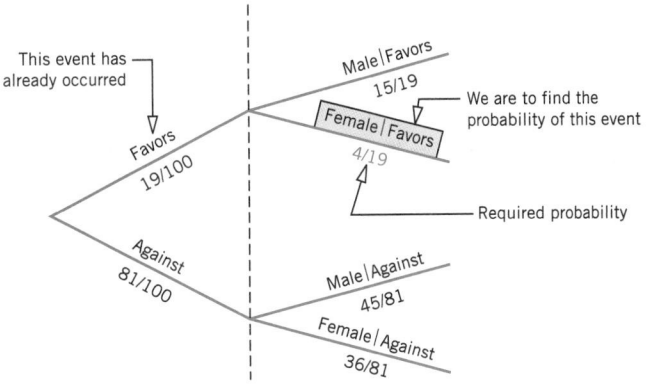

Figure 4.7 Tree diagram.

Case Study 4–2 illustrates conditional probability of home ownership, given marital status.

CASE STUDY 4-2 PROBABILITIES OF OWNING HOMES

The above chart, based on the U.S. Bureau of Labor Statistics survey for 1990, shows that 78% of the married couples, 46% of single persons, and 35% of single parents (with child at home under age 18) own homes in the United States. We can use these percentages to write the conditional probabilities. The Bureau of Labor Statistics surveys are based on samples of many thousands of households. Assume that these results hold true for the current population of all households. Suppose that one household is selected at random. Then, the fact that 78% of married couples own homes means that given that this selected household is a married couple, the probability that this married couple owns a home is .78. That is,

P(a household owns home | the household is a married couple) = .78

Similarly we can write the following two conditional probabilities using the above chart.

P(a household owns home | the household is a single person) = .46

P(a household owns home | the household is a single parent) = .35

Source: The chart is reproduced, with permission, from *USA TODAY*, January 17, 1992. Copyright © 1992, *USA TODAY*.

4.5 MUTUALLY EXCLUSIVE EVENTS

Events that cannot occur together are called **mutually exclusive events**. Such events do not have any common outcomes. If two or more events are mutually exclusive, then at most one of them will occur every time we repeat the experiment. Thus, the occurrence of one event excludes the occurrence of the other event or events.

> **MUTUALLY EXCLUSIVE EVENTS**
>
> Events that cannot occur together are said to be mutually exclusive events.

For any experiment, the final outcomes are always mutually exclusive because one and only one of these outcomes is expected to occur in one repetition of the experiment. For example, consider tossing a coin twice. This experiment has four outcomes: *HH, HT, TH,* and *TT*. These outcomes are mutually exclusive because one and only one of them will occur when we toss this coin twice.

Illustrating mutually exclusive and mutually nonexclusive events.

EXAMPLE 4–17   In a sample of 460 persons, 120 own stocks, 155 own bonds, and 225 own neither stocks nor bonds. Of the 120 persons who own stocks and 155 who own bonds, 40 own both stocks and bonds. Suppose one person is selected at random from these 460 persons. Let

A = the event that a randomly selected person owns stocks

B = the event that a randomly selected person owns bonds

C = the event that a randomly selected person owns neither stocks nor bonds

Are events A and B mutually exclusive? What about events A and C? What about events B and C?

Solution   Figure 4.8 shows the diagram of events *A*, *B*, and *C*.

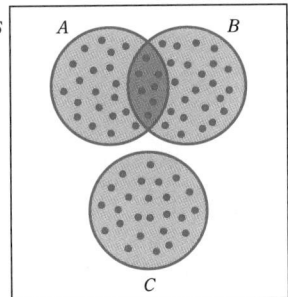

Figure 4.8   Events *A*, *B*, and *C*.

First consider events *A* and *B*. These two events have common outcomes because 40 persons own both stocks and bonds and, hence, belong to both events, *A* and *B*. This is shown by the common area that belongs to both events *A* and *B* in Figure 4.8. Consequently, if the selected person is one of the 40 who own both stocks and bonds, events *A* and *B* both happen. Therefore, events *A* and *B* are not mutually exclusive. Such events are called *mutually non-exclusive* or *overlapping events*.

Now consider events *A* and *C*. These events do not have any common outcome. This is obvious from Figure 4.8 as events *A* and *C* do not have any common area. A person who belongs to event *A* cannot belong to event *C* at the same time. Hence, events *A* and *C* are mutually exclusive events.

Similarly, because events *B* and *C* do not have any common outcome, they are also mutually exclusive events.

Illustrating mutually exclusive events.

EXAMPLE 4-18   Suppose an employee is selected at random from a large company. Consider the following two events.

$$B = \text{the employee selected is a blue-collar worker}$$

$$W = \text{the employee selected is a white-collar worker}$$

Are events *B* and *W* mutually exclusive?

Solution   Event *B* consists of all blue-collar workers at this company and event *W* includes all white-collar workers. These two events are shown in Figure 4.9.

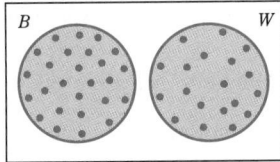

Figure 4.9   Mutually exclusive events *B* and *W*.

As we can see from the definition of events *B* and *W* and from Figure 4.9, events *B* and *W* have no common outcome. They represent two distinct sets of employees: blue-collar workers and white-collar workers. Hence, these two events are mutually exclusive events.

4.6 INDEPENDENT VERSUS DEPENDENT EVENTS

In the case of two **independent events**, the occurrence of one event does not change the probability of the occurrence of the other event.

INDEPENDENT EVENTS

Two events are said to be independent if the occurrence of one does not affect the probability of the occurrence of the other. In other words, A and B are independent events if

$$\text{either} \quad P(A \mid B) = P(A) \quad \text{or} \quad P(B \mid A) = P(B)$$

It can be shown that if one of these two conditions is true, then the second will also be true, and if one is not true then the second will also not be true.

If the occurrence of one event affects the probability of the occurrence of the other event, then the two events are said to be **dependent events**. Using probability notation, the two events will be dependent if either $P(A \mid B) \neq P(A)$ or $P(B \mid A) \neq P(B)$.

Illustrating two dependent events: two-way table.

EXAMPLE 4-19   Refer to the information on 100 employees given in Table 4.4 in Section 4.4. Are the two events "female (F)" and "in favor (A)" independent?

Solution   Events F and A will be independent if

$$P(F) = P(F \mid A)$$

Otherwise they will be dependent.

Using the information given in Table 4.4, we compute the following two probabilities.

$$P(F) = 40/100 = \textbf{.40} \quad \text{and} \quad P(F \mid A) = 4/19 = \textbf{.211}$$

Because these two probabilities are not equal, the two events are dependent. Here, dependence of events means that the percentage of males who are in favor of and against paying high salaries to CEOs is different from the percentage of females who are in favor and against.

In this example, the dependence of A and F can also be proved by showing that the probabilities $P(A)$ and $P(A \mid F)$ are not equal. ■

Illustrating two independent events.

EXAMPLE 4-20   A box contains a total of 100 cassettes that were manufactured on two machines. Of them, 60 were manufactured on Machine I. Of the total cassettes, 15 are defective. Of the 60 cassettes that were manufactured on Machine I, 9 are defective. Let D be an event that a randomly selected cassette is defective and A be an event that a randomly selected cassette was manufactured on Machine I. Are events D and A independent?

Solution   From the given information,

$$P(D) = 15/100 = .15 \quad \text{and} \quad P(D \mid A) = 9/60 = .15$$

Hence,

$$P(D) = P(D \mid A)$$

Consequently, the two events, D and A, are independent.

Independence, in this case, means that the probability for any cassette to be defective is the same, .15, irrespective of the machine on which it is manufactured. In other words, the two machines are producing the same percentage of defective items. For example, 9 of the 60 cassettes manufactured on Machine I are defective and 6 of the 40 cassettes manufactured on Machine II are defective. Thus, for each of the two machines, 15% of the cassettes produced are defective.

Actually, using the given information, we can prepare Table 4.6 for Example 4–20. The numbers in the shaded cells are given to us. The remaining numbers are calculated by doing some arithmetic.

Table 4.6 Two-Way Classification Table

	Defective (D)	Good (G)	Total
Machine I (A)	9	51	60
Machine II (B)	6	34	40
Total	15	85	100

Using this table, we can find the following probabilities.

$$P(D) = 15/100 = .15$$

and

$$P(D \mid A) = 9/60 = .15$$

Since these two probabilities are the same, the two events are independent.

☞ TWO IMPORTANT OBSERVATIONS

The following are two important observations about mutually exclusive, independent, and dependent events.

1. Two events are either mutually exclusive or independent.[3] In other words,
    a. mutually exclusive events are always dependent
    b. independent events are never mutually exclusive
2. Dependent events may or may not be mutually exclusive.

4.7 COMPLEMENTARY EVENTS

Two mutually exclusive events that taken together include all the outcomes for an experiment are called **complementary events**. Note that two complementary events are always mutually exclusive.

COMPLEMENTARY EVENTS

The complement of event A, denoted by $\overline{A}$ and read as "A bar" or "A complement," is the event that includes all the outcomes for an experiment that are not in A.

---

[3]The exception to this rule occurs when at least one of the two events has a zero probability.

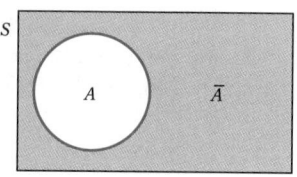

Figure 4.10 Venn diagram of two complementary events.

Events A and $\overline{A}$ are complements of each other. The Venn diagram in Figure 4.10 shows the complementary events A and $\overline{A}$.

Because two complementary events, taken together, include all the outcomes for an experiment and because the sum of the probabilities of all outcomes is 1, it is obvious that

$$P(A) + P(\overline{A}) = 1$$

From this equation we can deduce that

$$P(A) = 1 - P(\overline{A}) \quad \text{and} \quad P(\overline{A}) = 1 - P(A)$$

Thus, if we know the probability of an event, we can find the probability of its complementary event by subtracting the given probability from 1.0.

Calculating probabilities of complementary events.

EXAMPLE 4–21 In a lot of five washing machines, two are defective. If one machine is randomly selected, what are the two complementary events for this experiment and what are their probabilities?

Solution The two complementary events for this experiment are

$$A = \text{the machine selected is defective}$$
$$\overline{A} = \text{the machine selected is not defective}$$

Since there are two defective and three nondefective machines, the probabilities of events A and $\overline{A}$ are

$$P(A) = 2/5 = \textbf{.40} \quad \text{and} \quad P(\overline{A}) = 3/5 = \textbf{.60}$$

As we can observe, the sum of these two probabilities is 1. Figure 4.11 shows a Venn diagram for this example.

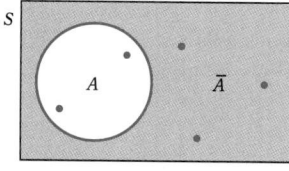

Figure 4.11 Venn diagram.

Calculating probability of the complement of an event.

EXAMPLE 4–22 The probability that a machine used to make computer chips at Rahming Inc. is out of control at any time is .04. What is the complement of this event and what is its probability?

Solution The given event is

$$A = \text{the machine is out of control}$$

Then, the complement of this event is

$$\overline{A} = \text{the machine is within control}$$

We are given that

$$P(A) = .04$$

Hence, $$P(\overline{A}) = 1 - .04 = .96$$ ■

EXERCISES

Concepts and Procedures

4.38 Briefly explain the difference between the marginal and conditional probabilities of events. Give one example of each.

4.39 What is meant by two mutually exclusive events? Give one example of two mutually exclusive events and another example of two mutually nonexclusive events.

4.40 Briefly explain the meaning of independent and dependent events. Suppose A and B are two events. What formula will you use to prove whether A and B are independent or dependent?

4.41 What is the complement of an event? What is the sum of the probabilities of two complementary events?

4.42 How many different outcomes are possible for four rolls of a die?

4.43 How many different outcomes are possible for 10 tosses of a coin?

4.44 A statistical experiment has eight equally likely outcomes that are denoted by 1, 2, 3, 4, 5, 6, 7, and 8. Let event $A = \{2, 5, 7\}$ and event $B = \{2, 4, 8\}$.
   a.   Are events A and B mutually exclusive events?
   b.   Are events A and B independent events?
   c.   What are the complements of events A and B and their probabilities?

4.45 A statistical experiment has 10 equally likely outcomes that are denoted by 1, 2, 3, 4, 5, 6, 7, 8, 9, and 10. Let event $A = \{3, 4, 6, 9\}$ and event $B = \{1, 2, 5\}$.
   a.   Are events A and B mutually exclusive events?
   b.   Are events A and B independent events?
   c.   What are the complements of events A and B and their probabilities?

Applications

4.46 A specific model of a car comes in five exterior colors and three interior colors. All exterior colors can be combined with any of the interior colors. How many different selections of one exterior and one interior color are possible? m·n   5·3=15cars

4.47 A manager just bought four suits, eight shirts, and nine ties. All of these suits, shirts, and ties coordinate with each other. If he is to randomly select one suit, one shirt, and one tie to wear on a certain day, how many different outcomes (selections) are possible?

4.48 A manager is refurbishing her office. She can choose one of three kinds of paints, one of four kinds of furniture, and one of three kinds of telephones. If she randomly selects one paint, one kind of furniture, and one telephone, how many different outcomes are possible? 3·4·3= 36outcomes

4.49 A restaurant menu has four kinds of soups, eight kinds of main courses, five kinds of desserts, and six kinds of drinks. If a customer randomly selects one item from each of these four categories, how many different outcomes are possible? 4·8·5·6= 960

4.50 All the 420 employees of a company were asked if they smoke or not and whether they are college graduates or not. Based on this information, the following two-way classification table was prepared.

	College Graduate	Not a College Graduate
Smoker	35	80
Nonsmoker	130	175

If one employee is selected at random from this company, find the probability that this employee is a

a. college graduate
b. nonsmoker
c. smoker given the employee is not a college graduate
d. college graduate given the employee is a nonsmoker

4.51 The following table gives a two-way classification of 200 corporations based on whether they earned profits or incurred losses in 1991 and 1992.

		1992	
		Earned Profits	Incurred Losses
1991	Earned Profits	110	40
	Incurred Losses	20	30

If one corporation is selected at random from these 200 corporations, find the probability that this corporation

a. earned profits in 1992
b. incurred losses in 1991
c. earned profits in 1992 given it earned profits in 1991
d. incurred losses in 1992 given it incurred losses in 1991

4.52 The following table gives a two-way classification of all 1000 employees of a large company based on whether they are single or married and whether or not they own stocks and bonds.

	Own Stocks and Bonds	
	Yes	No
Single	120	280
Married	180	420

a. If one employee is selected at random from these 1000 employees, find the probability that this employee is
  i. married
  ii. owns stocks and bonds
  iii. single given he/she owns stocks and bonds
  iv. does not own stocks and bonds given he/she is married
b. Are the events "single" and "yes" mutually exclusive? What about the events "yes" and "no"? Why or why not?
c. Are the events "married" and "yes" independent? Why or why not?

4.53 Five hundred employees were selected from a city's large private companies, and they were asked whether or not they have any retirement benefits provided by their companies. Based on this information, the following two-way classification table was prepared.

	Have Retirement Benefits	
	Yes	No
Men	225	75
Women	150	50

a. If one employee is selected at random from these 500 employees, find the probability that this employee
  i. is a woman $\frac{200}{500}$ .4
  ii. has retirement benefits $\frac{375}{500}$ .75
  iii. has retirement benefits given the employee is a man $\frac{225}{300}$ .75
  iv. is a woman given that she does not have retirement benefits $\frac{60}{125}$ .4

b. Are the events "men" and "yes" mutually exclusive? What about the events "yes" and "no"? Why or why not?

c. Are the events "women" and "yes" independent? Why or why not?

4.54 The following table gives the two-way classification of 2000 randomly selected employees from a city based on gender and commuting time from home to work.

	Commuting Time from Home to Work			Total
	Less Than 30 Minutes <30	30 Minutes to One Hour $30-1$	More Than One Hour >1	
Men	524	455	221	1200
Women	413	263	124	800
	937	718	345	2000

a. If one employee is selected at random from these 2000 employees, find the probability that this employee
  i. commutes for more than one hour $\frac{345}{2000}$ .1725
  ii. commutes for less than 30 minutes $\frac{937}{2000}$ .4685
  iii. is a man given that he commutes for 30 minutes to one hour $\frac{455}{718}$ .6337
  iv. commutes for more than one hour given the employee is a woman $\frac{124}{800}$ .155

b. Are the events "man" and "commutes for more than one hour" mutually exclusive? What about the events "less than 30 minutes" and "more than one hour"? Why or why not?

c. Are the events "woman" and "commutes for 30 minutes to one hour" independent? Why or why not?

4.55 Two thousand randomly selected adults were asked if they think they are financially better off than their parents. The following table gives the two-way classification of the responses based on the education levels of the persons included in the survey and whether they are better off, the same, or worse off financially than their parents.

	Education Level			Total
	Less Than High School $<$	High School	More Than High School $>$	
Better Off	140	450	420	1010
Same	60	250	110	420
Worse Off	200	300	70	570
	400	1000	600	2000

a. If one adult is selected at random from these 2000 adults, find the probability that this adult is
  i. financially better off than his/her parents $\frac{1010}{2000} = .51$
  ii. financially better off than his/her parents given he/she has less than high school education $\frac{140}{400} = .35$
  iii. financially worse off than his/her parents given he/she has high school education
  iv. financially the same as his/her parents given he/she has more than high school education

b. Are the events "better off" and "high school education" mutually exclusive? What about the events "less than high school" and "more than high school"? Why or why not?

c. Are the events "worse off" and "more than high school" independent? Why or why not?

$P(W|HS)$ $\frac{300}{1000} = .30$

$P(S|>HS)$ $\frac{110}{600} = .18$

4.56   Of a total of 100 diskettes manufactured on two machines, 20 are defective. Sixty of the total diskettes were manufactured on Machine I and 10 of these 60 are defective. Are the events "machine type" and "defective diskettes" independent? (*Note:* Compare this exercise with Example 4–20.)

4.57   A company hired 30 new graduates last week. Of these, 16 are female and 11 are business majors. Of the 16 females, 7 are business majors. Are the events "female" and "business major" independent? Are they mutually exclusive? Explain why or why not.

4.58   Define the following two events for two tosses of a coin.

$$A = \text{at least one head is obtained}$$

$$B = \text{both tails are obtained}$$

a.   Are A and B mutually exclusive events? Are they independent? Explain why or why not.
b.   Are A and B complementary events? If yes, first calculate the probability of B and then calculate the probability of A using the complementary event rule.

4.59   Let A be the event that a number less than 3 is obtained if we roll a die once. What is the probability of A? What is the complementary event of A, and what is its probability?

4.60   According to the U.S. Bureau of Labor Statistics, a total of 5729 thousand persons in the United States have more than one job and 3537 thousand of them are male. If one person is selected at random from these 5729 thousand persons, what are the two complementary events and their probabilities?

4.61   The probability that a business professor reads *The Wall Street Journal* is .65. What is its complementary event? What is the probability of this complementary event?

4.8   INTERSECTION OF EVENTS AND THE MULTIPLICATION RULE

This section discusses the intersection of two events and the application of the multiplication rule to compute the probability of the intersection of events.

4.8.1   INTERSECTION OF EVENTS

The **intersection of two events** is given by the outcomes that are common to both events.

INTERSECTION OF EVENTS

Let A and B be two events defined in a sample space. The intersection of A and B represents the collection of all outcomes that are common to both A and B and is denoted by

$$A \text{ and } B$$

The intersection of events A and B is also denoted by either $A \cap B$ or AB. Let

$$A = \text{the event that a family owns a VCR}$$

$$B = \text{the event that a family owns a telephone answering machine}$$

Figure 4.12 illustrates the intersection of events A and B. The shaded area in this figure gives the intersection of events A and B, and it includes all the families who own both a VCR and a telephone answering machine.

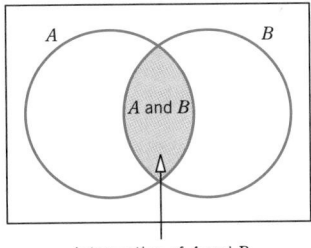

Intersection of A and B

Figure 4.12 Intersection of events A and B.

4.8.2 MULTIPLICATION RULE

The probability that events A and B happen together is called the **joint probability** of A and B and is written as $P(A \text{ and } B)$.

JOINT PROBABILITY

The probability of the intersection of two events is called their joint probability. It is written as

$$P(A \text{ and } B)$$

The probability of the intersection of two events is obtained by multiplying the marginal probability of one event by the conditional probability of the second event. This rule is called the **multiplication rule**.

MULTIPLICATION RULE

The probability of the intersection of two events A and B is

$$P(A \text{ and } B) = P(A)\, P(B \mid A)$$

The joint probability of events A and B can also be denoted by $P(A \cap B)$ or $P(AB)$.

Calculating joint probability of two events: two-way table.

EXAMPLE 4–23 The following table gives the classification of all employees of a company by sex and college degree.

	College Graduate (G)	Not a College Graduate (N)	Total
Male (M)	7	20	27
Female (F)	4	9	13
Total	11	29	40

If one of these employees is selected at random for membership on the employee-management committee, what is the probability that this employee is a female and a college graduate?

Solution   We are to calculate the probability of the intersection of events "female" (denoted by F) and "college graduate" (denoted by G). This probability will be computed using the formula

$$P(F \text{ and } G) = P(F)\, P(G|F)$$

The shaded area in Figure 4.13 gives the intersection of events "female" and "college graduate."

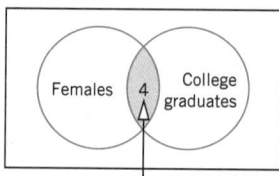

Female and college graduate   **Figure 4.13**

Notice that there are 13 females among 40 employees. Hence, the probability that a female is selected is

$$P(F) = 13/40$$

To calculate the probability $P(G|F)$, we know that F has already occurred. Consequently, the employee selected is one of the 13 females. In the table, there are 4 college graduates among 13 female employees. Hence, the conditional probability of G given F is

$$P(G|F) = 4/13$$

The joint probability of F and G is

$$P(F \text{ and } G) = P(F)\, P(G|F) = (13/40)\,(4/13) = \mathbf{.100}$$

Thus, the probability is .100 that a randomly selected employee is a female and a college graduate.

The probability in this example can also be calculated without using the multiplication rule. As we can notice from Figure 4.13 and from the table, there are four employees out of a total of 40 who are female and college graduates. Hence, if any of these four employees is selected, events "female" and "college graduate" both happen. Hence, the required probability is

$$P(F \text{ and } G) = 4/40 = \mathbf{.100}$$

Similarly, we can compute three other joint probabilities for the table as follows.

$$P(M \text{ and } G) = P(M)\, P(G|M) = (27/40)\,(7/27) = \mathbf{.175}$$
$$P(M \text{ and } N) = P(M)\, P(N|M) = (27/40)\,(20/27) = \mathbf{.500}$$
$$P(F \text{ and } N) = P(F)\, P(N|F) = (13/40)\,(9/13) = \mathbf{.225}$$

The tree diagram in Figure 4.14 shows all four joint probabilities for this example. The joint probability of F and G is highlighted in the tree diagram.

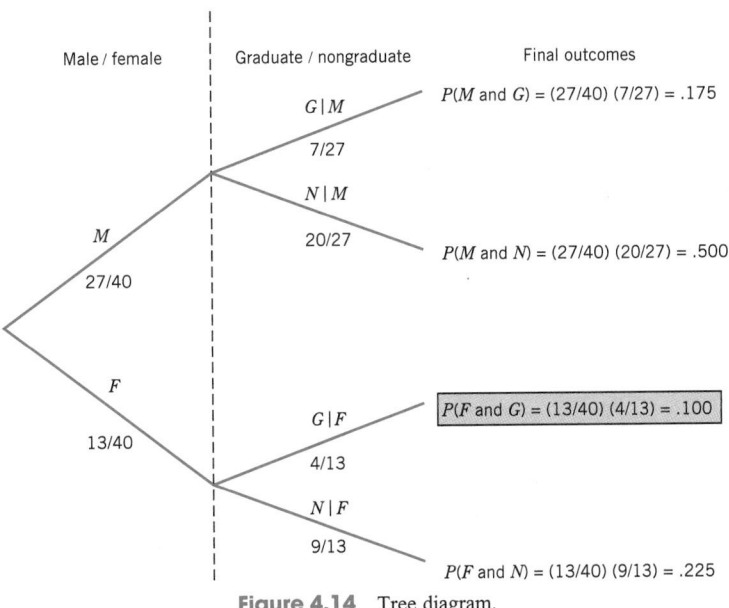

Figure 4.14 Tree diagram.

Calculating joint probability of two events.

EXAMPLE 4–24 A box contains 20 cassettes, 4 of which are defective. If 2 cassettes are selected at random (without replacement) from this box, what is the probability that both are defective?

Solution Let us define the following events for this experiment.

$$G_1 = \text{event that the first cassette selected is good}$$

$$D_1 = \text{event that the first cassette selected is defective}$$

$$G_2 = \text{event that the second cassette selected is good}$$

$$D_2 = \text{event that the second cassette selected is defective}$$

We are to calculate the joint probability of D_1 and D_2, which is given by

$$P(D_1 \text{ and } D_2) = P(D_1)\, P(D_2 | D_1)$$

As we know, there are 4 defective cassettes in 20. Consequently, the probability of selecting a defective cassette at the first selection is

$$P(D_1) = 4/20$$

To calculate the probability $P(D_2 | D_1)$, we know that the first cassette selected is defective because D_1 has already occurred. Because the selections are made without replacement, there are 19 total cassettes and 3 of them are defective at the time of the second selection. Therefore,

$$P(D_2 | D_1) = 3/19$$

Hence, the required probability is

$$P(D_1 \text{ and } D_2) = P(D_1)\, P(D_2 | D_1) = (4/20)(3/19) = \textbf{.032}$$

The tree diagram in Figure 4.15 shows the selection procedure and the final four outcomes for this experiment together with their probabilities. The joint probability of D_1 and D_2 is highlighted in the tree diagram.

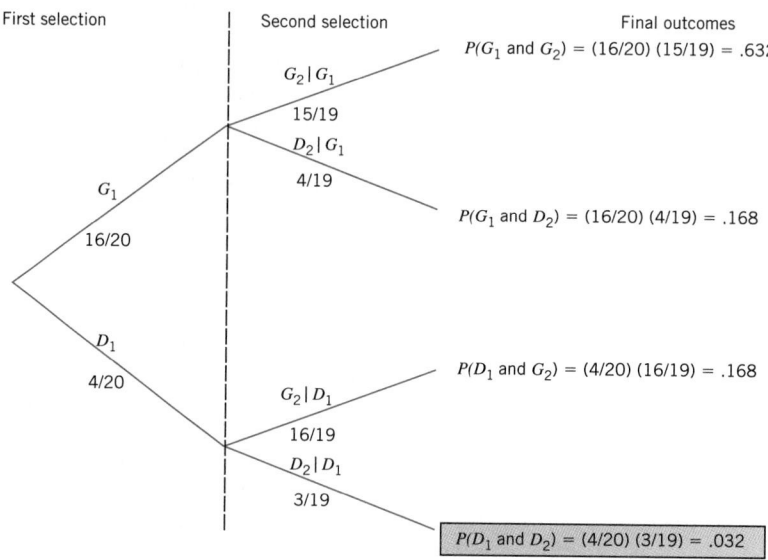

Figure 4.15   Selecting two cassettes.

Conditional probability was discussed in Section 4.4. It is obvious from the formula for joint probability that if we know the probability of an event A and the joint probability of events A and B, then we can calculate the conditional probability of B given A.

CONDITIONAL PROBABILITY

If A and B are two events, then,

$$P(B\,|\,A) = \frac{P(A \text{ and } B)}{P(A)} \quad \text{and} \quad P(A\,|\,B) = \frac{P(A \text{ and } B)}{P(B)}$$

given that $P(A) \neq 0$ and $P(B) \neq 0$.

Calculating conditional
probability of an event.

EXAMPLE 4–25   According to a survey, 60% of all homeowners owe money on home mortgages. Thirty-six percent of the homeowners owe money on both home mortgages and car loans. Find the conditional probability that a homeowner selected at random owes money on a car loan given that this homeowner owes money on a home mortgage.

Solution   Let us define the following two events.

A = the homeowner family selected owes money on a home mortgage

B = the homeowner family selected owes money on a car loan

From the given information,

$$P(A) = .60 \quad \text{and} \quad P(A \text{ and } B) = .36$$

Hence,

$$P(B\,|\,A) = \frac{P(A \text{ and } B)}{P(A)} = \frac{.36}{.60} = \textbf{.60}$$

Thus the (conditional) probability is .60 that a randomly selected homeowner family owes money on a car loan given that this family owes money on a home mortgage. ▄

MULTIPLICATION RULE FOR INDEPENDENT EVENTS

The foregoing discussion of the multiplication rule was based on the assumption that the two events are dependent. Now suppose that events A and B are independent. Then,

$$P(A) = P(A \mid B) \quad \text{and} \quad P(B) = P(B \mid A)$$

By substituting $P(B)$ for $P(B \mid A)$ into the formula for the joint probability of A and B, we obtain

$$P(A \text{ and } B) = P(A) \, P(B)$$

MULTIPLICATION RULE FOR INDEPENDENT EVENTS

The probability of the intersection of two independent events A and B is

$$P(A \text{ and } B) = P(A) \, P(B)$$

Calculating joint probability of two independent events.

EXAMPLE 4–26 Philmont Inc. makes computer chips that are supplied to many computer companies. The company has two quality control inspectors, Mr. Haines and Ms. Garcia, who independently inspect each computer chip manufactured in the company before it is shipped to a client. The probability that Mr. Haines fails to detect a defective computer chip is .02 and the probability that Ms. Garcia fails to detect a defective computer chip is .01. Find the probability that both Mr. Haines and Ms. Garcia will fail to detect a defective computer chip.

Solution   In this example, the two quality control inspectors are independent. This is so because whether or not one inspector fails to detect the defective chip, it has no effect on the second inspector's outcome. Let us define the following two events.

$A = $ the first inspector fails to detect a defective computer chip

$B = $ the second inspector fails to detect a defective computer chip

Then the joint probability of A and B is

$$P(A \text{ and } B) = P(A) \, P(B) = (.02) \, (.01) = \textbf{.0002} \quad ▄$$

The multiplication rule can be extended to calculate the joint probability of more than two events. Example 4–27 illustrates such a case for independent events.

Calculating joint probability of three events.

EXAMPLE 4–27 The probability that a patient is allergic to penicillin is .20. Suppose this drug is administered to three patients.

(a)   Find the probability that all three of them are allergic to it.

(b)   Find the probability that at least one of them is not allergic to it.

Solution

(a) Let A, B, and C denote the events that the first, second, and third patients are allergic to penicillin, respectively. We are to find the joint probability of A, B, and C. All three events are independent because whether or not one patient is allergic does not depend on whether or not any of the other patients is allergic. Hence,

$$P(A \text{ and } B \text{ and } C) = P(A)\,P(B)\,P(C) = (.20)\,(.20)\,(.20) = \textbf{.008}$$

The tree diagram in Figure 4.16 shows all the outcomes for this experiment. Events $\overline{A}$, $\overline{B}$, and $\overline{C}$ are the complementary events of A, B, and C, respectively. They represent the events that the respective patients are not allergic to penicillin. Note that the intersection of events A, B, and C is written as ABC in the tree diagram.

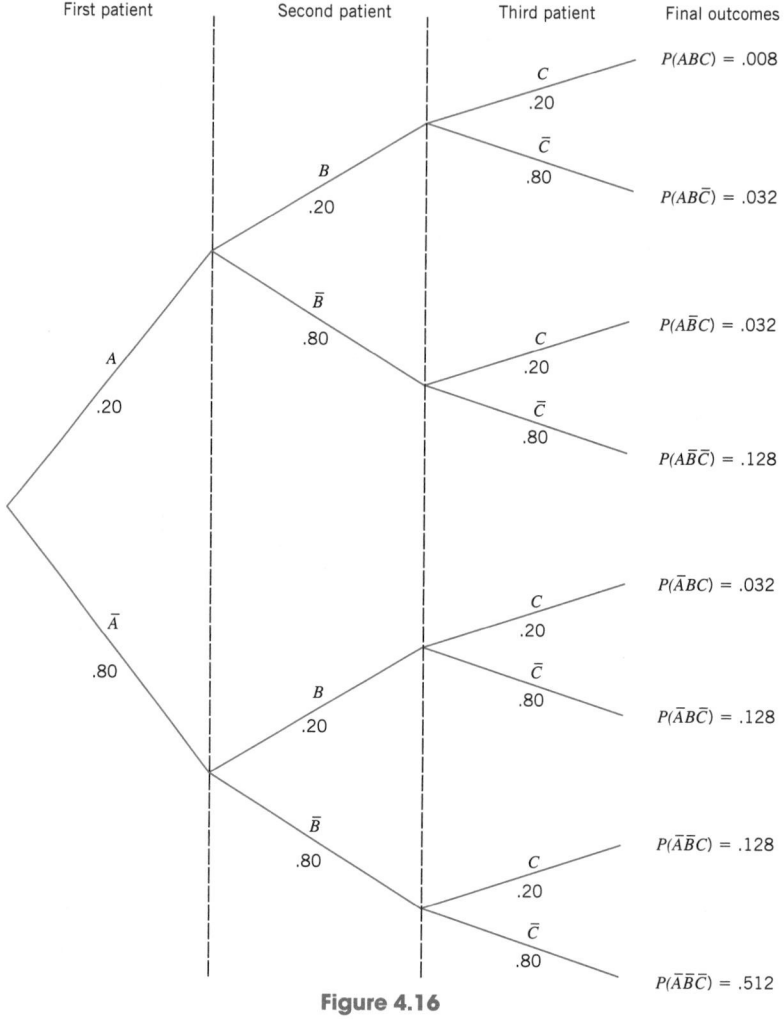

Figure 4.16

(b) Let us define the following events.

$$G = \text{all three patients are allergic}$$

$$H = \text{at least one patient is not allergic}$$

Events G and H are two complementary events. Event G consists of the intersection of events A, B, and C. Hence, from part (a),

$$P(G) = P(A \text{ and } B \text{ and } C) = .008$$

Therefore, using the complementary event rule, we obtain:

$$P(H) = 1 - P(G) = 1 - .008 = \mathbf{.992}$$

Case Study 4–3 calculates the probability of a hitless streak in baseball by using the multiplication rule.

CASE STUDY 4-3   BASEBALL PLAYERS HAVE "SLUMPS" AND "STREAKS"

Going '0 for July,' as former infielder Bob Aspromonte once put it, is enough to make a baseball player toss out his lucky bat or start seriously searching for flaws in his hitting technique. But the culprit is usually just simple mathematics.

Statistician Harry Roberts of the University of Chicago's Graduate School of Business studied the records of major-league baseball players and found that a batter is no more likely to hit worse when he is in a slump than when he is in a hot streak. The occurrences of hits followed the same pattern as purely random events such as pulling marbles out of a hat. If there were one white marble and three black ones in the hat, for example, then a white marble would come out about one quarter of the time—a .250 average. In the same way, a player who hits .250 will in the long run get a hit every four times at bat.

But that doesn't mean the player will hit the ball exactly every fourth time he comes to the plate—just as it's unlikely that the white marble will come out exactly every fourth time.

Even a batter who goes hitless 10 times in a row might safely be able to pin the blame on statistical fluctuations. The odds of pulling a black marble out of a hat 10 times in a row are about 6 percent—not a frequent occurrence, but not impossible, either. Only in the long run do these statistical fluctuations even out. . . .

If we assume a player hits .250 in the long run, the probability that this player does not hit during any visit to the plate is .75. Hence, we can calculate the probability that he goes hitless 10 times in a row as follows.

$$P(\text{hitless 10 times in a row}) = (.75)\,(.75)\,\ldots\,(.75) \text{ ten times}$$

$$= .056$$

Note that each visit to the plate is independent, and the probability that a player goes hitless 10 times in a row is given by the intersection of 10 hitless visits. This probability has been rounded off to "about 6%" in this illustration.

Source: U.S. News & World Report, July 11, 1988, p. 46. Copyright © 1988, by U.S. News & World Report, Inc. Excerpts reprinted with permission.

JOINT PROBABILITY OF MUTUALLY EXCLUSIVE EVENTS

We know from an earlier discussion that two mutually exclusive events cannot happen together. Consequently, their joint probability is zero.

JOINT PROBABILITY OF MUTUALLY EXCLUSIVE EVENTS

The joint probability of two mutually exclusive events is always zero. If A and B are two mutually exclusive events, then,

$$P(A \text{ and } B) = 0$$

Illustrating probability of two mutually exclusive events.

EXAMPLE 4-28   Consider the following two events for an application filed by a person to obtain a car loan.

$$A = \text{event that the car loan application is approved}$$

$$R = \text{event that the car loan application is rejected}$$

What is the joint probability of A and R?

Solution   The two events A and R are mutually exclusive. Either the loan application will be approved or it will be rejected. Hence,

$$P(A \text{ and } R) = \mathbf{0}$$

EXERCISES

Concepts and Procedures

4.62   Explain the meaning of the intersection of two events. Give one example.

4.63   What is meant by the joint probability of two or more events? Give one example.

4.64   How is the multiplication rule of probability for two dependent events different from the one for two independent events?

4.65   What is the joint probability of two mutually exclusive events? Give one example.

4.66   Find the joint probability of A and B for the following.
   a. $P(A) = .40$ and $P(B \mid A) = .32$
   b. $P(B) = .65$ and $P(A \mid B) = .36$

4.67   Find the joint probability of A and B for the following.
   a. $P(B) = .59$ and $P(A \mid B) = .77$
   b. $P(A) = .28$ and $P(B \mid A) = .15$

4.68   Given that A and B are two independent events, find their joint probability for the following.
   a. $P(A) = .61$ and $P(B) = .27$
   b. $P(A) = .39$ and $P(B) = .73$

4.69   Given that A and B are two independent events, find their joint probability for the following.
   a. $P(A) = .20$ and $P(B) = .86$
   b. $P(A) = .57$ and $P(B) = .32$

4.70   Given that A, B, and C are three independent events, find their joint probability for the following.
   a. $P(A) = .20$, $P(B) = .46$,   and $P(C) = .15$
   b. $P(A) = .44$, $P(B) = .27$,   and $P(C) = .33$

4.71 Given that A, B, and C are three independent events, find their joint probability for the following.

    a. $P(A) = .39$, $P(B) = .67$,   and $P(C) = .75$

    b. $P(A) = .71$, $P(B) = .34$,   and $P(C) = .41$

4.72 Given that $P(A) = .30$ and $P(A \text{ and } B) = .24$, find $P(B \mid A)$.

4.73 Given that $P(B) = .65$ and $P(A \text{ and } B) = .45$, find $P(A \mid B)$.

4.74 Given that $P(A \mid B) = .40$ and $P(A \text{ and } B) = .36$, find $P(B)$.

4.75 Given that $P(B \mid A) = .80$ and $P(A \text{ and } B) = .58$, find $P(A)$.

Applications

4.76 The following table gives a two-way classification of all 1000 employees of a large company based on whether they are single or married and whether or not they own stocks and bonds.

	Own Stocks and Bonds	
	Yes	**No**
Single	120	280
Married	180	420

    a. Suppose one employee is selected at random from these 1000 employees. Find the following probabilities.

        i. P(married and owns stocks & bonds)

        ii. P(owns stocks & bonds and single)

    b. Mention what other joint probabilities you can calculate for this table and then find all those joint probabilities. You may draw a tree diagram to find these probabilities.

4.77 Five hundred employees were selected from a city's large private companies, and they were asked whether or not they have any retirement benefits provided by their companies. Based on this information, the following two-way classification table was prepared.

	Have Retirement Benefits	
	Yes	**No**
Men	225	75
Women	150	50

    a. Suppose one employee is selected at random from these 500 employees. Find the following probabilities.

        i. Probability of the intersection of events "woman" and "has retirement benefits"

        ii. Probability of the intersection of events "does not have retirement benefits" and "man"

    b. Mention what other joint probabilities you can calculate for this table and then find them. You may draw a tree diagram to find these probabilities.

4.78 The following table gives the two-way classification of 2000 randomly selected employees from a city based on gender and commuting time from home to work.

	Commuting Time		
	Less Than 30 Minutes	**30 Minutes to One Hour**	**More Than One Hour**
Men	524	455	221
Women	413	263	124

a. Suppose one employee is selected at random from these 2000 employees. Find the following probabilities.
    i. *P*(commutes for more than one hour and man)
    ii. *P*(woman and commutes for less than 30 minutes)
b. Find the joint probability of events ''commutes for 30 minutes to one hour'' and ''commutes for more than one hour.'' Is this probability zero? Explain why or why not.

4.79 Two thousand randomly selected adults were asked if they think they are financially better off than their parents. The following table gives the two-way classification of the responses based on the education levels of the persons included in the survey and whether they are better off, the same, or worse off financially than their parents.

	Education Level		
	Less Than High School	High School	More Than High School
Better Off	140	450	420
Same	60	250	110
Worse Off	200	300	70

a. Suppose one adult is selected at random from these 2000 adults. Find the following probabilities.
    i. *P*(better off and high school)
    ii. *P*(more than high school and worse off)
b. Find the joint probability of events ''worse off'' and ''better off.'' Is this probability zero? Explain why or why not.

4.80 All 420 employees of a company were asked whether they smoke or not and whether they are college graduates or not. Based on this information, the following two-way classification table was prepared.

$P(C \cap N) = \dfrac{130}{420} = .31$

$P(S \cap \bar{C}) = \dfrac{80}{420} =$

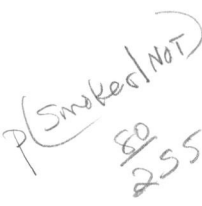
$P(\text{Smoker} | \text{NOT}) \dfrac{80}{255}$

	College Graduate	Not a College Graduate
Smoker	35	80
Nonsmoker	130	175

a. Suppose one employee is selected at random from this company. Find the following probabilities.
    i. *P*(college graduate and nonsmoker)
    ii. *P*(smoker and not a college graduate)
b. Find *P*(smoker and nonsmoker). Is this probability zero? Explain why or why not.
c. Mention what other joint probabilities you can calculate for this table and then find them. You may draw a tree diagram to find these probabilities.

4.81 The following table gives a two-way classification of 200 corporations based on whether they earned profits or incurred losses in 1991 and 1992.

		1992	
		Earned Profits	Incurred Losses
1991	Earned Profits	110	40
	Incurred Losses	20	30

a. If one corporation is selected at random from these 200 corporations, find the following probabilities.

      i. P(earned profits in 1991 and incurred losses in 1992)

     ii. P(earned profits in 1991 and earned profits in 1992)

  b. Find the joint probability of events that a randomly selected corporation "earned profits in 1992" and "incurred losses in 1992." Is this probability zero? Explain why or why not.

  c. Mention what other joint probabilities you can calculate for this table and then find them. You may draw a tree diagram to find these probabilities.

4.82 Of 20 economists, 14 hold the opinion that U.S. corporations do not spend enough money on research and development. If two economists are randomly selected from these 20, what is the probability that both of them hold the opinion that U.S. corporations do not spend enough money on research and development? Draw a tree diagram for this problem.

4.83 A company is to hire two new employees. They have prepared a final list of eight candidates, all of whom are equally qualified. Of these eight candidates, five are women. If the company decides to select two persons randomly from these eight candidates, what is the probability that both of them are women? Draw a tree diagram for this problem.

4.84 In a group of 10 managers, 4 have a type A personality and 6 have a type B personality. If 2 managers are selected at random from this group, what is the probability that the first of them has a type A personality and the second has a type B personality? Draw a tree diagram for this problem.

4.85 A small company has 20 employees and 12 of them are married. If 2 employees are randomly selected from this company, what is the probability that the first of them is married and the second is not? Draw a tree diagram for this problem.

4.86 The probability is .76 that a family owns a house. If two families are randomly selected, what is the probability that neither of them owns a house? (Note that the probability is .76 that any family owns a house. Hence, all families are independent.)

4.87 A contractor has submitted bids for two state construction projects. The probability that he will win any contract is .30 and it is the same for each of the two contracts.

  a. What is the probability that he will win both contracts?

  b. What is the probability that he will win neither contract?

Draw a tree diagram for this problem.

4.88 Based on past data, it is known that 2% of all calculators made by Calco Inc. malfunction and, hence, have to be replaced. What is the probability that two such calculators sold at a store on a given day will both have to be replaced?

4.89 Five percent of all items sold by a mail-order company are returned by customers for a refund. What is the probability that two items sold on a given day by this company will both be returned for a refund?

4.90 The probability that a farmer is in debt is .75. What is the probability that three randomly selected farmers are all in debt? Assume independence of events.

4.91 Five percent of all managers lose their jobs every year. What is the probability that three randomly selected managers will all lose jobs this year? Assume independence of events.

4.9 UNION OF EVENTS AND THE ADDITION RULE

This section discusses the union of events and the addition rule that is applied to compute the probability of the union of events.

4.9.1 UNION OF EVENTS

The **union of two events** A and B includes all outcomes that are either in A or in B or in both A and B.

> **UNION OF EVENTS**
>
> Let A and B be two events defined in a sample space. The union of events A and B is the collection of all outcomes that belong either to A or to B or to both A and B and is denoted by
>
> $$A \text{ or } B$$

The union of events A and B is also denoted by "$A \cup B$." Example 4–29 illustrates the union of events A and B.

Illustrating the union of two events.

EXAMPLE 4–29   A company has 1000 employees. Of them, 400 are female and 740 are labor union members. Of the 400 females, 250 are union members. Describe the union of events "female" and "union member."

Solution   The union of events "female" and "labor union member" for all employees of this company includes the employees who are either female or labor union members or both. The number of such employees is

$$400 + 740 - 250 = 890$$

Thus, there are a total of 890 employees who are either female or labor union members or both.

Why did we subtract 250 from the sum of 400 and 740? The reason is that 250 employees (who represent the intersection of events "female" and "labor union member") are common to both events "female" and "labor union member" and, hence, are counted twice. To avoid double counting, we subtracted 250 from the sum of the other two numbers. We can observe this double counting from Table 4.7, which is constructed using the given information. The sum of the numbers in the three shaded cells gives the employees who are either female or labor union members or both. However, if we add the numbers in the row labeled "Female" and the column labeled "Union Member," we count 250 twice.

Table 4.7

	Union Members	Not Union Members	Total
Male	490	110	600
Female	250	150	400
Total	740	260	1000

Counted twice

Figure 4.17 shows the diagram for the union of events "female" and "union member."

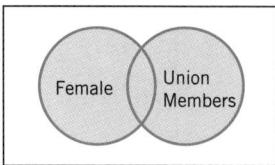

Shaded area gives the union of two events and includes 890 employees

Figure 4.17   Union of events "female" and "union member."

4.9.2 ADDITION RULE

The method used to calculate the probability of the union of events is called the **addition rule**. It is defined as follows.

ADDITION RULE

The probability of the union of two events A and B is

$$P(A \text{ or } B) = P(A) + P(B) - P(A \text{ and } B)$$

Thus, to calculate the probability of the union of two events A and B, we add their marginal probabilities and subtract their joint probability from this sum. We must subtract the joint probability of A and B from the sum of their marginal probabilities to avoid double counting due to common outcomes in A and B.

Calculating probability of union of two events: two-way table.

EXAMPLE 4–30 A university president has proposed that all students must take a course in ethics as a requirement for graduation. Three hundred faculty members and students from this university were asked about their opinion on this issue. The following table gives a two-way classification of the responses of these faculty members and students.

	Opinion			
	Favor	**Oppose**	**Neutral**	**Total**
Faculty	45	15	10	70
Student	90	110	30	230
Total	135	125	40	300

Find the probability that one person selected at random from these 300 persons is a faculty member or is in favor of this proposal.

Solution Let us define the following events.

$$A = \text{the person selected is a faculty member}$$

$$B = \text{the person selected is in favor of the proposal}$$

From the information given in the table,

$$P(A) = 70/300 = .233$$

$$P(B) = 135/300 = .450$$

and

$$P(A \text{ and } B) = P(A) \, P(B \mid A) = (70/300)(45/70) = .150$$

Using the addition rule

$$P(A \text{ or } B) = P(A) + P(B) - P(A \text{ and } B) = .233 + .450 - .150 = \textbf{.533}$$

Thus, the probability that a randomly selected person from these 300 persons is a faculty member or is in favor of this proposal is .533.

The probability in this example can also be calculated without using the addition rule. The total number of persons in the given table who are either faculty members or are in favor of this proposal is

$$45 + 15 + 10 + 90 = 160$$

Hence, the required probability is

$$P(A \text{ or } B) = 160/300 = \mathbf{.533}$$ ▬

EXAMPLE 4-31 There are a total of 7225 thousand persons with multiple jobs in the United States. Of them, 4115 thousand are male, 1742 thousand are single, and 905 thousand are male and single. (*Source:* U.S. Bureau of the Census.) What is the probability that a randomly selected person with multiple jobs is a male or single?

Solution Let us define the following two events.

$$M = \text{the randomly selected person is a male}$$

$$A = \text{the randomly selected person is single}$$

From the given information,

$$P(M) = 4115/7225 = .570$$

$$P(A) = 1742/7225 = .241$$

and

$$P(M \text{ and } A) = 905/7225 = .125$$

Hence,

$$P(M \text{ or } A) = P(M) + P(A) - P(M \text{ and } A) = .570 + .241 - .125 = \mathbf{.686}$$

Actually, using the given information, we can prepare Table 4.8 for this example. The numbers in the shaded cells are given to us. The remaining numbers are calculated by doing some arithmetic.

Table 4.8 Two-way Classification Table

	Single (A)	Married (B)	Total
Male (M)	905	3210	4115
Female (F)	837	2273	3110
Total	1742	5483	7225

Now, using this table, we can find the required probabilities as follows.

$$P(M) = 4115/7225 = .570$$

$$P(A) = 1742/7225 = .241$$

$$P(M \text{ and } A) = 905/7225 = .125$$

$$P(M \text{ or } A) = P(M) + P(A) - P(M \text{ and } A) = .570 + .241 - .125 = \mathbf{.686}$$ ▬

ADDITION RULE FOR MUTUALLY EXCLUSIVE EVENTS

We know from an earlier discussion that the joint probability of two mutually exclusive events is zero. When A and B are mutually exclusive events, the term $P(A \text{ and } B)$ in the addition rule becomes zero and is dropped from the formula. Thus, the probability of the union of two mutually exclusive events is given by the sum of their marginal probabilities.

> **ADDITION RULE FOR MUTUALLY EXCLUSIVE EVENTS**
>
> The probability of the union of two mutually exclusive events A and B is
>
> $$P(A \text{ or } B) = P(A) + P(B)$$

Calculating probability of union of two mutually exclusive events: two-way table.

EXAMPLE 4-32 A university president has proposed that all students must take a course in ethics as a requirement for graduation. Three hundred faculty members and students from this university were asked about their opinion on this issue. The following table, reproduced from Example 4–30, gives a two-way classification of the responses of these faculty members and students.

	Opinion			
	Favor	**Oppose**	**Neutral**	**Total**
Faculty	45	15	10	70
Student	90	110	30	230
Total	135	125	40	300

What is the probability that a randomly selected person from these 300 faculty members and students is in favor of the proposal or is neutral?

Solution Let us define the following events.

$$F = \text{the person selected is in favor of the proposal}$$

$$N = \text{the person selected is neutral}$$

As shown in Figure 4.18, events F and N are mutually exclusive because a person selected can be either in favor or neutral but not both.

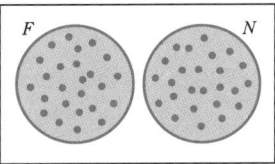

F *N*

Figure 4.18

From the given information,

$$P(F) = 135/300 = .450$$

and

$$P(N) = 40/300 = .133$$

Hence,

$$P(F \text{ or } N) = P(F) + P(N) = .450 + .133 = \mathbf{.583}$$

Calculating joint probability of three mutually exclusive events.

EXAMPLE 4-33 Eighteen percent of the working lawyers in the United States are female (*Self*, September 1992). Two lawyers are selected at random and it is observed whether they are male or female.

(a) Draw a tree diagram for this experiment.

(b) Find the probability that at least one of the two lawyers is a female.

Solution

(a)   Let

$$M = \text{event that the lawyer selected is a male}$$

$$F = \text{event that the lawyer selected is a female}$$

This experiment has four outcomes: both lawyers are male (*MM*), the first lawyer is a male and the second is a female (*MF*), the first lawyer is a female and the second is a male (*FM*), and both lawyers are female (*FF*). The tree diagram in Figure 4.19 shows these four outcomes and their probabilities.

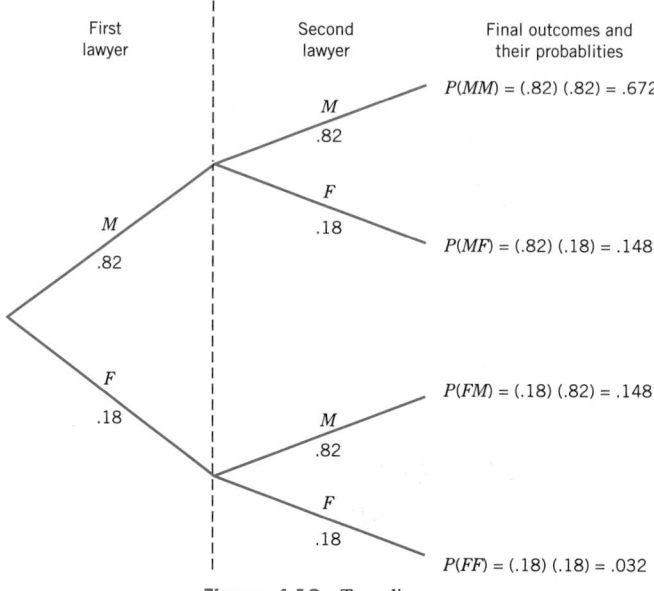

Figure 4.19   Tree diagram.

(b)   The probability that at least one lawyer is a female is given by the union of events *MF*, *FM*, and *FF*. These three outcomes are mutually exclusive. Hence,

$$P(\text{at least one lawyer is a female}) = P(MF \text{ or } FM \text{ or } FF)$$

$$= P(MF) + P(FM) + P(FF)$$

$$= .148 + .148 + .032 = \mathbf{.328}$$

EXERCISES

Concepts and Procedures

4.92   Explain the meaning of the union of two events. Give one example.

4.93   How is the addition rule of probability for two mutually exclusive events different from the one for two mutually nonexclusive events?

4.94   Consider the following addition rule to find the probability of the union of two events *A* and *B*.

$$P(A \text{ or } B) = P(A) + P(B) - P(A \text{ and } B)$$

When and why is the term $P(A$ and $B)$ subtracted from the sum of $P(A)$ and $P(B)$? Give one example where you may use this formula.

4.95 When is the following addition rule used to find the probability of the union of two events A and B?

$$P(A \text{ or } B) = P(A) + P(B)$$

Give one example where you may use this formula.

4.96 Find $P(A$ or $B)$ for the following.

    a. $P(A) = .58$, $P(B) = .66$, and $P(A$ and $B) = .47$
    b. $P(A) = .72$, $P(B) = .42$, and $P(A$ and $B) = .33$

4.97 Find $P(A$ or $B)$ for the following.

    a. $P(A) = .18$, $P(B) = .49$, and $P(A$ and $B) = .13$
    b. $P(A) = .83$, $P(B) = .71$, and $P(A$ and $B) = .68$

4.98 Given that A and B are two mutually exclusive events, find $P(A$ or $B)$ for the following.

    a. $P(A) = .57$ and $P(B) = .32$
    b. $P(A) = .16$ and $P(B) = .49$

4.99 Given that A and B are two mutually exclusive events, find $P(A$ or $B)$ for the following.

    a. $P(A) = .25$ and $P(B) = .17$
    b. $P(A) = .38$ and $P(B) = .09$

Applications

4.100 The following table gives a two-way classification of all 1000 employees of a large company based on whether they are single or married and whether or not they own stocks.

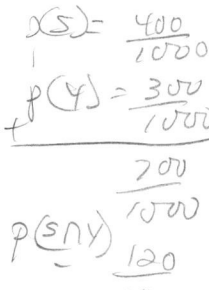

	Own Stocks	
	Yes	**No**
Single	120	280
Married	180	420

Suppose one employee is selected at random from these 1000 employees. Find the following probabilities.

    a. P(single or owns stocks)
    b. P(does not own stocks or single)

4.101 Five hundred employees were selected from a city's large private companies, and they were asked whether or not they have any retirement benefits provided by their companies. Based on this information, the following two-way classification table was prepared.

	Have Retirement Benefits	
	Yes	**No**
Men	225	75
Women	150	50

Suppose one employee is selected at random from these 500 employees. Find the following probabilities.

    a. The probability of the union of events "woman" and "has retirement benefits"
    b. The probability of the union of events "does not have retirement benefits" and "man"

4.102 The following table gives the two-way classification of 2000 randomly selected employees from a city based on gender and commuting time from home to work.

[handwritten top: symbol ∪ is OR, union / symbol ∩ is intersection]

	Commuting Time		
	Less Than 30 Minutes	30 Minutes to One Hour	More Than One Hour
Men	524	455	221
Women	413	263	124

If one employee is selected at random from these 2000 employees, find the following probabilities.

   a. *P*(commutes for more than one hour or man)
   b. *P*(woman or commutes for less than 30 minutes)
   c. *P*(man or woman)

4.103 Two thousand randomly selected adults were asked if they think they are financially better off than their parents. The following table gives the two-way classification of the responses based on the education levels of the persons included in the survey and whether they are financially better off, the same, or worse off than their parents.

[handwritten: pg 185 / pg 4.54]

	Education Level		
	Less Than High School	High School	More Than High School
Better Off	140	450	420
Same	60	250	110
Worse Off	200	300	70

[handwritten totals: 1010, 420, 570, 400, 1000, 600, 2000]

Suppose one adult is selected at random from these 2000 adults. Find the following probabilities.

   a. *P*(better off or high school) *[handwritten: P(B ∪ HS) = 1010/2000 + 1000/2000 − 450/2000 = .78]*
   b. *P*(more than high school or worse off) *[handwritten: P(>HS ∪ WO) = 600/2000 + 570/2000 − 70/2000 = 1100/2000]*
   c. *P*(better off or worse off) *[handwritten: P(B ∪ W)]*

[handwritten: mutually exclusive — 1010/2000 + 570/2000 = 1580/2000 = .79]

4.104 All the 420 employees of a company were asked whether or not they smoke and whether or not they are college graduates. Based on this information, the following two-way classification table was prepared.

	College Graduate	Not a College Graduate
Smoker	35	80
Nonsmoker	130	175

[handwritten: C, C̄ above columns; 115, 305 totals; 165, 255, 420]

Suppose one employee is selected at random from this company. Find the following probabilities.

   a. *P*(college graduate or smoker) *[handwritten: P(G ∪ S) = P(G) + P(S) − P(G ∩ S)]*
   b. *P*(smoker or not a college graduate) *[handwritten: P(S ∪ C̄) = P(S) + P(C̄) − P(S ∩ C̄)]*
   c. *P*(smoker or nonsmoker) *[handwritten: P(S ∪ S̄) = P(S) + P(S̄) − P(S ∩ S̄)]*

[handwritten: 165/420 + 115/420 − 35/420 = 245/420]

[handwritten: 115/420 + 305/420 − 0/420 = 420/420 = 1]

4.105 The following table gives a two-way classification of 200 corporations based on whether they earned profits or incurred losses in 1991 and 1992.

		1992	
		Earned Profits	Incurred Losses
1991	Earned Profits	110	40
	Incurred Losses	20	30

Suppose one corporation is selected at random from these 200 corporations. Find the following probabilities.

    a. *P*(earned profits in 1991 or incurred losses in 1992)
    b. *P*(earned profits in 1991 or earned profits in 1992)
    c. *P*(earned profits in 1992 or incurred losses in 1992)

4.106 The probability that a family owns a washing machine is .78, that it owns a VCR is .71, and that it owns both a washing machine and a VCR is .58. What is the probability that a randomly selected family owns a washing machine or a VCR?

4.107 The probability that a randomly selected student from a university is a senior is .18, a business major is .14, and a senior and a business major is .04. Find the probability that a student selected at random from this university is a senior or a business major.

4.108 The probability that a person has a checking account is .74, a savings account is .31, and both accounts is .22. Find the probability that a randomly selected person has a checking or a savings account.

4.109 The probability that a randomly selected elementary or secondary school teacher from a city is a female is .68, holds a second job is .42, and is a female and holds a second job is .29. Find the probability that an elementary or secondary school teacher selected at random from this city is a female or holds a second job.

4.110 The probability of a student getting an A in an economics class is .24 and that of getting a B is .28. What is the probability that a randomly selected student from this class will get an A or a B in this class?

4.111 Seventy-two percent of a town's voters favor the recycling issue, 12% oppose it, and 16% are indifferent. What is the probability that a randomly selected voter from this town will either favor recycling or be indifferent?

4.112 The probability that a corporation makes charitable contributions is .68. Two corporations are selected at random, and it is noted whether or not they make charitable contributions.

    a. Draw a tree diagram for this experiment.
    b. Find the probability that at most one corporation makes charitable contributions.

4.113 According to a survey, 22% of senior executives have been with the same firm for 25 years (*U.S. News & World Report*, June 8, 1992). Two senior executives are selected at random, and it is observed whether or not they have been with the same firm for 25 years.

    a. Draw a tree diagram for this experiment.
    b. Find the probability that at least one senior executive has been with the same firm for 25 years.

4.10 BAYES' THEOREM

Bayes' theorem, named after mathematician Thomas Bayes, is applied to revise probabilities of events after we obtain more information. Initially we know the probabilities of certain events, which are called the *prior probabilities* of those events. Then we obtain some additional information from a sample or experiment. Based on this information, we revise the prior probabilities of events. These revised probabilities of events are called the *posterior probabilities*.

    Suppose all loan applications in a bank are processed by two loan officers. The percentage of the applications processed by each of the two loan officers is known. Using these percentages, we can calculate the probability that a randomly selected application was processed by either of these two loan officers. These two probabilities will be called the prior probabilities. Now, suppose we know the percentage of applications that are approved and rejected by each of these two loan officers. Using this information, we can revise the probability that a randomly selected application was processed by a particular loan officer.

The following is Bayes' formula for a two-event case. This formula can be extended to more than two events.

BAYES' THEOREM

The revised probability of event A_1, after additional information on another event B is obtained, is calculated as follows.

$$P(A_1 \mid B) = \frac{P(A_1)\, P(B \mid A_1)}{P(A_1)\, P(B \mid A_1) \,+\, P(A_2)\, P(B \mid A_2)}$$

The probabilities $P(A_1)$ and $P(A_2)$ are called the prior probabilities of A_1 and A_2. The probability $P(A_1 \mid B)$ is called the revised or posterior probability of A_1. Similarly, $P(A_2 \mid B)$ is the revised or posterior probability of A_2.

Bayes' formula can also be written as

$$P(A_1 \mid B) = \frac{P(A_1 \text{ and } B)}{P(A_1 \text{ and } B) \,+\, P(A_2 \text{ and } B)}$$

Note that by using Bayes' rule, we are calculating the conditional probability $P(A_1 \mid B)$, which was discussed in Sections 4.4 and 4.8.2. In Section 4.8.2 the conditional probability $P(A_1 \mid B)$ was calculated as

$$P(A_1 \mid B) = \frac{P(A_1 \text{ and } B)}{P(B)}$$

When the probabilities $P(A_1 \text{ and } B)$ and $P(B)$ are not given, but instead the probabilities $P(A_1)$, $P(A_2)$, $P(B \mid A_1)$, and $P(B \mid A_2)$ are known, we use Bayes' rule to find the conditional probability $P(A_1 \mid B)$ instead of using the formula of Section 4.8.2. Note that the denominators in both formulas are the same. In other words,

$$P(B) = P(A_1)\, P(B \mid A_1) \,+\, P(A_2)\, P(B \mid A_2)$$

The denominator in Bayes' rule

$$P(A_1)\, P(B \mid A_1) \,+\, P(A_2)\, P(B \mid A_2)$$

is called the total probability of B, which is denoted by $P(B)$.

Applying Bayes' theorem to calculate revised probability of an event.

EXAMPLE 4–34 The Northeast Bank of Connecticut has two loan officers, Matt Sanger and Alison Terry, who process all loan applications. Of all the loan applications during the past five years, 55% were processed by Mr. Sanger and 45% were processed by Ms. Terry. It is further known that of all the loan applications processed by Mr. Sanger, 30% were rejected and 70% were approved. Of the loan applications processed by Ms. Terry, 20% were rejected and 80% were approved. One loan application is randomly selected from all the applications processed during the past five years and it is observed that this application was rejected. What is the probability that this application was processed by Mr. Sanger?

Solution   Let us define the following events.

$$A_1 = \text{a loan application is processed by Mr. Sanger}$$
$$A_2 = \text{a loan application is processed by Ms. Terry}$$
$$B = \text{a loan application is rejected}$$
$$C = \text{a loan application is approved}$$

Suppose one application is selected at random. The prior probabilities of A_1 and A_2 are

$$P(A_1) = .55 \quad \text{and} \quad P(A_2) = .45$$

We are given the following conditional probabilities.

$$P(B|A_1) = .30, \qquad P(C|A_1) = .70,$$
$$P(B|A_2) = .20, \quad \text{and} \quad P(C|A_2) = .80$$

The probability that a randomly selected loan application was processed by Mr. Sanger, when it is known that it was rejected, is obtained by using Bayes' rule as follows.

$$P(A_1|B) = \frac{P(A_1)\,P(B|A_1)}{P(A_1)\,P(B|A_1) + P(A_2)\,P(B|A_2)}$$

$$= \frac{(.55)\,(.30)}{(.55)\,(.30) + (.45)\,(.20)} = \textbf{.647}$$

Note that the numerator in this formula gives the joint probability of the events that a loan application was processed by Mr. Sanger and it was rejected. The denominator gives the total probability that an application was rejected.   ▄

There are three other revised probabilities that can be calculated for Example 4–34. These probabilities are the following.

1. The probability that a randomly selected application was processed by Ms. Terry given that this application was rejected.

$$P(A_2|B) = \frac{P(A_2)\,P(B|A_2)}{P(A_1)\,P(B|A_1) + P(A_2)\,P(B|A_2)}$$

$$= \frac{(.45)\,(.20)}{(.55)\,(.30) + (.45)\,(.20)} = .353$$

2. The probability that a randomly selected application was processed by Mr. Sanger given that it was approved.

$$P(A_1|C) = \frac{P(A_1)\,P(C|A_1)}{P(A_1)\,P(C|A_1) + P(A_2)\,P(C|A_2)}$$

$$= \frac{(.55)\,(.70)}{(.55)\,(.70) + (.45)\,(.80)} = .517$$

3. The probability that a randomly selected application was processed by Ms. Terry given that it was approved.

$$P(A_2 \mid C) = \frac{P(A_2)\,P(C \mid A_2)}{P(A_1)\,P(C \mid A_1) + P(A_2)\,P(C \mid A_2)}$$

$$= \frac{(.45)\,(.80)}{(.55)\,(.70) + (.45)\,(.80)} = .483$$

EXERCISES

Concepts and Procedures

4.114 Given that $P(A_1) = .33$, $P(A_2) = .67$, $P(B \mid A_1) = .40$, $P(C \mid A_1) = .60$, $P(B \mid A_2) = .25$, $P(C \mid A_2) = .75$, find the revised probabilities $P(A_1 \mid B)$, $P(A_2 \mid B)$, $P(A_1 \mid C)$, and $P(A_2 \mid C)$.

4.115 Given that $P(A_1) = .66$, $P(A_2) = .34$, $P(B \mid A_1) = .57$, $P(C \mid A_1) = .43$, $P(B \mid A_2) = .61$, $P(C \mid A_2) = .39$, find the revised probabilities $P(A_1 \mid B)$, $P(A_2 \mid B)$, $P(A_1 \mid C)$, and $P(A_2 \mid C)$.

Applications

4.116 At a department store, 20% of all customers spend $50 or less and 80% spend more than $50 per visit. Of those who spend $50 or less, 75% pay by cash or check and 25% pay by credit card. Of those who spend more than $50, 30% pay by cash or check and 70% pay by credit card. One randomly selected customer, who made a purchase at this store, paid by credit card. What is the probability that this customer spent more than $50 at this store?

4.117 Of all the cassettes manufactured at Kisling Inc., 60% are manufactured on Machine I and 40% are manufactured on Machine II. Five percent of the cassettes manufactured on Machine I are defective and 3% of the cassettes manufactured on Machine II are defective. One cassette is randomly selected from a lot that contains cassettes manufactured on both machines. This cassette is observed to be defective. What is the probability that this cassette was manufactured on Machine II?

4.118 A process that is used to make iron rods at a company is within control 95% of the time and out of control 5% of the time. When within control, it produces 3% of the rods with lengths outside the acceptable range and 97% within the acceptable range. When out of control, it produces 45% of the rods with lengths outside the acceptable range and 55% within the acceptable range. One randomly selected rod is found to have its length outside the acceptable range. What is the probability that the process is out of control?

4.119 Twenty-five percent of the customers who shop at a warehouse are small business owners and 75% of the customers shop for personal use. Of the small business owners, 85% spend more than $200 per visit and 15% spend $200 or less. Of the customers who shop for personal use, 20% spend more than $200 per visit and 80% spend $200 or less. A randomly selected purchase showed a total sale of more than $200. What is the probability that this purchase was made by a small business owner?

4.120 The probability that the price of a company's stock will be higher at the end of a given year is .55, and the probability that the price of this company's stock will be lower at the end of that year is .45. (Assume that the probability is zero that the stock price will be exactly the same at the end of the year as in the beginning of the year.) If the price of the stock for a company is higher at the end of a year, the probability is .90 that this company earned profits during that year and .10 that it incurred losses. If the price of the stock for a company is lower at the end of a year, the probability is .25 that this company earned profits during that year and .75 that it incurred losses. A company incurred losses during a given year. What is the probability that the price of this company's stock was higher at the end of this year?

4.121 Each day companies introduce thousands of new products in the market. Usually new products are test marketed before they are introduced for sale. The probability is .65 that a new product introduced by a company will be successful. For an eventually successful product, probability is .95 that 50% or more of the people included in the test like it. However, for an eventually unsuccessful product, probability is .20 that 50% or more of the people included in the test like it. A company recently introduced a new product. What is the probability that this product will be successful if less than 50% of the people included in the test liked it?

GLOSSARY

Bayes' theorem   A rule that is used to calculate the revised probabilities of events after more information has been obtained.

Classical probability rule   The method of assigning probabilities to outcomes or events of an experiment with equally likely outcomes.

Complementary events   Two events that taken together include all the outcomes for an experiment but do not contain any common outcome.

Compound event   An event that contains more than one outcome of an experiment. It is also called a *composite event.*

Conditional probability   The probability of an event subject to the condition that another event has already occurred.

Dependent events   Two events for which the occurrence of one changes the probability of the occurrence of the other.

Equally likely outcomes   Two (or more) outcomes or events that have the same probability of occurrence.

Event   A collection of one or more outcomes of an experiment.

Experiment   A process with well-defined outcomes that, when performed, results in one and only one of the outcomes per repetition.

Impossible event   An event that cannot occur.

Independent events   Two events for which the occurrence of one does not change the probability of occurrence of the other.

Intersection of events   Given by the outcomes that are common to two (or more) events.

Joint probability   The probability that two (or more) events occur together.

Law of Large Numbers   If an experiment is repeated again and again, the probability of an event obtained from the relative frequency approaches the actual or theoretical probability.

Marginal probability   The probability of one event or characteristic without consideration of any other event.

Mutually exclusive events   Two or more events that do not contain any common outcome and, hence, cannot occur together.

Outcome   The result of the performance of an experiment.

Probability   A numerical measure of the likelihood that a specific event will occur.

Relative frequency as an approximation of probability   Probability assigned to an event based on the results of an experiment or based on historical data.

Sample point   An outcome of an experiment.

Sample space   The collection of all sample points or outcomes of an experiment.

Simple event   An event that contains one and only one outcome of an experiment. It is also called an *elementary event.*

Subjective probability   The probability assigned to an event based on the information and judgment of a person.

Sure event   An event that is certain to occur.

Tree diagram   A diagram in which each outcome of an experiment is represented by a branch of a tree.

Union of two events   All outcomes that belong either to one or to both events.

Venn diagram   A picture that represents a sample space or certain events.

KEY FORMULAS

1. **Classical probability rule**

For a simple event E_i:

$$P(E_i) = \frac{1}{\text{Total number of outcomes for the experiment}}$$

For a compound event A:

$$P(A) = \frac{\text{Number of outcomes favorable to } A}{\text{Total number of outcomes for the experiment}}$$

2. **Relative frequency as an approximation of probability**

$$P(\text{an event}) = \frac{\text{Frequency of that event}}{\text{Sample size}} = \frac{f}{n}$$

3. **Counting rule**

If an experiment consists of three steps and if the first step can result in m outcomes, the second step in n outcomes, and the third step in k outcomes, then

$$\text{Total outcomes for the experiment} = m \cdot n \cdot k$$

4. **Conditional probability of an event**

$$P(B \mid A) = \frac{P(A \text{ and } B)}{P(A)} \qquad \text{and} \qquad P(A \mid B) = \frac{P(A \text{ and } B)}{P(B)}$$

5. **Independent events**

Two events A and B are independent if

$$P(A) = P(A \mid B) \qquad \text{and/or} \qquad P(B) = P(B \mid A)$$

6. **Complementary events**

For two complementary events A and $\overline{A}$:

$$P(A) + P(\overline{A}) = 1, \qquad P(A) = 1 - P(\overline{A}), \qquad \text{and} \qquad P(\overline{A}) = 1 - P(A)$$

7. **Multiplication rule for joint probability of events**

If A and B are dependent events, then

$$P(A \text{ and } B) = P(A)\, P(B \mid A)$$

If A and B are independent events, then

$$P(A \text{ and } B) = P(A)\, P(B)$$

8. **Joint probability of two mutually exclusive events**

For two mutually exclusive events A and B:

$$P(A \text{ and } B) = 0$$

9. **Addition rule for the probability of union of events**

If A and B are mutually nonexclusive events, then

$$P(A \text{ or } B) = P(A) + P(B) - P(A \text{ and } B)$$

If A and B are mutually exclusive events, then

$$P(A \text{ or } B) = P(A) + P(B)$$

10. Bayes' theorem

The revised probability of event A_1, after additional information on another event B is obtained, is calculated as follows.

$$P(A_1 \mid B) = \frac{P(A_1) \, P(B \mid A_1)}{P(A_1) \, P(B \mid A_1) + P(A_2) \, P(B \mid A_2)}$$

SUPPLEMENTARY EXERCISES

4.122 A statistical experiment has five equally likely outcomes that are denoted by E_1, E_2, E_3, E_4, and E_5. Let event $A = \{E_1, E_3, E_5\}$ and event $B = \{E_2, E_3, E_4\}$. Find the following probabilities.

  a. $P(A)$ **b.** $P(B)$ **c.** $P(A \mid B)$ **d.** $P(B \mid A)$ **e.** $P(A \text{ and } B)$ **f.** $P(A \text{ or } B)$

4.123 A statistical experiment has eight equally likely outcomes that are denoted by E_1, E_2, E_3, E_4, E_5, E_6, E_7, and E_8. Let event $A = \{E_1, E_4, E_5, E_7, E_8\}$ and event $B = \{E_2, E_3, E_4\}$. Find the following probabilities.

  a. $P(A)$ **b.** $P(B)$ **c.** $P(A \mid B)$ **d.** $P(B \mid A)$ **e.** $P(A \text{ and } B)$ **f.** $P(A \text{ or } B)$

4.124 Refer to Exercise 4.122. Using separate Venn diagrams, show the intersection and union of events A and B.

4.125 Refer to Exercise 4.123. Using separate Venn diagrams, show the intersection and union of events A and B.

4.126 For an experiment $P(A) = .65$, $P(B) = .48$, and $P(A \text{ or } B) = .78$. Find $P(A \text{ and } B)$ and $P(A \mid B)$. Are A and B mutually exclusive events? Are they independent events?

4.127 For an experiment $P(A) = .42$, $P(B) = .29$, and $P(A \text{ or } B) = .71$. Find $P(A \text{ and } B)$ and $P(A \mid B)$. Are A and B mutually exclusive events? Are they independent events?

4.128 Some of the workers at a company are union members and others are not. Two workers are selected at random from this company, and it is observed whether or not these workers are union members. How many total final outcomes are possible for this experiment? Draw a tree diagram for this experiment.

4.129 A list contains the names of a few companies. Some of these are manufacturing companies and the remaining are service companies. If two companies are selected at random from this list and it is observed whether the companies selected are manufacturing or service companies, how many total final outcomes are possible for this experiment? Draw a tree diagram for this experiment.

4.130 Refer to Exercise 4.128. List all the outcomes included in each of the following events and mention which of them are simple and which are compound events.

  a. Both persons selected are union members.
  b. The first person is a union member and the second is not.
  c. At least one person is a union member.

4.131 Refer to Exercise 4.129. List all the outcomes included in each of the following events and mention which of them are simple and which are compound events.

  a. At most one manufacturing company is selected.
  b. Not more than one service company is selected.
  c. Exactly one service company is selected.

4.132 A lawyers' association has 80 members. Of them, 12 are corporate lawyers. One lawyer is selected at random. Find the probability that this lawyer is

  a. a corporate lawyer   **b.** not a corporate lawyer

4.133   Of 100 company executives, 35 have a master's degree, 45 have a bachelor's degree, and 20 have a degree lower than a bachelor's. One executive is selected at random. Find the probability that this executive has a

   a.   master's degree   **b.**   degree lower than a bachelor's

4.134   A group of 150 randomly selected chief executive officers (CEOs) were tested for personality type. The following table gives the results of this survey.

	Type A	Type B
Men	78	42
Women	19	11

   a.   If one CEO is selected at random from this group, find the probability that this CEO
   - i.   has a type A personality
   - ii.   is a woman
   - iii.   is a man given he has a type A personality
   - iv.   has a type B personality given she is a woman
   - v.   has a type A personality and is a woman
   - vi.   is a man or has a type B personality
   b.   Are the events "woman" and "type A personality" mutually exclusive? What about the events "type A personality" and "type B personality"? Why or why not?
   c.   Are the events "type A personality" and "man" independent? Why or why not?

4.135   The following table gives a two-way classification of 1000 workers selected from a large city.

	Covered by Health Insurance	Not Covered by Health Insurance
Men	490	160
Women	220	130

   a.   If one worker is selected at random from this group, find the probability that this worker is
   - i.   not covered by health insurance
   - ii.   a woman
   - iii.   a man given he is covered by health insurance
   - iv.   not covered by health insurance given she is a woman
   - v.   not covered by health insurance and is a man
   - vi.   covered by health insurance or is a woman
   b.   Are the events "woman" and "not covered by health insurance" mutually exclusive? What about the events "covered by health insurance" and "not covered by health insurance"? Why or why not?
   c.   Are the events "covered by health insurance" and "man" independent? Why or why not?

4.136   The following table gives a two-way classification of 200 randomly selected purchases made at a department store.

	Paid by Cash/Check	Paid by Credit Card
Male	24	46
Female	77	53

   a.   If one of these 200 purchases is selected at random, find the probability that it is
   - i.   made by a female
   - ii.   paid by cash/check

    iii. paid by credit card given that the purchase is made by a male
    iv. made by a female given that it is paid by cash/check
    v. made by a female and paid by a credit card
    vi. paid by cash/check or made by a male
  b. Are the events "female" and "paid by credit card" independent? Are they mutually exclusive? Explain why or why not.

4.137 A random sample of 400 college students was asked if college athletes should be paid. The following table gives a two-way classification of the responses.

	Should Be Paid	Should Not Be Paid
Student Athlete	90	10
Student Nonathlete	210	90

  a. If one student is randomly selected from these 400 students, find the probability that this student
    i. is in favor of paying college athletes
    ii. favors paying college athletes given that the student selected is a nonathlete
    iii. is an athlete and favors paying student athletes
    iv. is a nonathlete or is against paying student athletes
  b. Are the events "student athlete" and "should be paid" independent? Are they mutually exclusive? Explain why or why not.

4.138 A survey conducted about job satisfaction showed that 20% of workers are not happy with their current jobs. Assume that this result is true for the population of all workers. Two workers are selected at random, and it is observed whether or not they are happy with their current jobs. Draw a tree diagram. Find the probability that in this sample of two workers
  a. both are not happy with their current jobs
  b. at least one of them is happy with the current job

4.139 Sixty-three percent of children who live with unmarried mothers live in poverty (*U.S. News & World Report*, June 8, 1992). Two children who live with unmarried mothers are selected at random and it is observed whether or not they live in poverty. Draw a tree diagram. Find the probability that in this sample of two children
  a. both live in poverty
  b. at most one lives in poverty

4.140 Refer to Exercise 4.132. Two lawyers are selected at random from this group of 80 lawyers. Find the probability that both of these lawyers are corporate lawyers.

4.141 Refer to Exercise 4.133. Two executives are selected at random from this group of 100 executives. Find the probability that the first of them has a master's degree and the second has a bachelor's degree.

4.142 Terry & Sons Inc. makes bearings for autos. The production system involves two independent processing machines so that each bearing passes through these two processes. The probability that the first processing machine will be out of control at any time is .08, and the probability that the second processing machine will be out of control at any time is .06. Find the probability that both processing machines will be out of control at any given time.

4.143 A company has installed a generator to back up the power in case there is a power failure. The probability that there will be a power failure during a snowstorm is .30. The probability that the generator will stop working during a snowstorm is .09. What is the probability that during a snowstorm the company will lose both sources of power?

4.144 Ross Computers Inc. buys 55% of all computer chips from Gill's Company and 45% from Dillon's Company. It is known that 1% of all the computer chips received from Gill's are defective, and 2% of all the computer chips received from Dillon's are defective. One computer is found to contain a defective chip. What is the probability that this chip came from Dillon's?

4.145   Sixty-five percent of the workers at Sanger's Inc. are men and 35% are women. The management of this company has proposed a new health care plan for all workers. Of the male workers, 74% are in favor of this new plan and 26% are against it. Of the female workers, 62% are in favor of this plan and 38% are against it. One randomly selected worker is observed to be in favor of this plan. What is the probability that this worker is a woman?

SELF-REVIEW TEST

1. The collection of all outcomes for an experiment is called
   a. a sample space   b.   intersection of events   c.   joint probability

2. A final outcome of an experiment is called
   a. a compound event   b.   a simple event   c.   a complementary event

3. A compound event includes
   a. all final outcomes   b.   exactly two outcomes
   c. more than one outcome for an experiment

4. Two equally likely events
   a. have the same probability of occurrence
   b. cannot occur together
   c. have no effect on the occurrence of each other

5. Which of the following probability approaches can be applied only to experiments with equally likely outcomes?
   a. Classical probability   b.   Empirical probability   c.   Subjective probability

6. Two mutually exclusive events
   a. have the same probability   b.   cannot occur together
   c. have no effect on the occurrence of each other

7. Two independent events
   a. have the same probability   b.   cannot occur together
   c. have no effect on the occurrence of each other

8. The probability of an event is always
   a. less than zero   b.   in the range zero to 1.0   c.   greater than 1.0

9. The sum of the probabilities of all final outcomes of an experiment is always
   a. 100   b.   1.0   c.   zero

10. The joint probability of two mutually exclusive events is always
    a. 1.0   b.   between 0 and 1   c.   zero

11. Two independent events are
    a. always mutually exclusive   b.   never mutually exclusive
    c. always complementary

12. Lucia graduated this year with an accounting degree from Eastern Connecticut State University. She has received job offers from an accounting firm, an insurance company, and an airline. She cannot decide which of the three job offers she should accept. Suppose she decides to select one of these three job offers randomly. Find the probability that the job offer selected is

    a. from the insurance company

    b. not from the accounting firm

13. A company has 500 employees. Of them, 300 are men and 280 are union members. Of the 300 men, 190 are union members.

    a. Are the events "man" and "union member" independent? Are they mutually exclusive? Explain why or why not.

    b. If one employee of this company is selected at random, what is the probability that this employee is

        i. a woman

        ii. a man given that he is a union member

14. Refer to Problem 13. If one employee is selected at random from this company, what is the probability that this employee is a man or a union member?

15. Refer to Problem 13. If two employees are selected at random from this company, what is the probability that both of them are union members?

16. In a survey conducted by Louis Harris & Associates for *Business Week*, 34% of adults said that they expect their children will have a better life than they have had (*Business Week*, April 6, 1992). Suppose this result holds true for the population of all adults. If two adults are randomly selected, what is the probability that both of them will hold this view?

17. The following two-way classification table gives information on whether or not a woman works outside the home and whether or not she has children living at home. The information was gathered on 500 women aged 24 to 60.

| | | Age of Children Living at Home | | |
		Under 6 Years	6 to 18 Years	No Children Living at Home
Works Outside Her Home	Full Time	30	95	105
	Part Time	60	50	25
	Does Not Work	50	55	30

    a. If one woman is selected at random from these 500 women, find the probability that this woman

        i. works full time outside her home

        ii. works part time given she has children aged 6 to 18 years of age living at home

        iii. does not work outside her home and has no children living at home

        iv. works full time outside her home or has children under 6 years of age living at home

        v. works full time outside her home and works part time outside her home

        vi. has children under 6 years of age living at home or has no children living at home

    b. Are the events "a woman works part time outside her home" and "she has children under 6 years of age" independent? Are they mutually exclusive? Explain why or why not.

18. A process that is used to make bearings at a company is within control 96% of the time and out of control 4% of the time. When within control, it produces 2% of the bearings with diameters outside the acceptable range. When out of control, it produces 55% of the bearings with diameters outside the acceptable range. One bearing is selected at random from the production line and its diameter is observed to be outside the acceptable range. What is the probability that the process is out of control?

5

DISCRETE RANDOM VARIABLES AND THEIR PROBABILITY DISTRIBUTIONS

Chapter 4 discussed the concepts and rules of probability. This chapter extends the concept of probability to explain probability distributions. As was seen in Chapter 4, any given statistical experiment has more than one outcome. It is impossible to predict which of those outcomes will occur if that experiment is performed. Consequently, decisions are made under uncertain conditions. For example, a lottery player does not know in advance whether or not he is going to win that lottery. If he knows that he is not going to win, he will definitely not play. It is the uncertainty about winning (some positive probability of winning) that makes him play. This chapter shows that if the outcomes and their probabilities for a statistical experiment are known, we can find out what will happen on average if that experiment is performed many times. For the lottery example, we can find out what a lottery player can expect to win (or lose) on average if he continues playing this lottery again and again.

First, random variables and types of random variables are explained in this chapter. Then, the concept of a probability distribution and its mean and standard deviation are discussed. Finally, three special probability distributions for a discrete random variable—the binomial probability distribution, the hypergeometric probability distribution, and the Poisson probability distribution—are developed.

5.1   RANDOM VARIABLES

Suppose Table 5.1 gives the frequency and relative frequency distributions of the number of vehicles owned by all 2000 families living in a small town.

Table 5.1   Frequency and Relative Frequency Distributions of the Number of Vehicles Owned by Families

Number of Vehicles Owned	Frequency	Relative Frequency
0	30	30/2000 = .015
1	470	470/2000 = .235
2	850	850/2000 = .425
3	490	490/2000 = .245
4	160	160/2000 = .080
	$N = 2000$	Sum = 1.0

Suppose one family is randomly selected from this population. The act of randomly selecting a family is called a *random* or *chance experiment*. Let x denote the number of vehicles owned by this family. Then x can assume any of the five possible values (0, 1, 2, 3, and 4) listed in the first column of Table 5.1. The value assumed by x depends on which family is selected. Thus, this value depends on the outcome of a random experiment. Consequently, x is called a **random variable** or a **chance variable**. In general, a random variable is denoted by x or y.

RANDOM VARIABLE

A random variable is a variable whose value is determined by the outcome of a random experiment.

As explained next, a random variable can be discrete or continuous.

5.1.1   DISCRETE RANDOM VARIABLE

A **discrete random variable** assumes values that can be counted. In other words, the consecutive values of a discrete random variable are separated by a certain gap.

DISCRETE RANDOM VARIABLE

A random variable that assumes countable values is called a discrete random variable.

In Table 5.1, *the number of vehicles owned by a family* is an example of a discrete random variable because the values of the random variable x are countable: 0, 1, 2, 3, and 4. Some other examples of discrete random variables are

1. The number of cars sold at a dealership during a given month
2. The number of employees working at a company
3. The number of suits a company executive owns
4. The number of complaints received at the office of an airline on a given day
5. The number of customers visiting a bank during any given hour
6. The number of heads obtained in three tosses of a coin

5.1.2 CONTINUOUS RANDOM VARIABLE

A random variable whose values are not countable is called a **continuous random variable**. A continuous random variable can assume any value over an interval or intervals.

CONTINUOUS RANDOM VARIABLE

A random variable that can assume any value contained in one or more intervals is called a continuous random variable.

Because the number of values contained in any interval is infinite, the possible number of values that a continuous random variable can assume is also infinite. Moreover, we cannot count these values. Consider the life of a battery. We can measure it as precisely as we want. For instance, the life of this battery may be 40 hours, or 40.25 hours, or 40.247 hours. Assume that the maximum life of such a battery is 200 hours. Let x denote the life of a randomly selected battery of this kind. Then x can assume any value in the interval 0 to 200. Consequently, x is a continuous random variable. As shown below, every point on the line representing the interval 0 to 200 gives a possible value of x.

Every point on this line represents a possible value of x that denotes the life of a battery. There are an infinite number of points on this line. The values represented by points on this line are uncountable.

The following are a few examples of continuous random variables.

1. Salaries of workers
2. Time taken by workers to learn a job
3. Amount of milk in a gallon (note that we do not expect a gallon to contain exactly one gallon of milk but either slightly more or slightly less than a gallon)
4. 1992 sales revenues of companies
5. Prices of houses

This chapter is limited to the discussion of discrete random variables and their probability distributions. Continuous random variables will be discussed in Chapter 6.

EXERCISES

Concepts and Procedures

5.1 Explain the meaning of a random variable, a discrete random variable, and a continuous random variable. Give one example each of a discrete random variable and a continuous random variable.

5.2 Classify the following random variables as discrete or continuous.
 a. The number of workers who carpool
 b. The amount of soda in a 12-oz can
 c. The number of cattle owned by a farmer
 d. The age of a house
 e. The number of CEOs of corporations who are older than 55 years of age
 f. The time spent by a technician to repair a machine

5.3 Indicate which of the following random variables are discrete and which are continuous.
 a. The number of new accounts opened at a bank during a certain month
 b. The time taken by a lawyer to write a real estate contract
 c. The price of a concert ticket
 d. The number of rotten eggs in a randomly selected box
 e. The number of workers employed at a randomly selected company
 f. The weight of a randomly selected package

Applications

5.4 A household can watch news on any of the three networks—ABC, CBS, or NBC. On a certain day, five households randomly and independently decide what channel to watch. Let x be the number of households in these five households that decide to watch news on ABC. Is x a discrete or a continuous random variable? Explain.

5.5 One of the four gas stations located at an intersection of two major roads is an Exxon station. Suppose the next six cars that stop at any of these four gas stations make the selections randomly and independently. Let x be the number of cars in these six that stop at the Exxon station. Is x a discrete or a continuous random variable? Explain.

5.2   PROBABILITY DISTRIBUTION OF A DISCRETE RANDOM VARIABLE

Let x be a discrete random variable. The **probability distribution** of x describes how the probabilities are distributed over all the possible values of x.

PROBABILITY DISTRIBUTION OF A DISCRETE RANDOM VARIABLE

The probability distribution of a discrete random variable lists all the possible values that the random variable can assume and their corresponding probabilities.

Example 5–1 illustrates the concept of the probability distribution of a discrete random variable.

Probability distribution of a discrete random variable.

EXAMPLE 5–1   Recall the frequency and relative frequency distributions of the number of vehicles owned by families given in Table 5.1. That table is reproduced as Table 5.2 on the next page. Let x be the number of vehicles owned by a randomly selected family. Write the probability distribution of x.

Table 5.2 Frequency and Relative Frequency Distributions of the Number of Vehicles Owned by Families

Number of Vehicles Owned	Frequency	Relative Frequency
0	30	.015
1	470	.235
2	850	.425
3	490	.245
4	160	.080
	$N = 2000$	Sum = 1.0

Solution   In Chapter 4 we learned that the relative frequencies obtained from an experiment or a sample can be used as approximate probabilities. However, when the relative frequencies are known for the population, they give the actual (theoretical) probabilities of outcomes. Using the relative frequencies of Table 5.2, we can write the *probability distribution* of the discrete random variable x in Table 5.3.

Table 5.3 Probability Distribution of the Number of Vehicles Owned by Families

Number of Vehicles Owned x	Probability $P(x)$
0	.015
1	.235
2	.425
3	.245
4	.080
	$\Sigma P(x) = 1.0$

The probability distribution of a discrete random variable possesses the following *two characteristics.*

1. The probability assigned to each value of the random variable x lies in the range 0 to 1, that is, $0 \leq P(x) \leq 1$ for each x.

2. The sum of the probabilities assigned to all possible values of x is equal to 1.0, that is, $\Sigma P(x) = 1$. (Remember, if the probabilities are rounded, the sum may not be exactly 1.0.)

TWO CHARACTERISTICS OF A PROBABILITY DISTRIBUTION

The probability distribution of a discrete random variable possesses the following two characteristics.

1. $0 \leq P(x) \leq 1$ for each value of x
2. $\Sigma P(x) = 1$

These two characteristics are also called the *two conditions* that a probability distribution must satisfy. Notice that in Table 5.3, each probability listed in the column labeled $P(x)$ is between 0 and 1. Also, $\Sigma P(x) = 1.0$. Because both conditions are satisfied, Table 5.3 represents the probability distribution of x.

From Table 5.3, the probability for any value of x can be read. For example, the probability that a randomly selected family from this town owns two vehicles is .425. This probability is written as

$$P(x = 2) = .425$$

The probability that the selected family owns more than two vehicles is given by the sum of the probabilities of three and four vehicles, respectively. This probability is .245 + .080 = .325. This probability is written as

$$P(x > 2) = P(x = 3) + P(x = 4) = .245 + .080 = .325$$

The probability distribution of a discrete random variable can be presented in the form of a *mathematical formula, a table, or a graph*. Table 5.3 presented the probability distribution in tabular form. Figure 5.1 shows the graphical presentation of the probability distribution of Table 5.3. In this figure, each value of x is marked on the horizontal axis. The probability for each value of x is exhibited by the height of the corresponding line. This section does not discuss the presentation of a probability distribution using a mathematical formula.

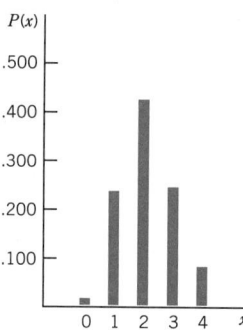

Figure 5.1 Graphical presentation of the probability distribution of Table 5.3.

Verifying conditions of a probability distribution.

EXAMPLE 5–2   Each of the following tables lists certain values of x and their probabilities. Determine whether or not each table represents a valid probability distribution.

(a)

x	$P(x)$
0	.08
1	.11
2	.39
3	.27

(b)

x	$P(x)$
2	.25
3	.34
4	.28
5	.13

(c)

x	$P(x)$
7	.70
8	.50
9	− .20

Solution

(a)   Because each probability listed in this table is in the range 0 to 1, it satisfies the first condition of a probability distribution. However, the sum of all probabilities is not equal to 1.0 because $\Sigma P(x) = .08 + .11 + .39 + .27 = .85$. Therefore, the second condition is not satisfied. Consequently, this table does not represent a valid probability distribution.

(b)   Each probability listed in this table is in the range 0 to 1. Also, $\Sigma P(x) = .25 + .34 + .28 + .13 = 1.0$. Consequently, this table represents a valid probability distribution.

(c) Although the sum of all probabilities listed in this table is equal to 1.0, one of the probabilities is negative. This violates the first condition of a probability distribution. Therefore, this table does not represent a valid probability distribution. ▬

(a) Graph of a probability distribution. (b) Probabilities of events for a discrete random variable.

EXAMPLE 5–3 The following table lists the probability distribution of the number of breakdowns per week for a machine based on past data.

Breakdowns per week	0	1	2	3
Probability	.15	.20	.35	.30

(a) Present this probability distribution graphically.

(b) Find the probability that the number of breakdowns for this machine during a given week is

    (i)   exactly 2   (ii)   zero to 2

    (iii) more than one   (iv) at most 1

Solution Let x denote the number of breakdowns for this machine during a given week. Table 5.4 lists the probability distribution of x.

Table 5.4 Probability Distribution of Breakdowns

x	$P(x)$
0	.15
1	.20
2	.35
3	.30
	$\Sigma P(x) = 1.0$

Graph of a probability distribution.

(a) Figure 5.2 shows the graphical presentation of the probability distribution of Table 5.4.

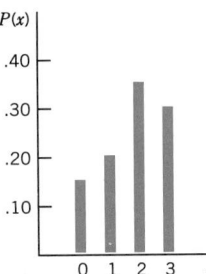

Figure 5.2 Graphical presentation of the probability distribution of Table 5.4.

Probabilities of events for a discrete random variable.

(b) Using Table 5.4, we can calculate the required probabilities as follows.

    (i) The probability of exactly two breakdowns is

$$P(\text{exactly 2 breakdowns}) = P(x = 2) = \textbf{.35}$$

(ii) The probability of zero to two breakdowns is given by the sum of the probabilities of 0, 1, and 2 breakdowns.

$$P(0 \text{ to } 2 \text{ breakdowns}) = P(0 \leq x \leq 2)$$

$$= P(x = 0) + P(x = 1) + P(x = 2)$$

$$= .15 + .20 + .35 = \mathbf{.70}$$

(iii) The probability of more than one breakdown is obtained by adding the probabilities of 2 and 3 breakdowns.

$$P(\text{more than } 1 \text{ breakdown}) = P(x > 1)$$

$$= P(x = 2) + P(x = 3)$$

$$= .35 + .30 = \mathbf{.65}$$

(iv) The probability of at most one breakdown is given by the sum of the probabilities of 0 and 1 breakdown.

$$P(\text{at most } 1 \text{ breakdown}) = P(x \leq 1)$$

$$= P(x = 0) + P(x = 1)$$

$$= .15 + .20 = \mathbf{.35}$$

EXERCISES

Concepts and Procedures

5.6 Explain the meaning of the probability distribution of a discrete random variable. Give one example of such a probability distribution. What are the three ways to present the probability distribution of a discrete random variable?

5.7 Briefly explain the two characteristics (conditions) of the probability distribution of a discrete random variable.

5.8 Each of the following tables lists certain values of x and their probabilities. Verify if each of them represents a valid probability distribution.

a.

x	$P(x)$
0	.10
1	.05
2	.45
3	.32

b.

x	$P(x)$
2	.35
3	.28
4	.23
5	.14

c.

x	$P(x)$
7	− .25
8	.85
9	.40

5.9 Each of the following tables lists certain values of x and their probabilities. Determine if each of them satisfies the two conditions required for a valid probability distribution.

a.

x	$P(x)$
5	− .36
6	.48
7	.62
8	.26

b.

x	$P(x)$
1	.16
2	.24
3	.49

c.

x	$P(x)$
0	.15
1	.00
2	.35
3	.50

Total probability = 1.0

5.10 The following table gives the probability distribution of a discrete random variable x.

x	0	1	2	3	4	5	6
$P(x)$	.11	.19	.28	.15	.12	.09	.06

= 1.0

.11 + .19 + .28 = .58

.12 + .09 + .06 = .27

Find the following probabilities.

a. $P(x = 3)$ *.15* b. $P(x \le 2)$ c. $P(x \ge 4)$ d. $P(1 \le x \le 4)$ *= .74*
e. Probability that x assumes a value less than 4   *$P(x < 4) = .73$*
f. Probability that x assumes a value greater than 2
g. Probability that x assumes a value in the interval 2 to 5

5.11 The following table gives the probability distribution of a discrete random variable x.

x	0	1	2	3	4	5
$P(x)$	.03	.13	.22	.31	.19	.12

Find the following probabilities.

a. $P(x = 1)$ b. $P(x \le 1)$ c. $P(x \ge 3)$ d. $P(0 \le x \le 2)$
e. Probability that x assumes a value less than 3
f. Probability that x assumes a value greater than 3
g. Probability that x assumes a value in the interval 2 to 4

Applications

5.12 Elmo's Sporting Goods sells exercise machines as well as other sporting goods. On different days, it sells different numbers of these machines. The following table, constructed using past data, lists the probability distribution of the number of exercise machines sold per day at Elmo's.

Number of exercise machines sold per day	4	5	6	7	8	9	10
Probability	.08	.11	.14	.19	.23	.16	.09

= 1.0

a. Graph the probability distribution.
b. Determine the probability that the number of exercise machines sold at Elmo's on a given day is
    i.   exactly 6   ii.   more than 8   *$P(x > 8) = .25$*
    iii. 5 to 8   iv.   at most 6   *$P(x \le 6) = .33$*

5.13 Despite all safety measures, accidents do happen at Brown's Manufacturing Corporation. Let x denote the number of accidents that occur during a month at this company. The following table lists the probability distribution of x.

x	0	1	2	3	4
$P(x)$	.25	.30	.20	.15	.10

a. Draw a graph of the probability distribution.
b. Determine the probability that the number of accidents that will occur during a given month at this company is
    i.   exactly 4   ii.   at least 2   *$P \ x \ge 2$*
    iii. less than 3   iv.   2 to 4

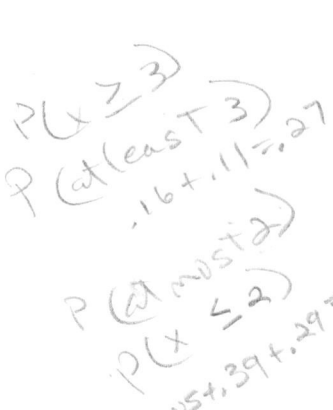

5.14 Lyle Corporation is a bicycle manufacturer that buys bearings from another company. The bearings are received in boxes containing 500 bearings each. Let x denote the number of defective bearings in a box. The following table lists the probability distribution of x.

x	0	1	2	3	4	5
$P(x)$	.17	.23	.29	.16	.11	.04

a. Draw a graph of the probability distribution.
b. Determine the probability that the number of defective bearings in a randomly selected box will be

   i.   exactly 3   ii.   at least 3
   iii. 2 or less   iv.   3 to 5

5.15 The following table, constructed using past data, lists the probability distribution of the number of employees absent per day at Young Clothing Manufacturers.

Number of employees absent per day	0	1	2	3	4	5
Probability	.18	.26	.31	.14	.08	.03

a. Graph the probability distribution.
b. Find the probability that the number of employees absent on a given day is

   i.   exactly 2   ii.   3 or more
   iii. 1 to 3   iv.   at most 2

5.16 A consumer agency surveyed all 2500 families living in a small town to collect data on the number of television sets owned by them. The following table lists the frequency distribution of the data collected by this agency.

rounded   .05   .39   .29   .16   .11

Number of TV sets owned	0	1	2	3	4
Number of families	120	970	730	410	270 = 2500

exact   .048   .388   .292   .164   .108

a. Construct a probability distribution table for the number of television sets owned by these families. Draw a graph of the probability distribution.
b. Let x denote the number of television sets owned by a randomly selected family from this town. Find the following probabilities.

   i. $P(x = 1)$ ii. $P(x > 2)$ .16 + .11 = .27
   iii. $P(x \leq 1)$ iv. $P(1 \leq x \leq 3)$

5.17 The Webster Mail Order Company sells expensive stereos by mail. The following table lists the frequency distribution of the number of orders received per day by this company during the past 100 days.

Number of orders received per day	2	3	4	5	6
Number of days	12	21	34	19	14

a. Construct a probability distribution table for the number of orders received per day. Draw a graph of the probability distribution.
b. Let x denote the number of orders received on any given day. Find the following probabilities.

   i. $P(x = 3)$ ii. $P(x \geq 3)$
   iii. $P(2 \leq x \leq 4)$ iv. $P(x < 4)$

$P(x \geq 3)$
$P(\text{at least } 3) = .27$
.16 + .11 = .27
$P(\text{at most } 2)$
$P(x \leq 2) = .73$
$P(x \leq 2)$
.05 + .39 + .29 = .73
$P(x < 2)$
.16 + .11 = .27

5.18 The Bank of Connecticut wants to determine how often its customers use the ATM installed outside its Main Street branch in Manchester. The following table gives the frequency distribution of the number of times its 500 customers used this ATM during the last one week.

Number of times ATM used	0	1	2	3	4	5
Number of customers	58	94	118	103	86	41

a. Prepare a probability distribution table for this frequency table. Draw a graph of the probability distribution.

b. Let x denote the number of times a customer uses this ATM in a week. Find the following probabilities.

    i. $P(x = 4)$ ii. $P(x \le 2)$

    iii. $P(3 \le x \le 5)$ iv. $P(x > 3)$

5.19 Let x denote the number of suits owned by a randomly selected CEO of a corporation. The following table lists the frequency distribution of x for 1000 CEOs.

x	4	5	6	7	8	9	10
f	70	180	240	210	170	90	40

a. Construct a probability distribution table for the number of suits owned by a CEO. Graph the probability distribution.

b. Find the following probabilities.

    i. $P(x = 5)$ ii. $P(x > 6)$

    iii. $P(4 \le x \le 7)$ iv. $P(x \le 6)$

5.3 MEAN OF A DISCRETE RANDOM VARIABLE

The **mean of a discrete random variable**, denoted by μ, is actually the mean of its probability distribution. The mean of a discrete random variable x is also called its *expected value* and is denoted by $E(x)$. The mean (or expected value) of a discrete random variable is the value that we expect to observe per repetition, on average, if we perform an experiment a large number of times. For example, we may expect a car salesperson to sell, on average, 2.4 cars per week. This does not mean that every week this salesperson will sell exactly 2.4 cars. (Actually she cannot sell exactly 2.4 cars.) This simply means that if we observe for many weeks, this salesperson will sell a different number of cars during different weeks. However, the average for all these weeks will be 2.4 cars.

To calculate the mean of a discrete random variable x, we multiply each value of x by the corresponding probability and sum the resulting products. This sum gives the mean (or expected value) of the discrete random variable x.

MEAN OF A DISCRETE RANDOM VARIABLE

The mean of a discrete random variable x is the value that is expected to occur per repetition, on average, if an experiment is repeated a large number of times. It is denoted by μ and calculated as

$$\mu = \Sigma x P(x)$$

> The mean of a discrete random variable x is also called its expected value and denoted by $E(x)$, that is,
>
> $$E(x) = \Sigma x P(x)$$

Example 5–4 illustrates the calculation of the mean of a discrete random variable.

Calculating and interpreting the mean of a discrete random variable.

EXAMPLE 5–4   Recall Example 5–3 of Section 5.2. The probability distribution Table 5.4 from that example is reproduced below. In this table, x represents the number of breakdowns for a machine during a given week and $P(x)$ is the probability of the corresponding value of x.

x	$P(x)$
0	.15
1	.20
2	.35
3	.30

Find the mean number of breakdowns per week for this machine.

Solution   To find the mean number of breakdowns per week for this machine, we multiply each value of x by its probability and add these products. This sum gives the mean of the probability distribution of x. The products $xP(x)$ are listed in the third column of Table 5.5.

Table 5.5   Calculating the Mean for the Probability Distribution of Breakdowns

x	$P(x)$	$xP(x)$
0	.15	$0(.15) = .00$
1	.20	$1(.20) = .20$
2	.35	$2(.35) = .70$
3	.30	$3(.30) = .90$
		$\Sigma xP(x) = 1.80$

The sum of these products gives $\Sigma xP(x)$, which is the mean of x. The mean is

$$\mu = \Sigma x P(x) = \mathbf{1.80}$$

Thus, on average, this machine is expected to break down 1.80 times per week over a period of time. In other words, if this machine is used for many weeks, then for certain weeks we will observe no breakdowns; for some other weeks we will observe 1 breakdown per week; and for still other weeks we will observe 2 or 3 breakdowns per week. The mean number of breakdowns is expected to be 1.80 per week for the entire period.   ■

Case Study 5–1 illustrates the calculation of the mean amount that an instant lottery player is expected to win.

CASE STUDY 5–1   CASH RICH INSTANT LOTTERY

EXAMPLE: Ticket As Printed

EXAMPLE: Scratched-off Ticket

Recently the state of Connecticut had in circulation an instant lottery game called ''cash rich.'' The cost of each ticket for this lottery was $1. A player could instantly win $500, $23, $13, $9, $7, $3, or $1. Each ticket had six erasable spots, which could reveal any of those amounts when rubbed off. If three of the six spots on a ticket showed the same amount, the player won that amount.

The following table lists the number of tickets with different prizes from a total of 6,000,000 tickets printed for this lottery. As is obvious from this table, out of a total of 6,000,000 tickets, 4,640,940 were nonwinning tickets (the ones with a prize of $0 in this table). Of the remaining tickets, 999,960 had a prize of $1 each, 222,000 had a prize of $3 each, and so on.

Prize (in dollars)	Number of Tickets
0	4,640,940
1	999,960
3	222,000
7	60,000
9	37,500
13	24,000
23	15,000
500	600
	Total = 6,000,000

The net gain to the players for each of the instant winning tickets is equal to the amount of the prize minus $1, which is the cost of the ticket. Thus, the net gain for each of the nonwinning tickets is $-\$1$, the cost of the ticket. Let

$$x = \text{the net amount a player won by playing this lottery}$$

The following table shows the probability distribution of x and all the calculations required to compute the mean of x for that probability distribution. The probability of an outcome (net winnings) is calculated by dividing the number of tickets with that outcome by the total number of tickets.

x (in dollars)	$P(x)$	$xP(x)$
-1	$4,640,940/6,000,000 = .77349$	$-.77349$
0	$999,960/6,000,000 = .16666$	$.00000$
2	$222,000/6,000,000 = .03700$	$.07400$
6	$60,000/6,000,000 = .01000$	$.06000$
8	$37,500/6,000,000 = .00625$	$.05000$
12	$24,000/6,000,000 = .00400$	$.04800$
22	$15,000/6,000,000 = .00250$	$.05500$
499	$600/6,000,000 = .00010$	$.04990$
		$\Sigma xP(x) = -.43659$

Hence, the mean of x is

$$\mu = \Sigma xP(x) = -\$.43659 \approx -\$.4366$$

This mean gives the expected value of the random variable x, that is,

$$E(x) = \Sigma xP(x) = -\$.4366$$

Thus, the mean of winnings for this lottery was $-\$.4366$. In other words, all players taken together lost an average of $\$.4366$ (or 43.66 cents) per ticket. This can also be interpreted as follows: Only 56.34% ($= 100 - 43.66$) of the total money spent by players on buying these lottery tickets was returned to them in the form of prizes and 43.66% was not returned. (The money that was not returned covered the costs of operating the lottery, the commission paid to agents, and revenue to the state of Connecticut.)

Source: The State of Connecticut Lottery Commission publications. Lottery ticket reproduced with permission.

5.4 STANDARD DEVIATION OF A DISCRETE RANDOM VARIABLE

The **standard deviation of a discrete random variable**, denoted by σ, measures the spread of its probability distribution. A higher value for the standard deviation of a discrete random variable indicates that x can assume values over a larger range about the mean. On the other hand, a smaller value for the standard deviation indicates that most of the values that x can assume are clustered closely about the mean. The basic formula to compute the standard deviation of a discrete random variable is

$$\sigma = \sqrt{\Sigma[(x - \mu)^2 \cdot P(x)]}$$

However, it is more convenient to use the following short-cut formula to compute the standard deviation of a discrete random variable.

STANDARD DEVIATION OF A DISCRETE RANDOM VARIABLE

The standard deviation of a discrete random variable x measures the spread of its probability distribution and is computed as

$$\sigma = \sqrt{\Sigma x^2 P(x) - \mu^2}$$

Note that the variance σ^2 of a discrete random variable is obtained by squaring its standard deviation.

Example 5–5 illustrates how to use the short-cut formula to compute the standard deviation of a discrete random variable.

Calculating standard deviation of a discrete random variable.

EXAMPLE 5–5   Baier's Electronics manufactures computer parts that are supplied to many computer companies. Despite the fact that two quality control inspectors at Baier's Electronics check every part for defects before it is shipped to another company, a few defective parts do pass through these inspections undetected. Let x denote the number of defective computer parts in a shipment of 400. The following table gives the probability distribution of x.

x	0	1	2	3	4	5
$P(x)$	.02	.20	.30	.30	.10	.08

Compute the standard deviation of x.

Solution   Table 5.6 shows all the calculations required for the computation of the standard deviation of x.

Table 5.6 Computations to Find the Standard Deviation

x	$P(x)$	$xP(x)$	x^2	$x^2P(x)$
0	.02	.00	0	.00
1	.20	.20	1	.20
2	.30	.60	4	1.20
3	.30	.90	9	2.70
4	.10	.40	16	1.60
5	.08	.40	25	2.00
		$\Sigma xP(x) = 2.50$		$\Sigma x^2P(x) = 7.70$

We perform the following steps to compute the standard deviation of x.

Step 1. Compute the mean of the discrete random variable.

The sum of the products $xP(x)$, recorded in the third column of Table 5.6, gives the mean of x.

$$\mu = \Sigma x P(x) = 2.50 \text{ defective computer parts in } 400$$

Step 2. Compute the value of $\Sigma x^2 P(x)$.

First we square each value of x and record it in the fourth column of Table 5.6. Then we multiply these values of x^2 by the corresponding values of $P(x)$. The resulting values of $x^2 P(x)$ are recorded in the fifth column of Table 5.6. The sum of this column gives

$$\Sigma x^2 P(x) = 7.70$$

Step 3. Substitute the values of μ and $\Sigma x^2 P(x)$ in the formula for the standard deviation of x and simplify.

By performing this step, we obtain:

$$\sigma = \sqrt{\Sigma x^2 P(x) - \mu^2} = \sqrt{7.70 - (2.50)^2} = \sqrt{1.45}$$

$$= \mathbf{1.20} \text{ defective computer parts}$$

Thus, a given shipment of 400 computer parts is expected to contain an average of 2.50 defective parts with a standard deviation of 1.20. ∎

☞ *Remember* Because the standard deviation of a discrete random variable is obtained by taking the positive square root, its value is never negative.

EXAMPLE 5–6 Loraine Corporation is planning to introduce a new makeup product. According to the analysis made by the financial department of the company, it will earn an annual profit of $4.5 million if this product has high sales, an annual profit of $1.2 million if the sales are mediocre, and it will lose $2.3 million a year if the sales are low. The probabilities of these three scenarios are .32, .51, and .17, respectively.

(a) Let x be the profits (in millions of dollars) earned per annum by the company from this product. Write the probability distribution of x.

(b) Calculate the mean and standard deviation of x.

Solution

Probability distribution of a discrete random variable.

(a) The following table lists the probability distribution of x. Note that since x denotes profits earned by the company, the loss is written as *negative profits* in the table.

x	$P(x)$
4.5	.32
1.2	.51
−2.3	.17

Calculating mean and standard deviation of a discrete random variable.

(b) Table 5.7 shows all the calculations needed for the computation of the mean and standard deviation of x.

Table 5.7 Computations to Find the Mean and Standard Deviation

x	$P(x)$	$xP(x)$	x^2	$x^2P(x)$
4.5	.32	1.440	20.25	6.4800
1.2	.51	.612	1.44	.7344
-2.3	.17	$-.391$	5.29	.8993
		$\Sigma xP(x) = 1.661$		$\Sigma x^2P(x) = 8.1137$

The mean of x is

$$\mu = \Sigma xP(x) = \textbf{\$1.661 million}$$

The standard deviation of x is

$$\sigma = \sqrt{\Sigma x^2P(x) - \mu^2} = \sqrt{8.1137 - (1.661)^2} = \textbf{\$2.314 million}$$

Thus, it is expected that Loraine Corporation will earn an average of $1.661 million profits a year with a standard deviation of $2.314 million from the new makeup product.

☞ **INTERPRETATION OF THE STANDARD DEVIATION**

The standard deviation of a discrete random variable can be interpreted or used the same way as the standard deviation of a data set in Section 3.4 of Chapter 3. In that section, we learned that according to Chebyshev's theorem, at least $[1 - (1/k^2)] \times 100\%$ of the total area under a curve lies within k standard deviations of the mean where k is any number greater than 1. Thus, if $k = 2$, then 75% of the area under a curve lies between $\mu - 2\sigma$ and $\mu + 2\sigma$. In Example 5–5,

$$\mu = 2.50 \quad \text{and} \quad \sigma = 1.20$$

Hence,

$$\mu - 2\sigma = 2.50 - 2(1.20) = .10$$

and

$$\mu + 2\sigma = 2.50 + 2(1.20) = 4.90$$

Using Chebyshev's theorem, we can state that at least 75% of the shipments (each containing 400 computer parts) are expected to contain .10 to 4.90 defective computer parts each.

EXERCISES

Concepts and Procedures

5.20 Briefly explain the concept of the mean and standard deviation of a discrete random variable.

5.21 Find the mean and standard deviation for each of the following probability distributions.

a.

x	$P(x)$
0	.12
1	.27
2	.43
3	.18

b.

x	$P(x)$
6	.36
7	.26
8	.21
9	.17

5.22 Find the mean and standard deviation for each of the following probability distributions.

a.	x	$P(x)$
	3	.09
	4	.21
	5	.34
	6	.23
	7	.13

b.	x	$P(x)$
	0	.43
	1	.31
	2	.17
	3	.09

Applications

5.23 The following table, reproduced from Exercise 5.12, lists the probability distribution of the number of exercise machines sold per day at Elmo's Sporting Goods store.

Number of exercise machines sold per day	4	5	6	7	8	9	10
Probability	.08	.11	.14	.19	.23	.16	.09

Calculate the mean and standard deviation for this probability distribution. Give a brief interpretation of the value of the mean.

5.24 Despite all safety measures, accidents do happen at Brown's Manufacturing Corporation. Let x denote the number of accidents that occur during a month at this company. The following table, reproduced from Exercise 5.13, lists the probability distribution of x.

x	0	1	2	3	4
$P(x)$	.25	.30	.20	.15	.10

Calculate the mean and standard deviation for the number of accidents per month at this company. Give a brief interpretation of the value of the mean.

5.25 Lyle Corporation is a bicycle manufacturer that buys bearings from another company. The bearings are received in boxes containing 500 bearings each. Let x denote the number of defective bearings in a box. The following table, reproduced from Exercise 5.14, lists the probability distribution of x.

x	0	1	2	3	4	5
$P(x)$	.17	.23	.29	.16	.11	.04

Calculate the mean and standard deviation for this probability distribution.

5.26 The following table, reproduced from Exercise 5.15, lists the probability distribution of the number of employees absent per day at Young Clothing Manufacturers.

Number of employees absent per day	0	1	2	3	4	5
Probability	.18	.26	.31	.14	.08	.03

Calculate the mean and standard deviation for this probability distribution.

5.27 Let x be the number of errors that a randomly selected page of a book contains. The following table lists the probability distribution of x.

x	0	1	2	3	4
$P(x)$	.73	.16	.06	.04	.01

Find the mean and standard deviation of x.

5.28 The following table gives the probability distribution of camcorders sold on a given day at an electronics store.

Camcorders sold	0	1	2	3	4	5	6
Probability	.05	.12	.23	.30	.16	.10	.04

Calculate the mean and standard deviation for this probability distribution.

5.29 Refer to Exercise 5.16. Find the mean and standard deviation for the probability distribution you developed for the number of television sets owned by all 2500 families living in a town. Give a brief interpretation of the values of the mean and standard deviation.

5.30 Refer to Exercise 5.17. Find the mean and standard deviation for the probability distribution you developed for the number of orders received per day for the past 100 days at the Webster Mail Order Company. Give a brief interpretation of the values of the mean and standard deviation.

5.31 Refer to Exercise 5.18. Find the mean and standard deviation for the probability distribution you developed for the number of times an ATM was used by 500 bank customers during the past one week. Give a brief interpretation of the values of the mean and standard deviation.

5.32 Refer to Exercise 5.19. Find the mean and standard deviation for the probability distribution you developed for the number of suits owned by 1000 CEOs. Give a brief interpretation of the values of the mean and standard deviation.

5.33 A farmer will earn a profit of $30 thousand in case of heavy rain next year, $60 thousand in case of moderate rain, and $15 thousand in case of little rain. A meteorologist forecasts that the probability is .35 for heavy rain, .40 for moderate rain, and .25 for little rain next year. Let x be the random variable that represents next year's profits in thousands of dollars for this farmer. Write the probability distribution of x. Find the mean and standard deviation of x. Give a brief interpretation of the values of the mean and standard deviation.

5.34 An instant lottery ticket costs $2. Out of a total of 10,000 tickets printed for this lottery, 1000 tickets contain a prize of $5 each, 100 tickets have a prize of $10 each, 5 tickets have a prize of $1000 each, and 1 ticket has a prize of $5000. Let x be the random variable that denotes the net amount a player wins by playing this lottery. Write the probability distribution of x. Determine the mean and standard deviation of x. How will you interpret the values of the mean and standard deviation of x?

5.5 FACTORIALS AND COMBINATIONS

This section introduces factorials and combinations, which will be used in the binomial formula discussed in Section 5.6.

5.5.1 FACTORIALS

The symbol "!" (read as *factorial*) is used to denote **factorials**. The value of the factorial of a number is obtained by multiplying all integers from that number to 1. For example, "7!" is read as *seven factorial* and it is evaluated by multiplying all integers from 7 to 1.

Permutations

have 4 digits

use only 2   4·3·

4P2

1 2   3   4¹   4²
1 2   31   32   42   43
1 3   21   22   31   24
1 4   12 numbers

FACTORIALS

The symbol $n!$, read as "n factorial," represents the product of all integers from n to 1. In other words,

$$n! = n(n - 1)(n - 2)(n - 3) \ldots 3 \cdot 2 \cdot 1$$

By definition,

$$0! = 1$$

Evaluating factorial.

EXAMPLE 5–7   Evaluate 7!.

Solution   To evaluate 7!, we multiply all integers from 7 to 1.

$$7! = 7 \cdot 6 \cdot 5 \cdot 4 \cdot 3 \cdot 2 \cdot 1 = \mathbf{5040}$$

Thus, the value of 7! is 5040.[1]

Evaluating factorial.

EXAMPLE 5–8   Evaluate 10!.

Solution   The value of 10! is given by the product of all integers from 10 to 1. Thus,

$$10! = 10 \cdot 9 \cdot 8 \cdot 7 \cdot 6 \cdot 5 \cdot 4 \cdot 3 \cdot 2 \cdot 1 = \mathbf{3,628,800}$$

Factorial of difference between two numbers.

EXAMPLE 5–9   Evaluate $(12 - 4)!$.

Solution   The value of $(12 - 4)!$ is

$$(12 - 4)! = 8! = 8 \cdot 7 \cdot 6 \cdot 5 \cdot 4 \cdot 3 \cdot 2 \cdot 1 = \mathbf{40,320}$$

Factorial of zero.

EXAMPLE 5–10   Evaluate $(5 - 5)!$.

Solution   As shown below, the value of $(5 - 5)!$ is 1.

$$(5 - 5)! = 0! = \mathbf{1}$$

Note that 0! is always equal to 1.

We can read the value of $n!$ for $n = 1$ through $n = 25$ from Table II of Appendix B. Example 5–11 illustrates how to read that table.

Using the table of factorials.

EXAMPLE 5–11   Find the value of 15! by using Table II of Appendix B.

Solution   To find the value of 15! from Table II, we locate 15 in the column labeled n. Then we read the value in the column for n! entered next to 15. Thus,

$$15! = \mathbf{1,307,674,368,000}$$

---

[1]**Using a calculator to evaluate $n!$:** Most calculators have an $n!$ or $x!$ function key. If your calculator has such a key, you can use it to evaluate 7! as follows: First enter 7 and then press the $n!$ or $x!$ key. The calculator screen will display 5040 as the answer.

5.5.2 COMBINATIONS

Quite often we face the problem of selecting a few elements from a large number of distinct elements. As an example, a company may have to select two managers out of four to attend a conference. As another example, a lottery player may have to pick 6 numbers from 49. The question arises: In how many ways can we make the selections in each of these examples? For instance, how many possible selections exist for the company that is to select any two managers out of four? Suppose we name the four managers A, B, C, and D. Then the possible selections are

$$(A,B) \quad (A,C) \quad (A,D) \quad (B,C) \quad (B,D) \quad (C,D)$$

The company can choose managers A and B, or A and C, or A and D, and so on. Thus, there are six possible ways to select two managers out of four. These six possible selections are called six combinations.

Each of the possible selections in this list is called a **combination**. All six combinations are distinct; that is, each combination contains a different set of managers. It is important to remember that the order in which the selections are made is not significant in the case of combinations. Thus, whether we write (A,B) or (B,A), both these arrangements represent only one combination.

COMBINATIONS NOTATION

Combinations give the number of ways x elements can be selected from n elements. The notation used to denote the total number of combinations is

$$\binom{n}{x}$$

p. 828 table

which is read as "the combinations of n elements selected x at a time."

Suppose there are a total of n elements from which we want to select x elements. Then,

20 stocks
5 picked

$$\binom{20}{5} = \frac{20 \cdot 19 \cdot 18 \cdot 17 \cdot 16}{5 \cdot 4 \cdot 3 \cdot 2 \cdot 1}$$

n denotes the total number of elements

$$\binom{n}{x} = \text{the number of combinations of } n \text{ elements selected } x \text{ at a time}$$

x denotes the number of elements selected per selection

NUMBER OF COMBINATIONS

The number of combinations for selecting x from n distinct elements is given by the formula

$$\binom{n}{x} = \frac{n!}{x! \, (n - x)!}$$

where $n!$, $x!$, and $(n - x)!$ are read as "n factorial," "x factorial," and "n minus x factorial," respectively.

In the combinations formula,

$$n! = n(n-1)(n-2)(n-3)\ldots 3 \cdot 2 \cdot 1$$

$$x! = x(x-1)(x-2)\ldots 3 \cdot 2 \cdot 1$$

and $$(n-x)! = (n-x)(n-x-1)(n-x-2)\ldots 3 \cdot 2 \cdot 1$$

Note that in combinations, n is always greater than or equal to x. If n is smaller than x, then we cannot select x distinct elements from n.

Finding the number of combinations by using formula.

EXAMPLE 5-12   Reconsider the example of a company that is to select two managers from four to attend a conference. Using the combinations formula, find the number of ways this company can select two managers from four.

Solution   For this example,

$$n = \text{total number of managers} = 4$$

$$x = \text{managers to be selected} = 2$$

Therefore, the number of ways this company can select two managers from four is

$$\binom{4}{2} = \frac{4!}{2!\,(4-2)!} = \frac{4!}{2!\,2!} = \frac{4 \cdot 3 \cdot 2 \cdot 1}{2 \cdot 1 \cdot 2 \cdot 1} = 6$$

We listed these six combinations earlier in this section.[2]

Finding the number of combinations and listing them.

EXAMPLE 5-13   Three members of a jury will be randomly selected from five persons. How many different combinations are possible?

Solution   There are a total of five persons and we are to select three of them. Hence,

$$n = 5 \quad \text{and} \quad x = 3$$

Applying the combinations formula,

$$\binom{5}{3} = \frac{5!}{3!\,(5-3)!} = \frac{5!}{3!\,2!} = \frac{120}{6 \cdot 2} = 10$$

If we assume that the five persons are A, B, C, D, and E, then the 10 possible combinations for the selection of three members of the jury are

ABC, ABD, ABE, ACD, ACE, ADE, BCD, BCE, BDE, CDE

Case Study 5–2 describes the number of ways a lottery player can select 6 numbers from 49 in a lotto game.

---

[2]**Using a calculator to evaluate** $\binom{n}{x}$: Most calculators have a Cn,r or Cn,x or $\binom{n}{r}$ or $\binom{n}{x}$ function key. (Note that all these notations are used to denote combinations.) If your calculator has such a key, read the manual that accompanies the calculator to find out how this key functions and then evaluate $\binom{4}{2}$ using that key.

CASE STUDY 5-2 PLAYING LOTTO

During the past few years, many states have initiated the popular lottery game called lotto. To play lotto, a player picks any 6 numbers from a list of numbers usually starting with 1, for example from 1 through 49. At the end of the lottery period, the state lottery commission randomly selects 6 numbers from the same list. If all 6 numbers picked by a player are the same as the ones randomly selected by the lottery commission, the player wins.

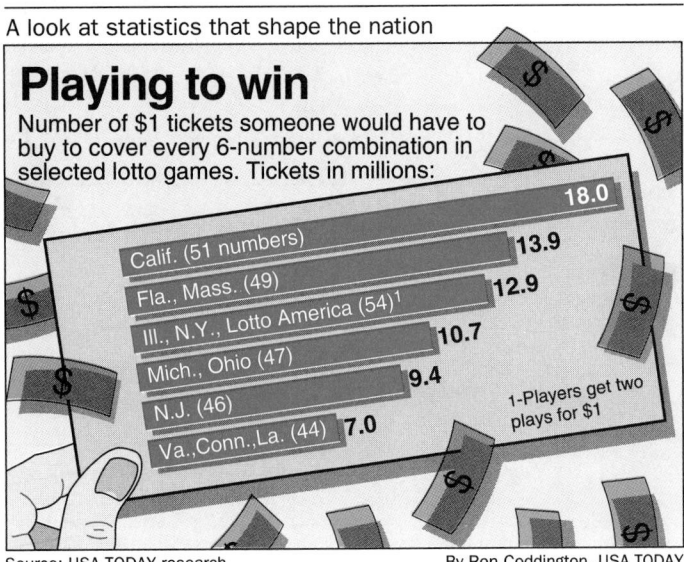

USA SNAPSHOTS ®

A look at statistics that shape the nation

Playing to win

Number of $1 tickets someone would have to buy to cover every 6-number combination in selected lotto games. Tickets in millions:

Calif. (51 numbers)	18.0
Fla., Mass. (49)	13.9
Ill., N.Y., Lotto America (54)[1]	12.9
Mich., Ohio (47)	10.7
N.J. (46)	9.4
Va.,Conn.,La. (44)	7.0

1-Players get two plays for $1

Source: USA TODAY research   By Ron Coddington, USA TODAY

The above chart shows the number of combinations (in millions) for picking 6 numbers for lotto games played in a few states. For example, in California a player has to pick 6 numbers from 1 through 51. As shown in the chart, there are approximately 18 million ways (combinations) to select 6 numbers from 1 through 51. In Florida and Massachusetts, a player has to pick 6 numbers from 1 through 49. For this lotto, there are approximately 13.9 million combinations.

Let us find the probability that a player who picks 6 numbers from 49 wins this game. The total combinations of selecting 6 numbers from 49 numbers are obtained as follows.

$$\binom{49}{6} = \frac{49!}{6! \, (49 - 6)!} = 13{,}983{,}816$$

Thus, there are a total of 13,983,816 different ways to select 6 numbers from 49 numbers. Hence, the probability that a player (who plays this lottery once) wins is

$$P(\text{player wins}) = 1/13{,}983{,}816 = .0000000715$$

Source: Chart reprinted with permission from *USA TODAY*, February 27, 1992. Copyright © 1992, *USA TODAY*.

5.5.3 THE TABLE OF COMBINATIONS

Table III in Appendix B lists the number of combinations of n elements selected x at a time. The following example illustrates how to read that table to find combinations.

Using the table of combinations.

EXAMPLE 5–14 Marv & Sons advertised to hire a financial analyst. The company has received applications from 10 candidates, who seem to be equally qualified. The company manager has decided to call only 3 of these candidates for an interview. If she selects 3 candidates randomly from the 10, how many total selections are possible?

Solution The total number of ways to select 3 applicants from 10 is given by $\begin{pmatrix} 10 \\ 3 \end{pmatrix}$. To find the value of $\begin{pmatrix} 10 \\ 3 \end{pmatrix}$ from Table III, we locate 10 in the column labeled n and 3 in the row labeled x. The relevant part of that table is reproduced here as Table 5.8.

Table 5.8 Determining the Value of $\begin{pmatrix} 10 \\ 3 \end{pmatrix}$

n	x	0	1	2	3	...	20
1		1	1				
2		1	2	1			
3		1	3	3	1		
⋮		⋮	⋮	⋮	⋮		
$n = 10 \rightarrow$ 10		1	10	45	120	...	...
⋮		⋮	⋮	⋮	⋮	...	...

$x = 3$

The value of $\begin{pmatrix} 10 \\ 3 \end{pmatrix}$

The number at the intersection of the row for $n = 10$ and the column for $x = 3$ gives the value of $\begin{pmatrix} 10 \\ 3 \end{pmatrix}$, which is

$$\begin{pmatrix} 10 \\ 3 \end{pmatrix} = 120$$

Thus, the company manager can select 3 applicants from 10 in 120 ways. ∎

If the total number of elements and the number of elements to be selected are the same, then there is only one combination. In other words,

$$\begin{pmatrix} n \\ n \end{pmatrix} = 1$$

Also, the number of combinations for selecting zero items from n is 1. That is,

$$\begin{pmatrix} n \\ 0 \end{pmatrix} = 1$$

For example,

$$\binom{5}{5} = \frac{5!}{5!\,(5-5)!} = \frac{5!}{5!\,0!} = \frac{120}{(120)\,(1)} = 1$$

and

$$\binom{8}{0} = \frac{8!}{0!\,(8-0)!} = \frac{8!}{0!\,8!} = \frac{40{,}320}{(1)\,(40{,}320)} = 1$$

EXERCISES

Concepts and Procedures

5.35 Determine the value of each of the following using the appropriate formulas.

$$3! \qquad (7-3)! \qquad 9! \qquad (14-12)! \qquad \binom{5}{3} \qquad \binom{7}{4} \qquad \binom{9}{3} \qquad \binom{6}{0} \qquad \binom{3}{3}$$

Verify the calculated values by using Tables II and III of Appendix B.

5.36 Find the value of each of the following using the appropriate formulas.

$$6! \qquad 11! \qquad (7-2)! \qquad (13-5)! \qquad \binom{8}{2} \qquad \binom{4}{0} \qquad \binom{5}{5} \qquad \binom{6}{4} \qquad \binom{11}{7}$$

Verify the calculated values by using Tables II and III of Appendix B.

Applications

5.37 An investor will randomly select 5 stocks from 20 for an investment. How many total combinations are possible? Use Table III of Appendix B.

5.38 A company employs a total of 16 workers. The management has asked these employees to select 2 workers who will negotiate a new contract with management. The employees have decided to select the 2 workers randomly. How many total selections are possible?

5.39 An inspector with the labor department plans to visit 3 of 12 businesses during next week to check whether or not these businesses comply with the labor safety laws. If she randomly selects 3 businesses from these 12, how many total selections are possible? Use the appropriate formula. Verify your answer by using Table III of Appendix B.

5.40 An environmental agency will randomly select 4 houses for a radon check from a block containing 25 houses. How many total selections are possible? Use Table III of Appendix B.

5.41 In how many ways can a sample (without replacement) of 8 items be selected from a population of 20 items?

5.42 In how many ways can a sample (without replacement) of 4 items be selected from a population of 15 items?

5.6 THE BINOMIAL PROBABILITY DISTRIBUTION

The **binomial probability distribution** is one of the most widely used discrete probability distributions. It is applied to find the probability that an outcome will occur x times in n performances of an experiment. For example, given that the probability is .05 that a VCR manufactured at a firm is defective, we may be interested in finding the probability that in a random sample of three VCRs manufactured at this firm, exactly one will be defective. As a second example, we may be interested in finding the probability that four out of nine customers who visit a department store will make a purchase when 25% of all customers who visit this store make a purchase.

To apply the binomial probability distribution, the random variable x must be a discrete dichotomous random variable. In other words, the variable must be a discrete random variable and each repetition of the experiment must result in one of two possible outcomes. The binomial distribution is applied to experiments that satisfy the four conditions of a *binomial experiment*. (These conditions are described in Section 5.6.1.) Each repetition of a binomial experiment is called a **trial** or a **Bernoulli trial** (after Jacob Bernoulli). For example, if an experiment is defined as one toss of a coin and this experiment is repeated 10 times, then each repetition (toss) is called a trial. Consequently, there are 10 total trials for this experiment.

5.6.1   THE BINOMIAL EXPERIMENT

An experiment that satisfies the following four conditions is called a **binomial experiment**.

1. There are n identical trials. In other words, the given experiment is repeated n times. All these repetitions are performed under identical conditions.
2. Each trial has two and only two outcomes. These outcomes are usually called a *success* and a *failure*.
3. The probability of success is denoted by p and that of failure by q, and $p + q = 1$. The probabilities p and q remain constant for each trial.
4. The trials are independent. In other words, the outcome of one trial does not affect the outcome of another trial.

CONDITIONS OF A BINOMIAL EXPERIMENT

A binomial experiment must satisfy the following four conditions.

1. There are n identical trials.
2. Each trial has only two possible outcomes.
3. The probabilities of the two outcomes remain constant.
4. The trials are independent.

Note that one of the two outcomes of a trial is called a *success* and the other a *failure*. Notice that a success does not mean that the corresponding outcome is considered favorable or desirable. Similarly, a failure does not necessarily refer to an unfavorable or undesirable outcome. Success and failure are simply the names used to denote the two possible outcomes of a trial. The outcome to which the question refers is usually called a success; the outcome to which it does not refer is called a failure.

Verifying conditions of a binomial experiment.

EXAMPLE 5–15   Consider the experiment consisting of 10 tosses of a coin. Determine if it is a binomial experiment.

Solution   As described below, the experiment consisting of 10 tosses of a coin satisfies all four conditions of a binomial experiment.

1. There are a total of 10 trials (tosses), and they are all identical. All 10 tosses are performed under identical conditions.
2. Each trial (toss) has only two possible outcomes: a head and a tail. Let a head be called a success and a tail be called a failure.
3. The probability of obtaining a head (a success) is 1/2 and that of a tail (a failure) is 1/2 for any toss. That is,

$$p = P(H) = 1/2 \quad \text{and} \quad q = P(T) = 1/2$$

The sum of these two probabilities is 1.0. Also, these probabilities remain the same for each toss.
4. The trials (tosses) are independent. The result of any preceding toss has no bearing on the result of any succeeding toss.

   Consequently, the experiment consisting of 10 tosses is a binomial experiment. ▄

Verifying conditions of a binomial experiment.

EXAMPLE 5-16 Five percent of all VCRs manufactured by a large electronics firm are defective. Three VCRs are randomly selected from the production line of this firm. The selected VCRs are inspected to determine if each of them is defective or good. Is this experiment a binomial experiment?

Solution

1. This example consists of three identical trials. A trial represents the selection of a VCR.
2. Each trial has two outcomes: A VCR is defective or a VCR is good. Let a defective VCR be called a success and a good VCR be called a failure.
3. Five percent of all VCRs are defective. So, the probability p that a VCR is defective is .05. As a result, the probability q that a VCR is good is .95. These two probabilities add up to 1.
4. Each trial (VCR) is independent. In other words, if one VCR is defective it does not affect the outcome of another VCR being defective or good. This is so because the size of the population is very large as compared to the sample size.

Since all four conditions of a binomial experiment are satisfied, this is an example of a binomial experiment. ▄

5.6.2 THE BINOMIAL PROBABILITY DISTRIBUTION AND BINOMIAL FORMULA

The random variable x that represents the number of successes in n trials for a binomial experiment is called a *binomial random variable*. The probability distribution of x in such experiments is called the **binomial probability distribution** or simply *binomial distribution*. Thus, the binomial probability distribution is applied to find the probability of x successes in n trials for a binomial experiment. The number of successes x in such an experiment is a discrete random variable. Consider Example 5–16. Let x be the number of defective VCRs in a sample of three. Since we can obtain any number of defective VCRs from zero to three in a sample of three, x can assume any of the values 0, 1, 2, and 3. Since the values of x are countable, it is a discrete random variable.

BINOMIAL FORMULA

For a binomial experiment, the probability of exactly x successes in n trials is given by the binomial formula

$$P(x) = \binom{n}{x} p^x q^{n-x}$$

where

$$n = \text{total number of trials}$$

$$p = \text{probability of success}$$

$$q = 1 - p = \text{probability of failure}$$

$$x = \text{number of successes in } n \text{ trials}$$

$$n - x = \text{number of failures in } n \text{ trials}$$

In the binomial formula, n is the total number of trials and x is the number of successes. The difference between the total number of trials and the total number of successes, $n - x$, gives the total number of failures in n trials. The value of $\binom{n}{x}$ gives the number of ways to obtain x successes in n trials. As mentioned earlier, p and q are the probabilities of success and failure, respectively. Again, although it does not matter which of the two outcomes is called a success and which one a failure, usually the outcome to which the question refers is called a success.

To solve a binomial problem, we determine the values of n, x, $n - x$, p, and q and then substitute these values in the binomial formula. To find the value of $\binom{n}{x}$, we can use either the combinations formula from Section 5.5.2 or the table of combinations (Table III of Appendix B).

To find the probability of x successes in n trials for a binomial experiment, the only values needed are those of n and p. These are called the *parameters of the binomial probability distribution* or simply the **binomial parameters**. The value of q is obtained by subtracting the value of p from 1.0. Thus, $q = 1 - p$.

Next we solve a binomial problem, first without using the binomial formula and then by using the binomial formula.

Calculating probability: using a tree diagram and the binomial formula.

EXAMPLE 5–17   Five percent of all VCRs manufactured by a large electronics firm are defective. A quality control inspector randomly selects three VCRs from the production line. What is the probability that exactly one of these three VCRs is defective?

Solution   Let

$$D = \text{a VCR is defective}$$

and

$$G = \text{a VCR is good}$$

As the tree diagram in Figure 5.3 shows, there are a total of eight outcomes and three of them contain exactly one defective VCR. These three outcomes are

$$DGG, \quad GDG, \quad \text{and} \quad GGD$$

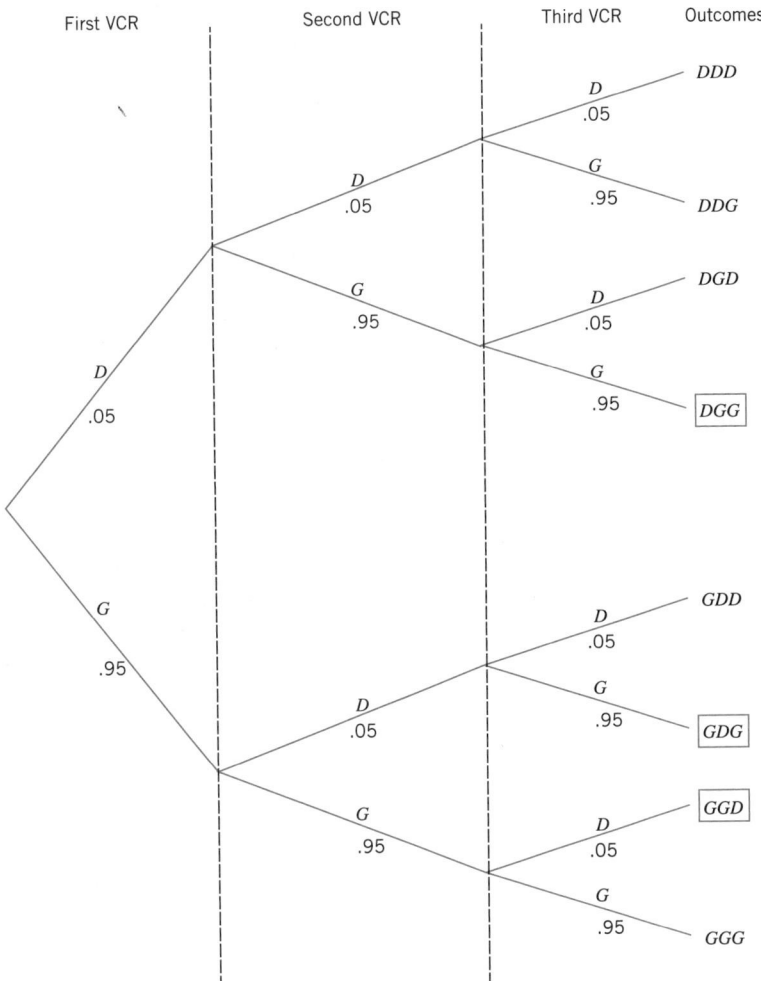

Figure 5.3   Tree diagram for selecting three VCRs.

We know that 5% of all VCRs manufactured at this firm are defective. As a result, 95% of all VCRs are good. So the probability that a randomly selected VCR is defective is .05 and the probability that it is good is .95.

$$P(D) = .05 \quad \text{and} \quad P(G) = .95$$

Because the size of the population is large (note that it is a large firm), the selections can be considered to be independent. The probability of each of the three outcomes, which give exactly one defective VCR, is calculated as follows.

$$P(DGG) = P(D) \cdot P(G) \cdot P(G) = (.05)(.95)(.95) = .0451$$

$$P(GDG) = P(G) \cdot P(D) \cdot P(G) = (.95)(.05)(.95) = .0451$$

$$P(GGD) = P(G) \cdot P(G) \cdot P(D) = (.95)(.95)(.05) = .0451$$

Note that *DGG* is simply the intersection of the three events *D*, *G*, and *G*. In other words, *P(DGG)* is the joint probability of three events: The first VCR selected is defective, the second is good, and the third is good. To calculate this probability, we use the multiplication rule for independent events learned in Chapter 4. The same is true about the probabilities of the other two outcomes: *GDG* and *GGD*.

Exactly one defective VCR will be selected if either *DGG* or *GDG* or *GGD* occurs. These are three mutually exclusive outcomes. Therefore, applying the addition rule of Chapter 4, the probability of the union of these three outcomes is simply the sum of their individual probabilities.

$$P(1 \text{ VCR is defective in 3}) = P(DGG \text{ or } GDG \text{ or } GGD)$$
$$= P(DGG) + P(GDG) + P(GGD)$$
$$= .0451 + .0451 + .0451 = \textbf{.1353}$$

Now let us use the binomial formula to compute this probability. Let us call the selection of a defective VCR a *success* and the selection of a good VCR a *failure*. The reason we have called a defective VCR a *success* is that the question refers to selecting exactly one defective VCR. Then,

$$n = \text{total number of trials} = 3 \text{ VCRs}$$
$$x = \text{number of successes} = \text{number of defective VCRs} = 1$$
$$n - x = \text{number of failures} = \text{number of good VCRs} = 3 - 1 = 2$$
$$p = P(\text{success}) = .05$$
$$q = P(\text{failure}) = 1 - p = .95$$

The probability of 1 success is denoted by $P(x = 1)$ or simply by $P(1)$. By substituting all the values in the binomial formula, we obtain:

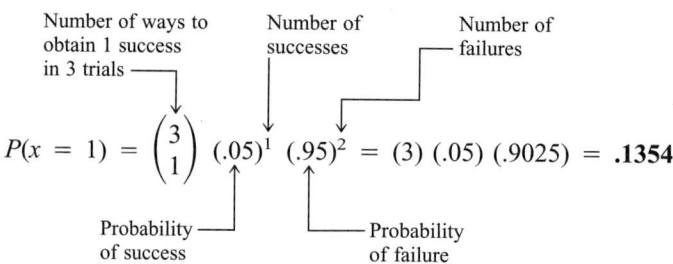

Note that the value of $\binom{3}{1}$ in the above formula can either be read from Table III of Appendix B or it can be computed as follows.

$$\binom{3}{1} = \frac{3!}{1! \, (3 - 1)!} = \frac{3 \cdot 2 \cdot 1}{1 \cdot 2 \cdot 1} = 3$$

In the above computation, $\binom{3}{1}$ gives the three ways to select one defective VCR in three selections. As listed earlier, these three ways to select one defective VCR are *DGG*, *GDG*, and *GGD*. The probability .1354 is slightly different from the earlier calculation (.1353) because of rounding.

*Calculating probability
by using the binomial
formula.*

EXAMPLE 5–18 At the Express House Delivery Service, providing high-quality service to its customers is the top priority of the management. The company guarantees a refund of all charges if a package it is delivering does not arrive at its destination by the specified time. It is known from past data that despite all efforts, 2% of the packages mailed through this company do not arrive at their destinations within the specified time. A corporation mailed 10 packages through Express House Delivery Service on Monday.

(a) Find the probability that exactly one of these 10 packages will not arrive at its destination within the specified time.

(b) Find the probability that at most one of these 10 packages will not arrive at its destination within the specified time.

Solution Let us call it a success if a package does not arrive at its destination within the specified time and a failure if it does arrive within the specified time. Then,

$$n = \text{total number of packages mailed} = 10$$

$$p = P(\text{success}) = .02$$

and $$q = P(\text{failure}) = 1 - .02 = .98$$

(a) For this part,

$$x = \text{number of successes} = 1$$

$$n - x = \text{number of failures} = 10 - 1 = 9$$

Substituting all values in the binomial formula, we obtain:

$$P(x = 1) = \binom{10}{1} (.02)^1 (.98)^9 = \frac{10!}{1! \, (10 - 1)!} (.02)^1 (.98)^9$$

$$= (10) (.02) (.83374776) = \textbf{.1667}$$

Thus, there is a .1667 probability that exactly one of the 10 packages mailed will not arrive at its destination within the specified time.[3]

(b) The probability that at most one of the 10 packages will not arrive at its destination within the specified time is given by the sum of the probabilities of $x = 0$ and $x = 1$. Thus,

$$P(x \le 1) = P(x = 0) + P(x = 1)$$

$$= \binom{10}{0} (.02)^0 (.98)^{10} + \binom{10}{1} (.02)^1 (.98)^9$$

$$= (1) (1) (.81707281) + (10) (.02) (.83374776)$$

$$= .8171 + .1667 = \textbf{.9838}$$

Thus, the probability that at most one of the 10 packages will not arrive at its destination within the specified time is .9838. ▬

*Constructing a binomial
probability distribution
and its graph.*

EXAMPLE 5–19 According to the U.S. Bureau of Labor Statistics, 56% of mothers with children under 6 years of age work outside their homes (*Newsweek*, June 4, 1990). A random sample of 3 mothers with children under 6 years of age is selected. Let x denote the number

---

[3]**Using a calculator to simplify the expression for $P(x = 1)$:** As explained in Chapter 4, you can evaluate $(.98)^9$ by using the y^x key on your calculator. To evaluate $(.98)^9$, first enter .98, then press y^x key, then enter 9, and finally press the " = " key. The calculator screen will display the answer.

of mothers in this sample who work outside their homes. Assuming that 56% of all mothers with children under 6 years of age work outside their homes, write the probability distribution of x and draw a graph of the probability distribution.

Solution   Let x denote the number of mothers in a sample of 3 who work outside their homes. Then $n - x$ is the number of mothers in the sample who do not work outside their homes. From the given information,

$$n = \text{total number of mothers in the sample} = 3$$

$$p = P(\text{a mother works outside her home}) = .56$$

and $q = P(\text{a mother does not work outside her home}) = 1 - .56 = .44$

The possible values that x can assume are 0, 1, 2, and 3. In other words, the number of mothers in a sample of 3 who work outside their homes can be 0, 1, 2, or 3. The probability of each of these four outcomes is calculated as follows.

If $x = 0$, then $n - x = 3$. Using the binomial formula, the probability of $x = 0$ is

$$P(x = 0) = \binom{3}{0} (.56)^0 (.44)^3 = (1)(1)(.085184) = \textbf{.0852}$$

Note that $\binom{3}{0}$ is equal to 1 by definition and $(.56)^0$ is equal to 1 because any number raised to the power zero is always 1.

If $x = 1$, then $n - x = 2$. The probability $P(x = 1)$ is

$$P(x = 1) = \binom{3}{1} (.56)^1 (.44)^2 = (3)(.56)(.1936) = \textbf{.3252}$$

If $x = 2$, then $n - x = 1$. The probability $P(x = 2)$ is

$$P(x = 2) = \binom{3}{2} (.56)^2 (.44)^1 = (3)(.3136)(.44) = \textbf{.4139}$$

If $x = 3$, then $n - x = 0$. The probability $P(x = 3)$ is

$$P(x = 3) = \binom{3}{3} (.56)^3 (.44)^0 = (1)(.175616)(1) = \textbf{.1756}$$

These probabilities are written in tabular form in Table 5.9. Figure 5.4 shows the graph of the probability distribution of Table 5.9.

Table 5.9 Probability Distribution of x

x	$P(x)$
0	.0852
1	.3252
2	.4139
3	.1756

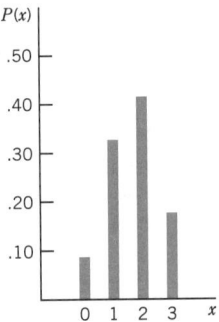

Figure 5.4 Graph of the probability distribution of *x*.

Case Study 5–3 shows how a probability calculation exposes gender bias in a jury selection system.

CASE STUDY 5-3 MISSING WOMEN

[This case study is based on] the 1968 trial of the pediatrician-author Dr. Benjamin Spock and others in the U.S. District Court in Boston for conspiracy to violate the Selective Service Act by encouraging resistance to the war in Vietnam. In that trial, the defense challenged the legality of the jury-selection method. Although more than half of all eligible jurors in Boston were women, there were no women on Dr. Spock's jury. Yet he, more than any defendant, would have wanted some because so many mothers have raised their children "according to Dr. Spock"; moreover, the opinion polls showed women in general to be more opposed to the Vietnam war than men.

The question was whether this total absence of women jurors was an accident of this particular jury or whether it had resulted from systematic discrimination. Statistical reasoning was to provide the answer.

In the Boston District Court, jurors are selected in three stages. The City Directory is used for the first stage; from it, the Clerk of the Court is supposed to select 300 names at random, that is, by a lotterylike method, and put a slip with each of these names into a box. The City Directory is renewed annually by censuslike household visits of the police, and it lists all adult individuals in the Boston area. The Directory lists slightly more women than men. The second selection stage occurs when a trial is about to begin. From the 300 names in the box, the names of 30 or more potential jurors are drawn. These people are ordered to appear in court on the morning of the trial. The subgroup of 30 or more is called a venire. In the third stage, the one that most of us think of as jury selection, 12 actual jurors are selected after interrogation by both the prosecutor and the defense counsel.

The average proportion of women drawn by the six judicial colleagues of the Spock trial judge was 29%, and furthermore, the averages of these six judges bunched closely around the group average. This suggests that the proportion of women among the names in the 300-name panels in the jury box was somewhere close to that 29% mark. But . . . the Spock judge's venires had consistently lower percentages of women, with an overall average of only 14.6% women, almost exactly half of that of his colleagues.

It is possible, of course, that the selection method used by the trial judge was the same as that of his colleagues. But what is the probability that a difference as large (or larger) as that between 14.6 and 29% could arise by chance? Statistical computation revealed the probability to be 1 in 1,000,000,000,000,000,000 that the ''luck of the draw'' would yield the distribution of women jurors obtained by the trial judge or a more extreme one. The conclusion, therefore, was virtually inescapable: the venires for the trial judge must have been drawn from the central jury lists in a fashion that somehow systematically reduced the proportion of women jurors.

Thus the proportion of women among the potential jurors twice suffered an improper reduction—first when the court clerk reduced their share from a majority in the City Directory to 29% in the jury lists and, second, when the judge managed to lower the 29% to his private average of 14.6%. In the Spock trial, only one potential woman juror came before the court, and she was easily eliminated in stage 3 by the prosecutor under his quota of peremptory challenges (for which he need not give any reasons).

For further discussion see H. Zeisel, ''Dr. Spock and the Case of the Vanishing Women Jurors.'' *University of Chicago Law Review*, 37, 1969, 1–18.

Source: From Judith M. Tanur et al., *Statistics: A Guide to the Unknown*, 2nd ed. Copyright © 1985 by Wadsworth, Inc. Reprinted by permission of Wadsworth & Brooks/Cole Advanced Books & Software, Pacific Grove, CA 93950.

Quiz

Fifty percent of the adult population in a large city are women. A court is to randomly select a jury of five adults from the population of all adults of this city. Using the binomial formula, find the probability that none of the five jury members is a woman.

5.6.3   USING THE TABLE OF BINOMIAL PROBABILITIES

The probabilities for a binomial experiment can also be read from Table IV, the table of binomial probabilities, given in Appendix B. That table lists the probabilities of x for $n = 1$ to $n = 25$ and for selected values of p. Example 5–20 illustrates how to read Table IV.

Using the binomial table to find probabilities and to construct the probability distribution and graph.

EXAMPLE 5–20   Based on data from a Peter D. Hart Research Associates' poll on consumer buying habits and attitudes, Peter Hart estimated that 5% of American shoppers are *status shoppers*, that is, shoppers who love to buy designer labels (*The American Way of Buying*, Dow Jones & Company, Inc., 1990). Assume that this result is true for the population of all American shoppers. A random sample of eight American shoppers is selected. Using Table IV of Appendix B, answer the following.

(a)   Find the probability that exactly three shoppers in this sample are status shoppers.
(b)   Find the probability that at most two shoppers in this sample are status shoppers.
(c)   Find the probability that at least three shoppers in this sample are status shoppers.
(d)   Find the probability that one to three shoppers in this sample are status shoppers.
(e)   Let x be the number of status shoppers in this sample. Write the probability distribution of x and draw a graph of this probability distribution.

Solution

(a)   To read the required probability from Table IV of Appendix B, first we determine the values of n, x, and p. These values are

$$n = \text{number of shoppers in the sample} = 8$$

$$x = \text{number of status shoppers in } 8 = 3$$

and $$p = P(\text{that a shopper is a status shopper}) = .05$$

Then we locate $n = 8$ in the column labeled n in Table IV. The relevant portion of Table IV with $n = 8$ is reproduced here as Table 5.10. Next, we locate 3 in the column for x in the portion of the table for $n = 8$ and locate $p = .05$ in the row for p at the top of the table. The entry at the intersection of the row for $x = 3$ and the column for $p = .05$ gives the probability of 3 successes in 8 trials when the probability of success is .05. From Table IV or Table 5.10,

$$P(x = 3) = \mathbf{.0054}$$

Table 5.10 Determining $P(x = 3)$ for $n = 8$ and $p = .05$

$p = .05$

					p		
n	x	**.05**	.10	.20	...	.95
$n = 8 \rightarrow$ 8	0	.6634	.4305	.1678	...	.0000
	1	.2793	.3826	.3355	...	.0000
	2	.0515	.1488	.2936	...	.0000
$x = 3 \rightarrow$	3	.0054 $\leftarrow$	.0331	.1468	...	.0000
	4	.0004	.0046	.0459	...	.0004
	5	.0000	.0004	.0092	...	.0054
	6	.0000	.0000	.0011	...	.0515
	7	.0000	.0000	.0001	...	.2793
	8	.0000	.0000	.0000	...	.6634

$P(x = 3) = .0054$

Using Table IV or Table 5.10, we write Table 5.11, which can be used to answer the remaining parts of this example.

Table 5.11 Portion of Table IV for $n = 8$ and $p = .05$

		p
n	x	**.05**
8	0	.6634
	1	.2793
	2	.0515
	3	.0054
	4	.0004
	5	.0000
	6	.0000
	7	.0000
	8	.0000

(b) The event that at most two shoppers in a sample of eight are status shoppers will occur if x is equal to 0, 1, or 2. Using Table IV or Table 5.11, the required probability is

$$P(\text{at most 2}) = P(0 \text{ or } 1 \text{ or } 2) = P(x = 0) + P(x = 1) + P(x = 2)$$

$$= .6634 + .2793 + .0515 = \mathbf{.9942}$$

(c) The probability that at least three shoppers in a sample of eight are status shoppers is given by the sum of the probabilities of 3, or 4, or 5, or 6, or 7, or 8 shoppers being status shoppers. Using Table IV of Appendix B or Table 5.11,

$$P(\text{at least } 3) = P(3 \text{ or } 4 \text{ or } 5 \text{ or } 6 \text{ or } 7 \text{ or } 8)$$

$$= P(x = 3) + P(x = 4) + P(x = 5)$$

$$+ P(x = 6) + P(x = 7) + P(x = 8)$$

$$= .0054 + .0004 + .0000 + .0000 + .0000 + .0000$$

$$= \textbf{.0058}$$

(d) The probability that one to three shoppers in a sample of eight are status shoppers is given by the sum of the probabilities that 1, or 2, or 3 shoppers are status shoppers. Using Table IV of Appendix B or Table 5.11,

$$P(1 \text{ to } 3) = P(x = 1) + P(x = 2) + P(x = 3)$$

$$= .2793 + .0515 + .0054 = \textbf{.3362}$$

(e) Using Table IV of Appendix B or Table 5.11, we list the probability distribution of x for $n = 8$ and $p = .05$ in Table 5.12. Figure 5.5 shows the graph of the probability distribution of x.

Table 5.12 Probability Distribution of x for $n = 8$ and $p = .05$

x	$P(x)$
0	.6634
1	.2793
2	.0515
3	.0054
4	.0004
5	.0000
6	.0000
7	.0000
8	.0000

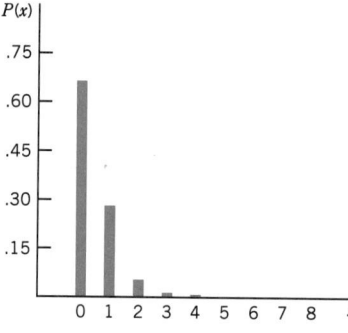

Figure 5.5 Graph of the probability distribution of x.

*Applying binomial
distribution to quality
control.*

EXAMPLE 5-21 Companies that receive large shipments of products use acceptance sampling methods to check if the products conform to a specified quality level. The quality control department selects a sample of n items from the shipment and observes the number of defective items in that sample. If the number of defective items is not more than a specified number k then the shipment is accepted, otherwise not. The sample of n items is called an acceptance sample.

Oaks Electronics Company makes food processors. The company buys motors for these food processors from New Haven Motors Inc. Oaks Electronics receives motors in shipments of 500 each. The quality control department at Oaks Electronics randomly selects 20 motors from each shipment and inspects them for being good or defective. If this sample contains more than two defective motors, the entire shipment is rejected. New Haven Motors promises that only 5% of its motors are defective.

(a) Using Table IV of Appendix B, find the probability that a given shipment of 500 motors received by Oaks Electronics will be accepted.

(b) Using Table IV of Appendix B, find the probability that a given shipment of 500 motors received by Oaks Electronics will not be accepted.

Solution Let p be the probability that a randomly selected motor from the shipment is defective. From the given information,

$$n = \text{number of motors in the sample} = 20$$

$$p = .05$$

The portion of Table IV corresponding to $n = 20$ and $p = .05$ is reproduced here as Table 5.13.

Table 5.13 Portion of Table IV for $n = 20$ and $p = .05$

n	x	p .05
20	0	.3585
	1	.3774
	2	.1887
	3	.0596
	4	.0133
	5	.0022
	6	.0003
	7	.0000
	⋮	⋮
	19	.0000
	20	.0000

(a) The shipment will be accepted if the sample of 20 motors contains 0, or 1, or 2 defective motors. Using Table 5.13, the required probability is

$$P(\text{the shipment is accepted}) = P(x \leq 2) = P(x = 0) + P(x = 1) + P(x = 2)$$

$$= .3585 + .3774 + .1887 = \textbf{.9246}$$

(b)   The shipment will not be accepted if the sample of 20 motors contains three or more defective motors. The two events ''shipment is accepted'' and ''shipment is not accepted'' are complementary events. Hence, using the complementary rule learned in Chapter 4, the required probability is

$$P(\text{the shipment is not accepted}) = 1 - P(\text{the shipment is accepted})$$

$$= 1 - .9246 = \textbf{.0754}$$

5.6.4   PROBABILITY OF SUCCESS AND THE SHAPE OF THE BINOMIAL DISTRIBUTION

For any number of trials n:

1.   The binomial probability distribution is symmetric if $p = .50$.
2.   The binomial probability distribution is skewed to the right if p is less than .50.
3.   The binomial probability distribution is skewed to the left if p is greater than .50.

These three cases are illustrated next with examples and graphs.

1.   Let $n = 4$ and $p = .50$. Using Table IV of Appendix B, we have written the probability distribution of x in Table 5.14 and plotted it in Figure 5.6. As we can observe from Table 5.14 and Figure 5.6, the probability distribution of x is symmetric.

Table 5.14   Probability Distribution of x for $n = 4$ and $p = .50$

x	$P(x)$
0	.0625
1	.2500
2	.3750
3	.2500
4	.0625

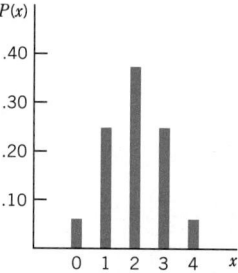

Figure 5.6   Graph of the probability distribution of Table 5.14.

2.   Let $n = 4$ and $p = .30$ (which is less than .50). Table 5.15, which is written by using Table IV of Appendix B, and the graph of the probability distribution in Figure 5.7 show that the probability distribution of x for $n = 4$ and $p = .30$ is skewed to the right.

Table 5.15 Probability Distribution of x for n = 4 and p = .30

x	P(x)
0	.2401
1	.4116
2	.2646
3	.0756
4	.0081

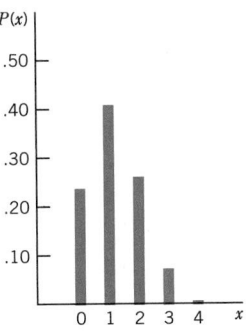

Figure 5.7 Graph of the probability distribution of Table 5.15.

3. Let n = 4 and p = .80 (which is greater than .50). Table 5.16, which is written by using Table IV of Appendix B, and the graph of the probability distribution in Figure 5.8 show that the probability distribution of x for n = 4 and p = .80 is skewed to the left.

Table 5.16 Probability Distribution of x for n = 4 and p = .80

x	P(x)
0	.0016
1	.0256
2	.1536
3	.4096
4	.4096

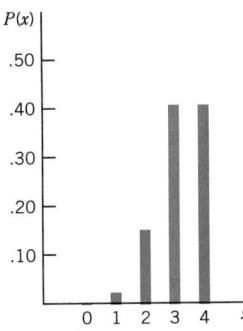

Figure 5.8 Graph of the probability distribution of Table 5.16.

5.6.5   MEAN AND STANDARD DEVIATION OF THE BINOMIAL DISTRIBUTION

Sections 5.3 and 5.4 explained how to compute the mean and standard deviation for a probability distribution of a discrete random variable. When a discrete random variable has a binomial distribution, the formulas learned in Sections 5.3 and 5.4 could still be used to compute its mean and standard deviation. However, it is more convenient and simpler to use the following formulas to find the mean and standard deviation in such cases.

MEAN AND STANDARD DEVIATION OF A BINOMIAL DISTRIBUTION

The mean and standard deviation for a binomial distribution are

$$\mu = np \quad \text{and} \quad \sigma = \sqrt{npq}$$

where n is the total number of trials, p is the probability of success, and q is the probability of failure.

Example 5–22 describes the calculation of the mean and standard deviation for a binomial distribution.

Mean and standard deviation of a binomial random variable.

EXAMPLE 5–22   In 1991, 18.6% of all the personal computers sold in the United States were manufactured by IBM (*Fortune*, September 21, 1992). Assume that of all the personal computers currently owned by all people in the United States, 18.6% are IBM computers. A sample of 25 computer owners is selected. Let x denote the number of IBM computers in this sample. Find the mean and standard deviation of the probability distribution of x.

Solution   This is a binomial experiment with 25 total trials. Each trial has two outcomes: A selected person owns either an IBM computer or a computer manufactured by another company. The probabilities p and q for these two outcomes are .186 and .814, respectively. Thus,

$$n = 25, \quad p = .186, \quad \text{and} \quad q = .814$$

Using the formulas for the mean and standard deviation of the binomial distribution, we obtain:

$$\mu = np = 25(.186) = \mathbf{4.65}$$
$$\sigma = \sqrt{npq} = \sqrt{(25)(.186)(.814)} = \mathbf{1.95}$$

Thus, the mean of the probability distribution of x is 4.65 and the standard deviation is 1.95. The value of the mean is what we expect to obtain, on average, per repetition of the experiment. In this example, if we take many samples of 25 personal computers owned by people, we expect that each sample will contain an average of 4.65 IBM personal computers with a standard deviation of 1.95.   ▬

EXERCISES

Concepts and Procedures

5.43   Briefly explain the following.

   a.   A binomial experiment   **b.**   A trial   **c.**   A binomial random variable

5.44 What are the parameters of the binomial probability distribution and what do they mean?

5.45 Which of the following are binomial experiments? Explain why.
a. Rolling a die many times and observing the number of spots
b. Rolling a die many times and observing whether the number obtained is even or odd
c. Selecting a few corporations from all corporations and observing whether or not each of them has a retirement plan for its employees when 65% of all corporations are known to have retirement plans

5.46 Which of the following are binomial experiments? Explain why.
a. Drawing 3 balls (with replacement) from a box that contains 10 balls, 6 of which are red and 4 blue, and observing the colors of the drawn balls
b. Drawing 3 balls (without replacement) from a box that contains 10 balls, 6 of which are red and 4 blue, and observing the colors of the drawn balls
c. Selecting a few households from New York City and observing whether or not they own stocks when it is known that 28% of all households in New York City own stocks

5.47 Let x be a discrete random variable that possesses a binomial distribution. Using the binomial formula, find the following probabilities.
a. $P(x = 5)$ for $n = 8$ and $p = .60$
b. $P(x = 3)$ for $n = 4$ and $p = .30$
c. $P(x = 2)$ for $n = 6$ and $p = .20$

Verify your answer by using Table IV of Appendix B.

5.48 Let x be a discrete random variable that possesses a binomial distribution. Using the binomial formula, find the following probabilities.
a. $P(x = 0)$ for $n = 5$ and $p = .10$
b. $P(x = 4)$ for $n = 7$ and $p = .80$
c. $P(x = 7)$ for $n = 10$ and $p = .40$

Verify your answer by using Table IV of Appendix B.

5.49 Let x be a discrete random variable that possesses a binomial distribution.
a. Using Table IV of Appendix B, write the probability distribution of x for $n = 7$ and $p = .30$ and graph it.
b. What are the mean and standard deviation of the probability distribution developed in part a?

5.50 Let x be a discrete random variable that possesses a binomial distribution.
a. Using Table IV of Appendix B, write the probability distribution of x for $n = 5$ and $p = .80$ and graph it.
b. What are the mean and standard deviation of the probability distribution developed in part a?

5.51 The binomial probability distribution is symmetric for $p = .50$, skewed to the right for $p < .50$, and skewed to the left for $p > .50$. Illustrate each of these three cases by writing a probability distribution table and by drawing a graph. Choose any values of n and p and use the table of binomial probabilities (Table IV of Appendix B) to write the probability distribution tables.

Applications

5.52 According to a survey conducted by Payment Systems Inc., 81% of consumers pay by cash at retail stores (*Business Week*, September 21, 1992).
a. Let x be a binomial random variable that denotes the number of customers in a sample of 10 who pay by cash when shopping at retail stores. What are the possible values that x can assume?
b. Find the probability that exactly 5 customers in a random sample of 10 will pay by cash when shopping at retail stores.

5.53 In a survey conducted for *Business Week* by Louis Harris & Associates, 1250 adults were asked, "Bearing in mind that they are paid out of taxes, do you think public school teachers are paid enough, too little, or too much?" Fifty-five percent of these adults said, "too little." Assume that this percentage is true for the population of all adults.

[handwritten margin notes:
$p = .05$
$n = 16$
$x \leq 5$
$p \ x = 0, 1, 2, 3, 4, 5$
$p(x=0) + p(x>1) +$
$p(x=2) + p(x=3) + p(x=4)$
$+ p(x=5)$ (use table p 832)*]*

a.   Let x be a binomial random variable that denotes the number of adults in a sample of 12 who think that public school teachers are paid too little. What are the possible values that x can assume?

b.   Find the probability that in a random sample of 12 adults, exactly 9 will say "too little" in response to the above question.

5.54   According to the U.S. Bureau of Labor Statistics, 5% of people working in the United States moonlight (i.e., they hold a second job). Using the binomial probabilities table (Table IV of Appendix B), find the probability that the number of workers who moonlight in a random sample of 16 workers is *[handwritten: (up to 16) $x \geq 1$ $p(x=0) = .44401$ $1 - .44401 = .5599$]*

a.   at most 5   b.   at least 1   c.   3 to 6

5.55   In a recent poll of chief executives of large companies conducted for *Fortune* magazine by the opinion research firm Clark Martire & Bartolomeo, 40% of CEOs said that it is *not likely at all* that their companies will have a female CEO within the next decade (*Fortune*, September 21, 1992). Assume that this result is true for the current population of CEOs of all large companies. Using the binomial probabilities table (Table IV of Appendix B), find the probability that in a random sample of 12 CEOs of large companies, the number who will hold this view is

a.   at least 8   b.   4 to 6   c.   at most 4

5.56   According to the U.S. Bureau of the Census, 66% of households headed by single women own autos. Find the probability that in a random sample of 9 households headed by single women

a.   exactly 8 own autos   b.   none own autos   c.   all 9 own autos   d.   exactly 3 own autos

5.57   According to the Motor Vehicle Manufacturers Association, 19% of cars in the United States were at least 12 years old in 1989 (*Business Week*, March 4, 1991). Assume that this result holds true for the current population of all cars in the United States. Find the probability that in a random sample of 10 cars

a.   exactly 4 are at least 12 years old

b.   exactly 2 are at least 12 years old

c.   none are at least 12 years old

d.   exactly 5 are at least 12 years old

5.58   According to Case Study 4–3 in Chapter 4, the probability that a baseball player will have no hits in 10 trips to the plate is .056, given that this player has a hitting percentage of 25%. Using the binomial formula, show that this probability is indeed .056.

5.59   As of May 1992, 32% of all sport-utility vehicles sold in the United States were manufactured by Ford Motor (*Fortune*, July 13, 1992). Assume that of all the sport-utility vehicles currently owned by all people in the United States, 32% are manufactured by Ford Motor. Find the probability that in a random sample of 12 sport-utility vehicles, exactly 6 are manufactured by Ford Motor.

5.60   The Food and Drug Administration approves 20% of drugs submitted for its approval (*U.S. News & World Report*, January 23, 1989). Using the binomial probabilities table, find the probability that in a random sample of 15 drugs submitted for the FDA's approval, the number of drugs that are approved is *[handwritten: $x < 4$ $x > 6$]*

a.   less than 4   b.   more than 6   c.   2 to 5

[handwritten margin notes:
$p = .20$
$n = 15$
$p(x=7) +$
$(x=8) +$
$(x=9) \ldots (x=15)$*]*

5.61   According to a survey, 30% of credit card holders pay off their balances in full each month (*The Wall Street Journal*, August 20, 1990). Assume that this result holds true for the current population of credit card holders. Using the binomial probabilities table, find the probability that in a random sample of 18 credit card holders, the number who pay off their balances in full each month is

a.   less than 3   b.   1 to 4   c.   more than 6

5.62   Johnson Electronics makes calculators. Consumer satisfaction is one of the top priorities of the company's management. The company guarantees a refund or a replacement for any calculator that malfunctions within two years from the date of purchase. It is known from past data that despite all efforts, 5% of the calculators manufactured by the company malfunction within a two-year period. The company mailed a package of 10 randomly selected calculators to a store.

a. Let x denote the number of calculators in this package of 10 that will be returned for refund or replacement within a two-year period. Using the binomial probabilities table, obtain the probability distribution of x and draw a graph of the probability distribution. Determine the mean and standard deviation of x.

b. Using the probability distribution of part a, find the probability that exactly 2 of the 10 calculators will be returned for refund or replacement within a 2-year period.

5.63 A fast food chain store conducted a taste survey before marketing a new hamburger. The results of the survey showed that 70% of the people who tried this hamburger liked it. Encouraged by this result, the company decided to market the new hamburger. Assume that 70% of all people like this hamburger. On a certain day, eight customers bought it.

a. Let x denote the number of customers in this sample of eight who will like this hamburger. Using the binomial probabilities table, obtain the probability distribution of x and draw a graph of the probability distribution. Determine the mean and standard deviation of x.

b. Using the probability distribution of part a, find the probability that exactly three of the eight customers will like this hamburger.

5.64 Hurbert Corporation makes font cartridges for laser printers that it sells to Alpha Electronics Inc. The cartridges are shipped to Alpha Electronics in lots of 400 each. The quality control department at Alpha Electronics randomly selects 15 cartridges from each shipment and inspects them for being good or defective. If this sample contains more than 2 defective cartridges, the entire shipment is rejected. Hurbert Corporation promises that only 5% of all cartridges are defective.

a. Using Table IV of Appendix B, find the probability that a given shipment of 400 cartridges received by Alpha Electronics will be accepted.

b. Using Table IV of Appendix B, find the probability that a given shipment of 400 cartridges received by Alpha Electronics will not be accepted.

5.7 THE HYPERGEOMETRIC PROBABILITY DISTRIBUTION

In Section 5.6, we learned that one of the conditions required to apply the binomial probability distribution is that the trials are independent so that the probabilities of the two outcomes (success and failure) remain constant. If the trials are not independent, we cannot apply the binomial probability distribution to find the probability of x successes in n trials. In such cases we replace the binomial by the **hypergeometric probability distribution**. Such a case occurs when a sample is drawn without replacement from a finite population.

As an example, suppose 20% of all auto parts manufactured at a company are defective. Four auto parts are selected at random. What is the probability that three of these four parts are good? Note that we are to find the probability that three of the four auto parts are good and one is defective. In this case, the population is very large and the probability of the first, second, third, and fourth auto parts being defective remains the same at .20. Similarly, the probability of any of the parts being good remains unchanged at .80. Consequently, we will apply the binomial probability distribution to find the probability of three good parts in four.

Now suppose this company shipped 25 auto parts to a dealer. Later on, it finds out that five of those parts were defective. By the time the company manager contacts the dealer, four auto parts from that shipment have already been sold. What is the probability that three of those four parts were good parts and one was defective? Here, because the four parts were selected without replacement from a small population, the probability of a part being good changes from the first selection to the second selection, to the third selection, and to the fourth selection. In this case we cannot apply the binomial probability distribution. In such instances, we use the hypergeometric probability distribution to find the required probability.

HYPERGEOMETRIC PROBABILITY DISTRIBUTION

Let

$$N = \text{total number of elements in the population}$$
$$r = \text{number of successes in the population}$$
$$N - r = \text{number of failures in the population}$$
$$n = \text{number of trials (sample size)}$$
$$x = \text{number of successes in } n \text{ trials}$$
$$n - x = \text{number of failures in } n \text{ trials}$$

The probability of x successes in n trials is given by

$$P(x) = \frac{\binom{r}{x}\binom{N-r}{n-x}}{\binom{N}{n}}$$

Examples 5–23 and 5–24 provide applications of the hypergeometric probability distribution.

Calculating probability by using hypergeometric distribution formula.

EXAMPLE 5–23   Brown Manufacturing Company makes auto parts that are sold to auto dealers. Last week the company shipped 25 auto parts to a dealer. Later on, it found out that five of those parts were defective. By the time the company manager contacted the dealer, four auto parts from that shipment had already been sold. What is the probability that three of those four parts were good parts and one was defective?

Solution   Let a good part be called a success and a defective part be called a failure. From the given information,

$$N = \text{total number of elements in the population} = 25$$
$$r = \text{number of successes in the population} = 20$$
$$N - r = \text{number of failures in the population} = 5$$
$$n = \text{number of trials (sample size)} = 4$$
$$x = \text{number of successes in four trials} = 3$$
$$n - x = \text{number of failures in four trials} = 1$$

Using the hypergeometric formula, the required probability is calculated as follows.

$$P(x = 3) = \frac{\binom{r}{x}\binom{N-r}{n-x}}{\binom{N}{n}} = \frac{\binom{20}{3}\binom{5}{1}}{\binom{25}{4}} = \frac{\dfrac{20!}{3!\,(20-3)!} \cdot \dfrac{5!}{1!\,(5-1)!}}{\dfrac{25!}{4!\,(25-4)!}}$$

$$= \frac{(1,140)\,(5)}{12,650} = \mathbf{.4506}$$

Thus, the probability that three of the four parts sold are good and one is defective is .4506.

In the above calculations, the values of combinations can either be calculated using the formula learned in Section 5.5.2 (as done here) or by using Table III of Appendix B. ■

Calculating probability by using hypergeometric distribution formula.

EXAMPLE 5–24 Dawn Corporation has 12 employees who hold managerial positions. Of them, 7 are female and 5 are male. The company is planning to send 3 of these 12 managers to a conference. If 3 managers are randomly selected out of 12,

(a) find the probability that all 3 of them are female

(b) find the probability that at most 1 of them is a female

Solution Let the selection of a female be called a success and the selection of a male be called a failure.

(a) From the given information,

$$N = \text{total number of managers in the population} = 12$$

$$r = \text{number of successes (females) in the population} = 7$$

$$N - r = \text{number of failures (males) in the population} = 5$$

$$n = \text{number of selections (sample size)} = 3$$

$$x = \text{number of successes (females) in three selections} = 3$$

$$n - x = \text{number of failures (males) in three selections} = 0$$

Using the hypergeometric formula, the required probability is calculated as follows.

$$P(x = 3) = \frac{\binom{r}{x}\binom{N - r}{n - x}}{\binom{N}{n}} = \frac{\binom{7}{3}\binom{5}{0}}{\binom{12}{3}} = \frac{(35)\,(1)}{220} = \mathbf{.1591}$$

Thus, the probability that all three of the managers selected are female is .1591.

(b) The probability that at most one of them is a female is given by the sum of the probabilities that either none or one of the selected managers is a female.

To find the probability that none of the selected managers is a female,

$$N = \text{total number of managers in the population} = 12$$

$$r = \text{number of successes (females) in the population} = 7$$

$$N - r = \text{number of failures (males) in the population} = 5$$

$$n = \text{number of selections (sample size)} = 3$$

$$x = \text{number of successes (females) in three selections} = 0$$

$$n - x = \text{number of failures (males) in three selections} = 3$$

Using the hypergeometric formula, the required probability is calculated as follows.

$$P(x = 0) = \frac{\binom{r}{x}\binom{N-r}{n-x}}{\binom{N}{n}} = \frac{\binom{7}{0}\binom{5}{3}}{\binom{12}{3}} = \frac{(1)(10)}{220} = .0455$$

To find the probability that one of the selected managers is a female,

$$N = \text{total number of managers in the population} = 12$$
$$r = \text{number of successes (females) in the population} = 7$$
$$N - r = \text{number of failures (males) in the population} = 5$$
$$n = \text{number of selections (sample size)} = 3$$
$$x = \text{number of successes (females) in three selections} = 1$$
$$n - x = \text{number of failures (males) in three selections} = 2$$

Using the hypergeometric formula, the required probability is calculated as follows.

$$P(x = 1) = \frac{\binom{r}{x}\binom{N-r}{n-x}}{\binom{N}{n}} = \frac{\binom{7}{1}\binom{5}{2}}{\binom{12}{3}} = \frac{(7)(10)}{220} = .3182$$

The probability that at most one of the three managers selected is a female is

$$P(x \leq 1) = P(x = 0) + P(x = 1) = .0455 + .3182 = \mathbf{.3637}$$

EXERCISES

Concepts and Procedures

5.65 Explain the hypergeometric probability distribution. Under what conditions is this probability distribution applied to find the probability of a discrete random variable x? Give one example of the application of the hypergeometric probability distribution.

5.66 Let $N = 8$, $r = 3$, and $n = 4$. Using the hypergeometric probability distribution formula, find
a. $P(x = 2)$ b. $P(x = 0)$ c. $P(x \leq 1)$

5.67 Let $N = 14$, $r = 6$, and $n = 5$. Using the hypergeometric probability distribution formula, find
a. $P(x = 4)$ b. $P(x = 5)$ c. $P(x \leq 1)$

5.68 Let $N = 11$, $r = 4$, and $n = 4$. Using the hypergeometric probability distribution formula, find
a. $P(x = 2)$ b. $P(x = 4)$ c. $P(x \leq 1)$

5.69 Let $N = 16$, $r = 10$, and $n = 5$. Using the hypergeometric probability distribution formula, find
a. $P(x = 5)$ b. $P(x = 0)$ c. $P(x \leq 1)$

$$\frac{6\cdot5\cdot4\cdot3\cdot2\cdot1}{3\cdot2\cdot1}$$

$n=12$ customers $\dfrac{6\cdot5\cdot4}{3\cdot2\cdot1} = \dfrac{120}{6} = 20$

$p=.07$ default

$q=.93$ not default $\dfrac{11\cdot10\cdot9\cdot8\cdot7\cdot6\cdot5}{7\cdot6\cdot5\cdot4\cdot3\cdot2\cdot1} = 330$

$x=3$

$\dfrac{12}{5}$ $\dfrac{13\cdot11\cdot12\cdot3\cdot f}{5\cdot4\cdot3\cdot2\cdot1}$ 792

.01393

n
p^x q^{n-x}
x

$12 \binom{.07^3}{.93^9}$
3

$230 \cdot .000343 \cdot .5204 = .0393$

$12\binom{.07^1}{.93^{11}}$ $12\binom{.07^0}{.93^{12}}$
1 0

$.1\cdot1$
.4485
.5817

$12(.07)(.4501) = .3780$
0

n	7	$7n$	$7\binom{n}{t}$
p	.30	.30p	$.30p^2$

5.59 32% SPORT VEHICLES SOLD
OF ALL 32% MANUFACTURED BY FORD
FIND 12 SAMPLES FIND 6 BY FORD

5.45 A. NO,
B. YES,
C. YES,

5.47

$$\frac{8!}{P} = \frac{8!}{40\,320} = \frac{40\,320}{.60 \times 7.4!}$$

$$\frac{}{.60} \quad \frac{}{.60} = \frac{}{.60 \times 7.4!} = .2787$$

5 / 10 0

5.49

$X = x$	$P(X = x)$
0	
1	
2	
3	
4	
5	
6	
7	

Applications

5.70 An Internal Revenue Service inspector is to select 3 corporations from a list of 15 for tax audit purposes. Of the 15 corporations, 6 earned profits and 9 incurred losses during the year for which the tax returns are to be audited. If the IRS inspector decides to select 3 corporations randomly, find the probability that the number of corporations in these 3 that incurred losses during the year for which the tax returns are to be audited is

    a. exactly 2   **b.** at most 1   **c.** none

5.71 GESCO Insurance Company has prepared a final list of eight candidates for two positions. Of the eight candidates, five are business majors and three are economics majors. If the company manager decides to select randomly two candidates from this list, find the probability that

    a. both candidates are business majors
    b. neither of the two candidates is a business major
    c. at most one of the candidates is a business major

5.72 The business department at a university has 18 faculty members. Of them, 11 are in favor of the proposition that all MBA students at the university be required to take a course in ethics and 7 are against this proposition. The department has agreed to form a committee of 5 faculty members who will survey the university administrators, MBA students, and the CEOs of major corporations on this issue. If 5 faculty members are randomly selected from 18, find the probability that the number of faculty members in this sample who are in favor of the proposition is

    a. exactly 2   **b.** exactly 4   **c.** exactly 5

5.73 Bender Electronics buys keyboards for its computers from another company. The keyboards are received in shipments of 100 boxes, each box containing 20 keyboards. The quality control department at Bender Electronics first randomly selects one box from each shipment and then randomly selects five keyboards from that box. The shipment is accepted if not more than one of the five keyboards is defective. The quality control inspector at Bender Electronics selected a box from a recently received shipment of keyboards. Unknown to the inspector, this box contains six defective keyboards.

    a. What is the probability that this shipment will be accepted?
    b. What is the probability that this shipment will not be accepted?

5.8 THE POISSON PROBABILITY DISTRIBUTION

The **Poisson probability distribution**, named after the French mathematician Simeon D. Poisson, is another important probability distribution of a discrete random variable that has a large number of applications. Suppose a washing machine in a laundromat breaks down an average of three times a month. We may want to find the probability of exactly two breakdowns during the next month. This is an example of a Poisson probability distribution problem. Each breakdown is called an *occurrence* in Poisson probability distribution terminology. The Poisson probability distribution is applied to experiments with random and independent occurrences. The occurrences are random in the sense that they do not follow any pattern and, hence, they are unpredictable. Independence of occurrences means that one occurrence (or nonoccurrence) of an event does not influence the successive occurrences or nonoccurrences of that event. The occurrences are always considered with respect to an interval. In the example of the washing machine, the interval represents one month. The interval may be a time interval, a space interval, or a volume interval. The actual number of occurrences within an interval is random and independent. If the average number of occurrences for a given interval is known, then by using the Poisson probability distribution we can compute the probability of a certain number of occurrences x in that interval. Note that the number of actual occurrences in an interval is denoted by x.

CONDITIONS TO APPLY POISSON PROBABILITY DISTRIBUTION

The following three conditions must be satisfied to apply the Poisson probability distribution.

1. x is a discrete random variable.
2. The occurrences are random.
3. The occurrences are independent.

The following are a few examples of discrete random variables for which the occurrences are random and independent. Hence, these are examples to which the Poisson probability distribution can be applied.

1. Consider the number of customers arriving at the Eagleville branch of the United National Bank during a one-hour interval. In this example, an occurrence is the arrival of a customer at this bank, the interval is one hour (an interval of time), and the occurrences are random. The total number of customers who may arrive at this bank during a one-hour interval may be 0, 1, 2, 3, 4, ... The independence of occurrences in this example means that the customers arrive individually and the arrival of any two (or more) customers is not related.

2. Consider the number of defective items in the next 100 items manufactured on a machine. In this case, the interval is a volume interval (100 items). The occurrences (number of defective items) are random because there may be 0, 1, 2, 3, . . . , 100 defective items in 100 items. We can assume the occurrence of defective items to be independent of one another.

3. Consider the number of defects in a 5-foot-long iron rod. The interval, in this example, is a space interval (5 feet). The occurrences (defects) are random because there may be any number of defects in a 5-foot iron rod. We can assume that these defects are independent of one another.

The following examples also qualify for the application of the Poisson probability distribution.

1. The number of accidents that occur at a company during a one-month period.
2. The number of customers coming to a grocery store during a one-hour interval.
3. The number of television sets sold at a department store during a given week.

On the other hand, arrival of patients at a physician's office will be nonrandom if the patients have to make appointments to see the doctor. The arrival of commercial airplanes at an airport is nonrandom because all planes are scheduled to arrive at certain times, and airport authorities know the exact number of arrivals for any period (although this number may change slightly because of late or early arrivals and cancellations).

In the Poisson probability distribution terminology, the average number of occurrences in an interval is denoted by λ (Greek letter lambda). The actual number of occurrences in that interval is denoted by x. Then, using the Poisson probability distribution, we find the probability of x occurrences during an interval given the mean occurrences are λ during that interval.

> **POISSON PROBABILITY DISTRIBUTION FORMULA**
>
> According to the Poisson probability distribution, the probability of x occurrences in an interval is
>
> $$P(x) = \frac{\lambda^x e^{-\lambda}}{x!}$$
>
> where λ (pronounced *lambda*) is the mean number of occurrences in that interval and the value of e is approximately 2.71828.

The mean number of occurrences in an interval, denoted by λ, is called the *parameter of the Poisson probability distribution* or the **Poisson parameter**. As is obvious from the Poisson probability distribution formula, we need to know only the value of λ to compute the probability of any given value of x. We can read the value of $e^{-\lambda}$ for a given λ from Table V of Appendix B. Examples 5–25 through 5–27 illustrate the use of the Poisson probability distribution formula.

Using Poisson formula: x equals a specific value.

EXAMPLE 5–25 The automatic teller machine (ATM) installed outside Mansfield Savings and Loan is used on average by five customers per hour. The bank closed this ATM for one hour for repairs. What is the probability that during that hour eight customers came to use this ATM?

Solution Let λ be the mean number of customers who use this ATM per hour. Then, $\lambda = 5$. Let x be the number of customers who came to use this ATM during the one-hour period when it was closed for repairs. We are to find the probability of $x = 8$. Substituting all the values in the Poisson probability distribution formula, we obtain:

$$P(x = 8) = \frac{\lambda^x e^{-\lambda}}{x!} = \frac{(5)^8 e^{-5}}{8!} = \frac{(390625)(.006738)}{40320} = .0653$$

Thus, the probability is .0653 that eight customers came to use this ATM during the one-hour period when it was closed for repairs.

In these calculations, we can find the value of 8! from Table II and the value of e^{-5} from Table V of Appendix B.[4]

To find the value of e^{-5} from Table V of Appendix B, we locate 5 in the column for λ and read the value across from 5 in the column for $e^{-\lambda}$. This value is .006738. ■

Calculating probabilities by using the Poisson formula.

EXAMPLE 5–26 A washing machine in a laundromat breaks down an average of three times per month. Using the Poisson probability distribution formula, find the probability that during the next month this machine will have

(a) exactly two breakdowns (b) at most one breakdown

Solution Let λ be the mean number of breakdowns per month and x be the actual number

---

[4]**Using a calculator to simplify the expression for $P(x = 8)$:** In the expression for $P(x = 8)$, you can evaluate $(5)^8$ and 8! by using the y^x and $n!$ keys respectively. If your calculator has the e^x key, you can evaluate e^{-5} as follows. First enter 5, then press the $+/-$ key. Finally press the e^x key.

of breakdowns observed during the next month for this machine. Then,

$$\lambda = 3$$

(a) The probability that exactly two breakdowns will be observed during the next month is

$$P(x = 2) = \frac{(3)^2 \, e^{-3}}{2!} = \frac{(9) \, (.049787)}{2} = \textbf{.2240}$$

(b) The probability that at most one breakdown will be observed during the next month is given by the sum of the probabilities of zero and one breakdown. Thus,

$$P(\text{at most 1 breakdown}) = P(0 \text{ or } 1 \text{ breakdown}) = P(x = 0) + P(x = 1)$$

$$= \frac{(3)^0 \, e^{-3}}{0!} + \frac{(3)^1 \, e^{-3}}{1!}$$

$$= \frac{(1) \, (.049787)}{1} + \frac{(3) \, (.049787)}{1}$$

$$= .0498 + .1494 = \textbf{.1992}$$

☞ *Remember*   One important point to remember is that *the intervals for λ and x must be equal.* If they are not, the mean λ should be redefined to make them equal. Example 5–27 illustrates this point.

Calculating probability by using the Poisson formula.

EXAMPLE 5–27   Cynthia's Mail Order Company provides free examination of its product for seven days. If not completely satisfied, a buyer can return the product within that period and get a full refund. According to past records of the company, an average of 2 of every 10 products sold by this company are returned for a refund. Using the Poisson probability distribution formula, find the probability that exactly 5 of the 20 products sold by this company on a given day will be returned for a refund.

Solution   Let x denote the number of products in 20 that will be returned for a refund. We are to find $P(x = 5)$. The given mean is defined per 10 products but x is defined for 20 products. As a result, we should first find the mean for 20 products. Because on average 2 out of 10 products are returned, the mean number of products returned out of 20 will be 4. Thus, $\lambda = 4$. Substituting $x = 5$ and $\lambda = 4$ in the Poisson probability distribution formula, we obtain:

$$P(x = 5) = \frac{\lambda^x \, e^{-\lambda}}{x!} = \frac{(4)^5 \, e^{-4}}{5!} = \frac{(1024) \, (.018316)}{120} = \textbf{.1563}$$

Thus, the probability is .1563 that exactly 5 products out of 20 sold on a given day will be returned.

5.8.1   USING THE TABLE OF POISSON PROBABILITIES

The probabilities for a Poisson distribution can also be read from Table VI, the table of Poisson probabilities, given in Appendix B. The following example describes how to read that table.

Using the table of Poisson probabilities.

EXAMPLE 5–28 On average, two new accounts are opened per day at an Imperial Savings Bank branch. Using Table VI of Appendix B, find the probability that on a given day the number of new accounts opened at this bank will be

(a) exactly 6 (b) at most 3 (c) at least 7

Solution Let

λ = the mean number of new accounts opened per day at this bank

x = the number of new accounts opened at this bank on a given day

(a) The values of λ and x are

$$\lambda = 2 \quad \text{and} \quad x = 6$$

In Table VI of Appendix B, we first locate the column that corresponds to $\lambda = 2$. In this column, we then read the value for $x = 6$. The relevant portion of that table is shown here as Table 5.17. The probability that exactly 6 new accounts will be opened on a given day is .0120. Therefore,

$$P(x = 6) = \mathbf{.0120}$$

Table 5.17 Portion of Table VI for $\lambda = 2.0$

			λ	
x	1.1	1.2	...	2.0 ← $\lambda = 2.0$
0				.1353
1				.2707
2				.2707
3				.1804
4				.0902
5				.0361
$x = 6$ → 6				.0120 ← $P(x = 6)$
7				.0034
8				.0009
9				.0002

Actually, Table 5.17 gives the probability distribution of x for $\lambda = 2.0$. Note that the sum of the 10 probabilities given in Table 5.17 is .9999 and not 1.0. This is so for two reasons. First, these probabilities are rounded to four decimal places. Second, on a given day more than 9 new accounts might be opened at this bank. However, the probabilities of 10, 11, 12, . . . new accounts are very small and they are not listed in the table.

(b) The probability that at most three new accounts are opened on a given day is obtained by adding the probabilities of 0, 1, 2, and 3 new accounts. Thus, using Table VI of Appendix B or Table 5.17, we obtain:

$$P(\text{at most } 3) = P(x = 0) + P(x = 1) + P(x = 2) + P(x = 3)$$
$$= .1353 + .2707 + .2707 + .1804 = \mathbf{.8571}$$

(c) The probability that at least 7 new accounts are opened on a given day is obtained by adding the probabilities of 7, 8, or 9 new accounts. Note that 9 is the last value of x for $\lambda = 2.0$ in Table VI of Appendix B or Table 5.17. Hence, 9 is the last value of x whose probability is included in the sum. However, this does not mean that on a given day more than 9 new accounts cannot be opened. It simply means that the probability of 10 or more accounts is close to zero.

$$P(\text{at least } 7) = P(x = 7) + P(x = 8) + P(x = 9)$$

$$= .0034 + .0009 + .0002 = \mathbf{.0045}$$ ▬

Constructing a Poisson probability distribution and graphing it.

EXAMPLE 5–29   An auto salesperson sells an average of .9 cars per day. Let x be the number of cars sold by this salesperson on any given day. Using the Poisson probability distribution table, write the probability distribution of x. Draw a graph of the probability distribution.

Solution   Let λ be the mean number of cars sold per day by this salesperson. Hence, $\lambda = .9$. Using the portion of Table VI corresponding to $\lambda = .9$, we write the probability distribution of x in Table 5.18. Figure 5.9 shows the line graph for the probability distribution of Table 5.18.

Table 5.18   Probability Distribution of x for $\lambda = .9$

x	$P(x)$
0	.4066
1	.3659
2	.1647
3	.0494
4	.0111
5	.0020
6	.0003

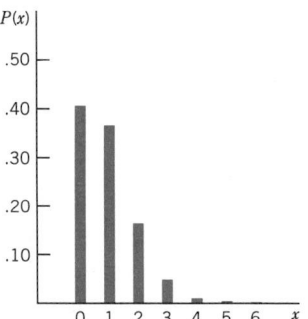

Figure 5.9   Graph of the probability distribution of Table 5.18.

Note that 6 is the largest value of x for $\lambda = .9$ listed in Table VI for which the probability is greater than zero. However, this does not mean that this salesperson cannot sell more than six cars on a given day. What this means is that the probability of selling 7 cars (or any higher number) is very small. Actually, the probability of $x = 7$ for $\lambda = .9$ calculated by using the Poisson formula is .000039. When rounded to four decimal places, this probability is .0000, as listed in Table VI. ▬

5.8.2   MEAN AND STANDARD DEVIATION OF THE POISSON PROBABILITY DISTRIBUTION

For the Poisson probability distribution, the mean and variance both are equal to λ and the standard deviation is equal to $\sqrt{\lambda}$. That is, for the Poisson probability distribution

$$\mu = \lambda$$
$$\sigma^2 = \lambda$$
$$\sigma = \sqrt{\lambda}$$

For Example 5–29, $\lambda = .9$. Therefore, for the probability distribution of x listed in Table 5.18, the mean, variance, and standard deviation are

$$\mu = \lambda = .9$$
$$\sigma^2 = \lambda = .9$$
$$\sigma = \sqrt{\lambda} = \sqrt{.9} = .949$$

EXERCISES

Concepts and Procedures

5.74 What are the conditions that must be satisfied to apply the Poisson probability distribution?

5.75 What is the parameter of the Poisson probability distribution, and what does it mean?

5.76 Using the Poisson formula, find the following probabilities.

    a. $P(x \leq 1)$ for $\lambda = 4$ **b.** $P(x = 2)$ for $\lambda = 2.2$

Verify these probabilities using Table VI of Appendix B.

5.77 Using the Poisson formula, find the following probabilities.

    a. $P(x < 2)$ for $\lambda = 3$ **b.** $P(x = 8)$ for $\lambda = 5.3$

Verify these probabilities using Table VI of Appendix B.

5.78 Let x be a Poisson random variable. Using the Poisson probabilities table, write the probability distribution of x for each of the following. Find the mean, variance, and standard deviation for each of these probability distributions. Draw a graph for each of these probability distributions.

    a. $\lambda = 1.3$ **b.** $\lambda = 2.1$

5.79 Let x be a Poisson random variable. Using the Poisson probabilities table, write the probability distribution of x for each of the following. Find the mean, variance, and standard deviation for each of these probability distributions. Draw a graph for each of these probability distributions.

    a. $\lambda = .6$ **b.** $\lambda = 1.8$

Applications

5.80 A mail-order company receives an average of 7.4 orders per day. Find the probability that it will receive exactly 10 orders on a certain day. Use the Poisson formula.

5.81 A commuter airline receives an average of 9.7 complaints per day from its passengers. Using the Poisson formula, find the probability that on a certain day this airline will receive exactly 7 complaints.

5.82 An average of 8.2 crimes are reported per day to police in a city. Find the probability that exactly 3 crimes will be reported to police on a certain day in this city. Use the Poisson formula.

5.83 On average, 3 households in 10 own answering machines. Using the Poisson formula, find the probability that in a random sample of 10 households, exactly 5 will own answering machines. (This is actually a binomial distribution problem. We are solving it by using the Poisson distribution. This is called *approximating binomial probability by using the Poisson distribution.*)

5.84 An average of 2.8 employees of a telephone company are absent per day.

    a. Using the Poisson formula, find the probability that at most one employee will be absent on a given day at this company.

  b. Using the Poisson probabilities table, find the probability that on a given day the number of employees who will be absent at this company is

    i. 1 to 5 ii. at least 7 iii. at most 3

5.85 A large proportion of small businesses in the United States fail during the first few years of operation. On average, 1.3 businesses file for bankruptcy per day in a large city.

  a. Using the Poisson formula, find the probability that 3 businesses will file for bankruptcy on a given day in this city.

  b. Using the Poisson probabilities table, find the probability that the number of businesses that will file for bankruptcy on a given day in this city is

    i. 2 to 3 ii. more than 3 iii. less than 3

5.86 One of the items made by Custom Clothiers Corporation is sweatshirts. The company mails these shirts to its retail customers in boxes of 200 sweatshirts each. Based on returns from the retailers, the quality control department at Custom Clothiers has estimated that each such box contains an average of 4 defective shirts.

  a. Using the Poisson formula, find the probability that a randomly selected box of 200 sweatshirts will contain exactly 7 defective shirts.

  b. Using the Poisson probabilities table, find the probability that the number of defective sweatshirts in a box of 200 will be

    i. 1 to 3 ii. at least 2 iii. at most 3

5.87 Despite all efforts by the quality control department, the fabric made at Benton Corporation always contains a few defects. A certain type of fabric made at this corporation contains an average of .4 defects per 500 yards.

  a. Using the Poisson formula, find the probability that a given piece of 500 yards of this fabric will contain exactly one defect.

  b. Using the Poisson probabilities table, find the probability that the number of defects in a given 500-yard piece of this fabric will be

    i. 2 to 4 ii. more than 3 iii. less than 2

5.88 The reception office at Tom's Building Corporation receives an average of 4.9 phone calls per half hour.

  a. Using the Poisson formula, find the probability that exactly 6 phone calls will be received at this office during a certain hour.

  b. Using the Poisson probabilities table, find the probability that the number of phone calls received at this office during a certain hour will be

    i. less than 8 ii. more than 12

5.89 An average of 4.5 customers come to Columbia Savings and Loan per half hour.

  a. Using the Poisson formula, find the probability that exactly 2 customers will come to this savings and loan during a given hour.

  b. Using the Poisson probabilities table, find the probability that during a given hour, the number of customers who will come to this savings and loan is

    i. at most 2 ii. at least 10

5.90 An insurance salesperson sells an average of 1.2 insurance policies per day.

  a. Using the Poisson formula, find the probability that this salesperson will sell no insurance policy on a certain day.

  b. Let x denote the number of insurance policies that this salesperson will sell on a given day. Using the Poisson probabilities table, write the probability distribution of x.

  c. Find the mean, variance, and standard deviation of the probability distribution developed in part b.

5.91 An average of .6 accidents occur per month at a large company.

  a. Using the Poisson formula, find the probability that no accident will occur at this company during a given month.

b. Let x denote the number of accidents that will occur at this company during a given month. Using the Poisson probabilities table, write the probability distribution of x.

c. Find the mean, variance, and standard deviation of the probability distribution developed in part b.

GLOSSARY

Bernoulli trial   One repetition of a binomial experiment. Also called a *trial*.

Binomial experiment   An experiment that contains n identical trials such that each of these n trials has only two possible outcomes, the probabilities of these two outcomes remain constant for each trial, and the trials are independent.

Binomial parameters   The total trials n and the probability of success p for the binomial probability distribution.

Binomial probability distribution   The probability distribution that gives the probability of x successes in n trials when the probability of success is p for each trial of a binomial experiment.

Combinations   The number of ways x elements can be selected from n elements.

Continuous random variable   A random variable that can assume any value in one or more intervals.

Discrete random variable   A random variable whose values are countable.

Factorial   Denoted by the symbol ''!.'' The product of all integers from a given number to 1. For example, ''$n!$'' (read as *n factorial*) represents the product of all integers from n to 1.

Hypergeometric probability distribution   The probability distribution that is applied to determine the probability of x successes in n trials when the trials are not independent.

Mean of a discrete random variable   The mean of a discrete random variable x is the value that is expected to occur per repetition, on average, if an experiment is performed a large number of times. The mean of a discrete random variable is also called its *expected value*.

Poisson parameter   The average occurrences, denoted by λ, during an interval for a Poisson probability distribution.

Poisson probability distribution   The probability distribution that gives the probability of x occurrences in an interval when the average occurrences in that interval are λ.

Probability distribution of a discrete random variable   A list of all the possible values that a discrete random variable can assume and their corresponding probabilities.

Random variable   A variable, denoted by x, whose value is determined by the outcome of a random experiment. Also called a *chance variable*.

Standard deviation of a discrete random variable   A measure of spread for the probability distribution of a discrete random variable.

KEY FORMULAS

1. **Mean of a discrete random variable x**

$$\mu = \Sigma x P(x)$$

The mean of a discrete random variable x is also called its expected value and is denoted by $E(x)$.

2. **Standard deviation of a discrete random variable x**

$$\sigma = \sqrt{\Sigma x^2 P(x) - \mu^2}$$

3. **Factorials**

$$n! = n\,(n-1)\,(n-2)\ldots 3 \cdot 2 \cdot 1$$

4. **Number of combinations of n items selected x at a time**

$$\binom{n}{x} = \frac{n!}{x!\,(n-x)!}$$

5. **Binomial probability formula**

$$P(x) = \binom{n}{x} p^x\, q^{n-x}$$

6. **Mean of the binomial probability distribution**

$$\mu = np$$

7. **Standard deviation of the binomial probability distribution**

$$\sigma = \sqrt{npq}$$

8. **Hypergeometric probability formula**

$$P(x) = \frac{\binom{r}{x}\binom{N-r}{n-x}}{\binom{N}{n}}$$

where

N = total number of elements in the population

r = number of successes in the population

$N - r$ = number of failures in the population

n = number of trials (sample size)

x = number of successes in n trials

$n - x$ = number of failures in n trials

9. **Poisson probability formula**

$$P(x) = \frac{\lambda^x\, e^{-\lambda}}{x!}$$

10. **Mean of the Poisson probability distribution**

$$\mu = \lambda$$

11. **Variance of the Poisson probability distribution**

$$\sigma^2 = \lambda$$

12. **Standard deviation of the Poisson probability distribution**

$$\sigma = \sqrt{\lambda}$$

SUPPLEMENTARY EXERCISES

5.92 The following table lists the probability distribution of the number of times the mainframe computer at a university breaks down per month.

Number of breakdowns	0	1	2	3
Probability	.55	.33	.08	.04

Let x denote the number of times the mainframe computer at this university will break down during any given month. Determine the following probabilities.

    a. $P(x = 2)$ **b.** $P(x \leq 1)$ **c.** $P(x \geq 1)$ **d.** $P(x < 3)$

5.93 The following table lists the probability distribution of the number of phone calls received per 10-minute period at an office of the Better Business Bureau.

Number of phone calls	0	1	2	3	4
Probability	.12	.26	.34	.18	.10

Let x denote the number of phone calls received during a certain 10-minute period at this office of the Better Business Bureau. Find the following probabilities.

    a. $P(x = 1)$ **b.** $P(x < 2)$ **c.** $P(x > 2)$ **d.** $P(1 \leq x \leq 3)$

5.94 Refer to Exercise 5.92. Find the mean and standard deviation of the probability distribution of the number of times the mainframe computer at this university breaks down per month.

5.95 Refer to Exercise 5.93. Find the mean and standard deviation of the probability distribution of the number of phone calls received during a certain 10-minute period at this office of the Better Business Bureau.

5.96 Let x be the number of cars that a randomly selected auto mechanic repairs on a given day. The following table lists the probability distribution of x.

x	2	3	4	5	6
$P(x)$	.05	.22	.35	.28	.10

Find the mean and standard deviation of x. Give a brief interpretation of the value of the mean.

5.97 Let x be the number of shopping trips made during a given week by a randomly selected family from a city. The following table lists the probability distribution of x.

x	0	1	2	3	4	5
$P(x)$	.08	.24	.39	.18	.07	.04

Calculate the mean and standard deviation of x. Give a brief interpretation of the value of the mean.

5.98 Based on its analysis of the future demand for its products, the financial department at Tipper Corporation has determined that there is a .17 probability that the company will lose \$1.2 million during the next year, a .21 probability that it will lose \$.7 million, a .37 probability that it will make a profit of \$.9 million, and a .25 probability that it will make a profit of \$2.3 million.

    a. Let x be a random variable that denotes the profit earned by this corporation during the next year. Write the probability distribution of x.

    b. Find the mean and standard deviation of the probability distribution of part a. Give a brief interpretation of the value of the mean.

5.99 GESCO Insurance Company charges a \$350 premium per annum for a \$100,000 life insurance policy for a 40-year-old female. The probability that a 40-year-old female will die within one year is .002.

    a. Let x be a random variable that denotes the expected gain for next year of the company from a \$100,000 life insurance policy sold to a 40-year-old female. Write the probability distribution of x.

b. Find the mean and standard deviation of the probability distribution of part a. Give a brief interpretation of the value of the mean.

5.100   Determine the value of each of the following using the appropriate formula.

$$6! \qquad 8! \qquad (10-10)! \qquad \binom{4}{4} \qquad \binom{10}{3} \qquad \binom{8}{6}$$

Verify the calculated values by using Tables II and III of Appendix B.

5.101   Determine the value of each of the following using the appropriate formula.

$$9! \qquad 7! \qquad (11-0)! \qquad \binom{6}{3} \qquad \binom{11}{7} \qquad \binom{9}{0}$$

Verify the calculated values by using Tables II and III of Appendix B.

5.102   An agency is to select 4 wells from a suburban subunit to test for water contamination. This subunit contains a total of 16 wells. How many total combinations are possible?

5.103   A financial consultant plans to subscribe to 5 of 12 business-related magazines. If he decides to select 5 magazines randomly from these 12, how many total combinations are possible?

5.104   Spoke Weaving Corporation has 8 weaving machines of the same kind and of the same age. The probability is .04 that any weaving machine will break down at any time. Find the probability that at any given time

   a.   all 8 weaving machines will be broken down
   b.   exactly 2 weaving machines will be broken down
   c.   none of the weaving machines will be broken down

5.105   At the Bank of California, past data show that 7% of all credit card holders default at some time in their lives. On one recent day, this bank issued 12 credit cards to new customers. Find the probability that of these 12 customers eventually

   a.   exactly 3 will default
   b.   exactly 1 will default
   c.   none will default

5.106   Cathy Lotus had $20,000 to invest in stocks. She randomly selected 10 stocks from *The Wall Street Journal* and invested all her money in those 10 stocks. Suppose one year after she made these investments, the prices of 60% of all stocks will be higher. Using the appropriate probabilities table from Appendix B, find the probability that of the 10 stocks in which Cathy Lotus invested her money

   a.   at least 5 will have higher prices one year later
   b.   less than 4 will have higher prices one year later
   c.   2 to 5 will have higher prices one year later

5.107   Brian Maroony owns an auto repair shop in Willingham. Past invoices show that 70% of the autos repaired at his shop have a repair bill of $100 or more each. Using the appropriate probabilities table from Appendix B, find the probability that of the 14 autos repaired at his shop on a given day

   a.   more than 9 will have a repair bill of $100 or more each
   b.   at most 5 will have a repair bill of $100 or more each
   c.   4 to 7 will have a repair bill of $100 or more each

5.108   Maine Corporation buys motors for electric fans from another company that guarantees that at most 5% of its motors are defective and that it will replace all defective motors at no cost to Maine Corporation. The motors are received in shipments of 500 each. The quality control department at Maine Corporation randomly selects 20 motors from each shipment and inspects them for being good or defective. If this sample contains more than 2 defective motors, the entire shipment is rejected.

   a.   Using the appropriate probabilities table from Appendix B, find the probability that a given shipment of 500 motors received by Maine Corporation will be accepted. Assume that 5% of all motors received by Maine Corporation are defective.

    b. Using the appropriate probabilities table from Appendix B, find the probability that a given shipment of 500 motors received by Maine Corporation will not be accepted.

5.109 One of the toys made by Dillon Corporation is called Speaking Joe, which is sold only by mail. Consumer satisfaction is one of the top priorities of the company's management. The company guarantees a refund or a replacement for any Speaking Joe toy if the chip that is installed inside becomes defective within a year from the date of purchase. It is known from past data that 10% of these chips become defective within a one-year period. The company sold 15 Speaking Joes on a given day.

    a. Let x denote the number of Speaking Joes in these 15 that will be returned for a refund or a replacement within a one-year period. Using the appropriate probabilities table from Appendix B, obtain the probability distribution of x and draw a graph of the probability distribution. Determine the mean and standard deviation of x.

    b. Using the probability distribution constructed in part a, find the probability that exactly 5 of the 15 Speaking Joes will be returned for a refund or a replacement within a one-year period.

5.110 In a list of 15 households, 9 own homes and 6 do not own homes. Four households are randomly selected from these 15 households. Find the probability that the number of households in these 4 who own homes is

    a. exactly 3   **b.** at most 1   **c.** exactly 4

5.111 Twenty corporations were asked whether or not they provide retirement benefits to their employees. Fourteen of the corporations said they do provide retirement benefits to their employees and six said they do not. Five corporations are randomly selected from these twenty. Find the probability that

    a. exactly two of them provide retirement benefits to their employees.

    b. none of them provides retirement benefits to its employees

    c. at most one of them provides retirement benefits to its employees

5.112 Uniroyal Electronics Company buys certain parts for its refrigerators from Bob's Corporation. The parts are received in shipments of 400 boxes, each box containing 16 parts. The quality control department at Uniroyal Electronics first randomly selects 1 box from each shipment and then randomly selects 4 parts from that box. The shipment is accepted if at most 1 of the 4 parts is defective. The quality control inspector at Uniroyal Electronics selected a box from a recently received shipment of such parts. Unknown to the inspector, this box contains 3 defective parts.

    a. What is the probability that this shipment will be accepted?

    b. What is the probability that this shipment will not be accepted?

5.113 Alison Bender works for an accounting firm. To make sure her work does not contain errors, her manager randomly checks on her work. Alison recently filled out 12 income tax returns for the company's clients. Unknown to anyone, 2 of these 12 returns have minor errors. Alison's manager randomly selects 3 returns from these 12 returns. Find the probability that

    a. exactly 1 of them contains errors

    b. none of them contains errors

    c. exactly 2 of them contain errors

5.114 A worker on an assembly line assembles an average of 5 products a day.

    a. Using the appropriate formula, find the probability that on a given day he will assemble exactly 7 products.

    b. Using the appropriate probabilities table from Appendix B, find the probability that on a given day the number of products assembled by this worker will be

        i. at least 6   **ii.** at most 4

5.115 An average of 8 videos are rented per day at a video-rental store.

    a. Using the appropriate formula, find the probability that on a given day exactly 5 videos will be rented at this store.

    b. Using the appropriate probabilities table from Appendix B, find the probability that on a given day the number of videos that will be rented at this store is

        i. at least 10   **ii.** at most 4

5.116 An average of 1.2 private airplanes arrive per hour at an airport.

    a. Using the appropriate formula, find the probability that during a given hour no private airplane will arrive at this airport.

    b. Let x denote the number of private airplanes that will arrive at this airport during a given hour. Using the appropriate probabilities table from Appendix B, write the probability distribution of x.

5.117 A machine produces an average of .8 items per minute.

    a. Using the appropriate formula, find the probability that during a certain minute this machine will produce exactly 3 items.

    b. Let x denote the number of items that this machine will produce during a given minute. Using the appropriate probabilities table from Appendix B, write the probability distribution of x.

SELF-REVIEW TEST

1. Briefly explain the meaning of a random variable, a discrete random variable, and a continuous random variable. Give one example each of a discrete and a continuous random variable.

2. What name is given to a table that lists all the values that a discrete random variable x can assume and their corresponding probabilities?

3. For the probability distribution of a discrete random variable, the probability of any single value of x is always

    a. in the range zero to 1   **b.** 1.0   **c.** less than 1.0

4. For the probability distribution of a discrete random variable, the sum of the probabilities of all possible values of x is always

    a. greater than 1   **b.** 1.0   **c.** less than 1.0

5. The number of combinations of 10 items selected 7 at a time is

    a. 120   **b.** 200   **c.** 80

6. State the four conditions of a binomial experiment. Give one example of such an experiment.

7. The parameters of the binomial probability distribution are

    a. n, p, and q **b.** n and p **c.** n, p, and x

8. The mean and standard deviation of a binomial probability distribution with $n = 25$ and $p = .20$ are

    a. 5 and 2   **b.** 8 and 4   **c.** 4 and 3

9. The binomial probability distribution is symmetric if

    a. $p < .5$ **b.** $p = .5$ **c.** $p > .5$

10. The binomial probability distribution is skewed to the right if

    a. $p < .5$ **b.** $p = .5$ **c.** $p > .5$

11. The binomial probability distribution is skewed to the left if

    a. $p < .5$ **b.** $p = .5$ **c.** $p > .5$

12. Briefly explain when a hypergeometric probability distribution is used. Give one example of a hypergeometric probability distribution.

13. The parameter/parameters of the Poisson probability distribution is/are

    a. λ **b.** λ and x **c.** λ and e

14. Describe the three conditions that must be satisfied to apply the Poisson probability distribution.

15. Let x be the number of homes sold per week by all four real estate agents working at a Century 21 office. The following table lists the probability distribution of x.

x	0	1	2	3	4	5
$P(x)$	.15	.24	.31	.14	.10	.06

Calculate the mean and standard deviation of x. Give a brief interpretation of the value of the mean.

16. In a nationwide poll of 21- to 29-year-old working people, 10% said that "the most important measure of living the good life is having a rewarding job/career" (*Fortune*, July 13, 1992). Assume that this result is true for the current population of all 21- to 29-year-old working people.

    a. Find the probability that the number in a random sample of ten 21- to 29-year-old working people who hold this view is

        i.   exactly 4 (use the appropriate formula)

        ii. at least 5 (use the appropriate table from Appendix B)

        iii. less than 3 (use the appropriate table from Appendix B)

    b. Let x be the number in a random sample of ten 21- to 29-year-old working people who hold the opinion that "the most important measure of living the good life is having a rewarding job/career." Using the appropriate table from Appendix B, write the probability distribution of x. Find the mean and standard deviation of this probability distribution.

17. A chamber of commerce has received 12 nominations for 4 positions of office-bearers for the next one year. Of these 12 persons nominated, 4 are female and 8 are male. All members have agreed to randomly select 4 office-bearers from these 12. Find the probability that

    a. exactly 2 of the office-bearers selected are female

    b. exactly 4 of the office-bearers selected are female

    c. at most 1 of the office-bearers selected is a female

18. A department store sells an average of 2 electric appliances per day.

    a. Find the probability that on a given day this store will sell

        i.   exactly 5 electric appliances (use the appropriate formula)

        ii. at most 4 electric appliances (use the appropriate table from Appendix B)

        iii. 5 to 9 electric appliances (use the appropriate table from Appendix B)

    b. Let x be the number of electric appliances sold at this store on a given day. Write the probability distribution of x. Use the appropriate table from Appendix B.

19. The binomial probability distribution is symmetric when $p = .50$; it is skewed to the right when $p < .50$; and it is skewed to the left when $p > .50$. Illustrate these three cases by writing three probability distributions and graphing them. Choose any values of n and p and use the table of binomial probabilities (Table IV of Appendix B).

USING MINITAB

Using MINITAB, we can find the probability of a single outcome for the binomial and Poisson probability distributions or we can list the probabilities for all outcomes of a binomial or a Poisson experiment.

THE BINOMIAL PROBABILITY DISTRIBUTION

The MINITAB command and subcommand given in Figure 5.10 help obtain the probability of a specific value of *x* for a binomial experiment.

Figure 5.10 MINITAB commands to find binomial probability.

MTB > PDF x=k; ⟵ ⎰ Note the semicolon at the end of this command. This semicolon instructs
  ⎱ MINITAB that a subcommand with the values of *n* and *p* is to follow.
  ⎱ Here, *k* is the value of *x* whose probability is to be determined.

SUBC > BINOMIAL n=a p=b. ⟵ ⎰ The period at the end of this subcommand indicates the
  ⎱ end of commands. Here, *n* = *a* is the total number of
  ⎱ trials and *p* = *b* is the probability of success.

In the first MINITAB command of Figure 5.10, **PDF** stands for the *probability density function*, which is another name for the probability of an outcome. The following illustration shows the use of these commands.

Illustration M5–1 Reconsider Example 5–17. Five percent of all VCRs manufactured by an electronics firm are defective. We need to find the probability that if three VCRs are randomly selected from the production line of that firm, exactly one of them will be defective. From the given information,

$$n = \text{total number of VCRs in the sample} = 3$$

$$x = \text{defective VCRs in the sample} = 1$$

$$p = P(\text{a VCR is defective}) = .05$$

MINITAB commands and the MINITAB solution for the probability of exactly one defective VCR in three are shown in Figure 5.11.

Figure 5.11 Finding binomial probability.

```
MTB  > NOTE: PROBABILITY OF x=1 FOR n=3 AND p=.05
MTB  > PDF x=1;
SUBC > BINOMIAL n=3 p=.05.

        k         P(x = k)   ⎤ ←── MINITAB output gives:
       1.00        0.1354    ⎦        P(x = 1) = 0.1354
```

Thus, the probability of one defective VCR in a sample of three VCRs is .1354.

If we put a semicolon after PDF in the second MINITAB command in Figure 5.11 and do not indicate the value of x, MINITAB will list the probabilities of all possible values of x for the binomial experiment with n and p values entered in the subcommand. Illustration M5–2 shows how we can obtain the probability distribution of x for a binomial experiment with $n = 3$ and $p = .05$ using MINITAB.

Illustration M5–2 Reconsider Illustration M5–1. Let x denote the number of defective VCRs in a random sample of 3. Suppose we want to list the probability distribution of x. As we know,

$$n = 3 \quad \text{and} \quad p = .05$$

The MINITAB commands and MINITAB output for this binomial problem are given in Figure 5.12.

Figure 5.12 Obtaining binomial probability distribution.

```
MTB  > NOTE: BINOMIAL PROBABILITY DISTRIBUTION FOR n=3 AND p=.05
MTB  > PDF;   ←──Note that we have not entered the value of x in this command.
SUBC > BINOMIAL n=3 p=.05.

      BINOMIAL WITH N = 3   P = 0.050000 ⎤
          k                 P(x = k)     ⎥
          0                  0.8574      ⎥ ←──MINITAB output
          1                  0.1354      ⎥
          2                  0.0071      ⎥
          3                  0.0001      ⎦
```

In the MINITAB output of Figure 5.12, the column labeled k lists the values of x and the column labeled $P(x = k)$ lists their probabilities. Using the MINITAB output, we can write the probability of any value of x for this example. For instance,

$$P(\text{at most 1 defective VCR}) = P(0) + P(1) = .8574 + .1354 = .9928$$

$$P(\text{at least 2 defective VCRs}) = P(2) + P(3) = .0071 + .0001 = .0072$$

THE POISSON PROBABILITY DISTRIBUTION

The MINITAB command and subcommand listed in Figure 5.13 help obtain the probability of a specific value of x for a Poisson problem.

Figure 5.13   MINITAB commands to find Poisson probability.

```
MTB  > PDF x=k;  ⟵  { This command instructs MINITAB that we need to determine the
                     { probability of x = k where k is a specific value of x.
SUBC > POISSON MEAN=a.  ⟵  { This command instructs MINITAB to use Poisson
                           { distribution with λ = a.
```

Again, note the semicolon at the end of the first command in Figure 5.13. That semicolon instructs MINITAB that a subcommand with a value of λ is to follow. The following illustration shows how to use these commands to find the probability of a specific value of x for a Poisson distribution.

Illustration M5-3   Refer to Example 5–25. The automatic teller machine (ATM) installed outside Mansfield Savings and Loan is used on average by five customers per hour. The bank closed this ATM for one hour for repairs. What is the probability that during that hour eight customers came to use the ATM?

Solution   From the given information,

$$\lambda = \text{mean number of customers who use ATM per hour} = 5$$

The MINITAB commands and MINITAB solution for this illustration are shown in Figure 5.14.

Figure 5.14   Finding Poisson probability.

```
MTB  > NOTE: PROBABILITY OF x = 8 FOR MEAN = 5
MTB  > PDF x=8;
SUBC > POISSON MEAN = 5.

        k       P(x = k) ⎤  ⟵  MINITAB output gives:
      8.00        0.0653 ⎦        P(x = 8) = .0653
```

Thus, from Figure 5.14,

$$P(x = 8) = .0653$$

That is, the probability is .0653 that during the one hour, when the ATM is closed, eight customers will come to use this ATM.   ▬

Again, if we put a semicolon after PDF in the second MINITAB command in Figure 5.14 and do not provide the value of x, MINITAB will list the probabilities of all possible values of x for the Poisson problem with the value of λ given in the subcommand. The following illustration shows how we can obtain the probability distribution of x for a Poisson problem.

Illustration M5-4   Reconsider Illustration M5–3. Let x be the number of customers who come to use this ATM during a given hour. Suppose we want to list the probability distribution of x. The MINITAB commands and MINITAB solution for this illustration are given in Figure 5.15.

Figure 5.15 Obtaining Poisson probability distribution.

```
MTB  > NOTE: POISSON DISTRIBUTION OF x FOR MEAN = 5
MTB  > PDF;
SUBC > POISSON MEAN = 5.

    POISSON WITH MEAN = 5.000
       k               P(x = k)
       0                0.0067
       1                0.0337
       2                0.0842        ←——MINITAB output
       3                0.1404
       4                0.1755
       5                0.1755
       6                0.1462
       7                0.1044
       8                0.0653
       9                0.0363
      10                0.0181
      11                0.0082
      12                0.0034
      13                0.0013
      14                0.0005
      15                0.0002
      16                0.0000
```

In the MINITAB output of Figure 5.15, the possible values that x can assume are listed in the column labeled k and the probabilities of those values of x are listed in the column labeled $P(x = k)$. Using this output, we can determine the probability of any value of x for this Poisson problem. For example,

$$P(x \leq 3) = P(0) + P(1) + P(2) + P(3)$$

$$= .0067 + .0337 + .0842 + .1404 = .2650$$

$$P(3 \leq x \leq 5) = P(3) + P(4) + P(5) = .1404 + .1755 + .1755 = .4914 \quad ■$$

COMPUTER ASSIGNMENTS

M5.1 In a poll of 201 chief executives of large companies conducted for *Fortune* magazine by the opinion research firm Clark Martire & Bartolomeo, 40% of CEOs said that it is *not likely at all* that their companies will have a female CEO within the next decade (*Fortune*, September 21, 1992). Assume that this result is true for the current population of CEOs of all large companies.
  a. Using MINITAB, find the probability that in a random sample of 12 CEOs of large companies, exactly 3 will hold this view.
  b. Let x denote the number of CEOs in a sample of 12 CEOs of large companies who hold this view. Using MINITAB, obtain the probability distribution of x.

M5.2 According to the Motor Vehicle Manufacturers Association, 19% of cars in the United States were at least 12 years old in 1989 (*Business Week*, March 4, 1991). Assume that this result holds true for the current population of all cars in the United States. Let x be the number of cars in a sample of 10 that are at least 12 years old.

    a.   Using MINITAB, find $P(x = 6)$.

    b.   Using MINITAB, construct the probability distribution of x.

M5.3   A mail-order company receives an average of 7.4 orders per day.

    a.   Using MINITAB, find the probability that it will receive exactly 10 orders on a certain day.

    b.   Let x denote the number of orders received by this company on a given day. Using MINITAB, obtain the probability distribution of x.

M5.4   A commuter airline receives an average of 3.7 complaints per day from its passengers. Let x denote the number of complaints received by this airline on a given day.

    a.   Using MINITAB, find $P(x = 0)$.

    b.   Using MINITAB, construct the probability distribution of x.

6

CONTINUOUS RANDOM VARIABLES AND THEIR PROBABILITY DISTRIBUTIONS

Discrete random variables and their probability distributions were presented in Chapter 5. Section 5.1 defined a continuous random variable as a variable that can assume any value in one or more intervals.

The possible values that a continuous random variable can assume are infinite and uncountable. For example, the variable representing the time taken by a worker to commute from home to work is a continuous random variable. Suppose 5 minutes is the minimum time and 130 minutes is the maximum time taken by all workers to commute from home to work. Let x be a continuous random variable that denotes the time taken to commute from home to work by a randomly selected worker. Then x can assume any value in the interval 5 to 130 minutes. This interval contains an infinite number of values that are uncountable.

A continuous random variable can possess one of many probability distributions. In this chapter, we discuss the normal probability distribution, the normal distribution as an approximation to the binomial distribution, the uniform probability distribution, and the exponential probability distribution.

6.1 CONTINUOUS PROBABILITY DISTRIBUTION

In Chapter 5 we defined a **continuous random variable** as a random variable whose values are not countable. A continuous random variable can assume any value over an interval or intervals. Because the number of values contained in any interval is infinite, the possible number of values that a continuous random variable can assume is also infinite. Moreover, we cannot count these values. In Chapter 5, it was stated that the life of a battery, salaries of workers, time taken by workers to learn a job, amount of milk in gallons, 1992 sales revenues of companies, and prices of houses are all examples of continuous random variables. Note that although money can be counted, usually all variables involving money are considered to be continuous random variables. This is so because a variable involving money often has a very large number of outcomes.

Jacinta Corporation manufactures auto batteries and Never Die is the top of the line auto battery it produces. Let *x* be a continuous random variable representing the life of a Never Die auto battery. Suppose Table 6.1 lists the frequency and relative frequency distributions of *x* for 1000 such batteries.

Table 6.1 Frequency and Relative Frequency Distributions of Lives of 1000 Never Die Auto Batteries

Life of a Battery (in months) x	f	Relative Frequency
55 to less than 56	20	.020
56 to less than 57	50	.050
57 to less than 58	90	.090
58 to less than 59	150	.150
59 to less than 60	190	.190
60 to less than 61	160	.160
61 to less than 62	130	.130
62 to less than 63	100	.100
63 to less than 64	70	.070
64 to less than 65	40	.040
	$\Sigma f = 1000$	Sum $= 1.0$

The relative frequencies listed in Table 6.1 can be used as approximate probabilities of respective classes.

Figure 6.1 displays the histogram and polygon for the relative frequency distribution of Table 6.1. Figure 6.2 shows the smoothed polygon for the data of Table 6.1. The smoothed polygon is an approximation of the *probability distribution curve* of the continuous random variable *x*. Note that each class in Table 6.1 has a width equal to 1 month. If the width of

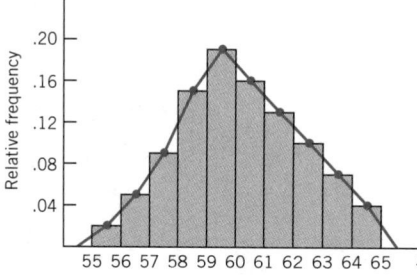

Figure 6.1 Histogram and polygon for Table 6.1.

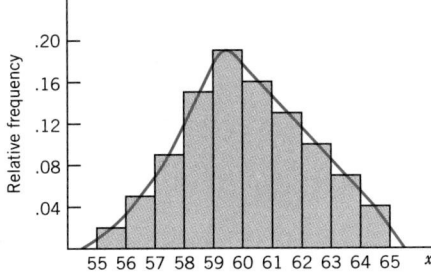

Figure 6.2 Probability distribution curve for lives of batteries.

classes is more than 1 unit, we first obtain the *relative frequency densities* and then graph these relative frequency densities to obtain the distribution curve. The relative frequency density of a class is obtained by dividing the relative frequency of that class by the class width. The relative frequency densities are calculated to make the sum of the areas of all rectangles in the histogram equal to 1.0. Case Study 6–1, which appears later in this section, illustrates this procedure. The probability distribution curve of a continuous random variable is also called its *probability density function.*

The probability distribution of a continuous random variable possesses the following *two characteristics.*

1. The probability that x assumes a value in any interval lies in the range 0 to 1.
2. The total probability of all the (mutually exclusive) intervals within which x can assume a value is 1.0.

The first characteristic states that the area under the probability distribution curve of a continuous random variable between any two points is between 0 and 1, as shown in Figure 6.3. The second characteristic indicates that the total area under the probability distribution curve of a continuous random variable is always 1.0 or 100%, as shown in Figure 6.4.

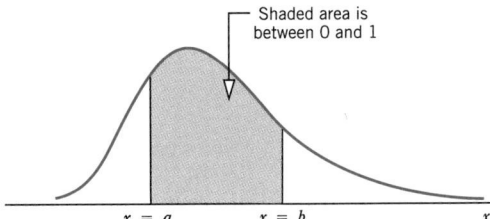

Figure 6.3 Area under a curve between two points.

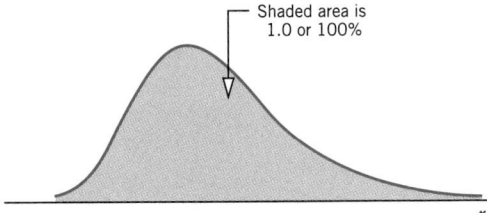

Figure 6.4 Total area under a probability distribution curve.

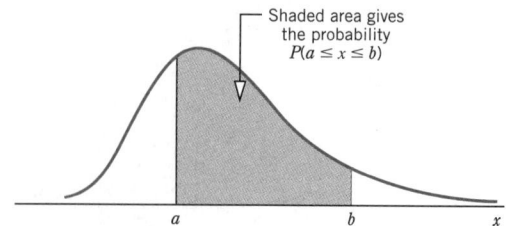

Figure 6.5   Area under the curve as probability.

The probability that a continuous random variable x assumes a value within a certain interval is given by the area under the curve between two limits of the interval, as shown in Figure 6.5. The shaded area under the curve from a to b in this figure gives the probability that x falls in the interval a to b. That is,

$$P(a \le x \le b) = \text{Area under the curve from } a \text{ to } b$$

Note that the interval $a \le x \le b$ states that x is greater than or equal to a but less than or equal to b.

Reconsider the example on the lives of Never Die auto batteries. The probability that the life of a randomly selected Never Die auto battery lies in the interval 60 to 63 months is given by the area under the distribution curve of the lives of these batteries from $x = 60$ to $x = 63$, as shown in Figure 6.6. This probability is written as

$$P(60 \le x \le 63)$$

which states that x is greater than or equal to 60 but less than or equal to 63.

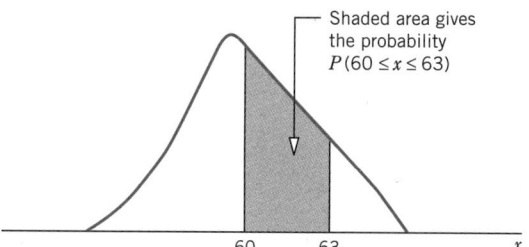

Figure 6.6   Probability that x lies in the interval 60 to 63.

For a continuous probability distribution, the probability is always calculated for an interval. For example, in Figure 6.6, the interval representing the shaded area is from 60 to 63. Consequently, the shaded area in that figure gives the probability for the interval $60 \le x \le 63$.

The probability that a continuous random variable x assumes a single value is always zero. This is so because the area of a line, which represents a single point, is zero. For example, if x is the life of a randomly selected Never Die auto battery, then the probability that the life of this battery is exactly 61 months is zero. That is,

$$P(x = 61) = 0$$

This probability is shown in Figure 6.7. Similarly, the probability for x to assume any other single value is zero.

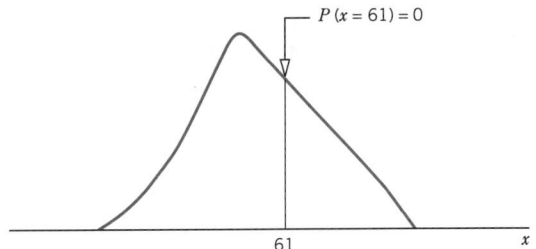

Figure 6.7 Probability of a single value of x is zero.

In general, if a and b are two of the values that x can assume, then,

$$P(a) = 0 \quad \text{and} \quad P(b) = 0$$

From this we can deduce that for a continuous random variable

$$P(a \leq x \leq b) = P(a < x < b)$$

In other words, the probability that x assumes a value in the interval a to b is the same whether or not the values a and b are included in the interval. For the example on the lives of Never Die auto batteries, the probability that the life of a randomly selected battery is between 60 and 63 months is the same as the probability that the life of this battery is 60 to 63 months. This is shown in Figure 6.8.

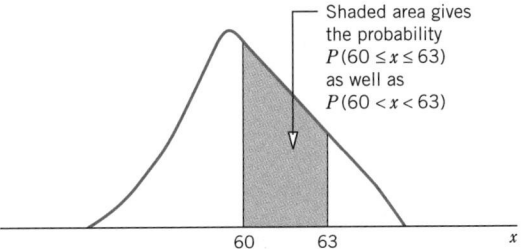

Figure 6.8 Probability "from 60 to 63" and "between 60 and 63."

Note that the interval "between 60 and 63" represents "$60 < x < 63$" and it does not include 60 and 63. On the other hand, the interval "from 60 to 63" represents "$60 \leq x \leq 63$" and it does include 60 and 63. However, as mentioned previously, in the case of a continuous random variable both these intervals contain the same probability or area under the curve.

Case Study 6–1 describes how we obtain the probability distribution curve of a continuous random variable.

CASE STUDY 6-1   DISTRIBUTION OF THE PRICE–EARNINGS RATIOS OF COMPANIES

Business Week publishes information on 1000 major U.S. companies each year. This case study is based on the data on price–earnings ratios for 871 U.S. companies published in the 1993 special issue of *Business Week*. (Although this issue of *Business Week* contained information on 1000 companies, the price–earnings ratios were not available for 129 companies.) As the name suggests, the price–earnings ratio gives the ratio of the stock price to the earnings per share of a company. The price–earnings ratio for a company is an indication of whether the stock of that company is undervalued or overvalued. If the price–earnings ratio is low, the price of the stock is low compared to the earnings of that company and, consequently, the stock of that company is undervalued. The converse is true if the price–earnings ratio is high for a company. Table 6.2 gives the frequency and relative frequency distributions for the 1993 price–earnings ratios for 871 companies. (These price–earnings ratios are based on 1992 earnings and March 5, 1993 stock prices.) The relative frequencies of Table 6.2 are used to construct the histogram and polygon in Figure 6.9.

Table 6.2   Frequency and Relative Frequency Distributions for the Price–Earnings Ratios

Price–Earnings Ratio	Frequency	Relative Frequency
0 to less than 10	31	.036
10 to less than 20	456	.524
20 to less than 30	255	.293
30 to less than 40	65	.075
40 to less than 50	31	.036
50 to less than 60	17	.020
60 to less than 70	10	.011
70 to less than 80	3	.003
80 to less than 90	3	.003
	$\Sigma f = 871$	Sum = 1.001

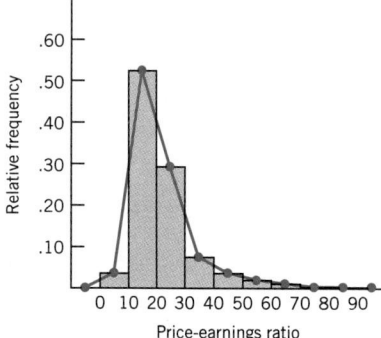

Figure 6.9   Histogram and polygon for the price–earnings ratio data.

To derive the probability distribution curve for these data, first the relative frequency densities are obtained by dividing the relative frequencies by the class widths. The width of each class in Table 6.2 is 10. The relative frequency densities, which are recorded in Table 6.3, are obtained by dividing the relative frequencies of Table 6.2 by 10. The histogram and smoothed polygon shown in Figure 6.10 are drawn using the relative frequency densities of Table 6.3. The curve in this figure gives the probability distribution curve for the price–earnings ratios of 871 companies.

Table 6.3 Relative Frequency Densities

Price–Earnings Ratio	Relative Frequency Density
0 to less than 10	.0036
10 to less than 20	.0524
20 to less than 30	.0293
30 to less than 40	.0075
40 to less than 50	.0036
50 to less than 60	.0020
60 to less than 70	.0011
70 to less than 80	.0003
80 to less than 90	.0003

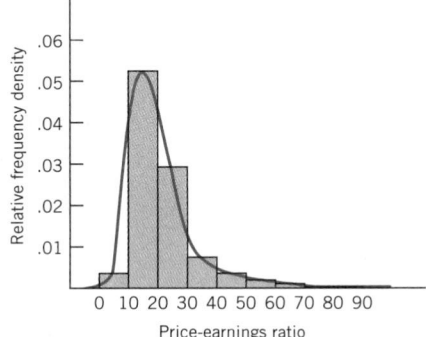

Figure 6.10 Probability distribution curve for price–earnings ratios.

Note that the areas of the rectangles in Figure 6.9 do not give probabilities (which are approximated by relative frequencies). Rather, it is the heights of these rectangles that give the probabilities. This is so because the base of each rectangle is 10 in this histogram. Consequently, the area of any rectangle is given by its height multiplied by 10. Thus, the total area of all rectangles in Figure 6.9 is 10.0 and not 1.0. However, in Figure 6.10 it is the areas and not the heights of rectangles that give the probabilities of respective classes. Thus, if we add the areas of all rectangles in Figure 6.10, we obtain the sum of all probabilities equal to 1.0. Consequently, the total area under the curve is equal to 1.0.

The probability distribution of a continuous random variable has a mean and a standard deviation, which are denoted by μ and σ, respectively. The mean and standard deviation of the probability distribution curve of Figure 6.10 are 21.433 and 11.440, respectively. These values of μ and σ were calculated by using the raw data on the price–earnings ratios of 871 companies.

Source: Data taken from *Business Week*, Special Issue 1993.

6.2   THE NORMAL DISTRIBUTION

As mentioned in Section 6.1, the normal distribution is one of the many probability distributions that a continuous random variable can possess. The normal distribution is the most important and most widely used of all the probability distributions. A large number of phenomena in the real world are normally distributed either exactly or approximately. The continuous random variables representing the weights of packages (e.g., cereal boxes, boxes of cookies), amount of milk in a gallon, life of an item (such as a light bulb or a television set), scores on an examination, time taken to complete a certain job, and heights and weights of people have all been observed to follow (approximately) a normal distribution.

The **normal probability distribution** or *normal curve* is given by a bell-shaped (symmetric) curve. Such a curve is shown in Figure 6.11. It has a mean of μ and a standard deviation of σ. A continuous random variable x that has a normal distribution is called a *normal random variable*. Note that not all bell-shaped curves represent a normal curve. Only a specific kind of bell-shaped curve represents a normal curve.

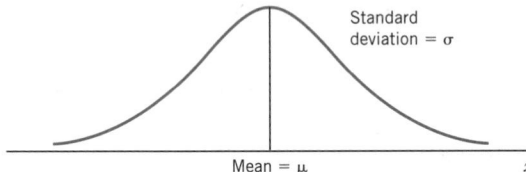

Figure 6.11   Normal distribution with mean μ and standard deviation σ.

NORMAL PROBABILITY DISTRIBUTION

A normal probability distribution, when plotted, gives a bell-shaped curve such that

1.   The total area under the curve is 1.0.
2.   The curve is symmetric about the mean.
3.   The two tails of the curve extend indefinitely.

A normal distribution possesses the following three characteristics.

1.   The total area under a normal curve is 1.0 or 100%, as shown in Figure 6.12.

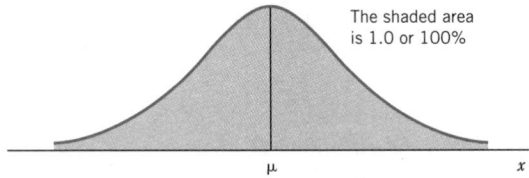

Figure 6.12   Total area under a normal curve.

2.   A normal curve is symmetric about the mean, as shown in Figure 6.13. Consequently, 1/2 of the total area under a normal curve lies on the left side of the mean and 1/2 lies on the right side of the mean.

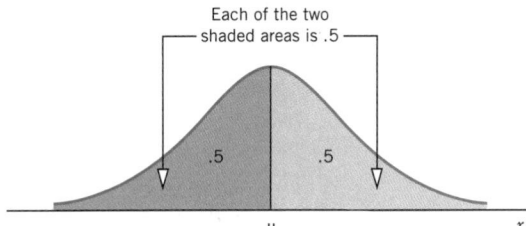

Figure 6.13 A normal curve is symmetric about the mean.

3. The tails of a normal distribution curve extend indefinitely in both directions without touching or crossing the horizontal axis. Although a normal curve never meets the horizontal axis, beyond the points represented by $\mu - 3\sigma$ and $\mu + 3\sigma$ it becomes so close to this axis that the area under the curve beyond these points in both directions can be taken as virtually zero. These areas are shown in Figure 6.14.

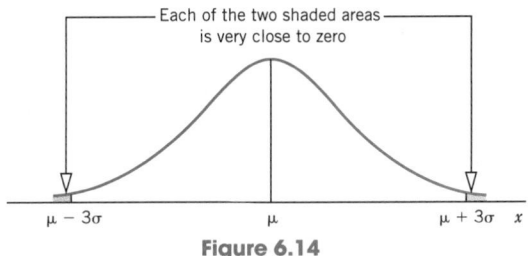

Figure 6.14

The mean μ and the standard deviation σ are the *parameters* of the normal distribution. Given the values of these two parameters, we can find the area under a normal curve for any interval. Remember, there is not just one normal distribution curve but rather a *family* of normal distribution curves. Each different set of values of μ and σ gives a different normal distribution. The value of μ determines the center of a normal distribution on the horizontal axis and the value of σ gives the spread of the normal distribution curve. The three normal distribution curves drawn in Figure 6.15 have the same mean but different standard deviations. By contrast, the three normal distribution curves in Figure 6.16 have different means but the same standard deviation.

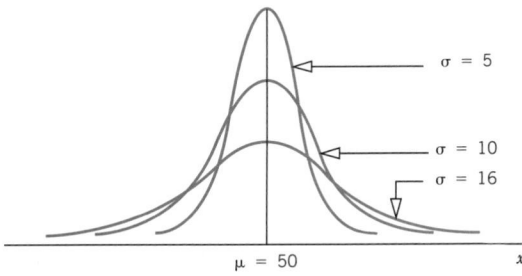

Figure 6.15 Three normal distribution curves with the same mean but different standard deviations.

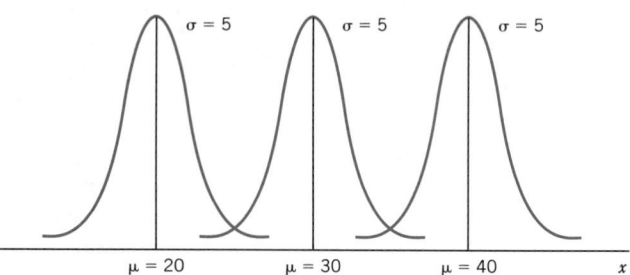

Figure 6.16   Three normal distribution curves with different means but the same standard deviation.

Like the binomial and Poisson probability distributions discussed in Chapter 5, the normal probability distribution can also be expressed by a mathematical equation.[1] However, we will not use this equation to find the area under a normal curve. Instead, we will use Table VII of Appendix B.

6.3   THE STANDARD NORMAL DISTRIBUTION

The **standard normal distribution** is a special case of the normal distribution. For the standard normal distribution, the value of the mean is equal to zero and the value of the standard deviation is equal to 1.

STANDARD NORMAL DISTRIBUTION

The normal distribution with $\mu = 0$ and $\sigma = 1$ is called the standard normal distribution.

Figure 6.17 displays the standard normal distribution curve. The random variable that possesses the standard normal distribution is denoted by z. In other words, the units for the standard normal distribution curve are denoted by z and are called the **z values** or **z scores**. They are also called *standard units* or *standard scores*.

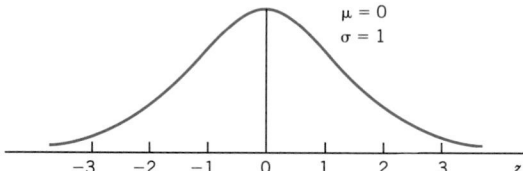

Figure 6.17   The standard normal distribution curve.

[1]The equation of the normal distribution is

$$f(x) = \frac{1}{\sigma\sqrt{2\pi}} e^{-(1/2)[(x-\mu)/\sigma]^2}$$

where $e = 2.71828$ and $\pi = 3.14159$ approximately; $f(x)$, called the probability density function, gives the vertical distance between the horizontal axis and the curve at point x. For the information of those who are familiar with integral calculus, the definite integral of this equation from a to b gives the probability that x assumes a value between a and b.

> ### z VALUES OR z SCORES
>
> The units marked on the horizontal axis of the standard normal curve are denoted by z and are called the z values or z scores. A specific value of z gives the distance between the mean and the point represented by z in terms of the standard deviation.

In Figure 6.17, the horizontal axis is labeled z. The z values on the right side of the mean are positive and those on the left side are negative. *The z value for a point on the horizontal axis gives the distance between the mean and that point in terms of the standard deviation.* For example, a point with a value of $z = 2$ is two standard deviations to the right of the mean. Similarly, a point with a value of $z = -2$ is two standard deviations to the left of the mean.

The standard normal distribution table, Table VII of Appendix B, lists the areas under the standard normal curve between $z = 0$ and the values of z from 0.00 to 3.09. To read the standard normal distribution table, we always start at $z = 0$, which represents the mean of the standard normal distribution. We learned earlier that the total area under a normal curve is 1.0. We also learned that, because of symmetry, the area on either side of the mean is .5. This is shown in Figure 6.18.

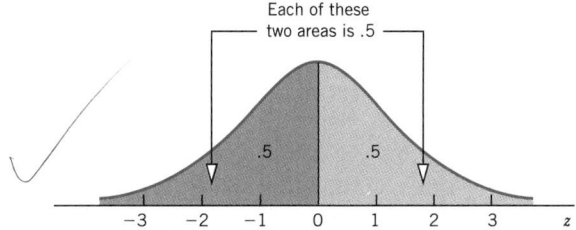

Figure 6.18 Area under the standard normal curve.

☞ *Remember* Although the values of z on the left side of the mean are negative, the area under the curve is always positive.

The area under the standard normal curve between any two points can be interpreted as the probability that z assumes a value within that interval. Examples 6–1 through 6–4 describe how to read Table VII of Appendix B to find areas under the standard normal curve.

Finding area between $z = 0$ and a positive z.

EXAMPLE 6-1 Find the area under the standard normal curve between $z = 0$ and $z = 1.95$.

Solution We divide the number 1.95 into two portions: 1.9 (the digit before the decimal and one digit after the decimal) and .05 (the second digit after the decimal). (Note that $1.9 + .05 = 1.95$.) To find the required area under the standard normal curve, we locate 1.9 in the column for z on the left side of Table VII and .05 in the row for z at the top of Table VII. The entry where the row for 1.9 and the column for .05 intersect gives the area under the standard normal curve between $z = 0$ and $z = 1.95$. The relevant portion of Table VII is reproduced below as Table 6.4. From Table VII or Table 6.4, the entry where the row for 1.9 and the column for .05 cross is .4744. Consequently, the area under the standard normal curve between $z = 0$ and $z = 1.95$ is .4744. This area is shown in Figure 6.19. (It is always helpful to sketch the curve and mark the area we are determining.)

Table 6.4   Area Under the Standard Normal Curve Between $z = 0$ and $z = 1.95$

z	**.00**	**.01**	**...**	**.05**	**...**	**.09**
0.0	.0000	.0040	...	.0199	...	.0359
0.1	.0398	.0438	...	.0596	...	.0753
0.2	.0793	.0832	...	.0987	...	.1141
.	.	.	...	.	...	.
.	.	.	...	.	...	.
1.9	.4713	.4719	...	.4744	...	.4767
.	.	.	...	.	...	.
.	.	.	...	.	...	.
3.0	.4987	.4987	...	.4989	...	.4990

Required area

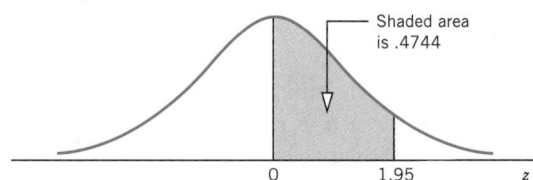

Figure 6.19   Area between $z = 0$ and $z = 1.95$.

The area between $z = 0$ and $z = 1.95$ can be interpreted as the probability that z assumes a value between 0 and 1.95. That is,

$$\text{Area between 0 and 1.95} = P(0 < z < 1.95) = \textbf{.4744}$$

As mentioned in Section 6.1, the probability that a continuous random variable assumes a single value is zero. Therefore,

$$P(z = 0) = 0 \quad \text{and} \quad P(z = 1.95) = 0$$

Hence,

$$P(0 < z < 1.95) = P(0 \leq z \leq 1.95) = .4744 \qquad \blacksquare$$

Finding area between a negative z and $z = 0$.

EXAMPLE 6–2   Find the area under the standard normal curve from $z = -2.17$ to $z = 0$.

Solution   Because the normal distribution is symmetric about the mean, the area from $z = -2.17$ to $z = 0$ is the same as the area from $z = 0$ to $z = 2.17$, as shown in Figure 6.20.

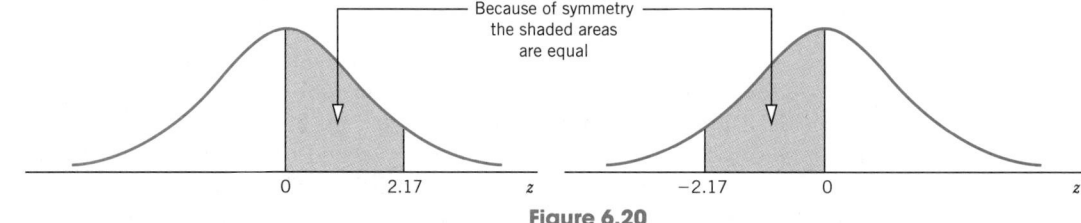

Figure 6.20

To find the area from $z = -2.17$ to $z = 0$, we look for the area from $z = 0$ to $z = 2.17$ in the standard normal distribution table (Table VII of Appendix B). To do so, first we locate 2.1 in the column for z and .07 in the row for z in that table. Then, we read the number at the intersection of the row for 2.1 and the column for .07. The relevant portion of Table VII is reproduced below as Table 6.5. As shown in Table 6.5 and Figure 6.21, this number is .4850.

Table 6.5 Finding Area Under the Standard Normal Curve from $z = 0$ to $z = 2.17$

z	.00	.01	...	.07	...	.09
0.0	.0000	.0040	...	.0279	...	.0359
0.1	.0398	.0438	...	.0675	...	.0753
0.2	.0793	.0832	...	.1064	...	.1141
.	.	.	...	.	...	.
.	.	.	...	.	...	.
2.1	.4821	.4826	...	.4850 ←	...	.4857
.	.	.	...	.	...	.
.	.	.	...	.	...	.
3.0	.4987	.4987	...	.4989	...	.4990

Required area

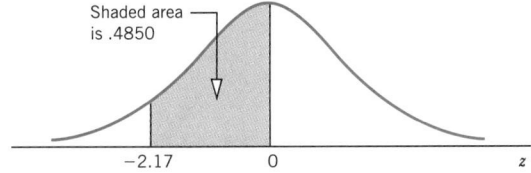

Shaded area is .4850

-2.17 0 z

Figure 6.21 Area from $z = -2.17$ to $z = 0$.

The area from $z = -2.17$ to $z = 0$ gives the probability that z lies in the interval -2.17 to 0. That is,

$$\text{Area from } -2.17 \text{ to } 0 = P(-2.17 \leq z \leq 0) = \mathbf{.4850}$$

Finding areas in the right and left tails.

EXAMPLE 6-3 Find the following areas under the standard normal curve.

(a) Area to the right of $z = 2.32$
(b) Area to the left of $z = -1.54$

Solution

(a) As mentioned earlier, to read the normal distribution table we must start with $z = 0$. To find the area to the right of $z = 2.32$, first we find the area between $z = 0$ and $z = 2.32$. Then we subtract this area from .5, which is the total area to the right of $z = 0$. From Table VII, the area between $z = 0$ and $z = 2.32$ is .4898. Consequently, the required area is $.5 - .4898 = .0102$, as shown in Figure 6.22.

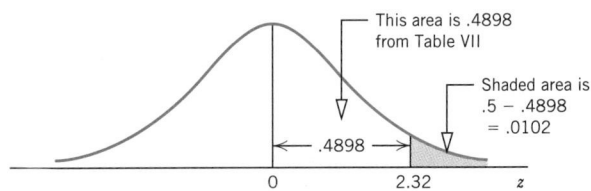

This area is .4898 from Table VII

Shaded area is .5 − .4898 = .0102

.4898

0   2.32 z

Figure 6.22 Area to the right of $z = 2.32$.

The area to the right of $z = 2.32$ gives the probability that z is greater than 2.32. Thus,

$$\text{Area to the right of } 2.32 = P(z > 2.32) = .5 - .4898 = \mathbf{.0102}$$

(b) To find the area under the standard normal curve to the left of $z = -1.54$, first we find the area between $z = -1.54$ and $z = 0$ and then we subtract this area from .5, which is the total area to the left of $z = 0$. From Table VII, the area between $z = -1.54$ and $z = 0$ is .4382. Hence, the required area is $.5 - .4382 = .0618$. This area is shown in Figure 6.23.

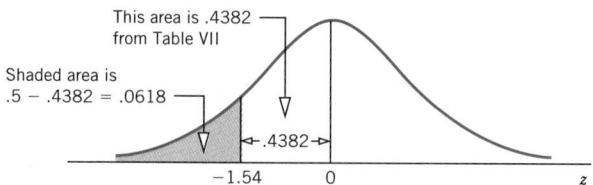

This area is .4382 from Table VII

Shaded area is
.5 − .4382 = .0618

.4382

−1.54　　0　　　　　　　　z

Figure 6.23　Area to the left of $z = -1.54$.

The area to the left of $z = -1.54$ gives the probability that z is less than -1.54. Thus,

$$\text{Area to the left of } -1.54 = P(z < -1.54) = .5 - .4382 = \mathbf{.0618}$$ ■

EXAMPLE 6–4　Find the following probabilities for the standard normal curve.

(a) $P(1.19 < z < 2.12)$　　(b) $P(-1.56 < z < 2.31)$　　(c) $P(z > -.75)$

Solution

Finding area between
two positive values of z.

(a) The probability $P(1.19 < z < 2.12)$ is given by the area under the standard normal curve between $z = 1.19$ and $z = 2.12$, which is the shaded area in Figure 6.24.

Both of the points, $z = 1.19$ and $z = 2.12$, are on the same (right) side of $z = 0$. To find the area between $z = 1.19$ and $z = 2.12$, first we find the areas between $z = 0$ and $z = 1.19$ and between $z = 0$ and $z = 2.12$. Then, we subtract the smaller area (the area between $z = 0$ and $z = 1.19$) from the larger area (the area between $z = 0$ and $z = 2.12$).

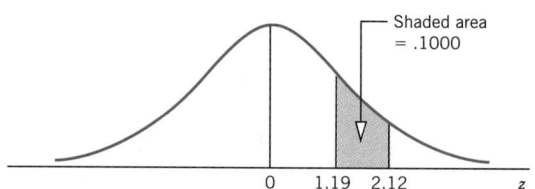

Shaded area
= .1000

0　　1.19　2.12　　　　z

Figure 6.24　Finding $P(1.19 < z < 2.12)$.

From Table VII, for the standard normal distribution,

$$\text{Area between } 0 \text{ and } 1.19 = .3830$$

and

$$\text{Area between } 0 \text{ and } 2.12 = .4830$$

The required probability is

$$P(1.19 < z < 2.12) = \text{Area between 1.19 and 2.12}$$

$$= .4830 - .3830 = \mathbf{.1000}$$

☞   *Remember*   As a general rule, when the two points are on the same side of the mean, first find the areas between the mean and each of the two points. Then, subtract the smaller area from the larger area.

Area between a positive and a negative value of z.

(b)   The probability $P(-1.56 < z < 2.31)$ is given by the area under the standard normal curve between $z = -1.56$ and $z = 2.31$, which is the shaded area in Figure 6.25.

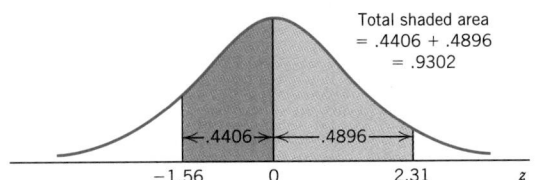

Figure 6.25   Finding $P(-1.56 < z < 2.31)$.

The two points, $z = -1.56$ and $z = 2.31$, are on different sides of $z = 0$. The area between $z = -1.56$ and $z = 2.31$ is obtained by adding the areas between $z = -1.56$ and $z = 0$ and between $z = 0$ and $z = 2.31$, respectively.
From Table VII, for the standard normal distribution,

$$\text{Area between } -1.56 \text{ and } 0 = .4406$$

and $$\text{Area between 0 and 2.31} = .4896$$

The required probability is

$$P(-1.56 < z < 2.31) = \text{Area between } -1.56 \text{ and } 2.31$$

$$= .4406 + .4896 = \mathbf{.9302}$$

☞   *Remember*   As a general rule, when the two points are on different sides of the mean, first find the areas between the mean and each of the two points. Then add these two areas.

Area to the right of a negative value of z.

(c)   The probability $P(z > -.75)$ is given by the area under the standard normal curve to the right of $z = -.75$, which is the shaded area in Figure 6.26.

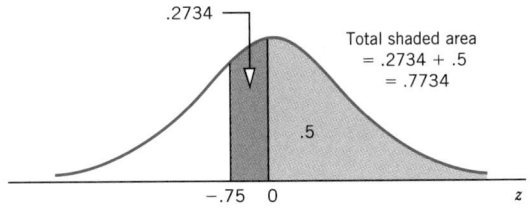

Figure 6.26   Finding $P(z > -.75)$.

The area to the right of $z = -.75$ is obtained by adding the area between $z = -.75$ and $z = 0$ and the area to the right of $z = 0$.

$$\text{Area to the right of } 0 = P(z > 0) = .5$$

From Table VII, for the standard normal distribution,

$$\text{Area between } -.75 \text{ and } 0 = .2734$$

The required probability is

$$P(z > -.75) = \text{Area to the right of } -.75 = .2734 + .5 = \mathbf{.7734} \quad \blacksquare$$

In the discussion in Section 3.4 of Chapter 3 on the use of the standard deviation, we also discussed the empirical rule for a bell-shaped curve. That empirical rule is based on the standard normal distribution table. By using the normal distribution table, we can now verify the empirical rule as follows.

1. The total area within one standard deviation of the mean is 68.26%. This area is given by the sum of the areas between $z = -1.0$ and $z = 0$ and between $z = 0$ and $z = 1.0$. As shown in Figure 6.27, each of these two areas is .3413 or 34.13%. Consequently, the total area between $z = -1.0$ and $z = 1.0$ is 68.26%.

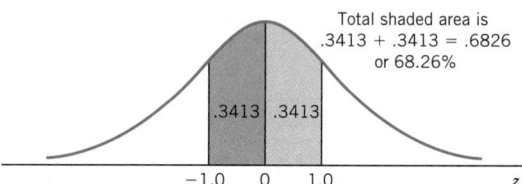

Figure 6.27   Area within one standard deviation of the mean.

2. The total area within two standard deviations of the mean is 95.44%. This area is given by the sum of the areas between $z = -2.0$ and $z = 0$ and between $z = 0$ and $z = 2.0$. As shown in Figure 6.28, each of these two areas is .4772 or 47.72%. Hence, the total area between $z = -2.0$ and $z = 2.0$ is 95.44%.

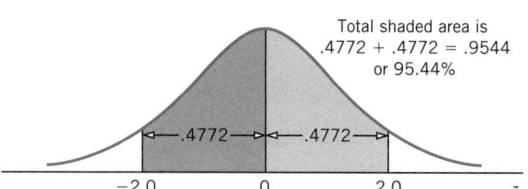

Figure 6.28   Area within two standard deviations of the mean.

3. The total area within three standard deviations of the mean is 99.74%. This area is given by the sum of the areas between $z = -3.0$ and $z = 0$ and between $z = 0$ and $z = 3.0$. As shown in Figure 6.29, each of these two areas is .4987 or 49.87%. Therefore, the total area between $z = -3.0$ and $z = 3.0$ is 99.74%.

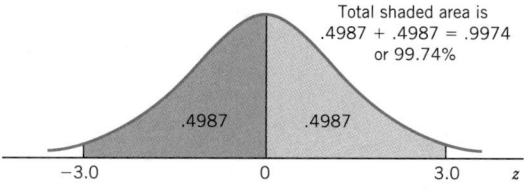

Figure 6.29   Area within three standard deviations of the mean.

Again, note that only a specific bell-shaped curve represents the normal distribution. Now we can state that a bell-shaped curve that contains (about) 68.26% of the total area within one standard deviation of the mean, (about) 95.44% of the total area within two standard deviations of the mean, and (about) 99.74% of the total area within three standard deviations of the mean represents a normal distribution curve.

The standard normal distribution table, Table VII of Appendix B, only goes up to $z = 3.09$. In other words, that table can be read only for $z = 0$ to $z = 3.09$ (or to $z = -3.09$). Consequently, if we need to find the area between $z = 0$ and a z value greater than 3.09 (or between a z value less than -3.09 and $z = 0$) under the standard normal curve, we cannot obtain it from the normal table because it does not contain a z value greater than 3.09. In such cases, the area under the normal curve between $z = 0$ and any z value greater than 3.09 (or less than -3.09) is approximated by .5. From the normal distribution table, the area between $z = 0$ and $z = 3.09$ is .4990. Hence, the area between $z = 0$ and any value of z greater than 3.09 is larger than .4990 and can be approximated by .5. Example 6–5 illustrates this procedure.

EXAMPLE 6–5   Find the following probabilities for the standard normal curve.

(a) $P(0 < z < 5.67)$ (b) $P(z < -5.35)$

Solution

Finding area between $z = 0$ and a value of z greater than 3.09.

(a)   The probability $P(0 < z < 5.67)$ is given by the area under the standard normal curve between $z = 0$ and $z = 5.67$. Because $z = 5.67$ is greater than 3.09 and is not in Table VII, the area under the standard normal curve between $z = 0$ and $z = 5.67$ can be approximated by .5. This area is shown in Figure 6.30.

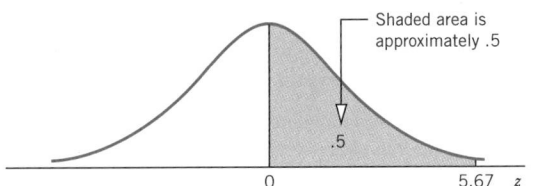

Figure 6.30   Area between $z = 0$ and $z = 5.67$.

The required probability is

$$P(0 < z < 5.67) = \text{Area between 0 and 5.67} = .5 \text{ approximately}$$

Note that the area between $z = 0$ and $z = 5.67$ is not exactly .5 but very close to .5.

Area to the left of a z that is less than -3.09.

(b)   The probability $P(z < -5.35)$ represents the area under the standard normal curve to the left of $z = -5.35$. The area between $z = -5.35$ and $z = 0$ is approximately .5. Consequently, the area under the standard normal curve to the left of $z = -5.35$ is approximately zero, as shown in Figure 6.31.

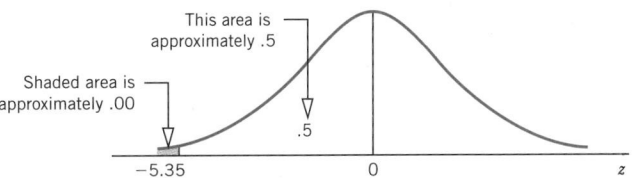

Figure 6.31   Area to the left of $z = -5.35$.

The required probability is

$$P(z < -5.35) = \text{Area to the left of } -5.35$$

$$= .5 - .5 = \mathbf{0.00} \text{ approximately}$$

Again, note that the area to the left of $z = -5.35$ is not exactly 0.00 but very close to 0.00.

EXERCISES

Concepts and Procedures

6.1 What is the difference between the probability distribution of a discrete random variable and that of a continuous random variable?

6.2 Let x be a continuous random variable. What is the probability that x assumes a single value, such as a?

6.3 For a continuous probability distribution, why is $P(a < x < b)$ equal to $P(a \leq x \leq b)$?

6.4 Briefly explain the main characteristics of a normal distribution. Illustrate with the help of graphs.

6.5 Briefly describe the standard normal distribution curve.

6.6 What are the parameters of the normal distribution?

6.7 How do the width and height of a normal distribution change when its mean remains the same but its standard deviation decreases?

6.8 Do the width and/or height of a normal distribution change when its standard deviation remains the same but its mean increases?

6.9 For a standard normal distribution, what does z represent?

6.10 For a standard normal distribution, find the area within one standard deviation of the mean, that is, the area between $\mu - \sigma$ and $\mu + \sigma$.

6.11 For a standard normal distribution, find the area within 1.5 standard deviations of the mean, that is, the area between $\mu - 1.5\sigma$ and $\mu + 1.5\sigma$.

6.12 For a standard normal distribution, what is the area within two standard deviations of the mean?

6.13 For a standard normal distribution, what is the area within 2.5 standard deviations of the mean?

6.14 For a standard normal distribution, what is the area within three standard deviations of the mean?

6.15 Find the area under the standard normal curve
 a. between $z = 0$ and $z = 1.90$ b. between $z = 0$ and $z = -1.75$
 c. between $z = 1.25$ and $z = 2.37$ d. from $z = -1.53$ to $z = -2.78$
 e. from $z = -1.67$ to $z = 2.34$

6.16 Find the area under the standard normal curve
 a. from $z = 0$ to $z = 2.34$ b. between $z = 0$ and $z = -2.78$
 c. from $z = .84$ to $z = 1.95$ d. between $z = -.57$ and $z = -2.39$
 e. between $z = -2.15$ and $z = 1.67$

6.17 Find the area under the standard normal curve
 a. to the right of $z = 1.56$ b. to the left of $z = -1.97$
 c. to the right of $z = -2.05$ d. to the left of $z = 1.86$

6.18 Obtain the area under the standard normal curve
 a. to the right of $z = 1.83$ b. to the left of $z = -1.65$
 c. to the right of $z = -.55$ d. to the left of $z = .79$

6.19 Find the area under the standard normal curve
  a. between $z = 0$ and $z = 4.28$ b. from $z = 0$ to $z = -3.75$
  c. to the right of $z = 7.43$ d. to the left of $z = -4.49$

6.20 Find the area under the standard normal curve
  a. from $z = 0$ to $z = 3.94$ b. between $z = 0$ and $z = -5.16$
  c. to the right of $z = 5.42$ d. to the left of $z = -3.68$

6.21 Determine the following probabilities for the standard normal distribution.
  a. $P(-1.83 \le z \le 2.67)$ b. $P(0 \le z \le 2.12)$
  c. $P(-1.89 \le z \le 0)$ d. $P(z \ge 1.38)$

6.22 Determine the following probabilities for the standard normal distribution.
  a. $P(-2.46 \le z \le 1.68)$ b. $P(0 \le z \le 1.86)$
  c. $P(-2.58 \le z \le 0)$ d. $P(z \ge .83)$

6.23 Find the following probabilities for the standard normal distribution.
  a. $P(z < -2.04)$ b. $P(.67 \le z \le 2.39)$
  c. $P(-2.07 \le z \le -.83)$ d. $P(z < 1.71)$

6.24 Find the following probabilities for the standard normal distribution.
  a. $P(z < -1.21)$ b. $P(1.03 \le z \le 2.79)$
  c. $P(-2.34 \le z \le -1.09)$ d. $P(z < 2.02)$

6.25 Obtain the following probabilities for the standard normal distribution.
  a. $P(z > -.78)$ b. $P(-2.47 \le z \le 1.09)$
  c. $P(0 \le z \le 4.25)$ d. $P(-5.36 \le z \le 0)$
  e. $P(z > 6.07)$ f. $P(z < -5.27)$

6.26 Obtain the following probabilities for the standard normal distribution.
  a. $P(z > -1.26)$ b. $P(-.68 \le z \le 1.74)$
  c. $P(0 \le z \le 3.85)$ d. $P(-4.34 \le z \le 0)$
  e. $P(z > 4.82)$ f. $P(z < -6.12)$

6.4   STANDARDIZING A NORMAL DISTRIBUTION

As was shown in the previous section, Table VII of Appendix B can be used to find areas under the standard normal curve. However, in real-world applications, a (continuous) random variable may have a normal distribution with values of the mean and standard deviation different from 0 and 1, respectively. The first step, in such a case, is to convert the given normal distribution to the standard normal distribution. This procedure is called *standardizing a normal distribution*. The units of a normal distribution (which is not the standard normal distribution) are denoted by x. We know from Section 6.3 that units of the standard normal distribution are denoted by z.

CONVERTING AN x VALUE TO A z VALUE

For a normal random variable x, a particular value of x can be converted to a z value by using the formula

$$z = \frac{x - \mu}{\sigma}$$

where μ and σ are the mean and standard deviation of the normal distribution of x.

Thus, to find the z value for an x value, we calculate the difference between the given x value and the mean μ and divide this difference by the standard deviation σ. If the value of x is equal to μ then its z value is equal to zero. Note that we will always round z values to two decimal places.

☞ **Remember** The z value for the mean of a normal distribution is always zero.

Examples 6–6 through 6–10 describe how to convert x values to the corresponding z values and how to find areas under a normal curve.

Converting x values to the corresponding z values.

EXAMPLE 6–6 Let x be a continuous random variable that has a normal distribution with a mean of 50 and a standard deviation of 10. Convert the following x values to z values.

(a) $x = 55$ (b) $x = 35$

Solution For the given normal distribution: $\mu = 50$ and $\sigma = 10$.

(a) The z value for $x = 55$ is computed as follows.

$$z = \frac{x - \mu}{\sigma} = \frac{55 - 50}{10} = \textbf{.50}$$

Thus, the z value for $x = 55$ is .50. The z values for $\mu = 50$ and $x = 55$ are shown in Figure 6.32. Note that the z value for $\mu = 50$ is zero. The value $z = .50$ for $x = 55$ indicates that the distance between the mean $\mu = 50$ of the given normal distribution and the point given by $x = 55$ is 1/2 of the standard deviation $\sigma = 10$. Consequently, we can state that the z value represents the distance between μ and x in terms of the standard deviation. Because $x = 55$ is greater than $\mu = 50$, its z value is positive.

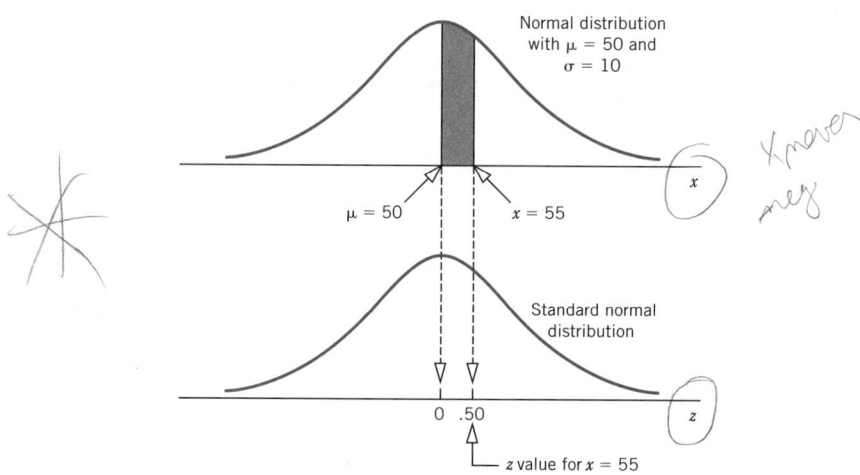

Figure 6.32 z value for $x = 55$.

From this point on, we will usually show only the z axis below the x axis, and not the standard normal curve itself.

(b) The z value for $x = 35$ is computed as follows and is shown in Figure 6.33.

$$z = \frac{x - \mu}{\sigma} = \frac{35 - 50}{10} = \textbf{-1.50}$$

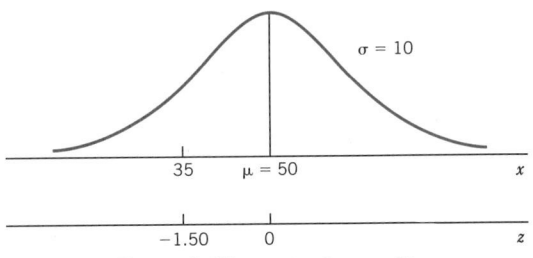

Figure 6.33 z value for $x = 35$.

Because $x = 35$ is on the left side of the mean (i.e., 35 is less than $\mu = 50$), its z value is negative. As a general rule, whenever an x value is less than the value of μ, its z value is negative. ∎

☞ *Remember* The z value for an x value greater than μ is positive; the z value for an x value equal to μ is zero, and the z value for an x value less than μ is negative.

To find the area between two values of x for a normal distribution, we first convert both values of x to their respective z values. Then we find the area under the standard normal curve between those two z values. The area between the two z values gives the area between the corresponding x values.

EXAMPLE 6–7 Let x be a continuous random variable that is normally distributed with a mean of 25 and a standard deviation of 4. Find the area

(a)   between $x = 25$ and $x = 32$ (b)   between $x = 18$ and $x = 34$

Solution   For the given normal distribution: $\mu = 25$ and $\sigma = 4$.

Area between the mean and a point to its right.

(a)   The first step in finding the required area is to standardize the given normal distribution by converting $x = 25$ and $x = 32$ to respective z values using the formula

$$z = \frac{x - \mu}{\sigma}$$

The z value for $x = 25$ is zero because it is the mean of the normal distribution. The z value for $x = 32$ is

$$z = \frac{32 - 25}{4} = 1.75$$

As shown in Figure 6.34, the area between $x = 25$ and $x = 32$ under the given normal curve is equivalent to the area between $z = 0$ and $z = 1.75$ under the standard normal curve. This area from Table VII is .4599. The area between $x = 25$ and $x = 32$ under the normal curve gives the probability that x assumes a value between 25 and 32. This probability can be written as

$$P(25 < x < 32) = P(0 < z < 1.75) = \textbf{.4599}$$

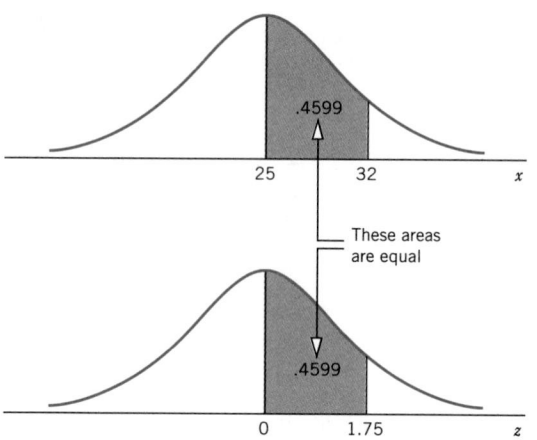

Figure 6.34   Area between $x = 25$ and $x = 32$.

Area between two points that are on different sides of the mean.

(b) First we calculate the z values for $x = 18$ and $x = 34$ as follows.

For $x = 18$: $z = \dfrac{18 - 25}{4} = -1.75$

For $x = 34$: $z = \dfrac{34 - 25}{4} = 2.25$

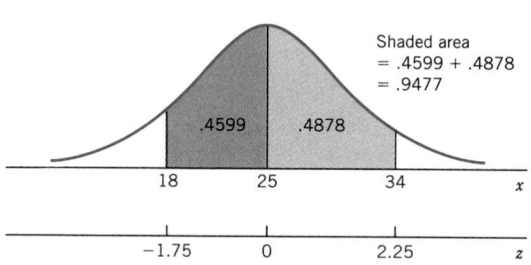

Figure 6.35   Area between $x = 18$ and $x = 34$.

The area under the given normal curve between $x = 18$ and $x = 34$ is given by the area under the standard normal curve between $z = -1.75$ and $z = 2.25$. This area is shown in Figure 6.35. The two values of z are on different sides of $z = 0$. Consequently, the total area is obtained by adding the areas between $z = -1.75$ and $z = 0$ and between $z = 0$ and $z = 2.25$. From Table VII, the area between $z = -1.75$ and $z = 0$ is .4599 and the area between $z = 0$ and $z = 2.25$ is .4878. Hence,

$$P(18 < x < 34) = P(-1.75 < z < 2.25) = .4599 + .4878 = \mathbf{.9477} \quad \blacksquare$$

EXAMPLE 6-8   Let x be a normal random variable with its mean equal to 40 and standard deviation equal to 5. Find the following probabilities for this normal distribution.

(a) $P(x > 55)$ (b) $P(x < 49)$

Solution

Probability of x falling in the right tail.

(a) The probability that x assumes a value greater than 55 is given by the area under the normal curve to the right of $x = 55$, as shown in Figure 6.36. This area is calculated by subtracting the area between $\mu = 40$ and $x = 55$ from .5, which is the total area to the right of the mean.

$$\text{For } x = 55: \quad z = \frac{55 - 40}{5} = 3.00$$

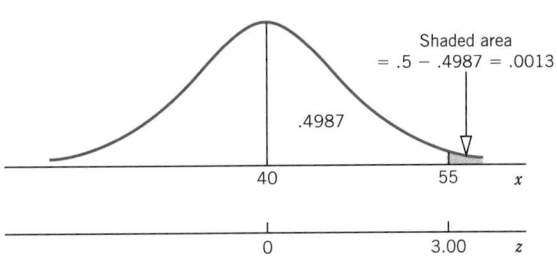

Figure 6.36 Finding $P(x > 55)$.

The required probability is given by the area to the right of $z = 3.00$. To find this area, first we find the area between $z = 0$ and $z = 3.00$, which is .4987. Then we subtract this area from .5. Thus,

$$P(x > 55) = P(z > 3.00) = .5 - .4987 = \mathbf{.0013}$$

Probability that x is less than a value that is to the right of the mean.

(b) The probability that x will assume a value less than 49 is given by the area under the normal curve to the left of 49, which is the shaded area in Figure 6.37. This area is given by the sum of the area to the left of $\mu = 40$ and the area between $\mu = 40$ and $x = 49$.

$$\text{For } x = 49: \quad z = \frac{49 - 40}{5} = 1.80$$

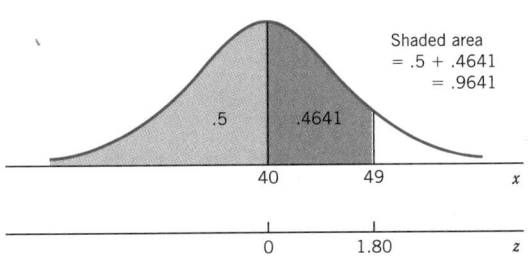

Figure 6.37 Finding $P(x < 49)$.

The required probability is given by the area to the left of $z = 1.80$. This area is obtained by adding the area to the left of $z = 0$ and the area between $z = 0$ and $z = 1.80$. The area to the left of $z = 0$ is .5. From Table VII, the area between $z = 0$ and $z = 1.80$ is .4641. Therefore, the required probability is

$$P(x < 49) = P(z < 1.80) = .5 + .4641 = \mathbf{.9641}$$

Area between two x values that are less than the mean.

EXAMPLE 6-9 Let x be a continuous random variable that has a normal distribution with $\mu = 50$ and $\sigma = 8$. Find the probability $P(30 \leq x \leq 39)$.

Solution For the given normal distribution: $\mu = 50$ and $\sigma = 8$. The probability $P(30 \leq x \leq 39)$ is given by the area from $x = 30$ to $x = 39$ under the normal curve.

As shown in Figure 6.38, this area is given by the difference between the area from $x = 30$ to $\mu = 50$ and the area from $x = 39$ to $\mu = 50$.

$$\text{For } x = 30: \qquad z = \frac{30 - 50}{8} = -2.50$$

$$\text{For } x = 39: \qquad z = \frac{39 - 50}{8} = -1.38$$

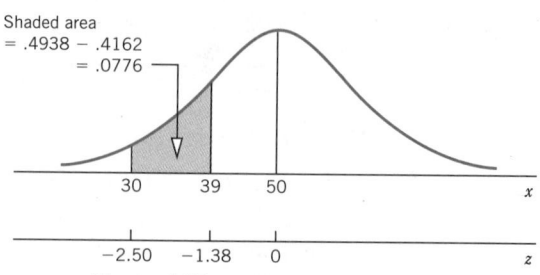

Figure 6.38   Finding $P(30 \leq x \leq 39)$.

To find the required area, we first find the areas from $z = -2.50$ to $z = 0$ and from $z = -1.38$ to $z = 0$ and then take the difference between these two areas. From Table VII, the area from $z = -2.50$ to $z = 0$ is .4938 and the area from $z = -1.38$ to $z = 0$ is .4162. Thus, the required probability is

$$P(30 \leq x \leq 39) = P(-2.50 \leq z \leq -1.38) = .4938 - .4162 = \mathbf{.0776} \qquad \blacksquare$$

EXAMPLE 6–10   Let x be a continuous random variable that has a normal distribution with a mean of 80 and a standard deviation of 12. Find the area under the normal curve

(a)   from $x = 70$ to $x = 135$ (b)   to the left of 27

Solution   For the given normal distribution: $\mu = 80$ and $\sigma = 12$.

Area between two values of x that are on different sides of the mean.

(a)   The area from $x = 70$ to $x = 135$ is obtained by adding the areas from $x = 70$ to $\mu = 80$ and from $\mu = 80$ to $x = 135$. This total area is given by the sum of the two shaded areas shown in Figure 6.39.

$$\text{For } x = 70: \qquad z = \frac{70 - 80}{12} = -.83$$

$$\text{For } x = 135: \qquad z = \frac{135 - 80}{12} = 4.58$$

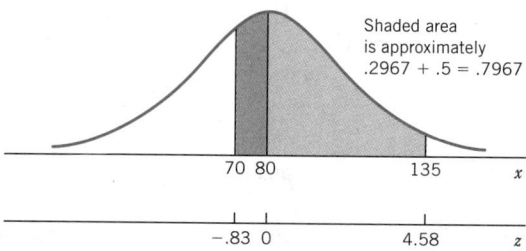

Figure 6.39   Area between $x = 70$ and $x = 135$.

Thus, the required area is obtained by adding the areas from $z = -.83$ to $z = 0$ and from $z = 0$ to $z = 4.58$ under the standard normal curve. From Table VII, the area from $z = -.83$ to $z = 0$ is .2967 and the area from $z = 0$ to $z = 4.58$ is approximately .5. Hence,

$$P(70 \leq x \leq 135) = P(-.83 \leq z \leq 4.58)$$

$$= .2967 + .5 = \mathbf{.7967} \text{ approximately}$$

Finding an area in the left tail.

(b) The area to the left of $x = 27$ is obtained by subtracting the area from $x = 27$ to $\mu = 80$ from .5, which is the total area to the left of the mean. This area is calculated as follows.

$$\text{For } x = 27: \qquad z = \frac{27 - 80}{12} = -4.42$$

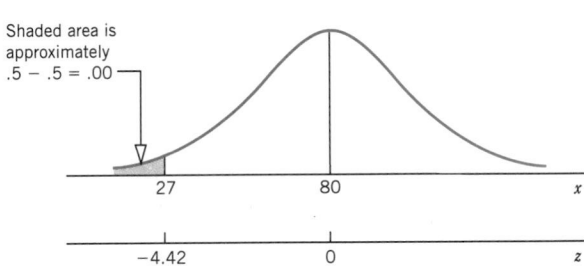

Shaded area is approximately
.5 − .5 = .00

Figure 6.40 Area to the left of $x = 27$.

As shown in Figure 6.40, the required area is given by the area under the normal curve to the left of $z = -4.42$. This area is

$$P(x < 27) = P(z < -4.42) = .5 - P(-4.42 < z < 0)$$

$$= .5 - .5 = \mathbf{.00} \text{ approximately}$$

EXERCISES

Concepts and Procedures

6.27 Find the z value for each of the following x values for a normal distribution with $\mu = 30$ and $\sigma = 5$.

    a. $x = 37$ **b.** $x = 19$ **c.** $x = 23$ **d.** $x = 44$

6.28 Determine the z value for each of the following x values for a normal distribution with $\mu = 16$ and $\sigma = 3$.

    a. $x = 11$ **b.** $x = 22$ **c.** $x = 18$ **d.** $x = 14$

6.29 Find the following areas under a normal curve with $\mu = 20$ and $\sigma = 4$.

    a. Area between $x = 20$ and $x = 27$
    b. Area from $x = 23$ to $x = 25$
    c. Area between $x = 9.5$ and $x = 17$

6.30 Find the following areas under a normal curve with $\mu = 12$ and $\sigma = 2$.

    a. Area between $x = 7.76$ and $x = 12$
    b. Area between $x = 14.48$ and $x = 16.34$
    c. Area from $x = 8.22$ to $x = 11.06$

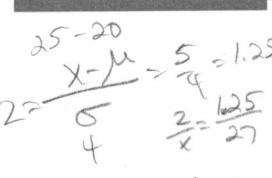
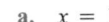
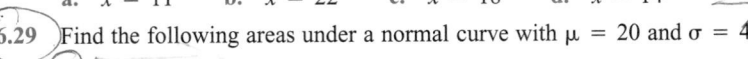
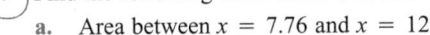
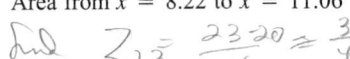
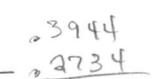

6.31 Determine the area under a normal curve with $\mu = 55$ and $\sigma = 7$
  a. to the right of $x = 58$
  b. to the right of $x = 43$
  c. to the left of $x = 67$
  d. to the left of $x = 24$

6.32 Find the area under a normal curve with $\mu = 37$ and $\sigma = 3$
  a. to the left of $x = 29$
  b. to the right of $x = 53$
  c. to the left of $x = 42$
  d. to the right of $x = 35$

6.33 Let x be a continuous random variable that is normally distributed with a mean of 25 and a standard deviation of 6. Find the probability that x assumes a value
  a. between 29 and 36   b. between 22 and 33

6.34 Let x be a continuous random variable that has a normal distribution with a mean of 40 and a standard deviation of 4. Find the probability that x assumes a value
  a. between 29 and 35   b. from 34 to 51

6.35 Let x be a continuous random variable that is normally distributed with a mean of 80 and a standard deviation of 12. Find the probability that x assumes a value
  a. greater than 70   b. less than 75
  c. greater than 100   d. less than 89

6.36 Let x be a continuous random variable that is normally distributed with a mean of 65 and a standard deviation of 15. Find the probability that x assumes a value
  a. less than 43   b. greater than 74
  c. greater than 56   d. less than 71

6.5 APPLICATIONS OF THE NORMAL DISTRIBUTION

Sections 6.2 through 6.4 discussed the normal distribution, how to convert a normal distribution to the standard normal distribution, and how to find the area under a normal distribution curve. This section presents examples that illustrate the applications of the normal distribution.

Application of the normal distribution: area between two points that are on different sides of the mean.

EXAMPLE 6–11   According to the U.S. Bureau of Census, the mean income of all U.S. families was $43,237 in 1991. Assume that the 1991 incomes of all U.S. families have a normal distribution with a mean of $43,237 and a standard deviation of $10,500. Find the probability that the 1991 income of a randomly selected U.S. family was between $30,000 and $50,000.

Solution   Let x denote the 1991 income of a randomly selected U.S. family. Then, x is normally distributed with

$$\mu = \$43,237 \quad \text{and} \quad \sigma = \$10,500$$

The probability that the 1991 income of a randomly selected family was between $30,000 and $50,000 is given by the area under the normal curve of x between $x = \$30,000$ and $x = \$50,000$. Because these two points are on different sides of the mean, the required probability is obtained by adding the two shaded areas shown in Figure 6.41.

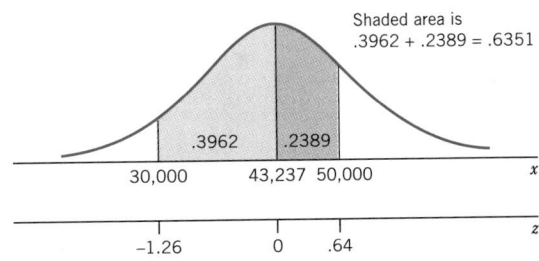

Figure 6.41   Area between $x = \$30,000$ and $x = \$50,000$.

For $x = \$30,000$: $\qquad z = \dfrac{30,000 - 43,237}{10,500} = -1.26$

For $x = \$50,000$: $\qquad z = \dfrac{50,000 - 43,237}{10,500} = .64$

The required probability is given by the area under the standard normal curve between $z = -1.26$ and $z = .64$, which is obtained by adding the area between $z = -1.26$ and $z = 0$ and the area between $z = 0$ and $z = .64$. From Table VII of Appendix B, the area between $z = -1.26$ and $z = 0$ is .3962 and the area between $z = 0$ and $z = .64$ is .2389. Hence, the required probability is

$$P(30,000 < x < 50,000) = P(-1.26 < z < .64) = .3962 + .2389 = \mathbf{.6351}$$

Thus, the probability is .6351 that the 1991 income of a randomly selected U.S. family was between \$30,000 and \$50,000. Converting this probability to a percentage, we can also state that the 1991 incomes of about 63.51% of U.S. families were between \$30,000 and \$50,000. ■

Application of the normal distribution: probability x is less than a value that is to the right of the mean.

EXAMPLE 6–12   A racing car is one of the many toys manufactured by Mack Corporation. The assembly time for this toy follows a normal distribution with a mean of 55 minutes and a standard deviation of 4 minutes. The company closes at 5 P.M. every day. If one worker starts assembling a racing car at 4 P.M., what is the probability that she will finish this job before the company closes for the day?

Solution   Let x denote the time taken by this worker to assemble a racing car. Then x is normally distributed with

$$\mu = 55 \text{ minutes} \quad \text{and} \quad \sigma = 4 \text{ minutes}$$

We are to find the probability that this worker can assemble this car in 60 minutes or less (between 4 and 5 P.M.). This probability is given by the area under the normal curve to the left of $x = 60$. This area is obtained by adding the two shaded areas shown in Figure 6.42.

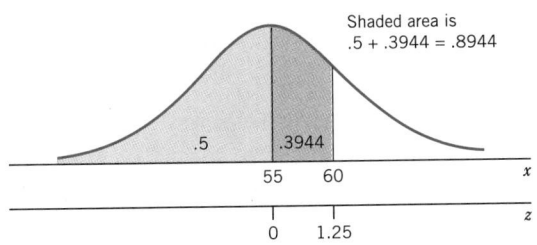

Figure 6.42   Area to the left of $x = 60$.

$$\text{For } x = 60: \quad z = \frac{60 - 55}{4} = 1.25$$

The required probability is given by the area under the standard normal curve to the left of $z = 1.25$, which is obtained by adding .5 and the area between $z = 0$ and $z = 1.25$. From Table VII of Appendix B, the area between $z = 0$ and $z = 1.25$ is .3944. Therefore, the required probability is

$$P(x \le 60) = P(z \le 1.25) = .5 + .3944 = \textbf{.8944}$$

Thus, the probability is .8944 that this worker will finish assembling this racing car before the company closes for the day.

Applications of the normal distribution.

EXAMPLE 6–13   Hupper Corporation produces many types of soft drinks, including Orange Cola. The filling machines are adjusted to pour 12 ounces of soda in each 12-ounce can of Orange Cola. However, the actual amount of soda poured into each can is not exactly 12 ounces; it varies from can to can. It is found that the net amount of soda in such a can has a normal distribution with a mean of 12 ounces and a standard deviation of .015 ounces.

(a)   What is the probability that a randomly selected can of Orange Cola contains 11.97 to 11.99 ounces of soda?

(b)   What percentage of the Orange Cola cans contain 12.02 to 12.07 ounces of soda?

Solution   Let x be the net amount of soda in a can of Orange Cola. Then, x has a normal distribution with $\mu = 12$ ounces and $\sigma = .015$ ounces.

Probability x is between two points that are to the left of the mean.

(a)   The probability that a randomly selected can contains 11.97 to 11.99 ounces of soda is given by the area under the normal curve from $x = 11.97$ to $x = 11.99$. This area is shown in Figure 6.43.

$$\text{For } x = 11.97: \quad z = \frac{11.97 - 12}{.015} = -2.00$$

$$\text{For } x = 11.99: \quad z = \frac{11.99 - 12}{.015} = -.67$$

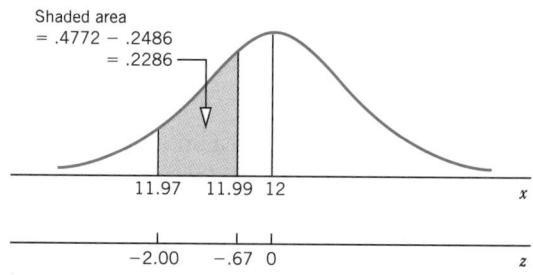

Figure 6.43   Area between $x = 11.97$ and $x = 11.99$.

The required probability is given by the area under the standard normal curve between $z = -2.00$ and $z = -.67$, which is obtained by subtracting the area from $z = -.67$ to $z = 0$ from the area from $z = -2.00$ to $z = 0$. From Table VII of Appendix B, the area from $z = -.67$ to $z = 0$ is .2486 and the area from $z = -2.00$ to $z = 0$ is .4772. Hence, the required probability is

$$P(11.97 \le x \le 11.99) = P(-2.00 \le z \le -.67) = .4772 - .2486 = \textbf{.2286}$$

Thus, the probability is .2286 that any randomly selected can of Orange Cola will contain 11.97 to 11.99 ounces of soda. We can also state that about 22.86% of the Orange Cola cans contain 11.97 to 11.99 ounces of soda.

Probability x is between two points that are to the right of the mean.

(b)   The percentage of Orange Cola cans that contain 12.02 to 12.07 ounces of soda is given by the area under the normal curve from $x = 12.02$ to $x = 12.07$, as shown in Figure 6.44.

$$\text{For } x = 12.02: \qquad z = \frac{12.02 - 12}{.015} = 1.33$$

$$\text{For } x = 12.07: \qquad z = \frac{12.07 - 12}{.015} = 4.67$$

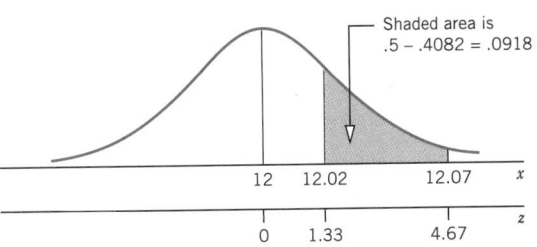

Figure 6.44   Area from $x = 12.02$ to $x = 12.07$.

The required probability is given by the area under the standard normal curve between $z = 1.33$ and $z = 4.67$, which is obtained by subtracting the area from $z = 0$ to $z = 1.33$ from the area from $z = 0$ to $z = 4.67$. From Table VII of Appendix B, the area from $z = 0$ to $z = 1.33$ is .4082. The area from $z = 0$ to $z = 4.67$ is approximately .5. Hence, the required probability is

$$P(12.02 \le x \le 12.07) = P(1.33 \le z \le 4.67) = .5 - .4082 = \mathbf{.0918}$$

Converting this probability to a percentage, we can state that approximately 9.18% of all the Orange Cola cans are expected to contain 12.02 to 12.07 ounces of soda.   ■

Area to the left of x that is less than the mean.

EXAMPLE 6–14   The life span of a calculator manufactured by Intal Corporation has a normal distribution with a mean of 54 months and a standard deviation of 8 months. The company guarantees that any calculator that starts malfunctioning within 36 months of the purchase will be replaced by a new one. About what percentage of such calculators made by this company are expected to be replaced?

Solution   Let x be the life span of such a calculator. Then x has a normal distribution with $\mu = 54$ and $\sigma = 8$ months. The probability that a randomly selected calculator will start malfunctioning within 36 months is given by the area under the normal curve to the left of $x = 36$, as shown in Figure 6.45.

$$\text{For } x = 36: \qquad z = \frac{36 - 54}{8} = -2.25$$

The required percentage is given by the area under the standard normal curve to the left of $z = -2.25$, which is obtained by subtracting the area between $z = -2.25$ and $z = 0$ from .5. From Table VII of Appendix B, the area between $z = -2.25$ and $z = 0$ is .4878. Hence,

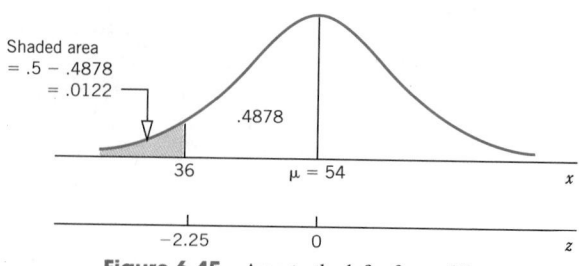

Figure 6.45   Area to the left of $x = 36$.

the required probability is

$$P(x < 36) = P(z < -2.25) = .5 - .4878 = \mathbf{.0122}$$

The probability that any randomly selected calculator manufactured by Intal Corporation will start malfunctioning within 36 months is .0122. Converting this probability to a percentage, we can state that approximately 1.22% of all such calculators manufactured by this company are expected to start malfunctioning within 36 months. Hence, 1.22% of the calculators are expected to be replaced.

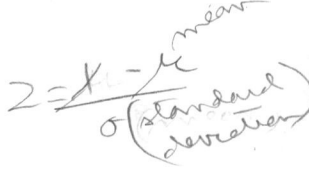

EXERCISES

Applications

6.37 The Bank of Connecticut issues Visa and Mastercard credit cards. It is estimated that the balances on all Visa credit cards issued by the Bank of Connecticut have a mean of $845 and a standard deviation of $270. Assume that the balances on all these Visa cards follow a normal distribution.

   a. What is the probability that a randomly selected Visa card issued by this bank has a balance between $1000 and $1400?

   b. What percentage of the Visa cards issued by this bank have a balance of $750 or more?

6.38 The U.S. Bureau of Labor conducts surveys to estimate the various kinds of expenditures incurred by households. According to a recent survey, the mean housing expenditure for households was $8886 in 1990. Assume that the 1990 housing expenditures for all households are (approximately) normally distributed with a mean of $8886 and a standard deviation of $1700.

   a. Find the percentage of households that had housing expenditures less than $9500 in 1990.

   b. What is the probability that a randomly selected household had housing expenditures of more than $12,000 in 1990?

6.39 According to a survey by Roth Young Personnel Service of Minneapolis, Inc., the mean annual salary of general managers of hotels and motels with 200 to 300 rooms is $55,000 (*Working Woman*, January 1993). Suppose that the annual salaries of all such managers of hotels and motels are (approximately) normally distributed with a mean of $55,000 and a standard deviation of $6500.

   a. Find the probability that the annual salary of a randomly selected such general manager is more than $65,000.

   b. What percentage of such general managers have an annual salary between $40,000 and $50,000?

6.40 The transmission on a model of a specific car has a warranty for 40,000 miles. It is known that the life of such a transmission has a normal distribution with a mean of 72,000 miles and a standard deviation of 12,000 miles.

   a. What percentage of the transmissions will fail before the end of the warranty period?

   b. What percentage of the transmissions will be good for more than 100,000 miles?

6.41 According to the records of the electric company serving the Boston area, the mean electric consumption for all households during winter is 1650 kilowatt hours per month. Assume that the monthly electric consumption during winter by all households in this area has a normal distribution with a mean of 1650 kilowatt hours and a standard deviation of 320 kilowatt hours.

    a. Find the probability that the monthly electric consumption during winter by a randomly selected household from this area is less than 1800 kilowatt hours.

    b. What percentage of the households in this area have a monthly electric consumption of 900 to 1300 kilowatt hours?

6.42 The management of a supermarket wants to adopt a new promotional policy of giving a free gift to every customer who spends more than a certain amount per visit at this supermarket. The expectation of the management is that after this promotional policy is advertised, the expenditures for all customers at this supermarket will be normally distributed with a mean of $95 and a standard deviation of $21. If the management decides to give free gifts to all those customers who spend more than $130 at this supermarket during a visit, what percentage of the customers are expected to get free gifts?

6.43 A health insurance company pays its members a maximum of $500 per year for prescription drugs. If a member spends more than $500 on prescription drugs during a year, any amount over $500 is paid either by the member or by the major medical insurance. The past data show that the annual expenses on prescription drugs for all members of this company have a normal distribution with a mean of $350 and a standard deviation of $65.

    a. What percentage of the members of this health insurance company exceed the maximum annual expenses allowed by the company on prescription drugs?

    b. What percentage of the members of this health insurance company spend less than $200 per year on prescription drugs?

6.44 The U.S. Bureau of Labor Statistics conducts surveys quite often and publishes results in many newsletters and magazines. According to one such recent survey, the mean weekly earnings of workers in U.S. manufacturing industries were $473 (*Monthly Labor Review*, November 1992). Assume that the current weekly earnings of all workers working in U.S. manufacturing industries have a normal distribution with a mean of $473 and a standard deviation of $50.

    a. Find the probability that a worker selected at random from the manufacturing industries earns more than $525 per week.

    b. What percentage of workers in the manufacturing industries earn between $400 and $450 per week?

6.45 The management at a large insurance company believes that the workers are more productive if they are happy with their jobs. To keep track of workers' satisfaction, the company regularly conducts surveys. According to a recent such survey, the mean job satisfaction score for all workers at this company was 13.10 (on a scale of 1 to 20) and the standard deviation was 1.95. Assume that the job satisfaction scores of workers are normally distributed.

    a. Find the probability that the job satisfaction score for a randomly selected worker from this company is less than 11.25.

    b. What percentage of the workers have a job satisfaction score between 14.50 and 18.70?

    c. A worker with a score of 8 or less is considered to be very unsatisfied with his/her job. What percentage of the workers are very unsatisfied with their jobs?

6.46 Most business schools require that every applicant for admission to a graduate degree program take the GMAT test. Suppose the GMAT scores of all students who took this test this year have a normal distribution with a mean of 550 and a standard deviation of 90.

    a. Sue Hopern scored 670 in this test. What percentage of the examinees scored higher than Sue?

    b. Joe Merck scored 610 in this test. What percentage of the examinees scored lower than Joe?

    c. A particular school requires that a student's GMAT score must be 500 or higher for consideration for admission. What percentage of the examinees are eligible for consideration for admission to this school?

6.47 The price–earnings ratio for a company is an indication of whether the stock of that company is undervalued or overvalued. If the price–earnings ratio is low, then the price of the stock is low compared to the earnings of that company and, consequently, the stock of that company is undervalued.

The converse is true if the price–earnings ratio is high for a company. Suppose the price–earnings ratios of all companies have a normal distribution with a mean of 20 and a standard deviation of 3.5.

    a.   If a price–earnings ratio of less than 11 is considered to be a relatively low ratio, what percentage of all companies have low price–earnings ratios?

    b.   If a price–earnings ratio of more than 28 is considered to be a relatively high ratio, what percentage of all companies have high price–earnings ratios?

6.48   Fast Auto Service guarantees that the maximum waiting time for its customers is 20 minutes for oil and lube service on their cars. It also guarantees that any customer who has to wait for more than 20 minutes for this service will receive a 50% discount on the charges. It is estimated that the mean time taken for oil and lube service at this garage is 15 minutes per car and the standard deviation is 2.4 minutes. Suppose the time taken for oil and lube service on a car follows a normal distribution.

    a.   What percentage of the customers will receive the 50% discount on their charges?

    b.   Is it possible that a car may take more than 25 minutes for oil and lube service? Explain.

6.49   The lengths of 3-inch nails manufactured on a machine are normally distributed with a mean of 3.0 inches and a standard deviation of .009 inches. The nails that are either less than 2.98 inches long or more than 3.02 inches long are unusable. What percentage of all the nails produced by this machine are unusable? (*Hint:* The required percentage is given by the sum of the areas in two tails of the normal curve, one to the left of $x = 2.98$ and the second to the right of $x = 3.02$.)

6.50   Lewis Corporation manufactures bicycles. The wheels of one type of bicycle are supposed to have a diameter of 30 inches. The processing system that makes these wheels does not produce every wheel with exactly 30-inch diameter. The diameters of wheels made by this system have a normal distribution with a mean of 30 inches and a standard deviation of .10 inches. The wheels that have a diameter of either less than 29.75 inches or more than 30.25 inches are discarded. What percentage of all the wheels produced by this processing system are discarded? (*Hint:* The required percentage is given by the sum of the areas in two tails of the normal curve, one to the left of $x = 29.75$ and the second to the right of $x = 30.25$.)

6.6   DETERMINING THE z AND x VALUES WHEN AN AREA UNDER THE NORMAL CURVE IS KNOWN

So far in this chapter we have discussed how to find the area under a normal curve for an interval of z or x. Now we reverse this procedure and learn how to find the corresponding value of z or x when an area under a normal curve is known. Examples 6–15 through 6–17 describe this procedure for finding the z value.

Finding z when area between the mean and z is known.

EXAMPLE 6–15   Find a point z such that the area under the standard normal curve between 0 and z is .4251 and the value of z is positive.

Solution   As shown in Figure 6.46, we are to find the z value such that the area between 0 and z is .4251.

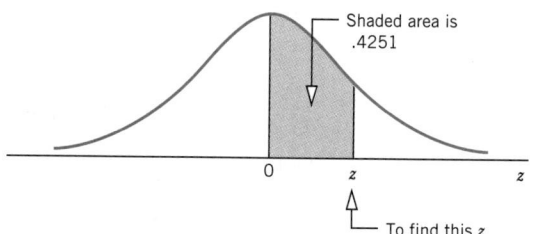

Figure 6.46   Finding the z value.

To find the required value of z, we locate .4251 in the body of the normal table, Table VII of Appendix B. The relevant portion of that table is reproduced below as Table 6.6. Next we read the numbers, in the column and row for z, which correspond to .4251. As shown in Table 6.6, these numbers are 1.4 and .04, respectively. Combining these two numbers, we obtain the required value of $z = \mathbf{1.44}$.

Table 6.6 Finding the z Value When Area Is Given

z	.00	.01	. . .	.04	. . .	.09
0.0	.0000	.0040	. . .	↑	. . .	.0359
0.1	.0398	.0438	. . .		. . .	.0753
0.2	.0793	.0832	. . .		. . .	.1141
.	.	.	. . .		. . .	.
.	.	.	. . .		. . .	.
1.4 ←				.4251 ←	. . .	. . .
.	.	.	. . .	.	. . .	.
.	.	.	. . .	.	. . .	.
3.0	.4987	.4987	. . .	.4988	. . .	.4990

We locate this
value in Table VII of
Appendix B

Finding z when area in the right tail is known.

EXAMPLE 6–16 Find the value of z such that the area under the standard normal curve in the right tail is .0050.

Solution To find the required value of z, first find the area between 0 and z. The total area to the right of $z = 0$ is .5. Hence,

$$\text{Area between 0 and } z = .5 - .0050 = .4950$$

This area is shown in Figure 6.47.

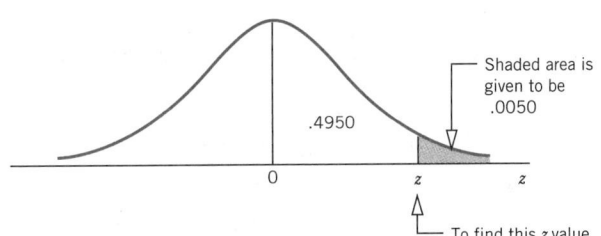

.4950

Shaded area is
given to be
.0050

To find this z value

Figure 6.47 Finding the z value.

Now we look for .4950 in the body of the normal table. Table VII does not contain .4950. So we find the value closest to .4950, which is either .4949 or .4951. We can use either of these two values. If we choose .4951, the corresponding z value is 2.58. Hence, the required value of z is **2.58** and the area to the right of $z = 2.58$ is approximately .0050. Note that there is no apparent reason to choose .4951 and not to choose .4949. We can use either of the two values. If we choose .4949, the corresponding z value will be 2.57.

Finding z when area in the left tail is known.

EXAMPLE 6–17 Find the value of z such that the area under the standard normal curve in the left tail is .05.

Solution Because .05 is smaller than .5 and it is the area in the left tail, the value of z is negative. To find the required value of z, first we find the area between 0 and $-z$. The total area to the left of $z = 0$ is .5. Hence,

$$\text{Area between 0 and } -z = .5 - .05 = .4500$$

This area is shown in Figure 6.48.

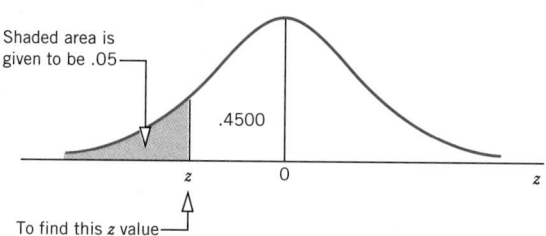

Shaded area is
given to be .05

.4500

z 0 z

To find this z value

Figure 6.48   Finding the z value.

Next, we look for .4500 in the body of the normal table. The value closest to .4500 in the normal table is either .4495 or .4505. Suppose we use the value .4505. The corresponding z value is 1.65. Because the value of z is negative in our example (see Figure 6.48), the required value of z is **−1.65** and the area to the left of $z = -1.65$ is approximately .05.   ■

To find an x value when an area under a normal distribution curve (that is not the standard normal distribution) is given, first we find the z value corresponding to that x value from the normal table. Then, to find the x value, we substitute the values of μ, σ, and z in the following formula, which is obtained from $z = (x - \mu)/\sigma$ by doing some algebraic manipulations. Actually, if we know the values of x, z, and σ, we can find μ using this same formula. Exercises 6.61 and 6.62 present such cases.

FINDING AN x VALUE FOR A NORMAL DISTRIBUTION

For a normal curve, with known values of μ and σ and for a given area under the curve between the mean and x, the x value is calculated as

$$z = \frac{x - \mu}{\sigma}$$

$$x = \mu + z\sigma$$

$$\mu = x \pm z\sigma$$

Examples 6–18 and 6–19 illustrate how to find an x value when an area under a normal curve is known.

Finding x when area in the left tail is known.

EXAMPLE 6–18   Recall Example 6–14. It is known that the life of a calculator manufactured by Intal Corporation has a normal distribution with a mean of 54 months and a standard deviation of 8 months. What should the warranty period be to replace a malfunctioning calculator if the company does not want to replace more than 1% of all the calculators sold?

Solution   Let x be the life of a calculator. Then, x follows a normal distribution with $\mu = 54$ months and $\sigma = 8$ months.

The calculators that would be replaced are the ones that start malfunctioning during the warranty period. The company's objective is to replace at most 1% of all the calculators sold. The shaded area in Figure 6.49 shows the proportion of calculators that are replaced. We are to find the value of x so that the area to the left of x under the normal curve is 1% or .01.

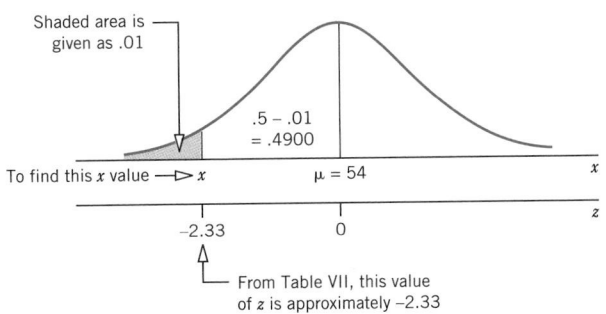

Figure 6.49 Finding an x value.

In the first step, we find the z value that corresponds to the required x value. To find this z value, we need to know the area between the mean and the x value. This area is calculated as follows.

$$\text{Area between the mean and the } x \text{ value} = .5 - .01 = .4900$$

Next, we find the z value from the normal table for .4900. Table VII of Appendix B does not contain a value that is exactly .4900. The value closest to .4900 in the table is .4901 and the z value for .4901 is 2.33. Since z is on the left side of the mean (see Figure 6.49), the value of z is negative. Hence,

$$z = -2.33$$

Substituting the values of μ, σ, and z in the formula $x = \mu + z\,\sigma$, we obtain the value of x as

$$x = \mu + z\,\sigma = 54 + (-2.33)\,(8) = 54 - 18.64 = \mathbf{35.36}$$

Thus, the company should replace all the calculators that start malfunctioning within 35.36 months (which can be rounded to 35 months) of the date of purchase so that they will not have to replace more than 1% of the calculators. ∎

Finding x when area in the right tail is known.

EXAMPLE 6-19 Most business schools require that every applicant for admission to a graduate degree program take the GMAT. Suppose the GMAT scores of all students have a normal distribution with a mean of 550 and a standard deviation of 90. Matt Sanger is planning to take this test soon. What should his score be on this test so that only 10% of all the examinees score higher than he does?

Solution: Let x represent the GMAT scores of examinees. Then, x follows a normal distribution with $\mu = 550$ and $\sigma = 90$. We are to find the value of x such that the area under the curve to the right of x is 10%, as shown in Figure 6.50.

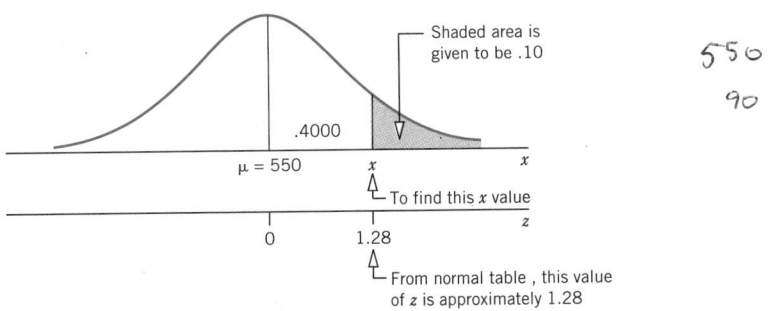

Figure 6.50 Finding an x value.

First we find the area under the normal curve between the mean and the x value.

$$\text{Area between } \mu \text{ and the } x \text{ value} = .5 - .10 = .4000$$

To find the z value that corresponds to the required x value, we look for .4000 in the body of the normal table. The value closest to .4000 in Table VII is .3997 and the corresponding z value is 1.28. Hence, the value of x is computed as

$$x = \mu + z\sigma = 550 + 1.28\,(90) = 550 + 115.2 = 665.2 \approx \mathbf{666}$$

Thus, if Matt Sanger scores 666 on the GMAT, only about 10% of the examinees are expected to score higher than he does. ▬

EXERCISES

Concepts and Procedures

6.51 Find the value of z (or $-z$) so that the area under the standard normal curve

    a. from 0 to z is .4772 and z is positive

    b. between 0 and z is (approximately) .4785 and z is negative

    c. in the left tail is (approximately) .3565

    d. in the right tail is (approximately) .1530

6.52 Find the value of z (or $-z$) so that the area under the standard normal curve

    a. from 0 to z is (approximately) .1965 and z is positive

    b. between 0 and z is (approximately) .2740 and z is negative

    c. in the left tail is (approximately) .2050

    d. in the right tail is (approximately) .1053

6.53 Determine the value of z (or $-z$) so that the area under the standard normal curve

    a. in the right tail is .0500

    b. in the left tail is .0250

    c. in the left tail is .0100

    d. in the right tail is .0050

6.54 Determine the value of z (or $-z$) so that the area under the standard normal curve

    a. in the right tail is .0250

    b. in the left tail is .0500

    c. in the left tail is .0010

    d. in the right tail is .0100

6.55 Let x be a continuous random variable that follows a normal distribution with a mean of 200 and a standard deviation of 25.

    a. Find the value of x so that the area under the normal curve to the left of x is approximately .6330.

    b. Find the value of x so that the area under the normal curve to the right of x is approximately .05.

    c. Find the value of x so that the area under the normal curve to the right of x is .8051.

    d. Find the value of x so that the area under the normal curve to the left of x is .015.

    e. Find the value of x so that the area under the normal curve between μ and x is .4525 and the value of x is smaller than μ.

    f. Find the value of x so that the area under the normal curve between μ and x is approximately .4800 and the value of x is greater than μ.

6.56 Let x be a continuous random variable that follows a normal distribution with a mean of 550 and a standard deviation of 75.

    a. Find the value of x so that the area under the normal curve to the left of x is .0250.

    b. Find the value of x so that the area under the normal curve to the right of x is .9345.

    c. Find the value of x so that the area under the normal curve to the right of x is approximately .0275.

    d. Find the value of x so that the area under the normal curve to the left of x is approximately .9600.

    e. Find the value of x so that the area under the normal curve between μ and x is approximately .4700 and the value of x is smaller than μ.

    f. Find the value of x so that the area under the normal curve between μ and x is approximately .4100 and the value of x is greater than μ.

Applications

6.57 Fast Auto Service provides oil and lube service for cars. It is known that the mean time taken for oil and lube service at this garage is 15 minutes per car and the standard deviation is 2.4 minutes. The management wants to promote the business by guaranteeing a maximum waiting time for its customers. If a customer's car is not serviced within that period, the customer will receive a 50% discount on the charges. The company wants to limit this discount to at most 5% of the customers. What should the maximum guaranteed waiting time be? Assume that the times taken for oil and lube service for all cars have a normal distribution.

6.58 The management of a supermarket wants to adopt a new promotional policy of giving a free gift to every customer who spends more than a certain amount per visit at this supermarket. The expectation of the management is that after this promotional policy is advertised, the expenditures for all customers at this supermarket will be normally distributed with a mean of $95 and a standard deviation of $21. If the management wants to give free gifts to at most 8% of the customers, what should the amount be above which a customer would receive a free gift?

6.59 According to the records of the electric company serving the Boston area, the mean electric consumption during winter for all households is 1650 kilowatt hours per month. Assume that the monthly electric consumption during winter by all households in this area has a normal distribution with a mean of 1650 kilowatt hours and a standard deviation of 320 kilowatt hours. The company sent a notice to Bill Johnson informing him that about 90% of the households use less electricity per month than he does. What is Bill Johnson's monthly electric consumption?

6.60 Rockingham Corporation makes electric shavers. The life (period before which a shaver does not need a major repair) of Model J795 of an electric shaver manufactured by this corporation has a normal distribution with a mean of 65 months and a standard deviation of 6 months. The company is to determine the warranty period for this shaver. Any shaver that will need a major repair during this warranty period will be replaced free by the company.

    a. What should the warranty period be if the company does not want to replace more than 1% of the shavers?

    b. What should the warranty period be if the company does not want to replace more than 5% of the shavers?

6.61 A study has shown that 20% of all college textbooks have a price of $70 or higher. It is known that the standard deviation of the prices of all college textbooks is $9.50. Suppose the prices of all college textbooks have a normal distribution. What is the mean price of all college textbooks?

6.62 A machine at Keats Corporation fills 64-ounce detergent jugs. The machine can be set to pour, on average, any amount of detergent into these jugs. However, the machine does not pour exactly the same amount of detergent in each jug; it varies from jug to jug. It is known that the net amount of detergent poured into each jug has a normal distribution with a standard deviation of .4 ounces. The quality control inspector wants to set the machine such that at least 95% of the jugs have more than 64 ounces of detergent. What should the mean amount of detergent poured by this machine into these jugs be?

6.7 THE NORMAL APPROXIMATION TO THE BINOMIAL DISTRIBUTION

Recall from Chapter 5 that

1. The binomial distribution is applied to a discrete random variable.
2. Each repetition, called a trial, of a binomial experiment results in one of two possible outcomes, either a success or a failure.
3. The probabilities of the two (possible) outcomes remain the same for each repetition of the experiment.
4. The trials are independent.

The binomial formula, which gives the probability of x successes in n trials, is

$$P(x) = \binom{n}{x} p^x \, q^{n-x}$$

However, the use of the binomial formula becomes very tedious when n is large. In such cases, the normal distribution can be used to approximate the binomial probability. Note that for a binomial problem, the exact probability is obtained by using the binomial formula. If we apply the normal distribution to solve a binomial problem, the probability that we obtain is an approximation to the exact probability. The approximation obtained by using the normal distribution is very close to the exact probability when n is large and p is very close to .50. However, this does not mean that we should not use the normal approximation when p is not close to .50. The reason for the approximation being closer to the exact probability when p is close to .50 is that the binomial distribution is symmetric when $p = .50$. The normal distribution is always symmetric. Hence, the two distributions are very close to each other when n is large and p is close to .50. However, this does not mean that whenever $p = .50$ the binomial distribution is the same as the normal distribution because not every symmetric bell-shaped curve is a normal curve.

NORMAL DISTRIBUTION AS AN APPROXIMATION TO BINOMIAL DISTRIBUTION

Usually, the normal distribution is used as an approximation to the binomial distribution when np and nq are both greater than 5, that is, when

$$np > 5 \quad \text{and} \quad nq > 5$$

Table 6.7 gives the binomial probability distribution of x for $n = 12$ and $p = .50$. This table is constructed using Table IV of Appendix B. Figure 6.51 shows the histogram and the smoothed polygon for the probability distribution of Table 6.7. As we can observe, the histogram in Figure 6.51 is symmetric and the curve obtained by joining the upper midpoints of the rectangles is approximately bell-shaped.

Examples 6–20 through 6–22 illustrate the application of the normal distribution as an approximation to the binomial distribution.

Table 6.7 The Binomial Probability Distribution for $n = 12$ and $p = .50$

x	$P(x)$
0	.0002
1	.0029
2	.0161
3	.0537
4	.1208
5	.1934
6	.2256
7	.1934
8	.1208
9	.0537
10	.0161
11	.0029
12	.0002

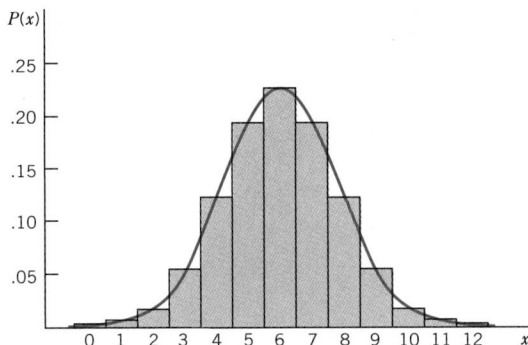

Figure 6.51 Histogram for the probability distribution of Table 6.7.

Normal approximation to binomial: x equals a specific value.

EXAMPLE 6-20 According to an estimate, 50% of the people in America have at least one credit card. If a random sample of 30 persons is taken, what is the probability that 19 of them will have at least one credit card?

Solution   Let n be the total number of persons in the sample, x be the number of persons in the sample who have at least one credit card, and p be the probability that a person has at least one credit card. Then, this is a binomial problem with

$$n = 30, \quad p = .50, \quad q = 1 - p = .50,$$
$$x = 19, \quad \text{and} \quad n - x = 30 - 19 = 11$$

Using the binomial formula, the exact probability that 19 persons in a sample of 30 have at least one credit card is

$$P(x = 19) = \binom{30}{19} (.50)^{19} (.50)^{11} = \mathbf{.0509}$$

Now let us solve this problem using the normal distribution as an approximation to the binomial distribution. For this example,

$$np = 30 \,(.50) = 15 \quad \text{and} \quad nq = 30 \,(.50) = 15$$

Since np and nq are both greater than 5, we can use the normal distribution as an approximation to solve this binomial problem.

Using the normal distribution as an approximation to the binomial involves the following three steps.

Step 1. *Compute* μ *and* σ *for the binomial distribution*

To use the normal distribution, we need to know the mean and standard deviation of the distribution. Hence, the first step in using the normal approximation to the binomial distribution is to compute the mean and standard deviation of the binomial distribution. As we know from Chapter 5, the mean and standard deviation of a binomial distribution are given by np and $\sqrt{npq}$, respectively. Using these formulas, we obtain:

$$\mu = np = 30 \,(.50) = 15$$

$$\sigma = \sqrt{npq} = \sqrt{30 \,(.50) \,(.50)} = 2.74$$

Step 2. *Convert the discrete random variable to a continuous random variable*

The normal distribution applies to a continuous random variable, whereas the binomial distribution applies to a discrete random variable. The second step in applying the normal approximation to the binomial distribution is to convert the discrete random variable to a continuous random variable by making the **correction for continuity**.

As shown in Figure 6.52, the probability of 19 successes in 30 trials is given by the area of the rectangle for $x = 19$. To make the correction for continuity, we use the interval 18.5 to 19.5 for 19 persons. This interval is actually given by the two boundaries of the rectangle for $x = 19$, which is obtained by subtracting .5 from 19 and by adding .5 to 19. Thus, $P(x = 19)$ for the binomial problem will be approximately equal to $P(18.5 \leq x \leq 19.5)$ for the normal distribution.

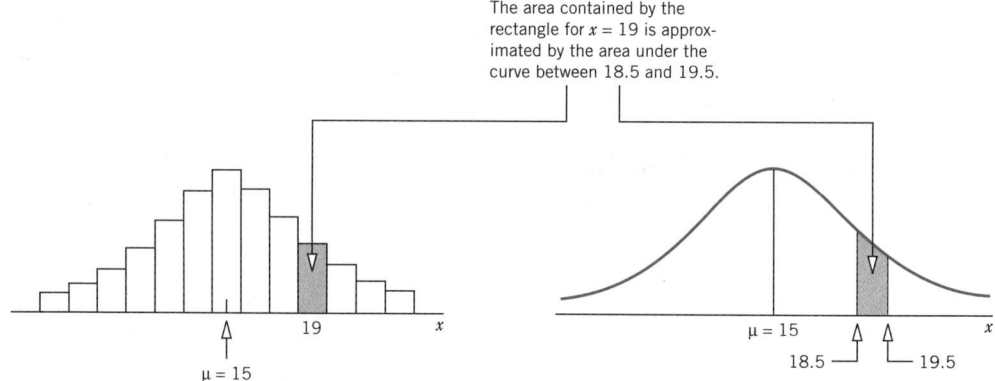

Figure 6.52

Step 3. *Compute the required probability using the normal distribution*

As shown in Figure 6.53, the area under the normal curve between $x = 18.5$ and $x = 19.5$ will give us the (approximate) probability that 19 persons possess at least one credit card. We calculate this probability as follows.

$$\text{For } x = 18.5: \quad z = \frac{18.5 - 15}{2.74} = 1.28$$

$$\text{For } x = 19.5: \quad z = \frac{19.5 - 15}{2.74} = 1.64$$

The required probability is given by the area under the standard normal curve between

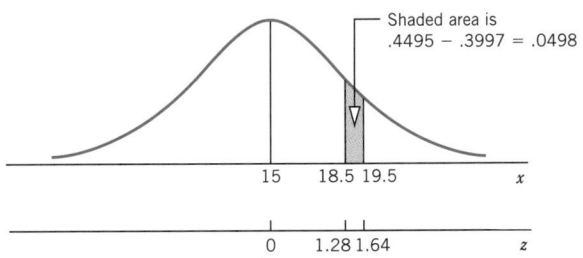

Figure 6.53 Area between $x = 18.5$ and $x = 19.5$.

$z = 1.28$ and $z = 1.64$. This area is obtained by subtracting the area between $z = 0$ and $z = 1.28$ from the area between $z = 0$ and $z = 1.64$. From Table VII of Appendix B, the area between $z = 0$ and $z = 1.28$ is .3997 and the area between $z = 0$ and $z = 1.64$ is .4495. Hence, the required probability is

$$P(18.5 \le x \le 19.5) = P(1.28 \le z \le 1.64) = .4495 - .3997 = \textbf{.0498}$$

Thus, based on the normal approximation, the probability that 19 persons in a sample of 30 will possess at least one credit card is approximately .0498. Earlier, using the binomial formula, we obtained the exact probability .0509. The error due to using the normal approximation is .0509 − .0498 = .0011. Thus, the exact probability is underestimated by .0011 if the normal approximation is used. ◼

☞ *Remember* When applying the normal distribution as an approximation to the binomial distribution, always make a correction for continuity. The continuity correction is made by subtracting .5 from the lower limit of the interval and/or by adding .5 to the upper limit of the interval. For example, the binomial probability $P(7 \le x \le 12)$ will be approximated by the probability $P(6.5 \le x \le 12.5)$ for the normal distribution; the binomial probability $P(x \ge 9)$ will be approximated by the probability $P(x \ge 8.5)$ for the normal distribution; and the binomial probability $P(x \le 10)$ will be approximated by the probability $P(x \le 10.5)$ for the normal distribution. Note that the probability $P(x \ge 9)$ has only the lower limit of 9 and no upper limit, and the probability $P(x \le 10)$ has only the upper limit of 10 and no lower limit.

Normal approximation to binomial: x assumes a value in an interval.

EXAMPLE 6–21 Due to tough competition from large companies, it has become very difficult for small businesses to survive and to be successful. According to the Small Business Administration, 23.7% of small businesses dissolve (that is, either they are voluntarily shut down or they file for bankruptcy) within 2 years of inception (*The Wall Street Journal*, October 16, 1992). What is the probability that out of a sample of 50 small businesses that were started recently, 7 to 9 will dissolve within 2 years?

Solution Let n be the total number of small businesses in the sample, x be the number of small businesses in the sample that will dissolve within two years, and p be the probability that a small business dissolves within 2 years. Then, this is a binomial problem with

$$n = 50, \quad p = .237, \quad \text{and} \quad q = 1 - .237 = .763$$

We are to find the probability of 7 to 9 successes in 50 trials. Because n is large, it is easier to apply the normal approximation than to use the binomial formula. We can check that np and nq are both greater than 5. The mean and standard deviation of the binomial distribution are

$$\mu = np = 50\,(.237) = 11.85$$
$$\sigma = \sqrt{npq} = \sqrt{50\,(.237)\,(.763)} = 3.007$$

For the continuity correction, we subtract .5 from 7 and add .5 to 9 to obtain the interval 6.5 to 9.5. Thus, the probability that 7 to 9 small businesses will dissolve within 2 years is approximated by the area under the normal curve from $x = 6.5$ to $x = 9.5$. This area is shown in Figure 6.54.

$$\text{For } x = 6.5: \qquad z = \frac{6.5 - 11.85}{3.007} = -1.78$$

$$\text{For } x = 9.5: \qquad z = \frac{9.5 - 11.85}{3.007} = -.78$$

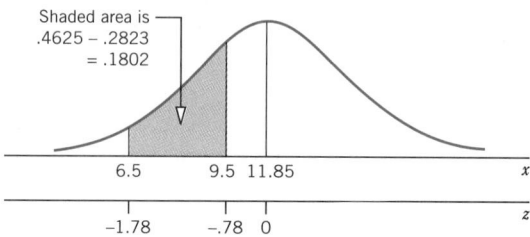

Figure 6.54   Area from $x = 6.5$ to $x = 9.5$.

The required probability is given by the area under the standard normal curve between $z = -1.78$ and $z = -.78$. This area is obtained by subtracting the area between $z = -.78$ and $z = 0$ from the area between $z = -1.78$ and $z = 0$. From Table VII of Appendix B, the area between $z = -.78$ and $z = 0$ is .2823 and the area between $z = -1.78$ and $z = 0$ is .4625. Hence,

$$P(6.5 \le x \le 9.5) = P(-1.78 \le z \le -.78) = .4625 - .2823 = \mathbf{.1802}$$

Thus, the probability that 7 to 9 small businesses in a sample of 50 would dissolve within 2 years is approximately .1802.   ■

Normal approximation to binomial: x greater than or equal to a value.

EXAMPLE 6–22 Due to cuts in federal funds and increases in expenses to maintain the provision of services to residents, more and more cities and states are facing ever-increasing deficit budget problems. As a result, almost all states have started lotteries and many cities either have resorted to casino gambling to cover the budget deficits or are considering doing so. In a Gallup Organization poll, 54% of adults polled approved casino gambling. Assume that this percentage is true for the current population of all adults. What is the probability that in a random sample of 100 adults, 60 or more will approve casino gambling?

Solution Let n be the total number of adults in the sample, x be the number of adults in the sample who approve casino gambling, and p be the probability that an adult approves casino gambling. Then, this is a binomial problem with

$$n = 100, \qquad p = .54, \qquad \text{and} \qquad q = 1 - .54 = .46$$

We are to find the probability of 60 or more successes in 100 trials. The mean and standard deviation of the binomial distribution are

$$\mu = np = 100\,(.54) = 54$$
$$\sigma = \sqrt{npq} = \sqrt{100\,(.54)\,(.46)} = 4.984$$

For the continuity correction, we subtract .5 from 60, which gives 59.5. Thus, the probability that 60 or more adults out of a random sample of 100 will approve casino gambling is approximated by the area under the normal curve to the right of $x = 59.5$, as shown in Figure 6.55.

$$\text{For } x = 59.5: \qquad z = \frac{59.5 - 54}{4.984} = 1.10$$

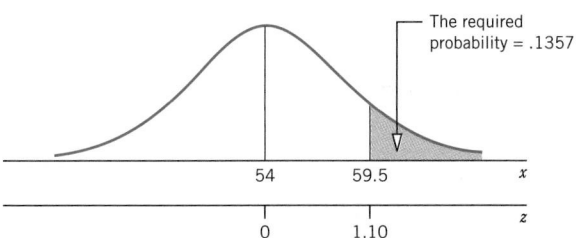

Figure 6.55 Area to the right of $x = 59.5$.

To find the required probability, we find the area between $z = 0$ and $z = 1.10$ and then subtract this area from .5, which is the total area to the right of $z = 0$. From Table VII of Appendix B, the area between $z = 0$ and $z = 1.10$ is .3643. Hence,

$$P(x \geq 59.5) = P(z \geq 1.10) = .5 - .3643 = \mathbf{.1357}$$

Thus, the probability that 60 or more adults in a random sample of 100 will approve casino gambling is approximately .1357.

EXERCISES

Concepts and Procedures

6.63 Under what conditions is the normal distribution usually used as an approximation to the binomial distribution?

6.64 For a binomial probability distribution, $n = 20$ and $p = .60$.
  a. Find the probability $P(x = 11)$ by using the table of binomial probabilities (Table IV of Appendix B).
  b. Find the probability $P(x = 11)$ by using the normal distribution as an approximation to the binomial distribution. What is the difference between this approximation and the exact probability calculated in part a?

6.65 For a binomial probability distribution, $n = 25$ and $p = .40$.
  a. Find the probability $P(8 \leq x \leq 12)$ by using the table of binomial probabilities (Table IV of Appendix B).
  b. Find the probability $P(8 \leq x \leq 12)$ by using the normal distribution as an approximation to the binomial distribution. What is the difference between this approximation and the exact probability calculated in part a?

6.66 For a binomial probability distribution, $n = 80$ and $p = .50$. Let x be the number of successes in 80 trials.
  a. Find the mean and standard deviation of the binomial distribution.
  b. Find $P(x \geq 37)$ using the normal approximation.
  c. Find $P(41 \leq x \leq 44)$ using the normal approximation.

6.67 For a binomial probability distribution, $n = 120$ and $p = .60$. Let x be the number of successes in 120 trials.

    a. Find the mean and standard deviation of the binomial distribution.
    b. Find $P(x \leq 70)$ using the normal approximation.
    c. Find $P(67 \leq x \leq 71)$ using the normal approximation.

6.68 Find the following binomial probabilities using the normal approximation.

    a. $n = 140,$ $p = .45,$ $P(x = 67)$
    b. $n = 100,$ $p = .55,$ $P(52 \leq x \leq 60)$
    c. $n = 90,$ $p = .42,$ $P(x \geq 40)$
    d. $n = 104,$ $p = .75,$ $P(x \leq 72)$

6.69 Find the following binomial probabilities using the normal approximation.

    a. $n = 70,$ $p = .30,$ $P(x = 18)$
    b. $n = 200,$ $p = .70,$ $P(133 \leq x \leq 145)$
    c. $n = 85,$ $p = .40,$ $P(x \geq 30)$
    d. $n = 150,$ $p = .38,$ $P(x \leq 62).$

Applications

6.70 Based on data from a Peter D. Hart Research Associates' poll on consumer buying habits and attitudes, Peter Hart estimated that 21% of American shoppers are *practical shoppers*, that is, the "smart shoppers who research their purchases and look for the best deal" (*The American Way of Buying*, Dow Jones & Company, Inc., 1990). Assume that this result is true for the current population of all American shoppers. Find the probability that in a random sample of 200 American shoppers, 37 to 46 will be practical shoppers.

6.71 According to the U.S. Bureau of Labor Statistics, 5% of people working in the United States moonlight (that is, they hold a second job). Find the probability that in a random sample of 400 workers, 12 to 16 workers moonlight.

6.72 In a poll of chief executives of large companies conducted by the opinion research firm Clark Martire & Bartolomeo for *Fortune* magazine, 40% of CEOs said that it is *not likely at all* that their companies will have a female CEO within the next decade (*Fortune*, September 21, 1992). Assume that this result is true for the current population of CEOs of all large companies.

    a. Find the probability that in a random sample of 250 CEOs of large companies, exactly 95 will hold this view.
    b. Find the probability that in a random sample of 250 CEOs of large companies, at most 85 will hold this view.
    c. What is the probability that in a random sample of 250 CEOs of large companies, 105 to 110 will hold this view?

6.73 According to the Motor Vehicle Manufacturers Association, 19% of cars in the United States were at least 12 years old in 1989 (*Business Week*, March 4, 1991). Assume that this result holds true for the current population of all cars in the United States.

    a. Find the probability that in a random sample of 500 cars, exactly 92 are at least 12 years old.
    b. Find the probability that in a random sample of 500 cars, 100 or more are at least 12 years old.
    c. What is the probability that in a random sample of 500 cars, 90 to 98 are at least 12 years old?

6.74 According to a survey, 30% of credit card holders pay off their balances in full each month (*The Wall Street Journal*, August 20, 1990). Assume that this result holds true for the current population of credit card holders.

    a. Find the probability that in a random sample of 400 credit card holders, exactly 125 pay off their balances in full each month.
    b. Find the probability that in a random sample of 400 credit card holders, at least 110 pay off their balances in full each month.

c.   What is the probability that in a random sample of 400 credit card holders, 115 to 130 pay off their balances in full each month?

6.75   According to a survey, 20.8% of the lawyers and judges in the United States are women (*Self*, January 1992). Assume that this result holds true for the current population of all lawyers and judges in the United States.

a.   Find the probability that in a random sample of 200 lawyers and judges, exactly 35 are women.

b.   Find the probability that in a random sample of 200 lawyers and judges, at most 45 are women.

c.   What is the probability that in a random sample of 200 lawyers and judges, 43 to 50 are women?

6.76   A fast food chain store conducted a taste survey before marketing a new hamburger. The results of the survey showed that 70% of the people who tried this hamburger liked it. Encouraged by this result, the company decided to market the new hamburger. Assume that 70% of all people like this hamburger. On a certain day, 100 customers bought this hamburger.

a.   Find the probability that exactly 65 of the 100 customers will like this hamburger.

b.   What is the probability that 60 or less of the 100 customers will like this hamburger?

c.   What is the probability that 75 to 80 of the 100 customers will like this hamburger?

6.77   Johnson Electronics makes calculators. Consumer satisfaction is one of the top priorities of the company's management. The company guarantees refund of money or a replacement for any calculator that malfunctions within two years from the date of purchase. It is known from past data that despite all efforts, 5% of the calculators manufactured by the company malfunction within a two-year period. The company recently mailed 500 such calculators to its customers.

a.   Find the probability that exactly 28 of the 500 calculators will be returned for refund or replacement within a 2-year period.

b.   What is the probability that 26 or more of the 500 calculators will be returned for refund or replacement within a 2-year period?

c.   What is the probability that 15 to 20 of the 500 calculators will be returned for refund or replacement within a 2-year period?

6.78   Hurbert Corporation makes font cartridges for laser printers that it sells to Alpha Electronics Inc. The cartridges are shipped to Alpha Electronics in lots of 2000 each. The quality control department at Alpha Electronics randomly selects 100 cartridges from each shipment and inspects them for being good or defective. If this sample contains 7 or more defective cartridges, the entire shipment is rejected. Hurbert Corporation promises that of all the cartridges, only 5% are defective.

a.   Find the probability that a given shipment of 2000 cartridges received by Alpha Electronics will be accepted.

b.   Find the probability that a given shipment of 2000 cartridges received by Alpha Electronics will not be accepted.

6.8   THE UNIFORM PROBABILITY DISTRIBUTION

Another probability distribution that a continuous random variable can follow is the uniform probability distribution. Let x be a continuous random variable. If the probability is the same for x to assume a value in any of the intervals that are of the same width, then x is said to be a **uniform random variable** and it possesses a **uniform probability distribution**. The graph of the uniform probability distribution has a rectangular shape, as shown in Figure 6.56. In this figure, x is a continuous random variable that is uniformly distributed over the range of values from $x = a$ to $x = b$. Here, x cannot assume a value less than a or greater than b.

Figure 6.56   The graph of the uniform probability distribution.

The area of a rectangle such as the one in Figure 6.56 is calculated as follows.

$$\text{Area of the rectangle} = \text{Width} \times \text{Height}$$

Note that the width of the rectangle in Figure 6.56 is $(b - a)$. The area under the curve in Figure 6.56 is given by the area of the rectangle. Since the total area of the rectangle must be 1.0, the height of the rectangle is

$$\text{Height} = \frac{\text{Area}}{\text{Width}} = \frac{1}{b - a}$$

Thus, for the uniform probability distribution, it is always true that

$$\text{Height} = \frac{1}{\text{Width}}$$

UNIFORM PROBABILITY DISTRIBUTION

Let x be a continuous random variable that possesses a uniform probability distribution in the interval $x = a$ to $x = b$. Then the probability density function of the uniform random variable x is

$$f(x) = \frac{1}{b - a} \quad \text{where } a \leq x \leq b$$

Here $f(x)$, called the probability density function, represents the height of the probability distribution curve at a point x. In the case of the uniform probability distribution, $f(x)$ gives the height of the rectangle.

The mean and standard deviation of the uniform probability distribution are

$$\mu = \frac{a + b}{2} \quad \text{and} \quad \sigma = \sqrt{\frac{(b - a)^2}{12}}$$

The probability that a uniformly distributed continuous random variable x assumes a value in an interval c to d is given by the rectangular area between c and d, as shown in Figure 6.57.

The width of the shaded rectangle in Figure 6.57 is $(d - c)$ and its height is $1/(b - a)$. Consequently, for Figure 6.57,

$$\text{Area of the shaded rectangle} = \text{Width} \times \text{Height} = \frac{d - c}{b - a}$$

This area gives the probability that a uniformly distributed continuous random variable x assumes a value between c and d.

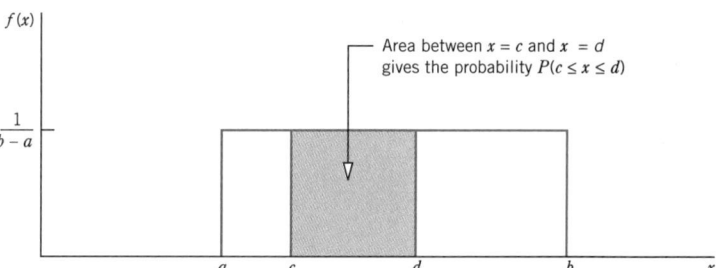

Figure 6.57   Area between two points for a uniform probability distribution.

THE PROBABILITY FOR THE UNIFORM PROBABILITY DISTRIBUTION

Let x be a continuous random variable that is uniformly distributed in an interval a to b. The probability that x assumes a value in an interval c to d is

$$P(c \leq x \leq d) = \frac{d - c}{b - a}$$

Example 6–23 illustrates the application of the uniform probability distribution.

Applications of the uniform probability distribution.

EXAMPLE 6–23   Sach Corporation manufactures toys. The time taken by a new worker to learn how to assemble a particular toy is uniformly distributed in the interval 20 to 27 hours.

(a)   Let x be the time taken by a new worker to learn how to assemble this toy. Draw the graph of the probability distribution of x.

(b)   Calculate the mean and standard deviation of the probability distribution of x.

(c)   What is the probability that a new worker will take 22 to 24 hours to learn how to assemble this toy?

(d)   What is the probability that a new worker will take 26 hours or more to learn how to assemble this toy?

(e)   What is the probability that a new worker will take at most 23 hours to learn how to assemble this toy?

Solution

Graphing a uniform probability distribution.

(a)   In this example, $a = 20$ and $b = 27$. The graph of x is given in Figure 6.58.

Figure 6.58   The graph of the uniform probability distribution.

The width of the rectangle extends from 20 to 27 hours. The height of the rectangle is

$$\text{Height} = \frac{1}{27 - 20} = \frac{1}{7}$$

Finding μ and σ for a uniform distribution.

(b) The mean and standard deviation of the probability distribution of x are as follows.

$$\mu = \frac{a + b}{2} = \frac{20 + 27}{2} = \mathbf{23.50}$$

$$\sigma = \sqrt{\frac{(b - a)^2}{12}} = \sqrt{\frac{(27 - 20)^2}{12}} = \mathbf{2.021}$$

Probability that x is in an interval for a uniform distribution.

(c) The probability that x assumes a value in the interval 22 to 24 is given by the shaded area in Figure 6.59.

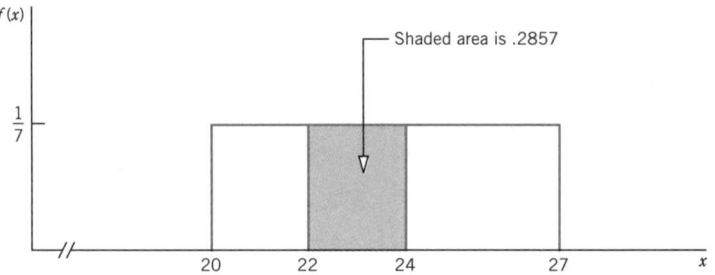

Figure 6.59 Area between $x = 22$ and $x = 24$.

The width of the rectangle made by the shaded area in Figure 6.59 is $24 - 22 = 2$ and its height is $1/7$. Hence, the required probability is

$$P(22 \le x \le 24) = \text{Width} \times \text{Height} = 2 \times \frac{1}{7} = \mathbf{.2857}$$

Thus, the probability is .2857 that a new worker will take 22 to 24 hours to learn the assembly procedure.

Probability that x is greater than or equal to a number.

(d) The probability that a worker will take 26 hours or more to learn the assembly procedure is given by the area of the rectangle from $x = 26$ to $x = 27$, as shown in Figure 6.60. Note that x cannot be greater than 27.

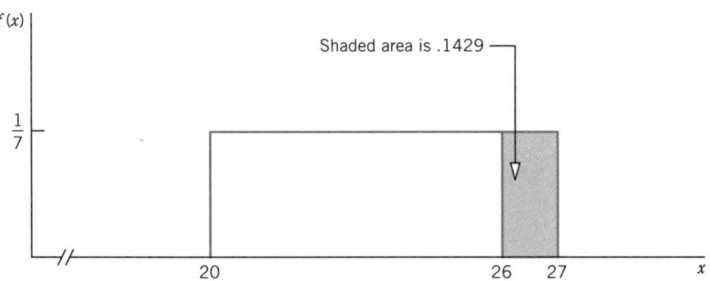

Figure 6.60 Area to the right of $x = 26$.

The width of the shaded rectangle in Figure 6.60 is $27 - 26 = 1$ and its height is $1/7$. The required probability is

$$P(x \ge 26) = P(26 \le x \le 27) = \text{Width} \times \text{Height} = 1 \times \frac{1}{7} = \mathbf{.1429}$$

Thus, the probability is .1429 that a new worker will take 26 or more hours to learn the assembly procedure.

Probability that x is less than or equal to a number.

(e) The probability that a new worker will take at most 23 hours to learn the assembly procedure is given by the area of the rectangle from $x = 20$ to $x = 23$, as shown in Figure 6.61. Note that x cannot be less than 20.

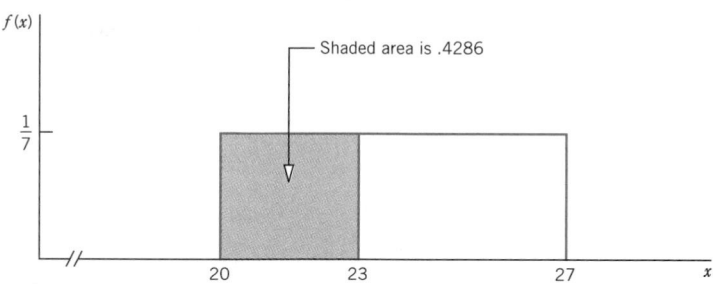

Figure 6.61 Area to the left of $x = 23$.

The width of the shaded rectangle in Figure 6.61 is $23 - 20 = 3$ and its height is $1/7$. The required probability is

$$P(x \leq 23) = P(20 \leq x \leq 23) = \text{Width} \times \text{Height} = 3 \times \frac{1}{7} = \textbf{.4286}$$

Thus, the probability that a new worker will take 23 hours or less to learn the assembly procedure is .4286.

EXERCISES

Concepts and Procedures

6.79 Briefly describe the uniform probability distribution. Give one example and illustrate it by drawing its graph.

6.80 Let x be a random variable that is uniformly distributed in the interval a to b. What are the width and height of the probability distribution of x?

6.81 Let x be a continuous random variable that is uniformly distributed in the interval 30 to 70.

    a. Draw the graph of the probability distribution of x.
    b. Calculate the mean and standard deviation of the probability distribution of x.
    c. Find the probability that x assumes a value in the interval 40 to 50.
    d. Find the probability that x assumes a value of 63 or greater.
    e. What is the probability that x assumes a value of 42 or lower?

6.82 Let x be a continuous random variable that is uniformly distributed in the interval 22 to 44.

    a. Draw the graph of the probability distribution of x.
    b. Calculate the mean and standard deviation of the probability distribution of x.
    c. Find the probability that x assumes a value in the interval 34 to 40.
    d. Find the probability that x assumes a value of 32 or greater.
    e. What is the probability that x assumes a value of 35 or lower?

Applications

6.83 Newcastle Mail Order Company claims that the delivery time for any order placed with the company is uniformly distributed in the interval 14 to 20 days.

    a. Let x be the delivery time for an order placed with this company. Draw the graph of the probability distribution of x.

b. Calculate the mean and standard deviation of the probability distribution of x.
c. What is the probability that a new order will be delivered in 15 to 18 days?
d. What is the probability that a new order will be delivered in 16 days or less?
e. What is the probability that a new order will take at least 18 days to be delivered?

6.84 Let x be a continuous random variable that denotes the time taken to service an auto at Eagleville Service Station. It is known that x is uniformly distributed in the interval 20 to 30 minutes.

a. Draw the graph of the probability distribution of x.
b. Calculate the mean and standard deviation of the probability distribution of x.
c. What is the probability that the service time for a randomly selected auto will be less than 24 minutes?
d. What is the probability that it will take more than 28 minutes to service an auto?
e. What is the probability that a randomly selected auto will take 21 to 24 minutes to be serviced?

6.85 Let x be a continuous random variable that denotes the weight of a letter package mailed through Federal Express. It is known that x is uniformly distributed in the interval 5 ounces to 20 ounces.

a. Draw the graph of the probability distribution of x.
b. Calculate the mean and standard deviation of the probability distribution of x.
c. What is the probability that the weight of a randomly selected letter package is 15 ounces or more?
d. What is the probability that the weight of a randomly selected letter package is 8 ounces or less?
e. What is the probability that the weight of a randomly selected letter package is between 12 and 16 ounces?

6.86 Becker Electronics Company manufactures computer chips. To check that the system is working properly, the quality control inspector takes a sample of chips from the production line every day and inspects them. The inspection time for a chip is uniformly distributed in the interval 10 to 16 minutes.

a. Let x be the inspection time for a chip. Draw the graph of the probability distribution of x.
b. Calculate the mean and standard deviation of the probability distribution of x.
c. What is the probability that a chip will take 11 to 12 minutes for inspection?
d. What is the probability that a chip will take at most 13 minutes for inspection?
e. What is the probability that a chip will take 14 minutes or more for inspection?

6.9 THE EXPONENTIAL PROBABILITY DISTRIBUTION

The **exponential probability distribution** is another important probability distribution that a continuous random variable can follow. This probability distribution is closely related to the Poisson probability distribution discussed in Chapter 5.

In the Poisson probability distribution, we deal with arrival or occurrence rates per interval, such as the number of customers arriving at a bank during a half-hour interval, the number of accidents at a factory during a month, the number of breakdowns of a machine per week, etc.

In the exponential probability distribution, we deal with the lapse of time between any two successive occurrences. For instance, in the exponential probability distribution, we can find the probability that the next customer at a bank will not arrive for another 10 minutes after the previous customer's arrival, that the next accident at a factory will not occur for three months after the previous accident's occurrence, or that a machine will not break down for five weeks after the occurrence of a previous breakdown. The exponential probability distribution has only one parameter λ, which denotes the average number of occurrences per unit of time. Then, we can find the probability that the lapse time between two successive occurrences is more than "a" units of time. Note that in the exponential probability distribution, the occurrences must follow the Poisson probability distribution. The probability density function for the exponential probability distribution is

$$f(x) = \lambda\, e^{-\lambda x}$$

The density function $f(x)$ gives the height of the exponential probability distribution curve above the horizontal axis at a point x.

Figure 6.62 shows the graphs of three exponential probability distributions. Note that each of these exponential distribution curves meets the vertical axis at the corresponding value of λ.

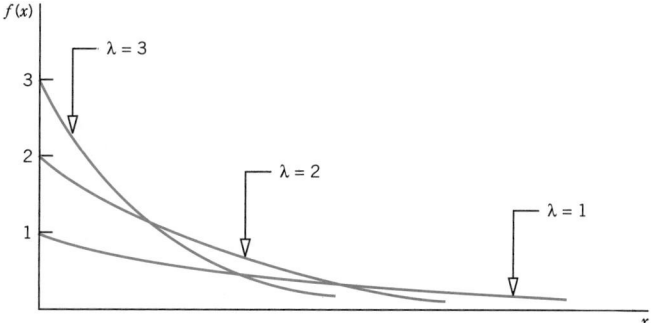

Figure 6.62 Three exponential probability distribution curves.

EXPONENTIAL PROBABILITY DISTRIBUTION

Let λ be the mean number of occurrences per unit of time and x be the lapse time between two successive occurrences. Then, the probability that x is more than or equal to "a" units of time is given by

$$P(x \geq a) = e^{-\lambda a} \qquad \text{where} \quad \lambda > 0 \quad \text{and} \quad a > 0$$

Note that λ is the mean number of occurrences per unit of time. For example, suppose an average of 10 customers come to a bank per 5-minute period. If we define a minute as a unit of time, then λ is 10/5 or 2, which states that on average two customers come to the bank per minute.

The probability $P(x \geq a)$ for the exponential probability distribution is given by the area in the tail of the exponential probability distribution curve beyond $x = a$, as shown in Figure 6.63.

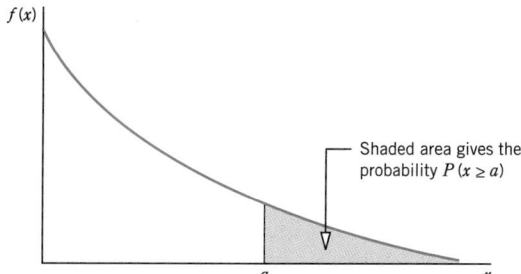

Figure 6.63 The probability $P(x \geq a)$ for the exponential probability distribution.

As we know from an earlier discussion in this chapter, for a continuous random variable x, $P(x > a)$ is equal to $P(x \geq a)$. Hence, for an exponential probability distribution,

$$P(x \geq a) = P(x > a) = e^{-\lambda a}$$

By using the complementary probability rule, we obtain:

$$P(x \leq a) = 1 - P(x > a) = 1 - e^{-\lambda a}$$

The probability that the lapse time between two successive occurrences is in the interval "a" to "b" is[2]

$$P(a \leq x \leq b) = P(x \geq a) - P(x \geq b) = e^{-\lambda a} - e^{-\lambda b}$$

PROBABILITIES FOR THE EXPONENTIAL PROBABILITY DISTRIBUTION

For the exponential probability distribution with the mean number of occurrences per unit of time equal to λ,

$$P(x \geq a) = e^{-\lambda a}$$
$$P(x \leq a) = 1 - e^{-\lambda a}$$
$$P(a \leq x \leq b) = e^{-\lambda a} - e^{-\lambda b}$$

Examples 6–24 and 6–25 illustrate the applications of the exponential probability distribution.

Applying the exponential distribution: probability $x \geq a$.

EXAMPLE 6–24   A processing machine at Wiley Inc. breaks down an average of once in four weeks. What is the probability that the next breakdown will not occur for at least six weeks after the previous breakdown? Assume that the time between breakdowns has an exponential distribution.

Solution   Let x denote the lapse time between any two successive breakdowns of this machine. We are to find the probability

$$P(x \geq 6 \text{ weeks})$$

Because the unit of time for x is in weeks, we must define the mean number of breakdowns λ per week. Since there is one breakdown in four weeks,

$$\lambda = \text{the mean number of breakdowns per week} = \frac{1}{4} = .25$$

The required probability is calculated using the formula $P(x \geq a) = e^{-\lambda a}$. In our example, $\lambda = .25$ and $a = 6$ weeks. The required probability is

$$P(x \geq 6 \text{ weeks}) = e^{-.25(6)} = e^{-1.5} = .223130 = \textbf{.2231}$$

The value of $e^{-1.5}$ can either be read from Table V of Appendix B or be calculated by using the e^x key on your calculator. Figure 6.64 shows the required probability.

---

[2]$P(a \leq x \leq b) = P(x \leq b) - P(x \leq a) = (1 - e^{-\lambda b}) - (1 - e^{-\lambda a})$

$\qquad = 1 - e^{-\lambda b} - 1 + e^{-\lambda a} = e^{-\lambda a} - e^{-\lambda b} = P(x \geq a) - P(x \geq b)$

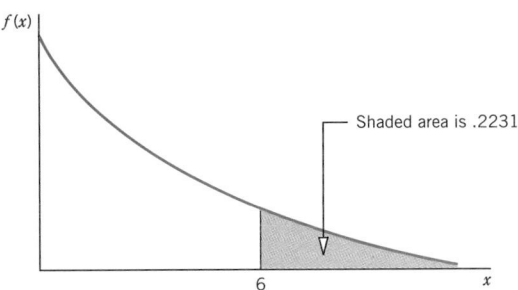

Figure 6.64 The probability $P(x \geq 6)$ for the exponential probability distribution.

Thus, the probability that the next breakdown will not occur for six weeks is .2231. ▬

Note that in Example 6–24 the number of breakdowns in any time interval has a Poisson probability distribution. However, the time between breakdowns has an exponential distribution. It is always true for an exponential probability distribution that the occurrences for an interval have a Poisson distribution.

Example 6–24 can also be solved using the Poisson probability distribution. According to this example, on average one breakdown occurs in a four-week period. Therefore, we can state that on average 1.5 breakdowns occur in a six-week period. We are to find the probability that no breakdown occurs in a six-week period. This is a Poisson probability with $\lambda = 1.5$. Hence, the probability of no breakdown in a six-week period is

$$P(x = 0) = \frac{\lambda^x e^{-\lambda}}{x!} = \frac{(1.5)^0 e^{-1.5}}{0!} = \frac{(1)(.223130)}{1} = \mathbf{.2231}$$

This probability is the same as calculated above using the exponential probability distribution.

Applying the exponential distribution.

EXAMPLE 6–25 A teller at New York Industrial Bank serves, on average, 30 customers per hour. Assume that the service time for a customer has an exponential probability distribution.

(a) What is the probability that the next customer will take five minutes or more to be served?

(b) Find the probability that the next customer will take two minutes or less to be served.

(c) What is the probability that the next customer will take two to four minutes to be served?

Solution Let x be the time taken by this teller to serve a customer. First we must find the mean number of customers served per minute by this teller to define λ per unit of time (minute). The teller serves on average 30 customers per 60 minutes. Hence,

$$\lambda = 30/60 = .5 \text{ customers served per minute}$$

Probability that x is greater than or equal to a value.

(a) We are to find the probability $P(x \geq 5)$. The probability $P(x \geq 5)$ will be calculated using the formula $P(x \geq a) = e^{-\lambda a}$. In this example, $a = 5$ minutes. The required probability is calculated below and shown in Figure 6.65.

$$P(x \geq 5) = e^{-\lambda a} = e^{-.5(5)} = e^{-2.5} = .082085 = \mathbf{.0821}$$

Thus, the probability is .0821 that a customer will take more than five minutes to be served.

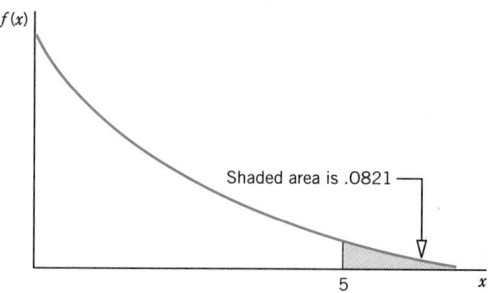

Figure 6.65   The probability $P(x \geq 5)$ for the exponential probability distribution.

Probability of x being less than or equal to a value.

(b)   We are to find the probability $P(x \leq 2)$. Note that this probability can also be written as $P(0 < x \leq 2)$. Remember that in the exponential probability distribution, x is always positive. The probability $P(x \leq 2)$ will be calculated using the formula $P(x \leq 2) = 1 - e^{-\lambda a}$. In this example, $a = 2$ minutes. The required probability is calculated below and shown in Figure 6.66.

$$P(x \leq 2) = 1 - P(x > 2) = 1 - e^{-\lambda a} = 1 - e^{-.5(2)}$$
$$= 1 - e^{-1.0} = 1 - .367879 = \mathbf{.6321}$$

Thus, the probability is .6321 that a customer will be served in two minutes or less.

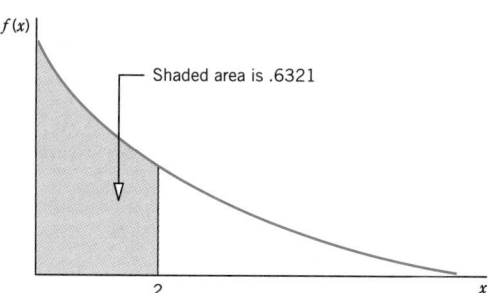

Figure 6.66   The probability $P(x \leq 2)$ for the exponential probability distribution.

Probability that value of x is in an interval.

(c)   We need to find the probability $P(2 \leq x \leq 4)$. In this case, we will use the formula $P(a \leq x \leq b) = P(x \geq a) - P(x \geq b)$. In this example, $a = 2$ minutes and $b = 4$ minutes. This probability is calculated as follows and shown in Figure 6.67.

$$P(2 \leq x \leq 4) = P(x \geq 2) - P(x \geq 4) = e^{-.5(2)} - e^{-.5(4)}$$
$$= e^{-1} - e^{-2} = .367879 - .135335 = \mathbf{.2325}$$

Thus, the probability that the teller will take two to four minutes to serve a customer is .2325.

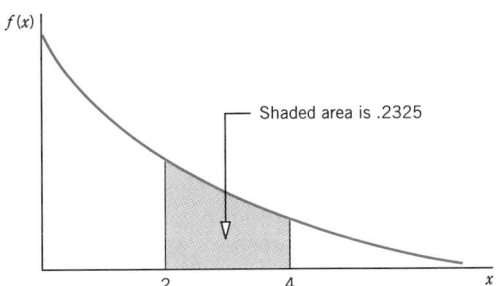

Figure 6.67 The probability $P(2 \le x \le 4)$ for the exponential probability distribution.

EXERCISES

Concepts and Procedures

6.87 Briefly describe the exponential probability distribution. What is its parameter?

6.88 Let x be a continuous random variable that possesses an exponential probability distribution with $\lambda = .40$. Find the following probabilities.

    a. $P(x \ge 7)$ **b.** $P(x \le 5)$ **c.** $P(4 \le x \le 9)$

6.89 Let x be a continuous random variable that possesses an exponential probability distribution with $\lambda = 1.0$. Find the following probabilities.

    a. $P(x \ge 3)$ **b.** $P(x \le 4)$ **c.** $P(2 \le x \le 6)$

Applications

6.90 The length of life of a light bulb manufactured by Standard Manufacturing Company is exponentially distributed with a mean of 700 hours. What is the probability that the life of a randomly selected such light bulb will be more than 800 hours?

6.91 The shelf life of a cake is exponentially distributed with a mean of three days. What is the probability that a cake that is baked just now will still be good after four days?

6.92 Delta Company sells household products by mail order. Its customers can order products using the company's toll free telephone number. On average, the company receives 10 orders per half hour. The time between orders is exponentially distributed.

    a. What is the probability that the company will receive no order during the next five minutes?
    b. What is the probability that the next order will be received within four minutes?

6.93 At a department store, a customer has to wait an average of four minutes in line before being served. The time a customer has to wait is exponentially distributed.

    a. Find the probability that a customer will have to wait for more than eight minutes in line.
    b. Find the probability that a customer will have to wait for three to seven minutes in line.

6.94 On average, two accidents per month occur at a large manufacturing company. The time between accidents is exponentially distributed.

    a. Find the probability that the next accident will not occur for at least three months.
    b. Find the probability that the next accident will occur in two to three months.

6.95 On average, 20 telephone calls are received per hour at an office of the Better Business Bureau. The time between calls received at this office is exponentially distributed.

    a. What is the probability that no call will come in during the next 10 minutes?
    b. What is the probability that the next call will come in within 4 minutes?

GLOSSARY

Continuity correction factor   The addition of .5 and/or the subtraction of .5 from x when the normal distribution is used as an approximation to the binomial distribution, where x is the number of successes in n trials.

Continuous random variable   A random variable that can assume any value in one or more intervals.

Exponential probability distribution   The probability distribution of a continuous random variable that gives the probability of lapse time between two successive occurrences when the average number of occurrences per unit of time is known.

Normal probability distribution   The probability distribution of a continuous random variable that, when plotted, gives a specific bell-shaped curve. The parameters of the normal distribution are the mean μ and the standard deviation σ.

Standard normal distribution   The normal distribution with a mean of zero and a standard deviation of one. The units of the standard normal distribution are denoted by z.

Uniform probability distribution   The probability distribution of a continuous random variable x that has a rectangular shape. For the uniform probability distribution, the probability that x assumes a value in any of the intervals (within the possible values of x) that are of equal width is the same.

z value or **z score**   The units of the standard normal distribution that are denoted by z.

KEY FORMULAS

1. **The z value for an x value to standardize a normal distribution**

$$z = \frac{x - \mu}{\sigma}$$

2. **The value of x for a normal distribution, when the values of μ, σ, and z are known**

$$x = \mu + z\,\sigma$$

3. **Normal approximation to the binomial distribution**

The normal distribution can be used as an approximation to the binomial distribution when

$$np > 5 \quad \text{and} \quad nq > 5$$

4. **Mean and standard deviation of the binomial distribution**

$$\mu = np \quad \text{and} \quad \sigma = \sqrt{npq}$$

5. **Uniform probability distribution**

Let x be a continuous random variable that is uniformly distributed in an interval a to b. The probability that x assumes a value in the interval c to d is

$$P(c \le x \le d) = \frac{d - c}{b - a}$$

The mean and standard deviation of the uniform probability distribution are

$$\mu = \frac{a + b}{2} \quad \text{and} \quad \sigma = \sqrt{\frac{(b - a)^2}{12}}$$

6. **Exponential probability distribution**

If λ is the mean number of occurrences per unit of time and x is the lapse time between two successive occurrences, then,

$$P(x \geq a) = e^{-\lambda a}$$
$$P(x \leq a) = 1 - e^{-\lambda a}$$
$$P(a \leq x \leq b) = e^{-\lambda a} - e^{-\lambda b}$$

where $\lambda > 0$, $a > 0$, and $b > a$.

SUPPLEMENTARY EXERCISES

6.96 Let x be a continuous random variable that has a normal distribution with a mean of 40 and a standard deviation of 3. Find the probability that x assumes a value

    a. between 32 and 37   **b.** between 38 and 46

    c. greater than 44   **d.** less than 42

6.97 Let x be a continuous random variable that has a normal distribution with a mean of 70 and a standard deviation of 12. Find the probability that x assumes a value

    a. between 52 and 63   **b.** between 65 and 86

    c. greater than 64   **d.** less than 76

6.98 The time taken by new employees at Shia Corporation to learn a packaging procedure is normally distributed with a mean of 24 hours and a standard deviation of 2.5 hours.

    a. What percentage of new employees can learn this procedure in less than 20 hours?

    b. Find the probability that a randomly selected new employee will take 26 to 30 hours to learn this procedure.

    c. Find the probability that a randomly selected new employee will take 21 to 28 hours to learn this procedure.

6.99 The print on the package of 100-watt General Electric soft-white light bulbs states that these bulbs have an average life of 750 hours. Assume that the lives of all such bulbs have a normal distribution with a mean of 750 hours and a standard deviation of 50 hours.

    a. What percentage of such bulbs will have a life of more than 840 hours?

    b. Find the probability that a randomly selected such bulb will have a life of 620 to 690 hours.

    c. Find the probability that a randomly selected such bulb will have a life of 710 to 830 hours.

6.100 The management at Ohio National Bank does not want its customers to wait in line for service for a long time. The manager of a branch of this bank estimated that the customers currently have to wait an average of eight minutes for service. Assume that the waiting time for all customers at this branch has a normal distribution with a mean of eight minutes and a standard deviation of two minutes.

    a. Find the probability that a randomly selected customer will have to wait for less than three minutes.

    b. What percentage of the customers have to wait for 10 to 13 minutes?

    c. What percentage of the customers have to wait for 6 to 12 minutes?

    d. Is it possible that a customer may have to wait for more than 16 minutes for service? Explain.

6.101 The research department at Colorado Bank estimated that the mean daily balance of all checking accounts at this bank is $870. Suppose the balances of all checking accounts at this bank have a normal distribution with a mean of $870 and a standard deviation of $350.

    a. Find the probability that on a certain day a randomly selected checking account will have a balance of more than $1500.

    b. What percentage of the checking accounts at this bank will have a balance of $200 to $300 on a certain day?

    c. What percentage of the checking accounts at this bank will have a balance of $1000 to $1400 on a certain day?

    d. Is it possible that a checking account will have a balance of more than $2500 on a certain day? Explain.

6.102 At Jen and Perry Ice Cream Company, the machine that fills one-pound cartons of Top Flavor ice cream is set to dispense 16 ounces of ice cream in every carton. However, some cartons contain slightly less than and some contain slightly more than 16 ounces of ice cream. The amounts of ice cream in all such cartons have a normal distribution with a mean of 16 ounces and a standard deviation of .18 ounces.

    a. Find the probability that a randomly selected such carton contains 16.25 to 16.50 ounces of ice cream.

    b. What percentage of such cartons contain less than 15.75 ounces of ice cream?

    c. Is it possible for a carton to contain less than 15.25 ounces of ice cream? Explain.

6.103 A machine at Kasem Steel Corporation makes iron rods that are supposed to be 50 inches long. However, the machine does not make all rods of exactly the same length. It is known that the probability distribution of the lengths of rods made on this machine is normal with a mean of 50 inches and a standard deviation of .06 inches. The rods that are either shorter than 49.85 inches or longer than 50.15 inches are discarded. What percentage of the rods made on this machine are discarded?

6.104 Find the value of z (or $-z$) so that the area under the standard normal curve

    a. from 0 to z is (approximately) .4205 and z is positive

    b. between 0 and z is .3485 and z is negative

    c. in the right tail is (approximately) .0130

    d. in the left tail is (approximately) .0300

    e. to the left of z is .8508

    f. to the right of z is .6700

6.105 Find the value of z (or $-z$) so that the area under the standard normal curve

    a. from 0 to z is .4975 and z is positive

    b. between 0 and z is (approximately) .2940 and z is negative

    c. in the left tail is (approximately) .0900

    d. in the right tail is .0080

    e. to the left of z is .6103

    f. to the right of z is .7088

6.106 Let x be a continuous random variable that follows a normal distribution with a mean of 270 and a standard deviation of 40.

    a. Find the value of x so that the area under the normal curve to the left of x is approximately .3050.

    b. Find the value of x so that the area under the normal curve to the right of x is approximately .2450.

    c. Find the value of x so that the area under the normal curve to the right of x is .7910.

    d. Find the value of x so that the area under the normal curve to the left of x is approximately .9100.

    e. Find the value of x so that the area under the normal curve between x and μ is .2910 and the value of x is smaller than μ.

    f. Find the value of x so that the area under the normal curve between μ and x is approximately .4870 and the value of x is greater than μ.

6.107 Let x be a continuous random variable that follows a normal distribution with a mean of 610 and a standard deviation of 55.

    a. Find the value of x so that the area under the normal curve to the left of x is .8340.

    b. Find the value of x so that the area under the normal curve to the right of x is .1075.

    c. Find the value of x so that the area under the normal curve to the right of x is approximately .8510.

    d. Find the value of x so that the area under the normal curve to the left of x is approximately .3300.

    e. Find the value of x so that the area under the normal curve between x and μ is approximately .3775 and the value of x is smaller than μ.

    f. Find the value of x so that the area under the normal curve between μ and x is approximately .2790 and the value of x is greater than μ.

6.108   The time taken by new employees at Shia Corporation to learn a packaging procedure is normally distributed with a mean of 24 hours and a standard deviation of 2.5 hours.

    a. Find the time beyond which only 1% of the new employees take to learn this procedure.

    b. What is the time below which only 5% of the new employees take to learn this procedure?

6.109   The print on the package of 100-watt General Electric soft-white light bulbs states that these bulbs have an average life of 750 hours. Assume that the lives of all such bulbs have a normal distribution with a mean of 750 hours and a standard deviation of 50 hours.

    a. Let x be the life of such a light bulb. Find x so that only 2.5% of such light bulbs have longer lives than this value.

    b. Let x be the life of such a light bulb. Find x so that about 80% of such light bulbs have shorter lives than this value.

6.110   It is known that 15% of all homeowners pay a monthly mortgage of more than $2400 and that the standard deviation of the monthly mortgage payments of all homeowners is $350. Suppose that the monthly mortgage payments of all homeowners have a normal distribution. What is the mean monthly mortgage paid by all homeowners?

6.111   At Jen and Perry Ice Cream Company, a machine fills one-pound cartons of Top Flavor ice cream. The machine can be set to dispense, on average, any amount of ice cream into these cartons. However, the machine does not put exactly the same amount of ice cream in each carton; it varies from carton to carton. It is known that the amount of ice cream put in each such carton has a normal distribution with a standard deviation of .18 ounces. The quality control inspector wants to set the machine such that at least 90% of the cartons have more than 16 ounces of ice cream. What should the mean amount of ice cream put by this machine into these cartons be?

6.112   Mong Corporation makes auto batteries. The company claims that 80% of its LL70 batteries are good for 70 months or more.

    a. What is the probability that in a sample of 100 such batteries, exactly 85 will be good for 70 months or more?

    b. Find the probability that in a sample of 100 such batteries, at most 74 will be good for 70 months or more.

    c. What is the probability that in a sample of 100 such batteries, 75 to 87 will be good for 70 months or more?

    d. Find the probability that in a sample of 100 such batteries, 72 to 77 will be good for 70 months or more.

6.113   Stress on the job is a major concern of a large number of people who go into managerial positions. It is estimated that 80% of managers of all companies suffer from job-related stress.

    a. What is the probability that in a sample of 200 managers of companies, exactly 150 suffer from job-related stress?

    b. Find the probability that in a sample of 200 managers of companies, at least 170 suffer from job-related stress.

    c. What is the probability that in a sample of 200 managers of companies, 165 or less suffer from job-related stress?

    d. Find the probability that in a sample of 200 managers of companies, 164 to 172 suffer from job-related stress.

6.114   Brooklyn Corporation manufactures computer diskettes. The machine that is used to make these diskettes produces 5% defective diskettes when working properly. The quality control inspector selects a sample of 75 diskettes every week and inspects them for being good or defective. If 6 or more of the diskettes in the sample are defective, the process is stopped and the machine is readjusted. What is the probability that a particular sample of 75 diskettes will contain 6 or more defective diskettes?

6.115 The K5 Corporation manufactures computer chips. The machine that is used to make these chips produces 7% defective chips when working properly. The quality control inspector selects a sample of 100 chips every week and inspects them for being good or defective. If 10 or more of the chips in the sample are defective, the process is stopped and the machine is readjusted. What is the probability that a particular sample of 100 computer chips will contain 10 or more defective chips?

6.116 The life of a television tube manufactured by Suburban Electronics Company has a uniform probability distribution with a minimum life of 8,000 hours and a maximum life of 10,000 hours.

    a. Let x be the life of a randomly selected such tube. Draw the graph of the probability distribution of x.

    b. Calculate the mean and standard deviation of the probability distribution of x.

    c. What is the probability that a randomly selected such tube will have a life of 8500 to 9000 hours?

    d. What is the probability that a randomly selected such tube will have a life of less than 8300 hours?

    e. What is the probability that a randomly selected such tube will have a life of more than 9500 hours?

6.117 The time taken to serve dinner to a customer at Angelino's Restaurant has a uniform probability distribution in the interval 20 minutes to 50 minutes.

    a. Let x be the time taken to serve dinner to a randomly selected customer at Angelino's. Draw the graph of the probability distribution of x.

    b. Calculate the mean and standard deviation of the probability distribution of x.

    c. What is the probability that the time taken to serve dinner to a randomly selected customer will be 44 minutes or more?

    d. Find the probability that the time taken to serve dinner to a randomly selected customer will be 34 to 42 minutes.

    e. What is the probability that the time taken to serve dinner to a randomly selected customer will be 24 minutes or less?

6.118 Private airplanes arrive at an airport at an average rate of 4 per 12-hour period. Assume that the time between successive arrivals of such airplanes at this airport follows an exponential distribution.

    a. What is the probability that the next airplane will arrive at this airport within one hour?

    b. Find the probability that the next airplane will not arrive at this airport for eight hours.

    c. What is the probability that the next airplane will arrive at this airport in four to seven hours?

6.119 Nicole Shooner works for a toy company and assembles five toys per hour on average. The assembly time for this toy follows an exponential distribution.

    a. Find the probability that the next toy will take more than 15 minutes to assemble.

    b. What is the probability that the next toy will take less than eight minutes to assemble?

    c. Find the probability that the next toy will take 10 to 16 minutes to assemble.

6.120 On average, five cars are rented per hour at a particular office of National Rental Inc. Assume that the time between successive car rentals at this office follows an exponential distribution.

    a. Find the probability that the next car will be rented at this office after 30 minutes.

    b. What is the probability that the next car will be rented within 5 minutes?

    c. Find the probability that the next car will be rented in 15 to 25 minutes.

6.121 Ken Sisco sells life insurance policies. The past data show that he sells, on average, 10 life insurance policies per 4-week period. Assume that the time between successive sales of life insurance policies by Ken Sisco has an exponential distribution.

    a. What is the probability that the next life insurance policy will not be sold for two weeks?

    b. Find the probability that the next life insurance policy will be sold within one week.

    c. What is the probability that the next life insurance policy will be sold in one to two weeks?

SELF-REVIEW TEST

1. The normal probability distribution is applied to

   a. a continuous random variable   **b.** a discrete random variable   **c.** any random variable

2. For a continuous random variable, the probability of a single value of x is always

   a. zero   **b.** 1.0   **c.** between 0 and 1

3. Which of the following is not a characteristic of the normal distribution?

   a. The total area under the curve is 1.0.
   b. The curve is symmetric about the mean.
   c. The two tails of the curve extend indefinitely.
   d. The value of the mean is always greater than the value of the standard deviation.

4. The parameters of a normal distribution are

   a. μ, z, and σ **b.** μ and σ **c.** μ, x, and σ

5. For the standard normal distribution

   a. $\mu = 0$ and $\sigma = 1$ **b.** $\mu = 1$ and $\sigma = 0$ **c.** $\mu = 100$ and $\sigma = 10$

6. The z value for μ for a normal curve is always

   a. positive   **b.** negative   **c.** zero

7. For a normal curve, the z value for an x value that is less than μ is always

   a. positive   **b.** negative   **c.** zero

8. Usually the normal distribution is used as an approximation to the binomial distribution when

   a. $n \geq 30$ **b.** $np > 5$ and $nq > 5$ **c.** $n > 20$ and $p = .50$

9. The parameter of the exponential probability distribution is

   a. x **b.** λ **c.** a

10. Find the following probabilities for the standard normal distribution.

    a. $P(.87 \leq z \leq 2.33)$ **b.** $P(-2.97 \leq z \leq 1.46)$ **c.** $P(z \leq -1.19)$ **d.** $P(z > -.71)$

11. Find the value of z for the standard normal curve such that the area

    a. in the left tail is .1000
    b. between 0 and z is .2291 and z is positive
    c. in the right tail is .0500
    d. between 0 and z is .3571 and z is negative

12. At Fleet Food Products Company, a machine is used to fill 18-ounce boxes of cornflake cereal. The net amount of cereal in all such boxes has a normal distribution with a mean of 18 ounces and a standard deviation of .24 ounces.

    a. What percentage of the boxes contain less than 17.5 ounces of cereal?
    b. Find the probability that a randomly selected box contains more than 18.4 ounces of cereal.
    c. What is the probability that a randomly selected box contains 17.8 to 18.3 ounces of cereal?
    d. What percentage of the boxes contain 18.2 to 18.5 ounces of cereal?
    e. Is it possible for a box to contain more than 19 ounces of cereal? Explain.

13. Refer to Problem 12.

    a. Find the net weight of a cereal box below which the net weights of 5% of all boxes fall.
    b. What is the net weight of a cereal box so that (about) 10% of all cereal boxes have their net weights more than this weight?

14. Refer to Problem 12. The machine that fills cereal boxes can be set to dispense, on average, any amount of cereal into these boxes. However, the machine does not put exactly the same amount of

cereal into each box; it varies from box to box. It is known that the amount of cereal put in each such box has a normal distribution with a standard deviation of .24 ounces. The quality control inspector wants to set the machine such that at least 90% of the boxes contain more than 18 ounces of cereal. What should the mean amount of cereal put by this machine into these boxes be?

15. The turnover rate at top-level management for U.S. companies is quite high. According to an estimate, only 22% of senior executives of U.S. companies have been with the same company for more than 25 years (*U.S. News & World Report*, June 8, 1992).

   a. What is the probability that in a sample of 200 senior executives, exactly 50 have been with the same company for more than 25 years?
   b. Find the probability that in a sample of 200 senior executives, at least 48 have been with the same company for more than 25 years.
   c. What is the probability that in a sample of 200 senior executives, 46 or less have been with the same company for more than 25 years?
   d. Find the probability that in a sample of 200 senior executives, 34 to 40 have been with the same company for more than 25 years.
   e. What is the probability that in a sample of 200 senior executives, 38 to 51 have been with the same company for more than 25 years?

16. The time taken by a nonstop flight from New York to Los Angeles has a uniform probability distribution with a minimum time of 5.0 hours and a maximum time of 5.50 hours.

   a. Let x be the time taken by a nonstop flight from New York to Los Angeles. Draw the graph of the probability distribution of x.
   b. Calculate the mean and standard deviation of the probability distribution of x.
   c. What is the probability that a nonstop flight from New York to Los Angeles will take 5.15 hours or less?
   d. Find the probability that a nonstop flight from New York to Los Angeles will take between 5.10 and 5.25 hours.
   e. What is the probability that a nonstop flight from New York to Los Angeles will take more than 5.30 hours?

17. Denise Baird works as a salesperson at an auto dealership. The past data show that she sells an average of five new cars in a two-week period. Assume that the time between successive sales of cars follows an exponential distribution.

   a. What is the probability that the next car will be sold within one-half week?
   b. Find the probability that the next car will not be sold for two weeks.
   c. What is the probability that the next car will be sold in one to 1.5 weeks?

USING MINITAB

THE NORMAL PROBABILITY DISTRIBUTION

To find a certain area under a normal curve by using MINITAB, we use the CDF command. Note that CDF stands for *cumulative density function*, which means cumulative probability. This command gives the total area under the normal curve to the left of an x or z value.

The MINITAB commands of Figure 6.68 will give the area under the standard normal curve to the left of a z value.

Figure 6.68 MINITAB commands to find an area under the standard normal distribution curve.

```
MTB  > CDF z=k;      ←—— This command provides information about the value of z.
SUBC > NORMAL MEAN=0 SD=1.  ←——  { This subcommand instructs MINITAB to use the
                                    normal distribution with μ = 0 and σ = 1.
```

In the first command in Figure 6.68, k is a specific value of z. The semicolon at the end of the first command instructs MINITAB that a subcommand with more information is to follow. Illustrations M6–1 and M6–2 describe how to find the area under the standard normal curve to the left of a point by using MINITAB.

Illustration M6–1 Suppose we want to find the area under the standard normal curve to the left of $z = -2.25$. The MINITAB commands to find this area and the resulting MINITAB output are given in the MINITAB display of Figure 6.69.

Figure 6.69 Finding area under the standard normal distribution curve.

```
MTB  > NOTE: FINDING AREA TO THE LEFT OF z=-2.25
MTB  > CDF z=-2.25;
SUBC > NORMAL MEAN=0 SD=1.

       -2.2500     0.0122   ←——This is the required area.
```

Thus, the area in the left tail of the standard normal curve to the left of $z = -2.25$ is .0122.

Illustration M6-2   Find the area under the standard normal distribution curve to the left of $z = 1.89$, to the right of $z = 1.89$, and between $z = 0$ and $z = 1.89$, respectively.

Solution   The MINITAB commands of Figure 6.70 will give the area under the standard normal curve to the left of $z = 1.89$. The MINITAB output shows that this area is .9706.

Figure 6.70   Finding area under the standard normal distribution curve.

```
MTB  > NOTE: FINDING AREA TO THE LEFT OF z=1.89
MTB  > CDF z=1.89;
SUBC > NORMAL MEAN=0 SD=1.

        1.8900    0.9706  ◀──This is the required area.
```

Note that MINITAB gives the whole area to the left of the z value whereas Table VII of Appendix B gives the area between the mean and a z value. The area obtained in the MINITAB display of Figure 6.70 is shown in Figure 6.71.

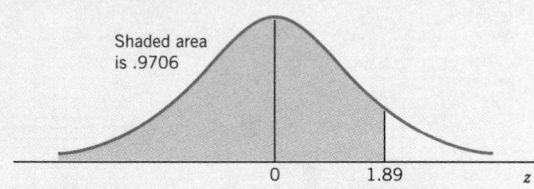

Figure 6.71   Area to the left of $z = 1.89$.

Now to find the area under the standard normal curve to the right of $z = 1.89$, we subtract the area to the left of $z = 1.89$ from 1.0, which is the total area under the normal curve. Thus, the area to the right of $z = 1.89$ is $1 - .9706 = .0294$.

To find the area under the standard normal curve between $z = 0$ and $z = 1.89$, we subtract .5 from .9706. This area is $.9706 - .5 = .4706$.   ■

The MINITAB command and subcommand given in Figure 6.72 will help us to find the area under any normal curve.

Figure 6.72   MINITAB commands to find an area under the normal curve.

```
MTB  > CDF x=k;  ◀──This command provides information about the value of x.
SUBC > NORMAL MEAN=b SD=c.  ◀──{ This subcommand instructs MINITAB to use the
                                 normal distribution with μ = b and σ = c.
```

The first command in Figure 6.72 lists the value of x. The semicolon at the end of the first command instructs MINITAB that a subcommand is to follow. This subcommand provides information about the mean and standard deviation of the normal distribution. In these MINITAB commands, k is a particular value of x, b is the value of the mean of the normal distribution, and c is the value of the standard deviation of the normal distribution. Again, note that MINITAB will give the area under the normal curve to the left of $x = k$. Illustration

M6–3 describes how we can find an area under a normal curve to the left of a point by using MINITAB.

Illustration M6–3   Refer to Example 6–12. According to that example, the assembly time for a racing car manufactured by Mack Corporation follows a normal distribution with a mean of 55 minutes and a standard deviation of 4 minutes. The company closes at 5 P.M. A worker starts assembling a racing car at 4 P.M. Find the probability that she will finish this job before the company closes for the day.

Solution   Let x denote the time taken to assemble this racing car. Then x is normally distributed with a mean of 55 minutes and a standard deviation of 4 minutes. We want to find the probability $P(x \leq 60)$.

The probability $P(x \leq 60)$ is given by the area under the normal curve to the left of $x = 60$. The MINITAB commands shown in Figure 6.73 give the area under the normal curve with $\mu = 55$ and $\sigma = 4$ to the left of $x = 60$.

Figure 6.73   Finding area under the normal curve.

```
MTB  > NOTE: FINDING AREA TO THE LEFT OF x=60
MTB  > CDF x=60;
SUBC > NORMAL MEAN=55  SD=4.

  60.0000        0.8944  ←——This is the required area.

     └—— This is the value of x.
```

Thus, the probability that this worker will finish assembling this racing car before the company closes for the day is .8944.

If we need to find an area under the normal curve to the right of a point using MINITAB, first we find the area to the left of that point and then subtract that area from 1.0. Suppose, for Illustration M6–3, we need to find the probability that this worker will not finish assembling this car within 60 minutes. This probability, $P(x > 60)$, is given by the area under the normal curve to the right of $x = 60$. To find this probability using MINITAB, first we find the area to the left of $x = 60$, as we did in Illustration M6–3. Then we subtract that area from 1.0. Thus,

$$P(x > 60) = 1 - .8944 = .1056$$

To find an area under a normal curve between two points using MINITAB, first we find the area to the left of each of the two points. Then we take the difference between these two areas. That gives the required probability.

THE UNIFORM PROBABILITY DISTRIBUTION

MINITAB can be used to find the probability $P(x \leq c)$ for a uniform probability distribution. The MINITAB commands of Figure 6.74 will help us find this probability.

Figure 6.74 MINITAB commands to find an area under the uniform probability distribution curve.

```
MTB  > NOTE: FINDING P(x ≤ c) FOR THE UNIFORM DISTRIBUTION
MTB  > CDF c;  ←—— c is the point to the left of which we need to find the probability.
SUBC > UNIFORM  a b.
                ↑
          These are the minimum and maximum values, respectively, of the uniform
          distribution.
```

Illustration M6–4 shows how we use MINITAB to find probabilities for the uniform distribution.

Illustration M6–4   Sach Corporation manufactures toys. The time taken by a new worker to learn how to assemble a particular toy is uniformly distributed in the interval 20 to 27 hours.

(a)   What is the probability that a new worker will take at most 23 hours to learn how to assemble this toy?

(b)   What is the probability that a new worker will take 22 to 24 hours to learn how to assemble this toy?

Solution   The uniform probability distribution in this example is defined for the interval 20 to 27. Hence, $a = 20$ and $b = 27$.

(a)   In this part, we are to find the probability $P(x \leq 23)$. The MINITAB display of Figure 6.75 shows the calculation of this probability.

Figure 6.75   Finding area under the uniform probability distribution curve.

```
MTB  > NOTE: FINDING P(x ≤ 23) FOR THE UNIFORM DISTRIBUTION
MTB  > CDF 23;
SUBC > UNIFORM  20 27.

   23.0000        0.4286  ←—— This is the required probability.
```

From this MINITAB display,

$$P(x \leq 23) = .4286$$

That is, the probability that a new worker will take at most 23 hours to learn how to assemble this toy is .4286.

     If we need to find the probability $P(x > 23)$ for the uniform distribution, we will first find the probability $P(x \leq 23)$ and then subtract this probability from 1.0.

(b)   In this part, we are to find the probability $P(22 \leq x \leq 24)$. To find this probability using MINITAB, first we find the two probabilities $P(x \leq 22)$ and $P(x \leq 24)$ and then take the difference between these two probabilities. The MINITAB display of Figure 6.76 shows the calculation of probabilities $P(x \leq 22)$ and $P(x \leq 24)$.

Figure 6.76 Finding area under the uniform probability distribution curve.

```
MTB  > NOTE: FINDING P(x ≤ 22) AND P(x ≤ 24)
MTB  > CDF 22;
SUBC > UNIFORM  20 27.

   22.0000        0.2857  ←——This is the probability P(x ≤ 22).

MTB  > CDF 24;
SUBC > UNIFORM  20 27.

   24.0000        0.5714  ←——This is the probability P(x ≤ 24).
```

Thus, from the MINITAB display, $P(x \leq 22) = .2857$ and $P(x \leq 24) = .5714$. Hence, the required probability is

$$P(22 \leq x \leq 24) = P(x \leq 24) - P(x \leq 22) = .5714 - .2857 = .2857$$

That is, the probability that a new worker will take 22 to 24 hours to learn how to assemble this toy is .2857. ▬

THE EXPONENTIAL PROBABILITY DISTRIBUTION

To find the probability $P(x \leq c)$ for an exponential probability distribution using MINITAB, we need to know the average time between two successive occurrences. Then, the set of MINITAB commands in Figure 6.77 will give us the probability to the left of a point c.

Figure 6.77 MINITAB commands to find an area under the exponential probability distribution curve.

```
MTB  > NOTE: FINDING P(x ≤ c) FOR THE EXPONENTIAL DISTRIBUTION
MTB  > CDF c;  ←——c is the point to the left of which we need to find the probability.
SUBC > EXPONENTIAL T.
              ↑
              T is the average time between two successive occurrences. Note that T is
              not the same as λ.
```

Illustration M6–5 describes how we use MINITAB to find probabilities for the exponential distribution.

Illustration M6–5 A teller at New York Industrial Bank serves on average 30 customers per hour. Assume that the service time for a customer has an exponential probability distribution.

(a) Find the probability that the next customer will take three minutes or less to be served.

(b) What is the probability that the next customer will take four to six minutes to be served?

Solution According to the given information, the teller serves an average of 30 customers per hour. Based on this information, we can state that this teller serves, on average, one customer in two minutes. Thus, the average time between any two customers served is two minutes. This is the number we will use as the value of T in MINITAB commands.

(a)   In this part, we are to find the probability $P(x \le 3)$. In the MINITAB commands, we will use $c = 3$ and $T = 2$. The MINITAB display of Figure 6.78 shows the calculation of $P(x \le 3)$.

Figure 6.78   Finding area under the exponential probability distribution curve.

```
MTB  > NOTE: FINDING P(x ≤ 3) FOR THE EXPONENTIAL DISTRIBUTION
MTB  > CDF 3;
SUBC > EXPONENTIAL  2.

     3.0000              0.7769  ←——This is the probability P(x ≤ 3).
```

From the MINITAB display,

$$P(x \le 3) = .7769$$

Thus, the probability that the next customer will take three minutes or less to be served is .7769.

(b)   In this part, we are to find the probability $P(4 \le x \le 6)$. To find this probability, first we find the two probabilities $P(x \le 4)$ and $P(x \le 6)$ and then take the difference between these two probabilities. The MINITAB display of Figure 6.79 shows the calculation of the probabilities $P(x \le 4)$ and $P(x \le 6)$.

Figure 6.79   Finding area under the exponential probability distribution curve.

```
MTB  > NOTE: FINDING P(x ≤ 4) and P(x ≤ 6)
MTB  > CDF 4;
SUBC > EXPONENTIAL  2.

     4.0000              0.8647  ←——This is the probability P(x ≤ 4).

MTB  > CDF 6;
SUBC > EXPONENTIAL  2.

     6.0000              0.9502  ←——This is the probability P(x ≤ 6).
```

From the MINITAB display, $P(x \le 4) = .8647$ and $P(x \le 6) = .9502$. Then, the required probability is

$$P(4 \le x \le 6) = P(x \le 6) - P(x \le 4) = .9502 - .8647 = .0855$$

Thus, the probability that the next customer will take four to six minutes to be served is .0855.   ▬

COMPUTER ASSIGNMENTS

M6.1   Find the area under the standard normal curve
   a.   to the left of $z = -1.94$
   b.   to the left of $z = .83$
   c.   to the right of $z = 1.45$
   d.   to the right of $z = -1.65$
   e.   between $z = .75$ and $z = 1.90$
   f.   between $z = -1.20$ and $z = 1.55$

M6.2 Find the following areas under a normal curve with $\mu = 86$ and $\sigma = 14$.
a. Area to the left of $x = 71$ b. Area to the left of $x = 96$
c. Area to the right of $x = 90$ d. Area to the right of $x = 75$
e. Area between $x = 65$ and $x = 75$ f. Area between $x = 72$ and $x = 95$

M6.3 Most business schools require that every applicant for admission to a graduate degree program take the GMAT test. Suppose the GMAT scores of all students who took this test this year have a normal distribution with a mean of 550 and a standard deviation of 90. Answer the following using MINITAB.
a. Sue Hopern scored 670 in this test. What percentage of the examinees scored higher than Sue?
b. Joe Merck scored 610 in this test. What percentage of the examinees scored lower than Joe?
c. What percentage of examinees scored 500 to 650?

M6.4 The transmission on a model of car has a warranty for 40,000 miles. It is known that the life of such a transmission has a normal distribution with a mean of 72,000 miles and a standard deviation of 12,000 miles. Answer the following using MINITAB.
a. What percentage of the transmissions will fail before the end of the warranty period?
b. What percentage of the transmissions will be good for more than 100,000 miles?
c. What percentage of the transmissions will be good for 80,000 to 100,000 miles?

M6.5 Newcastle Mail Order Company promises that the delivery time for any order placed with the company is uniformly distributed in the interval 14 to 20 days.
a. What is the probability that a new order will be delivered in 15 to 19 days?
b. What is the probability that a new order will be delivered in 16 days or less?
c. What is the probability that a new order will take at least 17 days to be delivered?

M6.6 Becker Electronics Company manufactures computer chips. To check that the system is working properly, the quality control inspector takes a sample of chips from the production line every day and inspects them. The inspection time for a chip is uniformly distributed in the interval 10 to 16 minutes.
a. What is the probability that a chip will take 11 to 12 minutes for inspection?
b. What is the probability that a chip will take at most 13 minutes for inspection?
c. What is the probability that a chip will take 14 minutes or more for inspection?

M6.7 Private airplanes arrive at an airport at an average rate of 4 per 12-hour period. Assume that the time between successive arrivals of such airplanes at this airport follows an exponential distribution.
a. What is the probability that the next airplane will arrive at this airport within one hour?
b. Find the probability that the next airplane will not arrive at this airport for eight hours.
c. What is the probability that the next airplane will arrive at this airport in four to seven hours?

M6.8 Nicole Shooner works for a toy company and assembles five toys per hour on average. The assembly time for this toy follows an exponential distribution.
a. Find the probability that the next toy will take more than 15 minutes to assemble.
b. What is the probability that the next toy will take less than 8 minutes to assemble?
c. Find the probability that the next toy will take 10 to 16 minutes to assemble.

1. The IRS would like to determine the proportion of taxpayers in a large group who cheated on their income tax returns. Since the results of directly asking each individual, ''Have you cheated on your income tax return?'' would be unreliable, the IRS might use the following randomized response scheme. Each taxpayer rolls a fair die once, the outcome of which only he or she knows. If a 1 or 2 is rolled, the taxpayer must answer the sensitive question truthfully. However, if a 3, 4, 5, or 6 is rolled, the taxpayer must answer the question with the opposite of the true answer. In this way, the IRS would not know whether a ''yes'' response means the taxpayer cheated on the tax return and answered truthfully, or the taxpayer did not cheat on the tax return and answered untruthfully. If 60% of the taxpayers in the group respond ''yes,'' what proportion of the group actually did cheat on their income tax returns? (Splitting the outcomes of the roll of the die as above represents only one possibility. The outcomes could have been split differently but not in such a way that the probability of answering truthfully is .5. In this case, it would not be possible to estimate the proportion of the group that actually cheated on their income tax returns.)

2. A local radio station plans to give away tickets as a promotional campaign for a singles riverboat cruise to random callers who are 21 years of age or older. From past experience, station managers know that 80% of the tickets given away are actually used. If the riverboat can hold at most 300 persons, how many tickets should the station give away so that there is a 99% chance that all ticket holders showing up for the cruise will be accommodated?

3. The print on the package of 100-watt General Electric soft-white light bulbs states that these light bulbs have an average life of 750 hours. Assume that the lives of 100-watt General Electric soft-white light bulbs have a normal distribution with the mean and standard deviation depending on the filament composition. A quality control manager at General Electric periodically checks to make sure the light bulbs conform to specifications by testing one hundred randomly selected light bulbs. If more than ten light bulbs in the sample last less than 670 hours he stops production and checks for problems in the manufacturing process.

   a. What is the probability that the quality control manager erroneously stops production when the mean life of all light bulbs is 750 hours with a standard deviation of 50 hours?

   b. What is the probability that the quality control manager stops production when the mean life of all light bulbs is 730 hours with a standard deviation of 55 hours?

4. The number of calls coming into a small mail order company follows a Poisson distribution. Currently, these calls are serviced by a single operator. The manager knows from past experience that an additional operator will be necessary if the rate of calls exceeds 20 per hour. The manager observes that 9 calls came into the mail order company during a randomly selected 15-minute period.

   a. If the rate of calls is actually 20 per hour, what is the probability that 9 or more calls will come in during a given 15-minute period?

   b. If the rate of calls is really 30 per hour, what is the probability that 9 or more calls will come in during a given 15-minute period?

   c. Would you advise the manager to hire a second operator? Explain.

5. Jeff Rumzis trades stocks in four companies on the U.S. stock exchange, each with an AAA rating. Jeff's strategy on buying and selling stocks follows two simple rules: (1) buy a company's stocks if he does not already hold any and the price of this stock falls below the first percentile of its prices over the past year, and (2) sell a company's stock if he owns any and the price of this stock exceeds the 95th percentile of its prices over the past year. The following table summarizes the mean and standard deviation of the prices of the stocks of the four companies over the past year. Assume that the stock prices for the past year follow a normal distribution.

	Stock			
	A	B	C	D
Mean	$29	68	47	24
Standard deviation	$2	5	6	2

Jeff currently has $10,000 worth of stocks in companies A and D, but no stocks in companies B and C. On checking into the stock prices, he finds the current price of stock in Company A to be $24, in Company B to be $55, in Company C to be $44, and in Company D to be $28. What trading, if any, should Jeff do at this time?

6. From her experience, an accountant knows that 20% of her clients file the 1040EZ form, 30% file the 1040A form, and 50% file the 1040 form. The accountant must decide to purchase one of two software packages for preparing this year's income tax returns. The software known as Accountant's Friend costs $1000, and the software called Uncle Sam's Enemy costs $3000. Accountant's Friend can handle 99% of all returns that require Form 1040EZ, 95% of all returns that require Form 1040A, and 90% of all returns that require Form 1040. On the other hand, Uncle Sam's Enemy can handle 100% of all returns that require Form 1040EZ, 98% of all returns that require Form 1040A, and 94% of all returns that require Form 1040. For tax returns having to be processed by hand, because they cannot be processed by the software package, the accountant figures a $30 loss when Form 1040EZ is required, a $50 loss when Form 1040A is required, and a $100 loss when Form 1040 is required.

    a.   If the accountant expects to do 1000 income tax returns this year, which software package would you suggest she buy?

    b.   If the accountant expects to do only 500 income tax returns this year, which software package would you suggest she buy?

    c.   At least how many tax returns must the accountant expect to do this year to make it worthwhile to purchase Uncle Sam's Enemy?

    d.   At least how many income tax returns must the accountant expect to do this year to make it worthwhile to purchase either of the software packages at all?

7. A clothing distributor is considering marketing a new line of fashions that she thinks has a 70% chance of being successful. The distributor can test market the fashions to gain insight into whether or not the new fashion line will be successful. Unfortunately, test marketing is not 100% accurate in predicting the success of a product. If the new fashion line will actually be successful, test marketing will predict a success 80% of the time. If the new fashion line will actually be unsuccessful, test marketing will predict it will be unsuccessful 75% of the time. The clothing distributor would not want to market the new fashion line unless she can be at least 85% certain it will be successful.

    a.   What decision would the clothing distributor make if no test marketing is done?

    b.   Should the clothing distributor test market the new fashion line?

    c.   What decision should the clothing distributor make if test marketing predicts a success?

    d.   How sure can the clothing distributor be that the new fashion line will be unsuccessful if test marketing predicts it will be unsuccessful?

8. Recall Example 5–3 of Section 5.2 of Chapter 5. The probability distribution of x, where x is the number of breakdowns per week for a machine, is reproduced below from that example.

x	0	1	2	3
$P(x)$	.15	.20	.35	.30

Suppose this machine is observed for a two-week period and let y represent the number of breakdowns during those two weeks. Find the probability distribution of y. Assume that breakdowns occur independently between weeks.

9. A casino offers two new games. Game I costs $36 to play and the player receives $10 times the outcome of the roll of a fair die. Game II costs $3 to play and depends on the spin of a wheel where half the spaces on the wheel pay $4 and the other half pay nothing.

a. How much would a player expect to win (or lose) on each play of Game I? What about Game II?

b. Which game would you rather play? Why?

c. The outcomes of 50 rolls of a fair die in Game I are

3	2	4	5	3	6	5	5	3	4
1	5	5	5	2	2	1	5	6	3
1	6	2	5	3	4	6	3	4	4
5	2	2	3	4	1	3	4	4	3
4	3	1	5	2	5	1	5	4	3

   i. Compute the amount you have won (or lost) at the end of each roll (i.e., -6, -22, -18, . . . , etc.).

   ii. Compute the average amount won (or lost) at the end of each roll (i.e., -6, -11, -6, . . . , etc.).

d. The outcomes of 50 spins of the wheel in Game II are

0	4	4	0	4	4	0	4	4	4
4	0	4	0	4	0	4	4	0	0
0	0	0	4	4	0	4	0	0	0
0	4	0	4	4	0	4	0	4	0
0	0	4	4	0	0	4	4	4	4

   i. Compute the amount you have won (or lost) at the end of each spin of the wheel (i.e., -3, -2, -1, . . . , etc.).

   ii. Compute the average amount you have won (or lost) at the end of each spin of the wheel (i.e., -3, -1, $-.33$, . . . , etc.).

e. What happens to the mean amount won (or lost) per play as the number of plays increases in part ii of parts c and d, respectively?

f. What do you think would happen to the mean amount won (or lost) per play if the number of plays was infinitely large?

7 | SAMPLING DISTRIBUTIONS

C hapter 1 had a brief discussion of sample surveys and a random sample. This chapter explains these concepts in more detail and, in addition, discusses the simple random sample and the use of a table of random numbers to select a sample. Chapters 5 and 6 discussed probability distributions of discrete and continuous random variables. This chapter extends the concept of probability distribution to that of a sample statistic. As we discussed in Chapter 3, a sample statistic is a numerical summary measure calculated for sample data. The mean, median, mode, and standard deviation calculated for sample data are called *sample statistics*. On the other hand, the same numerical summary measures calculated for population data are called *population parameters*. A population parameter is always a constant, whereas a sample statistic is always a random variable. Since every random variable must possess a probability distribution, each sample statistic possesses a probability distribution. The probability distribution of a sample statistic is more commonly called its *sampling distribution*. This chapter discusses the sampling distributions of the sample mean and the sample proportion. The concepts covered in this chapter are the foundation of the inferential statistics discussed in succeeding chapters.

7.1 A SAMPLE SURVEY

Usually, to conduct research, we select a portion of the target population. This portion of the population is called a sample. Then we collect the required information from the elements included in the sample.

A SAMPLE SURVEY

The technique of collecting information from a portion of the population is called a **sample survey**.

A survey can be conducted by personal interviews, by telephone, or by mail. The personal interview technique has the advantage of having a high response rate and a high quality of answers obtained. However, it is the most expensive and time-consuming technique. The telephone survey also gives a high response rate. Unlike personal interviews, it is less expensive and less time consuming. Nonetheless, a problem with this technique is that many people do not like to be called at home, and those who do not have a phone are left out of the survey. A survey conducted by mail is the least expensive method but the response rate is usually very low. Many people included in such a survey do not return the questionnaires.

Conducting a survey with accurate and reliable results is not an easy task. To quote Warren Mitofsky, director of Elections and Surveys for CBS News, "Any damn fool with 10 phones and a typewriter thinks he can conduct a poll."[1] The preparation of a questionnaire is probably the most difficult part of a survey. The way a question is phrased can affect the results of the survey. Case Study 7–1, which is excerpted from an article published in *Psychology Today*, shows that writing questions for a questionnaire is a much more complex task than is usually thought.

CASE STUDY 7–1   IS IT A SIMPLE QUESTION?

Even the seemingly simplest of questions can yield complex answers. "Do you own a car?" asks Stanley Presser, a sociologist at the National Science Foundation in Washington, D.C. "That sounds like an awfully simple question. But is it really? What does 'you' mean? Suppose a wife is answering the poll, and the car is registered in her husband's name. How is she supposed to answer? What does 'own' mean? What if the car is on a long-term lease? What does 'car' mean? What if they have one of those new little vans, or a four-wheel-drive vehicle? My God, that sounds like a simple question! You can imagine how diverse the factors become in a more complicated one."

Suppose, however, that the question about car ownership had been preceded by a series of related questions: "Are you married? Does your spouse drive an automotive vehicle? Is it a car, a van or some other sort of vehicle? Is it leased, or does your spouse own it? Now about you—do you own a car?" Such a series of questions would serve to clarify the intended meaning of the one about car ownership.

Source: Rich Jaroslovsky: "What's on Your Mind, America?" *Psychology Today*, July/August 1988, 54–59. Copyright © 1988 Sussex Publishers Inc. Reprinted with permission.

---

[1]"The Numbers Racket: How Polls and Statistics Lie," *U.S. News & World Report*, July 11, 1988.

Even polls taken at the same time can produce dramatically different results depending on how a question is phrased. The excerpts in Case Study 7–2 from *U.S. News & World Report* show how the structure of the questions can change the results of a poll.

CASE STUDY 7–2   HOW TO SKEW A POLL: LOADED QUESTIONS AND OTHER TRICKS

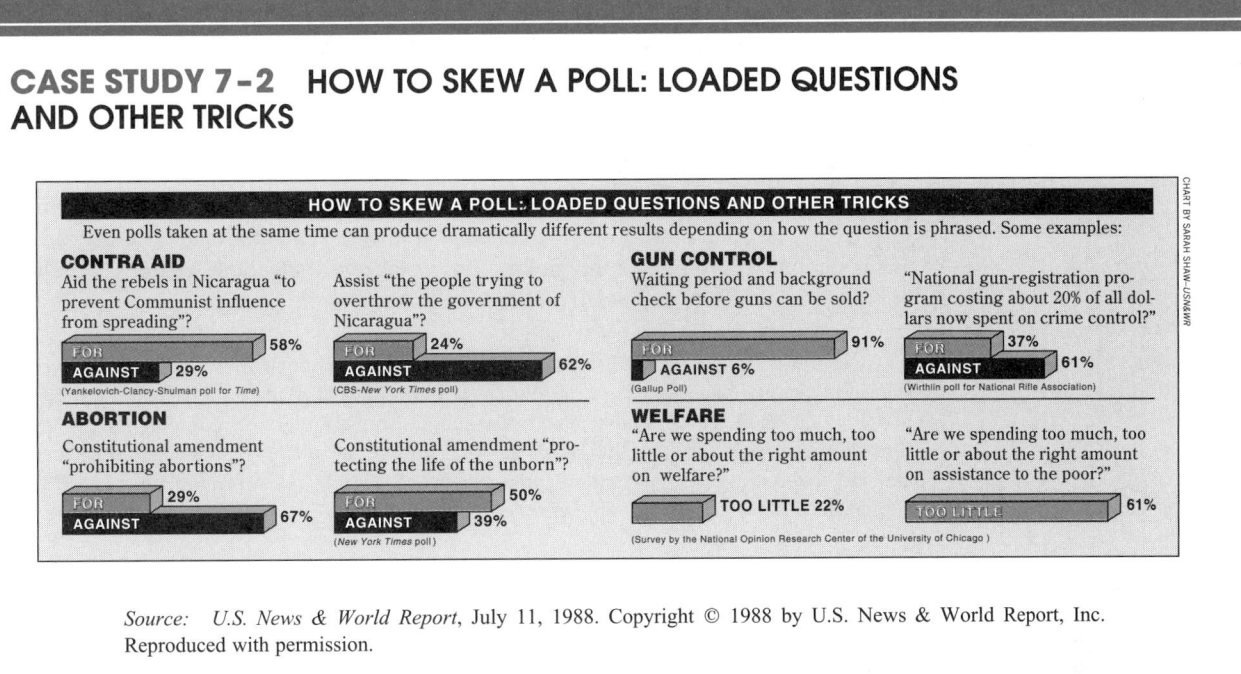

HOW TO SKEW A POLL: LOADED QUESTIONS AND OTHER TRICKS

Even polls taken at the same time can produce dramatically different results depending on how the question is phrased. Some examples:

CONTRA AID
Aid the rebels in Nicaragua "to prevent Communist influence from spreading"?
FOR 58%
AGAINST 29%
(Yankelovich-Clancy-Shulman poll for *Time*)

Assist "the people trying to overthrow the government of Nicaragua"?
FOR 24%
AGAINST 62%
(CBS-*New York Times* poll)

GUN CONTROL
Waiting period and background check before guns can be sold?
FOR 91%
AGAINST 6%
(Gallup Poll)

"National gun-registration program costing about 20% of all dollars now spent on crime control?"
FOR 37%
AGAINST 61%
(Wirthlin poll for National Rifle Association)

ABORTION
Constitutional amendment "prohibiting abortions"?
FOR 29%
AGAINST 67%

Constitutional amendment "protecting the life of the unborn"?
FOR 50%
AGAINST 39%
(*New York Times* poll)

WELFARE
"Are we spending too much, too little or about the right amount on welfare?"
TOO LITTLE 22%

"Are we spending too much, too little or about the right amount on assistance to the poor?"
TOO LITTLE 61%
(Survey by the National Opinion Research Center of the University of Chicago)

CHART BY SARAH SHAW-*USN&WR*

Source: U.S. News & World Report, July 11, 1988. Copyright © 1988 by U.S. News & World Report, Inc. Reproduced with permission.

7.2   RANDOM AND NONRANDOM SAMPLES

Depending on how a sample is drawn, it may be a **random sample** or a **nonrandom sample**.

> **RANDOM AND NONRANDOM SAMPLES**
>
> A random sample is a sample drawn in such a way that each member of the population has some chance for being selected in the sample. In a nonrandom sample, some members of the population may not have any chance of being selected in the sample.

Suppose we have a list of 100 companies and we want to select 10 of them. If we write the names of all 100 companies on pieces of paper, put them in a hat, mix them, and then draw 10 names, the result will be a random sample of 10 companies. However, if we arrange these 100 companies alphabetically and pick the first 10 names, it will be a nonrandom sample because the companies that are not among the first 10 have no chance of being selected in the sample.

A random sample is usually a representative sample. Note that for a random sample, each member of the population may or may not have the same chance for being included in the sample.

Two types of nonrandom samples are a *convenience sample* and a *judgment sample*. In a **convenience sample**, the most accessible members of the population are selected to obtain the results quickly. For example, an opinion poll may be conducted in a few hours by collecting information from certain shoppers at a single shopping mall. In a **judgment sample**, the members are selected from the population based on the judgment and prior knowledge of an expert. Although such a sample may happen to be a representative sample, the chances of it being so are small. If the population is large, it is not an easy task to select a representative sample based on judgment.

The so-called *pseudo polls* are examples of nonrepresentative samples. For instance, a survey conducted by a magazine that includes only its readers does not usually involve a representative sample. Similarly, a poll conducted by a television station giving two separate 900 telephone numbers for *yes* and *no* votes is not based on a representative sample. In these two examples, respondents will only be those people who read that magazine or watch that television station, who do not mind paying the postage and telephone charges, or who feel emotionally compelled to respond. To quote Larry King on this subject:

> All over the board . . . the 900 telephone number is very popular these days, but viewers should be warned that in the case of political polling it has absolutely no basis in fact. Poor people in the audience can't contribute to the survey, so it's faulty to begin with. . . . So next time you see a poll based on 900 numbers, treat it as some sort of middle-class amusement and forget about it. ("Larry King's People, News and Views," *USA TODAY*, July 17, 1989. Copyright © 1989, *USA TODAY*. Reprinted with permission.)

7.3 SELECTING A SIMPLE RANDOM SAMPLE

A sample that assigns the same probability to each member of the population for being selected is called a **simple random sample**.

SIMPLE RANDOM SAMPLE

A simple random sample is a sample that is selected in such a way that each member of the population has the same chance of being included in the sample.

One way to select a simple random sample is by a lottery or drawing. For example, if we need to select 5 students from a class of 50, we write each of the 50 names on separate pieces of paper. Then, we place all 50 names in a hat and mix them thoroughly. Next, we draw one name randomly from the hat. We repeat this experiment four more times. The five drawn names comprise a simple random sample.

The second procedure to select a simple random sample is to use a table of random numbers. Table I in Appendix B lists random numbers. These numbers are generated by a random process. Suppose we have a group of 400 persons and we need to select 30 persons randomly from this group. To select a simple random sample, we arrange the names of all 400 persons in alphabetic order and assign a three-digit number, from 001 to 400, to each person.

Next, we use the table of random numbers to select 30 persons. The random numbers in Table I are recorded in blocks of five digits. To use this table, we can start anywhere. One way to do so is to close our eyes and put a finger anywhere on the page and start at that point. From there, we can move in any direction. We need to pick three-digit numbers from the table because we have assigned three-digit numbers to the 400 persons in our population.

Suppose we start at the first block of the 31st row from the top of Table I. The five rows starting with the 31st row from that table are reproduced as Table 7.1 here. The first block of five numbers in Table 7.1 is 13049. We use the first three digits of this block to select the first person from the population. Hence, the first person selected is the one with a number 130. Suppose we move along the row to the right to make the next selection. The second block of five numbers in Table 7.1 is 85293. The first three digits of this block give 852. However, we have only 400 persons in the population with assigned numbers of 001 to 400. Consequently, we cannot use 852 to select a person. Therefore, we move to the next block of five numbers without making a selection. The third block of numbers is 32747. The first three digits of this block are 327. Consequently, the second person selected is the one with a number 327. We continue this process until all 30 required persons are selected. This gives us a simple random sample of 30 persons.

Table 7.1

13049	85293	32747	17728	50495	34617	73707	33976	86177
86544	52703	74990	98288	61833	48803	75258	83382	79099
77295	70694	97326	35430	53881	94007	70471	66815	73042
54637	32831	59063	72353	87365	15322	33156	40331	93942
50938	12004	18585	23896	62559	44470	27701	66780	56157

Although the table of random numbers given in Appendix B contains only 1485 blocks of five-digit numbers, we can easily construct a table of as many random numbers as we want using a computer software package such as MINITAB.

If we have access to a computer, we can use a statistical package, such as MINITAB, to select a simple random sample. The MINITAB sections at the end of Chapter 1 and this chapter explain and illustrate how we can draw such a sample by using MINITAB.

7.4 POPULATION AND SAMPLING DISTRIBUTIONS

This section introduces the concept of population distribution and sampling distribution. Subsection 7.4.1 explains the population distribution, and Subsection 7.4.2 describes the sampling distribution of $\bar{x}$.

7.4.1 POPULATION DISTRIBUTION

The **population distribution** is the probability distribution derived from the information on all elements of a population.

POPULATION DISTRIBUTION

The population distribution is the probability distribution of the population data.

Suppose there are only five employees working for a small company. The following data give the annual salaries (in thousands of dollars) of these five employees.

$$17 \quad 24 \quad 35 \quad 35 \quad 43$$

Let x denote the annual salary (in thousands of dollars) of an employee. Using single-valued classes (as there are only five data values, there is no need to group them), we can write the frequency distribution of annual salaries as in Table 7.2. Dividing the frequencies of classes by the population size we obtain the relative frequencies, which can be used as probabilities of those classes. Table 7.3, which lists the probabilities of various x values, presents the probability distribution of the population.

Table 7.2   Population Frequency Distribution

x	f
17	1
24	1
35	2
43	1
	$N = 5$

Table 7.3   Population Probability Distribution

x	$P(x)$
17	$1/5 = .20$
24	$1/5 = .20$
35	$2/5 = .40$
43	$1/5 = .20$
	Sum $= 1.0$

The values of the mean and standard deviation calculated for the probability distribution of Table 7.3 give the values of the population parameters μ and σ. These values are $\mu = 30.80$ and $\sigma = 9.174$. The values of μ and σ for the probability distribution of Table 7.3 can be calculated using the formulas given in Sections 5.3 and 5.4 of Chapter 5 (see Exercise 7.8).

7.4.2   SAMPLING DISTRIBUTION

As mentioned at the beginning of this chapter, the value of a population parameter is always constant. For example, for any population data set, there is only one value of the population mean μ. However, we cannot say the same about the sample mean $\bar{x}$. We would expect different samples of the same size drawn from the same population to yield different values of the sample mean $\bar{x}$. The value of the sample mean for any one sample will depend on the elements included in that sample. Consequently, *the sample mean $\bar{x}$ is a random variable.* Therefore, like other random variables, the sample mean $\bar{x}$ possesses a probability distribution, which is more commonly called the **sampling distribution of $\bar{x}$**. Other sample statistics such as the median, mode, and standard deviation also possess sampling distributions.

SAMPLING DISTRIBUTION OF $\bar{x}$

The probability distribution of $\bar{x}$ is called its sampling distribution. It lists the various values that $\bar{x}$ can assume and the probability for each value of $\bar{x}$.

In general, the probability distribution of a sample statistic is called its sampling distribution.

Reconsider the population of the annual salaries of five employees given in Table 7.2. Consider all possible samples of three salaries each that can be selected, without replacement, from that population. The total number of possible samples, given by the combinations formula discussed in Chapter 5, is 10, that is,

$$\text{Total number of samples} = \binom{5}{3} = \frac{5!}{3!\,(5-3)!} = \frac{5 \cdot 4 \cdot 3 \cdot 2 \cdot 1}{3 \cdot 2 \cdot 1 \cdot 2 \cdot 1} = 10$$

Suppose we assign letters A, B, C, D, and E to the salaries of five employees so that

$$A = 17, \quad B = 24, \quad C = 35, \quad D = 35, \quad E = 43$$

Then the 10 possible samples of three salaries each are

$$\text{ABC, ABD, ABE, ACD, ACE, ADE, BCD, BCE, BDE, CDE}$$

These 10 samples and their respective means are listed in Table 7.4. Note that the first two samples have the same three salaries. The reason for this is that two of the employees (C and D) have the same salaries and, hence, the samples ABC and ABD contain the same values. The mean of each sample is obtained by dividing the sum of the three salaries included in that sample by 3. For instance, the mean of the first sample is $(17 + 24 + 35)/3 = 25.33$. Note that the values of means of samples in Table 7.4 are rounded to two decimal places.

Table 7.4 All Possible Samples and Their Means When the Sample Size Is 3

Sample	Salaries in the Sample	$\bar{x}$
ABC	17, 24, 35	25.33
ABD	17, 24, 35	25.33
ABE	17, 24, 43	28.00
ACD	17, 35, 35	29.00
ACE	17, 35, 43	31.67
ADE	17, 35, 43	31.67
BCD	24, 35, 35	31.33
BCE	24, 35, 43	34.00
BDE	24, 35, 43	34.00
CDE	35, 35, 43	37.67

Table 7.5   Frequency
Distribution of $\bar{x}$ When
the Sample Size Is 3

$\bar{x}$	f
25.33	2
28.00	1
29.00	1
31.33	1
31.67	2
34.00	2
37.67	1

By using the values of $\bar{x}$ given in Table 7.4, we record the frequency distribution of $\bar{x}$ in Table 7.5. By dividing the frequencies of various values of $\bar{x}$ by the sum of all frequencies, we obtain the relative frequencies of classes, which can be used as probabilities of classes. These probabilities are listed in Table 7.6. This table gives the sampling distribution of $\bar{x}$.

Table 7.6   Sampling
Distribution of $\bar{x}$ When the
Sample Size Is 3

$\bar{x}$	$P(\bar{x})$
25.33	$2/10 = .20$
28.00	$1/10 = .10$
29.00	$1/10 = .10$
31.33	$1/10 = .10$
31.67	$2/10 = .20$
34.00	$2/10 = .20$
37.67	$1/10 = .10$
	$\Sigma P(\bar{x}) = 1.0$

If we draw just one sample of three salaries from the population of five salaries, we may draw any of the 10 possible samples. Hence, the sample mean $\bar{x}$ can assume any of the values listed in Table 7.6 with the corresponding probability. For instance, the probability that the mean of a randomly drawn sample of three salaries is 31.67 is .20. This can be written as

$$P(\bar{x} = 31.67) = .20$$

7.5   SAMPLING AND NONSAMPLING ERRORS

Usually, different samples selected from the same population will give different results because they contain different elements. This is obvious from Table 7.4, which shows that the mean of a sample of three salaries depends on which three of the five salaries are included in the sample. The result obtained from any one sample will generally be different from the one obtained from the corresponding population. The difference between the value of a sample statistic obtained from a sample and the value of the corresponding population parameter obtained from the population is called the **sampling error**. Note that this difference represents the sampling error only if the sample is random and no nonsampling error has been made. Otherwise only a part of this difference will be due to the sampling error.

SAMPLING ERROR

Sampling error is the difference between the value of a sample statistic and the value of the corresponding population parameter. In the case of the mean,

$$\text{Sampling error} = \bar{x} - \mu$$

assuming that the sample is random and no nonsampling error has been made.

It is important to remember that *a sampling error occurs because of chance.* The errors that occur for other reasons, such as errors made during collection, recording, and tabulation of data, are called **nonsampling errors**. Such errors occur because of human mistakes and not chance. Note that there is only one kind of sampling error—the error that occurs due to chance. However, there is not just one nonsampling error but many nonsampling errors that may occur due to different reasons.

NONSAMPLING ERRORS

The errors that occur in the collection, recording, and tabulation of data are called nonsampling errors.

The following paragraph, reproduced from the *Current Population Reports* of the U.S. Bureau of the Census, explains how the nonsampling errors can occur.

> Nonsampling errors can be attributed to many sources, e.g., inability to obtain information about all cases in the sample, definitional difficulties, differences in the interpretation of questions, inability or unwillingness on the part of the respondents to provide correct information, inability to recall information, errors made in collection such as in recording or coding the data, errors made in processing the data, errors made in estimating values for missing data, biases resulting from the differing recall periods caused by the interviewing pattern used, and failure of all units in the universe to have some probability of being selected for the sample (undercoverage).

The following are the main reasons for the occurrence of nonsampling errors.

1. If a sample is nonrandom (and, hence, nonrepresentative), the sample results may be too different from the census results. The following quote from *U.S. News & World Report* describes how even a randomly selected sample can become nonrandom if some of the members included in the sample cannot be contacted.

> A test poll conducted in the 1984 presidential election found that if the poll were halted after interviewing only those subjects who could be reached on the first try, Reagan showed a 3-percentage-point lead over Mondale. But when interviewers made a determined effort to reach everyone on their lists of randomly selected subjects—calling some as many as 30 times before finally reaching them—Reagan showed a 13 percent lead, much closer to the actual election result. As it turned out, people who were planning to vote Republican were simply less likely to be at home. ("The Numbers Racket: How Polls and Statistics Lie," *U.S. News & World Report*, July 11, 1988. Copyright © 1988 by U.S. News & World Report, Inc. Reprinted with permission.)

2. The questions may be phrased in such a way that they are not fully understood by the members of the sample or population. As a result, the answers obtained are not accurate.

3. The respondents may intentionally give false information in response to some sensitive questions. For example, people may not tell the truth about drinking habits, incomes, or opinions about minorities. Sometimes the respondents may give wrong answers because of ignorance. For example, a person may not remember the exact amount he spent on clothes during the last year. If asked in a survey, he may give an inaccurate answer.

4. The poll taker may make a mistake and enter a wrong number in the records or make an error while entering the data on a computer.

Note that nonsampling errors can occur both in a sample survey and in a census, whereas the sampling error occurs only when a sample survey is conducted. Nonsampling errors can be minimized by preparing the survey questionnaire carefully and handling the data cautiously. However, it is impossible to avoid the sampling error.

Example 7–1 illustrates the sampling and nonsampling errors using the mean.

Illustrating sampling and nonsampling errors.

EXAMPLE 7-1 Reconsider the population of five salaries given in Table 7.2. The salaries of five employees are $17, 24, 35, 35, and 43 thousand. The population mean is

$$\mu = (17 + 24 + 35 + 35 + 43)/5 = \$30.80 \text{ thousand}$$

Now suppose we take a random sample of three salaries from this population. Assume that this sample includes the salaries $17, 35, and 43 thousand. The mean for this sample is

$$\bar{x} = (17 + 35 + 43)/3 = \$31.67 \text{ thousand}$$

Consequently,

$$\text{Sampling error} = \bar{x} - \mu = 31.67 - 30.80 = \textbf{\$.87 thousand}$$

That is, the mean salary estimated from the sample is $.87 thousand higher than the mean salary of the population. Note that this difference occurred due to chance, that is, because we used a sample instead of the population.

Now suppose, when we select the above mentioned sample, we mistakenly record the second salary as $37 thousand instead of $35 thousand. As a result, we calculate the sample mean as

$$\bar{x} = (17 + 37 + 43)/3 = \$32.33 \text{ thousand}$$

Consequently, the difference between this sample mean and the population mean is

$$\bar{x} - \mu = 32.33 - 30.80 = \$1.53 \text{ thousand}$$

However, this difference between the sample mean and the population mean does not represent the sampling error. As we calculated earlier, only $.87 thousand of this difference is due to the sampling error. The remaining portion, which is equal to $1.53 - .87 = \$.66$ thousand, represents the nonsampling error because it occurred due to the error we made in recording the second salary in the sample. Thus, in this case,

$$\text{Sampling error} = \textbf{\$.87 thousand}$$

$$\text{Nonsampling error} = \textbf{\$.66 thousand}$$

Figure 7.1 shows the sampling and nonsampling errors for these calculations.

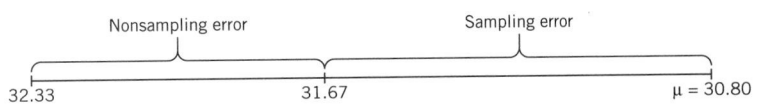

Figure 7.1 Sampling and nonsampling errors.

Note that in the real world we do not know the mean of a population. Hence, we select a sample to use the sample mean as an estimate of the population mean. Consequently, we never know the size of the sampling error.

EXERCISES

Concepts and Procedures

7.1 Explain the following.

    a. Random sample   **b.** Nonrandom sample
    c. Convenience sample   **d.** Judgment sample

7.2 Explain briefly what a simple random sample is and the various methods that can be used to select such a sample.

7.3 Under what kind of sample do all elements of a population have the same chance of being selected in a sample?

7.4 Briefly explain the meaning of a population distribution and a sampling distribution. Give an example of each.

7.5 Explain briefly the meaning of sampling and nonsampling errors. Give an example of each. Which occurs only in a sample survey and which occurs both in a sample survey and a census?

7.6 Consider the following population of six numbers.

$$15 \quad 12 \quad 8 \quad 16 \quad 9 \quad 10$$

    a. Find the population mean.
    b. Liza took one sample of four numbers from this population. The sample included the numbers 12, 8, 9, and 10. Calculate the sample mean and sampling error for this sample.
    c. Refer to part b. When Liza calculated the sample mean, she mistakenly used the numbers 12, 8, 6, and 10 to calculate the sample mean. Find the sampling and nonsampling errors in this case.
    d. List all samples of four numbers (without replacement) that can be selected from this population. Calculate the sample mean and sampling error for each of these samples.

7.7 Consider the following population of 10 numbers.

$$20 \quad 17 \quad 13 \quad 19 \quad 9 \quad 15 \quad 11 \quad 7 \quad 17 \quad 23$$

    a. Find the population mean.
    b. Rich took one sample of nine numbers from this population. The sample included the numbers 20, 17, 13, 9, 15, 11, 7, 17, and 23. Calculate the sample mean and sampling error for this sample.
    c. Refer to part b. When Rich calculated the sample mean, he mistakenly used the numbers 20, 17, 13, 9, 15, 11, 17, 17, and 23 to calculate the sample mean. Find the sampling and nonsampling errors in this case.
    d. List all samples of nine numbers (without replacement) that can be selected from this population. Calculate the sample mean and sampling error for each of these samples.

Applications

7.8 Using the formulas of Sections 5.3 and 5.4 of Chapter 5 for the mean and standard deviation of a discrete random variable, verify that the mean and standard deviation for the population probability distribution of Table 7.3 are 30.80 and 9.174, respectively.

7.9 The following data give the ages of all six employees of a company.

55   53   28   25   19   15

    a. Let x denote the age of an employee of this company. Write the population distribution of x.
    b. List all the possible samples of size five (without replacement) from this population. Calculate the mean for each of these samples. Write the sampling distribution of $\bar{x}$.
    c. Calculate the mean for the population data. Select one random sample of size five and calculate the sample mean $\bar{x}$. Compute the sampling error.

7.10 The following data give the years of experience for all five employees of a company.

7   8   12   7   5

    a. Let x denote the years of experience for an employee of this company. Write the population distribution of x.
    b. List all the possible samples of size four (without replacement) from this population. Calculate the mean for each of these samples. Write the sampling distribution of $\bar{x}$.
    c. Calculate the mean for the population data. Select one random sample of size four and calculate the sample mean $\bar{x}$. Compute the sampling error.

7.6 MEAN AND STANDARD DEVIATION OF $\bar{x}$

The mean and standard deviation calculated for the sampling distribution of $\bar{x}$ are called the **mean** and **standard deviation of $\bar{x}$**. Actually, the mean and standard deviation of $\bar{x}$ are, respectively, the mean and standard deviation of the means of all samples of the same size selected from a population. The standard deviation of $\bar{x}$ is also called the *standard error of $\bar{x}$*.

> **MEAN AND STANDARD DEVIATION OF $\bar{x}$**
>
> The mean and standard deviation of the sampling distribution of $\bar{x}$ are called the mean and standard deviation of $\bar{x}$ and are denoted by $\mu_{\bar{x}}$ and $\sigma_{\bar{x}}$, respectively.

If we calculate the mean and standard deviation of the 10 values of $\bar{x}$ listed in Table 7.4, we get the mean $\mu_{\bar{x}}$ and the standard deviation $\sigma_{\bar{x}}$ of $\bar{x}$. Alternatively, we can calculate the mean and standard deviation of the sampling distribution of $\bar{x}$ listed in Table 7.6. These will also be the values of $\mu_{\bar{x}}$ and $\sigma_{\bar{x}}$. From these calculations, we will obtain $\mu_{\bar{x}} = 30.80$ and $\sigma_{\bar{x}} = 3.747$ (see Exercise 7.27 at the end of this section).

The mean of the sampling distribution of $\bar{x}$ is always equal to the mean of the population.

> **MEAN OF THE SAMPLING DISTRIBUTION OF $\bar{x}$**
>
> The mean of the sampling distribution of $\bar{x}$ is equal to the mean of the population. Thus,
>
> $$\mu_{\bar{x}} = \mu$$

Hence, if we take all possible samples (of the same size) from a population and calculate their means, the mean $\mu_{\bar{x}}$ of all these sample means will be the same as the mean μ of the population. If we calculate the mean μ for the population probability distribution of Table 7.3 and the mean $\mu_{\bar{x}}$ for the sampling distribution of Table 7.6 by using the formula learned in Section 5.3 of Chapter 5, we get the same value of 30.80 for μ and $\mu_{\bar{x}}$ (see Exercise 7.27).

The sample mean $\bar{x}$ is called an **estimator** of the population mean μ. When the expected value (or mean) of a sample statistic is equal to the value of the corresponding population parameter, that sample statistic is said to be an **unbiased estimator**. For the sample mean $\bar{x}$, $\mu_{\bar{x}} = \mu$. Hence, $\bar{x}$ is an unbiased estimator of μ. This is a very important property that an estimator should possess.

However, the standard deviation $\sigma_{\bar{x}}$ of $\bar{x}$ is not equal to the standard deviation σ of the population distribution (unless $n = 1$). The standard deviation of $\bar{x}$ is equal to the standard deviation of the population divided by the square root of the sample size. That is,

$$\sigma_{\bar{x}} = \frac{\sigma}{\sqrt{n}}$$

This formula for the standard deviation of $\bar{x}$ holds true only when the sampling is done either with replacement from a finite population or with or without replacement from an infinite population. These two conditions can be replaced by the condition that the above formula holds true if the sample size is small in comparison to the population size. The sample size is considered to be small compared to the population size if the sample size is equal to or less than 5% of the population size, that is, if

$$n/N \le .05$$

If this condition is not satisfied, we use the following formula to calculate $\sigma_{\bar{x}}$.

$$\sigma_{\bar{x}} = \frac{\sigma}{\sqrt{n}} \sqrt{\frac{N - n}{N - 1}}$$

where the factor

$$\sqrt{\frac{N - n}{N - 1}}$$

is called the finite population correction factor.

In most practical applications, the sample size is usually small compared to the population size. Consequently, in most cases the formula used for calculating $\sigma_{\bar{x}}$ is $\sigma_{\bar{x}} = \sigma/\sqrt{n}$.

STANDARD DEVIATION OF THE SAMPLING DISTRIBUTION OF $\bar{x}$

The standard deviation of the sampling distribution of $\bar{x}$ is

$$\sigma_{\bar{x}} = \frac{\sigma}{\sqrt{n}}$$

where σ is the standard deviation of the population and n is the sample size. This formula is used when $n/N \leq .05$, where N is the population size.

Following are two important observations regarding the sampling distribution of $\bar{x}$.

1. *The spread of the sampling distribution of $\bar{x}$ is smaller than the spread of the corresponding population distribution.* In other words, $\sigma_{\bar{x}} < \sigma$. This is obvious from the formula for $\sigma_{\bar{x}}$. When n is greater than 1, which is usually true, the denominator in $\sigma/\sqrt{n}$ is greater than 1. Hence, $\sigma_{\bar{x}}$ is smaller than σ.

2. *The standard deviation of the sampling distribution of $\bar{x}$ decreases as the sample size increases.* This feature of the sampling distribution of $\bar{x}$ is also obvious from the formula

$$\sigma_{\bar{x}} = \frac{\sigma}{\sqrt{n}}$$

If the standard deviation of a sample statistic decreases as the sample size is increased, that statistic is said to be a **consistent estimator**. This is another important property that an estimator should possess. It is obvious from the above formula for $\sigma_{\bar{x}}$ that as n increases, the value of $\sqrt{n}$ also increases and, consequently, the value of $\sigma/\sqrt{n}$ decreases. Thus, the sample mean $\bar{x}$ is a consistent estimator of the population mean μ. Example 7–2 illustrates this feature.

Mean and standard deviation of $\bar{x}$.

EXAMPLE 7–2 The mean wage per hour for all 5000 employees working at a large company is $13.50 with a standard deviation of $2.90. Let $\bar{x}$ be the mean wage per hour for a random sample of certain employees selected from this company. Find the mean and standard deviation of $\bar{x}$ for a sample size of

(a)   30   (b)   75   (c)   200

Solution   From the given information, for the population of all employees,

$$N = 5000, \quad \mu = \$13.50, \quad \text{and} \quad \sigma = \$2.90$$

(a)   The mean $\mu_{\bar{x}}$ of the sampling distribution of $\bar{x}$ is

$$\mu_{\bar{x}} = \mu = \mathbf{\$13.50}$$

In this case, $n = 30$, $N = 5000$, and $n/N = 30/5000 = .006$. As n/N is less than .05, the standard deviation of $\bar{x}$ is obtained by using the formula $\sigma/\sqrt{n}$. Hence,

$$\sigma_{\bar{x}} = \frac{\sigma}{\sqrt{n}} = \frac{2.90}{\sqrt{30}} = \mathbf{\$.53}$$

Thus, we can state that if we take all possible samples of size 30 from the population of all employees of this company and prepare the sampling distribution of $\bar{x}$, the mean and standard deviation of this sampling distribution of $\bar{x}$ will be $13.50 and $.53, respectively.

(b) In this case, $n = 75$ and $n/N = 75/5000 = .015$, which is less than .05. The mean and standard deviation of $\bar{x}$ are

$$\mu_{\bar{x}} = \mu = \textbf{\$13.50} \qquad \text{and} \qquad \sigma_{\bar{x}} = \frac{\sigma}{\sqrt{n}} = \frac{2.90}{\sqrt{75}} = \textbf{\$.33}$$

(c) In this case, $n = 200$ and $n/N = 200/5000 = .04$, which is less than .05. Therefore, the mean and standard deviation of $\bar{x}$ are

$$\mu_{\bar{x}} = \mu = \textbf{\$13.50} \qquad \text{and} \qquad \sigma_{\bar{x}} = \frac{\sigma}{\sqrt{n}} = \frac{2.90}{\sqrt{200}} = \textbf{\$.21}$$

From the above calculations we observe that the mean of the sampling distribution of $\bar{x}$ is always equal to the mean of the population whatever the size of the sample. However, the value of the standard deviation of $\bar{x}$ decreases from $.53 to $.33 and then to $.21 as the sample size increases from 30 to 75 and then to 200. ▬

EXERCISES

Concepts and Procedures

7.11 Let $\bar{x}$ be the mean of a sample selected from a population.

a. What is the mean of $\bar{x}$ equal to?

b. What is the standard deviation of $\bar{x}$ equal to? Assume $n/N \leq .05$.

7.12 What is an estimator? When is an estimator unbiased? Is the sample mean $\bar{x}$ an unbiased estimator of μ? Explain.

7.13 When is an estimator said to be consistent? Is the sample mean $\bar{x}$ a consistent estimator of μ? Explain.

7.14 How does the value of $\sigma_{\bar{x}}$ change as the sample size increases? Explain.

7.15 Consider a large population with $\mu = 60$ and $\sigma = 12$. Assuming $n/N \leq .05$, find the mean and standard deviation of the sample mean $\bar{x}$ for a sample size of

a. 18   b. 90

7.16 Consider a large population with $\mu = 90$ and $\sigma = 16$. Assuming $n/N \leq .05$, find the mean and standard deviation of the sample mean $\bar{x}$ for a sample size of

a. 10   b. 35

7.17 A population of $N = 5000$ has a $\sigma = 20$. In each of the following cases which formula will you use to calculate $\sigma_{\bar{x}}$ and why? Using the appropriate formula, calculate $\sigma_{\bar{x}}$ for each of these cases.

a. $n = 300$ b. $n = 200$ c. $n = 500$ d. $n = 100$

7.18 A population of $N = 100,000$ has a $\sigma = 35$. In each of the following cases which formula will you use to calculate $\sigma_{\bar{x}}$ and why? Using the appropriate formula, calculate $\sigma_{\bar{x}}$ for each of these cases.

a. $n = 6000$ b. $n = 2500$ c. $n = 4000$ d. $n = 1000$

7.19 For a population $\mu = 125$ and $\sigma = 18$.

a. For a sample selected from this population, $\mu_{\bar{x}} = 125$ and $\sigma_{\bar{x}} = 3.6$. Find the sample size. Assume $n/N \leq .05$.

b. For a sample selected from this population, $\mu_{\bar{x}} = 125$ and $\sigma_{\bar{x}} = 2.25$. Find the sample size. Assume $n/N \leq .05$.

7.20 For a population $\mu = 46$ and $\sigma = 8$.

a. For a sample selected from this population, $\mu_{\bar{x}} = 46$ and $\sigma_{\bar{x}} = 2.0$. Find the sample size. Assume $n/N \leq .05$.

b. For a sample selected from this population, $\mu_{\bar{x}} = 46$ and $\sigma_{\bar{x}} = 1.6$. Find the sample size. Assume $n/N \leq .05$.

Applications

7.21 According to the U.S. Bureau of Labor Statistics, the mean number of hours worked per week by workers employed in the private (nonagricultural) industrial sector was 34.3 in 1991. Assume that for the hours worked per week by the current population of all workers employed in the private (non-agricultural) industrial sector $\mu = 34.3$ and $\sigma = 4$. Let $\bar{x}$ be the mean hours worked per week by a random sample of 40 such workers. Find the mean and standard deviation of the sampling distribution of $\bar{x}$.

7.22 According to the U.S. Bureau of Labor Statistics, the mean weekly earnings of workers in the manufacturing sector were \$455 in 1991. Assume that the current mean weekly earnings of all workers in the manufacturing sector are \$455 with a standard deviation of \$60. Let $\bar{x}$ be the mean weekly earnings of a sample of 400 workers selected from the manufacturing sector. Find the mean and standard deviation of the sampling distribution of $\bar{x}$.

7.23 According to McKinsey Global Institute, the mean annual output per worker is \$49,600 in the United States. Assume that the mean annual output for all workers in the United States is \$49,600 and its standard deviation is \$7000. Let $\bar{x}$ be the mean annual output for a sample of 900 workers selected from the United States. Find the mean and standard deviation of the sampling distribution of $\bar{x}$.

7.24 The mean annual salary for all 1050 professors at a university is \$47,600 and the standard deviation of their annual salaries is \$7200. Let $\bar{x}$ be the mean salary of a random sample of 16 professors selected from this university. Find the mean and standard deviation of the sampling distribution of $\bar{x}$.

7.25 The standard deviation of prices of all cars is known to be \$3600. Let $\bar{x}$ be the mean price of a sample of cars. What sample size will produce the standard deviation of $\bar{x}$ equal to \$180?

7.26 The standard deviation of the 1992 gross sales of all corporations is known to be \$139.50 million. Let $\bar{x}$ be the mean of the 1992 gross sales of a sample of corporations. What sample size will produce the standard deviation of $\bar{x}$ equal to \$15.50 million?

*7.27 Consider the sampling distribution of $\bar{x}$ given in Table 7.6.

a. Calculate the value of $\mu_{\bar{x}}$ using the formula: $\mu_{\bar{x}} = \Sigma \bar{x} \, P(\bar{x})$. Is the value of μ calculated in Exercise 7.8 the same as the value of $\mu_{\bar{x}}$ calculated here?

b. Calculate the value of $\sigma_{\bar{x}}$ by using the formula

$$\sigma_{\bar{x}} = \sqrt{\Sigma \bar{x}^2 P(\bar{x}) - (\mu_{\bar{x}})^2}$$

c. From Exercise 7.8, $\sigma = 9.174$. Also, our sample size is 3 so that $n = 3$. Therefore, $\sigma/\sqrt{n} = 9.174/\sqrt{3} = 5.297$. From part b, you should get $\sigma_{\bar{x}} = 3.747$. Why does $\sigma/\sqrt{n}$ not equal $\sigma_{\bar{x}}$ in this case?

d. In our example (given in the beginning of Section 7.4.1) on salaries, $N = 5$ and $n = 3$. Hence, $n/N = 3/5 = .60$. Because n/N is greater than .05, the appropriate formula to find $\sigma_{\bar{x}}$ will be

$$\sigma_{\bar{x}} = \frac{\sigma}{\sqrt{n}} \sqrt{\frac{N - n}{N - 1}}$$

Show that the value of $\sigma_{\bar{x}}$ calculated by using this formula gives the same value as the one calculated in part b.

7.7 SHAPE OF THE SAMPLING DISTRIBUTION OF $\bar{x}$

The shape of the sampling distribution of $\bar{x}$ relates to the following two cases.

1. The population from which samples are drawn has a normal distribution.
2. The population from which samples are drawn does not have a normal distribution.

7.7.1 SAMPLING FROM A NORMALLY DISTRIBUTED POPULATION

When the population from which samples are drawn is normally distributed with its mean equal to μ and standard deviation equal to σ, then

1. The mean of $\bar{x}$, $\mu_{\bar{x}}$, is equal to the mean of the population, μ.
2. The standard deviation of $\bar{x}$, $\sigma_{\bar{x}}$, is equal to $\sigma/\sqrt{n}$, assuming $n/N \leq .05$.
3. The shape of the sampling distribution of $\bar{x}$ is normal, whatever the value of n.

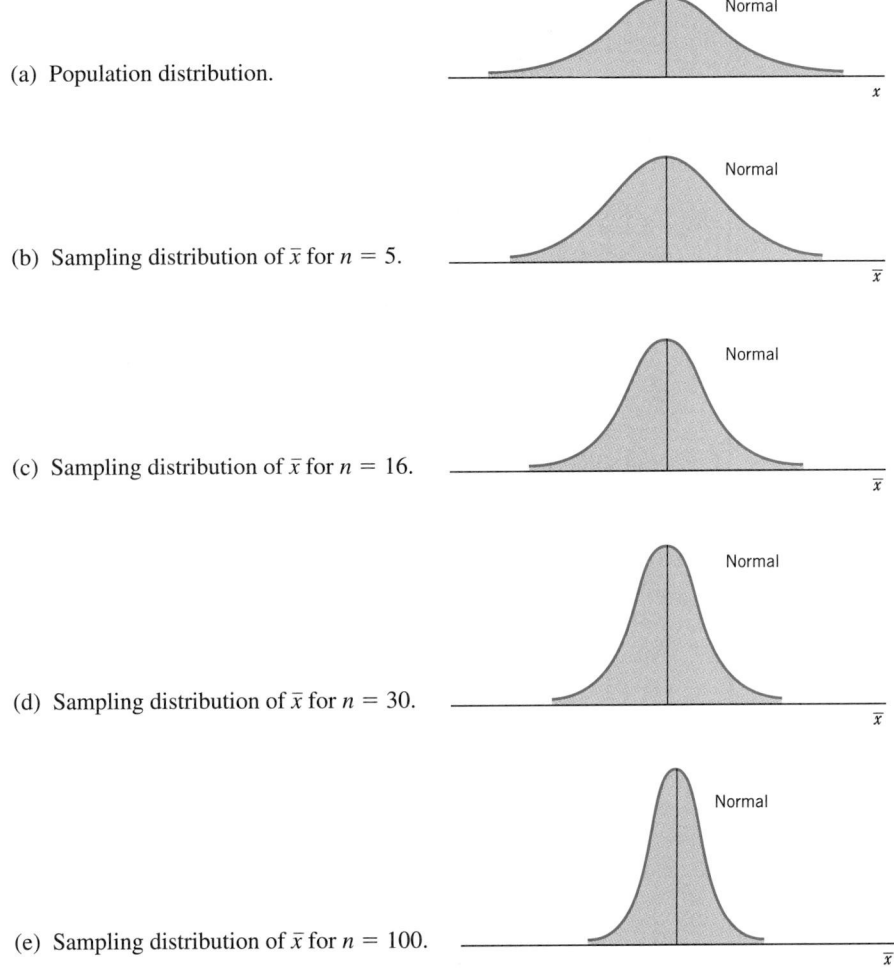

(a) Population distribution.

(b) Sampling distribution of $\bar{x}$ for $n = 5$.

(c) Sampling distribution of $\bar{x}$ for $n = 16$.

(d) Sampling distribution of $\bar{x}$ for $n = 30$.

(e) Sampling distribution of $\bar{x}$ for $n = 100$.

Figure 7.2 Population distribution and sampling distributions of $\bar{x}$.

SAMPLING DISTRIBUTION OF $\bar{x}$ WHEN THE POPULATION HAS A NORMAL DISTRIBUTION

If the population from which the samples are drawn is normally distributed with mean μ and standard deviation σ, then the sampling distribution of the sample mean $\bar{x}$ will also be normally distributed with the following mean and standard deviation, irrespective of the sample size.

$$\mu_{\bar{x}} = \mu \quad \text{and} \quad \sigma_{\bar{x}} = \frac{\sigma}{\sqrt{n}}$$

☞   *Remember*   For $\sigma_{\bar{x}} = \sigma/\sqrt{n}$ to be true, n/N must be less than or equal to .05.

Figure 7.2*a* shows the probability distribution curve for a population. The distribution curves in Figure 7.2*b* through Figure 7.2*e* show the sampling distributions of $\bar{x}$ for different sample sizes taken from the population of Figure 7.2*a*. As we can observe, the population has a normal distribution. Because of this, the sampling distribution of $\bar{x}$ is normal for each of the four cases illustrated in parts *b* through *e* of Figure 7.2. Also notice from Figure 7.2*b* through Figure 7.2*e* that the spread of the sampling distribution of $\bar{x}$ decreases as the sample size increases.

Example 7–3 illustrates the calculation of the mean and standard deviation of $\bar{x}$ and the description of the shape of its sampling distribution.

Mean, standard deviation, and the sampling distribution of $\bar{x}$: normal population.

EXAMPLE 7–3   According to the U.S. Bureau of Labor Statistics, the mean hourly wage for construction workers was $13.99 in 1991. Assume that the distribution of hourly wages of all construction workers is normal with a mean of $13.99 and a standard deviation of $2.25. Let $\bar{x}$ be the mean hourly wage of a random sample of certain construction workers. Calculate the mean and standard deviation of $\bar{x}$ and describe the shape of its sampling distribution when the sample size is

(a)   10   (b)   50   (c)   1000

Solution   Let μ and σ be the mean and standard deviation of hourly wages of all construction workers, and $\mu_{\bar{x}}$ and $\sigma_{\bar{x}}$ be the mean and standard deviation of the sampling distribution of $\bar{x}$.

(a)   The mean and standard deviation of $\bar{x}$ are

$$\mu_{\bar{x}} = \mu = \textbf{\$13.99}$$

$$\sigma_{\bar{x}} = \sigma/\sqrt{n} = 2.25/\sqrt{10} = \textbf{\$.712}$$

As the hourly wages of all construction workers are assumed to be normally distributed, the sampling distribution of $\bar{x}$ for samples of 10 construction workers will also be normal. Figure 7.3 shows the population distribution and the sampling distribution of $\bar{x}$. Note that because σ is greater than $\sigma_{\bar{x}}$, the population distribution has a wider spread and a smaller height than the sampling distribution of $\bar{x}$ in Figure 7.3.

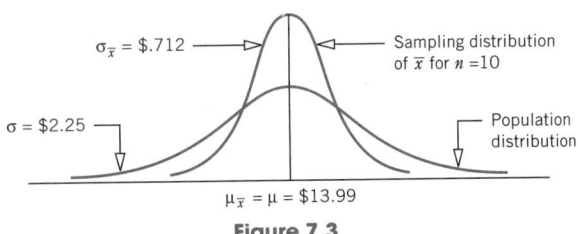

$\sigma_{\bar{x}} = \$.712$ ← Sampling distribution of $\bar{x}$ for $n = 10$

$\sigma = \$2.25$ ← Population distribution

$\mu_{\bar{x}} = \mu = \$13.99$

Figure 7.3

(b)   The mean and standard deviation of $\bar{x}$ are

$$\mu_{\bar{x}} = \mu = \mathbf{\$13.99}$$

$$\sigma_{\bar{x}} = \sigma/\sqrt{n} = 2.25/\sqrt{50} = \mathbf{\$.318}$$

Again, because the hourly wages of all construction workers are assumed to be normally distributed, the sampling distribution of $\bar{x}$ for samples of 50 construction workers will also be normal. The population distribution and the sampling distribution of $\bar{x}$ are shown in Figure 7.4.

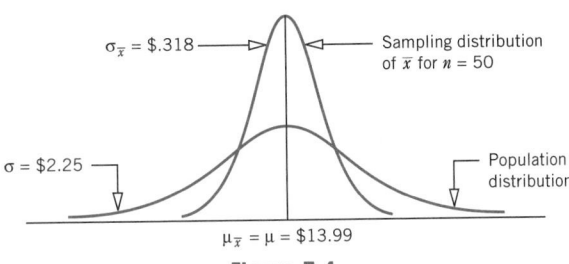

$\sigma_{\bar{x}} = \$.318$ ← Sampling distribution of $\bar{x}$ for $n = 50$

$\sigma = \$2.25$ ← Population distribution

$\mu_{\bar{x}} = \mu = \$13.99$

Figure 7.4

(c)   The mean and standard deviation of $\bar{x}$ are

$$\mu_{\bar{x}} = \mu = \mathbf{\$13.99}$$

$$\sigma_{\bar{x}} = \sigma/\sqrt{n} = 2.25/\sqrt{1000} = \mathbf{\$.071}$$

Again, because the hourly wages of all construction workers are assumed to be normally distributed, the sampling distribution of $\bar{x}$ for samples of 1000 construction workers will also be normal. The two distributions are shown in Figure 7.5.

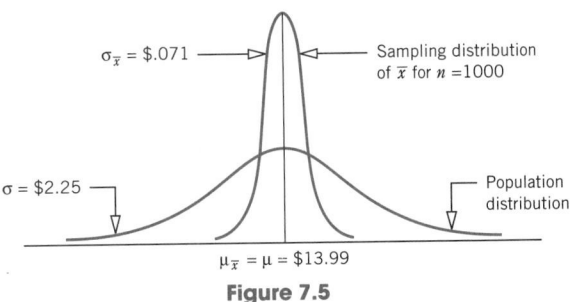

$\sigma_{\bar{x}} = \$.071$ ← Sampling distribution of $\bar{x}$ for $n = 1000$

$\sigma = \$2.25$ ← Population distribution

$\mu_{\bar{x}} = \mu = \$13.99$

Figure 7.5

Thus, whatever the sample size, the sampling distribution of $\bar{x}$ is normal when the population from which the samples are drawn is normally distributed.

7.7.2   SAMPLING FROM A POPULATION THAT IS NOT NORMALLY DISTRIBUTED

Most of the time the population from which the samples are taken is not normally distributed. In such cases, the shape of the sampling distribution of $\bar{x}$ is inferred from a very important theorem called the **central limit theorem**.

CENTRAL LIMIT THEOREM

According to the central limit theorem, for a large sample size, the sampling distribution of the sample mean $\bar{x}$ is approximately normal, irrespective of the shape of the population distribution. The mean and standard deviation of the sampling distribution of $\bar{x}$ are

$$\mu_{\bar{x}} = \mu \quad \text{and} \quad \sigma_{\bar{x}} = \frac{\sigma}{\sqrt{n}}$$

The sample size is usually considered to be large if $n \geq 30$.

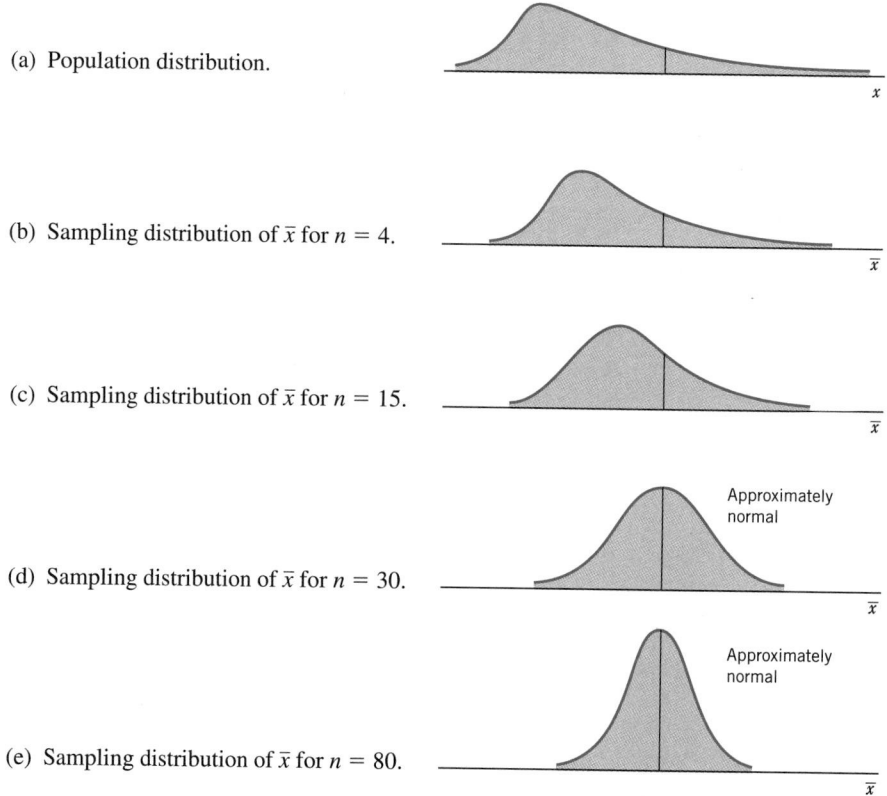

(a) Population distribution.

(b) Sampling distribution of $\bar{x}$ for $n = 4$.

(c) Sampling distribution of $\bar{x}$ for $n = 15$.

(d) Sampling distribution of $\bar{x}$ for $n = 30$.

(e) Sampling distribution of $\bar{x}$ for $n = 80$.

Figure 7.6   Population distribution and sampling distributions of $\bar{x}$.

Note that when the population does not have a normal distribution, the shape of the sampling distribution is not exactly normal but approximately normal for a large sample size. The approximation becomes more accurate as the sample size increases. Another point to remember is that the central limit theorem applies to *large* samples only. Usually, if the sample size is 30 or more, it is considered sufficiently large to apply the central limit theorem. Thus, according to the central limit theorem,

1. When $n \geq 30$, the shape of the sampling distribution of $\bar{x}$ is approximately normal irrespective of the shape of the population.
2. The mean of $\bar{x}$, $\mu_{\bar{x}}$, is equal to the mean of the population, μ.
3. The standard deviation of $\bar{x}$, $\sigma_{\bar{x}}$, is equal to $\sigma/\sqrt{n}$.

Again, remember that for $\sigma_{\bar{x}} = \sigma/\sqrt{n}$ to be true, n/N must be less than or equal to .05.

Figure 7.6*a* shows the probability distribution curve for a population. The distribution curves in Figure 7.6*b* through Figure 7.6*e* show the sampling distributions of $\bar{x}$ for different sample sizes taken from the population of Figure 7.6*a*. As we can observe, the population is not normally distributed. The sampling distributions of $\bar{x}$ shown in parts *b* and *c*, when $n < 30$, are not normal. However, the sampling distributions of $\bar{x}$ shown in parts *d* and *e*, when $n \geq 30$, are (approximately) normal. Also notice that the spread of the sampling distribution of $\bar{x}$ decreases as the sample size increases.

Example 7–4 illustrates the calculation of the mean and standard deviation of $\bar{x}$ and describes the shape of the sampling distribution of $\bar{x}$ when the sample size is large.

Mean, standard deviation, and the sampling distribution of $\bar{x}$: nonnormal population.

EXAMPLE 7–4 The mean rent paid by all tenants in a large city is $950 with a standard deviation of $225. However, the population distribution of rents for all tenants in this city is skewed to the right. Calculate the mean and standard deviation of $\bar{x}$ and describe the shape of its sampling distribution when the sample size is

(a) 30   (b) 100

Solution Although the population distribution of rents paid by all tenants is not normal, in each case the sample size is large ($n \geq 30$). Hence, the central limit theorem can be applied to infer the shape of the sampling distribution of $\bar{x}$.

(a) Let $\bar{x}$ be the mean rent paid by a sample of 30 tenants. Then, the sampling distribution of $\bar{x}$ is approximately normal with the values of the mean and standard deviation as

$$\mu_{\bar{x}} = \mu = \$950 \quad \text{and} \quad \sigma_{\bar{x}} = \sigma/\sqrt{n} = 225/\sqrt{30} = \$41.08$$

Figure 7.7 shows the population distribution and the sampling distribution of $\bar{x}$.

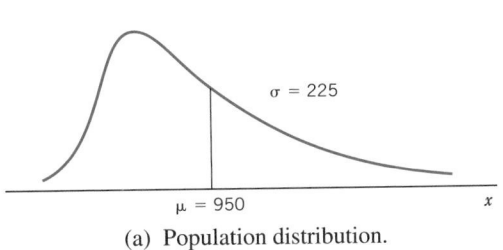

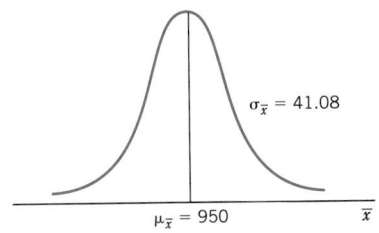

(a) Population distribution.   (b) Sampling distribution of $\bar{x}$ for $n = 30$.

Figure 7.7

(b) Let $\bar{x}$ be the mean rent paid by a sample of 100 tenants. Then, the sampling distribution of $\bar{x}$ is approximately normal with the values of the mean and standard deviation as

$$\mu_{\bar{x}} = \mu = \$950 \quad \text{and} \quad \sigma_{\bar{x}} = \sigma/\sqrt{n} = 225/\sqrt{100} = \$22.50$$

Figure 7.8 shows the population distribution and the sampling distribution of $\bar{x}$.

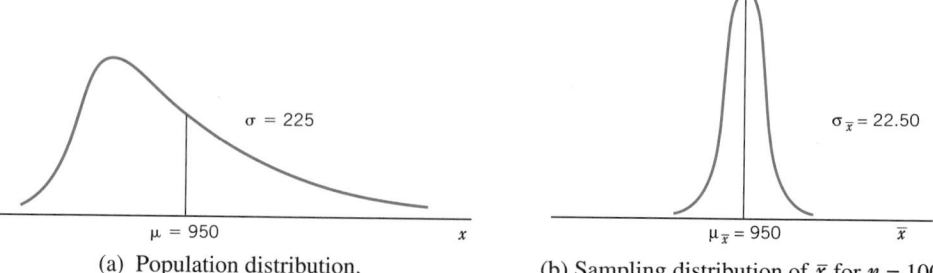

| (a) Population distribution. | (b) Sampling distribution of $\bar{x}$ for $n = 100$ |

Figure 7.8

EXERCISES

Concepts and Procedures

7.28   What condition or conditions must hold true for the sampling distribution of the sample mean to be normal when the sample size is less than 30?

7.29   Explain the central limit theorem.

7.30   A population has a distribution that is skewed to the left. Indicate in which of the following cases the central limit theorem will apply to describe the sampling distribution of the sample mean.

  a. $n = 400$ **b.** $n = 25$ **c.** $n = 16$ **d.** $n = 100$ **e.** $n = 36$

7.31   A population has a distribution that is skewed to the right. A sample of size n is selected from this population. Describe the shape of the sampling distribution of the sample mean for each of the following cases.

  a. $n = 25$ **b.** $n = 13$ **c.** $n = 80$ **d.** $n = 29$

7.32   A population has a normal distribution. A sample of size n is selected from this population. Describe the shape of the sampling distribution of the sample mean for each of the following cases.

  a. $n = 94$ **b.** $n = 11$ **c.** $n = 1900$ **d.** $n = 500$

7.33   A population has a normal distribution. A sample of size n is selected from this population. Describe the shape of the sampling distribution of the sample mean for each of the following cases.

  a. $n = 23$ **b.** $n = 450$ **c.** $n = 17$ **d.** $n = 27$

Applications

7.34   The time taken to learn an assembly job for a new worker hired in the assembly department of a company is normally distributed with a mean of 80 hours and a standard deviation of 6 hours. Let $\bar{x}$ be the mean time taken to learn this job by a random sample of 16 new workers. Calculate the mean and standard deviation of $\bar{x}$ and describe the shape of its sampling distribution.

7.35   The speeds of all cars traveling on a stretch of Interstate Highway I-95 are normally distributed with a mean of 68 miles per hour and a standard deviation of 3 miles per hour. Let $\bar{x}$ be the mean speed of a random sample of 20 cars traveling on this highway. Calculate the mean and standard deviation of $\bar{x}$ and describe the shape of its sampling distribution.

7.36   The amounts of electric bills for all households in a city have an approximate normal distribution with a mean of $42 and a standard deviation of $7. Let $\bar{x}$ be the mean amount of electric bills for a random sample of 25 households selected from this city. Find the mean and standard deviation of $\bar{x}$ and comment on the shape of its sampling distribution.

7.37 The ages of all CEOs of corporations have an approximate normal distribution with a mean of 54 years and a standard deviation of 5 years. Let $\bar{x}$ be the mean age of a random sample of 48 CEOs. Find the mean and standard deviation of $\bar{x}$ and comment on the shape of its sampling distribution.

7.38 The weights of all packages shipped from a post office have a distribution that is skewed to the right with a mean of 6 pounds and a standard deviation of 2 pounds. Let $\bar{x}$ be the mean weight of a random sample of 45 packages shipped from this post office. Find the mean and standard deviation of $\bar{x}$ and comment on the shape of its sampling distribution.

7.39 The amounts of telephone bills for all households in a large city have a distribution that is skewed to the right with a mean of $70 and a standard deviation of $25. Let $\bar{x}$ be the mean amount of telephone bills for a random sample of 90 households selected from this city. Calculate the mean and standard deviation of $\bar{x}$ and describe the shape of its sampling distribution.

7.40 The balances of checking accounts at a local bank have a distribution that is skewed to the right with its mean equal to $350 and standard deviation equal to $140. Let $\bar{x}$ be the mean balance of a random sample of 60 checking accounts selected from this bank. Calculate the mean and standard deviation of $\bar{x}$ and describe the shape of its sampling distribution.

7.41 According to the American Association of University Professors, the mean 1991–1992 salary of full-time instructional faculty at institutions of higher education was $45,360. Suppose the distribution of 1991–1992 salaries of these faculty is skewed to the right with its mean equal to $45,360 and standard deviation equal to $5900. Let $\bar{x}$ be the mean 1991–1992 salary of a random sample of 80 faculty members selected from institutions of higher education. Calculate the mean and standard deviation of $\bar{x}$ and describe the shape of its sampling distribution.

7.8   APPLICATIONS OF THE SAMPLING DISTRIBUTION OF $\bar{x}$

When conducting a survey, we usually take one sample and compute the value of $\bar{x}$ based on that sample. We never take all possible samples of the same size and then prepare the sampling distribution of $\bar{x}$. Rather, we are more interested in finding the probability that the value of $\bar{x}$ computed from one sample falls within a given interval. Examples 7–5 and 7–6 illustrate this procedure.

Calculating probability of $\bar{x}$ in an interval: normal population.

EXAMPLE 7–5 Assume that the weights of all packages of a certain brand of cookies are normally distributed with a mean of 32 ounces and a standard deviation of .3 ounces. Find the probability that the mean weight $\bar{x}$ of a random sample of 20 packages of this brand of cookies will be between 31.8 and 31.9 ounces.

Solution Although the sample size is small ($n < 30$) the shape of the sampling distribution of $\bar{x}$ is normal because the population is normally distributed. The mean and standard deviation of $\bar{x}$ are

$$\mu_{\bar{x}} = \mu \doteq 32 \text{ ounces} \quad \text{and} \quad \sigma_{\bar{x}} = \sigma/\sqrt{n} = .3/\sqrt{20} = .067$$

We are to compute the probability that the value of $\bar{x}$ calculated for one randomly drawn sample of 20 packages is between 31.8 and 31.9 ounces, that is,

$$P(31.8 < \bar{x} < 31.9)$$

This probability is given by the area under the normal curve for $\bar{x}$ between the points $\bar{x} = 31.8$ and $\bar{x} = 31.9$. The first step in finding this area is to convert the two $\bar{x}$ values to respective z values.

---

z VALUE FOR A VALUE OF $\bar{x}$

The z value for a value of $\bar{x}$ is calculated as

$$z = \frac{\bar{x} - \mu}{\sigma_{\bar{x}}}$$

The z values for $\bar{x} = 31.8$ and $\bar{x} = 31.9$ are computed below and they are shown on the z scale below the normal curve for $\bar{x}$ in Figure 7.9.

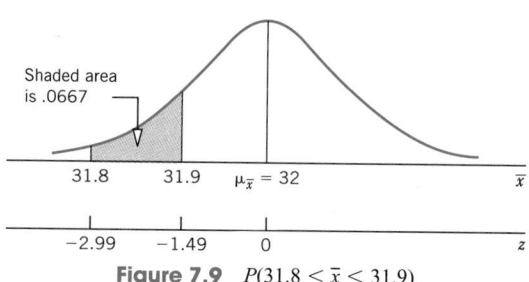

Figure 7.9 $P(31.8 < \bar{x} < 31.9)$

For $\bar{x} = 31.8$: $z = \dfrac{31.8 - 32}{.067} = -2.99$

For $\bar{x} = 31.9$: $z = \dfrac{31.9 - 32}{.067} = -1.49$

The probability that $\bar{x}$ is between 31.8 and 31.9 is given by the area under the standard normal curve between $z = -2.99$ and $z = -1.49$. Thus, the required probability is

$$P(31.8 < \bar{x} < 31.9) = P(-2.99 < z < -1.49)$$
$$= P(-2.99 < z < 0) - P(-1.49 < z < 0)$$
$$= .4986 - .4319 = \mathbf{.0667}$$

Therefore, the probability is .0667 that the mean weight of a sample of 20 packages will be between 31.8 and 31.9 ounces. ■

Calculating probability of $\bar{x}$ in an interval: $n > 30$.

EXAMPLE 7–6 The prices of all houses in New York State have a probability distribution that is skewed to the right with a mean of $157,000 and a standard deviation of $29,500. Let $\bar{x}$ be the mean price of a sample of 400 houses selected from New York State.

(a) What is the probability that the mean price obtained from this sample will be within $3000 of the population mean?

(b) What is the probability that the mean price obtained from this sample will be lower than the population mean by $2500 or more?

Solution Although the shape of the probability distribution of the population is skewed to the right, the sampling distribution of $\bar{x}$ is approximately normal because the sample size is large ($n > 30$). Remember that when the sample size is large, the central limit theorem applies.

The mean and standard deviation of the sampling distribution of $\bar{x}$ are

$$\mu_{\bar{x}} = \mu = \$157{,}000 \quad \text{and} \quad \sigma_{\bar{x}} = \sigma/\sqrt{n} = 29{,}500/\sqrt{400} = \$1475$$

(a) We need to find the probability

$$P(154{,}000 \leq \bar{x} \leq 160{,}000)$$

This probability is given by the area under the normal curve for $\bar{x}$ between $\bar{x} = \$154{,}000$ and $\bar{x} = \$160{,}000$, as shown in Figure 7.10. We find this area as follows.

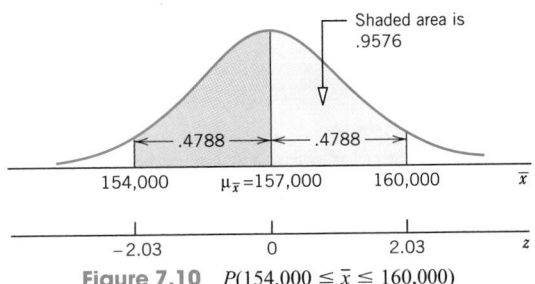

Figure 7.10 $P(154{,}000 \leq \bar{x} \leq 160{,}000)$

$$\text{For } \bar{x} = 154{,}000: \quad z = \frac{154{,}000 - 157{,}000}{1475} = -2.03$$

$$\text{For } \bar{x} = 160{,}000: \quad z = \frac{160{,}000 - 157{,}000}{1475} = 2.03$$

Hence, the required probability is

$$P(154{,}000 \leq \bar{x} \leq 160{,}000) = P(-2.03 \leq z \leq 2.03)$$
$$= P(-2.03 \leq z \leq 0) + P(0 \leq z \leq 2.03)$$
$$= .4788 + .4788 = \textbf{.9576}$$

Therefore, the probability that the mean price of a sample of 400 houses selected from New York State is within \$3000 of the population mean is .9576.

(b) The probability that the mean price obtained from a sample of 400 houses will be lower than the population mean by \$2500 or more is written as

$$P(\bar{x} \leq 154{,}500)$$

This probability is given by the area under the normal curve for $\bar{x}$ to the left of $\bar{x} = \$154{,}500$, as shown in Figure 7.11. We find this area as follows.

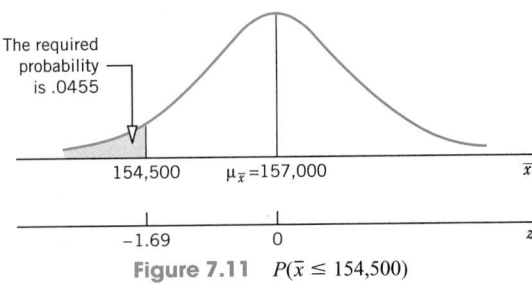

Figure 7.11 $P(\bar{x} \leq 154{,}500)$

$$\text{For } \bar{x} = 154{,}500: \quad z = \frac{154{,}500 - 157{,}000}{1475} = -1.69$$

Hence, the required probability is

$$P(\bar{x} \leq 154{,}500) = P(z \leq -1.69)$$
$$= .5 - P(-1.69 < z < 0)$$
$$= .5 - .4545 = \mathbf{.0455}$$

Therefore, the probability that the mean price of a sample of 400 houses selected from New York State is lower than the population mean by $2500 or more is .0455. ■

EXERCISES

Concepts and Procedures

7.42 If all possible samples of the same (large) size are selected from a population, what percent of all the sample means will be within 2.5 standard deviations of the population mean?

7.43 If all possible samples of the same (large) size are selected from a population, what percent of all the sample means will be within 1.5 standard deviations of the population mean?

7.44 For a population, $N = 10{,}000$, $\mu = 124$, and $\sigma = 18$. Find the z value for each of the following for $n = 36$.

    a. $\bar{x} = 128.60$ **b.** $\bar{x} = 119.30$ **c.** $\bar{x} = 116.88$ **d.** $\bar{x} = 132.05$

7.45 For a population, $N = 205{,}000$, $\mu = 66$, and $\sigma = 7$. Find the z value for each of the following for $n = 49$.

    a. $\bar{x} = 68.44$ **b.** $\bar{x} = 58.75$ **c.** $\bar{x} = 62.35$ **d.** $\bar{x} = 71.82$

7.46 A population has a mean of 80 and a standard deviation of 14. Assuming $n/N \leq .05$, find the following probabilities for a sample size of 110.

    a. $P(81.4 \leq \bar{x} \leq 83.6)$ **b.** $P(\bar{x} \geq 82.3)$

7.47 A population has a mean of 58 and a standard deviation of 12. Assuming $n/N \leq .05$, find the following probabilities for a sample size of 50.

    a. $P(53.7 \leq \bar{x} \leq 56.3)$ **b.** $P(\bar{x} \leq 59.2)$

Applications

7.48 According to the U.S. Bureau of Labor Statistics, the mean weekly wage for manufacturing workers was $455 in 1991. Assume that the weekly wages for all manufacturing workers for 1991 are normally distributed with a mean of $455 and a standard deviation of $60. Find the probability that the mean weekly wage for a random sample of 25 construction workers taken from this population would be

    a.   between $469 and $480
    b.   within $15 of the population mean weekly wage
    c.   lower than the population mean by $20 or more

7.49 The time taken to learn an assembly job for a new worker hired in the assembly department of a company is normally distributed with a mean of 80 hours and a standard deviation of 6 hours. Find the probability that the mean time taken to learn this job by a random sample of 16 new workers would be

    a.   between 76 and 78 hours
    b.   within 4 hours of the population mean
    c.   more than the population mean by at least 3.5 hours

7.50 Let x be a continuous random variable that has a distribution skewed to the right with $\mu = 60$ and $\sigma = 10$. Assuming $n/N \leq .05$, find the probability that the sample mean $\bar{x}$ for a random sample of 40 taken from this population will be

    a. less than 62.20   **b.** between 61.4 and 64.2

7.51 Let x be a continuous random variable that follows a distribution skewed to the left with $\mu = 90$ and $\sigma = 16$. Assuming $n/N \leq .05$, find the probability that the sample mean $\bar{x}$ for a random sample of 64 taken from this population will be

    a. less than 82.3   **b.** more than 86.7

7.52 According to the U.S. Bureau of the Census, the mean annual income of households in 1991 was \$37,922. Assume that the 1991 incomes of all U.S. households have a distribution that is skewed to the right with a mean of \$37,922 and a standard deviation of \$12,500. Find the probability that the 1991 mean income for a random sample of 400 households will be

    a. less than \$36,800   **b.** between \$38,500 and \$39,700
    c. within \$1100 of the population mean
    d. lower than the population mean by \$1250 or more

7.53 The amounts of electric bills for all households in a city have a skewed probability distribution with a mean of \$65 and a standard deviation of \$25. Find the probability that the mean amount of electric bills for a random sample of 75 households selected from this city will be

    a. more than \$70   **b.** between \$58 and \$63
    c. within \$6 of the population mean
    d. more than the population mean by at least \$5

7.54 The balances of all savings accounts at a local bank have a distribution that is skewed to the right with its mean equal to \$12,450 and standard deviation equal to \$4160. Find the probability that the mean of a sample of 50 savings accounts selected from this bank will be

    a. more than \$11,500   **b.** between \$12,000 and \$13,800
    c. within \$1500 of the population mean
    d. more than the population mean by at least \$1000

7.55 According to the American Association of University Professors, the mean 1991–1992 salary of full-time instructional faculty at institutions of higher education was \$45,360. Suppose the distribution of 1991–1992 salaries of these faculty is skewed to the right with a mean of \$45,360 and a standard deviation of \$5900. Find the probability that the mean salary of a sample of 80 such faculty will be

    a. less than \$46,500   **b.** between \$40,000 and \$49,200
    c. within \$1800 of the population mean
    d. lower than the population mean by \$1200 or more

7.56 Johnson Electronics Corporation makes electric tubes. It is known that the standard deviation of the lives of these tubes is 150 hours. The company's research department takes a sample of 100 such tubes and finds that the mean life of these tubes is 2250 hours. What is the probability that this sample mean is within 25 hours of the mean life of all tubes produced by this company?

7.57 A machine at Katz Steel Corporation makes 3-inch-long nails. The probability distribution of the lengths of these nails is normal with a mean of 3 inches and a standard deviation of .1 inches. The quality control inspector takes a sample of 25 nails once a week and calculates the mean length of these nails. If the mean of this sample is either less than 2.95 inches or greater than 3.05 inches, the inspector concludes that the machine needs an adjustment. What is the probability that based on a sample of 25 nails the inspector will conclude that the machine needs an adjustment?

7.9 POPULATION AND SAMPLE PROPORTIONS

The concept of proportion is the same as the concept of relative frequency discussed in Chapter 2 and the concept of probability of success in a binomial problem. The relative frequency of a category or class gives the proportion of the sample or population that belongs to that category or class. Similarly, the probability of success in a binomial problem represents the proportion of the sample or population that possesses a given characteristic.

The **population proportion**, denoted by *p*, is obtained by taking the ratio of the number of elements in a population with a specific characteristic to the total number of elements in the population. The **sample proportion**, denoted by $\hat{p}$ (read *p hat*), gives a similar ratio for a sample.

POPULATION AND SAMPLE PROPORTIONS

The population and sample proportions, denoted by p and $\hat{p}$, respectively, are calculated as

$$p = \frac{x}{N} \quad \text{and} \quad \hat{p} = \frac{x}{n}$$

where

$$N = \text{Total number of elements in the population}$$

$$n = \text{Total number of elements in the sample}$$

$$x = \text{Number of elements in the population or sample} \\ \text{that possess a specific characteristic}$$

Example 7–7 illustrates the calculation of the population and sample proportions.

Calculating population and sample proportions.

EXAMPLE 7–7   Suppose a total of 789,654 families live in a city and 563,282 of them own homes. Then,

$$N = \text{Population size} = 789,654$$

$$x = \text{Families in the population who own homes} = 563,282$$

The proportion of all families in this city who own homes is

$$p = x/N = 563,282/789,654 = \textbf{.71}$$

Now, suppose a sample of 240 families is taken from this city and 158 of them are homeowners. Then,

$$n = \text{Sample size} = 240$$

$$x = \text{Families in the sample who own homes} = 158$$

The sample proportion is

$$\hat{p} = x/n = 158/240 = \textbf{.66} \qquad \blacksquare$$

As in the case of the mean, the difference between the sample proportion and the corresponding population proportion gives the sampling error, assuming that the sample is random and no nonsampling error has been made. That is, in case of the proportion,

$$\text{Sampling error} = \hat{p} - p$$

7.10   MEAN, STANDARD DEVIATION, AND SHAPE OF THE SAMPLING DISTRIBUTION OF $\hat{p}$

This section discusses the sampling distribution of the sample proportion, and the mean, standard deviation, and shape of this sampling distribution.

7.10.1   SAMPLING DISTRIBUTION OF $\hat{p}$

Just like the sample mean $\bar{x}$, the sample proportion $\hat{p}$ is also a random variable. Hence, it possesses a probability distribution, which is called its **sampling distribution**.

SAMPLING DISTRIBUTION OF THE SAMPLE PROPORTION $\hat{p}$

The probability distribution of the sample proportion $\hat{p}$ is called its sampling distribution. It gives the various values that $\hat{p}$ can assume and their probabilities.

The value of $\hat{p}$ calculated for a particular sample depends on what elements of the population are included in that sample. Example 7–8 illustrates the concept of the sampling distribution of $\hat{p}$.

Illustrating the sampling distribution of $\hat{p}$.

EXAMPLE 7–8   Boe Consultant Associates has five employees. Table 7.7 gives the names of these five employees and information concerning their knowledge of statistics.

Table 7.7   Information on the Five Employees of Boe Consultant Associates

Name	Knows Statistics
Ally	yes
John	no
Susan	no
Lee	yes
Tom	yes

If we define the population proportion p as the proportion of employees who know statistics, then,

$$p = 3/5 = .60$$

Now, suppose we draw all possible samples of three employees each and compute the proportion of employees, for each sample, who know statistics. The total number of samples of size three that can be drawn from the population of five employees is

$$\text{Total number of samples} = \binom{5}{3} = \frac{5!}{3!\,(5-3)!} = \frac{5\cdot 4\cdot 3\cdot 2\cdot 1}{3\cdot 2\cdot 1\cdot 2\cdot 1} = 10$$

Table 7.8 lists these 10 possible samples and the proportion of employees who know statistics for each of those samples. Note that we have rounded the values of $\hat{p}$ to two decimal places.

Table 7.8   All Possible Samples of Size 3 and the Value of $\hat{p}$ for Each Sample

Sample	Proportion Who Know Statistics $\hat{p}$
Ally, John, Susan	$1/3 = .33$
Ally, John, Lee	$2/3 = .67$
Ally, John, Tom	$2/3 = .67$
Ally, Susan, Lee	$2/3 = .67$
Ally, Susan, Tom	$2/3 = .67$
Ally, Lee, Tom	$3/3 = 1.00$
John, Susan, Lee	$1/3 = .33$
John, Susan, Tom	$1/3 = .33$
John, Lee, Tom	$2/3 = .67$
Susan, Lee, Tom	$2/3 = .67$

Using Table 7.8, we prepare the frequency distribution of $\hat{p}$ as recorded in Table 7.9. Dividing the frequencies of classes by the sum of the frequencies, we obtain the relative frequencies, which can be used as probabilities of classes. These probabilities are listed in Table 7.10, which gives the sampling distribution of $\hat{p}$.

Table 7.9   Frequency Distribution of $\hat{p}$

$\hat{p}$	f
.33	3
.67	6
1.00	1
	$\Sigma f = 10$

Table 7.10   Sampling Distribution of $\hat{p}$

$\hat{p}$	$P(\hat{p})$
.33	$3/10 = .30$
.67	$6/10 = .60$
1.00	$1/10 = .10$
	$\Sigma P(\hat{p}) = 1.0$

7.10.2   MEAN AND STANDARD DEVIATION OF $\hat{p}$

The **mean of $\hat{p}$**, which is the same as the mean of the sampling distribution of $\hat{p}$, is always equal to the population proportion p just as the mean of the sampling distribution of $\bar{x}$ is always equal to the population mean μ.

MEAN OF THE SAMPLE PROPORTION

The mean of the sample proportion $\hat{p}$ is denoted by $\mu_{\hat{p}}$ and is equal to the population proportion p. Thus,

$$\mu_{\hat{p}} = p$$

The sample proportion $\hat{p}$ is called an **estimator** of the population proportion p. As mentioned earlier in this chapter, when the expected value (or mean) of a sample statistic is equal to the value of the corresponding population parameter, that sample statistic is said to be an **unbiased estimator**. Since for the sample proportion $\hat{p}$, $\mu_{\hat{p}} = p$, $\hat{p}$ is an unbiased estimator of p.

The **standard deviation of $\hat{p}$**, denoted by $\sigma_{\hat{p}}$, is given by the following formula. This formula is true only when the sample size is small as compared to the population size. As we know from Section 7.6, the sample size is said to be small compared to the population size if $n/N \leq .05$.

STANDARD DEVIATION OF THE SAMPLE PROPORTION

The standard deviation of the sample proportion $\hat{p}$ is denoted by $\sigma_{\hat{p}}$ and is given by the formula

$$\sigma_{\hat{p}} = \sqrt{\frac{pq}{n}}$$

where p is the population proportion, $q = 1 - p$, and n is the sample size.

This formula is used when $n/N \leq .05$ where N is the population size.

However, if n/N is greater than .05, then $\sigma_{\hat{p}}$ is calculated as follows.

$$\sigma_{\hat{p}} = \sqrt{\frac{pq}{n}} \sqrt{\frac{N - n}{N - 1}}$$

where the factor

$$\sqrt{\frac{N - n}{N - 1}}$$

is called the finite population correction factor.

In almost all cases, the sample size is small compared to the population size and, consequently, the formula used to calculate $\sigma_{\hat{p}}$ is $\sqrt{pq/n}$.

As mentioned earlier in this chapter, if the standard deviation of a sample statistic decreases as the sample size is increased, that statistic is said to be a **consistent estimator**. It is obvious from the above formula for $\sigma_{\hat{p}}$, that as n increases, the value of $\sqrt{pq/n}$ decreases. Thus, the sample proportion $\hat{p}$ is a consistent estimator of the population proportion p.

7.10.3 SHAPE OF THE SAMPLING DISTRIBUTION OF $\hat{p}$

The shape of the sampling distribution of $\hat{p}$ is inferred from the central limit theorem.

CENTRAL LIMIT THEOREM FOR SAMPLE PROPORTION

According to the central limit theorem, the sampling distribution of $\hat{p}$ is approximately normal for a sufficiently large sample size. In the case of proportion, the sample size n is considered to be sufficiently large if np and nq are both greater than 5, that is, if

$$np > 5 \quad \text{and} \quad nq > 5$$

Note that the sampling distribution of $\hat{p}$ will be approximately normal if $np > 5$ and $nq > 5$. This is the same condition that was required for the application of the normal approximation to the binomial probability distribution in Chapter 6.

Example 7–9 shows the calculation of the mean and standard deviation of $\hat{p}$ and describes the shape of its sampling distribution.

Mean, standard deviation, and the sampling distribution of $\hat{p}$.

EXAMPLE 7-9 Eighteen percent of the working people polled by the Roper Organization for Shearson Lehman Brothers said their careers are both personally and financially rewarding. Assume that this result is true for the population of all working people. Let $\hat{p}$ be the proportion in a random sample of 100 working people who hold this view. Find the mean and standard deviation of $\hat{p}$ and describe the shape of its sampling distribution.

Solution Let p be the proportion of all working people who think their careers are both personally and financially rewarding. Then,

$$p = .18 \quad \text{and} \quad q = 1 - p = 1 - .18 = .82$$

The mean of the sampling distribution of $\hat{p}$ is

$$\mu_{\hat{p}} = p = \mathbf{.18}$$

The standard deviation of $\hat{p}$ is

$$\sigma_{\hat{p}} = \sqrt{\frac{pq}{n}} = \sqrt{\frac{(.18)(.82)}{100}} = \mathbf{.038}$$

The values of np and nq are

$$np = 100\ (.18) = 18 \quad \text{and} \quad nq = 100\ (.82) = 82$$

As np and nq are both greater than 5, we can apply the central limit theorem to make an inference about the shape of the sampling distribution of $\hat{p}$. Therefore, the sampling distribution of $\hat{p}$ is approximately normal with a mean of .18 and a standard deviation of .038, as shown in Figure 7.12.

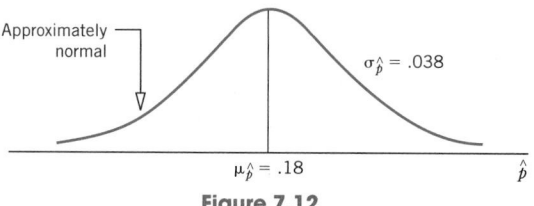

Figure 7.12

EXERCISES

Concepts and Procedures

7.58 In a population of 1000 subjects, 640 possess a certain characteristic. A sample of 40 subjects selected from this population has 24 subjects who possess the same characteristic. What are the values of the population and sample proportions?

7.59 In a population of 5000 subjects, 600 possess a certain characteristic. A sample of 120 subjects selected from this population contains 18 subjects who possess the same characteristic. What are the values of the population and sample proportions?

7.60 In a population of 18,700 subjects, 30% possess a certain characteristic. In a sample of 250 subjects selected from this population, 25% possess the same characteristic. How many subjects in the population and sample, respectively, possess this characteristic?

7.61 In a population of 9500 subjects, 75% possess a certain characteristic. In a sample of 400 subjects selected from this population, 78% possess the same characteristic. How many subjects in the population and sample, respectively, possess this characteristic?

7.62 Let $\hat{p}$ be the proportion of elements in a sample that possess a characteristic.

    a. What is the mean of $\hat{p}$?

    b. What is the standard deviation of $\hat{p}$? Assume $n/N \leq .05$.

7.63 For a population, $N = 12{,}000$ and $p = .72$. A random sample of 900 elements selected from this population gave $\hat{p} = .66$. Find the sampling error.

7.64 For a population, $N = 2800$ and $p = .28$. A random sample of 80 elements selected from this population gave $\hat{p} = .33$. Find the sampling error.

7.65 What is the estimator of the population proportion? Is this estimator an unbiased estimator of p? Explain why or why not.

7.66 Is the sample proportion a consistent estimator of the population proportion? Explain why or why not.

7.67 How does the value of $\sigma_{\hat{p}}$ change as the sample size increases? Explain. Assume $n/N \leq .05$.

7.68 Consider a large population with $p = .65$. Assuming $n/N \leq .05$, find the mean and standard deviation of the sample proportion $\hat{p}$ for a sample size of

    a. 100   **b.** 900

7.69 Consider a large population with $p = .18$. Assuming $n/N \leq .05$, find the mean and standard deviation of the sample proportion $\hat{p}$ for a sample size of

    a. 400   **b.** 750

7.70 A population of $N = 4000$ has a population proportion equal to .10. In each of the following cases which formula will you use to calculate $\sigma_{\hat{p}}$ and why? Using the appropriate formula, calculate $\sigma_{\hat{p}}$ for each of these cases.

    a. $n = 800$ **b.** $n = 30$ **c.** $n = 45$ **d.** $n = 240$

7.71 A population of $N = 1400$ has a population proportion equal to .45. In each of the following cases which formula will you use to calculate $\sigma_{\hat{p}}$ and why? Using the appropriate formula, calculate $\sigma_{\hat{p}}$ for each of these cases.

    a. $n = 90$ **b.** $n = 50$ **c.** $n = 120$ **d.** $n = 65$

7.72 According to the central limit theorem, the sampling distribution of $\hat{p}$ is approximately normal when the sample is large. What is considered a large sample in the case of the proportion? Briefly explain.

7.73 Indicate in which of the following cases the central limit theorem will apply to describe the sampling distribution of the sample proportion.

    a. $n = 400$ and $p = .28$ **b.** $n = 80$ and $p = .05$

    c. $n = 60$ and $p = .12$ **d.** $n = 100$ and $p = .035$

7.74 Indicate in which of the following cases the central limit theorem will apply to describe the sampling distribution of the sample proportion.

    a. $n = 20$ and $p = .45$ **b.** $n = 75$ and $p = .22$

    c. $n = 350$ and $p = .01$ **d.** $n = 200$ and $p = .022$

Applications

7.75 A company manufactured 6 television sets on a given day and these television sets were inspected for being good or defective. The results of the inspection are as follows.

Good   Good   Defective   Defective   Good   Good

    a. What proportion of these television sets are good?

    b. How many total samples (without replacement) of size 5 can be selected from this population?

    c. List all the possible samples of size 5 that can be selected from this population and calculate the sample proportion $\hat{p}$ of television sets that are good for each sample. Prepare the sampling distribution of $\hat{p}$.

    d. For each sample listed in part c, calculate the sampling error.

7.76 The following data give the information on all 5 employees of a company.

Male   Female   Female   Male   Female

    a. What proportion of employees of this company are female?

    b. How many total samples (without replacement) of size 3 can be drawn from this population?

    c. List all the possible samples of size 3 that can be selected from this population and calculate the sample proportion $\hat{p}$ of the employees who are female for each sample. Prepare the sampling distribution of $\hat{p}$.

    d. For each sample listed in part c, calculate the sampling error.

7.77 In a Harris poll, 31% of adults said that business executives have good moral and ethical standards (*USA TODAY*, September 3, 1992). Assume that this result holds true for the current population of all adults. Let $\hat{p}$ be the proportion of adults in a random sample of 500 who hold this view. Find the mean and standard deviation of $\hat{p}$ and describe the shape of its sampling distribution.

7.78 According to the U.S. Department of Education, (about) 69% of teachers in all public schools are female. Let $\hat{p}$ be the proportion in a random sample of 200 public school teachers who are female. Calculate the mean and standard deviation of $\hat{p}$ and comment on the shape of its sampling distribution.

7.79 A survey of all medium-sized and large corporations showed that 65% of them offer retirement plans to their employees. Let $\hat{p}$ be the proportion in a random sample of 50 such corporations that offer retirement plans to their employees. Calculate the mean and standard deviation of $\hat{p}$ and comment on the shape of its sampling distribution.

7.80 Dartmouth Distribution Warehouse makes deliveries of a large number of products to its customers. It is known that 85% of all the orders it receives from its customers are delivered on time. Let $\hat{p}$ be the proportion of orders in a random sample of 100 that are delivered on time. Calculate the mean and standard deviation of $\hat{p}$ and describe the shape of its sampling distribution.

7.11 APPLICATIONS OF THE SAMPLING DISTRIBUTION OF $\hat{p}$

As mentioned in Section 7.8, when we conduct a study we usually take only one sample and make all decisions or inferences on the basis of the results of that one sample. We use the concepts of the mean, standard deviation, and shape of the sampling distribution of $\hat{p}$ to determine the probability that the value of $\hat{p}$ computed from one sample falls within a given interval. Examples 7–10 and 7–11 illustrate this application.

Calculating probability that $\hat{p}$ is in an interval.

EXAMPLE 7–10 According to the U.S. Bureau of the Census, approximately 66% of the households headed by single women own autos. Assume that this percentage is true for the population of all households headed by single women. Let $\hat{p}$ be the proportion of households who own autos in a random sample of 120 households headed by single women. Find the probability that the value of $\hat{p}$ is between .68 and .72.

Solution From the given information,

$$p = .66 \quad \text{and} \quad q = 1 - p = 1 - .66 = .34$$

where p is the proportion of all households headed by single women who own autos.

The mean of the sample proportion $\hat{p}$ is

$$\mu_{\hat{p}} = p = .66$$

The standard deviation of $\hat{p}$ is

$$\sigma_{\hat{p}} = \sqrt{\frac{pq}{n}} = \sqrt{\frac{(.66)\,(.34)}{120}} = .043$$

The values of np and nq are

$$np = 120\,(.66) = 79.2 \quad \text{and} \quad nq = 120\,(.34) = 40.8$$

As np and nq are both greater than 5, we can infer from the central limit theorem that the sampling distribution of $\hat{p}$ is approximately normal. The probability that $\hat{p}$ is between .68 and .72 is given by the area under the normal curve for $\hat{p}$ between $\hat{p} = .68$ and $\hat{p} = .72$, as shown in Figure 7.13.

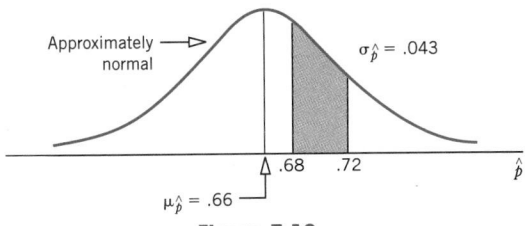

Figure 7.13

The first step in finding the area under the normal curve between $\hat{p} = .68$ and $\hat{p} = .72$ is to convert these two values to respective z values. The z value for $\hat{p}$ is computed using the following formula.

---

z VALUE FOR A VALUE OF $\hat{p}$

The z value for a value of $\hat{p}$ is calculated as

$$z = \frac{\hat{p} - p}{\sigma_{\hat{p}}}$$

Next, the two values of $\hat{p}$ are converted to their respective z values and then the area under the normal curve between these two points is found using the normal distribution table.

$$\text{For } \hat{p} = .68: \quad z = \frac{.68 - .66}{.043} = .47$$

$$\text{For } \hat{p} = .72: \quad z = \frac{.72 - .66}{.043} = 1.40$$

The probability that $\hat{p}$ is between .68 and .72 is given by the area under the standard normal distribution between $z = .47$ and $z = 1.40$. This area is shown in Figure 7.14. The required probability is

$$P(.68 < \hat{p} < .72) = P(.47 < z < 1.40)$$

$$= P(0 < z < 1.40) - P(0 < z < .47)$$

$$= .4192 - .1808 = \mathbf{.2384}$$

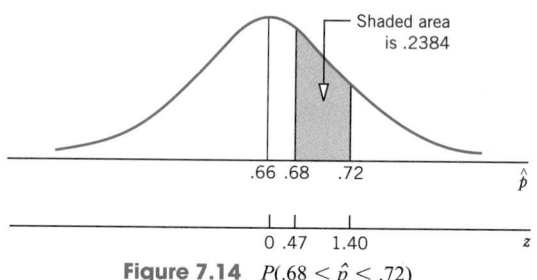

Figure 7.14 $P(.68 < \hat{p} < .72)$

Thus, the probability is .2384 that between 68% and 72% of households headed by single women own autos. ∎

Probability of $\hat{p}$ being greater than or equal to a certain value.

EXAMPLE 7–11 McMahn Corporation manufactures wheels for bicycles. The wheels that have diameters outside a predetermined range are useless and, hence, discarded. The machine that makes these wheels produces 5% of the wheels with diameters outside the acceptable range. The quality control inspector selects a sample of 110 wheels every week and measures their diameters. If 8% or more of the wheels in the sample have diameters outside the acceptable range, the process is stopped and the machine is readjusted. What is the probability that at least 8% of the wheels in a sample of 110 will have diameters in the unacceptable range?

Solution Let p be the proportion of all wheels that have diameters of unacceptable range. Then,

$$p = .05 \quad \text{and} \quad q = 1 - p = 1 - .05 = .95$$

Then, the mean of the sampling distribution of the sample proportion $\hat{p}$ is

$$\mu_{\hat{p}} = p = .05$$

The population of all wheels is large and the sample size is small as compared to the population. Consequently, we can assume that $n/N \leq .05$. Hence, the standard deviation of $\hat{p}$ is calculated as

$$\sigma_{\hat{p}} = \sqrt{\frac{pq}{n}} = \sqrt{\frac{(.05)\,(.95)}{110}} = .021$$

From the central limit theorem, the shape of the sampling distribution of $\hat{p}$ is approximately normal. (This is so because np and nq both are greater than 5.) The probability that $\hat{p}$ is greater than or equal to .08 is given by the area under the normal curve for $\hat{p}$ to the right of $\hat{p} = .08$, as shown in Figure 7.15.

The z value for $\hat{p} = .08$ is

$$z = \frac{\hat{p} - p}{\sigma_{\hat{p}}} = \frac{.08 - .05}{.021} = 1.43$$

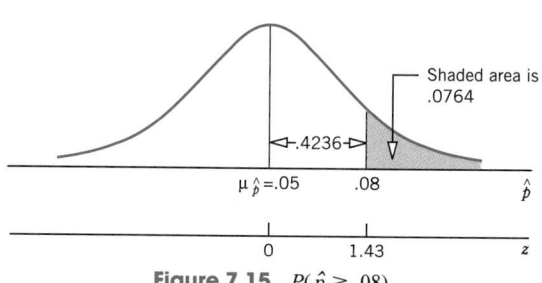

Figure 7.15 $P(\hat{p} \geq .08)$

Thus, the required probability is

$$P(\hat{p} \geq .08) = P(z \geq 1.43) = .5 - P(0 < z < 1.43)$$
$$= .5 - .4236 = .0764$$

Hence, the probability that at least 8% of the wheels in a random sample of 110 will have diameters in the unacceptable range is .0764.

EXERCISES

Concepts and Procedures

7.81 If all possible samples of the same (large) size are selected from a population, what percent of all sample proportions will be within 2.0 standard deviations of the population proportion?

7.82 If all possible samples of the same (large) size are selected from a population, what percent of all sample proportions will be within 3.0 standard deviations of the population proportion?

7.83 For a population, $N = 30,000$ and $p = .60$. Find the z value for each of the following for $n = 100$.

    a. $\hat{p} = .56$ **b.** $\hat{p} = .68$ **c.** $\hat{p} = .53$ **d.** $\hat{p} = .65$

7.84 For a population, $N = 18,000$ and $p = .24$. Find the z value for each of the following for $n = 70$.

    a. $\hat{p} = .26$ **b.** $\hat{p} = .32$ **c.** $\hat{p} = .17$ **b.** $\hat{p} = .20$

Applications

7.85 In a Harris poll, 31% of adults said that business executives have good moral and ethical standards (*USA TODAY*, September 3, 1992). Assume that this result holds true for the current population of all adults. Let $\hat{p}$ be the proportion of adults in a random sample of 500 who hold this view. Find the probability that the value of $\hat{p}$ will be

    a. between .27 and .34   **b.** less than .29

7.86 A survey of all medium-sized and large corporations showed that 65% of them offer retirement plans to their employees. Let $\hat{p}$ be the proportion in a random sample of 50 such corporations that offer retirement plans to their employees. Find the probability that the value of $\hat{p}$ will be

    a. between .54 and .61   **b.** more than .71

7.87 According to a 1992 Link Resources survey, 83% of households own VCRs (*The Wall Street Journal*, August 12, 1992). Assume that this percentage is true for the current population of all households and let $\hat{p}$ be the proportion of households who own VCRs in a random sample of 400 households. Find the probability that the value of $\hat{p}$ will be

    a. between .85 and .88   **b.** more than .80

7.88 Dartmouth Distribution Warehouse makes deliveries of a large number of products to its customers. It is known that 85% of all the orders it receives from its customers are delivered on time. Let $\hat{p}$ be the proportion of orders in a random sample of 100 that are delivered on time. Find the probability that the value of $\hat{p}$ will be

    a. between .81 and .88   **b.** less than .87

7.89 Brooklyn Corporation manufactures computer diskettes. The machine that is used to make these diskettes is known to produce 6% defective diskettes. The quality control inspector selects a sample of 100 diskettes every week and inspects them for being good or defective. If 8% or more of the diskettes in the sample are defective, the process is stopped and the machine is readjusted. What is the probability that based on a sample of 100 diskettes the process will be stopped to readjust the machine?

7.90 Mong Corporation makes auto batteries. The company claims that 80% of its LL70 batteries are good for 70 months or more. Let $\hat{p}$ be the proportion in a sample of 100 such batteries that are good for 70 months or more.

    a.   What is the probability that this sample proportion is within .05 of the population proportion?

    b.   What is the probability that this sample proportion is lower than the population proportion by .06 or more?

    c.   What is the probability that this sample proportion is greater than the population proportion by .07 or more?

GLOSSARY

Central limit theorem   The theorem from which it is inferred that for a large sample size ($n \geq 30$), the shape of the sampling distribution of $\bar{x}$ is approximately normal. Also, by the same theorem, the shape of the sampling distribution of $\hat{p}$ is approximately normal for a sample for which $np > 5$ and $nq > 5$.

Consistent estimator   A sample statistic with a standard deviation that decreases as the sample size increases.

Convenience sample   A sample that includes the most accessible members of the population.

Estimator   The sample statistic that is used to estimate a population parameter.

Judgment sample   A sample that includes the elements of the population selected based on the judgment and prior knowledge of an expert.

Mean of $\hat{p}$ The mean of the sampling distribution of $\hat{p}$, denoted by $\mu_{\hat{p}}$, is equal to the population proportion p.

Mean of $\bar{x}$ The mean of the sampling distribution of $\bar{x}$, denoted by $\mu_{\bar{x}}$, is equal to the population mean μ.

Nonrandom sample   A sample selected in such a way that some members of the population have no chance of being included in the sample.

Nonsampling errors   The errors that occur during the collection, recording, and tabulation of data.

Population distribution   The probability distribution of the population data.

Population proportion p   The ratio of the number of elements in a population with a specific characteristic to the total number of elements in the population.

Random sample   A sample that assigns some chance to each member of the population to be selected in the sample.

Sample proportion $\hat{p}$ The ratio of the number of elements in a sample with a specific characteristic to the total number of elements in that sample.

Sample survey   The technique of collecting information from a portion of the population.

Sampling distribution of $\hat{p}$ The probability distribution of all the values of $\hat{p}$ calculated from all possible samples of the same size selected from a population.

Sampling distribution of $\bar{x}$ The probability distribution of all the values of $\bar{x}$ calculated from all possible samples of the same size selected from a population.

Sampling error   The difference between the value of a sample statistic calculated from a random sample and the value of the corresponding population parameter. This type of error occurs due to chance.

Simple random sample   A sample chosen in such a way that each element of the population has the same probability of being included in the sample.

Standard deviation of $\hat{p}$ The standard deviation of the sampling distribution of $\hat{p}$, denoted by $\sigma_{\hat{p}}$, is equal to $\sqrt{pq/n}$ when $n/N \leq .05$.

Standard deviation of $\bar{x}$ The standard deviation of the sampling distribution of $\bar{x}$, denoted by $\sigma_{\bar{x}}$, is equal to $\sigma/\sqrt{n}$ when $n/N \leq .05$.

Unbiased estimator   An estimator with an expected value (or mean) that is equal to the value of the corresponding population parameter.

KEY FORMULAS

1. **Mean of $\bar{x}$**

$$\mu_{\bar{x}} = \mu$$

2. **Standard deviation of $\bar{x}$ when $n/N \leq .05$**

$$\sigma_{\bar{x}} = \frac{\sigma}{\sqrt{n}}$$

3. **The z value for $\bar{x}$**

$$z = \frac{\bar{x} - \mu}{\sigma_{\bar{x}}}$$

4. **Population proportion**

$$p = \frac{x}{N}$$

where

N = Total number of elements in the population

x = Number of elements in the population that possess a specific characteristic

5. **Sample proportion**

$$\hat{p} = \frac{x}{n}$$

where

n = Total number of elements in the sample

x = Number of elements in the sample that possess a specific characteristic

6. **Mean of $\hat{p}$**

$$\mu_{\hat{p}} = p$$

7. **Standard deviation of $\hat{p}$ when $n/N \leq .05$**

$$\sigma_{\hat{p}} = \sqrt{\frac{pq}{n}}$$

8. **The z value for $\hat{p}$**

$$z = \frac{\hat{p} - p}{\sigma_{\hat{p}}}$$

SUPPLEMENTARY EXERCISES

7.91 The print on the package of 100-watt General Electric soft-white light bulbs claims that these bulbs have an average life of 750 hours. Assume that the lives of all such bulbs have a normal distribution with a mean of 750 hours and a standard deviation of 50 hours. Let $\bar{x}$ be the mean life of a random sample of 25 such bulbs. Find the mean and standard deviation of $\bar{x}$ and describe the shape of its sampling distribution.

7.92 The mean of the net weights of all Top Taste cereal boxes is 18 ounces. Assume that the net weights of all such boxes have a normal distribution with a mean of 18 ounces and a standard deviation of .4 ounces. Let $\bar{x}$ be the mean net weight of a random sample of 16 such boxes. Find the mean and standard deviation of $\bar{x}$ and describe the shape of its sampling distribution.

7.93 According to the Securities Industry Association, retail stockbrokers earned an average of $98,401 in 1991. Assume that the standard deviation of their 1991 incomes was $12,500. Let $\bar{x}$ be the mean income for 1991 of a sample of 100 retail stockbrokers. Find the mean and standard deviation of $\bar{x}$. Comment on the shape of the sampling distribution of $\bar{x}$.

7.94 According to a survey by *American Banker*, households in the United States owe (including mortgages) an average of $71,500. Assume that this result is true for the current population of all U.S. households and that the standard deviation of the money owed by all households is $9800. Let $\bar{x}$ be the mean amount owed by a random sample of 200 U.S. households. Find the mean and standard deviation of $\bar{x}$. Comment on the shape of the sampling distribution of $\bar{x}$.

7.95 Refer to Exercise 7.91. The print on the package of 100-watt General Electric soft-white light bulbs says that these bulbs have an average life of 750 hours. Assume that the lives of all such bulbs have a normal distribution with a mean of 750 hours and a standard deviation of 50 hours. Find the probability that the mean life of a random sample of 25 such bulbs will be

    a. greater than 735 hours   **b.** between 725 and 740 hours
    c. within 15 hours of the population mean
    d. lower than the population mean by 20 hours or more

7.96 Refer to Exercise 7.92. The mean of the net weights of all Top Taste cereal boxes is 18 ounces. Assume that the net weights of all such boxes have a normal distribution with a mean of 18 ounces and a standard deviation of .4 ounces. Find the probability that the mean net weight of a random sample of 16 such boxes will be

    a. less than 17.75 ounces   **b.** between 18.15 and 18.30 ounces
    c. within .25 ounces of the population mean
    d. greater than the population mean by .20 ounces or more

7.97 Refer to Exercise 7.93. According to the Securities Industry Association, retail stockbrokers earned an average of $98,401 in 1991. Assume that the standard deviation of their 1991 incomes was $12,500. Find the probability that the mean income for 1991 of a sample of 100 retail stockbrokers is

    a. less than $100,000   **b.** between $97,000 and $99,000
    c. within $1500 of the population mean
    d. greater than the population mean by $1000 or more

7.98 Refer to Exercise 7.94. According to a survey by *American Banker*, households in the United States owe (including mortgages) an average of $71,500. Assume that this result is true for the current population of all U.S. households and that the standard deviation of the money owed by all households is $9800. Find the probability that the mean amount owed by a random sample of 200 U.S. households is

    a. greater than $70,500   **b.** between $70,000 and $72,000
    c. within $900 of the population mean
    d. lower than the population mean by $1000 or more

7.99 According to a Priority Management Survey, adults spend an average of 10 hours a day at work and commuting. Let the daily work and commute times for all adults have a mean of 10 hours and a standard deviation of 1.8 hours. Find the probability that the mean of the daily work and commute times for a random sample of 80 adults will be

    a. greater than 10.45 hours   **b.** between 9.75 and 10.50 hours
    c. within .25 hours of the population mean
    d. lower than the population mean by .50 hours or more

7.100 Smith Consultants conducted a sample survey to find the mean monthly mortgage payment made by home owners. The survey of 400 home owners showed that they pay an average of $1540 per month on their home loans. It is known that the standard deviation of the monthly payments on all home loans is $270.

    a. What is the probability that this sample mean is within $30 of the mean monthly mortgage payment made by all home owners?
    b. What is the probability that this sample mean is not within $30 of the mean monthly mortgage payment made by all home owners?

7.101   A machine at Keats Corporation fills 64-ounce detergent jugs. The probability distribution of the amount of detergent in these jugs is normal with a mean of 64 ounces and a standard deviation of .4 ounces. The quality control inspector takes a sample of 16 jugs once a week and measures the amount of detergent in these jugs. If the mean of this sample is either less than 63.75 ounces or greater than 64.25 ounces, the inspector concludes that the machine needs an adjustment. What is the probability that based on a sample of 16 jugs the inspector will conclude that the machine needs an adjustment when actually it does not?

7.102   Ten percent of all items produced on a machine are defective. Let $\hat{p}$ be the proportion of defective items in a random sample of 80 items selected from the production line. Calculate the mean and standard deviation of $\hat{p}$ and describe the shape of the sampling distribution of $\hat{p}$.

7.103   Seventy percent of adults favor some kind of government control on the prices of medicines. Assume that this percentage is true for the current population of all adults. Let $\hat{p}$ be the proportion of adults in a random sample of 400 who favor government control on the prices of medicines. Calculate the mean and standard deviation of $\hat{p}$ and describe the shape of the sampling distribution of $\hat{p}$.

7.104   Refer to Exercise 7.103. Seventy percent of adults favor some kind of government control on the prices of medicines. Assume that this percentage is true for the current population of all adults.

    a. Find the probability that the proportion of adults in a random sample of 400 who favor some kind of government control on the prices of medicines is
        i. less than .65   ii. between .73 and .76

    b. What is the probability that the proportion of adults in a random sample of 400 who favor some kind of government control is within .06 of the population proportion?

    c. What is the probability that the sample proportion is greater than the population proportion by .05 or more?

7.105   In a Roper Organization survey, 48% of adults favored banning all advertisements for beer and wine from television (*The American Way of Buying*, Dow Jones & Co., 1990). Assume that this percentage is true for the current population of all adults.

    a. Find the probability that the proportion of adults in a random sample of 100 who will favor banning all advertisements for beer and wine from television is
        i. more than .53   ii. between .41 and .45

    b. What is the probability that the proportion of adults in a random sample of 100 who will favor banning all advertisements for beer and wine from television is within .07 of the population proportion?

    c. What is the probability that the sample proportion is lower than the population proportion by .06 or more?

7.106   K5 Corporation manufactures computer chips. The machine that is used to make these chips is known to produce 7% defective chips. The quality control inspector selects a sample of 100 chips every week and inspects them for being good or defective. If 10% or more of the chips in the sample are defective, the process is stopped and the machine is readjusted. What is the probability that based on a sample of 100 chips the process will be stopped to readjust the machine?

7.107   Stress on the job is a major concern of a large number of people who go into managerial positions. Eighty percent of all managers of companies suffer from stress. Let $\hat{p}$ be the proportion in a sample of 100 managers of companies who suffer from stress.

    a. What is the probability that this sample proportion is within .08 of the population proportion?

    b. What is the probability that this sample proportion is not within .08 of the population proportion?

    c. What is the probability that this sample proportion is lower than the population proportion by .10 or more?

    d. What is the probability that this sample proportion is greater than the population proportion by .11 or more?

SELF-REVIEW TEST

1. A sampling distribution is the probability distribution of
   a. a population parameter   **b.** a sample statistic   **c.** any random variable

2. Nonsampling errors are
   a. the errors that occur because the sample size is too large in relation to the population size
   b. the errors made while collecting, recording, and tabulating data
   c. the errors that occur because an untrained person conducts the survey

3. A sampling error is
   a. the difference between the value of a sample statistic based on a random sample and the value of the corresponding population parameter
   b. the error made while collecting, recording, and tabulating data
   c. the error that occurs because the sample is too small

4. The mean of the sampling distribution of $\bar{x}$ is always equal to
   a. μ **b.** $\mu - 5$ **c.** $\sigma/\sqrt{n}$

5. The condition for the standard deviation of the sample mean to be $\sigma/\sqrt{n}$ is that
   a. $np > 5$ **b.** $n/N \leq .05$ **c.** $n > 30$

6. The standard deviation of the sampling distribution of the sample mean decreases when
   a. x increases   **b.** n increases   **c.** n decreases

7. When samples are taken from a normally distributed population, the sampling distribution of the sample mean has a normal distribution
   a. if $n \geq 30$ **b.** if $n/N \leq .05$ **c.** all the time

8. When samples are taken from a nonnormally distributed population, the sampling distribution of the sample mean has a normal distribution
   a. if $n \geq 30$ **b.** if $n/N \leq .05$ **c.** always

9. In a sample of 200 customers of a mail-order company, 148 are found to be satisfied with the service they receive from the company. The proportion of customers in this sample who are satisfied with the company's service is
   a. .26   **b.** .74   **c.** .148

10. The mean of the sampling distribution of $\hat{p}$ is always equal to
    a. p **b.** μ **c.** $\hat{p}$

11. The condition for the standard deviation of the sampling distribution of the sample proportion to be $\sqrt{pq/n}$ is
    a. $np > 5$ and $nq > 5$ **b.** $n > 30$ **c.** $n/N \leq .05$

12. The sampling distribution of $\hat{p}$ is (approximately) normal if
    a. $np > 5$ and $nq > 5$ **b.** $n > 30$ **c.** $n/N \leq .05$

13. Briefly state and explain the central limit theorem.

14. The weights of apples grown at a Washington farm have a normal distribution with a mean of 4.1 ounces and a standard deviation of .45 ounces. Let $\bar{x}$ be the mean weight of a random sample of certain apples grown at this farm. Calculate the mean and standard deviation of $\bar{x}$ and describe the shape of its sampling distribution for a sample size of
    a. 10   **b.** 25   **c.** 100

15. The market values of all large corporations in the United States have a mean of $12,570 million and a standard deviation of $6840 million. Let $\bar{x}$ be the mean market value of a random sample of

certain large U.S. corporations. Calculate the mean and standard deviation of $\bar{x}$ and describe the shape of its sampling distribution for a sample size of

    a. 20   **b.** 36   **c.** 80

16. At Jen and Perry Ice Cream Company, the machine that fills one-pound cartons of Top Flavor ice cream is set to dispense 16 ounces of ice cream in every carton. However, some cartons contain slightly less than and some contain slightly more than 16 ounces of ice cream. The amounts of ice cream in all such cartons have a normal distribution with a mean of 16 ounces and a standard deviation of .18 ounces.

    a. Find the probability that the mean amount of ice cream in a random sample of 16 such cartons will be

        i. between 15.90 and 15.95 ounces

        ii. less than 15.95 ounces

        iii. more than 15.97 ounces

    b. What is the probability that the mean amount of ice cream in a random sample of 16 such cartons will be within .10 ounces of the population mean?

    c. What is the probability that the mean amount of ice cream in a random sample of 16 such cartons will be lower than the population mean by .135 ounces or more?

17. Refer to Problem 15. As mentioned in that problem, the market values of all large corporations in the United States have a mean of $12,570 million and a standard deviation of $6840 million.

    a. Find the probability that the mean market value of a random sample of 36 such corporations will be

        i. between $11,600 and $14,000

        ii. more than $14,500

        iii. less than $13,000

    b. What is the probability that the mean market value of a random sample of 36 such corporations will be within $2000 of the population mean?

    c. What is the probability that the mean market value of a random sample of 36 such corporations will be greater than the population mean by $2500 or more?

18. According to a Gallup Organization poll, 78% of adults approve of state-sponsored lotteries (*USA TODAY*, April 2, 1992). Assume that this percentage is true for the current population of all adults. Let $\hat{p}$ be the proportion of adults in a random sample who approve of state-sponsored lotteries. Calculate the mean and standard deviation of $\hat{p}$ and describe the shape of the sampling distribution of $\hat{p}$ when the sample size is

    a. 50   **b.** 200   **c.** 900

19. Refer to Problem 18. According to a Gallup Organization poll, 78% of adults approve of state-sponsored lotteries (*USA TODAY*, April 2, 1992). Assume that this percentage is true for the current population of all adults.

    a. Find the probability that the proportion of adults in a random sample of 400 who will approve of state-sponsored lotteries is

        i. more than .81   ii. between .75 and .82

        iii. less than .80   iv. between .73 and .76

    b. What is the probability that the proportion of adults in a random sample of 400 who will approve of state-sponsored lotteries is within .05 of the population proportion?

    c. What is the probability that the sample proportion for a random sample of 400 adults is lower than the population proportion by .04 or more?

    d. What is the probability that the sample proportion for a random sample of 400 adults is greater than the population proportion by .03 or more?

USING MINITAB

We can use MINITAB to construct random number tables. For example, to construct a table of 75 three-digit random numbers, we will use the MINITAB commands given in Figure 7.16.

Figure 7.16 Constructing a table of 75 three-digit random numbers.

```
MTB  > RANDOM 75 OBSERVATIONS IN C1;
SUBC > INTEGERS BETWEEN 100 AND 999.
MTB  > PRINT C1
```

The first command shown in Figure 7.16 instructs MINITAB to generate 75 random numbers and put them in column C1. Note the semicolon at the end of the first MINITAB command. The subcommand tells MINITAB to select integers that are between 100 and 999. Because these selected numbers will be between 100 and 999, they will automatically be three-digit numbers. Note that 100 is the smallest three-digit number and 999 is the largest three-digit number. Also, note the period at the end of the subcommand. The last MINITAB command will print the random numbers of column C1 on the computer screen.

To generate two-digit random numbers, replace ''100 AND 999'' by ''10 AND 99'' in the subcommand; to generate four-digit random numbers, replace ''100 AND 999'' by ''1000 AND 9999'' in the subcommand.

Illustration M7–1 Suppose we want to generate 16 three-digit random numbers. Figure 7.17 gives the MINITAB commands and MINITAB output.

Figure 7.17 Generating 16 three-digit random numbers.

```
MTB  > NOTE: GENERATING 16 THREE-DIGIT RANDOM NUMBERS
MTB  > RANDOM 16 OBSERVATIONS IN C1;
SUBC > INTEGERS BETWEEN 100 AND 999.
MTB  > PRINT C1

C1
    465   838   628   524   115   176   699   322   967   539   169
    639   983   582   140   630
```

MINITAB can also be used to select a sample from a population. To do so, first enter the population data in MINITAB. Then use the MINITAB SAMPLE command to take a sample of any required size. Suppose we only have the population data on one variable. First, we enter these data in column C1 using the SET command. Then, we use the MINITAB command given in Figure 7.18 to take a sample of 9 elements from the data of column C1. The sample data will be recorded in column C2. MINITAB command PRINT C2 will print the sample data.

Figure 7.18

```
MTB  > SAMPLE 9 FROM C1 PUT IN C2
MTB  > PRINT C2
```

Now suppose the population data are on four variables and we want to take a sample of size 12 from these data. First, we enter the population data in columns C1 to C4. Then, we use the MINITAB command given in Figure 7.19 to take a sample of size 12 from the data of columns C1–C4. The sample data will be recorded in columns C5–C8. MINITAB command PRINT C5–C8 will print the sample data.

Figure 7.19

```
MTB  > SAMPLE 12 FROM C1—C4 PUT IN C5—C8
MTB  > PRINT C5—C8
```

The MINITAB section at the end of Chapter 1 illustrates how we can select a sample from a population using MINITAB.

CONSTRUCTING THE SAMPLING DISTRIBUTION OF THE SAMPLE MEAN

By using MINITAB, we can construct the sampling distribution of the sample mean by simulating the sampling procedure. Illustration M7–2 describes this procedure.

Illustration M7–2   Consider the experiment of rolling a die. Each roll has six possible outcomes: 1 spot, 2 spots, 3 spots, 4 spots, 5 spots, and 6 spots. Rolling a die can be simulated by randomly selecting a number from 1, 2, 3, 4, 5, and 6. The MINITAB commands given in Figure 7.20 will produce 1000 samples, each sample containing the results of 40 rolls. Note that a roll of the die is simulated by randomly selecting a number from 1 through 6. The set of values in the first row of the 40 columns gives the first sample, the set of values in the second row of the 40 columns produces the second sample, and so on. Thus, we obtain 1000 samples, each containing the results of 40 rolls of the die.

Figure 7.20   Taking 1000 samples, each containing 40 rolls of a die.

```
MTB  > NOTE: 1000 SAMPLES, EACH CONTAINING 40 ROLLS OF A DIE
MTB  > RANDOM 1000 C1—C40;
SUBC > INTEGER 1 6.
```

To calculate the sample means for 1000 samples obtained above, we use the RMEANS command, as shown in Figure 7.21. The RMEANS command calculates the means of rows. The command given in Figure 7.21 will calculate the means of 1000 samples given in 1000 rows of the 40 columns and will put these means in column C41.

Figure 7.21

```
MTB  > RMEANS C1–C40 PUT IN C41
```

Now, by using the MINITAB commands given in Figure 7.22, we can construct the histogram of the sample means given in column C41 and calculate the mean and standard deviation of these sample means.

Figure 7.22

```
MTB  > HISTOGRAM C41
MTB  > MEAN C41
MTB  > STDEV C41
```

By combining all these commands, we produce the histogram, the mean, and the standard deviation of the sample means of 1000 samples, each sample containing the results of 40 rolls of a die. These commands and the MINITAB output are shown in Figure 7.23.

Figure 7.23 Histogram, mean, and standard deviation of $\bar{x}$.

```
MTB  > NOTE: 1000 SAMPLES, EACH CONTAINING 40 ROLLS OF A DIE
MTB  > RANDOM 1000 C1–C40;
SUBC > INTEGER 1 6.
MTB  > RMEANS C1–C40 PUT IN C41
MTB  > MEAN C41
    MEAN    =        3.4952
MTB  > STDEV C41
    ST.DEV. =        0.26997
MTB  > HISTOGRAM C41

Histogram of C41    N = 1000
Each * represents 10 obs.

Midpoint      Count
     2.6         1     *
     2.8        10     **
     3.0        55     ******
     3.2       151     ****************
     3.4       268     ***************************
     3.6       280     ****************************
     3.8       155     ****************
     4.0        67     *******
     4.2        12     **
     4.4         1     *
```

The frequency histogram of Figure 7.23 can be used to construct the probability distribution of the sample mean $\bar{x}$. The shape of the sampling distribution of $\bar{x}$ for these 1000 samples will look exactly like the above histogram. The mean and standard deviation of this sampling distribution of $\bar{x}$ are 3.4952 and .26997 respectively. That is,

$$\mu_{\bar{x}} = 3.4952 \qquad \text{and} \qquad \sigma_{\bar{x}} = .26997$$

Note that each time we repeat this experiment, we will obtain different values of $\mu_{\bar{x}}$ and $\sigma_{\bar{x}}$ and a different histogram. ■

COMPUTER ASSIGNMENTS

M7.1 Using MINITAB, construct a table of 25 two-digit random numbers.

M7.2 Using MINITAB, construct a table of 60 four-digit random numbers.

M7.3 Using MINITAB, take a sample of 15 observations from Data Set II (City Data) of Appendix A.

M7.4 Using MINITAB, create 200 samples, each containing the results of 30 rolls of a die. Calculate the means of these 200 samples. Construct the histogram and calculate the mean and standard deviation of these 200 sample means.

M7.5 Using the following MINITAB commands, create 150 samples each containing the results of selecting 35 numbers from 1 through 100.

```
MTB > RANDOM 150 C1–C35;
SUBC > INTEGER 1 100.
```

Calculate the means of these 150 samples. Construct the histogram and calculate the mean and standard deviation of these 150 sample means.

8 ESTIMATION OF THE MEAN AND PROPORTION

Now we are entering that part of statistics called *inferential statistics*. In Chapter 1 inferential statistics was defined as the part of statistics that helps us to make decisions about some characteristics of a population based on sample information. In other words, inferential statistics uses the sample results to make decisions and draw conclusions about the population from which the sample is taken. Estimation is the first topic to be considered in our discussion of inferential statistics. Estimation and hypothesis testing (discussed in Chapter 9) taken together are usually referred to as inference making. This chapter explains how to estimate the population mean and population proportion for a single population.

8.1   ESTIMATION: AN INTRODUCTION

Estimation is a procedure by which numerical value or values are assigned to a population parameter based on the information collected from a sample.

> **ESTIMATION**
>
> The assignment of value(s) to a population parameter based on a value of the corresponding sample statistic is called estimation.

In inferential statistics, μ is called the *true population mean* and p is called the *true population proportion*. There are many other population parameters such as the median, mode, variance, and standard deviation.

An auto company may want to estimate the mean fuel consumption for a particular model of a car; a manager may want to estimate the average time taken by new employees to learn a job; the U.S. Census Bureau may want to find the mean housing expenditure per month incurred by households in the United States; and a company that plans to introduce a new product may want to estimate the proportion (or percentage) of all people who will like this product. All these examples relate to estimation.

The examples about estimating the average time taken to learn a job by new employees and estimating the mean housing expenditure per month incurred by households in the United States are illustrations of estimating the *true population mean* μ. The example about estimating the proportion (or percentage) of all people who will like a new product that a company plans to introduce is an illustration of estimating the *true population proportion p*.

If we can conduct a *census* (a survey that includes the entire population) each time we want to find the value of a population parameter, then the estimation procedures explained in this and subsequent chapters are not needed. For example, if the U.S. Census Bureau can contact every household living in the United States to find the mean housing expenditure incurred by households, the result of the survey (which will actually be a census) will give the value of μ and the procedures learned in this chapter will not be needed. However, it is too expensive, very time consuming, or virtually impossible to contact every member of a population to collect information to find the true value of a population parameter. Therefore, we usually take a sample from the population and calculate the value of the appropriate sample statistic. Then we assign a value or values to the corresponding population parameter based on the value of the sample statistic. This chapter (and subsequent chapters) explains how to assign values to population parameters based on the values of sample statistics.

For example, to estimate the mean time taken to learn a certain job by new employees, the manager will take a sample of new employees and record the time taken by these employees to learn the job. Using this information, she will calculate the sample mean $\bar{x}$. Then, based on the value of $\bar{x}$, she will assign certain values to μ. As another example, to estimate the mean housing expenditure per month incurred by all households in the United States, the U.S. Census Bureau will take a sample of certain households, collect the information on the housing expenditure that each of these households incurs per month, and compute the value of the sample mean $\bar{x}$. Based on this value of $\bar{x}$, the bureau will then assign values to the population mean μ. Similarly, the company that wants to find the proportion (or percentage) of all people who will like a new product will take a sample of certain people and determine the value of the sample proportion $\hat{p}$, which represents the proportion of the people in the sample who like the new product. Then, using this value of the sample proportion $\hat{p}$, the company will assign values to the population proportion p.

The value(s) assigned to a population parameter based on the value of a sample statistic is called an **estimate** of the population parameter. For example, suppose the manager takes a sample of 40 new employees and finds that the mean time $\bar{x}$ taken to learn this job for these employees is 5.5 hours. If she assigns this value to the population mean, then 5.5 hours will be called an estimate of μ. The sample statistic used to estimate a population parameter is called an **estimator**. Thus, the sample mean $\bar{x}$ is an estimator of the population mean μ, and the sample proportion $\hat{p}$ is an estimator of the population proportion p.

ESTIMATE AND ESTIMATOR

The value(s) assigned to a population parameter based on the value of a sample statistic is called an estimate. The sample statistic used to estimate a population parameter is called an estimator.

The estimation procedure involves the following steps.

1. Select a sample.
2. Collect the required information from the members of the sample.
3. Calculate the value of the sample statistic.
4. Assign value(s) to the corresponding population parameter.

8.2 POINT AND INTERVAL ESTIMATES

An estimate may be a point estimate or an interval estimate. These two types of estimates are described in this section.

8.2.1 A POINT ESTIMATE

If we select a sample and compute the value of the sample statistic for this sample, this value gives the **point estimate** of the corresponding population parameter.

POINT ESTIMATE

The value of a sample statistic that is used to estimate a population parameter is called a point estimate.

Thus, the value computed for the sample mean $\bar{x}$ from a sample is a point estimate of the corresponding population mean μ. For the example mentioned earlier, suppose the U.S. Census Bureau takes a sample of 10,000 households and determines that the mean housing expenditure per month $\bar{x}$ for this sample is $874. Then, using $\bar{x}$ as a point estimate of μ, the bureau can state that the mean housing expenditure per month μ for all households is about $874. This procedure is called **point estimation**.

Usually, whenever we use point estimation, we calculate the **margin of error** associated with that point estimation. For the estimation of the population mean, the margin of error is calculated as follows.

$$\text{Margin of error} = \pm 1.96\, \sigma_{\bar{x}} \quad \text{or} \quad \pm 1.96\, s_{\bar{x}}$$

That is, we find the standard deviation of the sample mean and multiply it by 1.96. Here $s_{\bar{x}}$ is a point estimator of $\sigma_{\bar{x}}$ (it will be discussed later in this chapter).

Each sample taken from a population is expected to yield a different value of the sample statistic. Thus, the value assigned to a population mean μ based on a point estimate depends on which of the samples is drawn. Consequently, the point estimate assigns a value to μ that almost always differs from the true value of the population mean.

8.2.2 AN INTERVAL ESTIMATE

In the case of **interval estimation**, instead of assigning a single value to a population parameter, an interval is constructed around the point estimate and then a probabilistic statement that this interval contains the corresponding population parameter is made.

INTERVAL ESTIMATE

In interval estimation, an interval is constructed around the point estimate, and it is stated that this interval is likely to contain the corresponding population parameter.

For the example about the mean housing expenditure (Section 8.2.1), instead of saying that the mean housing expenditure per month for all households is $874, we obtain an interval by subtracting a number from $874 and adding the same number to $874. Then we state that this interval contains the population mean μ. For purposes of illustration, suppose we subtract $110 from $874 and add $110 to $874. Consequently, we obtain the interval ($874 − $110) to ($874 + $110) or $764 to $984. Then we state that the interval $764 to $984 is likely to contain the population mean μ and that the mean housing expenditure per month for all households in the United States is between $764 and $984. This procedure is called *interval estimation*. The value $764 is called the *lower limit* of the interval and $984 is called the *upper limit* of the interval. Figure 8.1 illustrates the concept of interval estimation.

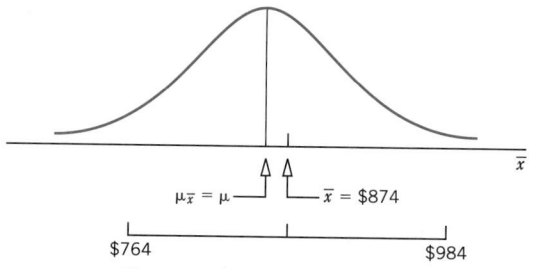

Figure 8.1 Interval estimation.

The question arises, what number should we subtract from and add to a point estimate to obtain an interval estimate? The answer to this question depends on two considerations:

1. The standard deviation $\sigma_{\bar{x}}$ of the sample mean $\bar{x}$
2. The level of confidence to be attached to the interval

First, the larger the standard deviation of $\bar{x}$, the greater the number subtracted from and added to the point estimate. Thus, it is obvious that if the range over which $\bar{x}$ can assume values is larger, the interval constructed around $\bar{x}$ must be wider to include μ.

Second, the quantity subtracted and added must be larger if we want to have a higher confidence in our interval. We always attach a probabilistic statement to the interval estimation. This probabilistic statement is given by the **confidence level**. An interval that is constructed based on this confidence level is called a **confidence interval**.

> **CONFIDENCE LEVEL AND CONFIDENCE INTERVAL**
>
> Each interval is constructed with regard to a given confidence level and is called a confidence interval. The confidence level associated with a confidence interval states how much confidence we have that this interval contains the true population parameter. The confidence level is denoted by $(1 - \alpha)100\%$.

The confidence level is denoted by $(1 - \alpha)100\%$ (α is the Greek letter *alpha*). When expressed as probability, it is called the *confidence coefficient* and is denoted by $1 - \alpha$. In passing, note that α is called the *significance level*, which will be explained in detail in Chapter 9.

Although any value of the confidence level can be chosen to construct a confidence interval, the more common values are 90%, 95%, and 99%. The corresponding confidence coefficients are .90, .95, and .99. The next section describes how to actually construct a confidence interval for the population mean for a large sample.

8.3 INTERVAL ESTIMATION OF A POPULATION MEAN: LARGE SAMPLES

This section explains how to construct a confidence interval for the population mean μ when the sample size is large.[1] Recall from the discussion in Chapter 7 that in the case of $\bar{x}$, the sample size is considered to be large when n is 30 or larger. According to the central limit theorem, for a large sample the sampling distribution of the sample mean $\bar{x}$ is (approximately) normal irrespective of the shape of the population from which the sample is drawn. Therefore, *when the sample size is 30 or larger, we will use the normal distribution to construct a confidence interval for* μ. We also know from Chapter 7 that the standard deviation of $\bar{x}$ is $\sigma_{\bar{x}} = \sigma/\sqrt{n}$. However, if the population standard deviation σ is not known, then we use the sample standard deviation s for σ. Consequently, we use

$$s_{\bar{x}} = \frac{s}{\sqrt{n}}$$

for $\sigma_{\bar{x}} = \sigma/\sqrt{n}$. Note that the value of $s_{\bar{x}}$ is a point estimate of $\sigma_{\bar{x}}$.

> **CONFIDENCE INTERVAL FOR μ FOR LARGE SAMPLES**
>
> The $(1 - \alpha)100\%$ confidence interval for μ is
>
> $$\bar{x} \pm z\sigma_{\bar{x}} \qquad \text{if } \sigma \text{ is known}$$
>
> $$\bar{x} \pm zs_{\bar{x}} \qquad \text{if } \sigma \text{ is not known}$$
>
> where $\sigma_{\bar{x}} = \sigma/\sqrt{n}$ and $s_{\bar{x}} = s/\sqrt{n}$.
>
> The value of z used here is read from the standard normal distribution table for the given confidence level.

[1]Some statisticians prefer to discuss estimation and tests of hypotheses based on whether or not the population standard deviation σ is known. However, we prefer to use the large sample and small sample criteria. The reason is that σ is almost always unknown. Hence, discussing estimation and tests of hypotheses based on large and small samples makes more sense than whether or not σ is known.

Appendix 8.1 (at the end of this chapter) explains how to obtain these formulas for the confidence interval for μ. An interested reader may refer to that appendix at this point.

The quantity $z\sigma_{\bar{x}}$ (or $zs_{\bar{x}}$ when σ is not known) in the confidence interval formula is called the **maximum error of estimate** and is denoted by E.

MAXIMUM ERROR OF ESTIMATE FOR μ

The maximum error of estimate for μ, denoted by E, is the quantity that is subtracted from and added to the value of $\bar{x}$ to obtain a confidence interval for μ. Thus,

$$E = z\sigma_{\bar{x}} \quad \text{or} \quad zs_{\bar{x}}$$

The value of z in the confidence interval formula is obtained from the standard normal distribution table (Table VII of Appendix B) for the given confidence level. To illustrate, suppose we want to construct a 95% confidence interval for μ. A 95% confidence level means that the total area under the normal curve for $\bar{x}$ between two points (at the same distance) on different sides of μ is 95% or .95, as shown in Figure 8.2. To find the value of z, we first divide the given confidence coefficient by 2. Then we look for this number in the body of the normal table. The corresponding value of z is the value we use in the confidence interval. Thus, to find the z value for a 95% confidence level, we perform the following two steps.

1. First we divide .95 by 2, which gives .4750.
2. Then we locate .4750 in the body of the normal distribution table and record the corresponding value of z. This value of z is 1.96.

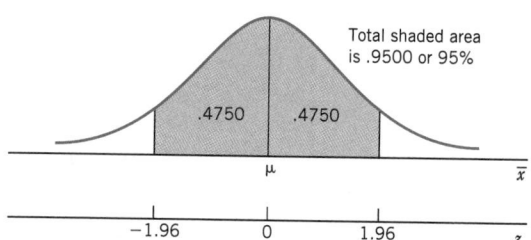

Figure 8.2   Finding z for a 95% confidence level.

For a $(1 - \alpha)100\%$ confidence level, the area between $-z$ and z is $1 - \alpha$. Because the total area under the normal curve is 1.0, the total area under the curve in the two tails is α. This, as mentioned earlier, is called the significance level. In the example of Figure 8.2, $\alpha = 1 - .95 = .05$. Therefore, as shown in Figure 8.3, the area under the curve in each of the two tails is $\alpha/2$. Thus, the value of z associated with a $(1 - \alpha)100\%$ confidence level is sometimes denoted by $z_{\alpha/2}$. However, this text will denote this value simply by z.

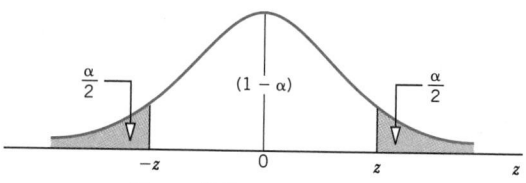

Figure 8.3   Area in the tails.

Example 8–1 describes the procedure to construct a confidence interval for μ for a large sample.

Point estimate and confidence interval for μ: σ known and $n \geq 30$.

EXAMPLE 8–1 A publishing company has just published a new college textbook. Before the company decides the price at which to sell this textbook, it wants to know the average price of all such textbooks in the market. The research department at the company took a sample of 36 such textbooks and collected information on their prices. This information produced a mean price of $48.40 for this sample. It is known that the standard deviation of the prices of all such textbooks is $4.50.

(a) What is the point estimate of the mean price of all such college textbooks? What is the margin of error for this estimate?

(b) Construct a 90% confidence interval for the mean price of all such college textbooks.

Solution From the given information,

$$n = 36, \quad \bar{x} = \$48.40, \quad \text{and} \quad \sigma = \$4.50$$

The standard deviation of $\bar{x}$ is

$$\sigma_{\bar{x}} = \sigma/\sqrt{n} = 4.50/\sqrt{36} = \$.75$$

(a) The point estimate of the mean price of all such college textbooks is $48.40, that is,

$$\text{Point estimate of } \mu = \bar{x} = \textbf{\$48.40}$$

The margin of error associated with this point estimate of μ is

$$\text{Margin of error} = \pm 1.96 \, \sigma_{\bar{x}} = \pm 1.96 \, (.75) = \textbf{\pm 1.47}$$

The margin of error states that the mean price of all such college textbooks is $48.40, give or take $1.47. Note that the margin of error is simply the maximum error of estimate for a 95% confidence interval.

(b) The confidence level is 90% or .90. First we find the z value for a 90% confidence level. To do so, we divide .90 by 2 to obtain .4500. Then we locate .4500 in the body of the normal distribution table (Table VII of Appendix B). Because .4500 is not in the normal table, we can use the number closest to .4500, which is either .4495 or .4505. If we use .4505 as an approximation for .4500, the value of z for this number is 1.65.[2]

Next, we substitute all the values in the confidence interval formula for μ. The 90% confidence interval for μ is

$$\bar{x} \pm z\sigma_{\bar{x}} = 48.40 \pm 1.65 \, (.75) = 48.40 \pm 1.24$$

$$= (48.40 - 1.24) \text{ to } (48.40 + 1.24) = \textbf{\$47.16 to \$49.64}$$

Thus, we are 90% confident that the mean price of all such college textbooks is between $47.16 and $49.64. Note that we cannot say for sure whether the interval $47.16 to $49.64 contains the true population mean or not. Since μ is a constant, we cannot say that the probability is .90 that this interval contains μ because either it contains μ or it does not. Consequently, the probability is either 1.0 or zero that this interval contains μ. All we can say is that we are 90% confident that the mean price of all such college textbooks is between $47.16 and $49.64. ∎

---

[2]Note that there is no apparent reason for choosing .4505 and not choosing .4495. If we choose .4495, the z value will be 1.64. An alternative is to use the average of 1.64 and 1.65, 1.645, which we will not do in this text.

How do we interpret a 90% confidence level? In terms of Example 8–1, if we take all possible samples of 36 such college textbooks each and construct a 90% confidence interval for μ around each sample mean, we can expect that 90% of these intervals will include μ and 10% will not. In Figure 8.4 we show means $\bar{x}_1$, $\bar{x}_2$, and $\bar{x}_3$ of three different samples of the same size drawn from the same population. Also shown in this figure are the 90% confidence intervals constructed around these three sample means. As we observe, the 90% confidence intervals constructed around $\bar{x}_1$ and $\bar{x}_2$ include μ, but the one constructed around $\bar{x}_3$ does not. We can state for a 90% confidence level that if we take many samples of the same size from a population and construct 90% confidence intervals around the means of these samples, then 90% of these confidence intervals will be like the ones around $\bar{x}_1$ and $\bar{x}_2$ in Figure 8.4, which include μ, and 10% will be like the one around $\bar{x}_3$, which does not include μ.

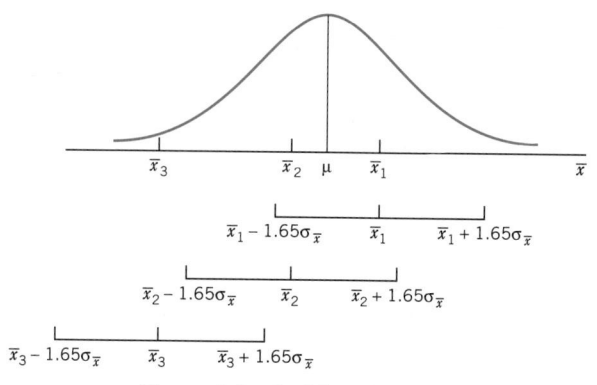

Figure 8.4 Confidence intervals.

In Example 8–1, the value of the population standard deviation σ was known. However, more often we do not know the value of σ. In such cases, we estimate the population standard deviation σ by the sample standard deviation s and estimate the standard deviation $\sigma_{\bar{x}}$ of $\bar{x}$ by $s_{\bar{x}}$. Then we use $s_{\bar{x}}$ for $\sigma_{\bar{x}}$ in the formula for the confidence interval for μ. Owing to the central limit theorem, as long as the sample size is large ($n \geq 30$), even if we do not know σ, we can use the normal distribution.

Example 8–2 illustrates the construction of a confidence interval for μ when σ is not known.

Constructing confidence interval for μ: σ not known and $n \geq 30$.

EXAMPLE 8–2 The U.S. Bureau of Labor conducts surveys quite often to collect information on the labor market. According to one such recent survey, the workers employed in manufacturing industries in the United States earned an average of $466.42 per week in July 1992 (*Monthly Labor Review*, November 1992). Assume that this mean is based on a random sample of 1000 workers selected from the manufacturing industries and that the standard deviation of weekly earnings for this sample is $70. Find a 99% confidence interval for the mean weekly earnings of all workers employed in manufacturing industries in July 1992.

Solution From the given information,

$$n = 1000, \qquad \bar{x} = \$466.42, \qquad s = \$70,$$

and

$$\text{Confidence level} = 99\% \text{ or } .99$$

First we find the standard deviation of $\bar{x}$. Because σ is not known, we will use $s_{\bar{x}}$ as an estimator of $\sigma_{\bar{x}}$. The value of $s_{\bar{x}}$ is

$$s_{\bar{x}} = s/\sqrt{n} = 70/\sqrt{1000} = 2.2136$$

Note that we have rounded $s_{\bar{x}}$ to four decimal places. We will continue with this rounding-off policy for all intermediary calculations.

Because the sample size is large ($n > 30$), we will use the normal distribution to determine the confidence interval for μ. To find z for a 99% confidence level, we divide .99 by 2 to obtain .4950. From the normal distribution table, the z value for .4950 is approximately 2.58. Substituting all the values in the formula, the 99% confidence interval for μ is

$$\bar{x} \pm zs_{\bar{x}} = 466.42 \pm 2.58\,(2.2136) = 466.42 \pm 5.71 = \textbf{\$460.71 to \$472.13}$$

Thus, we can state with 99% confidence that the average earnings of workers employed in manufacturing industries in July 1992 were between \$460.71 and \$472.13 per week. ■

The *width of a confidence interval* depends on the size of the maximum error $z\sigma_{\bar{x}}$, which depends on the values of z, σ, and n because $\sigma_{\bar{x}} = \sigma/\sqrt{n}$. However, the value of σ is not within the control of the investigator. Hence, the width of a confidence interval depends on

1. The value of z, which depends on the confidence level
2. The sample size n

The confidence level determines the value of z, which in turn determines the size of the maximum error. The value of z increases as the confidence level increases, and it decreases as the confidence level decreases. For example, the value of z is approximately 1.65 for a 90% confidence level, 1.96 for a 95%, and approximately 2.58 for a 99% confidence level. Hence, the higher the confidence level, the larger the width of the confidence interval, other things remaining the same.

For the same value of σ, an increase in the sample size decreases the value of $\sigma_{\bar{x}}$, which in turn decreases the size of the maximum error when the confidence level remains unchanged. Therefore, an increase in the sample size decreases the width of the confidence interval.

Thus, if we want to decrease the width of a confidence interval, we have two choices:

1. Lower the confidence level
2. Increase the sample size

However, lowering the confidence level is not a good choice because a lower confidence level may give less reliable results. Therefore, we should always prefer to increase the sample size if we want to decrease the width of a confidence interval. Next we illustrate, using Example 8–2, how either a decrease in the confidence level or an increase in the sample size decreases the width of the confidence interval.

1. CONFIDENCE LEVEL AND THE WIDTH OF CONFIDENCE INTERVAL

Reconsider Example 8–2. Suppose all the information given in that example remains the same. First, let us decrease the confidence level to 95%. From the normal distribution table, $z = 1.96$ for a 95% confidence level. Then, using $z = 1.96$ in the confidence interval for Example 8–2, we obtain:

$$\bar{x} \pm zs_{\bar{x}} = 466.42 \pm 1.96\,(2.2136) = 466.42 \pm 4.34 = \textbf{\$462.08 to \$470.76}$$

$$\bar{X} - E < \mu < \bar{X} + E$$

Comparing this confidence interval to the one obtained in Example 8–2, we observe that the width of the confidence interval for a 95% confidence level is smaller than the one for a 99% confidence level.

2. SAMPLE SIZE AND THE WIDTH OF CONFIDENCE INTERVAL

Consider Example 8–2 again. Now suppose the information given in that example is based on a sample size of 1600. Further assume that all other information given in that example, including the confidence level, remains the same. First we calculate the standard deviation of the sample mean using $n = 1600$.

$$s_{\bar{x}} = s/\sqrt{n} = 70/\sqrt{1600} = \$1.75$$

Then, the 99% confidence interval for μ is

$$\bar{x} \pm z s_{\bar{x}} = 466.42 \pm 2.58\,(1.75) = 466.42 \pm 4.52 = \textbf{\$461.90 to \$470.94}$$

Comparing this confidence interval to the one obtained in Example 8–2, we observe that the width of the confidence interval for $n = 1600$ is smaller than the one for $n = 1000$.

CASE STUDY 8-1　YOUNG FAMILIES IN DEBT

USA SNAPSHOTS®

A look at statistics that shape our lives

Young families in debt

Spending habits of young married couples[1] with children:

$19,783　　$21,401

Average after-tax income　　Average annual spending

1 – Both spouses ages 18 to 25

Source: *Family Economics Review*　　By Rod Little, USA TODAY

The above chart shows that young married couples (aged 18 to 25) with children have an average annual after-tax income of $19,783 and an average annual expenditure of $21,401.

Usually the data published in *Family Economics Review* are based on surveys conducted by the U.S. Bureau of Labor.

From the chart, one gets the impression that the average income and the average expenditure mentioned there are for the population of all young married couples with children. However, that is not the case. These two averages are based on a sample of young married couples with children. These averages are the point estimates of the corresponding population means.

Now assume that these sample means are based on a sample of 4000 young married couples with children. Further assume that the standard deviation of the annual incomes of these couples is $2600 and the standard deviation of the annual expenditures is $2200. Now let us make a 97% confidence interval for each of the two population means.

Confidence Interval for the Average Annual After-tax Income

For a 97% confidence interval, the value of z from Table VII of Appendix B is 2.17. From the given information,

$$n = 4000, \quad \bar{x} = \$19{,}783, \quad \text{and} \quad s = \$2600$$

The standard deviation of the sample mean is

$$s_{\bar{x}} = 2600/\sqrt{4000} = \$41.1096$$

The 97% confidence interval for the mean annual after-tax income of all young married couples with children is

$$\bar{x} \pm z s_{\bar{x}} = 19{,}783 \pm 2.17\,(41.1096)$$

$$= 19{,}783 \pm 89.21 = \$19{,}693.79 \text{ to } \$19{,}872.21$$

Confidence Interval for the Average Annual Spending

From the given information,

$$n = 4000, \quad \bar{x} = \$21{,}401, \quad \text{and} \quad s = \$2200$$

The standard deviation of the sample mean is

$$s_{\bar{x}} = 2200/\sqrt{4000} = \$34.7851$$

The 97% confidence interval for the mean annual spending by all young married couples with children is

$$\bar{x} \pm z s_{\bar{x}} = \$21{,}401 \pm 2.17\,(34.7851)$$

$$= \$21{,}401 \pm 75.48 = \$21{,}325.52 \text{ to } \$21{,}476.48$$

Source: The chart reprinted with permission from *USA TODAY*, May 20, 1992. Copyright © *USA TODAY* 1992.

EXERCISES

Concepts and Procedures

8.1 Briefly explain the meaning of an estimator and an estimate.

8.2 Explain the meaning of a point estimate and an interval estimate.

8.3   What is the point estimator of the population mean μ? How is the margin of error for a point estimate of μ calculated?

8.4   Explain the various alternatives for decreasing the width of a confidence interval. Which of them is the best alternative?

8.5   Briefly explain how the width of a confidence interval decreases with an increase in the sample size. Give an example.

8.6   Briefly explain how the width of a confidence interval decreases with a decrease in the confidence level. Give an example.

8.7   Briefly explain the difference between a confidence level and a confidence interval.

8.8   What is the maximum error of estimate for μ for a large sample? How is it calculated?

8.9   How will you interpret a 99% confidence interval for μ for a large sample? Explain.

8.10   Find z for each of the following confidence levels.
   a. 90%   b. 95%   c. 96%   d. 97%   e. 98%   f. 99%

8.11   For a data set obtained from a sample, $n = 64$, $\bar{x} = 22.5$, and $s = 3.4$.
   a.   What is the point estimate of μ?
   b.   What is the margin of error associated with the point estimate of μ?
   c.   Make a 99% confidence interval for μ.
   d.   What is the maximum error of estimate for part c?

8.12   For a data set obtained from a sample, $n = 81$, $\bar{x} = 44.25$, and $s = 4.5$.
   a.   What is the point estimate of μ?
   b.   What is the margin of error associated with the point estimate of μ?
   c.   Make a 95% confidence interval for μ.
   d.   What is the maximum error of estimate for part c?

8.13   The standard deviation for a population is $\sigma = 12.6$. A sample of 36 observations selected from this population gave a mean equal to 74.8.
   a.   Make a 90% confidence interval for μ.
   b.   Construct a 95% confidence interval for μ.
   c.   Determine a 99% confidence interval for μ.
   d.   Does the width of the confidence intervals constructed in parts a through c increase as the confidence level increases? Explain your answer.

8.14   The standard deviation for a population is $\sigma = 16.4$. A sample of 100 observations selected from this population gave a mean equal to 143.72.
   a.   Make a 99% confidence interval for μ.
   b.   Construct a 95% confidence interval for μ.
   c.   Determine a 90% confidence interval for μ.
   d.   Does the width of the confidence intervals constructed in parts a through c decrease as the confidence level decreases? Explain your answer.

8.15   The standard deviation for a population is $\sigma = 6.30$. A random sample selected from this population gave a mean equal to 78.90.
   a.   Make a 99% confidence interval for μ assuming $n = 36$.
   b.   Construct a 99% confidence interval for μ assuming $n = 81$.
   c.   Determine a 99% confidence interval for μ assuming $n = 100$.
   d.   Does the width of the confidence intervals constructed in parts a through c decrease as the sample size increases? Explain.

8.16   The standard deviation for a population is $\sigma = 7.14$. A random sample selected from this population gave a mean equal to 55.63.
   a.   Make a 95% confidence interval for μ assuming $n = 196$.
   b.   Construct a 95% confidence interval for μ assuming $n = 100$.
   c.   Determine a 95% confidence interval for μ assuming $n = 49$.

$z_{90\%} = 1.645$

$z_{95\%} = 1.96$

$z_{99\%} = 2.58$

    d. Does the width of the confidence intervals constructed in parts a through c increase as the sample size decreases? Explain.

8.17 For a sample data set, $\bar{x} = 16$ and $s = 5.3$.

    a. Construct a 95% confidence interval for μ assuming $n = 50$.

    b. Construct a 90% confidence interval for μ assuming $n = 50$. Is the width of the 90% confidence interval smaller than the width of the 95% confidence interval calculated in part a? If yes, why is it so?

    c. Find a 95% confidence interval for μ assuming $n = 100$. Is the width of the 95% confidence interval for μ with $n = 100$ smaller than the width of the 95% confidence interval for μ with $n = 50$ calculated in part a? If so, why?

8.18 For a sample data set, $\bar{x} = 18.5$ and $s = 3.7$.

    a. Construct a 90% confidence interval for μ assuming $n = 81$.

    b. Construct a 99% confidence interval for μ assuming $n = 81$. Is the width of the 99% confidence interval larger than the width of the 90% confidence interval calculated in part a? If yes, why is it so?

    c. Find a 90% confidence interval for μ assuming $n = 60$. Is the width of the 90% confidence interval for μ with $n = 60$ larger than the width of the 90% confidence interval for μ with $n = 81$ calculated in part a? If so, why?

8.19 **a.** A sample of 100 observations taken from a population produced a sample mean equal to 55.32 and a standard deviation equal to 8.4. Make a 90% confidence interval for μ.

    b. Another sample of 100 observations taken from the same population produced a sample mean equal to 57.40 and a standard deviation equal to 7.5. Make a 90% confidence interval for μ.

    c. A third sample of 100 observations taken from the same population produced a sample mean equal to 56.25 and a standard deviation equal to 7.9. Make a 90% confidence interval for μ.

    d. The true population mean for this population is 55.80. How many of the confidence intervals constructed in parts a through c cover this population mean and how many do not?

8.20 **a.** A sample of 400 observations taken from a population produced a sample mean equal to 92.45 and a standard deviation equal to 12.20. Make a 97% confidence interval for μ.

    b. Another sample of 400 observations taken from the same population producd a sample mean equal to 91.75 and a standard deviation equal to 14.50. Make a 97% confidence interval for μ.

    c. A third sample of 400 observations taken from the same population produced a sample mean equal to 89.63 and a standard deviation equal to 13.40. Make a 97% confidence interval for μ.

    d. The true population mean for this population is 90.65. How many of the confidence intervals constructed in parts a through c cover this population mean and how many do not?

8.21 For a population, the value of the standard deviation is 2.45. A sample of 35 observations taken from this population produced the following data.

42	51	42	31	28	36	49
29	46	37	32	27	33	41
44	41	28	46	34	39	48
26	35	37	38	46	48	37
29	31	44	41	37	38	46

    a. What is the point estimate of μ?

    b. What is the margin of error associated with the point estimate of μ?

    c. Make a 97% confidence interval for μ.

    d. What is the maximum error of estimate for part c?

8.22 For a population, the value of the standard deviation is 4.56. A sample of 32 observations taken from this population produced the following data.

74	85	72	73	86	81	77	80
83	78	79	88	76	73	84	78
81	72	82	81	79	83	88	86
78	83	87	82	80	84	76	74

a. What is the point estimate of μ?
b. What is the margin of error associated with the point estimate of μ?
c. Make a 99% confidence interval for μ.
d. What is the maximum error of estimate for part c?

Applications

8.23 According to a recent study by Professor Peter Hammerschmidt, the mean annual salary of mid- to upper-level managers at major companies is $80,722 for males and $65,258 for females (*Working Woman*, September 1992). Assume that these mean salaries are based on samples of 400 male managers and 200 female managers. Further assume that the standard deviation of the annual salaries of male managers is $11,500 and the standard deviation of the annual salaries of female managers is $8,400.

a. Make a 95% confidence interval for the mean annual salary of mid- to upper-level male managers at major companies.
b. Make a 95% confidence interval for the mean annual salary of mid- to upper-level female managers at major companies.

8.24 According to a recent survey by *American Banker*, the average amount owed by a household in the United States is $71,500 (*USA TODAY*, June 10, 1992). Assume that this mean is based on a random sample of 900 households and that the standard deviation for this sample is $21,760. Construct a 99% confidence interval for the average amount owed by all households in the United States.

8.25 A consumer agency that proposes that lawyers' rates are too high wanted to estimate the mean hourly rate for all lawyers in New York City. A sample of 70 lawyers taken from New York City showed that the mean hourly rate charged by them is $203 and the standard deviation of hourly charges is $55.

a. What is the point estimate of the mean hourly charges for all lawyers in New York City? What is the margin of error of this estimate?
b. Construct a 90% confidence interval for the mean hourly charges for all lawyers in New York City.

8.26 The American Federation of Teachers conducts surveys every year to estimate the salaries of teachers. According to one such recent survey, the average salary of teachers for 1991–1992 was $34,213 (*The Wall Street Journal*, August 28, 1992). Assume that this result is based on a sample of 400 teachers and that the standard deviation of the 1991–1992 salaries of teachers in the sample was $4800.

a. What is the point estimate of the mean 1991–1992 salaries of all teachers? What is the margin of error of this estimate?
b. Make a 98% confidence interval for the mean 1991–1992 salaries of all teachers.

8.27 Although the weekly hours spent at work by people in the United States are among the lowest in the world, it is still argued that if we include commuting time, the total time spent by American workers on work and commuting is quite high. According to a Priority Management Systems survey of 1400 adults, the mean time spent on work and commuting is 10 hours a day for these workers (*USA TODAY*, July 22, 1992). Assume that the standard deviation of the time spent on work and commuting per day by these workers is 1.25 hours.

a. Make a 97% confidence interval for the mean time spent on work and commuting per day by all workers in the United States.
b. Explain why we need to make the confidence interval. Why can we not say that the mean time spent on work and commuting is 10 hours a day for all workers?

8.28 Computer Action Company sells computers and computer parts by mail. The company assures its customers that products are mailed as soon as possible after an order is placed with the company. A sample of 50 recent orders showed that the mean time taken to mail products for these orders was 70 hours and the standard deviation was 14 hours.

    a. Construct a 95% confidence interval for the mean time taken to mail products for all orders received at the office of this company.

    b. Explain why we need to make the confidence interval. Why can we not say that the mean time taken to mail products for all orders received at the office of this company is 70 hours?

8.29 Lazurus Steel Corporation produces iron rods that are supposed to be 36 inches long. The machine that makes these rods does not produce each rod exactly 36 inches long. The lengths of these rods vary slightly. It is known that when the machine is working properly, the mean length of the rods made on this machine is 36 inches. The standard deviation of the lengths of all rods produced on this machine is always equal to .10 inches. The quality control department takes a sample of 40 such rods every week, calculates the mean length of these rods, and makes a 99% confidence interval for the population mean. If either the upper limit of this confidence interval is greater than 36.05 inches or the lower limit of this confidence interval is less than 35.95 inches, the machine is stopped and adjusted. A recent such sample of 40 rods produced a mean length of 36.02 inches. Based on this sample, will you conclude that the machine needs an adjustment?

8.30 At Farmer's Dairy, a machine is set to fill 32-ounce milk cartons. However, this machine does not put exactly 32 ounces of milk in each carton; the amount varies slightly from carton to carton. It is known that when the machine is working properly, the mean net weight of these cartons is 32 ounces. The standard deviation of the milk in all such cartons is always equal to .15 ounces. The quality control department takes a sample of 35 such cartons every week, calculates the mean net weight of these cartons, and makes a 99% confidence interval for the population mean. If either the upper limit of this confidence interval is greater than 32.15 ounces or the lower limit of this confidence interval is less than 31.85 ounces, the machine is stopped and adjusted. A recent sample of 35 such cartons produced a mean net weight of 31.94 ounces. Based on this sample, will you conclude that the machine needs an adjustment?

8.31 Harrods' Consultants, Inc., advises small businesses on energy saving tips and procedures. A business manager wants to know the average money saved per month by all clients of the company who adopt its tips and procedures. The manager of the consulting firm took a sample of 35 clients and found that these clients saved an average of $85 per month by adopting the tips provided by Harrods. The standard deviation of the savings for these 35 clients was $12.

    a. Make a 99% confidence interval for the population mean.

    b. Suppose the confidence interval obtained in part a is too wide. How can the width of this interval be reduced? Discuss all possible alternatives. Which of these alternatives is the best?

8.32 A bank manager wants to know the mean amount of mortgage paid per month by homeowners in an area. A random sample of 40 homeowners selected from this area showed that they pay an average of $1350 per month for their mortgage with a standard deviation of $215.

    a. Find a 97% confidence interval for the mean amount of mortgage paid per month by all homeowners in this area.

    b. Suppose the confidence interval obtained in part a is too wide. How can the width of this interval be reduced? Discuss all possible alternatives. Which of these alternatives is the best?

8.33 You are interested in estimating the mean rent paid per month by all tenants in your area. Briefly explain the procedure you will follow to conduct this study. Collect the required data from a sample of 30 or more such tenants and then estimate the population mean at a 99% confidence level. (*Hint:* Randomly select 30 or more tenants from your area [these may be your friends if they make a representative sample], collect information on the rent each of them pays per month, and then find the mean and standard deviation for these data using the formulas learned in Chapter 3. You may use MINITAB to calculate the mean and standard deviation for the sample. Finally, make a confidence interval for μ.)

8.34 You are interested in estimating the mean age of cars owned by all people in the United States. Briefly explain the procedure you will follow to conduct this study. Collect the required data on a sample of 30 or more cars and then estimate the population mean at a 95% confidence level. (*Hint:* Randomly select 30 or more people who own cars [these may be your friends if they make a representative sample], collect information on the ages of cars they own, and then find the mean and standard deviation for these data using the formulas learned in Chapter 3. You may use MINITAB to calculate the mean and standard deviation for the sample. Finally, make a confidence interval for μ.)

8.4   INTERVAL ESTIMATION OF A POPULATION MEAN: SMALL SAMPLES

Recall from Section 8.3 that for large samples ($n \geq 30$), whether or not σ is known, the normal distribution is used to estimate the population mean μ. We use the normal distribution in such cases because, according to the central limit theorem, the sampling distribution of $\bar{x}$ is approximately normal for large samples irrespective of the shape of the population distribution.

However, many times we can select only small samples. This may be due either to the nature of the experiment or to the cost involved in taking a sample. For example, to test a new drug on patients, research may have to be based on a small sample either because there are not many patients available or willing to participate or because it is too expensive to include enough patients in the research to have a large sample.

If the sample size is small, the normal distribution can still be used to construct a confidence interval for μ if (1) the population from which the sample is drawn is normally distributed, and (2) the value of σ is known. But more often we do not know σ and, consequently, we have to use the sample standard deviation s as an estimator of σ. In such cases, the normal distribution cannot be used to make confidence intervals about μ. When (1) the population from which the sample is selected is (approximately) normally distributed, (2) the sample size is small (that is, $n < 30$), and (3) the population standard deviation σ is not known, the normal distribution is replaced by the *t distribution* to construct confidence intervals about μ. The t distribution is described in the next subsection.

CONDITIONS UNDER WHICH THE *t* DISTRIBUTION IS USED TO MAKE A CONFIDENCE INTERVAL ABOUT μ.

The t distribution is used to make a confidence interval about μ if

1. The population from which the sample is drawn is (approximately) normally distributed
2. The sample size is small (that is, $n < 30$)
3. The population standard deviation σ is not known

8.4.1   THE *t* DISTRIBUTION

The ***t* distribution** was developed by W. S. Gossett in 1908 and published under the pseudonym Student. As a result, the t distribution is also called *Student's t distribution*. The t distribution is similar to the normal distribution in some respects. Like the normal distribution curve, the t distribution curve is symmetric (bell shaped) about the mean and it never meets the horizontal axis. The total area under a t distribution curve is 1.0 or 100%. However, the t distribution curve is flatter than the standard normal distribution curve. In other words, the t distribution curve has a lower height and wider spread (or, we can say, larger standard deviation) than the standard normal distribution. However, as the sample size increases, the t distribution approaches the standard normal distribution. The units of a t distribution are denoted by t.

The shape of a particular t distribution curve depends on the number of **degrees of freedom (df)**. For the purpose of Chapters 8 and 9, the number of degrees of freedom for a t distribution is equal to the sample size minus one, that is,

$$df = n - 1$$

The number of degrees of freedom is the only parameter of the t distribution. There is a different t distribution for each number of degrees of freedom. Like the standard normal distribution, the mean of the t distribution is zero. But unlike the standard normal distribution, whose standard deviation is 1, the standard deviation of a t distribution is $\sqrt{df/(df - 2)}$, which is always greater than 1. Thus, the standard deviation of a t distribution is larger than the standard deviation of the standard normal distribution.

THE t DISTRIBUTION

The t distribution is a specific type of bell-shaped distribution with a lower height and wider spread than the standard normal distribution. As the sample size becomes larger, the t distribution approaches the standard normal distribution. The t distribution has only one parameter, called the degrees of freedom (df). The mean of the t distribution is equal to zero and its standard deviation is $\sqrt{df/(df - 2)}$.

Figure 8.5 shows the standard normal distribution and the t distribution for 9 degrees of freedom. The standard deviation of the standard normal distribution is 1.0, and the standard deviation of the t distribution is $\sqrt{df/(df - 2)} = \sqrt{9/(9 - 2)} = 1.136$.

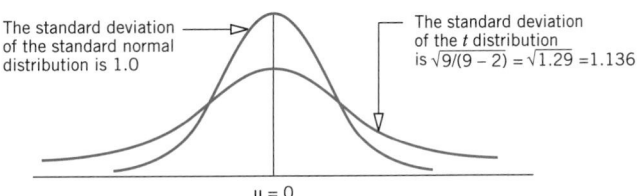

Figure 8.5   The t distribution for $df = 9$ and the standard normal distribution.

As stated earlier, the number of degrees of freedom for a t distribution for the purpose of this chapter is $n - 1$. **The number of degrees of freedom is defined as the number of observations that can be chosen freely.** As an example, suppose we know that the mean of 4 values is 20. Consequently, the sum of these 4 values is $20(4) = 80$. Now, how many values out of 4 can we choose freely so that the sum of these 4 values is 80? The answer is that we can freely choose $4 - 1 = 3$ values. Suppose we choose 27, 8, and 19 as the 3 values. Given these 3 values and the information that the mean of the 4 values is 20, the fourth value is $80 - 27 - 8 - 19 = 26$. Thus, once we have chosen 3 values, the fourth value is automatically determined. Consequently, the number of degrees of freedom for this example is

$$df = n - 1 = 4 - 1 = 3$$

We subtract 1 from n because we lose one degree of freedom to calculate the mean.

Table VIII of Appendix B lists the values of *t* for the given number of degrees of freedom and areas in the right tail of a *t* distribution. Because the *t* distribution is symmetric, these are also the values of −*t* for the same number of degrees of freedom and the same areas in the left tail of the *t* distribution. Example 8–3 describes how to read Table VIII of Appendix B.

Reading the t distribution table.

EXAMPLE 8–3 Find the value of *t* for 16 degrees of freedom and .05 area in the right tail of a *t* distribution curve.

Solution   In Table VIII, we locate 16 in the column of degrees of freedom (labeled *df*) and .05 in the row of *area in the right tail* at the top of the table. The entry at the intersection of the row of 16 and the column of .05, which is 1.746, gives the required value of *t*. The relevant portion of Table VIII is shown here as Table 8.1. The value of *t* read from the *t* distribution table is shown in Figure 8.6.

Table 8.1   Determining *t* for 16 *df* and .05 Area in the Right Tail

		—Area in the right tail			
		Area in the Right Tail			
df	**.10**	**.05**	**.025**	**. . .**	**.001**
1	3.078	6.314	12.706	. . .	318.309
2	1.886	2.920	4.303	. . .	22.327
3	1.638	2.353	3.182	. . .	10.215
.	. . .	. . .	. . .	. . .	. . .
.	. . .	. . .	. . .	. . .	. . .
df → **16**	1.337	**1.746** ←	2.120	. . .	3.686
.	. . .	. . .	. . .	. . .	. . .
.	. . .	. . .	. . .	. . .	. . .
75	1.293	1.665	1.992	. . .	3.202
∞	1.282	1.645	1.960	. . .	3.090

The required value of *t* for 16 *df* and .05 area in the right tail

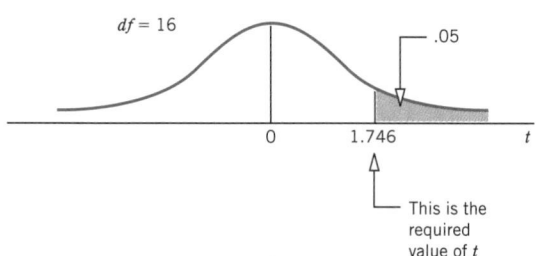

Figure 8.6   The value of *t* for 16 *df* and .05 area in the right tail.

Because of the symmetric shape of the *t* distribution curve, the value of *t* for 16 degrees of freedom and .05 area in the left tail is −1.746. Figure 8.7 illustrates this case.

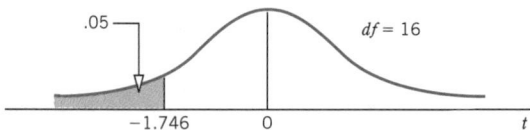

Figure 8.7   The value of *t* for 16 *df* and .05 area in the left tail.

8.4.2 CONFIDENCE INTERVAL FOR μ USING THE *t* DISTRIBUTION

To reiterate, when the following three conditions hold true, we use the *t* distribution to construct a confidence interval for the population mean μ.

1. The population from which the sample is drawn is (approximately) normally distributed
2. The sample size is small (that is, $n < 30$)
3. The population standard deviation σ is not known

CONFIDENCE INTERVAL FOR μ FOR SMALL SAMPLES

The $(1 - \alpha)100\%$ confidence interval for μ is

$$\bar{x} \pm ts_{\bar{x}} \qquad \text{where} \qquad s_{\bar{x}} = \frac{s}{\sqrt{n}}$$

The value of *t* is obtained from the *t* distribution table for $n - 1$ degrees of freedom and the given confidence level.

Examples 8–4 and 8–5 describe the procedure of constructing a confidence interval for μ using the *t* distribution.

Constructing a 95% confidence interval for μ using the t distribution.

EXAMPLE 8–4 The management of a supermarket is considering an area in which to open a new store. However, it is the policy of the management to open new stores only in those areas that have an average household income of at least $30,000 per year. The research department of the company took a sample of 25 households from this area and collected information on their annual incomes. This sample produced a mean annual household income of $29,300 and a standard deviation of $4500. Assume that the incomes of all households in this area have an approximate normal distribution. Construct a 95% confidence interval for the mean annual income of all households in this area. Does this interval include $30,000? Based on this result, should the management decide to open a new store in this area?

Solution From the given information,

$$n = 25, \qquad \bar{x} = \$29{,}300, \qquad s = \$4500,$$

and $\qquad\qquad\qquad$ Confidence level = 95% or .95

The value of $s_{\bar{x}}$ is

$$s_{\bar{x}} = s/\sqrt{n} = 4500/\sqrt{25} = \$900$$

To find the value of *t*, we need to know the degrees of freedom and the area under the *t* distribution curve in each tail.

$$\text{Degrees of freedom} = n - 1 = 25 - 1 = 24$$

To find the area in each tail, we divide the confidence level by 2 and subtract the number obtained from .5. Thus,

$$\text{Area in each tail} = .5 - .95/2 = .5 - .4750 = .025$$

From the *t* distribution table, Table VIII of Appendix B, the value of *t* for 24 *df* and .025 area in the right tail is 2.064. The value of *t* is shown in Figure 8.8.

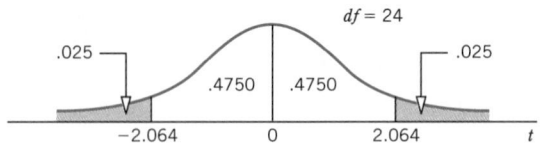

Figure 8.8   The value of t.

Substituting all values in the formula for the confidence interval for μ, the 95% confidence interval is

$$\bar{x} \pm ts_{\bar{x}} = \$29,300 \pm 2.064\,(900)$$

$$= \$29,300 \pm 1857.60 = \textbf{\$27,442.40 to \$31,157.60}$$

Thus, we can state with 95% confidence that the mean annual income of all households in this area is between \$27,442.40 and \$31,157.60. Since this interval includes \$30,000, the management should decide in favor of opening a new store in this area.   ▬

Constructing a 99% confidence interval for μ using the t distribution.

EXAMPLE 8–5   Each soap bar manufactured by Becker Corporation is supposed to have a net weight of 4.5 ounces. The machine that makes these soap bars does not produce each soap bar with exactly 4.5 ounces. The net weights of these bars vary slightly. It is known that when the machine is working properly, the mean net weight of these bars is 4.5 ounces. The quality control department at the company takes a sample of 16 such bars every week, calculates the mean net weight of these bars, and makes a 99% confidence interval for the population mean. If either the upper limit of this confidence interval is greater than 4.52 ounces or the lower limit of this confidence interval is less than 4.48 ounces, the machine is stopped and adjusted. A recent sample of 16 such bars produced a mean net weight of 4.496 ounces and a standard deviation of .03 ounces. Based on this sample, will you conclude that the machine needs an adjustment? Assume that the net weights of all such soap bars are approximately normally distributed.

Solution   From the given information,

$$n = 16, \quad \bar{x} = 4.496 \text{ ounces}, \quad s = .03 \text{ ounces},$$

and   Confidence level $= 99\%$ or .99

First we calculate the standard deviation of $\bar{x}$, the number of degrees of freedom, and the area in each tail of the t distribution as follows.

$$s_{\bar{x}} = s/\sqrt{n} = .03/\sqrt{16} = .0075$$

$$df = n - 1 = 16 - 1 = 15$$

$$\text{Area in each tail} = .5 - .99/2 = .5 - .4950 = .005$$

From the t distribution table, $t = 2.947$ for 15 degrees of freedom and .005 area in the right tail. The 99% confidence interval for μ is

$$\bar{x} \pm ts_{\bar{x}} = 4.496 \pm 2.947\,(.0075)$$

$$= 4.496 \pm .022 = \textbf{4.474 to 4.518 ounces}$$

Thus, we can state with 99% confidence that based on this sample the mean net weight of all such soap bars is between 4.474 and 4.518 ounces. The upper limit of this confidence interval is not greater than 4.52 ounces but the lower limit is less than 4.48 ounces. Consequently, the machine needs an adjustment.   ▬

Again, we can decrease the width of a confidence interval for μ either by lowering the confidence level or by increasing the sample size, as was done in Section 8.3. However, increasing the sample size is the better alternative.

EXERCISES

Concepts and Procedures

8.35   Briefly explain the similarities and the differences between the standard normal distribution and the t distribution.

8.36   What are the parameters of a normal distribution and a t distribution? Explain.

8.37   Briefly explain the meaning of the degrees of freedom for a t distribution. Give one example.

8.38   What assumptions must hold true to use the t distribution to make a confidence interval for μ?

8.39   Find the value of t for the t distribution for each of the following.
   a.   Area in the right tail $= .05$ and $df = 13$
   b.   Area in the left tail $= .025$ and $n = 22$
   c.   Area in the left tail $= .001$ and $df = 17$
   d.   Area in the right tail $= .005$ and $n = 26$

8.40   a.   Find the value of t for a t distribution with a sample size of 20 and area in the left tail equal to .10.
   b.   Find the value of t for a t distribution with a sample size of 12 and area in the right tail equal to .025.
   c.   Find the value of t for a t distribution with 14 degrees of freedom and .001 area in the right tail.
   d.   Find the value of t for a t distribution with 24 degrees of freedom and .005 area in the left tail.

8.41   For each of the following, find the area in the appropriate tail of the t distribution.
   a. $t = 2.060$ and $df = 25$ b. $t = -3.686$ and $df = 16$
   c. $t = -2.650$ and $n = 14$ d. $t = 2.845$ and $n = 21$

8.42   For each of the following, find the area in the appropriate tail of the t distribution.
   a. $t = -1.323$ and $df = 21$ b. $t = 2.467$ and $n = 29$
   c. $t = 3.250$ and $df = 9$ d. $t = -2.947$ and $df = 15$

8.43   Find the value of t from the t distribution table for each of the following.
   a.   Confidence level $= 99\%$ and $df = 16$
   b.   Confidence level $= 95\%$ and $n = 24$
   c.   Confidence level $= 90\%$ and $df = 19$

8.44   a.   Find the value of t from the t distribution table for a sample size of 20 and a confidence level of 95%.
   b.   Find the value of t from the t distribution table for 11 degrees of freedom and a 90% confidence level.
   c.   Find the value of t from the t distribution table for a sample size of 26 and a confidence level of 99%.

8.45   A sample of 12 observations taken from a normally distributed population produced the following data.

13	15	9	11	8	16
14	9	10	14	16	12

   a.   What is the point estimate of μ?
   b.   Make a 99% confidence interval for μ.
   c.   What is the maximum error of estimate for part b?

8.46   A sample of 10 observations taken from a normally distributed population produced the following data.

44   52   31   48   46   39   43   36   41   49

   a.   What is the point estimate of μ?
   b.   Make a 95% confidence interval for μ.
   c.   What is the maximum error of estimate for part b?

8.47   Suppose, for a sample selected from a normally distributed population, $\bar{x} = 65.50$ and $s = 8.6$.
   a.   Construct a 95% confidence interval for μ assuming $n = 16$.
   b.   Construct a 90% confidence interval for μ assuming $n = 16$. Is the width of the 90% confidence interval smaller than the width of the 95% confidence interval calculated in part a? If yes, explain why.
   c.   Find a 95% confidence interval for μ assuming $n = 25$. Is the width of the 95% confidence interval for μ with $n = 25$ smaller than the width of the 95% confidence interval for μ with $n = 16$ calculated in part a? If so, why? Explain.

8.48   Suppose, for a sample selected from a normally distributed population, $\bar{x} = 21.5$ and $s = 3.9$.
   a.   Construct a 95% confidence interval for μ assuming $n = 27$.
   b.   Construct a 99% confidence interval for μ assuming $n = 27$. Is the width of the 99% confidence interval larger than the width of the 95% confidence interval calculated in part a? If yes, explain why.
   c.   Find a 95% confidence interval for μ assuming $n = 20$. Is the width of the 95% confidence interval for μ with $n = 20$ larger than the width of the 95% confidence interval for μ with $n = 27$ calculated in part a? If so, why? Explain.

Applications

8.49   A tool manufacturing company wants to estimate the mean number of bolts produced per hour on a specific machine. The manager observed the machine for 20 randomly selected hours. The mean number of bolts produced per hour during these 20 hours was found to be 47 with a standard deviation of 2.4. Assume that the number of bolts produced per hour on this machine has a normal distribution. Construct a 90% confidence interval for the population mean μ.

8.50   A company wants to estimate the mean net weight of its Top Taste cereal boxes. A sample of 16 such boxes produced the mean net weight of 31.98 ounces with a standard deviation of .26 ounces. Make a 95% confidence interval for the mean net weight of all Top Taste cereal boxes. Assume that the net weights of all such cereal boxes have a normal distribution.

8.51   The high cost of health care is a matter of major concern for a large number of families. A random sample of 25 families selected from an area showed that they spend an average of $120 per month on health care with a standard deviation of $28. Make a 99% confidence interval for the mean health care expenditure per month incurred by all families in this area. Assume that the monthly health care expenditures of all families in this area have a normal distribution.

8.52   A random sample of 27 auto claims filed with an insurance company produced a mean amount for these claims equal to $2100 with a standard deviation of $460. Construct a 98% confidence interval for the corresponding population mean. Assume that the amounts of all such claims filed with this company have a normal distribution.

8.53   A random sample of 16 mid-sized cars, which were tested for fuel consumption, gave a mean of 26.4 miles per gallon with a standard deviation of 2.3 miles per gallon.
   a.   Assuming that the miles per gallon given by all mid-sized cars have a normal distribution, find a 99% confidence interval for the population mean μ.
   b.   Suppose the confidence interval obtained in part a is too wide. How can the width of this

interval be reduced? Describe all possible alternatives. Which alternative is the best and why?

8.54 The mean time taken to design a home plan by 20 architects was found to be 248 minutes with a standard deviation of 26 minutes.

    a. Assume that the time taken by all architects to design this home plan is normally distributed. Construct a 95% confidence interval for the population mean μ.

    b. Suppose the confidence interval obtained in part a is too wide. How can the width of this interval be reduced? Describe all possible alternatives. Which alternative is the best and why?

8.55 A sample of 10 customers who visited a supermarket was taken. The following data give the money (in dollars) they spent at this supermarket during that visit.

$df = 9\ (n-1)$
$ci = \bar{X}_{90\%}$

$74	89	121	63	146
47	91	28	84	76

Assuming that the money spent by all customers at this supermarket has a normal distribution, construct a 90% confidence interval for the population mean. (*Hint:* First, calculate the sample mean and the sample standard deviation for these data using the formulas from Sections 3.1.1 and 3.2.2 of Chapter 3. Then make the confidence interval for μ.)

8.56 A sample of eight adults was taken, and these adults were asked about the time they spend per week on leisure activities. Their responses (in hours) are as follows.

45	12	31	16	28	14	18	26

Assuming that the times spent on leisure activities by all adults are normally distributed, make a 95% confidence interval for the mean time spent per week by all adults on leisure activities. (*Hint:* First, calculate the sample mean and the sample standard deviation for these data using the formulas from Sections 3.1.1 and 3.2.2 of Chapter 3. Then make the confidence interval for μ.)

8.57 You are working for a supermarket. The manager has asked you to estimate the mean time taken by a cashier to serve customers at this supermarket. Unfortunately, he has asked you to take a small sample. Briefly explain how you will conduct this study. Collect data on time taken by any supermarket cashier to serve 10 customers. Then estimate the population mean. Choose your own confidence level.

8.58 You are working for a bank. The bank manager wants to know the mean waiting time for all customers who visit this bank. She has asked you to estimate this mean by taking a small sample. Briefly explain how you will conduct this study. Collect data on waiting times for 15 customers who visit this bank. Then estimate the population mean. Choose your own confidence level.

8.5 INTERVAL ESTIMATION OF A POPULATION PROPORTION: LARGE SAMPLES

Often we want to estimate the population proportion or percentage. *(Recall that a percentage is obtained by multiplying the proportion by 100.)* For example, the production manager of a company may want to estimate the proportion of defective items produced on a machine. A bank manager may want to find the percentage of customers who are satisfied with the service provided by the bank.

Again, if we can conduct a census each time we want to find the value of a population proportion, there is no need to learn the procedures discussed in this section. However, we usually derive our results from sample surveys. Hence, to take into account the variability in the results obtained from different sample surveys, we need to know the procedures, discussed in this section, for estimating a population proportion.

Recall from Chapter 7 that the population proportion is denoted by p and the sample proportion is denoted by $\hat{p}$. This section explains how to estimate the population proportion p using the sample proportion $\hat{p}$. The sample proportion $\hat{p}$ is a sample statistic, and it possesses a sampling distribution. From Chapter 7, we know that for large samples:

1. The sampling distribution of the sample proportion $\hat{p}$ is (approximately) normal
2. The mean $\mu_{\hat{p}}$ of the sampling distribution of $\hat{p}$ is equal to the population proportion p
3. The standard deviation $\sigma_{\hat{p}}$ of the sampling distribution of the sample proportion $\hat{p}$ is $\sqrt{pq/n}$ where $q = 1 - p$

Remember that in the case of proportion, a sample is considered to be large if np and nq are both greater than 5. If p and q are not known, then $n\hat{p}$ and $n\hat{q}$ should be greater than 5 for the sample to be large.

When estimating the value of a population proportion, we do not know the values of p and q. Consequently, we cannot compute $\sigma_{\hat{p}}$. Therefore, in the estimation of a population proportion, we use the value of $s_{\hat{p}}$ as an estimate of $\sigma_{\hat{p}}$. The value of $s_{\hat{p}}$ is calculated using the following formula.

ESTIMATOR OF THE STANDARD DEVIATION OF $\hat{p}$

The value of $s_{\hat{p}}$, which is a point estimate of $\sigma_{\hat{p}}$, is calculated as

$$s_{\hat{p}} = \sqrt{\frac{\hat{p}\,\hat{q}}{n}}$$

The sample proportion $\hat{p}$ is the point estimator of the corresponding population proportion p. As in the case of the mean, whenever we use point estimation for the population proportion, we calculate the **margin of error** associated with that point estimation. For the estimation of the population proportion, the margin of error is calculated as follows.

$$\text{Margin of error} = \pm 1.96\, s_{\hat{p}}$$

That is, we find the standard deviation of the sample proportion and multiply it by 1.96.

The $(1 - \alpha)100\%$ confidence interval for p is constructed using the following formula.

CONFIDENCE INTERVAL FOR THE POPULATION PROPORTION p

The $(1 - \alpha)100\%$ confidence interval for the population proportion p is

$$\hat{p} \pm zs_{\hat{p}}$$

The value of z used here is obtained from the standard normal distribution table for the given confidence level.

The rationale for using this formula for the estimation of p is the same as the one for using the formula for the estimation of μ, which is discussed in Appendix 8.1. Examples 8–6 and 8–7 illustrate the procedure for constructing a confidence interval for p.

Point estimate and a 99% confidence interval for p: large sample.

EXAMPLE 8-6   A food company is planning to market a new kind of cereal. However, before marketing this product, the company wants to find what percentage of people will like it. The company's research department selected a random sample of 500 persons and asked them to taste this cereal. Of these 500 persons, 290 said they liked it.

(a)   What is the point estimate of the population proportion? What is the margin of error of this estimate?

(b)   Find, with a 99% confidence level, what percentage of all people will like this cereal.

Solution   Let p be the proportion of all people who like this cereal, and let $\hat{p}$ be the corresponding sample proportion. From the given information,

$$n = 500, \qquad \hat{p} = 290/500 = .58, \qquad \hat{q} = 1 - \hat{p} = 1 - .58 = .42$$

First, we calculate the value of the standard deviation $s_{\hat{p}}$ of the sample proportion as follows.

$$s_{\hat{p}} = \sqrt{\frac{\hat{p}\,\hat{q}}{n}} = \sqrt{\frac{(.58)\,(.42)}{500}} = .0221$$

(a)   The point estimate of the proportion of all people who will like this cereal is equal to .58, that is,

$$\text{Point estimate of } p = \hat{p} = \textbf{.58}$$

The margin of error associated with this point estimate of p is

$$\text{Margin of error} = \pm 1.96\, s_{\hat{p}} = \pm 1.96\,(.0221) = \pm\textbf{.043}$$

The margin of error states that the proportion of all people who will like this cereal is .58 (or 58%), give or take .043 (or 4.3%). As in the case of the mean, the margin of error is simply the maximum error of estimate for a 95% confidence interval.

(b)   The confidence level is 99% or .99. To find z, we divide the confidence level by 2 and look for this number in the body of the normal distribution table and record the corresponding value of z.

$$.99/2 = .4950$$

The z value for .4950 is approximately 2.58 from the normal distribution table. Note that $n\hat{p}$ and $n\hat{q}$ are both greater than 5. Consequently, the sampling distribution of $\hat{p}$ is (approximately) normal.

Substituting all the values in the confidence interval formula for p, we obtain:

$$\hat{p} \pm z s_{\hat{p}} = .58 \pm 2.58\,(.0221) = .58 \pm .057$$

$$= \textbf{.523 to .637 or 52.3\% to 63.7\%}$$

Thus, we can state with 99% confidence that .523 to .637 or 52.3% to 63.7% of all people will like this cereal.   ■

Constructing a 95% confidence interval for p: large sample.

EXAMPLE 8-7   Besides being paid an hourly wage, the workers in the assembly line at Kozika Corporation are also paid a bonus based on the number of clocks they assemble per week. The quality control department of the company takes samples of clocks quite often to check if the assembly work is properly done because they fear that the quality of the work may deteriorate as workers attempt to assemble more clocks to earn a higher bonus. A recent sample of 120 such clocks taken by the quality control department showed that 7% of them were not properly assembled. Based on this sample, construct a 95% confidence interval for the proportion of all clocks that are not properly assembled.

Solution   Let p be the proportion of all clocks that are not properly assembled, and let $\hat{p}$ be the corresponding sample proportion. From the given information,

$$n = 120, \qquad \hat{p} = .07, \qquad \hat{q} = 1 - \hat{p} = 1 - .07 = .93$$

and   Confidence level = 95% or .95

The standard deviation $s_{\hat{p}}$ of the sample proportion $\hat{p}$ is

$$s_{\hat{p}} = \sqrt{\frac{\hat{p}\,\hat{q}}{n}} = \sqrt{\frac{(.07)\,(.93)}{120}} = .0233$$

From the normal distribution table, the value of z for $.95/2 = .4750$ is 1.96. The 95% confidence interval for p is

$$\hat{p} + zs_{\hat{p}} = .07 \pm 1.96\,(.0233) = .07 \pm .046 = \textbf{.024 to .116}$$

Thus, we can state with 95% confidence that the proportion of all clocks that are not properly assembled is between .024 and .116. The confidence interval can be converted to a percentage as 2.4% to 11.6%.   ▬

Again, we can decrease the width of a confidence interval for p either by lowering the confidence level or by increasing the sample size. However, lowering the confidence level is not a good choice because it simply decreases the likelihood that the confidence interval contains p. Hence, to decrease the width of a confidence interval for p, we should always increase the sample size.

Case Study 8–2 discusses the philosophy and meaning of confidence intervals and margin of errors.

CASE STUDY 8-2   ASK MR. STATISTICS

Fortune magazine frequently publishes a column titled *Ask Mr. Statistics*, which contains questions on statistics and answers to those questions. The following excerpts are reprinted from one such column. The Washington Post–ABC poll refers to 1992 Presidential election.

Dear Mr. Statistics: As a chap who took the liberal arts course and still keeps forgetting what a standard deviation is I was recently unnerved to read in the *Journal of Educational Psychology* (Vol. 74, No. 3) that Japanese school children study statistical sampling in the fifth grade. Not having a Niponese 10-year-old standing in the closet, I am writing for your help with my latest problem: explaining those mysterious formulations in the political news about margins of sampling error and whether it is or isn't a big deal that Ross Perot leads George Bush among registered voters by 34% to 31%, as stated in a recent Washington Post–ABC poll (whose interpreters in the Post disparaged this gap as ''statistically insignificant'' because the margin of sampling error was 4%).

                                                                    —Numbed by Numbers

Dear Numnum: To understand political survey data, you need to know several things the poll-persons never tell you. Well, almost never. The first is that they are aspiring to a "confidence level" of 95%. This means that if the sample of people polled is representative of voters generally, and the candidates are separated by more than the margin of error, then it is 19 to 1 (i.e., 95 to 5) that voters will in fact go for the candidate rated highest. The 95% confidence level is entirely arbitrary and somewhat unsatisfactory, since it means that the poll-persons can do everything right and still lose to a longshot, just like all the favorite players who glumly watched Lil E. Tee win the Kentucky Derby at 17 to 1 not long ago.

The pollsters could in principle offer 99% confidence levels (i.e., odds of 99 to 1). But this would typically require them either to (a) increase the sampling levels from the usual 1,300 or so to around 2,250, which is expensive, or else (b) increase their margins of error from the usual 3% to around 4%, which would result in more matchups in which the differences between candidates were within the margin of error.

Now about the Post–ABC poll you have cited: It is not at all an insignificant matter that Perot led Bush by three percentage points. In this case, the sample size was smaller than usual (784 registered voters instead of the usual 1,300 or so) and the margin of error was therefore larger than usual (4% instead of 3%). But it is fallacious to say that leads within the margin of error are meaningless. The correct statement would have been that Perot's advantage could be postulated only at a confidence level lower than 95%—in this case, it would have been about 91%. Even if the poll had shown Perot ahead by only two percentage points, he still would have been better off than Bush. The confidence level would then have been around 74%, meaning he was still a 3-to-1 favorite.

Now for the main thing the pollsters never tell you: It is very, very hard to produce a sample that talks the way voters vote. The pros are pretty good at producing samples that are kosher with respect to age, race, sex, income, and registration status; what they don't know how to control for is "response meaningfulness." Many people have no more ideas than a Valley Girl about the pending election, but are ashamed to admit this to pollsters. Far more serious, many folks limply broach whatever views they think the pollster will approve of. Since pollsters look to many citizens like members of the liberal media elite, there is a special tendency for conservative views to be understated. A week before Reagan's landslide election in 1980, the AP reported that "national polls say the race between Jimmy Carter and Ronald Reagan is too close to call." In the wake of the Tory upset in the April elections, Britain's Market Research Society has demonstrated that support for Labour has been overstated in polls in eight of the past ten national elections, a finding we rate significant at the 99% confidence level.

Source: Daniel Seligman, "Ask Mr. Statistics," *Fortune*, July 13, 1992. Copyright © 1992, Time Inc. Reprinted with permission. All Rights Reserved.

EXERCISES

Concepts and Procedures

8.59 What assumption(s) must hold true to use the normal distribution to make a confidence interval for the population proportion p?

8.60 What is the point estimator of the population proportion p? How is the margin of error for a point estimate of p calculated?

8.61 Check if the sample size is large enough to use the normal distribution to make a confidence interval for p for each of the following cases.

a.	$n = 50$	and	$\hat{p} = .25$	b. $n = 160$	and	$\hat{p} = .03$
c.	$n = 400$	and	$\hat{p} = .65$	d. $n = 75$	and	$\hat{p} = .06$

8.62 Check if the sample size is large enough to use the normal distribution to make a confidence interval for p for each of the following cases.

    a. $n = 120$ and $\hat{p} = .04$ **b.** $n = 60$ and $\hat{p} = .08$

    c. $n = 40$ and $\hat{p} = .50$ **d.** $n = 900$ and $\hat{p} = .15$

8.63 A sample of 900 observations taken from a population produced a sample proportion of .47.

    a. What is the point estimate of p?

    b. What is the margin of error associated with the point estimate of p?

    c. Make a 99% confidence interval for p.

    d. What is the maximum error of estimate for part c?

8.64 A sample of 200 observations taken from a population produced a sample proportion of .22.

    a. What is the point estimate of p?

    b. What is the margin of error associated with the point estimate of p?

    c. Make a 95% confidence interval for p.

    d. What is the maximum error of estimate for part c?

8.65 **a.** A sample of 400 observations taken from a population produced a sample proportion of .63. Make a 95% confidence interval for p.

    b. Another sample of 400 observations taken from the same population produced a sample proportion of .59. Make a 95% confidence interval for p.

    c. A third sample of 400 observations taken from the same population produced a sample proportion of .67. Make a 95% confidence interval for p.

    d. The true population proportion for this population is .65. How many of the confidence intervals constructed in parts a through c cover this population proportion and how many do not?

8.66 **a.** A sample of 900 observations taken from a population produced a sample proportion of .32. Make a 90% confidence interval for p.

    b. Another sample of 900 observations taken from the same population produced a sample proportion of .36. Make a 90% confidence interval for p.

    c. A third sample of 900 observations taken from the same population produced a sample proportion of .30. Make a 90% confidence interval for p.

    d. The true population proportion for this population is .34. How many of the confidence intervals constructed in parts a through c cover this population proportion and how many do not?

8.67 A sample of 500 observations selected from a population gave a sample proportion equal to .72.

    a. Make a 90% confidence interval for p.

    b. Construct a 95% confidence interval for p.

    c. Make a 99% confidence interval for p.

    d. Does the width of the confidence intervals constructed in parts a through c increase as the confidence level increases? If yes, explain why.

8.68 A sample of 200 observations selected from a population gave a sample proportion equal to .25.

    a. Make a 99% confidence interval for p.

    b. Construct a 97% confidence interval for p.

    c. Make a 90% confidence interval for p.

    d. Does the width of the confidence intervals constructed in parts a through c decrease as the confidence level decreases? If yes, explain why.

8.69 A sample selected from a population gave a sample proportion equal to .67.

    a. Make a 99% confidence interval for p assuming $n = 100$.

    b. Construct a 99% confidence interval for p assuming $n = 600$.

    c. Make a 99% confidence interval for p assuming $n = 1500$.

    d. Does the width of the confidence intervals constructed in parts a through c decrease as the sample size increases? If yes, explain why.

8.70 A sample selected from a population gave a sample proportion equal to .34.

    a. Make a 95% confidence interval for p assuming $n = 1200$.

    b. Construct a 95% confidence interval for p assuming $n = 500$.

    c. Make a 95% confidence interval for p assuming $n = 80$.

    d. Does the width of the confidence intervals constructed in parts a through c increase as the sample size decreases? If yes, explain why.

8.71 Suppose for a sample data set, $\hat{p} = .36$.

    a. Construct a 95% confidence interval for p assuming $n = 100$.

    b. Construct a 99% confidence interval for p assuming $n = 100$. Is the width of the 99% confidence interval larger than the 95% confidence interval calculated in part a? If yes, explain why.

    c. Find a 95% confidence interval for p assuming $n = 200$. Is the width of the 95% confidence interval for p with $n = 200$ smaller than the 95% confidence interval for p with $n = 100$ calculated in part a? If so, why? Explain.

8.72 Suppose for a sample data set, $\hat{p} = .76$.

    a. Construct a 95% confidence interval for p assuming $n = 900$.

    b. Construct a 90% confidence interval for p assuming $n = 900$. Is the width of the 90% confidence interval smaller than the 95% confidence interval calculated in part a? If yes,

    c. explain why.

    Find a 95% confidence interval for p assuming $n = 500$. Is the width of the 95% confidence interval for p with $n = 500$ larger than the 95% confidence interval for p with $n = 900$ calculated in part a? If so, why? Explain.

Applications

8.73 Recently an economic literacy test was given to a sample of high school students, 34% of whom defined profits correctly as "revenues minus costs" (*Test of Economic Literacy*, Joint Council on Economic Education). Assume that this test was given to 2500 high school students.

    a. What is the point estimate of the population proportion? What is the margin of error associated with this point estimate?

    b. Construct a 99% confidence interval for the proportion of all high school students who could correctly define profits.

8.74 A food company is planning to market a new ice cream. However, before marketing this ice cream, the company wants to find what percentage of the people like it. The company's research department selected a random sample of 400 persons and asked them to taste this ice cream. Of these 400 persons, 224 said they like it.

    a. What is the point estimate of the population proportion? What is the margin of error associated with this point estimate?

    b. Find, with a 95% confidence level, what percentage of all people like this ice cream.

8.75 According to a recent report by the U.S. Bureau of the Census, 26% of the households headed by single men own stocks, bonds, and mutual funds. Although Census Bureau estimates are based on very large samples, for convenience assume that this result is based on a random sample of 2000 households headed by single men.

    a. What is the point estimate of the proportion of all households headed by single men who own stocks, bonds, and mutual funds? What is the margin of error associated with this point estimate?

    b. Find a 95% confidence interval for the proportion of all households headed by single men who own stocks, bonds, and mutual funds.

8.76 It is said that happy and healthy workers are more efficient and productive. A company that manufactures exercising machines wanted to know the percentage of large companies that provide on-site health club facilities. A sample of 240 such companies showed that 80 of them provide such facilities on site.

    a. What is the point estimate of the percentage of all such companies that provide such facilities on site? What is the margin of error associated with this point estimate?

b. Construct a 98% confidence interval for the percentage of all such companies that provide such facilities on site.

8.77 A mail-order company promises its customers that the products ordered will be mailed within 72 hours after an order is placed. The quality control department at the company checks from time to time if this promise is fulfilled. Recently the quality control department took a sample of 50 orders and found that 42 of them were mailed within 72 hours of the placement of the orders. Construct a 97% confidence interval for the percentage of all orders that are mailed within 72 hours of their placement.

8.78 One of the major problems faced by department stores is a high percentage of returns. The manager of a department store wanted to estimate the percentage of all sales that result in returns. A sample of 40 sales showed that 8 of the products were returned within the time allowed for returns. Make a 99% confidence interval for the percentage of all sales that result in returns.

8.79 A researcher wanted to know the percentage of executives of companies who hold a degree beyond a bachelor's. He took a random sample of 15 executives of companies and asked them whether or not they hold such a degree. The responses of these executives are given below.

| Yes | No | Yes | Yes | No | No | No | Yes |
| Yes | No | Yes | Yes | Yes | No | Yes | |

Make a 90% confidence interval for the percentage of all executives of companies who hold a degree beyond a bachelor's.

8.80 The manager of a supermarket wanted to know the percentage of shoppers who prefer to buy name brand products. A random sample of 20 shoppers who shopped at this store were asked this question. The following are the responses of these shoppers.

No	No	No	Yes	Yes	No	No
Yes	No	Yes	No	No	Yes	No
No	No	No	Yes	No	Yes	

Construct a 99% confidence interval for the percentage of all shoppers who prefer to buy name brand products.

8.81 You want to estimate the proportion of students at your college who hold off-campus (part-time or full-time) jobs. Briefly explain how you will make such an estimate. Collect data from 40 students at your college on whether or not they hold off-campus jobs. Then, calculate the proportion of students in this sample who hold off-campus jobs. Using this information, estimate the population proportion. Select your own confidence level.

8.82 You want to estimate the percentage of people who are satisfied with the services provided by their banks. Briefly explain how you will make such an estimate. Collect data from 30 people on whether or not they are satisfied with the services provided by their banks. Then, calculate the percentage of people in this sample who are satisfied. Using this information, estimate the population percentage. Select your own confidence level.

8.6 SAMPLE SIZE DETERMINATION FOR THE ESTIMATION OF MEAN

One reason why we usually conduct a sample survey and not a census is that almost always we have limited resources at our disposal. In light of this, if a smaller sample can serve our purpose, then we will be wasting our resources by taking a larger sample. For instance, suppose we want to estimate the mean life of a certain auto battery. If a sample of 40 batteries can give us the type of confidence interval that we are looking for, then we will be wasting money and time if we take a sample of a much larger size, say 500 batteries. In such cases, if we know the confidence level and the width of the confidence interval that we want, then we can find the (approximate) size of the sample that will produce the required result.

In Section 8.3 we learned that $E = z\sigma_{\bar{x}}$ is called the maximum error of estimate for μ. As we know, the standard deviation $\sigma_{\bar{x}}$ of the sample mean $\bar{x}$ is equal to $\sigma/\sqrt{n}$. Therefore, we can write the maximum error of estimate for μ as

$$E = z \cdot \frac{\sigma}{\sqrt{n}}$$

Suppose we predetermine the size of the maximum error E and want to find the size of the sample that will yield this maximum error. From the above expression, the following formula is obtained that determines the required sample size n.

DETERMINING THE SAMPLE SIZE FOR THE ESTIMATION OF μ

Given the confidence level and the standard deviation of the population, the sample size that will produce a predetermined maximum error E of the confidence interval estimate of μ is

$$n = \frac{z^2 \sigma^2}{E^2}$$

If we do not know σ, we can take a preliminary sample (of any arbitrarily determined size) and find the sample standard deviation s. Then we can use s for σ in the formula. However, note that using s for σ may give a sample size that eventually may produce an error much larger (or smaller) than the predetermined maximum error. This will depend on how close s and σ are.

Example 8–8 illustrates how we determine the sample size that will produce the maximum error of estimate for μ within a certain limit.

Determining sample size for the estimation of μ.

EXAMPLE 8–8 Suppose the U.S. Bureau of the Census wants to estimate the mean family size for all U.S. families at a 99% confidence level. It is known that the standard deviation σ for the sizes of all families in the United States is .6. How large a sample should the bureau select if it wants its estimate to be within .01 of the population mean?

Solution The Bureau of the Census wants the 99% confidence interval for the mean family size to be

$$\bar{x} \pm .01$$

Hence, the maximum size of the error of estimate is to be .01, that is

$$E = .01$$

The value of z for a 99% confidence level is 2.58. The value of σ is given to be .6. Therefore, substituting all values in the formula and simplifying

$$n = \frac{z^2 \sigma^2}{E^2} = \frac{(2.58)^2 (.6)^2}{(.01)^2} = \frac{(6.6564)(.36)}{(.0001)} = 23963.04 \approx \mathbf{23,964}$$

Thus, the required sample size is 23,964. If the Bureau of the Census takes a sample of 23,964 families, computes the mean family size for this sample, and then constructs a 99% confidence interval around this sample mean, the maximum error of the estimate will be approximately .01. Note that we have rounded the final answer for the sample size to the next higher integer. This is always the case when determining the sample size. ■

EXERCISES

Concepts and Procedures

8.83 For a population, $\sigma = 10.5$.

    a. How large a sample should be selected so that the maximum error of estimate for a 99% confidence interval for μ is 2.50?

    b. How large a sample should be selected so that the maximum error of estimate for a 96% confidence interval for μ is 3.20?

8.84 For a population, $\sigma = 16.42$.

    a. What should the sample size be for a 98% confidence interval for μ to have a maximum error of estimate equal to 5.50?

    b. What should the sample size be for a 95% confidence interval for μ to have a maximum error of estimate equal to 4.25?

8.85 Determine the sample size for the estimate of μ for the following.

    a. $E = 2.3$, $\sigma = 15.40$,   confidence level = 99%

    b. $E = 4.1$, $\sigma = 23.45$,   confidence level = 95%

    c. $E = 25.9$, $\sigma = 122.25$,   confidence level = 90%

8.86 Determine the sample size for the estimate of μ for the following.

    a. $E = .17$, $\sigma = .90$,   confidence level = 99%

    b. $E = 1.45$, $\sigma = 5.82$,   confidence level = 95%

    c. $E = 5.65$, $\sigma = 18.20$,   confidence level = 90%

Applications

8.87 A researcher wants to determine a 95% confidence interval for the mean number of hours that college students who hold jobs spend at work per week. She knows that the standard deviation for hours spent at work per week by all college students who hold jobs is 3. How large a sample should the researcher select so that the estimate is within 1.5 hours of the population mean?

8.88 A company that produces detergents wants to estimate the mean amount of detergent in 64-ounce boxes at a 99% confidence level. The company knows that the standard deviation of amounts of detergent in such boxes is .2 ounces. How large a sample should the company take so that the estimate is within .04 ounces of the population mean?

8.89 A department store manager wants to estimate at a 90% confidence level the mean amount spent by all customers at this store. From an earlier study, the manager knows that the standard deviation of amounts spent by customers at this store is $27. What sample size should he choose so that the estimate is within $3 of the population mean?

8.90 A U.S. government agency wants to estimate at a 95% confidence level the mean speed for all cars traveling on Interstate Highway I-95. From a previous study, the agency knows that the standard deviation of speeds of cars traveling on this highway is 3.5 miles per hour. What sample size should the agency choose for the estimate to be within 1.5 miles per hour of the population mean?

8.7   SAMPLE SIZE DETERMINATION FOR THE ESTIMATION OF PROPORTION

Just as we did with the mean, we can also determine the sample size for estimating the population proportion p. This sample size will yield an error of estimate that may not be larger than a predetermined maximum error. By knowing the sample size that can give us the required results, we can save our scarce resources by not taking an unnecessary large sample. From Section 8.5, the maximum error E of the interval estimation of the population proportion is

$$E = z\,\sigma_{\hat{p}} = z \times \sqrt{\frac{pq}{n}}$$

By manipulating this expression algebraically, we obtain the following formula to find the required sample size given E, p, q, and z.

DETERMINING THE SAMPLE SIZE FOR THE ESTIMATION OF p

Given the confidence level and the values of p and q, the sample size that will produce a predetermined maximum error E of the confidence interval estimate of p is

$$n = \frac{z^2\,p\,q}{E^2}$$

We can observe from this formula that to find n, we need to know the values of p and q. However, the values of p and q are not known to us. In such a situation, we can choose one of the following alternatives.

1. We make the *most conservative estimate* of the sample size n by using $p = .50$ and $q = .50$. For a given E, these values of p and q will give us the largest sample size by comparison to any other pair of values of p and q because the product of $p = .50$ and $q = .50$ is greater than the product of any other pair of values for p and q.

2. We take a *preliminary sample* (of arbitrarily determined size) and calculate $\hat{p}$ and $\hat{q}$ for this sample. Then, we use these values of $\hat{p}$ and $\hat{q}$ to find n.

Examples 8–9 and 8–10 illustrate how to determine the sample size that will produce the error of estimation for the population proportion within a predetermined maximum value. Example 8–9 gives the most conservative estimate of n and Example 8–10 uses the results from a preliminary sample to determine the required sample size.

Most conservative estimate of n for the estimation of p.

EXAMPLE 8–9 Lombard Electronics Company has just installed a new machine that makes a part that is used in clocks. The company wants to estimate the proportion of these parts produced by this machine that are defective. The company manager wants this estimate to be within .02 of the population proportion for a 95% confidence level. What is the most conservative estimate of the sample size that will limit the maximum error to within .02 of the population proportion?

Solution The company manager wants the 95% confidence interval to be

$$\hat{p} \pm .02$$

Therefore,
$$E = .02$$

The value of z for a 95% confidence level is 1.96. For the most conservative estimate of the sample size, we will use $p = .50$ and $q = .50$. Hence, the required sample size is

$$n = \frac{z^2\,p\,q}{E^2} = \frac{(1.96)^2\,(.50)\,(.50)}{(.02)^2} = \mathbf{2401}$$

Thus, if the company takes a sample of 2401 parts, the estimate of p will be within .02 of the population proportion. ∎

Determining n for the estimation of p using the preliminary sample results.

EXAMPLE 8-10 Consider Example 8–9 again. Suppose a preliminary sample of 200 parts produced by this machine showed that 7% of them are defective. How large a sample should the company select so that the 95% confidence interval for p is within .02 of the population proportion?

Solution Again, the company wants the 95% confidence interval for p to be

$$\hat{p} \pm .02$$

Hence,

$$E = .02$$

The value of z for a 95% confidence level is 1.96. From the preliminary sample,

$$\hat{p} = .07 \quad \text{and} \quad \hat{q} = 1 - .07 = .93.$$

Using these values of $\hat{p}$ and $\hat{q}$ as estimates of p and q, we obtain:

$$n = \frac{z^2\,\hat{p}\,\hat{q}}{E^2} = \frac{(1.96)^2\,(.07)\,(.93)}{(.02)^2} = \frac{(3.8416)\,(.07)\,(.93)}{.0004} = 625.22 \approx \mathbf{626}$$

Thus, if the company takes a sample of 626 items, the estimate of p will be within .02 of the population proportion. However, we should note that this sample size will produce the maximum error within .02 points only if $\hat{p}$ is .07 or less for the new sample. But if $\hat{p}$ for the new sample happens to be much higher than .07, the maximum error will not be within .02. Therefore, to avoid such a situation, we may be more conservative and take a much larger sample than 626 items. ∎

EXERCISES

Concepts and Procedures

8.91 **a.** How large a sample should be selected so that the maximum error of estimate for a 99% confidence interval for p is .035 when the value of the sample proportion obtained from a preliminary sample is .26?

   b. Find the most conservative sample size that will produce the maximum error for a 99% confidence interval for p equal to .035.

8.92 **a.** How large a sample should be selected so that the maximum error of estimate for a 98% confidence interval for p is .045 when the value of the sample proportion obtained from a preliminary sample is .57?

   b. Find the most conservative sample size that will produce the maximum error for a 98% confidence interval for p equal to .045.

8.93 Determine the most conservative sample size for the estimation of the population proportion for the following.

   a. $E = .03$,   confidence level $= 99\%$
   b. $E = .04$,   confidence level $= 95\%$
   c. $E = .01$,   confidence level $= 90\%$

8.94 Determine the sample size for the estimation of the population proportion for the following where $\hat{p}$ is the sample proportion based on a preliminary sample.

   a. $E = .03$, $\hat{p} = .32$,   confidence level $= 99\%$
   b. $E = .04$, $\hat{p} = .78$,   confidence level $= 95\%$
   c. $E = .02$, $\hat{p} = .64$,   confidence level $= 90\%$

Applications

8.95 Tony's Pizza guarantees all pizza deliveries within 30 minutes of the placement of orders. An agency wants to estimate the proportion of all pizzas delivered within 30 minutes by Tony's. What is the most conservative estimate of the sample size that would limit the maximum error to within .02 of the population proportion for a 99% confidence interval?

8.96 Refer to Exercise 8.95. Assume that a preliminary study has shown that 93% of all Tony's pizzas are delivered within 30 minutes. How large should the sample size be so that the 99% confidence interval for the population proportion has a maximum error of .02?

8.97 Karasika Bakery sells cookies in boxes of 16 ounces. However, not all boxes contain exactly 16 ounces of cookies, with the weight varying slightly from box to box. The quality control department wants to know what percentage of such boxes contain less than 16 ounces of cookies. A preliminary sample of certain boxes showed that 25% of them contain less than 16 ounces of cookies. How large a sample should be taken so that the maximum error for a 95% confidence interval of the population proportion is .025?

8.98 Refer to Exercise 8.97. What is the most conservative estimate of the sample size that would limit the maximum error to be within 2.5% of the population percentage for a 95% confidence interval?

GLOSSARY

Confidence interval An interval constructed around the value of a sample statistic to estimate the corresponding population parameter.

Confidence level Confidence level, denoted by $(1 - \alpha)100\%$, states how much confidence we have that a confidence interval contains the true population parameter.

Degrees of freedom (df) The number of observations that can be chosen freely. For the estimation of μ using the t distribution, the degrees of freedom are $n - 1$.

Estimate The value of a sample statistic that is used to find the corresponding population parameter.

Estimation A procedure by which numerical value or values are assigned to a population parameter based on the information collected from a sample.

Estimator The sample statistic that is used to estimate a population parameter.

Interval estimate An interval constructed around the point estimate that is likely to contain the corresponding population parameter. Each interval estimate has a confidence level.

Maximum error of estimate The quantity that is subtracted from and added to the value of a sample statistic to obtain a confidence interval for the corresponding population parameter.

Point estimate The value of a sample statistic assigned to the corresponding population parameter.

t distribution A continuous distribution with a specific type of bell-shaped curve with its mean equal to zero and standard deviation equal to $\sqrt{df/(df - 2)}$.

KEY FORMULAS

1. **Margin of error associated with the point estimation of μ**

$$\pm 1.96 \, \sigma_{\bar{x}} \quad \text{or} \quad \pm 1.96 \, s_{\bar{x}}$$

where $\qquad \sigma_{\bar{x}} = \sigma/\sqrt{n} \quad$ and $\quad s_{\bar{x}} = s/\sqrt{n}$

2. **The $(1 - \alpha)100\%$ confidence interval for μ for a large sample ($n \geq 30$)**

$$\bar{x} \pm z\sigma_{\bar{x}} \quad \text{if } \sigma \text{ is known}$$
$$\bar{x} \pm zs_{\bar{x}} \quad \text{if } \sigma \text{ is not known}$$

where $\qquad \sigma_{\bar{x}} = \sigma/\sqrt{n} \quad$ and $\quad s_{\bar{x}} = s/\sqrt{n}$

442

CHAPTER 8 ESTIMATION OF THE MEAN AND PROPORTION

3. The $(1 - \alpha)100\%$ confidence interval for μ for a small sample ($n < 30$) when population is (approximately) normally distributed and σ is not known

$$\bar{x} \pm ts_{\bar{x}} \quad \text{where} \quad s_{\bar{x}} = s/\sqrt{n}$$

4. Margin of error associated with the point estimation of p

$$\pm 1.96\, s_{\hat{p}} \quad \text{where} \quad s_{\hat{p}} = \sqrt{\frac{\hat{p}\,\hat{q}}{n}}$$

5. The $(1 - \alpha)100\%$ confidence interval for p for a large sample

$$\hat{p} \pm zs_{\hat{p}} \quad \text{where} \quad s_{\hat{p}} = \sqrt{\frac{\hat{p}\,\hat{q}}{n}}$$

6. Maximum error E of the estimate for μ

$$E = z\sigma_{\bar{x}} \quad \text{or} \quad zs_{\bar{x}}$$

7. Required sample size for a predetermined maximum error for estimating μ

$$n = \frac{z^2\sigma^2}{E^2}$$

8. Maximum error E of the estimate for the population proportion

$$E = zs_{\hat{p}} \quad \text{where} \quad s_{\hat{p}} = \sqrt{\frac{\hat{p}\,\hat{q}}{n}}$$

9. Required sample size for a predetermined maximum error for estimating p

$$n = \frac{z^2\,p\,q}{E^2}$$

Use $p = .50$ and $q = .50$ for the most conservative estimate and the values of $\hat{p}$ and $\hat{q}$ if the estimate is to be based on a preliminary sample.

SUPPLEMENTARY EXERCISES

8.99 A company opened a new movie theater. Before setting the price of a movie ticket, the company wants to find the average price charged by other movie theaters. A random sample of 100 movie theaters taken by the company showed that the mean price of a movie ticket is $6.75 with a standard deviation of $.80.

    a. What is the point estimate of the mean price of movie tickets for all theaters? What is the margin of error associated with this estimate?

    b. Make a 95% confidence interval for the population mean μ.

8.100 A bank manager wants to know the mean amount owed on credit card accounts that become delinquent. A random sample of 100 delinquent credit card accounts taken by the manager produced a mean amount owed on these accounts equal to $2130 with a standard deviation of $578.

    a. What is the point estimate of the mean amount owed on all delinquent credit card accounts at this bank? What is the margin of error associated with this estimate?

    b. Construct a 97% confidence interval for the mean amount owed on all delinquent credit card accounts for this bank.

8.101 York Steel Corporation produces iron rings that are supplied to other companies. These rings are supposed to have a diameter of 24 inches. The machine that makes these rings does not produce each ring with a diameter of exactly 24 inches. The diameter of each of these rings varies slightly. It is known that when the machine is working properly, the rings made on this machine have a mean diameter of 24 inches. The standard deviation of the diameters of all rings produced on this machine is always equal to .06 inches. The quality control department takes a sample of 36 such rings every week, calculates the mean of diameters for these rings, and makes a 99% confidence interval for the population mean. If either the lower limit of this confidence interval is less than 23.975 inches or the upper limit of this confidence interval is greater than 24.025 inches, the machine is stopped and adjusted. A recent such sample of 36 rings produced a mean diameter of 24.015 inches. Based on this sample, can you conclude that the machine needs an adjustment? Explain.

8.102 Yunan Corporation produces bolts that are supplied to other companies. These bolts are supposed to be 4 inches long. The machine that makes these bolts does not produce each bolt exactly 4 inches long. It is known that when the machine is working properly, the mean length of the bolts made on this machine is 4 inches. The standard deviation of the lengths of all bolts produced on this machine is always equal to .04 inches. The quality control department takes a sample of 50 such bolts every week, calculates the mean length of these bolts, and makes a 98% confidence interval for the population mean. If either the upper limit of this confidence interval is greater than 4.02 inches or the lower limit of this confidence interval is less than 3.98 inches, the machine is stopped and adjusted. A recent such sample of 50 bolts produced a mean length of 3.99 inches. Based on this sample, will you conclude that the machine needs an adjustment?

8.103 A magazine wanted to estimate the mean daily cost for a room for all major hotels in the United States. The research department at the magazine took a sample of 33 hotels and obtained the following data on the daily charges.

$124	157	105	185	86	210	130	120	195	230	160
145	177	180	205	153	240	210	75	115	165	189
205	143	125	112	254	142	179	190	224	90	125

a. What is the point estimate of the mean daily cost per room for all hotels in the United States? What is the margin of error associated with this estimate?

b. Construct a 98% confidence interval for the mean daily cost per room for all hotels in the United States.

8.104 A business magazine wanted to estimate the mean leisure time per week enjoyed by managers. The research department at the magazine took a sample of 36 managers and obtained the following data on the weekly leisure time (in hours).

15	12	18	23	11	21	16	13	9	19	26	11
7	18	11	15	23	26	10	8	17	21	12	7
19	17	11	13	21	16	14	9	15	12	10	14

a. What is the point estimate of the mean leisure time per week enjoyed by all managers? What is the margin of error associated with this estimate?

b. Construct a 99% confidence interval for the mean leisure time per week enjoyed by all managers.

8.105 You are interested in estimating the mean weekly salary of all workers in your area. Briefly explain the procedure you will follow to conduct this study.

8.106 The manager of a large company wants to estimate the mean time taken by new employees to learn a process. Briefly explain the procedure he will follow to conduct this study.

8.107 A researcher wanted to estimate the mean time spent on housework per week by housewives. A random sample of 25 wives taken by this researcher showed that the mean time spent on housework is 28.9 hours a week with a standard deviation of 6.7 hours. Find a 99% confidence interval for the mean time spent on housework per week by all wives. Assume that the time spent on housework per week by all wives is (approximately) normally distributed.

8.108 According to the Metropolitan Life Insurance Company's claims data for 24 cases, the mean hospital and physician's charge for coronary bypass surgeries is $30,690 in Massachusetts (*Statistical Bulletin*, January–March 1989). Assume that these claims are representative of all such surgeries done in Massachusetts and that the standard deviation for charges for these 24 cases is $2350. Further assume that the charges for all such surgeries are normally distributed. Construct a 90% confidence interval for the mean charge for all such surgeries done in Massachusetts.

8.109 A random sample of 25 life insurance policyholders showed that the average premium they pay on their life insurance policies is $340 per year with a standard deviation of $62. Assuming that the life insurance policy premiums for all life insurance policyholders have a normal distribution, make a 95% confidence interval for the population mean μ.

8.110 A random sample of 20 major companies showed that they spent an average of $8.7 million on R&D (research and development) last year. Assuming that the expenditures incurred on R&D last year by all major companies have a normal distribution, make a 99% confidence interval for the population mean μ.

8.111 A researcher wanted to estimate the mean length of service of managers of large companies. A random sample of 16 managers produced the following data on length of service.

13	9	17	5	9	11	27	16
4	18	12	21	3	7	11	6

Make a 99% confidence interval for the mean length of service of managers of all large companies. Assume that the lengths of services of all managers of large companies have a normal distribution.

8.112 A researcher wanted to estimate the mean contributions made to charitable causes by major companies. A random sample of 18 companies produced the following data on the contributions (in millions of dollars) made by them.

1.8	.6	1.2	.3	2.6	1.9	3.4	2.6	.2
2.4	1.4	2.5	3.1	.9	1.2	2.0	.8	1.1

Make a 98% confidence interval for the mean contributions made to charitable causes by all major companies. Assume that the contributions made to charitable causes by all major companies have a normal distribution.

8.113 A computer company that recently introduced a new software product wanted to estimate the mean time taken to learn how to use this software by people who are somewhat familiar with computers. A random sample of 12 such persons was selected. The following data give the time taken (in hours) by these persons to learn how to use this software.

1.75	2.25	2.40	1.90	1.50	2.75
2.15	2.25	1.80	2.20	3.25	2.60

Construct a 95% confidence interval for the population mean. Assume that the time taken by all persons who are somewhat familiar with computers to learn how to use this software is approximately normally distributed.

8.114 A company that produces 8-ounce low-fat yogurt cups wanted to estimate the mean number of calories for such cups. A random sample of 10 such cups produced the following data on calories.

147	159	153	146	144	161	163	153	143	158

Construct a 90% confidence interval for the population mean. Assume that the number of calories for such cups of yogurt produced by this company has an approximate normal distribution.

8.115 An insurance company selected a sample of 50 auto claims filed with it and investigated those claims carefully. The company found that 18% of those claims were fraudulent.

    a. What is the point estimate of the percentage of all auto claims filed with this company that are fraudulent? What is the margin of error associated with this estimate?

    b. Make a 99% confidence interval for the percentage of all auto claims filed with this company that are fraudulent.

8.116 An auto company wanted to know the percentage of people who prefer to own safer cars (that is, cars that possess more safety features) even if they have to pay a few thousand dollars more. A random sample of 500 persons showed that 64% of them will not mind paying a few thousand dollars more to have safer cars.

  a. What is the point estimate of the percentage of all people who will not mind paying a few thousand dollars more to have safer cars? What is the margin of error associated with this estimate?

  b. Construct a 90% confidence interval for the percentage of all people who will not mind paying a few thousand dollars more to have safer cars.

8.117 A sample of 20 managers was taken and they were asked whether or not they usually take work home. The responses of these managers are given below where *yes* indicates they usually take work home and *no* means they do not.

Yes	Yes	No	No	No	Yes	No	No
No	No	Yes	Yes	No	Yes	Yes	No
No	No	No	Yes				

Make a 99% confidence interval for the percentage of all managers who take work home.

8.118 A sample of 16 CEOs was taken and they were asked whether or not they have ever been fired from their jobs. The following are the responses of these CEOs.

No	No	Yes	No	Yes	No	Yes	Yes
No	No	No	No	Yes	No	Yes	No

Construct a 97% confidence interval for the percentage of all CEOs who have ever been fired from their jobs.

8.119 A researcher wants to determine a 99% confidence interval for the mean number of hours that adults spend per week doing community service. How large a sample should the researcher select so that the estimate is within 1 hour of the population mean? Assume that the standard deviation for time spent per week doing community service by all adults is 3 hours.

8.120 An economist wants to find a 90% confidence interval for the mean sale price of houses in a state. How large a sample should she select so that the estimate is within $3500 of the population mean? Assume that the standard deviation for the sale prices of all houses in this state is $21,500.

8.121 A telephone company wants to estimate the proportion of all households who own telephone answering machines. What is the most conservative estimate of the sample size that would limit the maximum error to be within .03 of the population proportion for a 95% confidence interval?

8.122 Refer to Exercise 8.121. Assume that a preliminary sample has shown that 33% of the households in the sample own telephone answering machines. How large should the sample size be so that the 95% confidence interval for the population proportion has a maximum error of .03?

APPENDIX 8.1

RATIONALE BEHIND THE CONFIDENCE INTERVAL FORMULA FOR μ

We know from Chapter 7 that for a large sample ($n \geq 30$), the sampling distribution of $\bar{x}$ is approximately normal irrespective of the shape of the population distribution. Based on the discussion of Chapter 7, we can state that 95% of all means calculated for all possible (large) samples of the same size taken from a population are expected to fall within $1.96\sigma_{\bar{x}}$ of μ as shown in Figure 8.9.

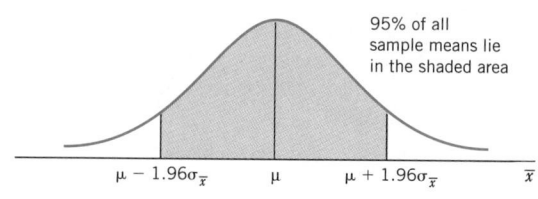

Figure 8.9

Therefore, we can state that for 95% of the (large) samples (of the same size) taken from a population, the sample means will lie in the interval $\mu - 1.96\sigma_{\bar{x}}$ to $\mu + 1.96\sigma_{\bar{x}}$. That is, for 95% of the samples

$$\mu - 1.96\,\sigma_{\bar{x}} \leq \bar{x} \leq \mu + 1.96\,\sigma_{\bar{x}}$$

By manipulating this expression algebraically, we can state that for 95% of the samples, the population mean μ will be contained in the interval $\bar{x} - 1.96\,\sigma_{\bar{x}}$ to $\bar{x} + 1.96\,\sigma_{\bar{x}}$. That is, for 95% of the samples

$$\bar{x} - 1.96\,\sigma_{\bar{x}} \leq \mu \leq \bar{x} + 1.96\,\sigma_{\bar{x}}$$

Generalizing this concept and replacing 95% by $(1 - \alpha)100\%$, we can state that for $(1 - \alpha)100\%$ of the samples the population mean μ will be contained in the interval $\bar{x} - z\sigma_{\bar{x}}$ to $\bar{x} + z\sigma_{\bar{x}}$. In other words, we can say that for $(1 - \alpha)100\%$ of the samples, the population mean μ will be contained in the interval

$$(\bar{x} - z\sigma_{\bar{x}}) \text{ to } (\bar{x} + z\sigma_{\bar{x}}) \qquad \text{or} \qquad \bar{x} \pm z\sigma_{\bar{x}}$$

This gives the $(1 - \alpha)100\%$ confidence interval for the population mean μ.

SELF-REVIEW TEST

1. Complete the following sentences using the terms *population parameter* and *sample statistic.*

    a. Estimation means assigning values to a _____ based on the value of a _____.
    b. An estimator is the _____ used to estimate a _____.
    c. The value of a _____ is called the point estimate of the corresponding _____.

2. A 95% confidence interval for μ can be interpreted to mean that if we take 100 samples of the same size and construct 100 such confidence intervals for μ then

    a. 95 of them will not include μ **b.** 95 will include μ **c.** 95 will include $\bar{x}$

3. The confidence level is denoted by

    a. $(1 - \alpha)100\%$ **b.** $100\alpha\%$ **c.** α

4. The maximum error of the estimate for μ is

    a. $z\sigma_{\bar{x}}$ (or $zs_{\bar{x}}$)   **b.** $\sigma/\sqrt{n}$ (or $s/\sqrt{n}$)   **c.** $\sigma_{\bar{x}}$ (or $s_{\bar{x}}$)

5. Which of the following assumptions is not required to use the t distribution to make a confidence interval for μ?

    a. The population from which the sample is taken is (approximately) normally distributed
    b. $n < 30$
    c. The population standard deviation σ is not known
    d. The sample size is at least 10

6. The parameter(s) of the *t* distribution is(are)

    a. *n*   **b.** degrees of freedom   **c.** μ and degrees of freedom

7. A sample of 50 packages mailed from a specific post office showed a mean mailing charge of $2.35 with a standard deviation of $.62.

    a. What is the point estimate of the population mean? What is the margin of error associated with this estimate?

    b. Construct a 99% confidence interval for the mean mailing charge for all packages mailed from this post office.

8. A sample of 25 malpractice lawsuits filed against doctors showed that the mean compensation awarded to the plaintiffs was $297,364 with a standard deviation of $74,820. Find a 99% confidence interval for the mean compensation awarded to plantiffs of all such lawsuits. Assume that the compensations awarded to plaintiffs of all such lawsuits are normally distributed.

9. More and more employees these days take their work home. According to a recent survey conducted by the U.S. Bureau of Labor Statistics, 15% of the wage and salary employees work at home at least some of the time. Assume that this survey is based on a sample of 500 employees.

    a. What is the point estimate of the population proportion? What is the margin of error associated with this estimate?

    b. Construct a 95% confidence interval for the proportion of all wage and salary employees who work at home at least some of the time.

10. A statistician is interested in estimating at a 95% confidence level the mean number of houses sold per month by all real estate agents in a large city. From an earlier study, it is known that the standard deviation of the number of houses sold per month by all real estate agents in this city is 2.1. How large a sample should be taken so that the estimate is within .65 of the population mean?

11. A company wants to estimate the proportion of all workers who hold more than one job. What is the most conservative estimate of the sample size that would limit the maximum error to be within .03 of the population proportion for a 99% confidence interval?

12. Refer to Problem 11. Assume that a preliminary study has shown that 12% of adults hold more than one job. How large a sample should be taken in this case so that the maximum error is within .03 of the population proportion for a 99% confidence interval?

13. Dr. Garcia estimated the mean stress score of managers of large companies for a random sample of 25 managers. She found the mean and standard deviation for this sample to be 6.8 (on a scale of 1 to 10) and 1.2, respectively. She used a 95% confidence level. However, she thinks that the confidence interval is too wide. How can she reduce the width of the confidence interval? Describe all possible alternatives. Which alternative do you think is best and why?

14. You want to estimate the mean number of hours that students at your college work per week. Briefly explain how you will conduct this study using a small sample. Take a sample of 12 students from your college who hold a job. Collect data on the number of hours that these students spent working last week. Then estimate the population mean. Choose your own confidence level. What assumptions will you make to estimate this population mean?

15. You want to estimate the proportion of people who are happy with their current jobs. Briefly explain how you will conduct this study. Take a sample of 35 persons and collect data on whether or not they are happy with their current jobs. Then estimate the population proportion. Choose your own confidence level.

USING MINITAB

This section describes how to use MINITAB commands to estimate the population mean for large and small samples. The current version of MINITAB does not have commands to construct a confidence interval for the population proportion directly. Although we can make such an interval for p by using the normal distribution as an approximation to the binomial distribution, we will not review that procedure here.

INTERVAL ESTIMATION OF A POPULATION MEAN: LARGE SAMPLES

The MINITAB command of Figure 8.10 gives the interval estimation of the population mean for a large sample.

Figure 8.10   MINITAB command to make a confidence interval for μ for large sample.

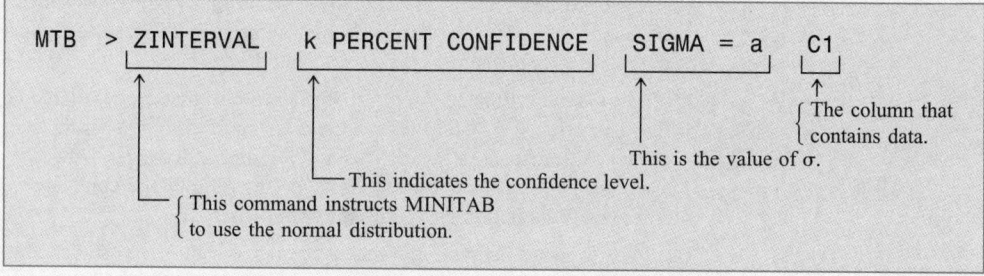

Illustration M8–1 describes the use of this procedure.

Illustration M8–1   The following data give the monthly grocery expenditures (in dollars) for 35 households selected from the 1990 Diary Survey conducted by the U.S. Bureau of Labor Statistics.

435	391	161	130	109	261	109
65	174	152	326	217	130	87
130	304	283	283	304	391	348
261	161	652	348	326	283	187
152	174	174	217	174	78	196

Using MINITAB, find a 95% confidence interval for the 1990 mean monthly grocery expenditure for all households.

(a) Assume that the standard deviation of the monthly grocery expenditures for 1990 for all households is $130.

(b) Assume that the population standard deviation is not known.

Solution

(a) For the MINITAB command ZINTERVAL 95 PERCENT CONFIDENCE SIGMA = 130 C1 in Figure 8.11, ZINTERVAL instructs MINITAB to use the normal distribution; 95 PERCENT CONFIDENCE indicates the confidence level in percentage; SIGMA = 130 represents the population standard deviation; and C1 instructs MINITAB to construct a confidence interval for μ using the value of $\bar{x}$ calculated for the data entered in column C1.

Figure 8.11   Confidence interval for μ for large sample, σ known.

```
MTB  > NOTE: MAKING CONFIDENCE INTERVAL FOR μ
MTB  > SET C1
DATA > 435       391      161      130      109      261      109
DATA >  65       174      152      326      217      130       87
DATA > 130       304      283      283      304      391      348
DATA > 261       161      652      348      326      283      187
DATA > 152       174      174    · 217      174       78      196
DATA > END
MTB  > ZINTERVAL  95 PERCENT CONFIDENCE  SIGMA = 130  C1

THE ASSUMED SIGMA = 130

          N      MEAN     STDEV    SE MEAN     95.0 PERCENT C.I.
C1       35     233.5     122.1       22.0    ( 190.4,      276.6)
```

Sample size $\quad$ $\bar{x}$ $\quad$ s $\quad$ $\sigma_{\bar{x}}$ $\quad$ 95% confidence interval for μ

In the MINITAB solution of Figure 8.11, 35 gives the number of values in the data entered in column C1; 233.5 is the sample mean $\bar{x}$ of the data in column C1; 122.1 is the standard deviation s of the sample data; 22.0 is the standard error (or standard deviation) of the sample mean $\bar{x}$ calculated as $\sigma/\sqrt{n} = 130/\sqrt{35} = 21.974 = 22.0$; 190.4 is the lower limit of the 95% confidence interval for μ; and 276.6 is the upper limit of that confidence interval.

Thus, the 95% confidence interval for the 1990 mean monthly grocery expenditure for all households based on this sample of 35 households is

$$\textbf{\$190.4 to \$276.6}$$

(b) Now, the population standard deviation σ is not known. As the sample size is large, we can use the normal distribution to make a confidence interval for μ. However, when we use the ZINTERVAL command in MINITAB, we must indicate the value of the standard deviation of the population. When σ is not known, we can first calculate the standard deviation for the data in column C1 and then use this value of s as an estimate of σ in the MINITAB command SIGMA = a. Assuming that the value of σ is not known in Illustration M8–1, we will proceed as shown in Figure 8.12 to find the confidence interval for μ.

Figure 8.12   Confidence interval for μ for large sample, σ not known.

```
MTB  > NOTE: MAKING CONFIDENCE INTERVAL FOR μ, σ UNKNOWN
MTB  > SET C1
DATA > 435       391       161       130       109       261       109
DATA >  65       174       152       326       217       130        87
DATA > 130       304       283       283       304       391       348
DATA > 261       161       652       348       326       283       187
DATA > 152       174       174       217       174        78       196
DATA > END
MTB  > STANDARD DEVIATION C1
       ST.DEV.=122.08
MTB  > ZINTERVAL  95 PERCENT CONFIDENCE  SIGMA = 122.08  C1

THE ASSUMED SIGMA = 122


         N       MEAN     STDEV    SE MEAN    95.0 PERCENT C.I.
C1      35       233.5    122.1     20.6     ( 193.0,      274.0)
```
      Sample $\bar{x}$ s $s_{\bar{x}}$ 95% confidence
      size   interval for μ

Thus, the 95% confidence interval for μ when we use s as an estimator of σ is

$193.0 to $274.0

INTERVAL ESTIMATION OF A POPULATION MEAN: SMALL SAMPLES

We know from the discussion in this chapter that when the sample size is small, the population is normally distributed, and the population standard deviation is not known, we use the t distribution to construct a confidence interval for μ. The MINITAB command for interval estimation of the population mean using the t distribution is shown in Figure 8.13.

Figure 8.13   MINITAB command for confidence interval for μ for small sample.

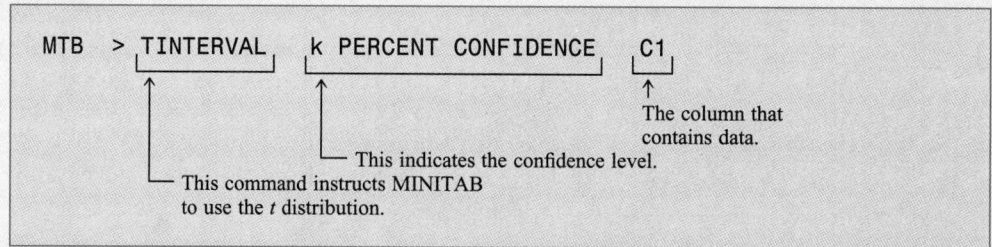

Illustration M8–2 explains the use of this procedure.

Illustration M8–2   The following data give the yearly earnings (in thousands of dollars) of 12 randomly selected households from an area.

36.75	52.43	18.82	28.45	39.50	22.65
14.30	46.75	24.48	31.70	17.25	40.27

Using MINITAB, construct a 99% confidence interval for the mean yearly earnings of all households in this area. Assume that the distribution of yearly earnings of all households in this area is approximately normal.

Solution　Figure 8.14 shows how a 99% confidence interval for μ is obtained for the given data using MINITAB.

Figure 8.14　Confidence interval for μ for small sample.

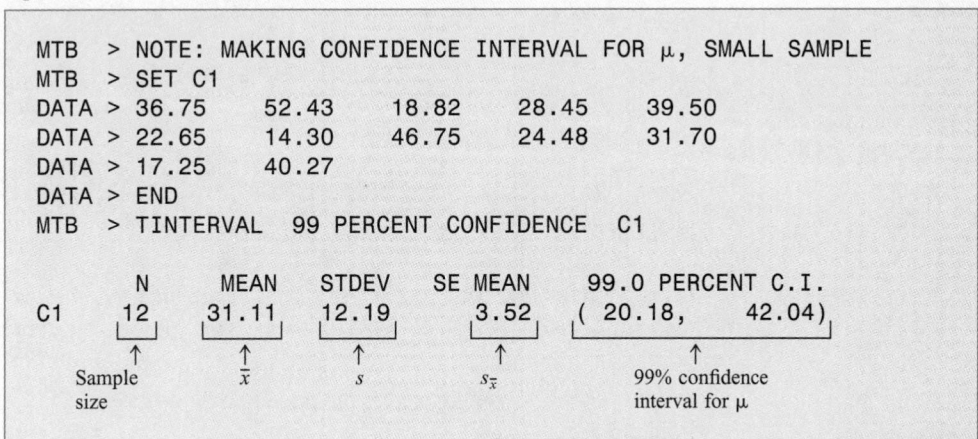

Thus, the 99% confidence interval for the yearly earnings of all households in this area based on the earnings of 12 households is

20.18 to 42.04

Because the given data are in thousands of dollars, the confidence interval can be written as

$20,180 to $42,040

COMPUTER ASSIGNMENTS

M8.1　Refer to Data Set I of Appendix A. Using MINITAB and data given in column C7, make a 95% confidence interval for the mean price–earnings ratio of all companies. Use the value of the sample standard deviation as an estimate of the population standard deviation to make the confidence interval.

M8.2　Refer to Data Set II of Appendix A on prices of various products in different cities across the country. Using the data on monthly telephone charges given in column C7, make a 99% confidence interval for the population mean μ.

M8.3　Refer to Data Set III of Appendix A. Column C3 of that data set contains incomes before taxes for 1990 for 200 households. Using MINITAB, make a 95% confidence interval for the mean income before taxes for 1990 for all households.

M8.4　Refer to Data Set III of Appendix A. Column C8 of that data set contains the entertainment expenditure incurred in 1990 by 200 households. Using MINITAB, construct a 95% confidence interval for the mean entertainment expenditure for 1990 for all households.

M8.5　The following table gives the market values (in billions of dollars) of 28 companies as of March 5, 1993 (*Source: Business Week, 1993 Special Issue*).

Company	Market Value (billions of dollars)	Company	Market Value (billions of dollars)
American Express	12.2	General Motors	27.6
Ameritech	20.0	H.J. Heinz	11.2
BankAmerica	17.9	IBM	31.5
Boeing	11.7	J.C. Penney	9.7
Caterpillar	5.9	Johnson & Johnson	26.5
CBS	2.8	McDonald's	18.7
Chase Manhattan	5.1	Microsoft	23.1
Chrysler	13.6	Nike	5.3
Cigna	4.3	Pepsico	33.1
Citicorp	9.6	Procter & Gamble	36.5
Coca-Cola	56.3	Travelers	3.6
Dow Jones	3.1	Walt Disney	24.3
Exxon	78.5	Wells Fargo	5.6
Ford Motor	24.1	Xerox	7.8

Using MINITAB, construct a 99% confidence interval for the mean market value of all companies as of March 5, 1993. Assume that the market values of all companies have a normal distribution.

M8.6   The following data give the prices (in thousands of dollars) of 16 recently sold houses in an area.

141	163	127	104	197	203	113	179
256	228	183	119	133	199	271	191

Using MINITAB, construct a 99% confidence interval for the mean price of all houses in this area. Assume that the distribution of prices of all houses in the given area is normal.

M8.7   A researcher wanted to estimate the mean contributions made to charitable causes by major companies. A random sample of 18 companies produced the following data on the contributions (in millions of dollars) made by them.

1.8	.6	1.2	.3	2.6	1.9	3.4	2.6	.2
2.4	1.4	2.5	3.1	.9	1.2	2.0	.8	1.1

Using MINITAB, make a 98% confidence interval for the mean contributions made to charitable causes by all major companies. Assume that the contributions made to charitable causes by all major companies have a normal distribution.

9

HYPOTHESIS TESTS ABOUT THE MEAN AND PROPORTION

This chapter introduces the second topic in inferential statistics: tests of hypotheses. In a test of hypothesis, we test a certain given theory or belief about a population parameter. We may want to find out, using some sample information, whether or not a given claim (or statement) about a population parameter is true. This chapter discusses how to make such tests of hypotheses about the population mean μ and the population proportion p.

As an example, a soft-drink company may claim that, on average, its cans contain 12 ounces of soda. A government agency may want to test whether or not such cans contain, on average, 12 ounces of soda. As another example, according to the U.S. Bureau of Labor Statistics, 57.3% of married women in the United States were working outside their homes in 1991. An economist may want to check if this percentage is still true for this year. In the first of these two examples we are to test a hypothesis about the population mean μ, and in the second example we are to test a hypothesis about the population proportion p.

9.1 HYPOTHESIS TESTS: AN INTRODUCTION

Why do we need to perform a test of hypothesis? Reconsider the example about soft-drink cans. Suppose we take a sample of 100 cans of the soft drink under investigation. We then find out that the mean amount of soda in these 100 cans is 11.89 ounces. Based on this result, can we state that, on average, all such cans contain less than 12 ounces of soda and that the company is lying to the public? Not until we perform a test of hypothesis can we make such an accusation. The reason is that the mean $\bar{x} = 11.89$ ounces is obtained from a sample. The difference between 12 ounces (the required average amount for the population) and 11.89 ounces (the observed average amount for the sample) may have occurred only because of the sampling error. Another sample of 100 cans may give us a mean of 12.04 ounces. Therefore, we make a test of hypothesis to find out how large the difference between 12 ounces and 11.89 ounces is and to investigate whether or not this difference has occurred as a result of chance alone. Now, if 11.89 ounces is the mean for all cans and not for only 100 cans, then we do not need to make a test of hypothesis. Instead, we can immediately state that the mean amount of soda in all such cans is less than 12 ounces. We perform a test of hypothesis only when we are making a decision about a population parameter based on the value of a sample statistic.

9.1.1 TWO HYPOTHESES

Consider a nonstatistical example of a person who has been indicted for committing a crime and is being tried in a court. Based on the available evidence, the judge or jury will make one of two possible decisions:

1. The person is not guilty.
2. The person is guilty.

At the outset of the trial, the person is presumed not guilty. The prosecutor's efforts are to prove that the person has committed the crime and, hence, is guilty. In statistics, *the person is not guilty* is called the **null hypothesis** and *the person is guilty* is called the **alternative hypothesis**. The null hypothesis is denoted by H_0 and the alternative hypothesis is denoted by H_1. In the beginning of the trial it is assumed that the person is not guilty. The null hypothesis is usually the hypothesis that is assumed to be true to begin with. The two hypotheses for the court case are written as follows (notice the colon after H_0 and H_1).

Null hypothesis: H_0: The person is not guilty

Alternative hypothesis: H_1: The person is guilty

In a statistics example, the null hypothesis states that a given claim (or statement) about a population parameter is true. Reconsider the example of the soft-drink company's claim that, on average, its cans contain 12 ounces of soda. In reality, this claim may or may not be true. However, we will initially assume that the company's claim is true (that is, the company is not guilty of cheating and lying). To test the claim of the soft-drink company, the null hypothesis will be that the company's claim is true. Let μ be the mean amount of soda in all cans. The company's claim will be true if $\mu = 12$ ounces. Thus, the null hypothesis will be written as

H_0: $\mu = 12$ ounces   (The company's claim is true)

In this example, the null hypothesis can also be written as $\mu \geq 12$ ounces because the claim of the company will still be true if the cans contain, on average, more than 12 ounces of soda.

The company will be accused of cheating the public only if the cans contain, on average, less than 12 ounces of soda. However, it will not affect the test whether we use an $=$ or a $\geq$ sign in the null hypothesis as long as the alternative hypothesis has a $<$ sign. Remember that in the null hypothesis (and in the alternative hypothesis also) we use the population parameter (such as μ or p) and not the sample statistic (such as $\bar{x}$ or $\hat{p}$).

NULL HYPOTHESIS

A null hypothesis is a claim (or statement) about a population parameter that is assumed to be true until it is declared false.

The alternative hypothesis in our statistics example will be that the company's claim is false and its soft-drink cans contain, on average, less than 12 ounces of soda, that is, $\mu < 12$ ounces. The alternative hypothesis will be written as

$$H_1: \mu < 12 \text{ ounces} \qquad \text{(The company's claim is false)}$$

ALTERNATIVE HYPOTHESIS

An alternative hypothesis is a claim about a population parameter that will be true if the null hypothesis is false.

Let us return to the example of the court trial. The trial begins with the assumption that the null hypothesis is true, that is, the person is not guilty. The prosecutor assembles all the possible evidence and presents it in the court to prove that the null hypothesis is false and the alternative hypothesis is true (that is, the person is guilty). In the case of our statistics example, the information obtained from a sample will be used as evidence to decide whether or not the claim of the company is true. In the court case, the decision made by the judge (or jury) depends on the amount of evidence presented by the prosecutor. At the end of the trial, the judge (or jury) will consider whether or not the evidence presented by the prosecutor is sufficient to declare the person guilty. The amount of evidence that will be considered to be sufficient to declare the person guilty depends on the discretion of the judge (or jury).

9.1.2   REJECTION AND NONREJECTION REGIONS

In Figure 9.1, which represents the court case, the point marked ''0'' indicates that there is no evidence against the person being tried. The farther the point is to the right on the horizontal axis, the more convincing the evidence is that the person has committed the crime. We have arbitrarily marked a point C on the horizontal axis. Let us assume that a judge (or jury) considers any amount of evidence to the right of point C to be sufficient and any amount of evidence to the left of C to be insufficient to declare the person guilty. Point C is called the **critical value** or **critical point** in statistics. If the amount of evidence presented by the prosecutor falls in the area to the left of point C, the verdict will reflect that there is not enough evidence to declare the person guilty. Consequently, the accused person will be de-clared not guilty. In statistics, this decision is stated as *do not reject H_0*. It is equivalent to saying that there is not enough evidence to declare the null hypothesis false. The area to the left of point C is called the *nonrejection region*, that is, this is the region where the null

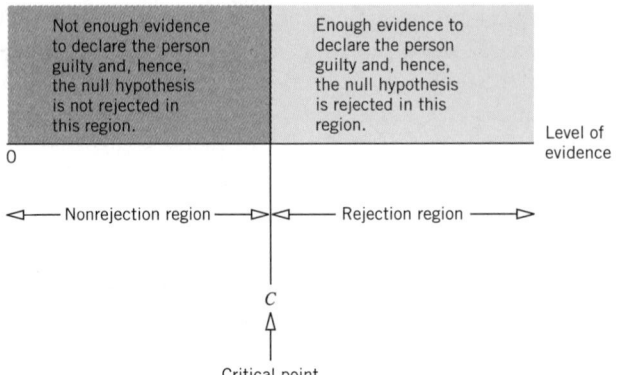

Figure 9.1 Nonrejection and rejection regions for the court case.

hypothesis is not rejected. However, if the amount of evidence falls in the area to the right of point C, the verdict will be that there is sufficient evidence to declare the person guilty. In statistics, this decision is stated as *reject H_0* or *the null hypothesis is false*. Rejecting H_0 is equivalent to saying that *the alternative hypothesis is true*. The area to the right of point C is called the *rejection region*, that is, this is the region where the null hypothesis is rejected.

9.1.3 TWO TYPES OF ERRORS

We all know that a court's verdict is not always correct. If a person is declared guilty at the end of a trial, there are two possibilities.

1. The person has *not* committed the crime but is declared guilty (because of what may be false evidence).
2. The person *has* committed the crime and is rightfully declared guilty.

In the first case, the court has made an error by punishing an innocent person. In statistics, this kind of error is called a **Type I** or an **α *(alpha)* error**. In the second case, because the guilty person has been punished, the court has made the correct decision. The second row in the shaded portion of Table 9.1 shows these two cases. The two columns of Table 9.1, corresponding to *the person is not guilty* and *the person is guilty*, give the two actual situations. Which one of these is true is known only to the person being tried. The two rows in this table, corresponding to *the person is not guilty* and *the person is guilty*, show the two possible court decisions.

Table 9.1

		Actual Situation	
		The Person Is Not Guilty	**The Person Is Guilty**
Court's decision	The person is not guilty	Correct decision	Type II or β error
	The person is guilty	Type I or α error	Correct decision

In our statistics example, a Type I error will occur when H_0 is actually true (that is, the cans do contain, on average, 12 ounces of soda), but it just happens that we draw a sample with a mean that is well below 12 ounces and we wrongfully reject the null hypothesis H_0.

The value of α, called the **significance level** of the test, represents the probability of making a Type I error. In other words, α is the probability of rejecting the null hypothesis H_0 when in fact it is true.

TYPE I ERROR

A Type I error occurs when a true null hypothesis is rejected. The value of α represents the probability of committing this type of error, that is,

$$\alpha = P(H_0 \text{ is rejected} \mid H_0 \text{ is true})$$

The value of α represents the significance level of the test.

The size of the rejection region in a statistics problem of a test of hypothesis depends on the value assigned to α. In a test of hypothesis, we usually assign a value to α before making the test. Although any value can be assigned to α, the commonly used values of α are .01, .025, .05, and .10. Usually the value assigned to α does not exceed .10 (or 10%).

Now, suppose that in the court trial case the person is declared not guilty at the end of the trial. Such a verdict does not indicate that the person has indeed *not* committed the crime. It is possible that the person is guilty but there is not enough evidence to prove the guilt. Consequently, in this situation there are again two possibilities.

1. The person has *not* committed the crime and is declared not guilty.
2. The person *has* committed the crime but, *because of the lack of enough evidence*, is declared not guilty.

In the first case, the court's decision is correct. But in the second case the court has committed an error by setting a guilty person free. In statistics, this type of error is called a **Type II** or a **β** (the Greek letter *beta*) **error**. These two cases are shown in the first row of the shaded portion of Table 9.1.

In our statistics example, a Type II error will occur when the null hypothesis H_0 is actually false (that is, the soda contained in all cans, on average, is less than 12 ounces), but it happens by chance that we draw a sample with a mean that is close to or larger than 12 ounces and we wrongfully conclude *do not reject H_0*. The value of β represents the probability of making a Type II error. It represents the probability that H_0 is not rejected when actually H_0 is false.

TYPE II ERROR

A Type II error occurs when a false null hypothesis is not rejected. The value of β represents the probability of committing a Type II error, that is,

$$\beta = P(H_0 \text{ is not rejected} \mid H_0 \text{ is false})$$

The value of $1 - \beta$ is called the **power of the test**. It represents the probability of not making a Type II error.

The two types of errors that occur in tests of hypotheses depend on each other. We cannot lower the values of α and β simultaneously for a test of hypothesis for a fixed sample size. Lowering the value of α will raise the value of β, and lowering the value of β will raise the value of α. However, we can decrease both α and β simultaneously by increasing the sample size. The computation of β and the relationship between α and β are discussed in Section 9.3.

Table 9.2, which is similar to Table 9.1, is written for the statistics problem of a test of hypothesis. In Table 9.2 *the person is not guilty* is replaced by H_0 *is true, the person is guilty* by H_0 *is false*, and the *court's decision* by *decision*.

Table 9.2

		Actual Situation	
		H_0 **is true**	H_0 **is false**
Decision	Do not reject H_0	Correct decision	Type II or β error
	Reject H_0	Type I or α error	Correct decision

9.1.4 TAILS OF A TEST

The statistical hypothesis-testing procedure is similar to the trial of a person in the court but with two major differences. The first major difference is that in a statistics test of hypothesis, the partition of the total region into rejection and nonrejection regions is not arbitrary. Instead, it depends on the value assigned to α (Type I error). As mentioned earlier, α is also called the significance level of the test.

The second major difference relates to the rejection region. In the court case, the rejection region is on the right side of the critical point, as shown in Figure 9.1. However, in statistics, the rejection region for a hypothesis-testing problem can be on both sides with the nonrejection region in the middle, or it can be on the left side or on the right side of the nonrejection region. These possibilities are explained in the next three parts of this section. A test with two rejection regions is called a **two-tailed test**, and a test with one rejection region is called a **one-tailed test**. The one-tailed test is called a **left-tailed test** if the rejection region is in the left tail of the distribution curve, and it is called a **right-tailed test** if the rejection region is in the right tail of the distribution curve.

TAILS OF THE TEST

A two-tailed test has rejection regions in both tails, a left-tailed test has the rejection region in the left tail, and a right-tailed test has the rejection region in the right tail of the distribution curve.

A Two-tailed Test

According to the U.S. Bureau of the Census, the mean family size in the United States was 3.17 in 1991. An economist wants to check whether or not this mean has changed since 1991. The key word here is *changed*. The mean family size has changed if it has either increased or decreased during the period since 1991. This is an example of a two-tailed test. Let μ be the current mean family size for all families. The two possible decisions are

1. The mean family size has not changed, that is, $\mu = 3.17$.
2. The mean family size has changed, that is, $\mu \neq 3.17$.

We write the null and alternative hypotheses for this test as

$$H_0: \mu = 3.17 \qquad \text{(The mean family size has not changed)}$$

$$H_1: \mu \neq 3.17 \qquad \text{(The mean family size has changed)}$$

Whether a test is two-tailed or one-tailed is determined by the sign in the alternative hypothesis. If the alternative hypothesis has a *not equal to* ($\neq$) sign, as in this example, it is a two-tailed test. As shown in Figure 9.2, a two-tailed test has two rejection regions, one in each tail of the distribution curve. Figure 9.2 shows the sampling distribution of $\bar{x}$ for a large sample. Assuming H_0 is true, $\bar{x}$ has a normal distribution with its mean equal to 3.17 (the value of μ in H_0). In Figure 9.2, the area of each of the two rejection regions is $\alpha/2$ and the total area of both rejection regions is α (the significance level). As shown in this figure, a two-tailed test of hypothesis has two critical values that separate the two rejection regions from the nonrejection region. We will reject H_0 if the value of $\bar{x}$ obtained from the sample falls in either of the two rejection regions. We will not reject H_0 if the value of $\bar{x}$ lies in the nonrejection region. By rejecting H_0, we are saying that the difference between the value of μ stated in H_0 and the value of $\bar{x}$ obtained from the sample is too large to have occurred because of the sampling error alone. Consequently, this difference is real. By not rejecting H_0, we are saying that the difference between the value of μ stated in H_0 and the value of $\bar{x}$ obtained from the sample is small and it may have occurred because of the sampling error alone.

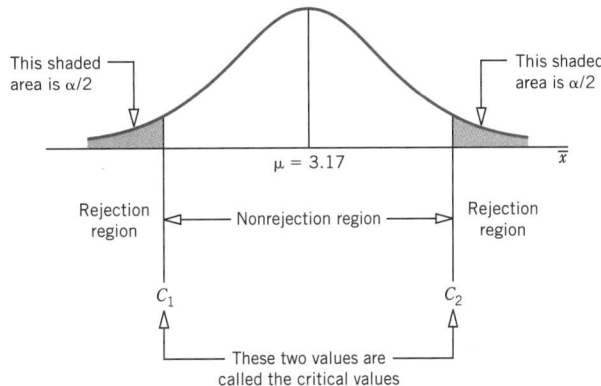

Figure 9.2 A two-tailed test.

A Left-tailed Test

Reconsider the example of mean amount of soda in all soft-drink cans produced by a company. The company claims that these cans, on average, contain 12 ounces of soda. However, if these cans contain less than the claimed amount of soda, then the company can be accused of cheating. Suppose a consumer agency wants to test whether the mean amount of soda per can is less than 12 ounces. Note that the key phrase this time is *less than*, which indicates a left-tailed test. Let μ be the mean amount of soda in all cans. The two possible decisions are

1. The mean amount of soda in all cans is not less than 12 ounces, that is, $\mu = 12$ ounces.
2. The mean amount of soda in all cans is less than 12 ounces, that is, $\mu < 12$ ounces.

The null and alternative hypotheses for this test are written as

$$H_0: \mu = 12 \text{ ounces} \qquad \text{(The mean is not less than 12 ounces)}$$

$$H_1: \mu < 12 \text{ ounces} \qquad \text{(The mean is less than 12 ounces)}$$

In this case, we can also write the null hypothesis as H_0: $\mu \geq 12$. This will not affect the result of the test as long as the sign in H_1 is *less than* ($<$).

When the alternative hypothesis has a *less than* ($<$) sign, as in this case, the test is always left-tailed. In a left-tailed test, the rejection region is always in the left tail of the distribution curve, as shown in Figure 9.3, and the area of this rejection region is equal to α (the significance level). We can observe from this figure that there is only one critical value in a left-tailed test.

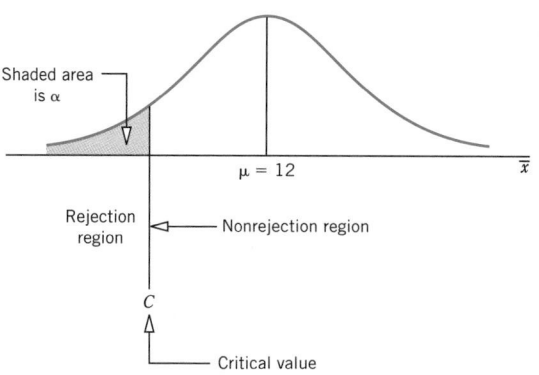

Figure 9.3   A left-tailed test.

Assuming H_0 is true, $\bar{x}$ has a normal distribution for a large sample with its mean equal to 12 ounces (the value of μ in H_0). We will reject H_0 if the value of $\bar{x}$ obtained from the sample falls in the rejection region; we will not reject H_0 otherwise.

A Right-tailed Test

To illustrate the third case, according to the U.S. Bureau of the Census, the mean income of all households in the United States was $37,922 in 1991. Suppose we want to test if the current mean income of all households in the United States is higher than $37,922. The key phrase in this case is *higher than*, which indicates a right-tailed test. Let μ be the current mean income of all households in the United States. The two possible decisions this time are

1.   The current mean income of all households is not higher than $37,922, that is, $\mu = \$37,922$.

2.   The current mean income of all households is higher than $37,922, that is, $\mu > \$37,922$.

We write the null and alternative hypotheses for this test as

H_0: $\mu = \$37,922$ (The current mean income is not higher than $37,922)

H_1: $\mu > \$37,922$ (The current mean income is higher than $37,922)

In this case, we can also write the null hypothesis as H_0: $\mu \leq \$37,922$, which states that the current mean income is either equal to or less than $37,922. Again, the result of the test will not be affected whether we use an *equal to* ($=$) or a *less than or equal to* ($\leq$) sign in H_0 as long as the alternative hypothesis has a *greater than* ($>$) sign.

When the alternative hypothesis has a *greater than* ($>$) sign, the test is always right-tailed. As shown in Figure 9.4, in a right-tailed test, the rejection region is in the right tail of the distribution curve. The area of this rejection region is equal to α, the significance level. Like a left-tailed test, a right-tailed test has only one critical value.

Again, assuming H_0 is true, $\bar{x}$ has a normal distribution for a large sample with its mean

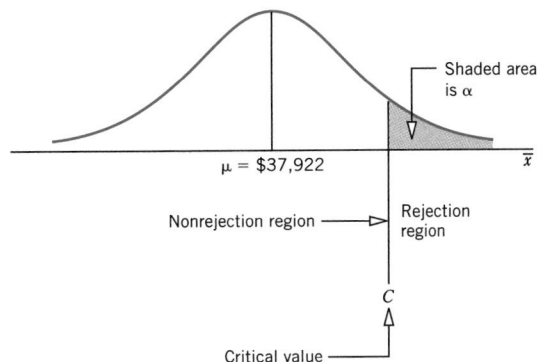

Figure 9.4   A right-tailed test.

equal to \$37,922 (the value of μ in H_0). We will reject H_0 if the value of $\bar{x}$ obtained from the sample falls in the rejection region. Otherwise, we will not reject H_0.

Table 9.3 summarizes the foregoing discussion about the relationship between the signs in H_0 and H_1 and the tails of a test.

Table 9.3

	Two-tailed Test	Left-tailed Test	Right-tailed Test
Sign in the null hypothesis H_0	$=$	$=$ or $\geq$	$=$ or $\leq$
Sign in the alternative hypothesis H_1	$\neq$	$<$	$>$
Rejection region	In both tails	In the left tail	In the right tail

Note that the null hypothesis always has an *equal to* ($=$) or a *less than or equal to* ($\leq$) or a *greater than or equal to* ($\geq$) sign and the alternative hypothesis always has a *not equal to* ($\neq$) or a *greater than* ($>$) or a *less than* ($<$) sign.

A test of hypothesis involves five steps, which are listed below.

STEPS OF A TEST OF HYPOTHESIS

A statistical test of hypothesis procedure contains the following five steps.

1. State the null and alternative hypotheses
2. Select the distribution to use
3. Determine the rejection and nonrejection regions
4. Calculate the value of the test statistic
5. Make a decision

With the help of examples, these steps will be described in the next section.

EXERCISES

Concepts and Procedures

9.1 Briefly explain the meaning of each of the following terms.

    a.   Null hypothesis   b.   Alternative hypothesis
    c.   Critical point(s)   d.   Significance level
    e.   Nonrejection region   f.   Rejection region
    g.   Tails of a test   h.   Two types of errors

9.2 What are the four possible outcomes for a test of hypothesis? Show these outcomes by writing a table. Briefly describe the Type I and Type II errors.

9.3 Explain how the tails of a test depend on the sign in the alternative hypothesis. Describe the signs in the null and alternative hypotheses for a two-tailed, a left-tailed, and a right-tailed test, respectively.

9.4 Explain which of the following is a two-tailed test, a left-tailed test, or a right-tailed test.

    a. $H_0: \mu = 45,$ $H_1: \mu > 45$
    b. $H_0: \mu = 23,$ $H_1: \mu \neq 23$
    c. $H_0: \mu \geq 75,$ $H_1: \mu < 75$

Show the rejection and nonrejection regions for each of these cases by drawing a sampling distribution curve for the sample mean, assuming that the sample size is large in each case.

9.5 Explain which of the following is a two-tailed test, a left-tailed test, or a right-tailed test.

    a. $H_0: \mu = 12,$ $H_1: \mu < 12$
    b. $H_0: \mu \leq 85,$ $H_1: \mu > 85$
    c. $H_0: \mu = 33,$ $H_1: \mu \neq 33$

Show the rejection and nonrejection regions for each of these cases by drawing a sampling distribution curve for the sample mean, assuming that the sample size is large in each case.

9.6 Which of the two hypotheses (null and alternative) is initially assumed to be true in a test of hypothesis?

9.7 Consider $H_0: \mu = 20$ versus $H_1: \mu < 20$.

    a.   What type of error would you make if the null hypothesis is actually false and you fail to reject it?
    b.   What type of error would you make if the null hypothesis is actually true and you reject it?

9.8 Consider $H_0: \mu = 55$ versus $H_1: \mu \neq 55$.

    a.   What type of error would you make if the null hypothesis is actually false and you fail to reject it?
    b.   What type of error would you make if the null hypothesis is actually true and you reject it?

Applications

9.9 Write the null and alternative hypotheses for each of the following examples. Determine if each is a case of a two-tailed, a left-tailed, or a right-tailed test.

    a.   To test whether or not the mean price of houses in Connecticut is greater than $143,000
    b.   To test if the mean number of hours spent working per week by college students who hold jobs is different from 15 hours
    c.   To test whether the mean life of a particular brand of auto batteries is less than 45 months
    d.   To test if the mean amount of time taken by all workers to do a certain assembly job is more than 35 minutes
    e.   To test if the mean age of all managers of companies is different from 50 years

9.10 Write the null and alternative hypotheses for each of the following examples. Determine if each is a case of a two-tailed, a left-tailed, or a right-tailed test.

    a.   To test if the mean amount of time spent per week watching sports on television by adult males is different from 9.5 hours
    b.   To test if the mean amount of money spent by all customers at a supermarket is less than $85

c.   To test whether the mean starting salary of college graduates is higher than $29,000 per year
d.   To test if the mean rent paid by all tenants in Boston is different from $1000 a month
e.   To test whether the mean net weight of all boxes of a certain brand of cereal is less than 20 ounces

9.2   HYPOTHESIS TESTS ABOUT A POPULATION MEAN: LARGE SAMPLES

From the central limit theorem discussed in Chapter 7, the sampling distribution of $\bar{x}$ is approximately normal for large samples ($n \geq 30$). Consequently, whether or not σ is known, the normal distribution is used to test hypotheses about the population mean when a sample size is large.

TEST STATISTIC

In tests of hypotheses about μ for large samples, the random variable

$$z = \frac{\bar{x} - \mu}{\sigma_{\bar{x}}} \quad \text{or} \quad \frac{\bar{x} - \mu}{s_{\bar{x}}}$$

is called the test statistic. The test statistic can be defined as a rule or criterion that is used to make the decision whether or not to reject the null hypothesis.

At the end of Section 9.1, it was mentioned that a test of hypothesis procedure involves the following five steps.

1.   State the null and alternative hypotheses
2.   Select the distribution to use
3.   Determine the rejection and nonrejection regions
4.   Calculate the value of the test statistic
5.   Make a decision

Examples 9–1 through 9–3 illustrate the use of these five steps to perform tests of hypotheses about the population mean μ. Example 9–1 is concerned with a two-tailed test and Examples 9–2 and 9–3 describe one-tailed tests.

Conducting a two-tailed test of hypothesis about μ for a large sample.

EXAMPLE 9–1   When a machine that is used to make bolts at Sabana Steel Corporation is working properly, the mean length of these bolts is 2.5 inches. However, from time to time this machine falls out of alignment and produces bolts that have a mean length of either less than 2.5 inches or more than 2.5 inches. When this happens, the process is stopped and the machine is adjusted. To check whether or not the machine is producing bolts with a mean length of 2.5 inches, the quality control department at the company takes a sample of bolts each week and makes a test of hypothesis. One such random sample of 49 bolts produced a mean length of 2.49 inches and a standard deviation of .021 inches. Using the 5% significance level, can we conclude that the machine needs to be adjusted?

Solution   Let μ be the mean length of all bolts made on this machine and $\bar{x}$ be the corresponding mean for the sample. From the given information,

$$n = 49, \quad \bar{x} = 2.49 \text{ inches}, \quad \text{and} \quad s = .021 \text{ inches}$$

The mean length of all bolts is supposed to be 2.5 inches. The significance level α is .05. That is, the probability of rejecting the null hypothesis when it actually is true should not exceed .05. This is the probability of making a Type I error. We perform the test of hypothesis using the five steps as follows.

Step 1. *State the null and alternative hypotheses*

Notice that we are testing to find whether or not the machine needs to be adjusted. The machine will not need an adjustment if the mean length of the bolts produced is equal to 2.5 inches, that is, $\mu = 2.5$ inches. The machine will need an adjustment if the mean length of these bolts is either less than 2.5 inches or more than 2.5 inches, which can be written as $\mu \neq 2.5$ inches. We write the null and alternative hypotheses as follows.

$$H_0: \mu = 2.5 \qquad \text{(The machine does not need an adjustment)}$$

$$H_1: \mu \neq 2.5 \qquad \text{(The machine needs an adjustment)}$$

Step 2. *Select the distribution to use*

Because the sample size is large ($n > 30$), the sampling distribution of $\bar{x}$ is (approximately) normal. Consequently, we use the normal distribution to make the test.

Step 3. *Determine the rejection and nonrejection regions*

The significance level is .05. The $\neq$ sign in the alternative hypothesis indicates that the test is two-tailed with two rejection regions, one in each tail of the normal distribution curve of $\bar{x}$. Because the total area of both rejection regions is .05 (the significance level), the area of the rejection region in each tail is .025, that is,

$$\text{Area in each tail} = \alpha/2 = .05/2 = .025$$

These areas are shown in Figure 9.5. Two critical points in this figure separate the two rejection regions from the nonrejection region. Next we find the z values for the two critical points using the area of the rejection region. To find the z values for these critical points, we first find the area between the mean and one of the critical points. We obtain this area by subtracting .025 (the area in each tail) from .5, which gives .4750. Next we look for .4750 in the standard normal distribution table, Table VII of Appendix B. The value of z for .4750 is 1.96. Hence, the z values of the two critical points, as shown in Figure 9.5, are -1.96 and 1.96.

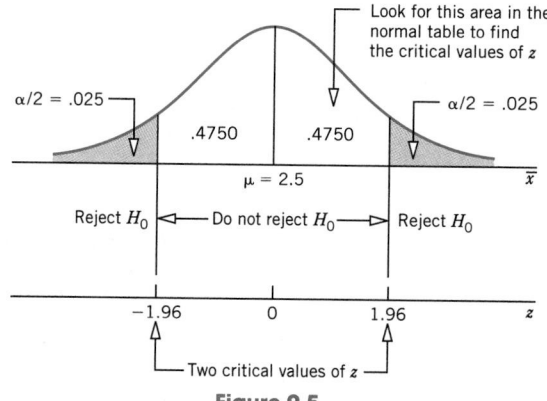

Figure 9.5

Step 4. *Calculate the value of the test statistic*

The decision to reject or not to reject the null hypothesis will depend on whether the evidence from the sample falls in the rejection or nonrejection region. If the value of the sample mean $\bar{x}$ falls in either of the two rejection regions, we reject H_0. Otherwise, we do not reject H_0. The value of $\bar{x}$ obtained from the sample is called the *observed value of* $\bar{x}$. To locate the position of $\bar{x} = 2.49$ on the sampling distribution curve of $\bar{x}$ in Figure 9.5, we first calculate the z value for $\bar{x} = 2.49$. This is called the *value of the test statistic*. Then, we compare the value of the test statistic with the two critical values of z, -1.96 and 1.96, shown in Figure 9.5. If the value of the test statistic is between -1.96 and 1.96, we do not reject H_0. If the value of the test statistic is either greater than 1.96 or less than -1.96, we reject H_0.

CALCULATING THE VALUE OF THE TEST STATISTIC

The value of the test statistic z for $\bar{x}$ for a test of hypothesis about μ for a large sample is computed as follows.

$$z = \frac{\bar{x} - \mu}{\sigma_{\bar{x}}} \qquad \text{if } \sigma \text{ is known}$$

$$z = \frac{\bar{x} - \mu}{s_{\bar{x}}} \qquad \text{if } \sigma \text{ is not known}$$

where $\qquad \sigma_{\bar{x}} = \sigma/\sqrt{n} \qquad$ and $\qquad s_{\bar{x}} = s/\sqrt{n}$

The value of z calculated for $\bar{x}$ using the above formula is also called the **observed value of** z.

The value of $\bar{x}$ from the sample is 2.49. As σ is not known, we calculate the z value using $s_{\bar{x}}$ as follows.

$$s_{\bar{x}} = s/\sqrt{n} = .021/\sqrt{49} = .003$$

From H_0

$$z = \frac{\bar{x} - \mu}{s_{\bar{x}}} = \frac{2.49 - 2.5}{.003} = -3.33$$

The value of μ in the calculation of the z value is substituted from the null hypothesis. The value of $z = -3.33$ calculated for $\bar{x}$ is called the *computed value of the test statistic z*. This is the value of z that corresponds to the value of $\bar{x}$ observed from the sample. It is also called the *observed value of z*.

Step 5. *Make a decision*

In the final step we make a decision based on the location of the value of the test statistic z computed for $\bar{x}$ in Step 4. This value of $z = -3.33$ is less than the critical value of $z = -1.96$, and it falls in the rejection region in the left tail. Hence, we reject H_0 and conclude that based on the sample information, it appears that the mean length of all such bolts produced on this machine is not equal to 2.5 inches. Therefore, the machine needs to be adjusted.

By rejecting the null hypothesis we are stating that the difference between the sample mean $\bar{x} = 2.49$ and the hypothesized value of the population mean $\mu = 2.5$ is too large and may not have occurred because of chance or sampling error alone. This difference seems to be real and, hence, the mean length of bolts is different from 2.5 inches. Note that the rejection of the null hypothesis does not necessarily indicate that the mean length of bolts is definitely different from 2.5 inches. It simply indicates that there is strong evidence (from the sample) that the mean length of bolts is not equal to 2.5 inches. There is a possibility that the mean length of bolts is equal to 2.5 inches but, by the luck of the draw, we selected a sample with a mean that is too far from the required mean of 2.5 inches. If so, we have wrongfully rejected the null hypothesis H_0. This is a Type I error and its probability is .05 in this example. ■

Making a right-tailed test of hypothesis about μ for a large sample.

EXAMPLE 9–2   According to the National Association of Realtors, the mean sales price of existing single-family homes in the United States was \$128,400 in 1991 (*Home Sales*, 6(9), September 1992). A random sample of 500 such homes that were recently sold gave a mean sales price of \$137,670 with a standard deviation of \$23,700. Test at the 1% significance level if the current mean sales price of such homes is greater than \$128,400.

Solution   Let μ be the current mean sales price of all existing single-family homes in the United States and $\bar{x}$ be the corresponding mean for the sample. From the given information,

$$n = 500, \qquad \bar{x} = \$137,670, \qquad \text{and} \qquad s = \$23,700$$

The significance level is $\alpha = .01$.

Step 1.   *State the null and alternative hypotheses*

We are to test if the current mean sales price of existing single-family homes is greater than \$128,400. The null and alternative hypotheses are

H_0: $\mu = \$128,400$ (The current mean is not greater than \$128,400)

H_1: $\mu > \$128,400$ (The current mean is greater than \$128,400)

Step 2.   *Select the distribution to use*

Because the sample size is large ($n > 30$), the sampling distribution of $\bar{x}$ is (approximately) normal. Consequently, we use the normal distribution to make the test.

Step 3.   *Determine the rejection and nonrejection regions*

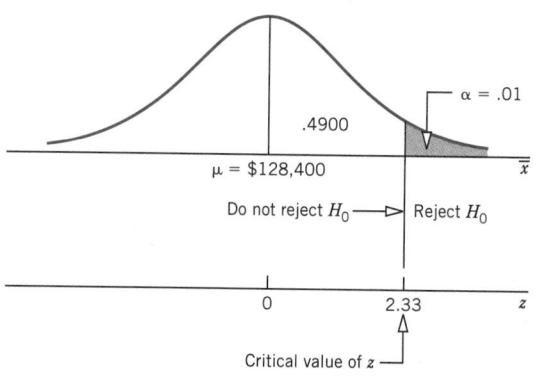

Figure 9.6

The significance level is .01. The $>$ sign in the alternative hypothesis indicates that the test is right-tailed with its rejection region in the right tail of the sampling distribution curve of $\bar{x}$. Because there is only one rejection region, its area is $\alpha = .01$. As shown in Figure 9.6, the critical value of z, obtained from Table VII of Appendix B for .4900, is approximately 2.33.

Step 4. *Calculate the value of the test statistic*

The value of the test statistic z for $\bar{x} = \$137,670$ is computed as follows.

$$s_{\bar{x}} = s/\sqrt{n} = 23,700/\sqrt{500} = \$1059.8962$$

$$z = \frac{\bar{x} - \mu}{s_{\bar{x}}} = \frac{137,670 - \overset{\text{From } H_0}{128,400}}{1059.8962} = 8.75$$

Step 5. *Make a decision*

Because the value of the test statistic $z = 8.75$ is larger than the critical value of $z = 2.33$ and it falls in the rejection region, we reject H_0. Consequently, we can state that the sample mean $\bar{x} = \$137,670$ is too far from the hypothesized population mean $\mu = \$128,400$. The difference between the two may not be attributed to chance or sampling error alone. Therefore, the current mean sales price of existing single-family homes in the United States is greater than $128,400. ■

Making a left-tailed test of hypothesis about μ for a large sample.

EXAMPLE 9–3 Because couples are deciding to have fewer children, the family size in the United States has declined continuously during the past few decades. According to the U.S. Bureau of the Census, the mean family size was 3.17 in 1991. An economist wanted to check if the current mean family size is less than 3.17. A sample of 900 families taken this year by this economist produced a mean family size of 3.13 with a standard deviation of .7. Using the .025 significance level, can we conclude that the mean family size has declined since 1991?

Solution Let μ be the current mean size of all families and $\bar{x}$ be the mean family size for the sample. From the given information,

$$n = 900, \quad \bar{x} = 3.13, \quad \text{and} \quad s = .7$$

The mean family size for 1991 is given to be 3.17. The significance level α is .025.

Step 1. *State the null and alternative hypotheses*

Notice that we are testing for a *decline* in the mean family size. The null and alternative hypotheses are written as follows.

$$H_0: \mu = 3.17 \quad \text{(The mean family size has not declined)}$$
$$H_1: \mu < 3.17 \quad \text{(The mean family size has declined)}$$

Step 2. *Select the distribution to use*

Because the sample size is large ($n > 30$), the sampling distribution of $\bar{x}$ is (approximately) normal. Consequently, we use the normal distribution to make the test.

Step 3. *Determine the rejection and nonrejection regions*

The significance level is .025. The $<$ sign in the alternative hypothesis indicates that the test is left-tailed with the rejection region in the left tail of the sampling distribution curve of $\bar{x}$. The critical value of z, obtained from the normal table for .4750 is -1.96, as shown in Figure 9.7.

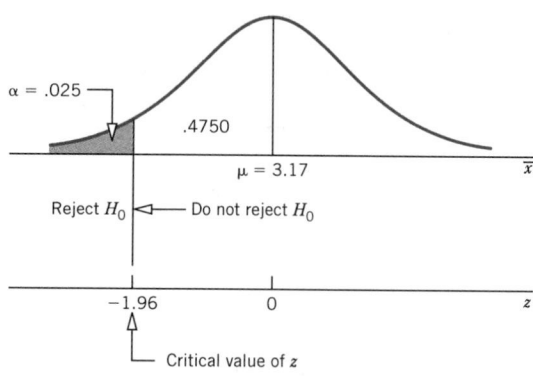

Figure 9.7

Step 4. *Calculate the value of the test statistic*

The value of the test statistic z for $\bar{x} = 3.13$ is calculated as follows.

$$s_{\bar{x}} = s/\sqrt{n} = .7/\sqrt{900} = .0233$$

$$z = \frac{\bar{x} - \mu}{s_{\bar{x}}} = \frac{3.13 - 3.17}{.0233} = -1.72 \quad \text{—From } H_0$$

Step 5. *Make a decision*

The value of the test statistic $z = -1.72$ is greater than the critical value of $z = -1.96$ and it falls in the nonrejection region. As a result, we fail to reject H_0. Consequently, we can state that based on the sample information, it appears that the mean family size has not declined since 1991. Note that we are not concluding that the mean family size has definitely not declined. By not rejecting the null hypothesis, we are saying that the information obtained from the sample is not strong enough to reject the null hypothesis and to conclude that the family size has declined since 1991. ■

In studies published in various journals, authors usually use the terms *significantly different* and *not significantly different* when deriving conclusions based on hypothesis tests. These terms are short versions of the terms *statistically significantly different* and *statistically not significantly different*. The statement *significantly different* means that the difference between the observed value of the sample mean $\bar{x}$ and the hypothesized value of the population mean μ is so large that it probably did not occur because of the sampling error alone. Consequently, the null hypothesis is rejected. In other words, the difference between $\bar{x}$ and μ is statistically significant. Thus, the statement *significantly different* is equivalent to saying that the *null hypothesis is rejected*. In Example 9–2, we can state as a conclusion that the observed value of $\bar{x} = \$137,670$ is significantly different from the hypothesized value of $\mu = \$128,400$. That is, the current mean sales price of existing single-family homes is significantly different from $128,400.

On the other hand, the statement *not significantly different* means that the difference between the observed value of the sample mean $\bar{x}$ and the hypothesized value of the population mean μ is so small that it may have occurred just because of chance. Consequently, the null hypothesis is not rejected. Thus, the statement *not significantly different* is equivalent to saying that we *fail to reject the null hypothesis*. In Example 9–3, we can state as a conclusion that the observed value of $\bar{x} = 3.13$ is not significantly different from the hypothesized value of $\mu = 3.17$. In other words, the current mean family size does not seem to be significantly different from 3.17.

EXERCISES

Concepts and Procedures

9.11 What are the five steps of a test of hypothesis? Explain briefly.

9.12 What does the level of significance represent in a test of hypothesis? Explain.

9.13 By rejecting the null hypothesis in a test of hypothesis example, are you stating that the alternative hypothesis is true?

9.14 What is the difference between the critical value of z and the observed value of z?

9.15 For each of the following examples of tests of hypotheses about μ, show the rejection and nonrejection regions on the sampling distribution of the sample mean.

    a.   A two-tailed test with $\alpha = .05$ and $n = 40$
    b.   A left-tailed test with $\alpha = .01$ and $n = 67$
    c.   A right-tailed test with $\alpha = .02$ and $n = 55$

9.16 For each of the following examples of tests of hypotheses about μ, show the rejection and nonrejection regions on the sampling distribution of the sample mean.

    a.   A two-tailed test with $\alpha = .01$ and $n = 100$
    b.   A left-tailed test with $\alpha = .005$ and $n = 60$
    c.   A right-tailed test with $\alpha = .025$ and $n = 36$

9.17 Consider the following null and alternative hypotheses.

$$H_0: \mu = 25 \quad \text{versus} \quad H_1: \mu \neq 25$$

Suppose you perform this test at $\alpha = .05$ and reject the null hypothesis. Would you state that the difference between the hypothesized value of the population mean and the observed value of the sample mean is "statistically significant" or would you state that this difference is "statistically not significant"? Explain.

9.18 Consider the following null and alternative hypotheses.

$$H_0: \mu = 60 \quad \text{versus} \quad H_1: \mu > 60$$

Suppose you perform this test at $\alpha = .01$ and fail to reject the null hypothesis. Would you state that the difference between the hypothesized value of the population mean and the observed value of the sample mean is "statistically significant" or would you state that this difference is "statistically not significant"? Explain.

9.19 For each of the following significance levels, what is the probability of making a Type I error?

    a. $\alpha = .025$ **b.** $\alpha = .05$ **c.** $\alpha = .01$

9.20 For each of the following significance levels, what is the probability of making a Type I error?

    a. $\alpha = .10$ **b.** $\alpha = .02$ **c.** $\alpha = .005$

9.21 A random sample of 100 observations produced a sample mean of 32 and a standard deviation of 6. Find the critical and observed values of z for each of the following tests of hypotheses using $\alpha = .05$.

    a. $H_0: \mu = 28$ versus $H_1: \mu > 28$
    b. $H_0: \mu = 28$ versus $H_1: \mu \neq 28$

472

9.22 A random sample of 80 observations produced a sample mean of 15 and a standard deviation of 4. Find the critical and observed values of z for each of the following tests of hypotheses using $\alpha = .01$.

    a. $H_0: \mu = 20$ versus $H_1: \mu < 20$
    b. $H_0: \mu = 20$ versus $H_1: \mu \neq 20$

9.23 Consider the null hypothesis $H_0: \mu = 50$. Suppose a random sample of 100 observations is taken to perform this test. Using $\alpha = .05$, show the rejection and nonrejection regions on the sampling distribution curve of the sample mean and find the critical value(s) of z when the alternative hypothesis is

    a. $H_1: \mu < 50$ b. $H_1: \mu \neq 50$ c. $H_1: \mu > 50$

9.24 Consider the null hypothesis $H_0: \mu = 35$. Suppose a random sample of 60 observations is taken to perform this test. Using $\alpha = .01$, show the rejection and nonrejection regions on the sampling distribution curve of the sample mean and find the critical value(s) of z for a

    a.   left-tailed test   b.   two-tailed test   c.   right-tailed test

9.25 Consider $H_0: \mu = 100$ versus $H_1: \mu \neq 100$.

    a.   A random sample of 64 observations produced a sample mean of 98 and a standard deviation of 12. Using $\alpha = .01$, would you reject the null hypothesis?
    b.   Another random sample of 64 observations taken from the same population produced a sample mean of 104 and a standard deviation of 10. Using $\alpha = .01$, would you reject the null hypothesis?

Comment on the results of parts a and b.

9.26 Consider $H_0: \mu = 45$ versus $H_1: \mu < 45$.

    a.   A random sample of 100 observations produced a sample mean of 43 and a standard deviation of 5. Using $\alpha = .025$, would you reject the null hypothesis?
    b.   Another random sample of 100 observations taken from the same population produced a sample mean of 43.8 and a standard deviation of 7. Using $\alpha = .025$, would you reject the null hypothesis?

Comment on the results of parts a and b.

9.27 Make the following tests of hypotheses.

    a. $H_0: \mu = 25,$ $H_1: \mu \neq 25,$ $n = 81,$ $\bar{x} = 28,$ $s = 3,$ $\alpha = .01$
    b. $H_0: \mu = 12,$ $H_1: \mu < 12,$ $n = 45,$ $\bar{x} = 11,$ $\sigma = 4.5,$ $\alpha = .05$
    c. $H_0: \mu = 40,$ $H_1: \mu > 40,$ $n = 100,$ $\bar{x} = 46,$ $s = 7,$ $\alpha = .10$

9.28 Make the following tests of hypotheses.

    a. $H_0: \mu = 80,$ $H_1: \mu \neq 80,$ $n = 33,$ $\bar{x} = 76,$ $s = 15,$ $\alpha = .10$
    b. $H_0: \mu = 32,$ $H_1: \mu < 32,$ $n = 75,$ $\bar{x} = 27,$ $s = 7.4,$ $\alpha = .01$
    c. $H_0: \mu = 55,$ $H_1: \mu > 55,$ $n = 40,$ $\bar{x} = 60,$ $s = 4,$ $\alpha = .05$

Applications

9.29 The U.S. Bureau of the Census often conducts surveys about households to collect information on a number of variables. One such survey showed that people with a bachelor's degree earned an average of $2116 a month in 1990. A sample of 900 persons with a bachelor's degree taken recently by a researcher showed that the persons in this sample earned an average of $2345 a month with a standard deviation of $210. Test at the 5% significance level whether people with a bachelor's degree currently earn more than $2116 a month.

9.30 According to the Hertz Corporation, the mean cost of owning and operating a car was $3002 in 1986. Suppose this estimate is true for the population of all cars for 1986. A random sample of 45 cars showed that the mean cost of owning and operating these cars was $3550 in 1993 with a standard deviation of $475. Using the 2.5% significance level, can you conclude that the mean cost of owning and operating a car in 1993 was greater than $3002?

9.31 The American Bar Association conducted a survey in 1986 that showed that the mean household income for lawyers was $120,000 in 1986. A researcher took a random sample of 64 lawyers recently that produced a mean household income of $140,500 with a standard deviation of $24,500. Test at the 2.5% significance level whether the current mean household income for all lawyers is greater than $120,000. Explain your conclusion in words.

9.32 The U.S. Bureau of Labor Statistics often conducts surveys to collect information on the labor market. According to the bureau, workers in the private sector earned an average of $10.33 an hour in 1991. A labor economist took a random sample of 1000 private sector workers recently that produced a mean hourly wage of $11.20 with a standard deviation of $1.90. Test at the 1% significance level if the current mean hourly wage for private sector workers is greater than $10.33. Explain your conclusion in words.

9.33 According to a study, the mean child support paid to custodial mothers by noncustodial fathers was $185 a month in 1985 (*The Forgotten Half: Pathways to Success for America's Youth and Young Families*, Washington, DC: Youth and America's Future: The William T. Grant Commission on Work, Family and Citizenship, 1988). A random sample of 340 custodial mothers taken recently by a researcher showed that the mean child support paid to these mothers is $236 per month with a standard deviation of $35.

a. Using the 1% significance level, can you conclude that the current mean child support paid to custodial mothers is higher than $185?

b. What is the Type I error in this case? Explain in words. What is the probability of making this error?

9.34 Are we enjoying more or less leisure time now than in the past? A survey conducted by Louis Harris and Associates showed that the mean time that Americans spent "to relax, watch TV, take part in sports or hobbies, go swimming or skiing, go to the movies, theater, concerts, or other forms of entertainment, get together with friends, and so forth" was 16.6 hours per week in 1988. A recent poll of 200 Americans showed that they spend an average of 17.2 hours a week on these leisure activities with a standard deviation of 3.9 hours.

a. Testing at the 5% significance level, do you think the mean number of hours spent per week on leisure activities by all Americans is now different from 16.6?

b. What is the Type I error in this case? Explain in words. What is the probability of making this error?

9.35 The U.S. Bureau of Labor Statistics often collects information on consumer expenditures on various items. According to one such survey, the mean housing expenditure incurred by households was $9252 in 1991. A recent sample of 400 households taken by an economist showed that these households had an average housing expenditure of $9358 in 1993 with a standard deviation of $1190.

a. Test at the 2.5% significance level whether the current mean housing expenditure incurred by all households is different from $9252.

b. What will your decision be in part a if the probability of making a Type I error is zero? Explain.

9.36 A restaurant franchise company has a policy of opening new restaurants only in those areas that have a mean household income of at least $35,000 per year. The company is currently considering an area to open a new restaurant. The company's research department took a sample of 150 households from this area and found that the mean income of these households is $33,124 per year with a standard deviation of $5400.

a. Using the 1% significance level, would you conclude that the company should not open a restaurant in this area?

b. What will your decision be in part a if the probability of making a Type I error is zero? Explain.

9.37 The manufacturer of a certain brand of auto batteries claims that the mean life of these batteries is 45 months. A consumer protection agency that wants to check this claim took a random sample of 36 such batteries and found that the mean life for this sample is 43.75 months with a standard deviation of 4 months.

a. Using the 2.5% significance level, would you conclude that the mean life of these batteries is less than 45 months?

    b. Make the test of part a using a 5% significance level. Is your decision different from the one in part a? Comment on the results of parts a and b.

9.38 A study claims that all adults spend an average of 14 hours or more on chores during a weekend. An economist wanted to check if this claim is true. A random sample of 200 adults taken by this economist showed that these adults spend an average of 13.55 hours on chores during a weekend with a standard deviation of 3.1 hours.

    a. Using the 1% significance level, can you conclude that the claim that all adults spend an average of 14 hours or more on chores during a weekend is false?

    b. Make the test of part a using a 2.5% significance level. Is your decision different from the one in part a? Comment on the results of parts a and b.

9.39 Lazurus Steel Corporation produces iron rods that are supposed to be 36 inches long. The machine that makes these rods does not produce each rod exactly 36 inches long. The lengths of these rods vary slightly. It is known that when the machine is working properly, the mean length of the rods is 36 inches. The standard deviation of the lengths of all rods produced on this machine is always equal to .05 inches. The quality control department at the company takes a sample of 40 such rods each week, calculates the mean length of these rods, and tests the null hypothesis $\mu = 36$ inches against the alternative hypothesis $\mu \neq 36$ inches using a 1% significance level. If the null hypothesis is rejected, the machine is stopped and adjusted. A recent such sample of 40 rods produced a mean length of 36.015 inches. Based on this sample, would you conclude that the machine needs to be adjusted?

9.40 At Farmer's Dairy, a machine is set to fill 32-ounce milk cartons. However, this machine does not put exactly 32 ounces of milk in each carton; the amount varies slightly from carton to carton. It is known that when the machine is working properly, the mean net weight of these cartons is 32 ounces. The standard deviation of the milk in all such cartons is always equal to .15 ounces. The quality control inspector at this dairy takes a sample of 35 such cartons each week, calculates the mean net weight of these cartons, and tests the null hypothesis $\mu = 32$ ounces against the alternative hypothesis $\mu \neq 32$ ounces using a 2% significance level. If the null hypothesis is rejected, the machine is stopped and adjusted. A recent sample of 35 such cartons produced a mean net weight of 31.90 ounces. Based on this sample, would you conclude that the machine needs to be adjusted?

9.41 A company claims that the mean net weight of the contents of its All Taste cereal boxes is at least 18 ounces. Suppose you want to test whether or not the claim of the company is true. Explain briefly how you would conduct this test using a large sample.

9.42 A researcher claims that college students spend an average of 45 minutes per week on community service. You want to test if the mean time spent per week on community service by college students is different from 45 minutes. Explain briefly how you would conduct this test using a large sample.

9.3   CALCULATING THE PROBABILITY OF A TYPE II ERROR

As discussed earlier in this chapter, the probability of making a Type II error is denoted by β. A Type II error is made when a false null hypothesis is not rejected, that is,

$$\beta = P(H_0 \text{ is not rejected} \mid H_0 \text{ is false})$$

Consider Example 9–2 of Section 9.2. The two hypotheses in that example are

$$H_0: \mu = \$128,400$$

$$H_1: \mu > \$128,400$$

Suppose, in that example, H_0 is false, that is, the current mean sales price of all single-family homes is greater than $128,400. However, we fail to reject H_0 based on the sample information. In this case we make a Type II error. We can calculate the probability of a Type II error if and only if the null hypothesis is false and we know the true (actual) population mean.

Therefore, the calculation of the probability of a Type II error is not possible in real cases. We would usually never know whether or not the null hypothesis is true. If the null hypothesis is false, we would not know the true value of the population mean. Example 9–4 illustrates how the probability of making a Type II error is calculated for a two-tailed test.

Calculating β for a two-tailed test of hypothesis.

EXAMPLE 9–4   Reconsider Example 9–1. Suppose the null hypothesis stated in that example is false and the true mean length of all bolts produced by the machine at the time of the selection of the sample was 2.498 inches. What is the probability of making a Type II error if $\alpha = .05$? What is the power of the test?

Solution   From the information given in Example 9–1,

$$n = 49 \quad \text{and} \quad s = .021 \text{ inches}$$

We are to calculate the probability of making a Type II error with $\alpha = .05$. The following four steps are performed to calculate this probability.

Step 1.   *State the null and alternative hypotheses*

From Example 9–1, the null and alternative hypotheses are

$$H_0: \mu = 2.5 \quad \text{(The machine does not need an adjustment)}$$

$$H_1: \mu \neq 2.5 \quad \text{(The machine needs an adjustment)}$$

Step 2.   *Select the distribution to use*

Because the sample size is large ($n > 30$), the sampling distribution of $\bar{x}$ is (approximately) normal. Consequently, we use the normal distribution to make the test or to calculate the probability of making a Type II error.

Step 3.   *Identify the area where the null hypothesis is not rejected and calculate the values of $\bar{x}$ that correspond to the two critical points*

The test is two-tailed and the significance level is .05. The rejection and nonrejection regions are shown in Figure 9.8. The two critical points are $z = -1.96$ and $z = 1.96$. In this step, we find the two values of $\bar{x}$ that correspond to these two critical points.

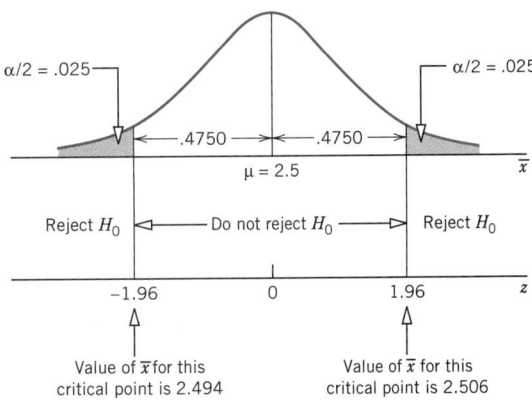

Figure 9.8

The standard deviation of $\bar{x}$ is

$$s_{\bar{x}} = s/\sqrt{n} = .021/\sqrt{49} = .003$$

THE VALUE OF $\bar{x}$ FOR A PARTICULAR VALUE OF z

The value of $\bar{x}$ for a particular value of z is calculated using the formula

$$\bar{x} = \mu + zs_{\bar{x}}$$

The value of $\bar{x}$ for $z = -1.96$ is

$$\bar{x} = 2.5 + (-1.96)(.003) = 2.494$$

The value of $\bar{x}$ for $z = 1.96$ is

$$\bar{x} = 2.5 + 1.96(.003) = 2.506$$

These two values of $\bar{x}$ are shown in Figure 9.8. Thus, we can state that with the null and alternative hypotheses as given in Step 1, we will not reject the null hypothesis if $\bar{x}$ is between 2.494 and 2.506.

Step 4. *Draw the distribution curve of $\bar{x}$ with the true population mean and calculate* β

Figure 9.9 combines Figure 9.8 with the distribution curve of $\bar{x}$ with the true population mean $\mu = 2.498$. The probability β of making a Type II error is given by the shaded area under the sampling distribution curve of $\bar{x}$ with the true population mean $\mu = 2.498$. Note that this area corresponds to the nonrejection region in Figure 9.8. This is always true.

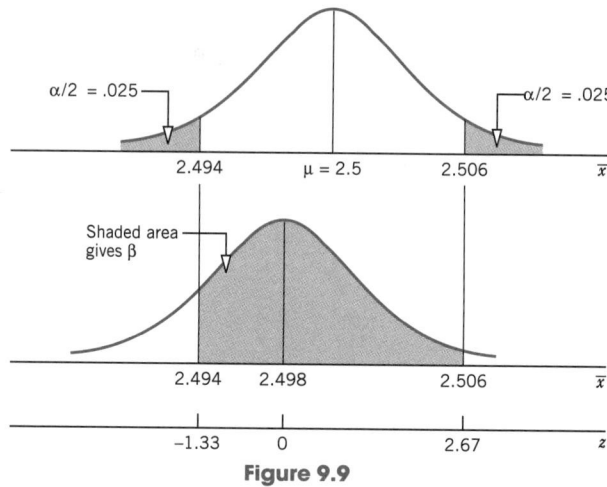

Figure 9.9

The shaded area in the bottom part of Figure 9.9 (which gives the probability of making a Type II error for this example) is calculated as follows.

For $\bar{x} = 2.494$: $z = \dfrac{2.494 - 2.498}{.003} = -1.33$

For $\bar{x} = 2.506$: $z = \dfrac{2.506 - 2.498}{.003} = 2.67$

Thus, the probability of making a Type II error is given by the area between $\bar{x} = 2.494$ and $\bar{x} = 2.506$ of the sampling distribution curve of $\bar{x}$ in the bottom part of Figure 9.9, which is equal to the area under the standard normal distribution curve between $z = -1.33$ and $z = 2.67$.

$$\beta = P(-1.33 < z < 2.67) = .4082 + .4962 = \mathbf{.9044}$$

The probability is .9044 that a Type II error will be made in this example when H_0 is false (that is, μ is not equal to 2.5 inches), the true population mean is 2.498 inches, and $\alpha = .05$. The power of the test for this example is

$$1 - \beta = 1 - .9044 = \mathbf{.0956}$$

That is, the probability is .0956 that we will reject the null hypothesis if it is false. This is the probability of not making a Type II error for this example. ∎

Example 9–5 describes the calculation of the Type II error for a one-tailed test.

Calculating β for a one-tailed test of hypothesis.

EXAMPLE 9-5   Refer to Example 9–2. Suppose the null hypothesis stated in that example is false and the current mean sales price of all single-family homes is $131,250. What is the probability of making a Type II error? Use $\alpha = .01$. What is the power of the test?

Solution   From the information given in Example 9–2,

$$n = 500 \quad \text{and} \quad s = \$23,700$$

The value of α is .01. We perform the following four steps to calculate the probability of making a Type II error.

Step 1.   *State the null and alternative hypotheses*

From Example 9–2, the two hypotheses are

H_0: $\mu = \$128,400$ (The current mean is not greater than $128,400)

H_1: $\mu > \$128,400$ (The current mean is greater than $128,400)

Step 2.   *Select the distribution to use*

Because the sample size is large ($n > 30$), the sampling distribution of $\bar{x}$ is (approximately) normal. Consequently, we use the normal distribution to make the test and to calculate the probability of making a Type II error.

Step 3.   *Identify the area where the null hypothesis is not rejected and calculate the value of $\bar{x}$ that corresponds to the critical point*

The test is right-tailed and the significance level is .01. The rejection and nonrejection regions are shown in Figure 9.10. The critical point is $z = 2.33$. We find the value of $\bar{x}$ that corresponds to this critical point.

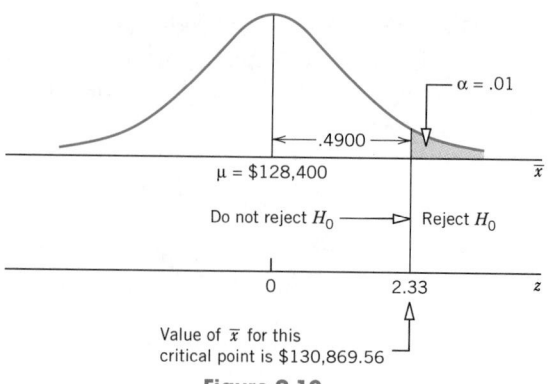

Figure 9.10

The standard deviation of $\bar{x}$ is

$$s_{\bar{x}} = s/\sqrt{n} = 23{,}700/\sqrt{500} = \$1059.8962$$

The value of $\bar{x}$ for $z = 2.33$ is

$$\bar{x} = \mu + zs_{\bar{x}} = 128{,}400 + (2.33)(1059.8962) = \$130{,}869.56$$

Thus, we can state that with the null and alternative hypotheses as stated in Step 1, we will not reject the null hypothesis if $\bar{x}$ is less than \$130,869.56.

Step 4.   *Draw the distribution curve of $\bar{x}$ with the true population mean and calculate β*

Figure 9.11 combines Figure 9.10 with the distribution curve of $\bar{x}$ with the true population mean $\mu = \$131{,}250$. The probability β of making a Type II error is given by the shaded area under the distribution curve of $\bar{x}$ with the true population mean $\mu = \$131{,}250$, as shown in the bottom part of Figure 9.11. Note that this area corresponds to the nonrejection region in Figure 9.10.

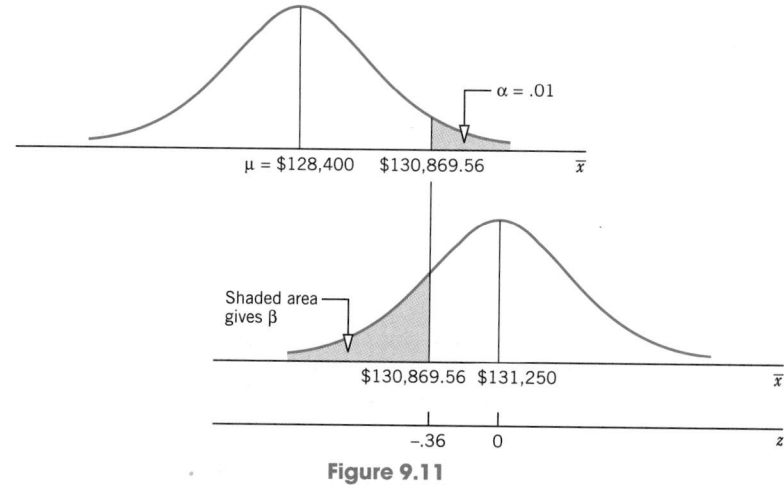

Figure 9.11

The shaded area in the bottom part of Figure 9.11 (which gives the probability of making a Type II error for this example) is calculated as follows.

$$\text{For } \bar{x} = \$130{,}869.56: \qquad z = \frac{130{,}869.56 - 131{,}250}{1059.8962} = -.36$$

Thus, the probability of making a Type II error is given by the area to the left of $z = -.36$ under the standard normal distribution curve, as shown in the bottom part of Figure 9.11.

$$\beta = P(z < -.36) = .5 - .1406 = \mathbf{.3594}$$

The probability is .3594 that a Type II error will be made in this example when H_0 is false, the true population mean is \$131,250, and $\alpha = .01$.

The power of the test for this example is

$$1 - \beta = 1 - .3594 = \mathbf{.6406}$$

That is, the probability is .6406 that we will reject the null hypothesis if it is false. This is the probability of not making a Type II error for this example. ■

EXERCISES

Concepts and Procedures

9.43 When is a Type II error made in a test of hypothesis? Explain briefly.

9.44 Consider the following null and alternative hypotheses.

$$H_0: \mu = 35 \quad \text{versus} \quad H_1: \mu \neq 35$$

A random sample of 100 observations taken from this population produced a standard deviation equal to 4.

   a. Calculate the probability of committing a Type II error for this test assuming that the null hypothesis is false and the true population mean is 37. Use a significance level of .05.
   b. Calculate the probability of committing a Type II error for this test assuming that the null hypothesis is false and the true population mean is 34. Use a significance level of .05.

9.45 Consider the following null and alternative hypotheses.

$$H_0: \mu = 55 \quad \text{versus} \quad H_1: \mu > 55$$

A random sample of 40 observations taken from this population produced a standard deviation equal to 8.

   a. Calculate the probability of committing a Type II error for this test assuming that the null hypothesis is false and the true population mean is 57.5. Use a significance level of .01.
   b. Calculate the probability of committing a Type II error for this test assuming that the null hypothesis is false and the true population mean is 56. Use a significance level of .01.

9.46 Consider the following null and alternative hypotheses.

$$H_0: \mu = 20 \quad \text{versus} \quad H_1: \mu < 20$$

A random sample of 50 observations taken from this population produced a standard deviation equal to 5. Suppose the null hypothesis mentioned above is false and the true population mean is 19. What is the probability that the test of hypothesis performed using a significance level of .025 will fail to reject the null hypothesis? What is the power of the test?

9.47 Consider the following null and alternative hypotheses.

$$H_0: \mu = 47 \quad \text{versus} \quad H_1: \mu \neq 47$$

A random sample of 36 observations taken from this population produced a standard deviation equal to 9. Suppose the null hypothesis mentioned above is false and the true population mean is 49. What is the probability that the test of hypothesis performed using a significance level of .01 will fail to reject the null hypothesis? What is the power of the test?

9.48 Consider $H_0: \mu = 50$ versus $H_1: \mu \neq 50$. Suppose this null hypothesis is false and the true population mean is 51. A random sample of 64 observations taken from this population produced a standard deviation equal to 12. Find the probability of committing a Type II error when

  a. $\alpha = .05$ b. $\alpha = .02$ c. $\alpha = .01$

9.49   Consider H_0: $\mu = 66$ versus H_1: $\mu < 66$. Suppose this null hypothesis is false and the true population mean is 64.5. A random sample of 49 observations taken from this population produced a standard deviation equal to 10.5. Find the probability of committing a Type II error when

  a. $\alpha = .005$ b. $\alpha = .01$ c. $\alpha = .025$

Applications

9.50   Refer to Exercise 9.29. Suppose the null hypothesis for that exercise is false and that the true population mean is $2250. What is the probability of committing a Type II error if $\alpha = .05$?

9.51   Refer to Exercise 9.30. Suppose the null hypothesis for that exercise is false and that the true population mean is $3100. What is the probability of committing a Type II error if $\alpha = .025$?

9.52   Refer to Exercise 9.34. Suppose the null hypothesis for that exercise is false and that the true population mean is 17 hours.

  a.   Find the probability of making a Type II error if $\alpha = .05$.
  b.   Find the power of the test.

9.53   Refer to Exercise 9.35. Suppose the null hypothesis for that exercise is false and that the true population mean is $9305.

  a.   Find the probability of making a Type II error if $\alpha = .025$.
  b.   Find the power of the test.

9.54   Refer to Exercise 9.37. Suppose the null hypothesis for that exercise is false and that the true population mean is 43.5 months.

  a.   Find the probability of making a Type II error if $\alpha = .01$.
  b.   Find the power of the test.

9.55   Refer to Exercise 9.39. Suppose the null hypothesis for that exercise is false and that the true population mean is 36.01 inches.

  a.   Find the probability of making a Type II error if $\alpha = .01$.
  b.   Find the power of the test.

9.4   HYPOTHESIS TESTS USING THE *p*-VALUE APPROACH

In the discussion of tests of hypotheses in Section 9.2, the value of the significance level α was selected before the test was performed. Sometimes we may prefer not to predetermine α. Instead, we may want to find a value such that a given null hypothesis will be rejected for any α greater than this value and it will not be rejected for any α smaller than this value. The *probability-value approach*, more commonly called the *p-value approach*, gives such a value. In this approach, we calculate the **p-value** for the test, which is defined as the smallest level of significance at which the given null hypothesis is rejected.

---

p-VALUE

The *p*-value is the smallest significance level at which the null hypothesis is rejected.

Using the *p*-value approach, we reject the null hypothesis if

$$p\text{-value} < \alpha$$

and we do not reject the null hypothesis if

$$p\text{-value} \geq \alpha$$

For a one-tailed test, the *p*-value is given by the area in the tail of the sampling distribution curve beyond the observed value of the sample statistic. Figure 9.12 shows the *p*-value for a left-tailed test about μ.

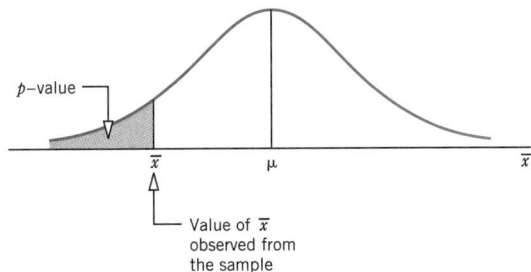

Figure 9.12   *p*-value for a left-tailed test.

For a two-tailed test, the *p*-value is twice the area in the tail of the sampling distribution curve beyond the observed value of the sample statistic. Figure 9.13 shows the *p*-value for a two-tailed test. Each of the areas in the two tails gives one-half the *p*-value.

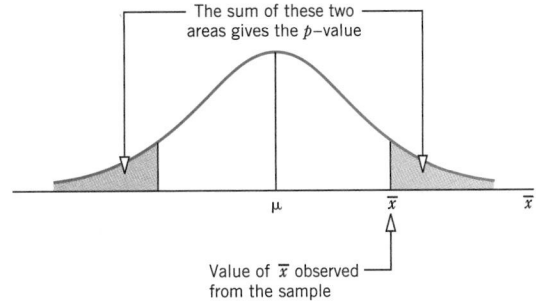

Figure 9.13   *p*-value for a two-tailed test.

Examples 9–6 and 9–7 illustrate the calculation and use of the *p*-value.

Calculating p-value
for a one-tailed test
of hypothesis.

EXAMPLE 9–6   The management of Priority Health Club claims that its members lose an average of 10 pounds or more within the first month after joining the club. A consumer agency that wanted to check this claim took a random sample of 36 members of this health club and found that they lost an average of 9.2 pounds within the first month of membership with a standard deviation of 2.4 pounds. Find the *p*-value for this test.

Solution   Let μ be the mean weight lost during the first month of membership by all members of this health club and $\bar{x}$ be the corresponding mean for the sample. From the given information,

$$n = 36, \qquad \bar{x} = 9.2 \text{ pounds}, \qquad \text{and} \qquad s = 2.4 \text{ pounds}$$

The claim of the club is that its members lose, on average, 10 pounds or more within the first month of membership. To calculate the *p*-value, we apply the following three steps.

Step 1.   *State the null and alternative hypotheses*

$$H_0: \mu \geq 10 \qquad \text{(The mean weight lost is 10 pounds or more)}$$

$$H_1: \mu < 10 \qquad \text{(The mean weight lost is less than 10 pounds)}$$

Step 2.   *Select the distribution to use*

Because the sample size is large, we use the normal distribution to make the test and to calculate the *p*-value.

Step 3.   *Calculate the p-value*

The $<$ sign in the alternative hypothesis indicates that the test is left-tailed. The *p*-value is given by the area in the left tail of the sampling distribution curve of $\bar{x}$ where $\bar{x}$ is less than 9.2, as shown in Figure 9.14. To find this area, we first find the *z* value for $\bar{x} = 9.2$ as follows.

$$s_{\bar{x}} = s/\sqrt{n} = 2.4/\sqrt{36} = .40$$

$$z = \frac{\bar{x} - \mu}{s_{\bar{x}}} = \frac{9.2 - 10}{.40} = -2.00$$

The area to the left of $\bar{x} = 9.2$ under the sampling distribution of $\bar{x}$ is equal to the area under the standard normal curve to the left of $z = -2.00$. From the normal distribution table, the area between the mean and $z = -2.00$ is .4772. Hence, the area to the left of $z = -2.00$ is $.5 - .4772 = .0228$. Consequently,

$$p\text{-value} = \textbf{.0228}$$

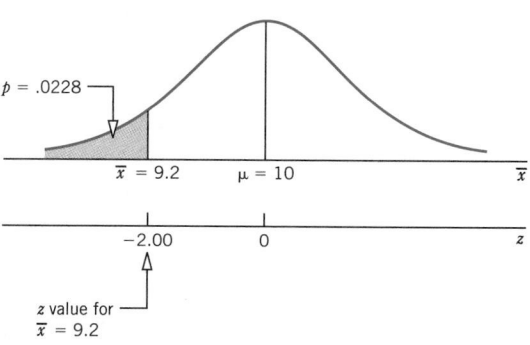

Figure 9.14   *p*-value for a left-tailed test.

Thus, based on the *p*-value of .0228 we can state that for any α (significance level) greater than .0228 we will reject the null hypothesis stated in Step 1 and for any α less than .0228 we will not reject the null hypothesis. Suppose we make the test for this example at $\alpha = .01$. Because $\alpha = .01$ is less than the *p*-value of .0228, we will not reject the null hypothesis. Now, suppose we make the test at $\alpha = .05$. This time, because $\alpha = .05$ is greater than the *p*-value of .0228, we will reject the null hypothesis.   ▬

The reader should make the test of hypothesis for Example 9–6 at $\alpha = .01$ and at $\alpha = .05$ by using the five steps learned in Section 9.2. The null hypothesis will not be rejected at $\alpha = .01$ (as .01 is less than $p = .0228$), and the null hypothesis will be rejected at $\alpha = .05$ (as .05 is greater than $p = .0228$).

Calculating p-value
for a two-tailed test
of hypothesis.

EXAMPLE 9–7   At Canon Food Corporation, it took, on average, 50 minutes for new workers to learn a food processing job. Recently the company installed a new food processing machine. The supervisor at the company wants to find if the mean time taken by new workers to learn the food processing procedure on this new machine is different from 50 minutes. A sample of 40 workers showed that it took, on average, 47 minutes for them to learn the food processing procedure on the new machine with a standard deviation of 7 minutes. Find the *p*-value for the test that the mean learning time for the food processing procedure on the new machine is different from 50 minutes.

Solution   Let μ be the mean time (in minutes) taken to learn the food processing procedure on the new machine by all workers and $\bar{x}$ be the corresponding sample mean. From the given information,

$$n = 40, \qquad \bar{x} = 47 \text{ minutes}, \qquad \text{and} \qquad s = 7 \text{ minutes}$$

To calculate the *p*-value, we apply the following three steps.

Step 1.   *State the null and alternative hypotheses*

$$H_0: \mu = 50 \text{ minutes}$$

$$H_1: \mu \neq 50 \text{ minutes}$$

Note that the null hypothesis states that the mean time for learning the food processing procedure on the new machine is 50 minutes and the alternative hypothesis states that this time is different from 50 minutes.

Step 2.   *Select the distribution to use*

Because the sample size is large, we use the normal distribution to make the test and to calculate the *p*-value.

Step 3.   *Calculate the p-value*

The $\neq$ sign in the alternative hypothesis indicates that the test is two-tailed. The *p*-value is equal to twice the area in the tail of the sampling distribution curve of $\bar{x}$ to the left of $\bar{x} = 47$, as shown in Figure 9.15. To find this area, we first find the *z* value for $\bar{x} = 47$ as follows.

$$s_{\bar{x}} = s/\sqrt{n} = 7/\sqrt{40} = 1.1068 \text{ minutes}$$

$$z = \frac{\bar{x} - \mu}{s_{\bar{x}}} = \frac{47 - 50}{1.1068} = -2.71$$

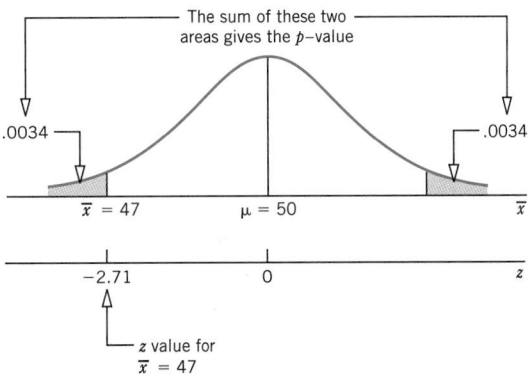

Figure 9.15   *p*-value for a two-tailed test.

The area to the left of $\bar{x} = 47$ is equal to the area under the standard normal curve to the left of $z = -2.71$. From the normal distribution table, the area between the mean and $z = -2.71$ is .4966. Hence, the area to the left of $z = -2.71$ is

$$.5 - .4966 = .0034$$

Consequently, the p-value is

$$p\text{-value} = 2 (.0034) = \textbf{.0068}$$

Thus, based on the p-value of .0068, we conclude that for any α (significance level) greater than .0068 we will reject the null hypothesis and for any α less than .0068 we will not reject the null hypothesis.   ■

EXERCISES

Concepts and Procedures

9.56 Briefly explain the procedure used to calculate the p-value for a two-tailed and for a one-tailed test, respectively.

9.57 Find the p-value for each of the following hypothesis tests.

a.	H_0: $\mu = 23$,	H_1: $\mu \neq 23$,	$n = 50$,	$\bar{x} = 21$,	$s = 5$
b.	H_0: $\mu = 15$,	H_1: $\mu < 15$,	$n = 80$,	$\bar{x} = 13.2$,	$s = 5.5$
c.	H_0: $\mu = 38$,	H_1: $\mu > 38$,	$n = 35$,	$\bar{x} = 40.6$,	$s = 7.2$

9.58 Find the p-value for each of the following hypothesis tests.

a.	H_0: $\mu = 46$,	H_1: $\mu \neq 46$,	$n = 40$,	$\bar{x} = 49.43$,	$s = 9.7$
b.	H_0: $\mu = 26$,	H_1: $\mu < 26$,	$n = 33$,	$\bar{x} = 24.2$,	$s = 4.3$
c.	H_0: $\mu = 18$,	H_1: $\mu > 18$,	$n = 55$,	$\bar{x} = 20.4$,	$s = 7.8$

9.59 Consider H_0: $\mu = 29$ versus H_1: $\mu \neq 29$. A random sample of 60 observations taken from this population produced a sample mean of 31.4 and a standard deviation of 8.

    a. Calculate the p-value.
    b. Considering the p-value of part a, would you reject the null hypothesis if the test were made at the significance level of .05?
    c. Considering the p-value of part a, would you reject the null hypothesis if the test were made at the significance level of .01?

9.60 Consider H_0: $\mu = 72$ versus H_1: $\mu > 72$. A random sample of 36 observations taken from this population produced a sample mean of 74.07 and a standard deviation of 6.

    a. Calculate the p-value.
    b. Considering the p-value of part a, would you reject the null hypothesis if the test were made at the significance level of .01?
    c. Considering the p-value of part a, would you reject the null hypothesis if the test were made at the significance level of .025?

Applications

9.61 According to the Oceanic and Atmospheric Administration, the mean consumption of seafood in the United States was 15.4 pounds per person in 1987. Assume that this result holds true for the population of all Americans for 1987. A random sample of 80 persons showed that they consumed an average of 16.1 pounds of seafood in 1993 with a standard deviation of 3.7 pounds. Find the p-value for the hypothesis test that the mean consumption of seafood for 1993 is different from 15.4 pounds.

9.62 According to Exercise 9.37, the manufacturer of a certain brand of auto batteries claims that the mean life of these batteries is 45 months. A consumer protection agency that wants to check this claim took a random sample of 36 such batteries and found that the mean life for this sample is 43.75 months with a standard deviation of 4 months. Find the *p*-value for the test of hypothesis with the alternative hypothesis that the mean life of these batteries is less than 45 months.

9.63 According to Exercise 9.38, a study claims that all adults spend an average of 14 hours or more on chores during a weekend. An economist wanted to check if this claim is true. A random sample of 200 adults taken by this economist showed that these adults spend an average of 13.55 hours on chores during a weekend with a standard deviation of 3.1 hours. Find the *p*-value for the hypothesis test with the alternative hypothesis that all adults spend less than 14 hours on chores during a weekend.

9.64 A survey conducted by the Hertz Corporation showed that cars driven by people in the United States were, on average, 7.6 years old in 1987. A sample of 50 cars taken recently showed that they are an average of 8.3 years old with a standard deviation of 2.1 years. Find the *p*-value for the test that the mean age of all current cars is greater than 7.6 years.

9.65 Refer to Exercise 9.39. Lazurus Steel Corporation produces iron rods that are supposed to be 36 inches long. The machine that makes these rods does not produce each rod exactly 36 inches long. The lengths of these rods vary slightly. It is known that when the machine is working properly, the mean length of the rods is 36 inches. The standard deviation of the lengths of all rods produced on this machine is always equal to .05 inches. The quality control department at the company takes a sample of 40 such rods every week, calculates the mean length of these rods, and tests the null hypothesis $\mu = 36$ inches against the alternative hypothesis $\mu \neq 36$ inches. If the null hypothesis is rejected, the machine is stopped and adjusted. A recent such sample of 40 rods produced a mean length of 36.015 inches.

    a. Calculate the *p*-value for this test of hypothesis.
    b. Based on the *p*-value calculated in part a, will the quality control inspector decide to stop the machine and adjust it if he chooses the maximum probability of a Type I error to be .02? What if the maximum probability of a Type I error is .05?

9.66 Refer to Exercise 9.40. At Farmer's Dairy, a machine is set to fill 32-ounce milk cartons. However, this machine does not put exactly 32 ounces of milk in each carton; the amount varies slightly from carton to carton. It is known that when the machine is working properly, the mean net weight of these cartons is 32 ounces. The standard deviation of the milk in all such cartons is always equal to .15 ounces. The quality control inspector at this company takes a sample of 35 such cartons every week, calculates the mean net weight of these cartons, and tests the null hypothesis $\mu = 32$ ounces against the alternative hypothesis $\mu \neq 32$ ounces. If the null hypothesis is rejected, the machine is stopped and adjusted. A recent sample of 35 such cartons produced a mean net weight of 31.90 ounces.

    a. Calculate the *p*-value for this test of hypothesis.
    b. Based on the *p*-value calculated in part a, will the quality control inspector decide to stop the machine and readjust it if she chooses the maximum probability of a Type I error to be .01? What if the maximum probability of a Type I error is .05?

9.67 According to the U.S. Bureau of the Census, the mean monthly salary of people with a professional degree was $4961 in 1990. Assume that this result holds true for the 1990 population of all people with a professional degree. A random sample of 400 people with a professional degree taken recently showed that their mean monthly salary is $5067 with a standard deviation of $985.

    a. Find the *p*-value for the test of hypothesis with the alternative hypothesis that the current mean monthly salary of all people with a professional degree is greater than $4961.
    b. If $\alpha = .01$, based on the *p*-value calculated in part a, would you reject the null hypothesis? Explain.
    c. If $\alpha = .025$, based on the *p*-value calculated in part a, would you reject the null hypothesis? Explain.

9.68 A telephone company claims that the mean duration of all long-distance phone calls made by its residential customers is 10 minutes. A random sample of 100 long-distance calls made by its residential customers taken from the records of this company showed that the mean duration of calls in this sample is 9.0 minutes with a standard deviation of 5.2 minutes.

a. Find the p-value for the test that the mean duration of all long-distance calls is less than 10 minutes.
b. If $\alpha = .02$, based on the p-value calculated in part a, would you reject the null hypothesis? Explain.
c. If $\alpha = .05$, based on the p-value calculated in part a, would you reject the null hypothesis? Explain.

9.5 HYPOTHESIS TESTS ABOUT A POPULATION MEAN: SMALL SAMPLES

Many times the size of a sample that is used to make a test of hypothesis about μ is small, that is, $n < 30$. This may be the case because we have limited resources and cannot afford to take a large sample or it may be because of the nature of the experiment itself. For example, to test a new model of a car for fuel efficiency (miles per gallon), the company may prefer to use a small sample. All cars included in such a test must be sold as used cars. In the case of a small sample, if the population from which the sample is drawn is (approximately) normally distributed and the population standard deviation σ is known, we can still use the normal distribution to make a test of hypothesis about μ. However, if the population is (approximately) normally distributed, the population standard deviation σ is not known, and the sample size is small ($n < 30$), then the normal distribution is replaced by the t distribution to make a test of hypothesis about μ. In such a case the random variable

$$t = \frac{\bar{x} - \mu}{s_{\bar{x}}} \qquad \text{where} \qquad s_{\bar{x}} = \frac{s}{\sqrt{n}}$$

has a t distribution. The t is called the **test statistic** to make a hypothesis test about a population mean for small samples.

CONDITIONS UNDER WHICH THE t DISTRIBUTION IS USED TO MAKE TESTS OF HYPOTHESIS ABOUT μ

The t distribution is used to conduct a test of hypothesis about μ if

1. The sample size is small ($n < 30$)
2. The population from which the sample is taken is (approximately) normally distributed
3. The population standard deviation σ is unknown

The procedure that is used to make hypothesis tests about μ in the case of small samples is similar to the one for large samples. We perform the same five steps with the only difference being the use of the t distribution in place of the normal distribution.

TEST STATISTIC

The value of the test statistic t for the sample mean $\bar{x}$ is computed as

$$t = \frac{\bar{x} - \mu}{s_{\bar{x}}} \qquad \text{where} \qquad s_{\bar{x}} = \frac{s}{\sqrt{n}}$$

The value of t calculated for $\bar{x}$ by using the above formula is also called the **observed value of t**.

Examples 9–8, 9–9, and 9–10 describe the procedure of testing hypotheses about the population mean using the t distribution.

Conducting a two-tailed test of hypothesis about μ: $n < 30$.

EXAMPLE 9–8   The mean age of all CEOs (chief executive officers) for major corporations in the United States was 48 years in 1991. A random sample of 25 CEOs taken recently from major corporations showed a mean age of 46 years with a standard deviation of 5 years. Assume that the ages of all CEOs of major corporations have an approximate normal distribution. Using the 1% significance level, would you conclude that the current mean age of all CEOs of major corporations is different from that in 1991?

Solution   Let μ be the current mean age of all CEOs of major corporations and $\bar{x}$ be the mean age of the CEOs included in the sample. Then, from the given information,

$$n = 25, \qquad \bar{x} = 46 \text{ years}, \qquad \text{and} \qquad s = 5 \text{ years}$$

The mean age of all CEOs for 1991 is given to be 48 years. The significance level is $\alpha = .01$.

Step 1.   State the null and alternative hypotheses

We are to test whether or not the current mean age of CEOs is different from 48 years. The null and alternative hypotheses are

$$H_0\text{: } \mu = 48 \qquad \text{(The current mean age is not different from 48 years)}$$
$$H_1\text{: } \mu \neq 48 \qquad \text{(The current mean age is different from 48 years)}$$

Step 2.   Select the distribution to use

The sample size is small ($n < 30$) and the population has an approximate normal distribution. However, we do not know the population standard deviation σ. Hence, we use the t distribution to make the hypothesis test.

Step 3.   Determine the rejection and nonrejection regions

The $\neq$ sign in the alternative hypothesis indicates that the test is two-tailed. The significance level is .01. Consequently, the total area of the two rejection regions is .01 and the area of the rejection region in each tail of the t distribution curve is

$$\text{Area in each tail} = \alpha/2 = .01/2 = .005$$

To find the critical values of t, we also need to know the degrees of freedom (df). The degrees of freedom for the t distribution for a test of hypothesis about μ are $n - 1$. Thus,

$$df = n - 1 = 25 - 1 = 24$$

From the t distribution table (Table VIII of Appendix B), the critical values of t for 24 degrees of freedom and .005 area in each tail are -2.797 and 2.797. These critical values are shown in Figure 9.16.

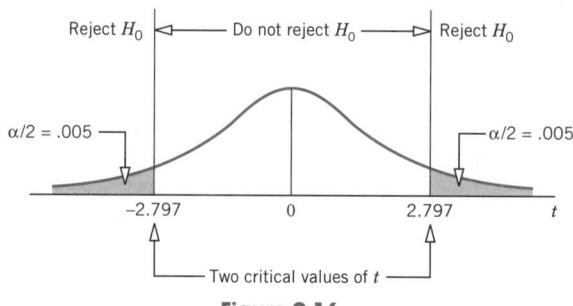

Figure 9.16

Step 4. *Calculate the value of the test statistic*

The value of the test statistic t for $\bar{x} = 46$ is calculated as follows.

$$s_{\bar{x}} = s/\sqrt{n} = 5/\sqrt{25} = 1.0$$

$$t = \frac{\bar{x} - \mu}{s_{\bar{x}}} = \frac{46 - 48}{1.0} = -2.000$$

From H_0

Step 5. *Make a decision*

The value of the test statistic $t = -2.00$ is between the two critical points, -2.797 and 2.797, which is the nonrejection region. Consequently, we fail to reject H_0. Therefore, we can state that the difference between the population mean for 1991 and the current sample mean is so small that it may have occurred because of sampling error. The mean age of the current CEOs of major corporations is not different from the mean age of CEOs of major corporations in 1991.   ■

Conducting a left-tailed test of hypothesis about **μ:** *$n < 30$.*

EXAMPLE 9-9   Grand Auto Corporation produces auto batteries. The company claims that its top of the line Never Die batteries are good on average, for at least 65 months. A consumer protection agency tested 15 such batteries to check this claim. It found the mean life of these 15 batteries to be 63 months with a standard deviation of 2 months. At the 5% significance level, can you conclude that the claim of the company is true? Assume that the life of such a battery has an approximate normal distribution.

Solution   Let μ be the mean life of all Never Die batteries and $\bar{x}$ be the corresponding mean for the sample. Then, from the given information,

$$n = 15, \quad \bar{x} = 63 \text{ months}, \quad \text{and} \quad s = 2 \text{ months}$$

The significance level is $\alpha = .05$. The company's claim is that the mean life of these batteries is at least 65 months.

Step 1.   *State the null and alternative hypotheses*

We are to test whether or not the mean life of Never Die batteries is at least 65 months. The null and alternative hypotheses are as follows.

$$H_0\text{: } \mu \geq 65 \quad \text{(The mean life is at least 65 months)}$$

$$H_1\text{: } \mu < 65 \quad \text{(The mean life is less than 65 months)}$$

Step 2. *Select the distribution to use*

The sample size is small, and the life of a battery is approximately normally distributed. However, the population standard deviation σ is not known. Hence, we use the t distribution to make the test.

Step 3. *Determine the rejection and nonrejection regions*

The significance level is .05. The $<$ sign in the alternative hypothesis indicates that the test is left-tailed with the rejection region in the left tail of the t distribution curve. To find the critical value of t, we need to know the area in the left tail and the degrees of freedom.

$$\text{Area in the left tail} = \alpha = .05$$

$$df = n - 1 = 15 - 1 = 14$$

From the t distribution table, the critical value of t for 14 degrees of freedom and an area of .05 in the left tail is -1.761. This value is shown in Figure 9.17.

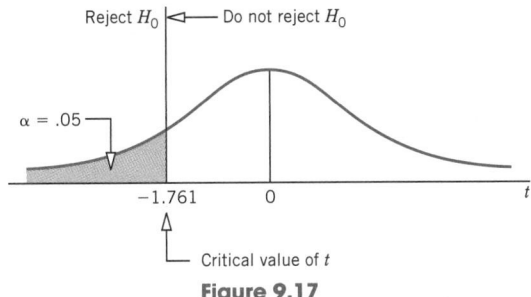

Figure 9.17

Step 4. *Calculate the value of the test statistic*

The value of the test statistic t for $\bar{x} = 63$ is calculated as follows.

$$s_{\bar{x}} = s/\sqrt{n} = 2/\sqrt{15} = .5164$$

$$t = \frac{\bar{x} - \mu}{s_{\bar{x}}} = \frac{63 - 65}{.5164} = -3.873 \qquad \text{From } H_0$$

Step 5. *Make a decision*

The value of the test statistic $t = -3.873$ is less than the critical value of $t = -1.761$, and it falls in the rejection region. Therefore, we reject H_0 and conclude that the sample mean is too small compared to 65 (company's claimed value of μ) and the difference between the two may not be attributed to chance alone. We can conclude that the mean life of the company's Never Die batteries is less than 65 months. ■

Making a right-tailed test of hypothesis about μ: $n < 30$.

EXAMPLE 9–10 The management at Massachusetts Savings Bank is always concerned about the quality of service provided to its customers. With the old computer system, a teller at this bank could serve, on average, 22 customers per hour. The management noticed that with this service rate, the waiting time for customers was too long. Recently the management of the bank installed a new computer system in the bank expecting that it would increase the service rate and consequently make the customers happier by reducing the waiting time. To check if the new computer system is more efficient than the old system, the management of the bank took a random sample of 18 hours and found that during these hours the mean number of customers served by tellers was 28 per hour with a standard deviation of 2.5. Testing at the 1% significance level, would you conclude that the new computer system is more efficient than the old computer system? Assume that the number of customers served per hour by a teller on this computer system has an approximate normal distribution.

Solution   Let μ be the mean number of customers served per hour by a teller and $\bar{x}$ be the corresponding mean for the sample. Then, from the given information,

$$n = 18, \quad \bar{x} = 28 \text{ customers}, \quad s = 2.5 \text{ customers}, \quad \text{and} \quad \alpha = .01$$

Step 1.   *State the null and alternative hypotheses*

We are to test whether or not the new computer system is more efficient than the old system. The new computer system will be more efficient than the old system if the mean number of customers served per hour by using the new computer system is significantly more than 22; otherwise, it will not be more efficient. The null and alternative hypotheses are

$$H_0: \mu = 22 \quad \text{(The new computer system is not more efficient)}$$
$$H_1: \mu > 22 \quad \text{(The new computer system is more efficient)}$$

Step 2.   *Select the distribution to use*

The sample size is small, and the population is approximately normally distributed. However, we do not know the population standard deviation σ. Hence, we use the t distribution to make the test.

Step 3.   *Determine the rejection and nonrejection regions*

The significance level is .01. The $>$ sign in the alternative hypothesis indicates that the test is right-tailed and the rejection region lies in the right tail of the t distribution curve.

$$\text{Area in the right tail} = \alpha = .01$$
$$df = n - 1 = 18 - 1 = 17$$

From the t distribution table, the critical value of t for 17 degrees of freedom and .01 area in the right tail is 2.567. This value is shown in Figure 9.18.

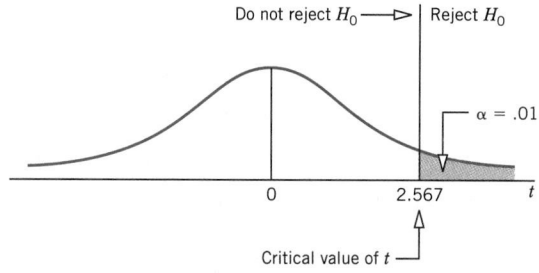

Figure 9.18

Step 4.   *Calculate the value of the test statistic*

The value of the test statistic t for $\bar{x} = 28$ is calculated as follows.

$$s_{\bar{x}} = s/\sqrt{n} = 2.5/\sqrt{18} = .5893$$

$$t = \frac{\bar{x} - \mu}{s_{\bar{x}}} = \frac{28 - 22}{.5893} = 10.182 \qquad \text{From } H_0$$

Step 5. *Make a decision*

The value of the test statistic $t = 10.182$ is larger than the critical value of $t = 2.567$, and it falls in the rejection region. Consequently, we reject H_0. As a result, we conclude that the value of the sample mean is too large compared to the hypothesized value of the population mean and the difference between the two may not be attributed to chance alone. The mean number of customers served per hour using the new computer system is more than 22. The new computer system is more efficient than the old computer system. ▬

EXERCISES

Concepts and Procedures

9.69 Briefly explain the conditions that must hold true to use the t distribution to make a test of hypothesis about the population mean.

9.70 For each of the following examples of tests of hypotheses about μ, show the rejection and nonrejection regions on the t distribution curve.

    a. A two-tailed test with $\alpha = .02$ and $n = 20$
    b. A left-tailed test with $\alpha = .01$ and $n = 16$
    c. A right-tailed test with $\alpha = .05$ and $n = 18$

9.71 For each of the following examples of tests of hypotheses about μ, show the rejection and nonrejection regions on the t distribution curve.

    a. A two-tailed test with $\alpha = .01$ and $n = 15$
    b. A left-tailed test with $\alpha = .005$ and $n = 25$
    c. A right-tailed test with $\alpha = .025$ and $n = 22$

9.72 A random sample of 25 observations taken from a population that is normally distributed produced a sample mean of 58.5 and a standard deviation of 7.5. Find the critical and observed values of t for each of the following tests of hypotheses using $\alpha = .01$.

    a. $H_0: \mu = 55$ versus $H_1: \mu > 55$
    b. $H_0: \mu = 55$ versus $H_1: \mu \neq 55$

9.73 A random sample of 16 observations taken from a population that is normally distributed produced a sample mean of 42.4 and a standard deviation of 8. Find the critical and observed values of t for each of the following tests of hypotheses using $\alpha = .05$.

    a. $H_0: \mu = 46$ versus $H_1: \mu < 46$
    b. $H_0: \mu = 46$ versus $H_1: \mu \neq 46$

9.74 Consider the null hypothesis $H_0: \mu = 70$ about the mean of a population that is normally distributed. Suppose a random sample of 20 observations is taken from this population to make this test. Using $\alpha = .01$, show the rejection and nonrejection regions and find the critical value(s) of t for a

    a. left-tailed test   **b.** two-tailed test   **c.** right-tailed test

9.75 Consider the null hypothesis $H_0: \mu = 35$ about the mean of a population that is normally distributed. Suppose a random sample of 22 observations is taken from this population to make this test. Using $\alpha = .05$, show the rejection and nonrejection regions and find the critical value(s) of t for a

    a. left-tailed test   **b.** two-tailed test   **c.** right-tailed test

9.76 Consider H_0: $\mu = 80$ versus H_1: $\mu \neq 80$ for a population that is normally distributed.

    a. A random sample of 25 observations taken from this population produced a sample mean of 77 and a standard deviation of 8. Using $\alpha = .01$, would you reject the null hypothesis?

    b. Another random sample of 25 observations taken from the same population produced a sample mean of 86 and a standard deviation of 6. Using $\alpha = .01$, would you reject the null hypothesis?

Comment on the results of parts a and b.

9.77 Consider H_0: $\mu = 40$ versus H_1: $\mu > 40$ for a population that is normally distributed.

    a. A random sample of 16 observations taken from this population produced a sample mean of 45 and a standard deviation of 5. Using $\alpha = .025$, would you reject the null hypothesis?

    b. Another random sample of 16 observations taken from the same population produced a sample mean of 41.9 and a standard deviation of 7. Using $\alpha = .025$, would you reject the null hypothesis?

Comment on the results of parts a and b.

9.78 Assuming that the respective populations are normally distributed, make the following hypothesis tests.

    a. H_0: $\mu = 24$, H_1: $\mu \neq 24$, $n = 25$, $\bar{x} = 28$, $s = 4.9$, $\alpha = .01$

    b. H_0: $\mu = 30$, H_1: $\mu < 30$, $n = 16$, $\bar{x} = 27$, $s = 6.6$, $\alpha = .025$

    c. H_0: $\mu = 18$, H_1: $\mu > 18$, $n = 20$, $\bar{x} = 22$, $s = 8$, $\alpha = .10$

9.79 Assuming that the respective populations are normally distributed, make the following hypothesis tests.

    a. H_0: $\mu = 60$, H_1: $\mu \neq 60$, $n = 14$, $\bar{x} = 56$, $s = 9$, $\alpha = .05$

    b. H_0: $\mu = 35$, H_1: $\mu < 35$, $n = 24$, $\bar{x} = 29$, $s = 5.4$, $\alpha = .005$

    c. H_0: $\mu = 47$, H_1: $\mu > 47$, $n = 18$, $\bar{x} = 51$, $s = 6$, $\alpha = .001$

Applications

9.80 According to the U.S. Bureau of the Census, people with a high school diploma earned an average of $1077 a month in 1990. A random sample of 28 high school diploma holders taken recently showed that their mean monthly income is $1112 with a standard deviation of $160. Assuming that the monthly incomes of all high school diploma holders are approximately normally distributed, test at the 1% significance level whether the mean monthly income of such persons has increased since 1990.

9.81 A soft-drink manufacturer claims that its 12-ounce cans do not contain, on average, more than 30 calories. A random sample of 16 cans of this soft drink, which were checked for calories, contained a mean of 31.8 calories with a standard deviation of 3 calories. Assume that the number of calories in 12-ounce soda cans is normally distributed. Does the sample information support the alternative hypothesis that the manufacturer's claim is false? Use a significance level of 1%.

9.82 According to the National Agricultural Statistics Service, the mean yield of potatoes per acre was 292 cwt. (a cwt. is equal to 100 pounds) in 1986. A random sample of 20 acres gave a mean yield of potatoes to be 301 cwt. for 1993 with a standard deviation of 22 cwt. Assume that the yield of potatoes per acre is normally distributed. Using the 5% significance level, can you conclude that the mean yield of potatoes for 1993 is different from 292 cwt.? Explain your conclusion in words.

9.83 The mean balance of all checking accounts at a bank on December 31, 1992 was $850. A random sample of 25 checking accounts taken recently from this bank gave a mean balance of $775 with a standard deviation of $230. Assume that the balances of all checking accounts at this bank are normally distributed. Using the 5% significance level, can you conclude that the mean balance of such accounts has decreased during this period? Explain your conclusion in words.

9.84 A paint manufacturing company claims that the mean drying time for its paints is not more than 45 minutes. A random sample of 20 gallons of paints selected from the production line of this company showed that the mean drying time for this sample is 50 minutes with a standard deviation of 3 minutes. Assume that the drying times for these paints have a normal distribution.

a. Using the 5% significance level, would you conclude that the company's claim is true?

b. What is the Type I error in this exercise? Explain in words. What is the probability of making such an error?

9.85 The Metropolitan Life Insurance Company calculates the mean charges for various medical procedures by using its own claims data. According to one such estimate, the mean hospital and physician's charges for a coronary artery bypass graft for patients aged 35–64 were $43,370 in 1990 for all states. The mean hospital and physician's charges for 21 such procedures performed in Minnesota were $36,180 in 1990 (*Statistical Bulletin*, 73(3), July–September 1992). Assume that the standard deviation for the hospital and physician's charges for these 21 procedures is $12,680. Assume that the hospital and physician's charges for all coronary artery bypass grafts for the state of Minnesota have a normal distribution.

a. Using the 1% significance level, can you conclude that the mean hospital and physician's charges for a coronary artery bypass graft for patients aged 35–64 performed in Minnesota in 1990 were lower than $43,370?

b. What is the Type I error in this exercise? Explain in words. What is the probability of making such an error?

9.86 *Harper's* magazine claims that the mean household income of its readers is $79,600 a year. A researcher wanted to test whether or not the mean income of readers of *Harper's* magazine is equal to $79,600. She took a sample of 25 readers of *Harper's* magazine and found that their mean annual income is $76,445 with a standard deviation of $9,864. Assume that the annual household incomes of all readers of *Harper's* magazine have an approximate normal distribution.

a. Suppose the probability of making a Type I error is selected to be zero. What is your decision about the claim of *Harper's* magazine? Answer without performing the five steps of a test of hypothesis.

b. Using the 1% significance level, can you conclude that the claim of *Harper's* magazine is true?

9.87 A business school claims that students who complete a three-month typing course can type, on average, at least 1200 words an hour. A random sample of 25 students who completed this course typed, on average, 1130 words an hour with a standard deviation of 85 words. Assume that the typing speeds for all students who complete this course have an approximate normal distribution.

a. Suppose the probability of making a Type I error is selected to be zero. Can you conclude that the claim of the business school is true? Answer without performing the five steps of a test of hypothesis.

b. Using the 5% significance level, can you conclude that the claim of the business school is true?

9.88 The past records of a supermarket show that its customers spend an average of $65 per visit at this store. Recently the management of the store initiated a promotional campaign according to which each customer receives points based on the total money spent at the store and these points can be used to buy products at the store. The management expects that as a result of this campaign, the customers should be encouraged to spend more at the store. To check whether this is true, the manager of the store took a sample of 12 customers who visited the store. The following data give the money (in dollars) spent by these customers at this supermarket during their visits.

$88	69	141	28	106	45
32	51	78	54	110	83

Assume that the money spent by all customers at this supermarket has a normal distribution. Using the 1% significance level, can you conclude that the mean amount of money spent by all customers at this supermarket after the campaign was started is higher than $65? (*Hint:* First, calculate the sample mean and the sample standard deviation for these data using the formulas learned in Sections 3.1.1 and 3.2.2 of Chapter 3. Then make the test of hypothesis about μ.)

9.89   A past study claims that adults in America spend an average of 18 hours a week on leisure activities. A researcher wanted to test this claim. She took a sample of 10 adults and asked them about the time they spend per week on leisure activities. Their responses (in hours) are as follows.

14   25   22   38   16   26   19   23   41   33

Assume that the time spent on leisure activities by all adults is normally distributed. Using the 5% significance level, can you conclude that the claim of the earlier study is true? (*Hint:* First, calculate the sample mean and the sample standard deviation for these data using the formulas learned in Sections 3.1.1 and 3.2.2 of Chapter 3. Then make the test of hypothesis about μ.)

9.90   The manager of a service station claims that the mean amount spent on gas by its customers is $10.90. You want to test if the mean amount spent on gas at this station is different from $10.90. Briefly explain how you would conduct this test by taking a small sample.

9.91   A tool manufacturing company claims that its top of the line machine that is used to manufacture bolts produces an average of 88 or more bolts per hour. A company that is interested in buying this machine wants to check this claim. Suppose you are asked to conduct this test. Briefly explain how you would do so by taking a small sample.

9.6   HYPOTHESIS TESTS ABOUT A POPULATION PROPORTION: LARGE SAMPLES

Often we want to conduct a test of hypothesis about a population proportion. For example, a mail-order company claims that 90% of all orders it receives are shipped within 72 hours. The quality control department of the company may want to check from time to time whether or not this claim is true. As another example, a company claims that 25% of all males use its after-shave products. A rival company may want to check whether or not this claim is true.

This section presents the procedure to perform tests of hypotheses about the population proportion p for large samples. The procedure to make such tests is similar in many respects to the one for the population mean μ. The procedure includes the same five steps. Again, the test can be two-tailed or one-tailed. We know from Chapter 7 that when the sample size is large, the sample proportion $\hat{p}$ is approximately normally distributed with its mean equal to p and standard deviation equal to $\sqrt{pq/n}$. Hence, we use the normal distribution to perform a test of hypothesis about the population proportion p for a large sample. As was mentioned in Chapters 7 and 8, in the case of a proportion, the sample size is considered to be large when np and nq are both greater than 5.

TEST STATISTIC

The value of the test statistic z for the sample proportion $\hat{p}$ is computed as

$$z = \frac{\hat{p} - p}{\sigma_{\hat{p}}} \quad \text{where} \quad \sigma_{\hat{p}} = \sqrt{\frac{pq}{n}}$$

The value of p used in this formula is the one used in the null hypothesis. The value of q is equal to $1 - p$.

The value of z calculated for $\hat{p}$ using the above formula is also called the **observed value of z.**

Examples 9–11, 9–12, and 9–13 describe the procedure to make tests of hypotheses about the population proportion p.

Conducting a two-tailed test of hypothesis about p: large sample.

EXAMPLE 9–11 According to Information Resources Inc., based on the sales of teeth-cleaning products in supermarkets and drugstores during the period from October 1991 to September 1992, Crest toothpaste controlled a 31.2% share of the market. For convenience, assume that a 31.2% share of the market means that 31.2% of all people in the United States used Crest toothpaste. A researcher from a rival company wants to find whether or not the current market share controlled by Crest is different from 31.2%. She took a sample of 400 persons and found that 29% of them use Crest toothpaste. Using the .01 significance level, can you conclude that the current market share of Crest toothpaste is different from that for 1991–1992?

Solution Let p be the proportion of all people in the United States who currently use Crest toothpaste and $\hat{p}$ be the corresponding sample proportion. Then, from the given information,

$$n = 400, \quad \hat{p} = .29, \quad \text{and} \quad \alpha = .01$$

Based on 1991–1992 sales data, 31.2% of all people use Crest toothpaste. Assuming this claim is true,

$$p = .312 \quad \text{and} \quad q = 1 - p = 1 - .312 = .688$$

Note that we have changed all percentages to proportions.

Step 1. *State the null and alternative hypotheses*

Crest toothpaste still controls the same market share if $p = .312$ and the current market share is different if $p \neq .312$. The null and alternative hypotheses are as follows.

$$H_0: p = .312 \quad \text{(The current market share of Crest is the same)}$$
$$H_1: p \neq .312 \quad \text{(The current market share of Crest is different)}$$

Step 2. *Select the distribution to use*

The values of np and nq are

$$np = 400 \,(.312) = 124.8 \quad \text{and} \quad nq = 400 \,(.688) = 275.20$$

Because both np and nq are greater than 5, the sample size is large. Consequently, we use the normal distribution to make the hypothesis test about p.

Step 3. *Determine the rejection and nonrejection regions*

The $\neq$ sign in the alternative hypothesis indicates that the test is two-tailed. The significance level is .01. Therefore, the total area of the two rejection regions is .01 and the rejection region in each tail of the sampling distribution of $\hat{p}$ is $\alpha/2 = .01/2 = .005$. The critical values of z, obtained from the standard normal distribution table for .4950, are -2.58 and 2.58, as shown in Figure 9.19.

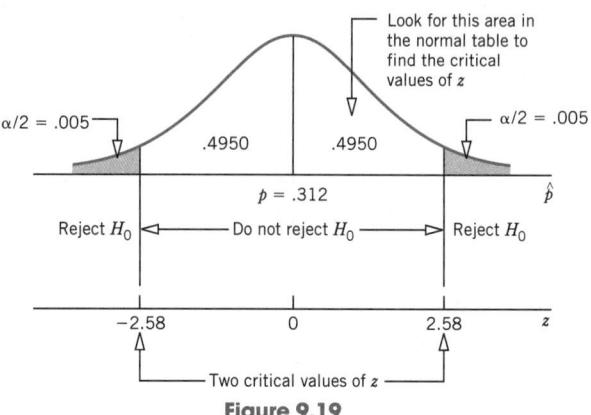

Figure 9.19

Step 4.   *Calculate the value of the test statistic*

The value of the test statistic z for $\hat{p} = .29$ is calculated as follows.

$$\sigma_{\hat{p}} = \sqrt{\frac{pq}{n}} = \sqrt{\frac{(.312)\,(.688)}{400}} = .0232$$

$$z = \frac{\hat{p} - p}{\sigma_{\hat{p}}} = \frac{.29 - .\overset{\text{From } H_0}{312}}{.0232} = -.95$$

Step 5.   *Make a decision*

The value of the test statistic $z = -.95$ for $\hat{p}$ lies between -2.58 and 2.58, and it falls in the nonrejection region. Consequently, we fail to reject H_0. Therefore, we can state that the sample proportion is not too far from the hypothesized value of the population proportion and the difference between the two can be attributed to chance. We conclude that the current market share of Crest toothpaste is not different from 31.2%.   ■

Making a right-tailed test of hypothesis about p: large sample.

EXAMPLE 9–12   When working properly, a machine that is used to make chips for calculators does not produce more than 4% defective chips. Whenever the machine produces more than 4% defective chips, it needs an adjustment. To check if the machine is working properly, the quality control department at the company often takes samples of chips and inspects them to determine if they are good or defective. One such random sample of 200 chips taken recently from the production line contained 14 defective chips. Test at the 5% significance level whether or not the machine needs an adjustment.

Solution   Let p be the proportion of defective chips in all chips produced by this machine and $\hat{p}$ be the corresponding sample proportion. Then, from the given information,

$$n = 200 \qquad \hat{p} = 14/200 = .07, \qquad \text{and} \qquad \alpha = .05$$

When the machine is working properly it produces 4% defective chips. Consequently, assuming that the machine is working properly,

$$p = .04 \qquad \text{and} \qquad q = 1 - p = 1 - .04 = .96$$

Step 1.   *State the null and alternative hypotheses*

The machine will not need an adjustment if the percentage of defective chips is 4% or less, and it will need an adjustment if this percentage is greater than 4%. Hence, the null and alternative hypotheses are

$$H_0: p \leq .04 \qquad \text{(The machine does not need an adjustment)}$$

$$H_1: p > .04 \qquad \text{(The machine needs an adjustment)}$$

Step 2. *Select the distribution to use*

The values of np and nq are

$$np = 200\ (.04) = 8 > 5 \qquad \text{and} \qquad nq = 200\ (.96) = 192 > 5$$

Because the sample size is large, we use the normal distribution to make the hypothesis test about p.

Step 3. *Determine the rejection and nonrejection regions*

The significance level is .05. The $>$ sign in the alternative hypothesis indicates that the test is right-tailed and the rejection region lies in the right tail of the sampling distribution of $\hat{p}$ with its area equal to .05. As shown in Figure 9.20, the critical value of z, obtained from the normal distribution table for .4500, is approximately 1.65.

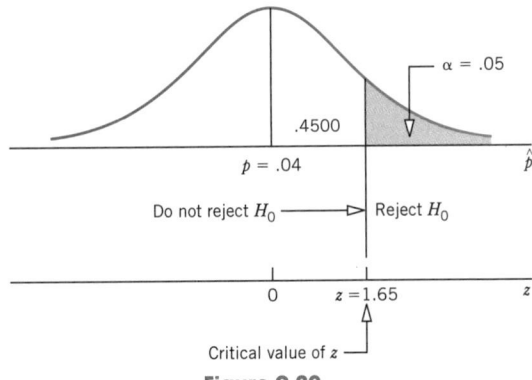

Figure 9.20

Step 4. *Calculate the value of the test statistic*

The value of the test statistic z for $\hat{p} = .07$ is calculated as follows.

$$\sigma_{\hat{p}} = \sqrt{\frac{pq}{n}} = \sqrt{\frac{(.04)\ (.96)}{200}} = .0139$$

$$z = \frac{\hat{p} - p}{\sigma_{\hat{p}}} = \frac{.07 - .04}{.0139} = 2.16 \qquad \text{—From } H_0$$

Step 5. *Make a decision*

Because the value of the test statistic $z = 2.16$ is greater than the critical value of $z = 1.65$ and it falls in the rejection region, we reject H_0. We conclude that the sample proportion is too far from the hypothesized value of the population proportion and the difference between the two cannot be attributed to chance alone. Therefore, based on the sample information, we conclude that the machine needs an adjustment. ∎

Conducting a left-tailed test of hypothesis about p: large sample.

EXAMPLE 9–13 Direct Mailing Company sells computers and computer parts by mail. The company claims that at least 90% of all orders are mailed within 72 hours after they are received. The quality control department at the company often takes samples to check if this claim is valid. A recently taken sample of 150 orders showed that 129 of them were mailed within 72 hours. Do you think the company's claim is true? Use a 2.5% significance level.

Solution Let p be the proportion of all orders that are mailed by the company within 72 hours and $\hat{p}$ be the corresponding sample proportion. Then, from the given information,

$$n = 150 \quad \text{and} \quad \hat{p} = 129/150 = .86$$

The company claims that at least 90% of all orders are mailed within 72 hours. Assuming that this claim is true, the values of p and q are

$$p = .90 \quad \text{and} \quad q = 1 - p = 1 - .90 = .10$$

The significance level is $\alpha = .025$.

Step 1. *State the null and alternative hypotheses*

The null and alternative hypotheses are

$$H_0: p \geq .90 \quad \text{(The company's claim is true)}$$
$$H_1: p < .90 \quad \text{(The company's claim is false)}$$

Step 2. *Select the distribution to use*

We first check whether both np and nq are greater than 5.

$$np = 150 \ (.90) = 135 > 5 \quad \text{and} \quad nq = 150 \ (.10) = 15 > 5$$

Consequently, the sample size is large. Therefore, we use the normal distribution to make the hypothesis test about p.

Step 3. *Determine the rejection and nonrejection regions*

The significance level is .025. The $<$ sign in the alternative hypothesis indicates that the test is one-tailed and the rejection region lies in the left tail of the sampling distribution of $\hat{p}$ with its area equal to .025. As shown in Figure 9.21, the critical value of z, obtained from the normal distribution table for .4750, is (approximately) -1.96.

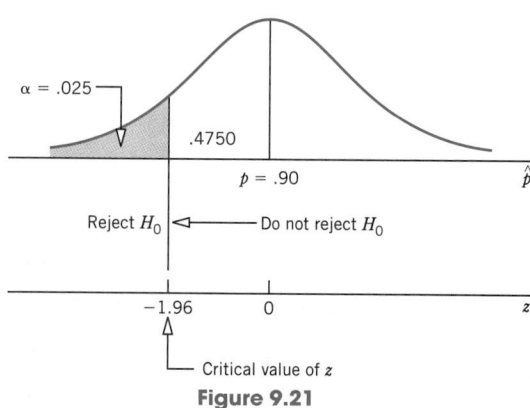

Figure 9.21

Step 4. *Calculate the value of the test statistic*

The value of the test statistic z for $\hat{p} = .90$ is calculated as follows.

$$\sigma_{\hat{p}} = \sqrt{\frac{pq}{n}} = \sqrt{\frac{(.90)\,(.10)}{150}} = .0245$$

—From H_0

$$z = \frac{\hat{p} - p}{\sigma_{\hat{p}}} = \frac{.86 - .90}{.0245} = -1.63$$

Step 5. *Make a decision*

The value of the test statistic $z = -1.63$ is greater than the critical value of $z = -1.96$, and it falls in the nonrejection region. Therefore, we fail to reject H_0. We can state that the difference between the sample proportion and the hypothesized value of the population proportion is small and this difference may have occurred owing to chance alone. Therefore, the proportion of all orders that are mailed within 72 hours is at least 90% and the company's claim is true. ∎

We can also use the *p*-value approach to make tests of hypotheses about the population proportion *p*. The procedure to calculate the *p*-value for the sample proportion is similar to the one applied to the sample mean in Section 9.4.

EXERCISES

Concepts and Procedures

9.92 Explain when a sample is large enough to use the normal distribution to make a test of hypothesis about the population proportion.

9.93 In each of the following cases, do you think the sample size is large enough to use the normal distribution to make a test of hypothesis about the population proportion? Explain why or why not.

a. $n = 40$ and $p = .011$ b. $n = 100$ and $p = .73$
c. $n = 80$ and $p = .05$ d. $n = 50$ and $p = .14$

9.94 In each of the following cases, do you think the sample size is large enough to use the normal distribution to make a test of hypothesis about the population proportion? Explain why or why not.

a. $n = 30$ and $p = .65$ b. $n = 70$ and $p = .05$
c. $n = 60$ and $p = .06$ d. $n = 900$ and $p = .17$

9.95 For each of the following examples of tests of hypotheses about the population proportion, show the rejection and nonrejection regions on the graph of the sampling distribution of the sample proportion.

a. A two-tailed test with $\alpha = .10$
b. A left-tailed test with $\alpha = .01$
c. A right-tailed test with $\alpha = .05$

9.96 For each of the following examples of tests of hypotheses about the population proportion, show the rejection and nonrejection regions on the graph of the sampling distribution of the sample proportion.

a. A two-tailed test with $\alpha = .05$
b. A left-tailed test with $\alpha = .02$
c. A right-tailed test with $\alpha = .025$

9.97 A random sample of 500 observations produced a sample proportion equal to .37. Find the critical and observed values of z for each of the following tests of hypotheses using $\alpha = .05$.

    a. $H_0: p = .30$ versus $H_1: p > .30$
    b. $H_0: p = .30$ versus $H_1: p \neq .30$

9.98 A random sample of 200 observations produced a sample proportion equal to .59. Find the critical and observed values of z for each of the following tests of hypotheses using $\alpha = .01$.

    a. $H_0: p = .63$ versus $H_1: p < .63$
    b. $H_0: p = .63$ versus $H_1: p \neq .63$

9.99 Consider the null hypothesis $H_0: p = .65$. Suppose a random sample of 1000 observations is taken to make this test about the population proportion. Using $\alpha = .05$, show the rejection and non-rejection regions and find the critical value(s) of z for a

    a. left-tailed test   **b.** two-tailed test   **c.** right-tailed test

9.100 Consider the null hypothesis $H_0: p = .25$. Suppose a random sample of 400 observations is taken to make this test about the population proportion. Using $\alpha = .01$, show the rejection and non-rejection regions and find the critical value(s) of z for a

    a. left-tailed test   **b.** two-tailed test   **c.** right-tailed test

9.101 Consider $H_0: p = .70$ versus $H_1: p \neq .70$.

    a. A random sample of 600 observations produced a sample proportion equal to .67. Using $\alpha = .01$, would you reject the null hypothesis?
    b. Another random sample of 600 observations taken from the same population produced a sample proportion equal to .76. Using $\alpha = .01$, would you reject the null hypothesis?

Comment on the results of parts a and b.

9.102 Consider $H_0: p = .45$ versus $H_1: p < .45$.

    a. A random sample of 400 observations produced a sample proportion equal to .41. Using $\alpha = .025$, would you reject the null hypothesis?
    b. Another random sample of 400 observations taken from the same population produced a sample proportion of .395. Using $\alpha = .025$, would you reject the null hypothesis?

Comment on the results of parts a and b.

9.103 Make the following hypothesis tests about p.

    a. $H_0: p = .45,$ $H_1: p \neq .45,$ $n = 100,$ $\hat{p} = .48,$ $\alpha = .10$
    b. $H_0: p = .72,$ $H_1: p < .72,$ $n = 700,$ $\hat{p} = .65,$ $\alpha = .05$
    c. $H_0: p = .30,$ $H_1: p > .30,$ $n = 200,$ $\hat{p} = .34,$ $\alpha = .01$

9.104 Make the following hypothesis tests about p.

    a. $H_0: p = .57,$ $H_1: p \neq .57,$ $n = 800,$ $\hat{p} = .51,$ $\alpha = .05$
    b. $H_0: p = .26,$ $H_1: p < .26,$ $n = 400,$ $\hat{p} = .22,$ $\alpha = .01$
    c. $H_0: p = .84,$ $H_1: p > .84,$ $n = 125,$ $\hat{p} = .86,$ $\alpha = .025$

Applications

9.105 According to a 1992 survey conducted by the Roper Organization for Shearson Lehman Brothers, 45% of the persons polled said that they would change careers if they could (*The Wall Street Journal*, September 1, 1992). Suppose this percentage is true for the 1992 population of all workers. A researcher wanted to check whether this percentage is still true. A sample of 350 workers taken recently by this researcher showed that (approximately) 39% of them would change careers if they could. Using the 2.5% significance level, can you conclude that the current percentage of workers who would change careers if they could is less than 45%?

9.106 Providing health insurance coverage to all people was one of the major issues in the recent presidential election. According to a 1992 survey, 45% of the American people would support higher taxes to pay for health insurance for people who cannot afford it (*U.S. News & World Report*, August 10, 1992). A consumer agency wanted to check whether or not this percentage is true. The agency recently took a sample of 400 people and found that 50% of them would support higher taxes to pay for health insurance for people who cannot afford it. Using the 1% significance level, can you conclude that the percentage of people who would support higher taxes to pay for health insurance for people who cannot afford it is greater than 45%?

9.107 According to the National Education Association, 12% of school teachers have a second job to supplement their incomes (*USA TODAY*, July 7, 1992). A random sample of 400 teachers taken this year showed that 19% of them hold a second job. Testing at the 1% significance level, can you conclude that the current percentage of all teachers who hold a second job to supplement their incomes is higher than 12%? Explain your conclusion.

9.108 In a 1985 study conducted by the Roper Organization, 45% of women aged 18 and older said they would "prefer to stay home and take care of a house and family" than "to have a job outside the home" (*The 1985 Virginia Slims American Women's Opinion Poll*). Assume that this percentage is true for the 1985 population of all women aged 18 and older. An economist wanted to check if this result still holds true. A random sample of 500 women aged 18 and older taken recently by this researcher showed that 39% of them hold this view. Test at the 5% significance level if the current percentage of women aged 18 and older who hold this view is less than 45%. Explain your conclusion.

9.109 In a 1987 study conducted by the Roper Organization, 66% of American adults said that factors within their personal control have determined their lot in life (*The American Dream*, A National Survey Conducted for *The Wall Street Journal* by the Roper Organization). Assume that this percentage is true for the 1987 population of all American adults. A recent study based on a random sample of 500 American adults showed that 71% of them hold this view.

    a. Test at the 2% significance level whether the current percentage of American adults who hold this view is different from 66%.

    b. What is the Type I error in this case? What is the probability of making this error?

9.110 Due to the high turnover rate, only a small percentage of senior executives of companies have been with the same company for a long time. According to an estimate, 22% of senior executives had been with the same company for 25 years or more in 1990 (*U.S. News & World Report*, June 8, 1992). Assume that this percentage is true for the 1990 population of all CEOs of major companies. An employment agency wanted to check whether this percentage has changed since then. A random sample of 200 senior executives taken recently by this agency showed that 17% of them have been with the same company for 25 years or more.

    a. Test at the 5% significance level whether the current percentage of senior executives who have been with the same company for 25 years or more is different from 22%.

    b. What is the Type I error in this case? What is the probability of making this error?

9.111 A food company is planning to market a new type of frozen yogurt. However, before marketing this yogurt, the company wants to find what percentage of the people like it. The company's management has decided that it will market this yogurt only if at least 35% of the people like it. The company's research department selected a random sample of 400 persons and asked them to taste this yogurt. Of these 400 persons, 128 said they liked it.

    a. Testing at the 2.5% significance level, can you conclude that the company should market this yogurt?

    b. What will your decision be in part a if the probability of making a Type I error is zero? Explain.

9.112 A mail-order company claims that at least 60% of all orders are mailed within 48 hours. From time to time the quality control department at the company checks if this promise is fulfilled. Recently the quality control department at this company took a sample of 400 orders and found that 224 of them were mailed within 48 hours of the placement of the orders.

    a. Testing at the 1% significance level, can you conclude that the company's claim is true?

    b. What will your decision be in part a if the probability of making a Type I error is zero? Explain.

9.113 Brooklyn Corporation manufactures computer diskettes. The machine that is used to make these diskettes is known to produce not more than 5% defective diskettes. The quality control inspector selects a sample of 200 diskettes each week and inspects them for being good or defective. Using the sample proportion, the quality control inspector tests the null hypothesis $p \leq .05$ against the alternative hypothesis $p > .05$, where p is the proportion of diskettes that are defective. She always uses a 2.5% significance level. If the null hypothesis is rejected, the production process is stopped to make any necessary adjustments. A recent such sample of 200 diskettes contained 17 defective diskettes.

    a. Using the 2.5% significance level, would you conclude that the production process should be stopped to make necessary adjustments?

    b. Perform the test of part a using a 1% significance level. Is your decision different from the one in part a?

Comment on the results of parts a and b.

9.114   Shulman Steel Corporation makes bearings that are supplied to other companies. One of the machines makes bearings that are supposed to have a diameter of 4 inches. The bearings that have a diameter of either more or less than 4 inches are considered defective and are discarded. When working properly, the machine does not produce more than 7% of bearings that are defective. The quality control inspector selects a sample of 200 bearings each week and inspects them for the size of their diameters. Using the sample proportion, the quality control inspector tests the null hypothesis $p \leq .07$ against the alternative hypothesis $p > .07$, where p is the proportion of bearings that are defective. He always uses a 2% significance level. If the null hypothesis is rejected, the machine is stopped to make any necessary adjustments. One such sample of 200 bearings taken recently contained 22 defective bearings.

    a. Using the 2% significance level, will you conclude that the machine should be stopped to make necessary adjustments?

    b. Perform the test of part a using a 1% significance level. Is your decision different from the one in part a?

Comment on the results of parts a and b.

9.115   Two years ago, 75% of the customers of a bank said that they were satisfied with the services provided by the bank. The manager of the bank wants to know if this percentage of satisfied customers has changed since then. She assigns this responsibility to you. Briefly explain how you would conduct such a test.

9.116   A study claims that 65% of students at all colleges and universities hold off-campus (part-time or full-time) jobs. You want to check if the percentage of students at your school who hold off-campus jobs is different from 65%. Briefly explain how you would conduct such a test. Collect data from 40 students at your school on whether or not they hold off-campus jobs. Then, calculate the proportion of students in this sample who hold off-campus jobs. Using this information, test the hypothesis. Select your own significance level.

GLOSSARY

α The significance level of a test of hypothesis that denotes the probability of rejecting a null hypothesis when it actually is true. (The probability of committing a Type I error.)

Alternative hypothesis   A claim about a population parameter that will be true if the null hypothesis is false.

β The probability of not rejecting a null hypothesis when it actually is false. (The probability of committing a Type II error.)

Critical value or **critical point**   One or two values that divide the whole region under the sampling distribution of a sample statistic into rejection and nonrejection regions.

Left-tailed test   A test in which the rejection region lies in the left tail of the distribution curve.

Null hypothesis   A claim about a population parameter that is assumed to be true until proven otherwise.

Observed value of z or t   The value of z or t calculated for a sample statistic such as the sample mean or the sample proportion.

One-tailed test A test in which there is only one rejection region, either in the left tail or in the right tail of the distribution curve.

***p*-value** The smallest significance level at which a null hypothesis can be rejected.

Right-tailed test A test in which the rejection region lies in the right tail of the distribution curve.

Significance level The value of α that gives the probability of committing a Type I error.

Test statistic The value of z or t calculated for a sample statistic such as the sample mean or the sample proportion.

Two-tailed test A test in which there are two rejection regions, one in each tail of the distribution curve.

Type I error An error that occurs when a true null hypothesis is rejected.

Type II error An error that occurs when a false null hypothesis is not rejected.

KEY FORMULAS

1. **Value of the test statistic z for $\bar{x}$ in a test of hypothesis about μ for a large sample**

$$z = \frac{\bar{x} - \mu}{\sigma_{\bar{x}}} \quad \text{if } \sigma \text{ is known, where } \sigma_{\bar{x}} = \frac{\sigma}{\sqrt{n}}$$

or

$$z = \frac{\bar{x} - \mu}{s_{\bar{x}}} \quad \text{if } \sigma \text{ is not known, where } s_{\bar{x}} = \frac{s}{\sqrt{n}}$$

2. **Value of the test statistic t for $\bar{x}$ in a test of hypothesis about μ for a small sample**

$$t = \frac{\bar{x} - \mu}{s_{\bar{x}}} \quad \text{where} \quad s_{\bar{x}} = \frac{s}{\sqrt{n}}$$

3. **Value of the test statistic z for $\hat{p}$ in a test of hypothesis about p for a large sample**

$$z = \frac{\hat{p} - p}{\sigma_{\hat{p}}} \quad \text{where} \quad \sigma_{\hat{p}} = \sqrt{\frac{pq}{n}}$$

SUPPLEMENTARY EXERCISES

9.117 Consider the following null and alternative hypotheses.

$$H_0:\ \mu = 120 \text{ versus } H_1:\ \mu > 120$$

A random sample of 81 observations taken from this population produced a sample mean of 123.5 and a sample standard deviation of 15.

    a. If this test is made at the 2.5% significance level, would you reject the null hypothesis?

    b. What is the probability of making a Type I error in part a?

    c. Calculate the *p*-value for the test. Based on this *p*-value, would you reject the null hypothesis if $\alpha = .01$? What if $\alpha = .05$?

    d. Suppose the null hypothesis mentioned above is false and the true population mean is 121. What is the probability of making a Type II error in part a? What is the power of the test?

9.118 Consider the following null and alternative hypotheses.

$$H_0:\ \mu = 40 \text{ versus } H_1:\ \mu \neq 40$$

A random sample of 64 observations taken from this population produced a sample mean of 38.4 and a sample standard deviation of 6.

a. If this test is made at the 5% significance level, would you reject the null hypothesis?
b. What is the probability of making a Type I error in part a?
c. Calculate the p-value for the test. Based on this p-value, would you reject the null hypothesis if $\alpha = .01$? What if $\alpha = .02$?
d. Suppose the null hypothesis mentioned above is false and the true population mean is 39. What is the probability of making a Type II error in part a? What is the power of the test?

9.119 Consider the following null and alternative hypotheses.

$$H_0: p = .82 \text{ versus } H_1: p \neq .82$$

A random sample of 600 observations taken from this population produced a sample proportion of .855.

a. If this test is made at the 2% significance level, would you reject the null hypothesis?
b. What is the probability of making a Type I error in part a?
c. Calculate the p-value for the test. Based on this p-value, would you reject the null hypothesis if $\alpha = .025$? What if $\alpha = .005$?
d. Suppose the null hypothesis mentioned above is false and the true population proportion is .84. What is the probability of making a Type II error in part a? What is the power of the test?

9.120 Consider the following null and alternative hypotheses.

$$H_0: p = .44 \text{ versus } H_1: p < .44$$

A random sample of 450 observations taken from this population produced a sample proportion of .39.

a. If this test is made at the 2% significance level, would you reject the null hypothesis?
b. What is the probability of making a Type I error in part a?
c. Calculate the p-value for the test. Based on this p-value, would you reject the null hypothesis if $\alpha = .01$? What if $\alpha = .025$?
d. Suppose the null hypothesis mentioned above is false and the true population proportion is .41. What is the probability of making a Type II error in part a? What is the power of the test?

9.121 A manufacturer of fluorescent light bulbs claims that the mean life of these bulbs is at least 2500 hours. A consumer agency wanted to check whether or not this claim is true. The agency took a random sample of 36 such bulbs and tested them. The mean life for the sample was found to be 2447 hours with a standard deviation of 180 hours.

a. Do you think the sample information supports the company's claim? Use $\alpha = 2.5\%$.
b. What is the Type I error in this case? Explain. What is the probability of making this error?
c. Will your conclusion of part a change if the probability of making a Type I error is zero?

9.122 According to the U.S. Bureau of the Census, the mean annual earnings of women 25 years of age or older with a master's degree were $30,781 in 1991. An economist wanted to find if the current mean annual earnings of such women are higher than $30,781. A recent random sample of 400 such women taken by this economist produced mean annual earnings of $31,249 with a standard deviation of $4,645.

a. Does the sample information support the alternative hypothesis that the current mean earnings of such women are greater than $30,781? Use $\alpha = .01$.
b. What is the Type I error in this case? Explain. What is the probability of making this error?
c. Will your conclusion of part a change if the probability of making a Type I error is zero?

9.123 According to the U.S. Bureau of Labor Statistics, the mean expenditure incurred on food by households was $4271 in 1991. A random sample of 500 households showed that they spent, on average, $4319 on food in 1993 with a standard deviation of $524.

a. Using $\alpha = .025$, can you conclude that the mean food expenditure for all households for 1993 is different from $4271?
b. Using $\alpha = .005$, can you conclude that the mean food expenditure for all households for 1993 is different from $4271?

Comment on the results of parts a and b.

9.124 During the last few years people have become more health conscious, especially in regard to the consumption of red meat. In 1990, the average consumption of red meat per person was 112.3 pounds in the United States. A sample of 100 persons showed that they consumed, on average, 106.8 pounds of red meat in 1993 with a standard deviation of 26.5 pounds.

    a. Using $\alpha = .05$, does the sample information support the alternative hypothesis that the 1993 mean consumption of red meat is different from 112.3 pounds?

    b. Using $\alpha = .01$, does the sample information support the alternative hypothesis that the 1993 mean consumption of red meat is different from 112.3 pounds?

Comment on the results of parts a and b.

9.125 Customers often complain about long waiting times at restaurants before the food is served. A restaurant claims that it serves food to its customers, on average, within 15 minutes after the order is placed. A local newspaper journalist wanted to check if the company's claim is true. A sample of 36 customers showed that the mean time taken to serve food to them was 15.9 minutes with a standard deviation of 2.4 minutes. Using the sample mean, the journalist says that the restaurant's claim is false. Do you think the journalist's conclusion is fair to the restaurant? Use the 1% significance level to answer this question.

9.126 The customers at a bank complained about long lines and the time they had to spend waiting for service. It is known that the customers at this bank had to wait 8 minutes, on average, before being served. The management made some changes to reduce the waiting time for its customers. A sample of 32 customers taken after these changes were made produced a mean waiting time of 7.4 minutes with a standard deviation of 2.1 minutes. Using this sample mean, the bank manager displayed a huge banner inside the bank mentioning that the mean waiting time for customers has been reduced by new changes. Do you think the bank manager's claim is justifiable? Use the 2.5% significance level to answer this question.

9.127 Refer to Exercise 9.121.

    a. Suppose the null hypothesis stated in that exercise is false and the true mean life of these bulbs is 2450 hours. Find the probability of making a Type II error when $\alpha = .025$.

    b. Suppose the null hypothesis stated in that exercise is false and the true mean life of these bulbs is 2400 hours. Find the probability of making a Type II error when $\alpha = .025$.

9.128 Refer to Exercise 9.124.

    a. Suppose the null hypothesis stated in that exercise is false and the true mean consumption of red meat is 105 pounds. Find the probability of making a Type II error when $\alpha = .05$.

    b. Suppose the null hypothesis stated in that exercise is false and the true mean consumption of red meat is 116 pounds. Find the probability of making a Type II error when $\alpha = .05$.

9.129 According to the U.S. Department of Labor, private sector workers earned, on average, $354.32 a week in 1991. A recently taken random sample of 400 private sector workers showed that they earn, on average, $362.50 a week with a standard deviation of $72. Find the p-value for the test with an alternative hypothesis that the current mean weekly salary of private sector workers is different from $354.32.

9.130 The mean consumption of water per household in a city was 1245 cubic feet per month. Due to a water shortage because of a drought, the city council campaigned for water usage conservation by households. A few months after the campaign was started, the mean consumption of water for a sample of 100 households was found to be 1175 cubic feet with a standard deviation of 250 cubic feet. Find the p-value for the hypothesis test that the mean consumption of water per household has decreased due to the campaign by the city council.

9.131 According to the Hertz Corporation, the mean repair and maintenance cost of a car in the United States was $1035 in 1986. A random sample of 27 cars showed that the mean repair and maintenance cost for these cars was $1446 in 1993 with a standard deviation of $244. Test at the 1% significance level if the 1993 mean repair and maintenance cost of a car is different from $1035. Assume that the 1993 repair and maintenance costs for all cars are approximately normally distributed.

9.132 The administrative office of a hospital claims that the mean waiting time for patients to get treatment in its emergency ward is 25 minutes. A random sample of 16 patients who received treatment in the emergency ward of this hospital produced a mean waiting time of 27.5 minutes with a standard deviation of 4.8 minutes. Using the 1% significance level, test whether the mean waiting time at the emergency ward is different from 25 minutes. Assume that the waiting times for all patients at this emergency ward have a normal distribution.

9.133 According to the American Medical Association, the mean annual income of surgeons was $236,400 in 1991 (*The Wall Street Journal*, August 5, 1992). A recently taken sample of 20 surgeons showed that their mean income for last year was $269,347 with a standard deviation of $45,372. Assume that the last year's incomes of all surgeons are approximately normally distributed.

    a. Using the 5% significance level, can you conclude that the last year's mean income of all surgeons is higher than $236,400?

    b. Suppose the probability of making a Type I error is zero. Can you make a decision for the test of part a without going through the five steps of hypothesis testing? If yes, what is your decision? Explain.

9.134 According to a 1992 survey by Priority Management Systems, adults spend an average of 114 minutes with their family per day (*USA TODAY*, July 22, 1992). Suppose this is true of all adults at the time of that survey. A recently taken sample of 25 adults showed that they spend an average of 109 minutes with their families. The sample standard deviation is 11 minutes. Assume that the time spent by adults with their families has an approximate normal distribution.

    a. Using the 1% significance level, test whether the mean time spent currently by all adults with their families is less than 114 minutes a day.

    b. Suppose the probability of making a Type I error is zero. Can you make a decision for the test of part a without going through the five steps of hypothesis testing? If yes, what is your decision? Explain.

9.135 A computer company that recently introduced a new software product claims that the mean time taken to learn how to use this software is not more than 2 hours for those people who are somewhat familiar with computers. A random sample of 12 such persons was selected. The following data give the time taken (in hours) by these persons to learn how to use this software.

| 1.75 | 2.25 | 2.40 | 1.90 | 1.50 | 2.75 |
| 2.15 | 2.25 | 1.80 | 2.20 | 3.25 | 2.60 |

Test at the 1% significance level if the company's claim is true. Assume that the time taken by all persons who are somewhat familiar with computers to learn how to use this software is approximately normally distributed.

9.136 A company claims that its 8-ounce low-fat yogurt cups contain, on average, at most 150 calories per cup. A consumer agency wanted to check whether or not this claim is true. A random sample of 10 such cups produced the following data on calories.

| 147 | 159 | 153 | 146 | 144 | 161 | 163 | 153 | 143 | 158 |

Test at the 2.5% significance level if the company's claim is true. Assume that the number of calories for such cups of yogurt produced by this company has an approximate normal distribution.

9.137 Economic reasons are compelling more and more women to participate in the labor force. According to the U.S. Bureau of Labor Statistics, 56% of mothers with children under the age of 6 were working outside their homes in 1988. A recently taken random sample of 900 mothers with children under the age of 6 showed that 64% of them work outside their homes.

    a. Test at the 1% significance level whether the current percentage of such mothers is higher than 56%.

    b. How will you explain the Type I error in this case? What is the probability of making this error?

9.138 The United States' increasing trade deficit with Japan during recent years has changed the attitude of a lot of Americans toward Japan. According to a survey conducted by the Daniel Yankelovich

Group and Marttila and Kiley Inc., 59% of the public see economic competitors like Japan as a greater danger than military adversaries, because economic competitors threaten their jobs and economic security (*Psychology Today*, September 1988). A random sample of 500 persons taken recently gave this percentage as 69%.

a. Test at the 5% significance level whether the current percentage of people who hold this view is different from 59%.

b. How will you explain the Type I error in this case? What is the probability of making this error?

9.139 More and more people are abandoning national brand products and buying store brand products to save money. The president of a company that produces national brand coffee claims that 40% of the people prefer to buy national brand coffee. A random sample of 700 people who buy coffee showed that 252 of them buy national brand coffee. Using $\alpha = .01$, can you conclude that the percentage of people who buy national brand coffee is different from 40%?

9.140 According to the U.S. Department of Labor, 57% of families had two or more wage earners in 1987. Assume that this percentage is true for the population of all 1987 families. A recent poll of 800 randomly selected families showed that 488 of them have two or more wage earners. Testing at the 2.5% significance level, can you conclude that the current percentage of families with two or more wage earners is higher than 57%?

9.141 Mong Corporation makes auto batteries. The company claims that 80% of its LL70 batteries are good for 70 months or more. A consumer agency wanted to check if this claim is true. The agency took a random sample of 40 batteries and found that 75% of them were good for 70 months or more.

a. Using the 1% significance level, can you conclude that the company's claim is false?

b. What will your decision be in part a if the probability of making a Type I error is zero? Explain.

9.142 Dartmouth Distribution Warehouse makes deliveries of a large number of products to its customers. To keep its customers happy and satisfied, the company's policy is to deliver on time at least 90% of all the orders it receives from its customers. The quality control inspector at the company quite often takes samples of orders delivered and checks if this policy is maintained. A recent such sample of 70 orders taken by this inspector showed that 59 of those were delivered on time.

a. Using the 2% significance level, can you conclude that the company's policy is maintained?

b. What will your decision be in part a if the probability of making a Type I error is zero? Explain.

∗9.143 Refer to Exercise 9.139. Find the *p*-value for the test of hypothesis mentioned in that exercise. Using this *p*-value, would you reject the null hypothesis at $\alpha = .05$? What if $\alpha = .02$?

∗9.144 Refer to Exercise 9.140. Find the *p*-value for the test of hypothesis mentioned in that exercise. Using this *p*-value, would you reject the null hypothesis at $\alpha = .005$? What if $\alpha = .02$?

∗9.145 Refer to Exercise 9.139. Find the probability of making a Type II error assuming that the null hypothesis mentioned in that exercise is false and that the true population proportion is .38. Use $\alpha = .01$.

∗9.146 Refer to Exercise 9.140. Find the probability of making a Type II error assuming that the null hypothesis mentioned in that exercise is false and the true population proportion is .59. Use $\alpha = .025$.

SELF-REVIEW TEST

1. A test of hypothesis is always about

a. a population parameter   **b.** a sample statistic   **c.** a test statistic

2. A Type I error is made when

    a. a null hypothesis is not rejected when it is actually false
    b. a null hypothesis is rejected when it is actually true
    c. an alternative hypothesis is rejected when it is actually true

3. A Type II error is made when

    a. a null hypothesis is not rejected when it is actually false
    b. a null hypothesis is rejected when it is actually true
    c. an alternative hypothesis is rejected when it is actually true

4. A critical value is the value

    a. calculated from sample data
    b. determined from a table (e.g., the normal distribution table)
    c. neither a nor b

5. The computed value of a test statistic is the value

    a. calculated for a sample statistic
    b. determined from a table (e.g., the normal distribution table)
    c. neither a nor b

6. The observed value of a test statistic is the value

    a. calculated for a sample statistic
    b. determined from a table (e.g., the normal distribution table)
    c. neither a nor b

7. The significance level, denoted by α, is

    a. the probability of committing a Type I error
    b. the probability of committing a Type II error
    c. neither a nor b

8. The value of β gives the

    a. probability of committing a Type I error
    b. probability of committing a Type II error
    c. power of the test

9. The value of $1 - \beta$ gives the

    a. probability of committing a Type I error
    b. probability of committing a Type II error
    c. power of the test

10. A two-tailed test is a test with

    a. two rejection regions   **b.** two nonrejection regions   **c.** two test statistics

11. A one-tailed test

    a. has one rejection region   **b.** has one nonrejection region   **c.** both a and b

12. The smallest level of significance at which a null hypothesis will be rejected is called

    a. α **b.** p-value   **c.** β

13. Which of the following is not required to apply the t distribution to make a test of hypothesis about μ?

    a. $n < 30$ **b.** Population is normally distributed   **c.** σ is unknown   **d.** β is known

14. The sign in the alternative hypothesis in a two-tailed test is always

    a. $<$ **b.** $>$ **c.** $\neq$

15. The sign in the alternative hypothesis in a left-tailed test is always

    a. $<$ **b.** $>$ **c.** $\neq$

16. The sign in the alternative hypothesis in a right-tailed test is always

    a. $<$ **b.** $>$ **c.** $\neq$

17. A bank loan officer claims that the mean monthly mortgage payment made by all home owners in a certain city is $1365. A housing magazine wanted to test this claim. A random sample of 100 home owners taken by this magazine produced the mean monthly mortgage of $1489 with a standard deviation of $278.

    a. Testing at the 1% significance level, would you conclude that the mean monthly mortgage payment made by all home owners in this city is different from $1365?

    b. What is the Type I error in part a? What is the probability of making this error?

    c. What will your decision be in part a if the probability of making a Type I error is zero? Explain.

18. An editor of a New York publishing company claims that the mean time taken to write a textbook is at least 15 months. A sample of 16 textbook authors showed that the mean time taken by them to write a textbook was 12.5 months with a standard deviation of 3.6 months.

    a. Using the 2.5% significance level, would you conclude that the editor's claim is true? Assume that the time taken to write a textbook is normally distributed for all textbook authors.

    b. What is the Type I error in part a? What is the probability of making this error?

    c. What will your decision be in part a if the probability of making a Type I error is .001?

19. In a *Time*/CNN poll of adult Americans, 73% said "there should be more government spending on educational and recreational facilities for teenagers to reduce teenage violence" (*Time*, June 12, 1989). Among a recent sample of 400 adult Americans, 77% hold this view.

    a. Test at the 1% significance level if the current percentage of adult Americans who hold this view is higher than 73%.

    b. What is the Type I error in part a? What is the probability of making this error?

    c. What will your decision be in part a if the probability of making a Type I error is zero? Explain.

20. According to an IRS study, it takes an average of 60 minutes to prepare, copy, and mail a 1040EZ tax form. A sample of 100 taxpayers who filed the 1040EZ form last year showed that they took, on average, 62.6 minutes to prepare, copy, and mail this tax form. The standard deviation for the sample was 11 minutes.

    a. Find the p-value for the test that the mean time taken to prepare, copy, and mail a 1040EZ tax form is different from 60 minutes.

    b. Using the p-value calculated in part a, will you reject the null hypothesis if $\alpha = .01$? What if $\alpha = .05$?

21. Refer to Problem 17.

    a. Suppose the null hypothesis stated in that problem is false and the true population mean is $1375. Find the probability of making a Type II error. What is the power of the test?

    b. Suppose the null hypothesis stated in that problem is false and the true population mean is $1400. Find the probability of making a Type II error. What is the power of the test?

*22. Refer to Problem 19.

    a. Find the p-value for the test of hypothesis mentioned in part a of that problem.

    b. Using this p-value, will you reject the null hypothesis if $\alpha = .01$? What if $\alpha = .05$?

*23. Refer to Problem 19.

    a. Suppose the null hypothesis stated in that problem is false and the true population proportion is .75. Find the probability of making a Type II error. What is the power of the test?

    b. Suppose the null hypothesis stated in that problem is false and the true population proportion is .79. Find the probability of making a Type II error. What is the power of the test?

USING MINITAB

The MINITAB commands ZTEST and TTEST are used to make hypothesis tests about the population mean using the normal and the t distributions, respectively. MINITAB does not have commands like ZTEST to perform tests of hypotheses about the population proportion. Although a program can be written to make such tests, we will not do so in this text.

HYPOTHESIS TESTS ABOUT A POPULATION MEAN: LARGE SAMPLES

When the sample size is large ($n \geq 30$), whether σ is known or not, we can use the normal distribution to make a hypothesis test about a population mean. The MINITAB commands to make such a test using the normal distribution are as shown in Figure 9.22.

Figure 9.22   MINITAB commands for a test of hypothesis about μ for large samples.

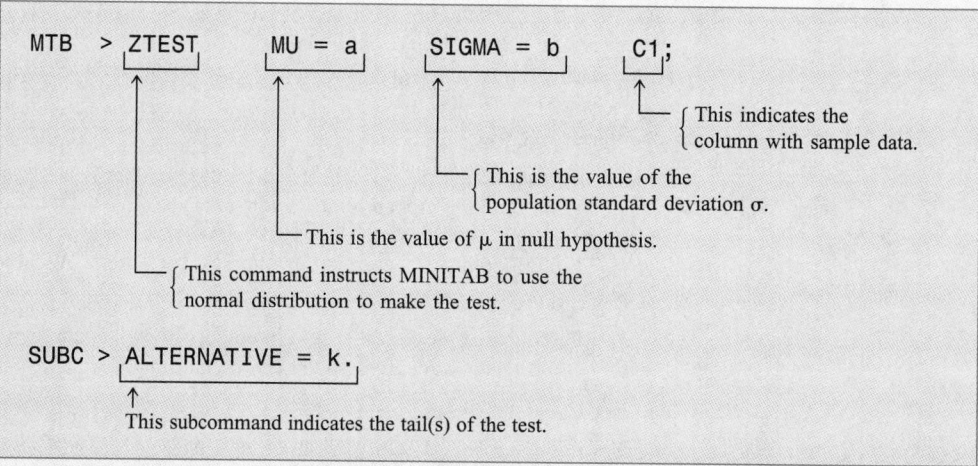

Note the semicolon at the end of the first MINITAB command in Figure 9.22. This instructs MINITAB that a subcommand is to follow. The period at the end of subcommand ALTERNATIVE = k indicates the end of subcommands. In the first MINITAB command, we replace a by the value of μ in the null hypothesis and b by the value of σ.

In the subcommand, k is replaced by either -1, 0, or 1 depending on whether the test is *left-tailed*, *two-tailed*, or *right-tailed*, respectively. In other words,

$$\text{ALTERNATIVE} = -1 \qquad \text{if the test is left-tailed}$$

$$\text{ALTERNATIVE} = 0 \qquad \text{if the test is two-tailed}$$

$$\text{ALTERNATIVE} = 1 \qquad \text{if the test is right-tailed}$$

Illustration M9–1 describes the use of these commands.

Illustration M9–1 An earlier study showed that the mean time spent by all college students on community service is 50 minutes a week with a standard deviation of 27 minutes. The following data give the time spent on community service by each of 34 college students based on a recently taken random sample.

34	56	74	23	12	89	87	56	48	13	9	85
76	56	17	28	66	38	46	81	29	33	41	78
11	57	17	22	91	54	19	35	65	47		

Using MINITAB, test at the 5% significance level whether the mean time spent doing community service by all college students is different from 50 minutes. Assume that the population standard deviation is 27.

Solution From the given information,

$$\sigma = 27 \qquad \text{and} \qquad \alpha = .05$$

We are to test whether the mean time spent doing community service by all college students is different from 50 minutes. The null and alternative hypotheses are

$$H_0: \mu = 50$$

$$H_1: \mu \neq 50$$

To perform this test, we first enter the given data on 34 college students in column C1 using the SET command. Then we write the MINITAB command that indicates that the test is to be made using the normal distribution with $\mu = 50$ and $\sigma = 27$. This MINITAB command also includes the information that the test is to be made using the sample data entered in column C1. Because the test is two-tailed, the value of ALTERNATIVE in the subcommand will be 0.

Figure 9.23 MINITAB output for a test of hypothesis about μ for a large sample.

```
MTB  > NOTE: MAKING A TEST OF HYPOTHESIS FOR ILLUSTRATION M9-1
MTB  > SET C1
DATA > 34    56    74    23    12    89    87    56    48
DATA > 13     9    85    76    56    17    28    66    38
DATA > 46    81    29    33    41    78    11    57    17
DATA > 22    91    54    19    35    65    47
DATA > END
MTB  > ZTEST    MU = 50    SIGMA = 27    C1;
SUBC > ALTERNATIVE = 0.

TEST OF MU = 50.000 VS MU N.E. 50.000
THE ASSUMED SIGMA = 27.0
```

Figure 9.23   *(Continued)*

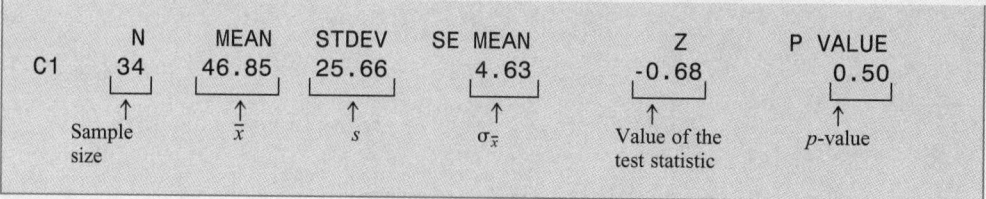

In the MINITAB solution given in Figure 9.23, TEST OF MU = 50.000 VS MU N.E. 50.000 indicates that the null hypothesis is that μ is equal to 50 and the alternative hypothesis is that μ is not equal to (N.E.) 50. In the row of C1 in MINITAB output, 34 is the sample size, 46.85 is the mean of the sample data, 25.66 is the standard deviation of the sample data, 4.63 is the standard deviation (or standard error) of the sample mean, -0.68 is the observed value of the test statistic z and 0.50 is the *p*-value discussed in Section 9.4 of this chapter.

To make a decision, we find the critical values of z from the normal distribution table for a two-tailed test with $\alpha = .05$. These values (for .4750 area between the mean and z) are approximately -1.96 and 1.96. Because the observed value of the test statistic, $z = -.68$, is between the two critical values of z and it falls in the nonrejection region, we fail to reject the null hypothesis. (The reader is advised to draw a graph that shows the rejection and nonrejection regions.) Based on this sample information, we can conclude that the mean time spent doing community service by college students is not different from 50 minutes a week.

We can also use the P VALUE, printed in the MINITAB solution, to make the decision. As we know from Section 9.4, we will reject H_0 if the value of α is larger than the *p*-value and we will fail to reject H_0 if the value of α is smaller than the *p*-value. In the MINITAB solution of Figure 9.23, the *p*-value is .50. As $\alpha = .05$ is smaller than the *p*-value of .50, we fail to reject H_0.   ▬

If the population standard deviation is not known but the sample size is large, we first compute the standard deviation of C1 and then use that value as an estimate of sigma in the ZTEST command as we did in the case of the ZINTERVAL command in Figure 8.12 of Chapter 8. An alternative to this is to use the t distribution to make a test of hypothesis about μ irrespective of the sample size when σ is not known.

HYPOTHESIS TESTS ABOUT A POPULATION MEAN: SMALL SAMPLES

We know from the discussion in this chapter that we apply the t distribution to make a hypothesis test about the population mean when

1.   The sample size is small, that is, $n < 30$
2.   The population is (approximately) normally distributed
3.   We do not know the population standard deviation

The MINITAB TTEST commands to make such a test using the t distribution are as shown in Figure 9.24.

Figure 9.24 MINITAB commands for a test of hypothesis about μ for small samples.

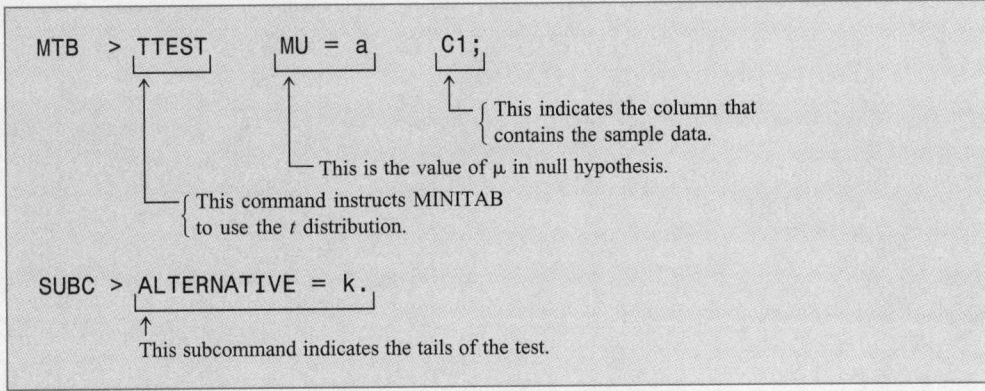

Note that because σ is not known, its value is not mentioned in these MINITAB commands. MINITAB will use the value of the sample standard deviation calculated for the data of column C1 as an estimate of σ to make a test of hypothesis. The value of k for ALTERNATIVE in the subcommand will again be −1, 0, or 1, depending on whether the test is left-tailed, two-tailed, or right-tailed. Illustration M9–2 describes the use of these commands.

Illustration M9–2   Customers do not like to wait for a long time in line at supermarkets. The management of a supermarket wants to keep the mean waiting time for its customers to at most 5 minutes. The manager of the store took a sample of 20 customers over a period of one week. The following data give the time that these 20 customers spent in line at this store before being served.

7	4	9	12	3	5	2	8	9	10
14	10	8	15	11	6	4	9	6	13

Based on this sample information, would you conclude that the mean waiting time for all customers visiting this supermarket is greater than 5 minutes? Use MINITAB and $\alpha = .025$ to make the test. Assume that the times that all customers spend in line waiting for service at this supermarket have a normal distribution.

Solution   We are to test whether the mean waiting time for all customers visiting this supermarket is greater than 5 minutes. The null and alternative hypotheses are

$$H_0: \mu = 5 \text{ minutes}$$

$$H_1: \mu > 5 \text{ minutes}$$

The test is right-tailed. We assign a value of 1 to ALTERNATIVE in the MINITAB subcommand. To make this test, first we enter the given data values in column C1 using the SET command. Then we write the MINITAB command that indicates that the test is to be made using the t distribution, that $\mu = 5$, and the test is to be made using the sample data entered in column C1.

Figure 9.25   MINITAB output for a test of hypothesis about μ for a small sample.

```
MTB  > NOTE: MAKING A TEST OF HYPOTHESIS FOR ILLUSTRATION M9-2
MTB  > SET C1
DATA >  7   4   9  12   3   5   2   8   9  10
DATA > 14  10   8  15  11   6   4   9   6  13
DATA > END
MTB  > TTEST   MU = 5    C1;
SUBC > ALTERNATIVE = 1.

TEST OF MU = 5.000 VS MU G.T. 5.000
```

	N	MEAN	STDEV	SE MEAN	T	P VALUE
C1	20	8.250	3.669	0.820	3.96	0.0004

       Sample $\bar{x}$ s $s_{\bar{x}}$ Value of the p-value
       size   test statistic

In the MINITAB solution of Figure 9.25, TEST OF MU = 5.000 VS MU G.T. 5.000 indicates that the null hypothesis is that the mean is equal to 5 and that the alternative hypothesis is that the mean is greater than (G.T.) 5. In the row of C1 in the MINITAB solution, 20 represents the sample size, 8.250 is the mean of the sample data, 3.669 is the standard deviation of the sample data, 0.820 is the standard deviation (or standard error) of the sample mean, 3.96 is the value of the test statistic t calculated as $(\bar{x} - \mu)/s_{\bar{x}} = (8.250 - 5)/0.820$, and 0.0004 is the p-value.

To make a decision, we find the critical value of t from the t distribution table for a right-tailed test with .025 area in the right tail and $df = 20 - 1 = 19$. This value of t is 2.093. Because the value of the test statistic $t = 3.96$ is greater than 2.093, it falls in the rejection region. (The reader is advised to draw a graph showing the rejection and nonrejection regions.) Consequently, we reject the null hypothesis. Thus, based on this sample information, we conclude that the mean waiting time for all customers visiting this supermarket is greater than 5 minutes.

We can reach the same conclusion by using the p-value. As $\alpha = .025$ is larger than the p-value of .0004, we reject the null hypothesis.

COMPUTER ASSIGNMENTS

M9.1   Refer to Data Set III of Appendix A. Column C3 of that data set contains 1990 incomes before taxes for 200 households. Based on these data, can you conclude that the mean income before taxes for 1990 for all households was different from $45,000? Use MINITAB and $\alpha = .05$.

M9.2   Refer to Data Set III of Appendix A. Column C4 of that data set contains the amount of social security paid in 1990 by 200 households. Based on these data, test if the mean social security payments in 1990 by all households were more than $2500. Use MINITAB and $\alpha = .01$.

M9.3   Refer to Data Set III of Appendix A. Column C7 of that data set contains annual food expenditures for 1990 for 200 households. Based on these data, can you conclude that the mean annual food expenditure for 1990 for all households was less than $3000? Use MINITAB and $\alpha = .025$.

M9.4 The past records of a supermarket show that its customers spend an average of $65 per visit at this store. Recently the management of the store initiated a promotional campaign according to which each customer receives points based on the total money spent at the store and these points can be used to buy products at the store. The management expects that as a result of this campaign, the customers should be encouraged to spend more at the store. To check whether this is true, the manager of the store took a sample of 12 customers who visited the store. The following data give the money (in dollars) spent by these customers at this supermarket during their visits.

$88	69	141	28	106	45
32	51	78	54	110	83

Assume that the money spent by all customers at this supermarket has a normal distribution. Using the 1% significance level, can you conclude that the mean money spent by all customers at this supermarket after the campaign was started is higher than $65? Use MINITAB to answer this question.

M9.5 A past study claims that adults in America spend an average of 18 hours a week on leisure activities. A researcher wanted to test this claim. She took a sample of 10 adults and asked them about the time they spend per week on leisure activities. Their responses (in hours) are as follows.

14	25	22	38	16	26	19	23	41	33

Assume that the times spent on leisure activities by all adults are normally distributed. Using the 2.5% significance level, can you conclude that the claim of the earlier study is true? Use MINITAB to answer this question.

1. Owners of two donut shops are independently considering moving their businesses to new locations. Each owner's business is currently in a location where an average of 2000 cars pass each day. The null hypothesis H_0: $\mu = 2000$ will be tested in each case for the new locations.

    a. What is the meaning of the unknown parameter μ in this problem? Explain.
    b. Write the alternative hypothesis H_1 that Owner I should use if he likes his current location and does not want to move to a new location, unless he is fairly confident it will be busier than the current location.
    c. Write the alternative hypothesis H_1 that Owner II should use if she is unhappy at her current location and she does not want to pass up a new location, unless she is fairly confident the new location will not be busier.

2. Reconsider Exercise 1. Owner I randomly selected 20 days and counted the number of cars passing by the new location he's considering. This survey produced the following data.

1830	2115	2085	2218	2057	1930	2408	1982	2035	2192
2322	2018	2263	1911	2178	2007	2521	1739	2289	1881

    a. Assuming the standard deviation of the number of cars passing the new location to be 180 (the same as his current location), would you recommend Owner I change locations?
    b. If Owner I does decide to move to the new location based on these data, what probability of moving erroneously must he allow?
    c. Owner I guesses that the standard deviation of the number of cars passing the new location is 180, but what if this value is incorrect? How far off can this guess be for Owner I to still decide to change locations based on these data if a .05 level of significance is used?

3. Consider Owner II from Exercise 1. Even though she does not like her current location, business there is profitable. In fact, she knows she needs at least 1800 cars to pass her business on an average day for the business to be profitable.

    a. How would you advise Owner II in her decision-making process? Can Owner II protect herself against passing up a busier location and also protect herself against moving to a location where her business will fail? If so, how? Be specific and include recommended levels of protection.
    b. Assuming the standard deviation of the number of cars passing the new location each day is known to be 250 (the same as her current location), on how many randomly selected days should she observe the number of cars passing the new location to provide the protection you recommend in part a?

4. A restaurant owner would like to estimate the amount of money an average household in his area spends on dining out each week. To do this, he decides to randomly select 100 of his customers and ask each how much money was spent by members of their household on dining out during the past week. As a point estimate of the mean weekly amount of money spent on dining out by all households in his area, he will compute $\bar{x}$ the average amount spent by the 100 customers. How would you critique the restaurant owner's estimation procedure? Is there anything wrong with this procedure that would result in a sampling error or bias? If so, identify the problems and suggest ways to alleviate them.

5. A thumbtack that is tossed on a desk can land in one of the following two ways.

Heads

Tails

Brad and Dan cannot agree on the likelihood of obtaining a head or a tail. Brad argues that obtaining a tail is more likely than obtaining a head because of the shape of the tack. If the tack had no point at all it would resemble a coin that has the same probability of coming up heads or tails when tossed. And, the longer the point, the less likely it is that the tack will stand up on its head when tossed. Dan believes that as the tack lands tails, the point causes the tack to jump around and come to rest in the heads position. Brad and Dan need you to settle their dispute. Do you think the tack is equally likely to land heads or tails? To investigate this question, find an ordinary thumbtack and toss it a large number of times (say 100 times).

    a. What is the meaning, in words, of the unknown parameter in this problem?

    b. Set up the null and alternative hypotheses and compute the p-value based on your results from tossing the tack.

    c. How would you answer the original question now? If you decide the tack is not fair, do you side with Brad or Dan?

    d. What would you estimate the value of the parameter in part a to be? Find a 90% confidence interval for this parameter.

    e. After doing this experiment, do you feel 100 tosses are enough to infer about the nature of your tack? Using your result as a preliminary estimate, determine how many tosses would be necessary to be 95% certain of having 4% accuracy; that is, the maximum error of estimate is $\pm 4\%$. Have you observed enough tosses?

6. A U.S. senator heading a committee preparing to propose changes in the federal income tax structure hires a statistician to estimate the proportion of all households with annual incomes before taxes exceeding $70,000. The statistician uses the 1991 Interview Survey data collected by the Bureau of Labor Statistics.

    a. If the statistician uses the data given in column C3 of Data Set III of Appendix A, what 99% confidence interval will she get for this unknown parameter?

    b. Are the 200 observations in Data Set III enough to be able to estimate the proportion of all households with annual incomes before taxes in excess of $70,000 to within 2% with a 99% confidence level?

    c. How many observations should the statistician use to be 99% confident of estimating the true proportion to within 2%?

7. When an insured individual is sued, the insurance company issuing the policy must hire legal counsel to defend the suit if the company feels the suit is unfounded. In fact, anticipated legal expenses account for much, if not most, of the premiums charged for certain types of insurance (for example, umbrella policies). Insurance companies must choose law firms based on a balance between the cost of the defense and the firm's success in defending suits. An insurance company has been using Miller and Associates to defend its catastrophic loss cases and knows that this firm either wins or favorably settles 80% of such suits, with an average cost of $75,000 per case. Cases that are either lost or not settled favorably cost the insurance company an average of $1.5 million each, in addition to the legal representation fees. Insurance company executives are reasonably happy with Miller and Associates. However, a vice-president at the insurance company has asked the claims supervisor to consider using the Rose and Powers law firm to defend its catastrophic loss suits in an effort to keep legal costs down. A sample of 15 recent catastrophic loss suits defended by the Rose and Powers law firm reveals that the average defense cost was $69,000 with a standard deviation of $20,000. Two of these 15 law suits were either lost or not settled favorably. How might this sample information be used to determine if the insurance company should switch law firms? What decision would you suggest the claims supervisor make?

8. Samples of employees from each of the 50 companies listed in Data Set I of Appendix A will be used to independently test the null hypothesis that the mean company salary equals the national average, against the alternative hypothesis that the mean company salary exceeds the national average. Each test will be performed using a .10 level of significance. It should be noted that each of these 50 companies has a mean salary of all employees equal to the national average.

    a. For how many of the 50 companies do you expect it will be incorrectly concluded that the mean salary of all employees exceeds the national average?

    b. What is the probability that it will be concluded that at least one of the 50 companies has a mean salary of all employees exceeding the national average?

9. The weights of the packages handled by a parcel delivery service have a mean of 20 pounds and a standard deviation of 6 pounds. The van that is used to pick up these parcels has a load limit of 1500 pounds. On a particular day, the van must pick up 68 packages from different locations.

    a. Find the probability the load limit of the van will not be exceeded.

    b. What is the highest number of packages the van can pick up so that the probability the load limit of the van will be exceeded is no more than .05?

10. Refer to Exercise 9 of the More Challenging Exercises for Chapters 4 to 6. What does the Central Limit Theorem say about the mean amount won (or lost) per play? What would it say if the number of plays was infinitely large?

10 ESTIMATION AND HYPOTHESIS TESTING: TWO POPULATIONS

C hapters 8 and 9 discussed the estimation and hypothesis-testing procedures for μ and p involving a single population. This chapter extends the discussion of estimation and hypothesis testing procedures to the difference between two population means and the difference between two population proportions. For example, we may want to make a confidence interval for the difference between mean prices of houses in California and in New York. Or we may want to test the hypothesis that the mean price of houses in California is different from that in New York. As another example, we may want to make a confidence interval for the difference between the proportions of defective items in all items manufactured on each of the two machines. Or we may want to test the hypothesis that the proportion of defective items in all items manufactured on Machine I is different from the proportion of defective items in all items manufactured on Machine II. Constructing confidence intervals and testing hypotheses about population parameters are referred to as *making inferences.*

10.1 INFERENCES ABOUT THE DIFFERENCE BETWEEN TWO POPULATION MEANS FOR LARGE AND INDEPENDENT SAMPLES

Let μ_1 be the mean of the first population and μ_2 be the mean of the second population. Suppose we want to make a confidence interval and test a hypothesis about the difference between these two population means, that is, $\mu_1 - \mu_2$. Let $\bar{x}_1$ be the mean of a sample taken from the first population and $\bar{x}_2$ be the mean of a sample taken from the second population. Then, $\bar{x}_1 - \bar{x}_2$ is the sample statistic that is used to make an interval estimate and to test a hypothesis about $\mu_1 - \mu_2$. This section discusses how to make confidence intervals and test hypotheses about $\mu_1 - \mu_2$ when the two samples are large and independent. As discussed in earlier chapters, in the case of μ, a sample is considered to be large if it contains 30 or more observations. The concept of independent and dependent samples is explained next.

10.1.1 INDEPENDENT VERSUS DEPENDENT SAMPLES

Two samples are **independent** if they are drawn from two different populations and the elements of one sample have no relationship to the elements of the second sample. If the elements of the two samples are somehow related, then the samples are said to be **dependent**. Thus, in two independent samples, the selection of one sample has no effect on the selection of the second sample.

> **INDEPENDENT VERSUS DEPENDENT SAMPLES**
>
> Two samples drawn from two populations are independent if the selection of one sample from one population does not affect the selection of the second sample from the second population. Otherwise, the samples are dependent.

Examples 10–1 and 10–2 illustrate independent and dependent samples, respectively.

Illustrating two independent samples.

EXAMPLE 10–1 Suppose we want to estimate the difference between the mean salaries of all male and all female executives. To do so, we draw two samples, one from the population of male executives and another from the population of female executives. These two samples are independent because they are drawn from two different populations and the samples have no effect on each other. ▬

Illustrating two dependent samples.

EXAMPLE 10–2 Suppose we want to estimate the difference between the mean number of customers served per hour by all tellers working at a bank before and after the tellers take a new training course. To accomplish this, suppose we take a sample of 20 tellers and observe the mean number of customers served per hour by them before and after this course. Note that these two samples include the same 20 tellers. This is an example of two dependent samples. Such samples are also called *paired* or *matched samples*. ▬

This section and Sections 10.2 and 10.4 discuss how to make confidence intervals and test hypotheses about the difference between two population parameters when samples are independent. Section 10.3 discusses how to make confidence intervals and test hypotheses about the difference between two population means when samples are dependent.

10.1.2 THE MEAN, STANDARD DEVIATION, AND SAMPLING DISTRIBUTION OF $\bar{x}_1 - \bar{x}_2$

Suppose we draw two (independent) large samples from two different populations that are referred to as population 1 and population 2. Let

μ_1 = the mean of population 1

μ_2 = the mean of population 2

σ_1 = the standard deviation of population 1

σ_2 = the standard deviation of population 2

n_1 = the size of the sample drawn from population 1 ($n_1 \geq 30$)

n_2 = the size of the sample drawn from population 2 ($n_2 \geq 30$)

$\bar{x}_1$ = the mean of the sample drawn from population 1

$\bar{x}_2$ = the mean of the sample drawn from population 2

Then, from the central limit theorem, $\bar{x}_1$ is approximately normally distributed with mean μ_1 and standard deviation $\sigma_1/\sqrt{n_1}$, and $\bar{x}_2$ is approximately normally distributed with mean μ_2 and standard deviation $\sigma_2/\sqrt{n_2}$.

Using these results, we can now make the following statements about the mean, the standard deviation, and the shape of the sampling distribution of $\bar{x}_1 - \bar{x}_2$. Figure 10.1 shows the sampling distribution of $\bar{x}_1 - \bar{x}_2$.

1. The mean of $\bar{x}_1 - \bar{x}_2$, denoted by $\mu_{\bar{x}_1 - \bar{x}_2}$, is

$$\mu_{\bar{x}_1 - \bar{x}_2} = \mu_1 - \mu_2$$

2. The standard deviation of $\bar{x}_1 - \bar{x}_2$, denoted by $\sigma_{\bar{x}_1 - \bar{x}_2}$, is[1]

$$\sigma_{\bar{x}_1 - \bar{x}_2} = \sqrt{\frac{\sigma_1^2}{n_1} + \frac{\sigma_2^2}{n_2}}$$

3. Regardless of the shapes of the two populations, the shape of the sampling distribution of $\bar{x}_1 - \bar{x}_2$ is approximately normal. This is so because the difference between two normally distributed random variables is also normally distributed. Note again that for this to hold true, both samples must be large.

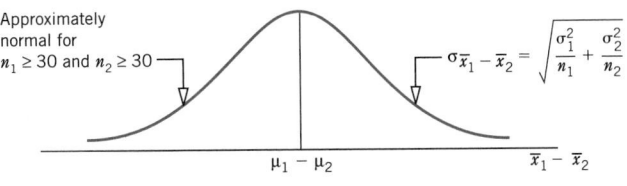

Figure 10.1

---

[1]The formula for the standard deviation of $\bar{x}_1 - \bar{x}_2$ can also be written as

$$\sigma_{\bar{x}_1 - \bar{x}_2} = \sqrt{\sigma_{\bar{x}_1}^2 + \sigma_{\bar{x}_2}^2}$$

where $\sigma_{\bar{x}_1} = \sigma_1/\sqrt{n_1}$ and $\sigma_{\bar{x}_2} = \sigma_2/\sqrt{n_2}$.

THE SAMPLING DISTRIBUTION, MEAN, AND STANDARD DEVIATION OF $\bar{x}_1 - \bar{x}_2$

For two large and independent samples, taken from two different populations, the sampling distribution of $\bar{x}_1 - \bar{x}_2$ is (approximately) normal with its mean and standard deviation as follows.

$$\mu_{\bar{x}_1 - \bar{x}_2} = \mu_1 - \mu_2 \quad \text{and} \quad \sigma_{\bar{x}_1 - \bar{x}_2} = \sqrt{\frac{\sigma_1^2}{n_1} + \frac{\sigma_2^2}{n_2}}$$

However, we usually do not know the standard deviations σ_1 and σ_2 of the two populations. In such cases, we replace $\sigma_{\bar{x}_1 - \bar{x}_2}$ by its point estimator $s_{\bar{x}_1 - \bar{x}_2}$, which is calculated as follows.

AN ESTIMATE OF THE STANDARD DEVIATION OF $\bar{x}_1 - \bar{x}_2$

The value of $s_{\bar{x}_1 - \bar{x}_2}$, which gives an estimate of $\sigma_{\bar{x}_1 - \bar{x}_2}$, is calculated as

$$s_{\bar{x}_1 - \bar{x}_2} = \sqrt{\frac{s_1^2}{n_1} + \frac{s_2^2}{n_2}}$$

where s_1 and s_2 are the standard deviations of the two samples taken from the two populations.

Thus, when both samples are large, the sampling distribution of $\bar{x}_1 - \bar{x}_2$ is approximately normal. Consequently, in such cases, we use the normal distribution to make a confidence interval and to test a hypothesis about $\mu_1 - \mu_2$.

10.1.3   INTERVAL ESTIMATION OF $\mu_1 - \mu_2$

By constructing a confidence interval for $\mu_1 - \mu_2$ we find the difference between the means of two populations. For example, we may want to find the difference between the mean salaries of male and female managers. The difference between the two sample means $\bar{x}_1 - \bar{x}_2$ is the point estimator of the difference between the two population means $\mu_1 - \mu_2$. Again, in this section we assume that the two samples are large and independent. When these assumptions hold true, we use the normal distribution to make a confidence interval for the difference between the two population means. The following formula gives the interval estimation for $\mu_1 - \mu_2$.

CONFIDENCE INTERVAL FOR $\mu_1 - \mu_2$

The $(1 - \alpha)100\%$ confidence interval for $\mu_1 - \mu_2$ is

$$(\bar{x}_1 - \bar{x}_2) \pm z\sigma_{\bar{x}_1 - \bar{x}_2} \qquad \text{if } \sigma_1 \text{ and } \sigma_2 \text{ are known}$$

$$(\bar{x}_1 - \bar{x}_2) \pm zs_{\bar{x}_1 - \bar{x}_2} \qquad \text{if } \sigma_1 \text{ and } \sigma_2 \text{ are not known}$$

The value of z is obtained from the normal distribution table for the given confidence level. The values of $\sigma_{\bar{x}_1 - \bar{x}_2}$ and $s_{\bar{x}_1 - \bar{x}_2}$ are calculated as explained earlier.

Examples 10–3 and 10–4 illustrate the procedure to construct a confidence interval for $\mu_1 - \mu_2$ for large samples. In Example 10–3 the population standard deviations are known, and in Example 10–4 they are not known.

Constructing a
confidence interval
for $\mu_1 - \mu_2$:
σ_1 and σ_2 known.

EXAMPLE 10-3   According to the Bureau of Labor Statistics, in 1992 construction workers earned an average of $538 per week and manufacturing workers earned an average of $470 per week. Assume that these mean earnings have been calculated for samples of 500 and 700 workers taken from the two populations, respectively. Further assume that the standard deviations of weekly earnings of the two populations are $66 and $60, respectively.

(a)   What is the point estimate of $\mu_1 - \mu_2$?

(b)   Construct a 95% confidence interval for the difference between the mean weekly earnings of the two populations.

Solution   Refer to all construction workers as population 1 and all manufacturing workers as population 2. The respective samples, then, are samples 1 and 2. Let μ_1 and μ_2 be the means of populations 1 and 2, and let $\bar{x}_1$ and $\bar{x}_2$ be the means of the respective samples. From the given information,

$$n_1 = 500, \quad \bar{x}_1 = \$538, \quad \sigma_1 = \$66$$

$$n_2 = 700, \quad \bar{x}_2 = \$470, \quad \sigma_2 = \$60$$

(a)   The point estimate of $\mu_1 - \mu_2$ is given by the value of $\bar{x}_1 - \bar{x}_2$. Thus,

$$\text{Point estimate of } \mu_1 - \mu_2 = \$538 - \$470 = \$68$$

(b)   The confidence level is $1 - \alpha = .95$.

First, we calculate the standard deviation of $\bar{x}_1 - \bar{x}_2$ as follows.

$$\sigma_{\bar{x}_1 - \bar{x}_2} = \sqrt{\frac{\sigma_1^2}{n_1} + \frac{\sigma_2^2}{n_2}} = \sqrt{\frac{(66)^2}{500} + \frac{(60)^2}{700}} = 3.7222$$

Next, we find the z value for the 95% confidence level. From the normal distribution table, this value of z is 1.96.

Finally, substituting all the values in the confidence interval formula, we obtain the 95% confidence interval for $\mu_1 - \mu_2$ as

$$(\bar{x}_1 - \bar{x}_2) \pm z\sigma_{\bar{x}_1 - \bar{x}_2} = (538 - 470) \pm 1.96 \, (3.7222)$$

$$= 68 \pm 7.30 = \textbf{\$60.70 to \$75.30}$$

Thus, with 95% confidence we can state that the difference in the mean weekly earnings of all construction workers and all manufacturing workers in 1992 was between $60.70 and $75.30.   ■

Constructing a
confidence interval
for $\mu_1 - \mu_2$:
σ_1 and σ_2 not known.

EXAMPLE 10-4   Ann Howard conducted a management progress study with AT&T management job holders to assess the performance of those who possessed a college degree and those who did not (Ann Howard, "College Experiences and Managerial Performance," *Journal of Applied Psychology*, 71(3), 1986, pp. 530–552). Let us refer to the two groups of participants as college and noncollege participants, respectively. The samples included 274 college and 148 noncollege participants. In the area of motivation for advancement, the mean scores were 2.89 for college participants and 2.70 for noncollege participants, with the two standard deviations being .57 and .64, respectively. Find a 99% confidence interval for the difference between the mean scores of the two respective populations in the area of motivation for advancement.

Solution   Let all the management job holders with a college degree be referred to as population 1 and the ones without a college degree be referred to as population 2. We can refer to the respective samples as samples 1 and 2. Let μ_1 and μ_2 be the means of populations 1 and 2, respectively, and let $\bar{x}_1$ and $\bar{x}_2$ be the means of the respective samples. From the given information,

$$n_1 = 274, \qquad \bar{x}_1 = 2.89, \qquad s_1 = .57$$
$$n_2 = 148, \qquad \bar{x}_2 = 2.70, \qquad s_2 = .64$$

The confidence level is $1 - \alpha = .99$.

Because σ_1 and σ_2 are not known, we replace $\sigma_{\bar{x}_1 - \bar{x}_2}$ by $s_{\bar{x}_1 - \bar{x}_2}$ in the confidence interval formula. The value of $s_{\bar{x}_1 - \bar{x}_2}$ is

$$s_{\bar{x}_1 - \bar{x}_2} = \sqrt{\frac{s_1^2}{n_1} + \frac{s_2^2}{n_2}} = \sqrt{\frac{(.57)^2}{274} + \frac{(.64)^2}{148}} = .0629$$

From the normal distribution table, the z value for a 99% confidence level is (approximately) 2.58. The 99% confidence interval for $\mu_1 - \mu_2$ is

$$(\bar{x}_1 - \bar{x}_2) \pm z s_{\bar{x}_1 - \bar{x}_2} = (2.89 - 2.70) \pm 2.58\,(.0629)$$
$$= .19 \pm .16 = \textbf{.03 to .35}$$

Thus, with 99% confidence we can state that the difference in the mean scores of the two populations of managers in the area of motivation for advancement is between .03 and .35.   ▪

10.1.4   HYPOTHESIS TESTING ABOUT $\mu_1 - \mu_2$

It is often necessary to compare the means of two populations. For example, we may want to know if the mean price of houses in Chicago is the same as that in Los Angeles. Similarly, we may be interested in knowing if, on average, the mean number of units of a product assembled by using Process I is less than those assembled by using Process II. In both these cases we will perform a test of hypothesis about $\mu_1 - \mu_2$. The alternative hypothesis in a test of hypothesis may be that the means of two populations are different, or that the mean of the first population is greater than the mean of the second population, or that the mean of the first population is smaller than the mean of the second population. These three situations are described below.

1.   Testing an alternative hypothesis that the means of two populations are different is equivalent to $\mu_1 \neq \mu_2$, which is the same as $\mu_1 - \mu_2 \neq 0$.
2.   Testing an alternative hypothesis that the mean of the first population is greater than the mean of the second population is equivalent to $\mu_1 > \mu_2$, which is the same as $\mu_1 - \mu_2 > 0$.
3.   Testing an alternative hypothesis that the mean of the first population is smaller than the mean of the second population is equivalent to $\mu_1 < \mu_2$, which is the same as $\mu_1 - \mu_2 < 0$.

The procedure followed to perform a test of hypothesis about the difference between two population means is similar to the one used to test hypotheses about single population parameters in Chapter 9. The procedure involves the same five steps that were used in Chapter 9 to test hypotheses about μ and p. Because we are dealing with large (and independent) samples in this section, we will use the normal distribution to conduct a test of hypothesis about $\mu_1 - \mu_2$.

TEST STATISTIC z FOR $\bar{x}_1 - \bar{x}_2$

The value of the test statistic z for $\bar{x}_1 - \bar{x}_2$ is computed as

$$z = \frac{(\bar{x}_1 - \bar{x}_2) - (\mu_1 - \mu_2)}{\sigma_{\bar{x}_1 - \bar{x}_2}}$$

The value of $\mu_1 - \mu_2$ is substituted from H_0. If the values of σ_1 and σ_2 are not known, we replace $\sigma_{\bar{x}_1 - \bar{x}_2}$ by $s_{\bar{x}_1 - \bar{x}_2}$ in the formula.

Making a two-tailed test of hypothesis about $\mu_1 - \mu_2$: large samples.

EXAMPLE 10–5 Reconsider Example 10–3 on the mean weekly earnings of construction workers and manufacturing workers. Test at the 1% significance level if the mean weekly earnings of the two groups of workers are different.

Solution From the information given in Example 10–3,

$$n_1 = 500, \qquad \bar{x}_1 = \$538, \qquad \sigma_1 = \$66$$

$$n_2 = 700, \qquad \bar{x}_2 = \$470, \qquad \sigma_2 = \$60$$

where the subscript 1 refers to construction workers and 2 to manufacturing workers. The significance level is $\alpha = .01$. Let

$$\mu_1 = \text{the mean weekly earnings of all construction workers}$$

$$\mu_2 = \text{the mean weekly earnings of all manufacturing workers}$$

Step 1. *State the null and alternative hypotheses*

We are to test if the two population means are different. The two possibilities are
(i) The mean weekly earnings of construction workers and manufacturing workers are not different. In other words, $\mu_1 = \mu_2$, which can be written as $\mu_1 - \mu_2 = 0$.
(ii) The mean weekly earnings of construction workers and manufacturing workers are different. That is, $\mu_1 \neq \mu_2$, which can be written as $\mu_1 - \mu_2 \neq 0$.

Considering these two possibilities, the null and alternative hypotheses are

$$H_0: \mu_1 - \mu_2 = 0 \qquad \text{(the mean weekly earnings are not different)}$$

$$H_1: \mu_1 - \mu_2 \neq 0 \qquad \text{(the mean weekly earnings are different)}$$

Step 2. *Select the distribution to use*

Because $n_1 > 30$ and $n_2 > 30$, both sample sizes are large. Therefore, the sampling distribution of $\bar{x}_1 - \bar{x}_2$ is approximately normal, and we use the normal distribution to make the hypothesis test.

Step 3. *Determine the rejection and nonrejection regions*

The significance level is given to be .01. The $\neq$ sign in the alternative hypothesis indicates that the test is two-tailed. The area in each tail of the normal distribution curve is

$\alpha/2 = .01/2 = .005$. The critical values of z for .005 area in each tail of the normal distribution are (approximately) 2.58 and -2.58. These values are shown in Figure 10.2.

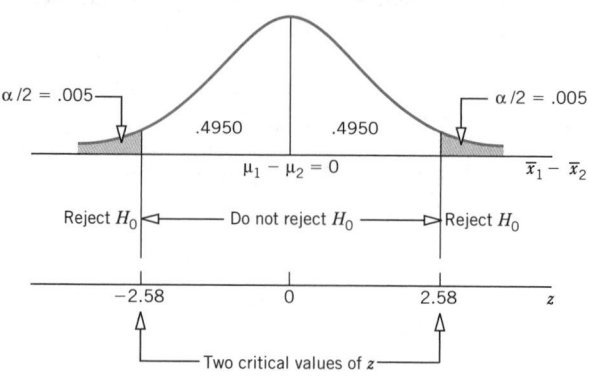

Figure 10.2

Step 4. *Calculate the value of the test statistic*

The value of the test statistic z for $\bar{x}_1 - \bar{x}_2$ is computed as follows.

$$\sigma_{\bar{x}_1 - \bar{x}_2} = \sqrt{\frac{\sigma_1^2}{n_1} + \frac{\sigma_2^2}{n_2}} = \sqrt{\frac{(66)^2}{500} + \frac{(60)^2}{700}} = 3.7222$$

$$z = \frac{(\bar{x}_1 - \bar{x}_2) - (\mu_1 - \mu_2)}{\sigma_{\bar{x}_1 - \bar{x}_2}} = \frac{(538 - 470) - 0}{3.7222} = 18.27$$

$-$From H_0

Step 5. *Make a decision*

Because the value of the test statistic $z = 18.27$ falls in the rejection region, we reject the null hypothesis H_0. Therefore, we conclude that the mean weekly earnings of the two groups of workers are different. Note that we cannot say for sure that the two means are different. All we can say is that the evidence from the two samples is very strong that the corresponding population means are different.   ▬

Conducting a right-tailed test of hypothesis about $\mu_1 - \mu_2$: large samples.

EXAMPLE 10-6   Refer to Example 10–4 on the mean scores of college and noncollege participants in the management progress study done by Ann Howard. Test at the 5% significance level if the mean score in the area of motivation for advancement is higher for college degree holders than for noncollege participants.

Solution   From the information given in Example 10–4,

$$n_1 = 274, \qquad \bar{x}_1 = 2.89, \qquad s_1 = .57$$
$$n_2 = 148, \qquad \bar{x}_2 = 2.70, \qquad s_2 = .64$$

where subscript 1 refers to college degree holders and 2 to no~ The significance level is $\alpha = .05$.

Step 1. *State the null and alternative hypotheses*

The two possibilities are
(i)   The mean score of college degree holders is not higher than that of the noncollege participants, which can be written as $\mu_1 = \mu_2$ or $\mu_1 - \mu_2 = 0$.

(ii) The mean score of college degree holders is higher than that of the noncollege partici-pants, which can be written as $\mu_1 > \mu_2$ or $\mu_1 - \mu_2 > 0$.

The null and alternative hypotheses are

$$H_0: \mu_1 - \mu_2 = 0 \qquad (\mu_1 \text{ is equal to } \mu_2)$$

$$H_1: \mu_1 - \mu_2 > 0 \qquad (\mu_1 \text{ is greater than } \mu_2)$$

Note that we can also write the null hypothesis as $\mu_1 - \mu_2 \le 0$, which states that the mean score of college participants is lower than or equal to the mean score of noncollege partici-pants.

Step 2. Select the distribution to use

Because $n_1 > 30$ and $n_2 > 30$, both sample sizes are large. As a result, we use the normal distribution to make the test.

Step 3. Determine the rejection and nonrejection regions

The significance level is .05. The $>$ sign in the alternative hypothesis indicates that the test is right-tailed. Consequently, the critical value of z is 1.65, as shown in Figure 10.3.

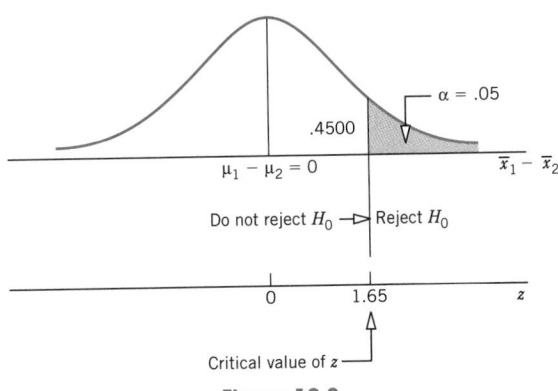

Figure 10.3

Step 4. Calculate the value of the test statistic

The value of the test statistic z for $\bar{x}_1 - \bar{x}_2$ is computed as follows.

$$s_{\bar{x}_1 - \bar{x}_2} = \sqrt{\frac{s_1^2}{n_1} + \frac{s_2^2}{n_2}} = \sqrt{\frac{(.57)^2}{274} + \frac{(.64)^2}{148}} = .0629$$

$$z = \frac{(\bar{x}_1 - \bar{x}_2) - (\mu_1 - \mu_2)}{s_{\bar{x}_1 - \bar{x}_2}} = \frac{(2.89 - 2.70) - 0}{.0629} = 3.02$$

From H_0

Step 5. Make a decision

Because the value of the test statistic $z = 3.02$ for $\bar{x}_1 - \bar{x}_2$ falls in the rejection region, we reject the null hypothesis H_0. Therefore, we conclude that the mean score in the area of motivation for advancement is higher for those who hold a college degree than for those who do not hold a college degree.

Case Study 10–1 further illustrates the application of the confidence interval and hypothesis testing for the difference between two population means.

CASE STUDY 10-1 INFERENCES MADE BY THE BUREAU OF THE CENSUS

The U.S. Bureau of the Census publishes a large number of reports every year that contain results of various surveys done by the bureau. Using these survey results, the bureau estimates many population parameters like the mean, median, and proportion. It also tests the hypotheses about the change in the values of these parameters over time. Following are two statements quoted from two of the bureau's recent publications. The figures in parentheses give the 90% confidence intervals of the estimates.

1. "Between 1990 and 1991, (real) per capita income declined 2.5 (± 1.0) percent to $14,617 ($\pm$ $123)."

2. "The average household size in 1992 of 2.62 (± 0.01) persons was significantly smaller than a decade or so ago."

The first statement gives two estimates. First, it gives a 90% confidence interval for the decline in real per capita income between 1990 and 1991. (Real per capita income is actually the mean income per person adjusted for inflation.) This confidence interval is 2.5% $\pm$ 1.0% = 1.5% to 3.5%. This indicates that the real per capita income in 1991 was somewhere between 1.5% and 3.5% lower than the real per capita income in 1990. This is simply a 90% confidence interval for $\mu_1 - \mu_2$ presented as a percentage of μ_1 where μ_1 is the per capita income for 1990 and μ_2 is the per capita income for 1991. The second estimate in the first statement is that the real per capita income in the United States in 1991 was $14,617 $\pm$ $123 = $14,494 to $14,740.

The first part of the second statement gives an estimate of the mean household size in the United States in 1992. From the information given in this statement, we can say with 90% confidence that the mean household size in 1992 was 2.62 $\pm$ .01 = 2.61 to 2.63. The second part of this statement gives a conclusion of a test of hypothesis. The conclusion is that the mean household size in 1992 was significantly smaller than that of a decade or so ago. Let μ_1 be the mean number of persons per household in 1992 and μ_2 be the mean number of persons per household a decade or so ago (say 1980). The bureau performed the following test of hypothesis.

$$H_0: \mu_1 = \mu_2$$

$$H_1: \mu_1 < \mu_2$$

Based on the information f

of the Census, April 1993.

EXERCISES

Concepts and Procedures

10.1 Briefly explain the meaning of independent and dependent samples. Give one example of each.

10.2 Describe the sampling distribution of $\bar{x}_1 - \bar{x}_2$ for large and independent samples. What are the mean and standard deviation of this sampling distribution?

10.3 The following information is obtained from two independent samples selected from two populations.

$$n_1 = 150 \qquad \bar{x}_1 = 5.56 \qquad s_1 = 1.65$$
$$n_2 = 170 \qquad \bar{x}_2 = 4.80 \qquad s_2 = 1.58$$

    a. What is the point estimate of $\mu_1 - \mu_2$?
    b. Construct a 99% confidence interval for $\mu_1 - \mu_2$.

10.4 The following information is obtained from two independent samples selected from two populations.

$$n_1 = 300 \qquad \bar{x}_1 = 25.0 \qquad s_1 = 4.9$$
$$n_2 = 250 \qquad \bar{x}_2 = 28.5 \qquad s_2 = 4.5$$

    a. What is the point estimate of $\mu_1 - \mu_2$?
    b. Construct a 95% confidence interval for $\mu_1 - \mu_2$.

10.5 Refer to the information given in Exercise 10.3. Test at the 5% significance level if the two population means are different.

10.6 Refer to the information given in Exercise 10.4. Test at the 1% significance level if the two population means are different.

10.7 Refer to the information given in Exercise 10.3. Test at the 1% significance level if μ_1 is greater than μ_2.

10.8 Refer to the information given in Exercise 10.4. Test at the 5% significance level if μ_1 is less than μ_2.

Applications

10.9 A company recently opened two supermarkets in two different areas. The management wants to know if the mean sales per day for these two supermarkets are different. A sample of 35 days for Supermarket A produced mean daily sales of $53.70 thousand with a standard deviation of $2.90 thousand. A sample of 30 days for Supermarket B produced mean daily sales of $58.45 thousand with a standard deviation of $3.10 thousand.

    a. Find the point estimate of $\mu_1 - \mu_2$ where μ_1 and μ_2 are the mean daily sales of supermarkets A and B, respectively.
    b. Construct a 99% confidence interval for the difference between the mean daily sales for these two supermarkets.
    c. Test at the 1% significance level if the mean daily sales for these two supermarkets are different.

10.10 According to the U.S. Bureau of the Census, the mean income of families was $38,608 in 1988 and $43,704 in 1991. Assume that these means are based on sample sizes of 1800 families for 1988 and 2200 families for 1991. Further assume that the standard deviations for the two populations are $9,569 and $9,885, respectively.

    a. Find the point estimate of $\mu_1 - \mu_2$ where μ_1 and μ_2 are the mean incomes of all families in 1988 and 1991, respectively.
    b. Construct a 95% confidence interval for the difference between the two population means.
    c. Test at the 5% significance level if the 1988 mean income for all families is different from that for 1991.

10.11   According to a management progress study done by Ann Howard (see Example 10–4), the mean score in motivation for advancement for a sample of 43 management job holders with a master's degree was 2.92 with a standard deviation of .46. The mean score in the same area for a sample of 112 management job holders with a bachelor's degree was 2.81 with a standard deviation of .57.

   a.   Let μ_1 and μ_2 be the mean scores in motivation for advancement for all management job holders with a master's degree and with a bachelor's degree, respectively. What is the point estimate of $\mu_1 - \mu_2$?
   b.   Make a 90% confidence interval for the difference between the mean scores of the two populations.
   c.   Test at the 1% significance level if the mean score of master's degree holders is higher than that for bachelor's degree holders.

10.12   According to the authors of a study about employee reliability, "Although theft is a major problem, it is only one factor of the larger construct of organizational delinquency. Excessive absences, tardiness, malingering, equipment damage, drug and alcohol abuse, grievances, suspensions from work, insubordination, and ordinary rule infractions are all components of the delinquency syndrome" (Joyce Hogan and Robert Hogan, "How to Measure Employee Reliability." *Journal of Applied Psychology*, 74(2), April 1989). The authors measured the employee reliability scores for male and female employees. A sample of 1637 male employees produced a mean score of 45.4 with a standard deviation of 8.0, and a sample of 590 female employees gave a mean score of 46.5 with a standard deviation of 8.3.

   a.   Let μ_1 and μ_2 be the mean reliability scores of male and female employees, respectively. What is the point estimate of $\mu_1 - \mu_2$?
   b.   Construct a 97% confidence interval for the difference between the mean reliability scores of all male and all female employees.
   c.   Test at the 2% significance level if the mean reliability scores for all male and all female employees are different.

10.13   A business consultant wanted to investigate if providing day care facilities on premises by companies reduces the absentee rate of working mothers with 6-year-old or younger children. She took a sample of 45 such mothers from companies that provide day care facilities on premises. These mothers missed an average of 6.4 days from work last year with a standard deviation of 1.20 days. Another sample of 50 such mothers taken from companies that do not provide day care facilities on premises showed that these mothers missed an average of 9.3 days last year with a standard deviation of 1.85 days.

   a.   Construct a 98% confidence interval for the difference between the two population means.
   b.   Test at the 2.5% significance level if the mean number of days missed per year by mothers working for companies that provide day care facilities on premises is less than the mean number of days missed per year by mothers working for companies that do not provide day care facilities on premises.
   c.   What is the Type I error and its probability for the test of hypothesis in part b? Explain.

10.14   According to Metropolitan Life Insurance Company's claims data, the mean hospital and physician's charges for coronary bypass surgery are $24,710 for New York State and $27,670 for Texas (*Statistical Bulletin*, January–March 1989). The two means are based on 55 claims for New York State and 186 claims for Texas. Assume that these claims are representative of all such surgeries performed in these two states. Further assume that the population standard deviations for such surgeries for these two states are $1300 and $1450, respectively.

wage in February 1993 was $13.62 for transportation and public utility workers and $11.61 for manufacturing workers. Assume that these two estimates are based on random samples of 1000 and 1200 workers taken, respectively, from the two populations. Further assume that the standard deviations of the two populations are $1.85 and $1.40, respectively.

a. Construct a 95% confidence interval for the difference between the mean hourly wages of the two populations.

b. Test at the 5% significance level if the mean hourly wage for transportation and public utility workers in February 1993 was higher than that of manufacturing workers.

c. What will your decision be in part b if the probability of making a Type I error is zero? Explain.

10.16 Professors Paul W. Kingston and Steven L. Nock studied the time spent together by single- and dual-earner couples. According to the records kept by wives in this study, the mean time spent together by a husband and wife watching television was 61.6 minutes per day for single-earner couples and 44.4 minutes per day for dual-earner couples. The respective sample sizes were 144 and 177 and the sample standard deviations were 79.0 and 56.8 minutes, respectively (Paul William Kingston and Steven L. Nock, "Time Together Among Dual-earner Couples." *American Sociological Review*, 52(3), June 1987, pp. 391–400).

a. Construct a 99% confidence interval for the difference between the two population means.

b. Test at the 1% significance level if the mean time spent together watching television by single-earner couples is higher than that of dual-earner couples.

c. What will your decision be in part b if the probability of making a Type I error is zero? Explain.

10.17 The management at the New Century Bank claims that the mean waiting time for all customers at its branches is less than that at the Public Bank, which is its main competitor. A business consulting firm took a sample of 200 customers from the New Century Bank and found that they waited an average of 4.60 minutes with a standard deviation of 1.2 minutes before being served. Another sample of 300 customers taken from the Public Bank showed that these customers waited an average of 4.85 minutes with a standard deviation of 1.5 minutes before being served.

a. Make a 97% confidence interval for the difference between the two population means.

b. Test at the 2.5% significance level if the claim of the management of the New Century Bank is true.

c. Calculate the *p*-value for the test of part b. Based on this *p*-value, would you reject the null hypothesis if $\alpha = .01$? What if $\alpha = .05$?

10.18 Maine Mountain Dairy claims that its 8-ounce low-fat yogurt cups contain, on average, fewer calories than the 8-ounce low-fat yogurt cups produced by a competitor. A consumer agency wanted to check this claim. A sample of 50 such yogurt cups produced by this company showed that they contained an average of 144 calories per cup with a standard deviation of 5.4 calories. A sample of 40 such yogurt cups produced by its competitor showed that they contained an average of 147 calories per cup with a standard deviation of 6.3 calories.

a. Make a 98% confidence interval for the difference between the mean number of calories in the 8-ounce low-fat yogurt cups produced by the two companies.

b. Test at the 1% significance level if Maine Mountain Dairy's claim is true.

c. Calculate the *p*-value for the test of part b. Based on this *p*-value, would you reject the null hypothesis if $\alpha = .005$? What if $\alpha = .025$?

10.2 INFERENCES ABOUT THE DIFFERENCE BETWEEN TWO POPULATION MEANS FOR SMALL AND INDEPENDENT SAMPLES: EQUAL STANDARD DEVIATIONS

Many times, due to either budget constraint or the nature of the populations, it may not be possible to take large samples to make inferences about the difference between two population means. This section discusses how to make a confidence interval and test a hypothesis about the difference between two population means when the samples are small ($n_1 < 30$ and $n_2 < 30$) and independent. Our main assumption in this case is that the two populations from which the two samples are drawn are (approximately) normally distributed. If this assumption is true, and we know the population standard deviations, we can still use the normal distribution to make inferences about $\mu_1 - \mu_2$ when samples are small and independent. However, we usually do not know the population standard deviations σ_1 and σ_2. In such cases, we replace the normal distribution by the t distribution to make inferences about $\mu_1 - \mu_2$ for small and independent samples. We will make one more assumption in this section that the standard deviations of the two populations are equal. In other words, we assume that although σ_1 and σ_2 are unknown, they are equal.[2]

WHEN TO USE THE t DISTRIBUTION TO MAKE INFERENCES ABOUT $\mu_1 - \mu_2$

The t distribution is used to make inferences about $\mu_1 - \mu_2$ when the following assumptions hold true.

1. The two populations from which the two samples are drawn are (approximately) normally distributed.
2. The samples are small ($n_1 < 30$ and $n_2 < 30$) and independent.
3. The standard deviations σ_1 and σ_2 of the two populations are unknown but they are equal, that is, $\sigma_1 = \sigma_2$.

When the standard deviations of the two populations are equal, we can use σ for both σ_1 and σ_2. Since σ is unknown, we replace it by its point estimator s_p, which is called the **pooled sample standard deviation** (hence, the subscript p). The value of s_p is computed by using the information from the two samples as follows.

---

[2]If the two population standard deviations are unknown and unequal (and other conditions of this section are satisfied), then the degrees of freedom of the t distribution are given by

$$df = \frac{\left(\dfrac{s_1^2}{n_1} + \dfrac{s_2^2}{n_2}\right)^2}{\dfrac{\left(\dfrac{s_1^2}{n_1}\right)^2}{n-1} + \dfrac{\left(\dfrac{s_2^2}{n_2}\right)^2}{}}$$

same as discussed in Subsections 10.2.1 and

THE POOLED STANDARD DEVIATION FOR TWO SAMPLES

The pooled standard deviation for two samples is computed as

$$s_p = \sqrt{\frac{(n_1 - 1)s_1^2 + (n_2 - 1)s_2^2}{n_1 + n_2 - 2}}$$

where n_1 and n_2 are the sizes of the two samples and s_1^2 and s_2^2 are the variances of the two samples.

In this formula, $n_1 - 1$ are the degrees of freedom for sample 1, $n_2 - 1$ are the degrees of freedom for sample 2, and $n_1 + n_2 - 2$ *are the degrees of freedom for the two samples taken together.*

When s_p is used as an estimator of σ, the standard deviation $\sigma_{\bar{x}_1 - \bar{x}_2}$ of $\bar{x}_1 - \bar{x}_2$ is estimated by $s_{\bar{x}_1 - \bar{x}_2}$. The value of $s_{\bar{x}_1 - \bar{x}_2}$ is calculated by using the following formula.

ESTIMATOR OF THE STANDARD DEVIATION OF $\bar{x}_1 - \bar{x}_2$

The estimator of the standard deviation of $\bar{x}_1 - \bar{x}_2$ is

$$s_{\bar{x}_1 - \bar{x}_2} = s_p \sqrt{\frac{1}{n_1} + \frac{1}{n_2}}$$

Now we are ready to discuss the procedures that are used to make confidence intervals and test hypotheses about $\mu_1 - \mu_2$ for small and independent samples taken from two populations with unknown but equal standard deviations.

10.2.1 INTERVAL ESTIMATION OF $\mu_1 - \mu_2$

As was mentioned earlier, the difference between the two sample means $\bar{x}_1 - \bar{x}_2$ is the point estimator of the difference between the two population means $\mu_1 - \mu_2$. The following formula gives the confidence interval for $\mu_1 - \mu_2$ when the t distribution is used.

CONFIDENCE INTERVAL FOR $\mu_1 - \mu_2$

The $(1 - \alpha)100\%$ confidence interval for $\mu_1 - \mu_2$ is

$$(\bar{x}_1 - \bar{x}_2) \pm t s_{\bar{x}_1 - \bar{x}_2}$$

where the value of t is obtained from the t distribution table for the given confidence level and $n_1 + n_2 - 2$ degrees of freedom, and $s_{\bar{x}_1 - \bar{x}_2}$ is calculated as explained earlier in Section 10.2.

Example 10–7 describes the procedure to make a confidence interval for $\mu_1 - \mu_2$ using the t distribution.

Constructing a confidence interval for $\mu_1 - \mu_2$: small and independent samples, and $\sigma_1 = \sigma_2$.

EXAMPLE 10–7   A consumer agency wanted to estimate the difference in the mean amounts of caffeine in two brands of coffee. The agency took a sample of 15 one-pound jars of Brand I coffee that showed the mean amount of caffeine in these jars to be 80 milligrams per jar with a standard deviation of 5 milligrams. Another sample of 12 one-pound jars of Brand II coffee gave a mean amount of caffeine equal to 77 milligrams per jar with a standard deviation of 6 milligrams. Construct a 95% confidence interval for the difference between the mean amounts of caffeine in one-pound jars of these two brands of coffee. Assume that the two populations are normally distributed and that the standard deviations of the two populations are equal.

Solution   Let μ_1 and μ_2 be the mean amounts of caffeine per jar in all one-pound jars of Brands I and II, respectively, and let $\bar{x}_1$ and $\bar{x}_2$ be the means of the two respective samples. From the given information,

$$n_1 = 15 \qquad \bar{x}_1 = 80 \text{ milligrams} \qquad s_1 = 5 \text{ milligrams}$$
$$n_2 = 12 \qquad \bar{x}_2 = 77 \text{ milligrams} \qquad s_2 = 6 \text{ milligrams}$$

The confidence level is $1 - \alpha = .95$.
First we calculate the standard deviation of $\bar{x}_1 - \bar{x}_2$ as follows.

$$s_p = \sqrt{\frac{(n_1 - 1)s_1^2 + (n_2 - 1)s_2^2}{n_1 + n_2 - 2}} = \sqrt{\frac{(15 - 1)(5)^2 + (12 - 1)(6)^2}{15 + 12 - 2}} = 5.4626$$

$$s_{\bar{x}_1 - \bar{x}_2} = s_p \sqrt{\frac{1}{n_1} + \frac{1}{n_2}} = (5.4626)\sqrt{\frac{1}{15} + \frac{1}{12}} = 2.1157$$

Next, to find the t value from the t distribution table, we need to know the area in each tail of the t distribution and the degrees of freedom.

$$\text{Area in each tail} = \alpha/2 = .5 - (.95/2) = .025$$

$$\text{Degrees of freedom} = n_1 + n_2 - 2 = 15 + 12 - 2 = 25$$

The t value for $df = 25$ and .025 area in the right tail of the t distribution is 2.060. The 95% confidence interval for $\mu_1 - \mu_2$ is

$$(\bar{x}_1 - \bar{x}_2) \pm t s_{\bar{x}_1 - \bar{x}_2} = (80 - 77) \pm 2.060 \, (2.1157)$$

10.2.2 HYPOTHESIS TESTING ABOUT $\mu_1 - \mu_2$

When the three assumptions mentioned in Section 10.2 are satisfied, then the t distribution is applied to make a hypothesis test about the difference between two population means. The test statistic in this case is t, which is calculated as follows.

TEST STATISTIC t FOR $\bar{x}_1 - \bar{x}_2$

The value of the test statistic t for $\bar{x}_1 - \bar{x}_2$ is computed as

$$t = \frac{(\bar{x}_1 - \bar{x}_2) - (\mu_1 - \mu_2)}{s_{\bar{x}_1 - \bar{x}_2}}$$

The value of $\mu_1 - \mu_2$ in this formula is substituted from the null hypothesis and $s_{\bar{x}_1 - \bar{x}_2}$ is calculated as explained in Section 10.2.

Examples 10–8 and 10–9 illustrate how a test of hypothesis about the difference between two population means is conducted using the t distribution.

Making a two-tailed test of hypothesis about $\mu_1 - \mu_2$: small and independent samples, and $\sigma_1 = \sigma_2$.

EXAMPLE 10–8 A sample of 14 cans of Brand 1 diet soda gave the mean number of calories of 23 per can with a standard deviation of 3 calories. Another sample of 16 cans of Brand 2 diet soda gave the mean number of calories of 25 per can with a standard deviation of 4 calories. At the 1% significance level, are the mean number of calories per can different for these two brands of diet soda? Assume that the calories per can of diet soda are normally distributed for each of the two brands and that the standard deviations for the two populations are equal.

Solution Let μ_1 and μ_2 be the mean number of calories per can for diet soda of Brand 1 and Brand 2, respectively, and let $\bar{x}_1$ and $\bar{x}_2$ be the means of the respective samples. From the given information,

$$n_1 = 14 \qquad \bar{x}_1 = 23 \qquad s_1 = 3$$
$$n_2 = 16 \qquad \bar{x}_2 = 25 \qquad s_2 = 4$$

The significance level is $\alpha = .01$.

Step 1. *State the null and alternative hypotheses*

We are to test for the difference in the mean number of calories per can for two brands. The null and alternative hypotheses are

$$H_0: \mu_1 - \mu_2 = 0 \qquad \text{(the mean number of calories are not different)}$$
$$H_1: \mu_1 - \mu_2 \neq 0 \qquad \text{(the mean number of calories are different)}$$

Step 2. *Select the distribution to use*

The two populations are normally distributed, the samples are small and independent, and the standard deviations of the two populations are unknown but equal. Consequently, we will use the t distribution.

Step 3. *Determine the rejection and nonrejection regions*

The $\neq$ sign in the alternative hypothesis indicates that the test is two-tailed. The significance level is .01. Hence,

$$\text{Area in each tail} = \alpha/2 = .01/2 = .005$$

$$\text{Degrees of freedom} = n_1 + n_2 - 2 = 14 + 16 - 2 = 28$$

The critical values of t for $df = 28$ and .005 area in each tail of the t distribution are -2.763 and 2.763, as shown in Figure 10.4.

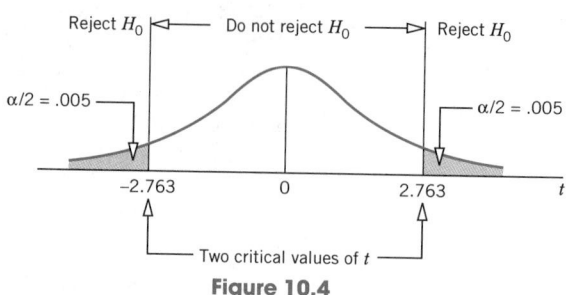

Figure 10.4

Step 4.   *Calculate the value of the test statistic*

The value of the test statistic t for $\bar{x}_1 - \bar{x}_2$ is computed as follows.

$$s_p = \sqrt{\frac{(n_1 - 1)s_1^2 + (n_2 - 1)s_2^2}{n_1 + n_2 - 2}} = \sqrt{\frac{(14 - 1)(3)^2 + (16 - 1)(4)^2}{14 + 16 - 2}} = 3.5707$$

$$s_{\bar{x}_1 - \bar{x}_2} = s_p \sqrt{\frac{1}{n_1} + \frac{1}{n_2}} = (3.5707)\sqrt{\frac{1}{14} + \frac{1}{16}} = 1.3067$$

$$t = \frac{(\bar{x}_1 - \bar{x}_2) - (\mu_1 - \mu_2)}{s_{\bar{x}_1 - \bar{x}_2}} = \frac{(23 - 25) - \overset{\text{From } H_0}{0}}{1.3067} = -1.531$$

Step 5.   *Make a decision*

Because the value of the test statistic $t = -1.531$ for $\bar{x}_1 - \bar{x}_2$ falls in the nonrejection region, we fail to reject the null hypothesis. Consequently we conclude that there is no difference in the mean number of calories per can for the two brands of diet soda. The difference in $\bar{x}_1$ and $\bar{x}_2$ observed for the two samples may have occurred due to sampling error only.

Making a right-tailed test of hypothesis about $\mu_1 - \mu_2$: small and independent samples, and $\sigma_1 = \sigma_2$.

EXAMPLE 10–9   The management at a supermarket wanted to investigate whether or not a promotional campaign increases the sales of a product. A sample of 28 days during the promotional campaign showed that an average of 316 units of this product were sold per day with a standard deviation of 18 units. A sample of 24 days before the promotional campaign showed that an average of 282 units of this product were sold per day with a standard deviation of 13 units. Test at the 5% significance level if the promotional campaign increases the mean number of units sold per day. Assume that the number of units sold per day has a normal distribution for both populations and that the standard deviations for the two populations are equal.

Solution Let the number of units sold per day during the promotional campaign be referred to as population 1 and the number of units sold per day before the promotional campaign be referred to as population 2. Let μ_1 and μ_2 be the means of populations 1 and 2, respectively, and let $\bar{x}_1$ and $\bar{x}_2$ be the means of the respective samples. From the given information,

$$\text{For population 1:} \quad n_1 = 28, \quad \bar{x}_1 = 316, \quad s_1 = 18$$

$$\text{For population 2:} \quad n_2 = 24, \quad \bar{x}_2 = 282, \quad s_2 = 13$$

The significance level is $\alpha = .05$.

Step 1. *State the null and alternative hypotheses*

To conclude that the promotional campaign increases the mean number of units sold per day, μ_1 must be greater than μ_2. The two possible decisions are

(i) The promotional campaign does not increase the mean number of units sold per day. This can be written as $\mu_1 = \mu_2$ or $\mu_1 - \mu_2 = 0$.

(ii) The promotional campaign increases the mean number of units sold per day. This can be written as $\mu_1 > \mu_2$ or $\mu_1 - \mu_2 > 0$.

Hence, the null and alternative hypotheses are

$$H_0: \mu_1 - \mu_2 = 0 \quad \text{(promotional campaign does not increase the mean daily sales)}$$

$$H_1: \mu_1 - \mu_2 > 0 \quad \text{(promotional campaign does increase the mean daily sales)}$$

Note that the null hypothesis can also be written as $\mu_1 - \mu_2 \leq 0$.

Step 2. *Select the distribution to use*

The two populations are normally distributed, the samples are small and independent, and the standard deviations of the two populations are unknown but equal. Consequently, we use the t distribution to make the test.

Step 3. *Determine the rejection and nonrejection regions*

The $>$ sign in the alternative hypothesis indicates that the test is right-tailed. The significance level is .05.

$$\text{Area in the right tail} = \alpha = .05$$

$$\text{Degrees of freedom} = n_1 + n_2 - 2 = 28 + 24 - 2 = 50$$

From the t distribution table, the critical value of t for $df = 50$ and .05 area in the right tail of the t distribution is 1.676. This value is shown in Figure 10.5.

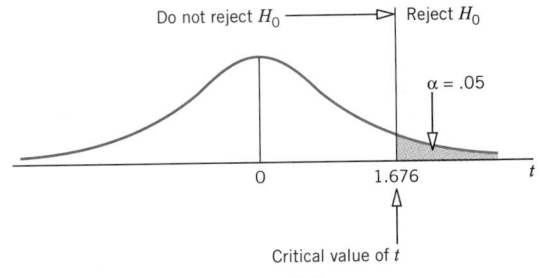

Figure 10.5

Step 4.   *Calculate the value of the test statistic*

The value of the test statistic t for $\bar{x}_1 - \bar{x}_2$ is computed as follows.

$$s_p = \sqrt{\frac{(n_1 - 1)s_1^2 + (n_2 - 1)s_2^2}{n_1 + n_2 - 2}} = \sqrt{\frac{(28 - 1)(18)^2 + (24 - 1)(13)^2}{28 + 24 - 2}} = 15.8965$$

$$s_{\bar{x}_1 - \bar{x}_2} = s_p \sqrt{\frac{1}{n_1} + \frac{1}{n_2}} = (15.8965)\sqrt{\frac{1}{28} + \frac{1}{24}} = 4.4220$$

$$t = \frac{(\bar{x}_1 - \bar{x}_2) - (\mu_1 - \mu_2)}{s_{\bar{x}_1 - \bar{x}_2}} = \frac{(316 - 282) - \overset{\text{From } H_0}{0}}{4.4220} = 7.689$$

Step 5.   *Make a decision*

Because the value of the test statistic $t = 7.689$ for $\bar{x}_1 - \bar{x}_2$ falls in the rejection region, we reject the null hypothesis H_0. Thus, we conclude that the promotional campaign increases the mean number of units sold per day.

EXERCISES

Concepts and Procedures

10.19   Explain what conditions must hold true to use the t distribution to make a confidence interval and to test a hypothesis about $\mu_1 - \mu_2$ for two independent samples selected from two populations with unknown but equal standard deviations.

10.20   The following information was obtained from two independent samples selected from two normally distributed populations with unknown but equal standard deviations.

$$n_1 = 22 \qquad \bar{x}_1 = 12.50 \qquad s_1 = 3.75$$
$$n_2 = 18 \qquad \bar{x}_2 = 14.60 \qquad s_2 = 3.10$$

    a.   What is the point estimate of $\mu_1 - \mu_2$?
    b.   Construct a 95% confidence interval for $\mu_1 - \mu_2$.

10.21   The following information was obtained from two independent samples selected from two normally distributed populations with unknown but equal standard deviations.

$$n_1 = 20 \qquad \bar{x}_1 = 33.75 \qquad s_1 = 5.25$$
$$n_2 = 23 \qquad \bar{x}_2 = 28.50 \qquad s_2 = 4.55$$

    a.   What is the point estimate of $\mu_1 - \mu_2$?
    b.   Construct a 99% confidence interval for $\mu_1 - \mu_2$.

10.22   Refer to the information given in Exercise 10.20. Test at the 5% significance level if the two population means are different.

10.23   Refer to the information given in Exercise 10.21. Test at the 1% significance level if the two population means are different.

10.24   Refer to the information given in Exercise 10.20. Test at the 1% significance level if μ_1 is less than μ_2.

10.25   Refer to the information given in Exercise 10.21. Test at the 5% significance level if μ_1 is greater than μ_2.

10.26 The following information was obtained from two independent samples selected from two normally distributed populations with unknown but equal standard deviations.

Sample 1:	27	31	25	33	21	35	30	26	25	31	33	30	28
Sample 2:	24	28	22	25	24	22	29	26	25	28	19	29	

    a. Let μ_1 be the mean of population 1 and μ_2 be the mean of population 2. What is the point estimate of $\mu_1 - \mu_2$?

    b. Construct a 98% confidence interval for $\mu_1 - \mu_2$.

    c. Test at the 1% significance level if μ_1 is greater than μ_2.

10.27 The following information was obtained from two independent samples selected from two normally distributed populations with unknown but equal standard deviations.

Sample 1:	13	11	9	12	8	10	5	10	9	12	13	
Sample 2:	16	14	11	19	14	17	13	16	17	18	19	12

    a. Let μ_1 be the mean of population 1 and μ_2 be the mean of population 2. What is the point estimate of $\mu_1 - \mu_2$?

    b. Construct a 99% confidence interval for $\mu_1 - \mu_2$.

    c. Test at the 2.5% significance level if μ_1 is lower than μ_2.

Applications

10.28 The management of a supermarket wanted to investigate if the male customers spend less money, on average, than the female customers. A sample of 25 male customers who shopped at this supermarket showed that they spent an average of $73 with a standard deviation of $17.50. Another sample of 20 female customers who shopped at the same supermarket showed that they spent an average of $87 with a standard deviation of $14.40. Assume that the amounts spent at this supermarket by all male and all female customers are normally distributed with equal but unknown population standard deviations.

    a. Construct a 99% confidence interval for the difference between the mean amounts spent by all male and all female customers at this supermarket.

    b. Test at the 5% significance level if the mean amount spent by all male customers at this supermarket is less than that of all female customers.

10.29 A manufacturing company is interested in buying one of two different kinds of machines. The company tested the two machines for production purposes. The first machine was run for 8 hours. It produced an average of 123 items per hour with a standard deviation of 9 items. The second machine was run for 10 hours. It produced an average of 114 items per hour with a standard deviation of 6 items. Assume that the production per hour for each machine is (approximately) normally distributed. Further assume that the standard deviations of the hourly productions of the two populations are equal.

    a. Make a 95% confidence interval for the difference between the two population means.

    b. Test at the 2.5% significance level if the mean number of items produced per hour by the first machine is higher than that of the second machine.

10.30 An insurance company wants to know if the average speed at which men drive cars is higher than that of women drivers. The company took a random sample of 27 cars driven by men on a highway and found the mean speed to be 68 miles per hour with a standard deviation of 2.2 miles. Another sample of 18 cars driven by women on the same highway gave a mean speed of 65 miles per hour with a standard deviation of 2.5 miles. Assume that the speeds at which all men and all women drive cars on this highway are both normally distributed with the same population standard deviation.

    a. Construct a 98% confidence interval for the difference between the mean speeds of cars driven by all men and all women drivers on this highway.

    b. Test at the 1% significance level if the mean speed of cars driven by all men drivers on this highway is higher than that of cars driven by all women drivers.

10.31 According to Metropolitan Life Insurance Company's claims data, the mean hospital and physician's charges for a typical Caesarean birth are $6400 for Florida and $6940 for California (*Statistical Bulletin*, July–September 1988). Assume that these two means are based on 24 claims for Florida and 28 claims for California and that the standard deviations for these two sets of claims for the two states are $980 and $1150, respectively. Further assume that such charges have a normal distribution with the same population standard deviation for each of these two states.

    a. Construct a 90% confidence interval for the difference between the mean charges for all Caesarean births for these two states.

    b. Test at the 5% significance level if the mean charges for all Caesarean births are different for these two states.

10.32 Quadro Corporation has two supermarket stores in a city. The company's quality control department wanted to check if the customers are equally satisfied with the service provided at these two stores. A sample of 25 customers selected from Supermarket I produced a mean satisfaction index of 7.8 (on a scale of 1 to 10, 1 being the lowest and 10 being the highest) with a standard deviation of .75. Another sample of 28 customers selected from Supermarket II produced a mean satisfaction index of 8.3 with a standard deviation of .59. Assume that the customer satisfaction index for each supermarket has a normal distribution with the same population standard deviation.

    a. Construct a 98% confidence interval for the difference between the mean satisfaction indexes for all customers for the two supermarkets.

    b. Test at the 1% significance level if the mean satisfaction indexes for all customers for the two supermarkets are different.

10.33 A company claims that its medicine, Brand A, provides faster relief from pain than another company's medicine, Brand B. A researcher tested both brands of medicine on two groups of randomly selected patients. The results of the test are given in the following table. The mean and standard deviation of relief times are in minutes.

Brand	Sample Size	Mean of Relief Times	Standard Deviation of Relief Times
A	25	44	13
B	22	49	11

    a. Construct a 99% confidence interval for the difference between the mean relief times for the two brands of medicines.

    b. Test at the 1% significance level if the mean relief time for Brand A is less than that for Brand B.

Assume that the two populations are normally distributed with equal standard deviations.

10.34 A professor took two samples, one of 21 males and another of 18 females, from college students who were enrolled in an introductory course in business statistics at the same school. The professor found that the mean score of male students in a mid-term examination in statistics was 76.2 with a standard deviation of 7.3, and the mean score of female students was 78.5 with a standard deviation of 6.7. Assume that the scores of all male and all female students are normally distributed with equal but unknown standard deviations.

    a. Construct a 95% confidence interval for the difference between the mean scores of all male and all female students.

    b. Test at the 2.5% significance level if the mean score in business statistics for all male students is lower than that for all female students.

10.35 According to the American Association of University Professors, the mean salary of male professors at American colleges and universities was $54,340 and that of female professors was $48,080 in 1989–1990. For convenience, assume that these two means are based on random samples of 28 male professors and 26 female professors. Further assume that the standard deviations for the two samples are $3100 and $2800, respectively.

    a. Construct a 90% confidence interval for the difference between the two population means.

b. Test at the 1% significance level if the mean salary of all male professors for 1989–1990 is higher than that of all female professors.

Assume that the salaries of all male and all female professors are both normally distributed with equal standard deviations.

10.3 INFERENCES ABOUT THE DIFFERENCE BETWEEN TWO POPULATION MEANS FOR PAIRED SAMPLES

Sections 10.1 and 10.2 were concerned with estimation and hypothesis testing about the difference between two population means when the two samples were drawn independently from two different populations. This section describes estimation and hypothesis-testing procedures for the difference between two population means when the samples are dependent.

In a case of two dependent samples, two data values—one in each sample—are collected from the same source (or element) and, hence, these are also called **paired** or **matched samples**. For example, we may want to make inferences about the mean number of units of a product sold by salespersons before and after they take a course in sales management. To do so, suppose we select a sample of 15 salespersons and record their sales before and after they take this course. In this example, both sets of data are collected from the same 15 salespersons, once before and once after the course. Thus, although there are two samples, they contain the same 15 salespersons. This is an example of paired (or dependent or matched) samples. The procedures to make confidence intervals and test hypotheses in the case of paired samples are different from the ones for independent samples discussed in earlier sections of this chapter.

PAIRED OR MATCHED SAMPLES

Two samples are said to be paired or matched samples when for each data value collected from one sample there is a corresponding data value collected from the second sample, and both these data values are collected from the same source.

As another example of paired samples, suppose an agronomist wants to measure the effect of a new brand of fertilizer on the yield of potatoes. To do so, he selects 10 pieces of land and divides each piece of land into two portions. Then he randomly assigns one of the two portions from each piece of land to grow potatoes without using fertilizer (or using some other brand of fertilizer). The second portion from each piece of land is used to grow potatoes using the new brand of fertilizer. Thus, he will have 10 pairs of data values. Then, using the procedure to be discussed in this section, he will make inferences about the difference in the mean yields of potatoes with the new fertilizer and without it.

The question arises, why does the agronomist not choose 10 pieces of land on which to grow potatoes without using the new brand of fertilizer and another 10 pieces of land to grow potatoes by using the new brand of fertilizer? If he does so, the effect of the fertilizer might be confused with the effects due to soil differences at different locations. Thus, he will not be able to identify the effect of only the new brand of fertilizer on the yield of potatoes. Consequently, the results will not be reliable. By choosing 10 pieces of land and then dividing each of them into two portions, the researcher decreases the possibility that the difference in the productivities of different pieces of land affects the results.

In paired samples, the difference between the two data values for each element of the two samples is denoted by **d**. This value of d is called the **paired difference**. We then treat all the values of d as one sample and make inferences applying procedures similar to the ones used for one-sample cases in Chapters 8 and 9. Note that as each source (or element) gives a pair of values (one for each of the two data sets), each sample contains the same number of values. That is, both samples are of the same size. Therefore, we denote the (common) **sample size** by n, which gives the number of paired difference d values. The **degrees of freedom** for the paired samples are $n - 1$. Let

μ_d = the mean of the paired differences for the population

σ_d = the standard deviation of the paired differences for the population

$\bar{d}$ = the sample mean of the paired differences

s_d = the sample standard deviation of the paired differences

n = the number of paired difference values

MEAN AND STANDARD DEVIATION OF THE PAIRED DIFFERENCES FOR SAMPLES

The values of $\bar{d}$ and s_d are calculated as[3]

$$\bar{d} = \frac{\Sigma d}{n}$$

$$s_d = \sqrt{\frac{\Sigma d^2 - \dfrac{(\Sigma d)^2}{n}}{n - 1}}$$

In paired samples, instead of using $\bar{x}_1 - \bar{x}_2$ as the sample statistic to make inferences about $\mu_1 - \mu_2$, we use the sample statistic $\bar{d}$ to make inferences about μ_d. Actually the value of $\bar{d}$ is always equal to $\bar{x}_1 - \bar{x}_2$, and the value of μ_d is always equal to $\mu_1 - \mu_2$.

SAMPLING DISTRIBUTION, MEAN, AND STANDARD DEVIATION OF $\bar{d}$

If the number of paired values is large ($n \geq 30$), because of the central limit theorem the sampling distribution of $\bar{d}$ is approximately normal with its mean and standard deviation as

$$\mu_{\bar{d}} = \mu_d \qquad \text{and} \qquad \sigma_{\bar{d}} = \frac{\sigma_d}{\sqrt{n}}$$

In cases when $n \geq 30$, the normal distribution can be used to make inferences about μ_d.

---

[3]The basic formula to calculate s_d is

$$s_d = \sqrt{\frac{\Sigma(d - \bar{d})^2}{n - 1}}$$

However, we will not use this formula to make calculations in this chapter.

However, in cases of paired samples, the sample sizes are usually small and σ_d is unknown. In such cases, assuming that the paired differences for the population are (approximately) normally distributed, the normal distribution is replaced by the t distribution to make inferences about μ_d. When σ_d is not known, the standard deviation of $\overline{d}$ is estimated by $s_{\overline{d}} = s_d / \sqrt{n}$.

ESTIMATE OF THE STANDARD DEVIATION OF PAIRED DIFFERENCES

If

1. n is less than 30
2. σ_d is not known
3. the population of paired differences is (approximately) normally distributed

then the t distribution is used to make inferences about μ_d. The standard deviation $\sigma_{\overline{d}}$ of $\overline{d}$ is estimated by $s_{\overline{d}}$, which is calculated as

$$s_{\overline{d}} = \frac{s_d}{\sqrt{n}}$$

Sections 10.3.1 and 10.3.2 describe the procedures to make a confidence interval and test a hypothesis about μ_d when σ_d is unknown and n is small. The inferences are made using the t distribution. However, if n is large, even if σ_d is unknown, the normal distribution can be used to make inferences about μ_d.

10.3.1 INTERVAL ESTIMATION OF μ_d

The mean $\overline{d}$ of paired differences for two paired samples is the point estimator of μ_d. The following formula is used to construct a confidence interval for μ_d in the case of (approximately) normally distributed populations.

CONFIDENCE INTERVAL FOR μ_d

The $(1 - \alpha)100\%$ confidence interval for μ_d is

$$\overline{d} \pm ts_{\overline{d}}$$

where the value of t is obtained from the t distribution table for the given confidence level and $n - 1$ degrees of freedom, and $s_{\overline{d}}$ is calculated as explained earlier.

Example 10–10 illustrates the procedure to construct a confidence interval for μ_d.

Constructing a confidence interval for μ_d: paired samples.

EXAMPLE 10–10 Sham Corporation is considering installing a new assembly procedure. The company randomly selected seven employees and gathered information on the times taken by them to assemble one unit of the product by using the old procedure. Then, the same employees were asked to assemble one unit of the product by using the new procedure. The following table gives the assembly times (in minutes) for these seven employees for each procedure.

Employee	Old Procedure	New Procedure
1	64	60
2	71	66
3	68	66
4	66	69
5	73	63
6	62	57
7	70	62

Let μ_d be the mean of the differences between the assembly times for the two procedures. Construct a 95% confidence interval for μ_d. Assume that the population of paired differences is (approximately) normally distributed.

Solution Because the information obtained is from paired samples, we will make the confidence interval for the paired difference mean μ_d of the population by using the paired difference mean $\bar{d}$ of the sample. Let d be the difference in the times taken by an employee to assemble one unit of the product by using the old procedure and by using the new procedure. The values of d are obtained by subtracting the assembly time for the new procedure from the assembly time for the old procedure for an employee. The fourth column of Table 10.1 lists the values of d for seven employees. The fifth column of the table records the values of d^2, which are obtained by squaring each of the d values.

Table 10.1

Employee	Old Procedure	New Procedure	Difference d	d^2
1	64	60	4	16
2	71	66	5	25
3	68	66	2	4
4	66	69	-3	9
5	73	63	10	100
6	62	57	5	25
7	70	62	8	64
			$\Sigma d = 31$	$\Sigma d^2 = 243$

The values of $\bar{d}$ and s_d are calculated as follows.

$$\bar{d} = \frac{\Sigma d}{n} = \frac{31}{7} = 4.43$$

$$s_d = \sqrt{\frac{\Sigma d^2 - \frac{(\Sigma d)^2}{n}}{n - 1}} = \sqrt{\frac{243 - \frac{(31)^2}{7}}{7 - 1}} = 4.1975$$

The standard deviation of $\bar{d}$ is

$$s_{\bar{d}} = \frac{s_d}{\sqrt{n}} = \frac{4.1975}{\sqrt{7}} = 1.5865$$

For a 95% confidence interval, the area in each tail of the t distribution is

$$\alpha/2 = .5 - (.95/2) = .025$$

The degrees of freedom are

$$df = n - 1 = 7 - 1 = 6$$

From the t distribution table, the t value for $df = 6$ and .025 area in the right tail of the t distribution is 2.447. The 95% confidence interval for μ_d is

$$\bar{d} \pm ts_{\bar{d}} = 4.43 \pm 2.447 \,(1.5865) = 4.43 \pm 3.88 = \mathbf{.55 \ to \ 8.31}$$

Thus, we can state with 95% confidence that the mean difference between the assembly times for the two procedures is between .55 and 8.31 minutes. In other words, the old procedure takes, on average, .55 to 8.31 minutes longer than the new procedure to assemble one unit of the product. ■

10.3.2 HYPOTHESIS TESTING ABOUT μ_d

A hypothesis about μ_d is tested by using the sample statistic $\bar{d}$. If n is 30 or larger, we can use the normal distribution to test a hypothesis about μ_d. However, if n is less than 30, we replace the normal distribution by the t distribution. To use the t distribution, we assume that the population of all paired differences is (approximately) normally distributed and that the population standard deviation σ_d of paired differences is not known. This section illustrates the case of the t distribution only. The following formula is used to calculate the value of the test statistic t when testing a hypothesis about μ_d.

TEST STATISTIC t FOR $\bar{d}$

The value of the test statistic t for $\bar{d}$ is computed as follows.

$$t = \frac{\bar{d} - \mu_d}{s_{\bar{d}}}$$

The critical value of t is found from the t distribution table for the given significance level and $n - 1$ degrees of freedom.

Examples 10–11 and 10–12 illustrate the hypothesis-testing procedure for μ_d.

Conducting a left-tailed test of hypothesis about μ_d: paired samples.

EXAMPLE 10–11 A company wanted to know if attending a course on "how to be a successful salesperson" can increase the average sales of its employees. The company sent six of its salespersons to attend this course. The following table gives the one-week sales of these salespersons before and after they attended this course.

Before	12	18	25	9	14	16
After	18	24	24	14	19	20

Using the 1% significance level, can you conclude that the mean weekly sales for all salespersons increase as a result of attending this course? Assume that the population of paired differences has a normal distribution.

Solution Because the data are for paired samples, we test a hypothesis about the paired difference mean μ_d of the population using the paired difference mean $\bar{d}$ of the sample. Let

$$d = (\text{weekly sales before the course}) - (\text{weekly sales after the course})$$

In Table 10.2, we calculate d for each of the six salespersons by subtracting the sales after the course from the sales before the course. The fourth column of the table lists the values of d^2.

Table 10.2

Before	After	Difference d	d^2
12	18	-6	36
18	24	-6	36
25	24	1	1
9	14	-5	25
14	19	-5	25
16	20	-4	16
		$\Sigma d = -25$	$\Sigma d^2 = 139$

The values of $\bar{d}$ and s_d are calculated as follows.

$$\bar{d} = \frac{\Sigma d}{n} = \frac{-25}{6} = -4.17$$

$$s_d = \sqrt{\frac{\Sigma d^2 - \frac{(\Sigma d)^2}{n}}{n-1}} = \sqrt{\frac{139 - \frac{(-25)^2}{6}}{6-1}} = 2.6394$$

The standard deviation of $\bar{d}$ is

$$s_{\bar{d}} = \frac{s_d}{\sqrt{n}} = \frac{2.6394}{\sqrt{6}} = 1.0775$$

Step 1. *State the null and alternative hypotheses*

We are to test if the mean weekly sales for all salespersons increase as a result of taking the course. Let μ_1 be the mean weekly sales for all salespersons before the course and μ_2 be the mean weekly sales for all salespersons after the course. Then, $\mu_d = \mu_1 - \mu_2$. The mean weekly sales for all salespersons will increase due to attending the course if μ_1 is less than μ_2, which can be written as $\mu_1 - \mu_2 < 0$ or $\mu_d < 0$. Consequently, the null and alternative hypotheses are

$H_0: \mu_d = 0$ ($\mu_1 - \mu_2 = 0$ or the mean weekly sales do not increase)

$H_1: \mu_d < 0$ ($\mu_1 - \mu_2 < 0$ or the mean weekly sales do increase)

Note that we can also write the null hypothesis as $\mu_d \geq 0$.

Step 2. *Select the distribution to use*

The sample size is small ($n < 30$), the population of paired differences is normal, and σ_d is unknown. Therefore, we use the t distribution to conduct the test.

Step 3. *Determine the rejection and nonrejection regions*

The $<$ sign in the alternative hypothesis indicates that the test is left-tailed. The significance level is .01. Hence,

Area in the left tail $= \alpha = .01$

Degrees of freedom $= n - 1 = 6 - 1 = 5$

The critical value of t for $df = 5$ and .01 area in the left tail of the t distribution is -3.365. This value is shown in Figure 10.6.

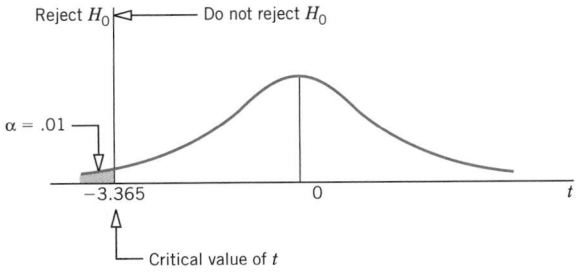

Figure 10.6

Step 4. *Calculate the value of the test statistic*

The value of the test statistic t for $\bar{d}$ is computed as follows.

$$t = \frac{\bar{d} - \mu_d}{s_{\bar{d}}} = \frac{-4.17 - \overset{\text{From } H_0}{0}}{1.0775} = -3.870$$

Step 5. *Make a decision*

Because the value of the test statistic $t = -3.870$ for $\bar{d}$ falls in the rejection region, we reject the null hypothesis. Consequently, we conclude that the mean weekly sales for all salespersons increase as a result of this course. ▬

Making a two-tailed test of hypothesis about μ_d: paired samples.

EXAMPLE 10-12 Refer to Example 10–10. The table that gives the assembly times for the two procedures is reproduced below.

Employee	Old Procedure	New Procedure
1	64	60
2	71	66
3	68	66
4	66	69
5	73	63
6	62	57
7	70	62

Test at the 5% significance level if the mean of the paired differences μ_d is different from zero. Assume that the population of paired differences is (approximately) normally distributed.

Solution Table 10.3 gives d and d^2 for each of the seven workers.

Table 10.3

Employee	Old Procedure	New Procedure	Difference d	d^2
1	64	60	4	16
2	71	66	5	25
3	68	66	2	4
4	66	69	−3	9
5	73	63	10	100
6	62	57	5	25
7	70	62	8	64
			$\Sigma d = 31$	$\Sigma d^2 = 243$

The values of $\overline{d}$ and s_d are calculated as follows.

$$\overline{d} = \frac{\Sigma d}{n} = \frac{31}{7} = 4.43$$

$$s_d = \sqrt{\frac{\Sigma d^2 - \frac{(\Sigma d)^2}{n}}{n - 1}} = \sqrt{\frac{243 - \frac{(31)^2}{7}}{7 - 1}} = 4.1975$$

The standard deviation of $\overline{d}$ is

$$s_{\overline{d}} = \frac{s_d}{\sqrt{n}} = \frac{4.1975}{\sqrt{7}} = 1.5865$$

Step 1. *State the null and alternative hypotheses*

H_0: $\mu_d = 0$ (mean of the paired differences is not different from zero)

H_1: $\mu_d \neq 0$ (mean of the paired differences is different from zero)

Step 2. *Select the distribution to use*

The sample size is small, the population of paired differences is (approximately) normal, and σ_d is not known. Therefore, we use the t distribution to make the test.

Step 3. *Determine the rejection and nonrejection regions*

The $\neq$ sign in the alternative hypothesis indicates that the test is two-tailed. The significance level is .05.

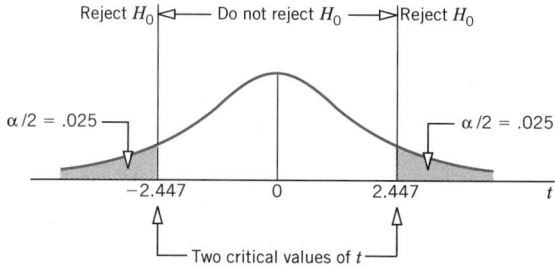

Figure 10.7

Area in each tail $= \alpha/2 = .05/2 = .025$

Degrees of freedom $= n - 1 = 7 - 1 = 6$

The two critical values of t for $df = 6$ and .025 area in each tail of the t distribution are -2.447 and 2.447. These values are shown in Figure 10.7.

Step 4. *Calculate the value of the test statistic*

The value of the test statistic t for $\bar{d}$ is computed as follows.

$$t = \frac{\bar{d} - \mu_d}{s_{\bar{d}}} = \frac{4.43 - \overset{\text{From } H_0}{0}}{1.5865} = 2.792$$

Step 5. *Make a decision*

Because the value of the test statistic $t = 2.792$ for $\bar{d}$ falls in the rejection region, we reject the null hypothesis. We conclude that the mean of the population paired differences is different from zero. In other words, we can state that the mean assembly time for the two procedures is not the same.

EXERCISES

Concepts and Procedures

10.36 Explain when you would use the paired samples procedure to make confidence intervals and test hypotheses.

10.37 Find the following confidence intervals for μ_d assuming that the populations of paired differences are normally distributed.

    a. $n = 9$, $\bar{d} = 25.4$, $s_d = 13.5$,   confidence level $= 99\%$
    b. $n = 26$, $\bar{d} = 13.2$, $s_d = 4.8$,   confidence level $= 95\%$
    c. $n = 12$, $\bar{d} = 34.6$, $s_d = 11.7$,   confidence level $= 90\%$

10.38 Find the following confidence intervals for μ_d assuming that the populations of paired differences are normally distributed.

    a. $n = 10$, $\bar{d} = 17.5$, $s_d = 6.3$,   confidence level $= 99\%$
    b. $n = 24$, $\bar{d} = 55.9$, $s_d = 14.7$,   confidence level $= 95\%$
    c. $n = 14$, $\bar{d} = 29.3$, $s_d = 8.3$,   confidence level $= 90\%$

10.39 Perform the following tests of hypotheses assuming that the populations of paired differences are normally distributed.

    a. $H_0: \mu_d = 0$, $H_1: \mu_d \neq 0$, $n = 9$, $\bar{d} = 6.7$, $s_d = 2.5$, $\alpha = .10$
    b. $H_0: \mu_d = 0$, $H_1: \mu_d > 0$, $n = 22$, $\bar{d} = 14.8$, $s_d = 6.4$, $\alpha = .05$
    c. $H_0: \mu_d = 0$, $H_1: \mu_d < 0$, $n = 17$, $\bar{d} = -9.3$, $s_d = 4.8$, $\alpha = .01$

10.40 Conduct the following tests of hypotheses assuming that the populations of paired differences are normally distributed.

    a. $H_0: \mu_d = 0$, $H_1: \mu_d \neq 0$, $n = 26$, $\bar{d} = 9.6$, $s_d = 3.9$, $\alpha = .05$
    b. $H_0: \mu_d = 0$, $H_1: \mu_d > 0$, $n = 15$, $\bar{d} = 8.8$, $s_d = 4.7$, $\alpha = .01$
    c. $H_0: \mu_d = 0$, $H_1: \mu_d < 0$, $n = 20$, $\bar{d} = -7.4$, $s_d = 2.3$, $\alpha = .10$

Applications

10.41 A company sent seven of its employees to attend a course in building self-confidence. These employees were evaluated for their self-confidence before and after attending this course. The following table gives the scores (on a scale of 1 to 15, 1 being the lowest and 15 being the highest score) of these employees before and after they attended the course.

Before	8	5	4	9	6	8	5
After	10	7	5	11	6	7	9

a. Construct a 95% confidence interval for the mean μ_d of the population paired differences where a paired difference is equal to the score of an employee before attending the course minus the score of the same employee after attending the course.

b. Test at the 1% significance level if attending this course increases the mean score of employees.

Assume that the population of paired differences has a normal distribution.

10.42 Many students suffer from math anxiety. A professor who teaches business statistics offered her students a two-hour lecture on math anxiety and ways to overcome it. The following table gives the scores of seven students in business statistics before and after they attended this lecture.

Before	56	69	48	74	65	71	58
After	62	73	44	85	71	70	69

a. Construct a 99% confidence interval for the mean μ_d of the population paired differences where a paired difference is equal to the score before attending this lecture minus the score after attending this lecture.

b. Test at the 2.5% significance level if attending this lecture increases the average score in business statistics.

Assume that the population of paired differences is (approximately) normally distributed.

10.43 A private agency claims that the crash course it offers significantly increases the writing speed of secretaries. The following table gives the scores of eight secretaries before and after they attended this course.

Before	81	75	89	91	65	70	90	69
After	97	72	93	110	78	69	115	75

a. Make a 90% confidence interval for the mean μ_d of the population paired differences where a paired difference is equal to the score before attending the course minus the score after attending the course.

b. Using the 5% significance level, can you conclude that attending this course increases the writing speed of secretaries?

Assume that the population of paired differences is (approximately) normally distributed.

10.44 A company claims that its 12-week special exercise program significantly reduces weight. A random sample of six persons were put on this exercise program for 12 weeks. The following table gives the weights (in pounds) of those six persons before and after the program.

Before	180	195	177	221	208	199
After	183	187	161	204	197	189

a. Make a 95% confidence interval for the mean μ_d of the population paired differences where a paired difference is equal to the weight before joining this exercise program minus the weight at the end of the 12-week program.

b. Test at the 1% significance level if the mean weight loss for all persons due to this special exercise program is greater than zero.

Assume that the population of all paired differences is (approximately) normally distributed.

10.45 The manufacturer of a gasoline additive claims that the use of this additive increases gasoline mileage. A random sample of six cars was selected and these cars were driven for one week without the gasoline additive and then for one week with the gasoline additive. The following table gives the miles per gallon for these cars without and with the gasoline additive.

Without	24.6	28.3	18.9	23.7	15.4	29.5
With	26.3	31.7	18.2	25.3	18.3	30.9

a. Construct a 99% confidence interval for the mean μ_d of the population paired differences where a paired difference is equal to the miles per gallon without the gasoline additive minus the miles per gallon with the gasoline additive.

b. Test at the 2.5% significance level if the use of the gasoline additive increases the gasoline mileage.

Assume that the population of paired differences is (approximately) normally distributed.

10.46 A company is considering installing new machines to assemble its products. The company is considering two types of machines, but it will buy only one type. The company selected eight assembly workers and asked them to use these two types of machines to assemble products. The following table gives the time taken (in minutes) to assemble one unit of the product on each type of machine for each of these eight workers.

Machine I	23	26	19	24	26	22	20	18
Machine II	21	24	23	25	24	25	24	23

a. Construct a 98% confidence interval for the mean μ_d of the population paired differences where a paired difference is equal to the time taken to assemble a unit of the product on Machine I minus the time taken to assemble a unit of the product on Machine II.

b. Test at the 5% significance level if the mean time taken to assemble a unit of the product is different for the two types of machines.

Assume that the population of paired differences is (approximately) normally distributed.

10.4   INFERENCES ABOUT THE DIFFERENCE BETWEEN TWO POPULATION PROPORTIONS FOR LARGE AND INDEPENDENT SAMPLES

Quite often we need to construct a confidence interval and test a hypothesis about the difference between two population proportions. For instance, we may want to estimate the difference between the proportion of defective items produced on two different machines. If p_1 and p_2 are the proportions of defective items produced on the first and second machine, respectively, then we are to make a confidence interval for $p_1 - p_2$. Or we may want to test the hypothesis that the proportion of defective items produced on Machine I is different from the proportion of defective items produced on Machine II. In this case, we are to test the null hypothesis $p_1 - p_2 = 0$ against the alternative hypothesis $p_1 - p_2 \neq 0$.

This section discusses how to make a confidence interval and test a hypothesis about $p_1 - p_2$ for two large and independent samples. The sample statistic used to make inferences about $p_1 - p_2$ is $\hat{p}_1 - \hat{p}_2$ where $\hat{p}_1$ and $\hat{p}_2$ are the proportions for two large and independent samples. As discussed in Section 7.9 of Chapter 7, we determine a sample proportion by dividing the number of elements in the sample with a given attribute by the sample size. Thus,

$$\hat{p}_1 = x_1/n_1 \quad \text{and} \quad \hat{p}_2 = x_2/n_2$$

where x_1 and x_2 are the number of elements with a given characteristic in the two samples and n_1 and n_2 are the sizes of the two samples, respectively.

10.4.1   THE MEAN, STANDARD DEVIATION, AND SAMPLING DISTRIBUTION OF $\hat{p}_1 - \hat{p}_2$

As discussed in Chapter 7, for a large sample the sample proportion $\hat{p}$ is (approximately) normally distributed with mean p and standard deviation $\sqrt{pq/n}$. Hence, for two large and independent samples of sizes n_1 and n_2, respectively, their sample proportions $\hat{p}_1$ and $\hat{p}_2$ are (approximately) normally distributed with means p_1 and p_2 and standard deviations $\sqrt{p_1q_1/n_1}$ and $\sqrt{p_2q_2/n_2}$, respectively. Using these results, we can make the following statements about the shape of the sampling distribution of $\hat{p}_1 - \hat{p}_2$ and its mean and standard deviation.

> **MEAN, STANDARD DEVIATION, AND SAMPLING DISTRIBUTION OF $\hat{p}_1 - \hat{p}_2$**
>
> For two large and independent samples, the sampling distribution of $\hat{p}_1 - \hat{p}_2$ is (approximately) normal with its mean and standard deviation as
>
> $$\mu_{\hat{p}_1 - \hat{p}_2} = p_1 - p_2$$
>
> and
> $$\sigma_{\hat{p}_1 - \hat{p}_2} = \sqrt{\frac{p_1q_1}{n_1} + \frac{p_2q_2}{n_2}}$$
>
> respectively, where $q_1 = 1 - p_1$ and $q_2 = 1 - p_2$.

Thus, to construct a confidence interval and test a hypothesis about $p_1 - p_2$ for large and independent samples, we use the normal distribution. As was indicated in Chapter 7, in the case of proportion the sample is large if np and nq are both greater than 5. In the case of two samples, both sample sizes will be large if n_1p_1, n_1q_1, n_2p_2, and n_2q_2 are all greater than 5.

10.4.2   INTERVAL ESTIMATION OF $p_1 - p_2$

The difference between two sample proportions $\hat{p}_1 - \hat{p}_2$ is the point estimator for the difference between two population proportions $p_1 - p_2$. Because we do not know p_1 and p_2 when we are making a confidence interval for $p_1 - p_2$, we cannot calculate the value of $\sigma_{\hat{p}_1 - \hat{p}_2}$. Therefore, we use $s_{\hat{p}_1 - \hat{p}_2}$ as the point estimator of $\sigma_{\hat{p}_1 - \hat{p}_2}$ in the interval estimation. We construct the confidence interval for $p_1 - p_2$ using the following formula.

> **CONFIDENCE INTERVAL FOR $p_1 - p_2$**
>
> The $(1 - \alpha)100\%$ confidence interval for $p_1 - p_2$ is
>
> $$(\hat{p}_1 - \hat{p}_2) \pm z s_{\hat{p}_1 - \hat{p}_2}$$
>
> where the value of z is read from the normal distribution table for the given confidence level, and $s_{\hat{p}_1 - \hat{p}_2}$ is calculated as
>
> $$s_{\hat{p}_1 - \hat{p}_2} = \sqrt{\frac{\hat{p}_1\hat{q}_1}{n_1} + \frac{\hat{p}_2\hat{q}_2}{n_2}}$$

Example 10–13 describes the procedure to make a confidence interval for the difference between two population proportions for large samples.

Constructing a confidence interval for $p_1 - p_2$: large samples.

EXAMPLE 10-13 A business consultant agency wanted to estimate the difference between the percentages of users of two toothpastes who will never switch to another toothpaste. In a sample of 500 users of Toothpaste A, 20% said that they will never switch to another toothpaste. In another sample of 400 users of Toothpaste B, 13% said that they will never switch to another toothpaste.

(a) Let p_1 and p_2 be the proportions of all users of Toothpastes A and B, respectively, who will never switch to another toothpaste. What is the point estimate of $p_1 - p_2$?

(b) Construct a 97% confidence interval for the difference between the proportions of users of the two toothpastes who will never switch.

Solution Let p_1 and p_2 be the proportions of all users of Toothpastes A and B, respectively, who will never switch to another toothpaste and let $\hat{p}_1$ and $\hat{p}_2$ be the respective sample proportions. From the given information,

For Toothpaste A: $n_1 = 500$ $\hat{p}_1 = .20$ $\hat{q}_1 = 1 - .20 = .80$

For Toothpaste B: $n_2 = 400$ $\hat{p}_2 = .13$ $\hat{q}_2 = 1 - .13 = .87$

(a) The point estimate of $p_1 - p_2$ is as follows.

$$\text{Point estimate of } p_1 - p_2 = \hat{p}_1 - \hat{p}_2 = .20 - .13 = \textbf{.07}$$

(b) The values of $n_1\hat{p}_1$, $n_1\hat{q}_1$, $n_2\hat{p}_2$, and $n_2\hat{q}_2$ are

$$n_1\hat{p}_1 = 500 \,(.20) = 100, \qquad n_1\hat{q}_1 = 500 \,(.80) = 400,$$

$$n_2\hat{p}_2 = 400 \,(.13) = 52, \quad \text{and} \quad n_2\hat{q}_2 = 400 \,(.87) = 348$$

Since each of these values is greater than 5, both sample sizes are large. Consequently we use the normal distribution to make a confidence interval for $p_1 - p_2$.
The standard deviation of $\hat{p}_1 - \hat{p}_2$ is

$$s_{\hat{p}_1 - \hat{p}_2} = \sqrt{\frac{\hat{p}_1\hat{q}_1}{n_1} + \frac{\hat{p}_2\hat{q}_2}{n_2}} = \sqrt{\frac{(.20)\,(.80)}{500} + \frac{(.13)\,(.87)}{400}} = .0246$$

The z value for a 97% confidence level, obtained from the normal distribution table for $.97/2 = .4850$, is 2.17. The 97% confidence interval for $p_1 - p_2$ is

$$(\hat{p}_1 - \hat{p}_2) \pm zs_{\hat{p}_1 - \hat{p}_2} = (.20 - .13) \pm 2.17 \,(.0246)$$

$$= .07 \pm .05 = \textbf{.02 to .12}$$

Thus, with 97% confidence we can state that the difference between the two population proportions is between .02 and .12. ▬

10.4.3 HYPOTHESIS TESTING ABOUT $p_1 - p_2$

In this section we learn how to test a hypothesis about $p_1 - p_2$ for two large and independent samples. The procedure involves the same five steps that we have used previously. Once again, we calculate the standard deviation of $\hat{p}_1 - \hat{p}_2$ as

$$\sigma_{\hat{p}_1 - \hat{p}_2} = \sqrt{\frac{p_1q_1}{n_1} + \frac{p_2q_2}{n_2}}$$

When a test of hypothesis about $p_1 - p_2$ is performed, usually the null hypothesis is $p_1 = p_2$ and the values of p_1 and p_2 are not known. Assuming that the null hypothesis is true and $p_1 = p_2$, a common value of p_1 and p_2, denoted by $\bar{p}$, is calculated by using one of the following formulas.

$$\bar{p} = \frac{x_1 + x_2}{n_1 + n_2} \quad \text{or} \quad \frac{n_1 \hat{p}_1 + n_2 \hat{p}_2}{n_1 + n_2}$$

Which of these formulas is used depends on whether the values of x_1 and x_2 or the values of $\hat{p}_1$ and $\hat{p}_2$ are known. Note that x_1 and x_2 are the number of elements in each of the two samples that possess a certain characteristic. This value of $\bar{p}$ is called the **pooled sample proportion**. Using the value of the pooled sample proportion, we compute an estimate of the standard deviation of $\hat{p}_1 - \hat{p}_2$ as

$$s_{\hat{p}_1 - \hat{p}_2} = \sqrt{\bar{p}\,\bar{q}\left(\frac{1}{n_1} + \frac{1}{n_2}\right)}$$

where $\bar{q} = 1 - \bar{p}$.

TEST STATISTIC z FOR $\hat{p}_1 - \hat{p}_2$

The value of the test statistic z for $\hat{p}_1 - \hat{p}_2$ is calculated as

$$z = \frac{(\hat{p}_1 - \hat{p}_2) - (p_1 - p_2)}{s_{\hat{p}_1 - \hat{p}_2}}$$

The value of $p_1 - p_2$ is substituted from H_0, which is usually zero.

Examples 10–14 and 10–15 illustrate the procedure to test hypotheses about the difference between two population proportions for large samples.

Making a right-tailed test of hypothesis about $p_1 - p_2$: large samples.

EXAMPLE 10–14　Reconsider Example 10–13 about the percentages of users of two toothpastes who will never switch to another toothpaste. At the 1% significance level, can we conclude that the proportion of users of Toothpaste A who will never switch to another toothpaste is greater than the proportion of users of Toothpaste B who will never switch to another toothpaste?

Solution　Let p_1 and p_2 be the proportions of all users of Toothpastes A and B, respectively, who will never switch to another toothpaste and let $\hat{p}_1$ and $\hat{p}_2$ be the corresponding sample proportions. From the given information,

For Toothpaste A:　　$n_1 = 500$　　$\hat{p}_1 = .20$　　$\hat{q}_1 = 1 - .20 = .80$

For Toothpaste B:　　$n_2 = 400$　　$\hat{p}_2 = .13$　　$\hat{q}_2 = 1 - .13 = .87$

The significance level is $\alpha = .01$.

Step 1.　*State the null and alternative hypotheses*

We are to test if the proportion of users of Toothpaste A who will never switch to another toothpaste is greater than the proportion of users of Toothpaste B who will never switch to another toothpaste. In other words, we are to test whether or not p_1 is greater than p_2. This can be written as $p_1 - p_2 > 0$. Thus, the two hypotheses are

$$H_0: p_1 - p_2 = 0$$
$$H_1: p_1 - p_2 > 0$$

Step 2. *Select the distribution to use*

As shown in Example 10–13, $n_1\hat{p}_1$, $n_1\hat{q}_1$, $n_2\hat{p}_2$, and $n_2\hat{q}_2$ are all greater than 5. Consequently both samples are large and we apply the normal distribution to make the test.

Step 3. *Determine the rejection and nonrejection regions*

The $>$ sign in the alternative hypothesis indicates that the test is right-tailed. From the normal distribution table, for .01 significance level, the critical value of z is 2.33. This is shown in Figure 10.8.

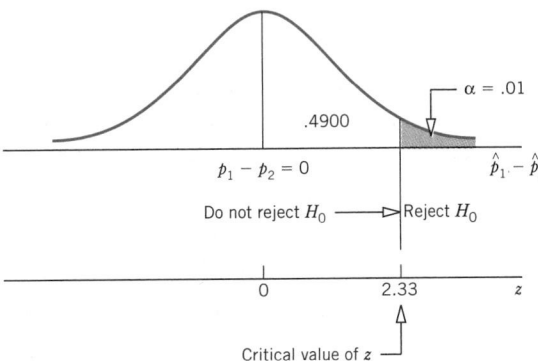

Figure 10.8

Step 4. *Calculate the value of the test statistic*

The pooled sample proportion is

$$\bar{p} = \frac{n_1\hat{p}_1 + n_2\hat{p}_2}{n_1 + n_2} = \frac{500\,(.20) + 400\,(.13)}{500 + 400} = .169$$

and
$$\bar{q} = 1 - \bar{p} = 1 - .169 = .831$$

The estimate of the standard deviation of $\hat{p}_1 - \hat{p}_2$ is

$$s_{\hat{p}_1 - \hat{p}_2} = \sqrt{\bar{p}\,\bar{q}\left(\frac{1}{n_1} + \frac{1}{n_2}\right)} = \sqrt{(.169)\,(.831)\left(\frac{1}{500} + \frac{1}{400}\right)} = .0251$$

The value of the test statistic z for $\hat{p}_1 - \hat{p}_2$ is

$$z = \frac{(\hat{p}_1 - \hat{p}_2) - (p_1 - p_2)}{s_{\hat{p}_1 - \hat{p}_2}} = \frac{(.20 - .13) - 0}{.0251} = 2.79$$

From H_0

Step 5. *Make a decision*

Since the value of the test statistic $z = 2.79$ for $\hat{p}_1 - \hat{p}_2$ falls in the rejection region, we reject the null hypothesis. Therefore, we conclude that the proportion of users of Toothpaste A who will never switch to another toothpaste is greater than the proportion of users of Toothpaste B who will never switch to another toothpaste. ∎

Conducting a two-tailed
test of hypothesis about
$p_1 - p_2$*: large samples.*

EXAMPLE 10-15 A company is planning to buy a few machines. It is considering two types of machines but will buy all of the same type. The company selected one machine of each type and used it for production for a few days. A sample of 800 items produced on Machine I showed that 48 of them were defective. Another sample of 900 items produced on Machine II showed that 45 of them were defective. Testing at the 1% significance level, can we conclude based on the information from these samples that the proportions of defective items produced on the two machines are different?

Solution Let p_1 be the proportion of defective items in all items produced on Machine I and p_2 be the proportion of defective items in all items produced on Machine II. Let $\hat{p}_1$ and $\hat{p}_2$ be the corresponding sample proportions. Let x_1 and x_2 be the number of defective items in two samples, respectively. From the given information,

$$\text{Machine I:} \qquad n_1 = 800 \qquad x_1 = 48$$
$$\text{Machine II:} \qquad n_2 = 900 \qquad x_2 = 45$$

The significance level is $\alpha = .01$.

The two sample proportions are calculated as follows.

$$\hat{p}_1 = x_1/n_1 = 48/800 = .06$$
$$\hat{p}_2 = x_2/n_2 = 45/900 = .05$$

Step 1. *State the null and alternative hypotheses*

The null and alternative hypotheses are

$$H_0: p_1 - p_2 = 0 \qquad \text{(the two proportions are not different)}$$
$$H_1: p_1 - p_2 \neq 0 \qquad \text{(the two proportions are different)}$$

Step 2. *Select the distribution to use*

Because the samples are large and independent, we apply the normal distribution to make the test. (The reader should check that $n_1\hat{p}_1$, $n_1\hat{q}_1$, $n_2\hat{p}_2$, and $n_2\hat{q}_2$ are all greater than 5.)

Step 3. *Determine the rejection and nonrejection regions*

The $\neq$ sign in the alternative hypothesis indicates that the test is two-tailed. For a 1% significance level, the critical values of z are -2.58 and 2.58. These values are shown in Figure 10.9.

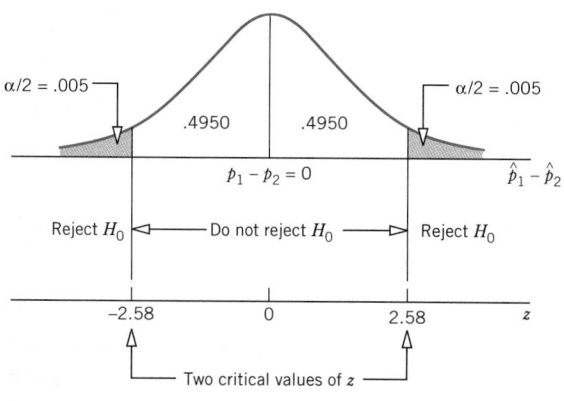

Figure 10.9

Step 4. *Calculate the value of the test statistic*

The pooled sample proportion is

$$\bar{p} = \frac{x_1 + x_2}{n_1 + n_2} = \frac{48 + 45}{800 + 900} = .055$$

and
$$\bar{q} = 1 - \bar{p} = 1 - .055 = .945$$

The estimate of the standard deviation of $\hat{p}_1 - \hat{p}_2$ is

$$s_{\hat{p}_1 - \hat{p}_2} = \sqrt{\bar{p}\,\bar{q}\left(\frac{1}{n_1} + \frac{1}{n_2}\right)} = \sqrt{(.055)(.945)\left(\frac{1}{800} + \frac{1}{900}\right)} = .0111$$

The value of the test statistic z for $\hat{p}_1 - \hat{p}_2$ is

$$z = \frac{(\hat{p}_1 - \hat{p}_2) - (p_1 - p_2)}{s_{\hat{p}_1 - \hat{p}_2}} = \frac{(.06 - .05) - \overset{\text{From } H_0}{0}}{.0111} = .90$$

Step 5. *Make a decision*

The value of the test statistic $z = .90$ for $\hat{p}_1 - \hat{p}_2$ falls in the nonrejection region. Consequently, we fail to reject the null hypothesis. As a result, we can conclude that the proportions of defective items produced by two machines are not different. ∎

CASE STUDY 10-2 MORE ON THE INFERENCES MADE BY THE BUREAU OF THE CENSUS

As we mentioned in Case Study 10–1, the U.S. Bureau of the Census publishes a large number of reports every year that give results of various surveys done by the bureau. The following statement is quoted from one of the bureau's publications. The figures in parentheses denote the 90% confidence interval of the estimate.

The percent of persons without health insurance coverage in 1991 was 14.1 (± 0.3) percent, unchanged from 1990.

The first part of this statement gives an estimate of the percentage of persons who were not covered by health insurance in the United States in 1991. The 90% confidence interval for this percentage is $14.1 \pm .3 = 13.8\%$ to 14.4%. Thus, we can state with 90% confidence that 13.8% to 14.4% of all persons in the United States were not covered by health insurance in 1991.

The second part of this statement gives the decision of a test of hypothesis. In other words, it states that the proportions of all persons who were covered by health insurance in 1990 and 1991 were the same (unchanged). Let p_1 be the proportion of all persons who were covered by health insurance in 1990 and p_2 be the proportion of all persons who were covered by health insurance in 1991. The bureau tested the following hypothesis.

$$H_0: p_1 = p_2$$

$$H_1: p_1 \neq p_2$$

Based on the information from samples for 1990 and 1991, the bureau failed to reject the null hypothesis.

Source: Money Income of Households, Families, and Persons in the United States: 1991. Series P-60, 180. U.S. Department of Commerce, Bureau of the Census, August 1992.

EXERCISES

Concepts and Procedures

10.47   What is the shape of the sampling distribution of $\hat{p}_1 - \hat{p}_2$ for two large samples? What are the mean and standard deviation of this sampling distribution?

10.48   When are the samples considered large enough for the sampling distribution of the difference between two sample proportions to be (approximately) normal?

10.49   Construct a 99% confidence interval for $p_1 - p_2$ for the following.

$$n_1 = 300 \qquad \hat{p}_1 = .53 \qquad n_2 = 200 \qquad \hat{p}_2 = .59$$

10.50   Construct a 95% confidence interval for $p_1 - p_2$ for the following.

$$n_1 = 100 \qquad \hat{p}_1 = .81 \qquad n_2 = 150 \qquad \hat{p}_2 = .76$$

10.51   Refer to the information given in Exercise 10.49. Test at the 1% significance level if the two population proportions are different.

10.52   Refer to the information given in Exercise 10.50. Test at the 5% significance level if $p_1 - p_2$ is different from zero.

10.53   Refer to the information given in Exercise 10.49. Test at the 1% significance level if p_1 is less than p_2.

10.54   Refer to the information given in Exercise 10.50. Test at the 2.5% significance level if p_1 is greater than p_2.

10.55   A sample of 500 observations taken from the first population gave $x_1 = 310$. Another sample of 600 observations taken from the second population gave $x_2 = 348$.

     a.   Find the point estimate of $p_1 - p_2$.
     b.   Make a 97% confidence interval for $p_1 - p_2$.
     c.   Show the rejection and nonrejection regions on the sampling distribution of $\hat{p}_1 - \hat{p}_2$ for $H_0: p_1 = p_2$ versus $H_1: p_1 > p_2$. Use a significance level of 2.5%.
     d.   Find the value of the test statistic z for the test of part c.
     e.   Will you reject the null hypothesis mentioned in part c at a significance level of 2.5%?

10.56   A sample of 1000 observations taken from the first population gave $x_1 = 280$. Another sample of 1200 observations taken from the second population gave $x_2 = 396$.

     a.   Find the point estimate of $p_1 - p_2$.
     b.   Make a 98% confidence interval for $p_1 - p_2$.

c. Show the rejection and nonrejection regions on the sampling distribution of $\hat{p}_1 - \hat{p}_2$ for $H_0: p_1 = p_2$ versus $H_1: p_1 < p_2$. Use a significance level of 1%.

d. Find the value of the test statistic z for the test of part c.

e. Will you reject the null hypothesis mentioned in part c at a significance level of 1%?

Applications

10.57 According to the U.S. Bureau of the Census, 66% of households headed by single women and 81.9% of households headed by single men own cars. Assume that these estimates are based on random samples of 1640 households headed by single women and 1800 households headed by single men.

a. Determine a 99% confidence interval for the difference between the two population proportions.

b. At the 1% significance level, can you conclude that the proportion of households headed by single women who own cars is less than the proportion of households headed by single men who own cars?

10.58 A company has two restaurants in two different areas of New York City. The company wants to estimate the percentages of patrons who think that the food and service at each of these restaurants is excellent. A sample of 200 patrons taken from the restaurant in area A showed that 114 of them think that the food and service are excellent at this restaurant. Another sample of 250 patrons taken from the restaurant in area B showed that 155 of them think that the food and service are excellent at this restaurant.

a. Construct a 97% confidence interval for the difference between the two population proportions.

b. Testing at the 2.5% significance level, can you conclude that the proportion of patrons at the restaurant in area A who think that the food and service is excellent is lower than the corresponding proportion at the restaurant in area B?

10.59 According to the U.S. Bureau of the Census, 48.8% of households headed by single women and 41.5% of households headed by single men own homes. Assume that these estimates are based on random samples of 1640 households headed by single women and 1800 households headed by single men.

a. Construct a 98% confidence interval for the difference between the two population proportions.

b. Test at the 5% significance level if the proportion of households headed by single women who own homes is higher than the proportion of households headed by single men who own homes.

10.60 According to Roper Organization surveys done in 1985 and 1990, 51% of women in 1985 and 42% of women in 1990 said that they would prefer to have a job than to stay at home (*The 1990 Virginia Slims Opinion Poll*, A Study Conducted by the Roper Organization, Inc.). The sample for each year included 3000 women.

a. Construct a 95% confidence interval for the difference between the proportions of all women for 1985 and 1990 who would prefer to have a job.

b. Test at the 1% significance level if the proportion of all women who would prefer to have a job is higher for 1985 than that for 1990.

10.61 The management of a supermarket wanted to investigate if the percentages of men and women who prefer to buy national brand products over the store brand products are different. A sample of 600 men shoppers at the company's supermarkets showed that 246 of them prefer to buy national brand products over the store brand products. Another sample of 700 women shoppers at the company's supermarkets showed that 266 of them prefer to buy national brand products over the store brand products.

a. What is the point estimate of the difference between the two population proportions?

b. Construct a 95% confidence interval for the difference between the proportions of all men and all women shoppers at these supermarkets who prefer to buy national brand products over the store brand products.

c. Testing at the 5% significance level, can you conclude that the proportions of all men and all women shoppers at these supermarkets who prefer to buy national brand products over the store brand products are different?

10.62 The lottery commissioner's office in a state wanted to find if the percentages of men and women who play the lottery often are different. A sample of 500 men taken by the commissioner's office showed that 165 of them play the lottery often. Another sample of 300 women showed that 69 of them play the lottery often.

a. What is the point estimate of the difference between the two population proportions?

b. Construct a 99% confidence interval for the difference between the proportions of all men and all women who play the lottery often.

c. Testing at the 1% significance level, can you conclude that the proportions of all men and all women who play the lottery often are different?

10.63 A mail-order company has two warehouses, one on the West Coast and the second on the East Coast. The company's policy is to mail all orders placed with it within 72 hours. The company's quality control department checks quite often whether or not this policy is maintained at the two warehouses. A recently taken sample of 400 orders placed with the warehouse on the West Coast showed that 368 of them were mailed within 72 hours. Another sample of 300 orders placed with the warehouse on the East Coast showed that 285 of them were mailed within 72 hours.

a. Construct a 97% confidence interval for the difference between the proportions of all orders placed at the two warehouses that are mailed within 72 hours.

b. Using the 2.5% significance level, can you conclude that the proportion of all orders placed at the warehouse on the West Coast that are mailed within 72 hours is lower than the corresponding proportion for the warehouse on the East Coast?

*c. Find the p-value for the test mentioned in part b.

10.64 A company that has many department stores in the southern states wanted to find the percentage of sales at two such stores for which at least one of the items was returned. A sample of 800 sales randomly selected from Store A showed that for 280 of them at least one item was returned. Another sample of 900 sales randomly selected from Store B showed that for 324 of them at least one item was returned.

a. Construct a 98% confidence interval for the difference between the proportions of all sales at the two stores for which at least one item is returned.

b. Using the 1% significance level, can you conclude that the proportions of all sales at the two stores for which at least one item is returned are different?

*c. Find the p-value for the test mentioned in part b.

GLOSSARY

d The difference between two matched values in two samples collected from the same source. It is called the paired difference.

$\bar{d}$ The mean of the paired differences for a sample.

Independent samples Two samples drawn from two populations such that the selection of one does not affect the selection of the other.

Paired or **matched samples** Two samples drawn in such a way that they include the same elements and two data values are obtained from each element, one for each sample. Also called **dependent samples**.

μ_d The mean of the paired differences for the population.

s_d The standard deviation of the paired differences for a sample.

σ_d The standard deviation of the paired differences for the population.

KEY FORMULAS

1. **Mean of the sampling distribution of $\bar{x}_1 - \bar{x}_2$**

$$\mu_{\bar{x}_1 - \bar{x}_2} = \mu_1 - \mu_2$$

2. **Standard deviation of $\bar{x}_1 - \bar{x}_2$ for two large and independent samples**

$$\sigma_{\bar{x}_1 - \bar{x}_2} = \sqrt{\frac{\sigma_1^2}{n_1} + \frac{\sigma_2^2}{n_2}}$$

3. **The $(1 - \alpha)100\%$ confidence interval for $\mu_1 - \mu_2$ for two large and independent samples**

$$(\bar{x}_1 - \bar{x}_2) \pm z\sigma_{\bar{x}_1 - \bar{x}_2}$$

If σ_1 and σ_2 are not known, then $\sigma_{\bar{x}_1 - \bar{x}_2}$ is replaced by its point estimator $s_{\bar{x}_1 - \bar{x}_2}$, which is calculated as

$$s_{\bar{x}_1 - \bar{x}_2} = \sqrt{\frac{s_1^2}{n_1} + \frac{s_2^2}{n_2}}$$

4. **Value of the test statistic z for $\bar{x}_1 - \bar{x}_2$ for two large and independent samples**

$$z = \frac{(\bar{x}_1 - \bar{x}_2) - (\mu_1 - \mu_2)}{\sigma_{\bar{x}_1 - \bar{x}_2}}$$

If σ_1 and σ_2 are not known, then $\sigma_{\bar{x}_1 - \bar{x}_2}$ is replaced by $s_{\bar{x}_1 - \bar{x}_2}$.

5. **Pooled standard deviation for two small and independent samples taken from two populations with equal standard deviations**

$$s_p = \sqrt{\frac{(n_1 - 1)s_1^2 + (n_2 - 1)s_2^2}{n_1 + n_2 - 2}}$$

6. **Estimate of the standard deviation of $\bar{x}_1 - \bar{x}_2$ for two small and independent samples taken from two populations with equal standard deviations**

$$s_{\bar{x}_1 - \bar{x}_2} = s_p \sqrt{\frac{1}{n_1} + \frac{1}{n_2}}$$

7. **The $(1 - \alpha)100\%$ confidence interval for $\mu_1 - \mu_2$ for two small and independent samples taken from two normally distributed populations with equal standard deviations**

$$(\bar{x}_1 - \bar{x}_2) \pm ts_{\bar{x}_1 - \bar{x}_2}$$

8. **Value of the test statistic t for $\bar{x}_1 - \bar{x}_2$ for two small and independent samples taken from two normally distributed populations with equal standard deviations**

$$t = \frac{(\bar{x}_1 - \bar{x}_2) - (\mu_1 - \mu_2)}{s_{\bar{x}_1 - \bar{x}_2}}$$

9. **Sample mean for paired differences**

$$\bar{d} = \frac{\Sigma d}{n}$$

10. **Sample standard deviation for paired differences**

$$s_d = \sqrt{\frac{\Sigma d^2 - \frac{(\Sigma d)^2}{n}}{n - 1}}$$

11. **Mean and standard deviation of the sampling distribution of $\bar{d}$**

$$\mu_{\bar{d}} = \mu_d \qquad \text{and} \qquad s_{\bar{d}} = \frac{s_d}{\sqrt{n}}$$

12. **The $(1 - \alpha)100\%$ confidence interval for μ_d**

$$\bar{d} \pm t s_{\bar{d}}$$

13. **Value of the test statistic t for $\bar{d}$**

$$t = \frac{\bar{d} - \mu_d}{s_{\bar{d}}}$$

14. **Mean of the sampling distribution of $\hat{p}_1 - \hat{p}_2$**

$$\mu_{\hat{p}_1 - \hat{p}_2} = p_1 - p_2$$

15. **Estimate of the standard deviation of $\hat{p}_1 - \hat{p}_2$ for two large and independent samples**

$$s_{\hat{p}_1 - \hat{p}_2} = \sqrt{\frac{\hat{p}_1 \hat{q}_1}{n_1} + \frac{\hat{p}_2 \hat{q}_2}{n_2}}$$

16. **The $(1 - \alpha)100\%$ confidence interval for $p_1 - p_2$ for two large and independent samples**

$$(\hat{p}_1 - \hat{p}_2) \pm z s_{\hat{p}_1 - \hat{p}_2}$$

17. **Pooled sample proportion for two samples**

$$\bar{p} = \frac{x_1 + x_2}{n_1 + n_2} \qquad \text{or} \qquad \frac{n_1 \hat{p}_1 + n_2 \hat{p}_2}{n_1 + n_2}$$

18. **Estimator of the standard deviation of $\hat{p}_1 - \hat{p}_2$ using the pooled sample proportion**

$$s_{\hat{p}_1 - \hat{p}_2} = \sqrt{\bar{p}\,\bar{q} \left(\frac{1}{n_1} + \frac{1}{n_2}\right)}$$

19. **Value of the test statistic z for $\hat{p}_1 - \hat{p}_2$ for large and independent samples**

$$z = \frac{(\hat{p}_1 - \hat{p}_2) - (p_1 - p_2)}{s_{\hat{p}_1 - \hat{p}_2}}$$

SUPPLEMENTARY EXERCISES

10.65   A mathematics proficiency test was given to 905 randomly selected 13-year-old American students. The following table gives the mean scores of male and female students along with the standard deviations of the sample means (*A World of Differences: An International Assessment of Mathematics and Science*, Educational Testing Service, January 1989).

Male students	$\bar{x}_1 = 474.6$	$s_{\bar{x}_1} = 6.4$
Female students	$\bar{x}_2 = 473.2$	$s_{\bar{x}_2} = 5.1$

a.   Construct a 99% confidence interval for the difference between the two population means. Assume that the two samples are large.

b. Test at the 5% significance level if the mean scores in the mathematics proficiency test are different for all male and all female 13-year-old students.

(*Note:* The standard deviation of $\bar{x}_1 - \bar{x}_2$ will be calculated using the formula given in the footnote on page 523, except that we substitute s for σ.)

10.66 According to the Bureau of Labor Statistics, the mean hourly wage for mine workers was $14.51 and that for transportation and public utility workers was $13.49 in 1992. Assume that these mean hourly wages are based on samples of 1100 mine workers and 1400 transportation and public utility workers. Further assume that the standard deviations of the two samples are $1.55 and $1.78, respectively.

    a. Construct a 98% confidence interval for the difference between the two population means.

    b. Test at the 1% significance level if the 1992 mean hourly wage for all mine workers was higher than that for all transportation and public utility workers.

10.67 A consulting firm was asked by a large insurance company to investigate if business majors were better salespersons. A sample of 40 salespersons with a business degree showed that they sold an average of 10 insurance policies per week with a standard deviation of 1.80. Another sample of 45 salespersons with a degree other than business showed that they sold an average of 8.5 insurance policies per week with a standard deviation of 1.35.

    a. Construct a 99% confidence interval for the difference between the two population means.

    b. Using the 1% significance level, can you conclude that persons with a business degree are better salespersons than those who have a degree in another area?

10.68 According to the U.S. Department of Labor, the average weekly pay of teachers was $470 and that of registered nurses was $482 in 1987–1988. Suppose these averages are calculated based on random samples of 1300 teachers and 1500 registered nurses and that the standard deviations for these two samples are $30 and $35, respectively.

    a. Construct a 95% confidence interval for the difference between the two population means.

    b. Test at the 2.5% significance level if the mean weekly pay of all teachers is less than that of all registered nurses.

10.69 A researcher wants to test if the mean GPAs (grade point averages) of all male and all female college students who hold jobs are different. She took a random sample of 28 male students and 24 female students who hold jobs. She found the mean GPAs of the two groups to be 2.62 and 2.74, respectively, with the corresponding standard deviations equal to .43 and .38.

    a. Test at the 5% significance level if the mean GPAs of the two populations are different.

    b. Construct a 99% confidence interval for the difference between the two population means.

Assume that the GPAs of all male and all female students who hold jobs are normally distributed with the same standard deviation.

10.70 According to the National Agricultural Statistics Service, the average yield of corn per acre was 121 bushels for Ohio and 126 bushels for Iowa in 1990. Assume that these two results are based on random samples of 25 acres from Ohio and 28 acres from Iowa. Further assume that the sample standard deviations for the two states are 6 and 7 bushels, respectively.

    a. Construct a 95% confidence interval for the difference between the two population means.

    b. Test at the 2.5% significance level if the mean yield of corn per acre for Ohio is lower than that for Iowa.

Assume that the yields of corn for all acres in Ohio and Iowa are normally distributed with the same standard deviation.

10.71 An agency wanted to estimate the difference between the auto insurance premiums paid by drivers insured with two different insurance companies. A random sample of 25 drivers insured by insurance company A showed that they paid an average monthly insurance premium of $83 with a standard deviation of $14. Another random sample of 20 drivers insured by insurance company B showed that these drivers paid an average monthly insurance premium of $76 with a standard deviation of $12. Assume that the insurance premiums paid by all drivers insured with companies A and B are normally distributed with equal standard deviations.

    a.   Construct a 90% confidence interval for the difference between the two population means.
    b.   Test at the 1% significance level if the mean monthly insurance premium paid by drivers insured with company A is higher than that of drivers insured with company B.

10.72   A business consultant wanted to find the difference between the mean tolerance levels for men and women managers. A random sample of 28 men managers selected by this consultant gave a mean tolerance level of 4.2 (on a scale of 1 to 8, with 1 being the lowest and 8 being the highest level) with a standard deviation of .72. Another random sample of 25 women managers produced a mean tolerance level of 4.9 with a standard deviation of .57.

    a.   Construct a 99% confidence interval for the difference between the two population means.
    b.   Test at the 5% significance level if the mean tolerance level for men managers is lower than that for women managers.

Assume that the tolerance levels for men and women managers each has a normal distribution with the same standard deviation.

10.73   A random sample of eight employees was selected to test for the effectiveness of hypnosis on their job performances. The following table gives the job performance ratings (on a scale of 1 to 4, with 1 being the lowest and 4 being the highest) before and after these employees tried hypnosis.

Before	2.3	2.8	3.1	2.7	3.4	2.6	2.8	2.5
After	2.6	3.2	3.0	3.5	3.7	2.4	2.9	2.9

    a.   Construct a 99% confidence interval for the mean μ_d of the population paired differences where a paired difference is equal to the job performance rating of an employee before trying hypnosis minus the job performance rating of an employee after trying hypnosis.
    b.   Test at the 5% significance level if there is an improvement in the job performances of employees due to hypnosis.

Assume that the population of paired differences is (approximately) normally distributed.

10.74   A random sample of nine employees was selected to test for the effectiveness of a special course designed to improve memory. The following table gives the results of a memory test given to these employees before and after this course.

Before	43	57	48	65	71	49	38	69	58
After	49	56	55	77	79	57	36	64	69

    a.   Construct a 95% confidence interval for the mean μ_d of the population paired differences where a paired difference is equal to the memory test score of an employee before attending this course minus the memory test score of the same employee after attending this course.
    b.   Test at the 1% significance level if this course makes any statistically significant improvement in the memory of the employees.

Assume that the population of the paired differences has a normal distribution.

10.75   In a *USA TODAY* poll of 1010 adults conducted by R.H. Bruskin Associates in 1989, 74.2% of men and 88.8% of women said that they are concerned about living near a nuclear power plant. Assume that there were 520 men and 490 women in this sample.

    a.   Construct a 99% confidence interval for the difference between the proportions of all men and all women who are concerned about living near a nuclear power plant.
    b.   Test at the 1% significance level if the proportion of all men who are concerned about living near a nuclear power plant is lower than that of all women.

10.76   According to a Gallup poll, 58% of adults in 1988 and 45% in 1983 said they would like their child to ''take up teaching in the public schools as a career'' (''The 20th Annual Gallup Poll of the Public's Attitude Towards the Public Schools,'' *Phi Delta Kappa*, September 1988). The sample size for the 1988 poll was 2118. Assume that the sample size for the 1983 poll was 1940.

    a. Construct a 97% confidence interval for the difference between the population proportions for 1988 and 1983.

    b. Test at the 2% significance level if the population proportion of adults who would like their child to take up teaching in the public schools as a career is higher for 1988 than for 1983.

10.77 According to a 1990 survey of CEOs (chief executive officers) of major corporations conducted by Korn/Ferry International and UCLA's Anderson Graduate School of Management, 48% "would choose the same career if they were starting over again". In a similar survey conducted 10 years ago, 60% of the CEOs said that they would choose the same career if they were starting over again (*The Wall Street Journal*, July 3, 1990). Assume that the 1990 survey is based on a sample of 800 CEOs and the one done 10 years ago included 600 CEOs.

    a. Construct a 95% confidence interval for the difference between the two population proportions.

    b. Test at the 5% significance level if the proportion of all CEOs who would choose the same career if they were starting over again has decreased during the past 10 years.

SELF-REVIEW TEST

1. To test the hypothesis that the mean blood pressure of university professors is lower than that of company executives, which of the following would you use?

    a. A left-tailed test   **b.** A two-tailed test   **c.** A right-tailed test

2. Briefly explain the meaning of independent and dependent samples. Give one example of each of these cases.

3. A company psychologist wanted to test if company executives have job-related stress scores higher than those of university professors. He took a sample of 40 executives and 50 professors and tested them for job-related stress. The sample of 40 executives gave a mean stress score of 7.6 with a standard deviation of .8. The sample of 50 professors produced a mean stress score of 5.4 with a standard deviation of 1.3.

    a. Construct a 99% confidence interval for the difference between the mean stress scores of all executives and all professors.

    b. Test at the 5% significance level if the mean stress score of all executives is higher than that of professors.

4. A sample of 20 fathers who were company executives showed that they spend an average of 2.3 hours per week playing with their children with a standard deviation of .54 hours. A sample of 25 fathers who were medical professionals gave a mean of 4.6 hours per week with a standard deviation of .8 hours.

    a. Construct a 95% confidence interval for the difference between the mean time spent per week playing with their children by all fathers who are executives and all fathers who are medical professionals.

    b. Test at the 1% significance level if the mean time spent per week playing with their children by all fathers who are executives is less than that of all fathers who are medical professionals.

Assume that the times spent per week playing with their children by all fathers who are executives and all fathers who are medical professionals have normal distributions with equal but unknown standard deviations.

5. The following table gives the number of items made in one hour by seven randomly selected workers on two different machines.

Worker	1	2	3	4	5	6	7
Machine I	15	18	14	20	16	18	21
Machine II	16	20	13	23	19	18	20

Let μ_d be the mean of differences between the number of items made in one hour on the two machines by all workers.

  a. Construct a 99% confidence interval for the mean μ_d of the population paired differences where a paired difference is equal to the number of items made by an employee in one hour on Machine I minus the number of items made by the same employee in one hour on Machine II.

  b. Test at the 5% significance level if the mean μ_d of the population paired differences is different from zero.

Assume that the population of paired differences is (approximately) normally distributed.

6.   A sample of 500 male registered voters showed that 57% of them are in favor of higher taxes on wealthy people. Another sample of 400 female registered voters showed that 54% of them are in favor of higher taxes on wealthy people.

  a. Construct a 97% confidence interval for the difference between the proportion of all male and all female registered voters who are in favor of higher taxes on wealthy people.

  b. Test at the 1% significance level if the proportion of all male voters who are in favor of higher taxes on wealthy people is different from that of all female voters.

USING MINITAB

INFERENCES ABOUT THE DIFFERENCE BETWEEN TWO POPULATION MEANS FOR LARGE AND INDEPENDENT SAMPLES

MINITAB does not have a direct set of commands that can be used to make a confidence interval and test a hypothesis about the difference between two population means for large and independent samples using the normal distribution. The simplest way to make such inferences with MINITAB is to use the t distribution irrespective of the sample sizes. This procedure is explained next.

INFERENCES ABOUT THE DIFFERENCE BETWEEN TWO POPULATION MEANS FOR SMALL AND INDEPENDENT SAMPLES: EQUAL POPULATION STANDARD DEVIATIONS

The first step in making a confidence interval and testing a hypothesis about the difference between two population means for small and independent samples is to enter data for two samples in columns C1 and C2 using the SET C1 and SET C2 commands. Note that if the sample sizes are the same, we can use the READ command. After the data are entered, the MINITAB command and subcommand given in Figure 10.10 will give a confidence interval for $\mu_1 - \mu_2$ assuming that the standard deviations of the two populations are equal.

Figure 10.10

```
MTB  > TWOSAMPLE (1 - α)100% CONFIDENCE INTERVAL C1 C2;
SUBC > POOLED.
             ↑
       This subcommand instructs MINITAB that σ₁ = σ₂.
```

The first command in Figure 10.10 instructs MINITAB that we are to make a $(1 - \alpha)100\%$ confidence interval for the difference between the two population means using the data on two samples entered in columns C1 and C2. In this command, we will replace $(1 - \alpha)100\%$ by the given confidence level. The subcommand POOLED instructs MINITAB that the standard deviations of the two populations are equal. MINITAB will not only give a $(1 - \alpha)100\%$ confidence interval for $\mu_1 - \mu_2$ but it will also give the results for a test of hypothesis (using the t distribution) for H_0: $\mu_1 - \mu_2 = 0$ against H_1: $\mu_1 - \mu_2 \neq 0$.

This test of hypothesis is based on the assumption that the standard deviations of the two populations are equal. Illustration M10–1 describes how to use the above commands.

Illustration M10-1 An insurance company wanted to find the difference between the mean speeds of cars driven by all men and all women drivers on a highway. A random sample of 16 men who were driving on this highway produced the following data on the speeds of their cars at the time of the survey.

70	67	65	72	71	54	74	69
63	57	64	76	60	55	63	69

Another random sample of 14 women who were driving on the same highway produced the following data on the speeds of their cars at the time of the survey.

61	55	58	66	70	54	57
60	63	72	65	63	59	67

Construct a 99% confidence interval for the difference between the mean speeds of cars driven by all men and all women drivers on this highway. Assume that the speeds at which all men and all women drive cars on this highway are normally distributed with equal but unknown population standard deviations.

Solution Let μ_1 and μ_2 be the mean speeds of cars driven by all men and all women on this highway and let $\bar{x}_1$ and $\bar{x}_2$ be the means of the respective samples. The MINITAB commands and the MINITAB solution are as shown in Figure 10.11.

Figure 10.11

```
MTB  > NOTE: CONFIDENCE INTERVAL FOR ILLUSTRATION M10-1
MTB  > SET C1
DATA > 70 67 65 72 71 54 74 69
DATA > 63 57 64 76 60 55 63 69
DATA > END
MTB  > SET C2
DATA > 61 55 58 66 70 54 57 60 63 72 65 63 59 67
DATA > END
MTB  > TWOSAMPLE 99% CONFIDENCE INTERVAL C1 C2;
SUBC > POOLED.

TWOSAMPLE T FOR C1 VS C2
       N       MEAN      STDEV     SE MEAN
C1   16        65.56      6.64       1.66
C2   14        62.14      5.43       1.45

99 PCT CI FOR MU C1  -  MU C2:  (-2.761, 9.601)
```
99% confidence interval for $\mu_1 - \mu_2$
```
TTEST MU C1 = MU C2 (VS NE):  T = 1.53  P = 0.14  DF = 28

POOLED STDEV = 6.11
```

In the MINITAB solution of Figure 10.11, the row of C1 gives

$$n_1 = 16, \qquad \bar{x}_1 = 65.56, \qquad s_1 = 6.64, \qquad \text{and} \qquad s_{\bar{x}} = 1.7$$

The row of C2 gives

$$n_2 = 14, \qquad \bar{x}_2 = 62.14, \qquad s_2 = 5.43, \qquad \text{and} \qquad s_{\bar{x}} = 1.5$$

The following portion of the MINITAB solution in Figure 10.11 gives the 99% confidence interval for $\mu_1 - \mu_2$.

$$\text{99 PCT CI FOR MU C1} - \text{MU C2: } (-2.8, 9.6)$$

From this, the 99% confidence interval for $\mu_1 - \mu_2$ is -2.8 to 9.6.

The second line from the bottom in the MINITAB solution of Figure 10.11 gives the test of hypothesis (using the t distribution) for H_0: $\mu_1 = \mu_2$ against H_1: $\mu_1 \neq \mu_2$ assuming that the standard deviations of the two populations are equal. The last line in this solution gives the value of the pooled standard deviation s_p, which is 6.11. ▬

The MINITAB commands given in Figure 10.12 are used to make a test of hypothesis about $\mu_1 = \mu_2$ for small and independent samples assuming that the two samples are taken from two populations that are normally distributed with equal (but unknown) standard deviations.

Figure 10.12

```
MTB  > TWOSAMPLE T C1 C2;
SUBC > POOLED;
SUBC > ALTERNATIVE = k.
```

The first command in Figure 10.12, TWOSAMPLE T C1 C2, instructs MINITAB that we are to make a test of hypothesis about the difference between the two population means using the t distribution for the data from two samples entered in columns C1 and C2. The subcommand POOLED in Figure 10.12 indicates that the test is to be made using the pooled standard deviation (see Section 10.2) assuming that the standard deviations of the two populations are equal. The subcommand ALTERNATIVE = k gives the alternative hypothesis. The value of k will be -1, 0, or 1 depending on whether the alternative hypothesis is $\mu_1 - \mu_2 < 0$, $\mu_1 - \mu_2 \neq 0$, or $\mu_1 - \mu_2 > 0$.

Illustration M10–2 shows how to make a test of hypothesis about $\mu_1 - \mu_2$ for small and independent samples when the population standard deviations are unknown but equal.

Illustration M10-2 Refer to Illustration M10–1. Test at the 1% significance level if the mean speed of cars driven by all male drivers on this highway is greater than that of cars driven by all female drivers.

Solution The null and alternative hypotheses are

$$H_0: \mu_1 = \mu_2 \qquad \text{or} \qquad \mu_1 - \mu_2 = 0$$
$$H_1: \mu_1 > \mu_2 \qquad \text{or} \qquad \mu_1 - \mu_2 > 0$$

Because we already entered data from two samples in columns C1 and C2 in Figure 10.11, we do not need to repeat that step here. The MINITAB commands and the MINITAB solution for the test of hypothesis appear in Figure 10.13. Since the test is right-tailed, the value of k will be 1 in the subcommand ALTERNATIVE = k.

Figure 10.13

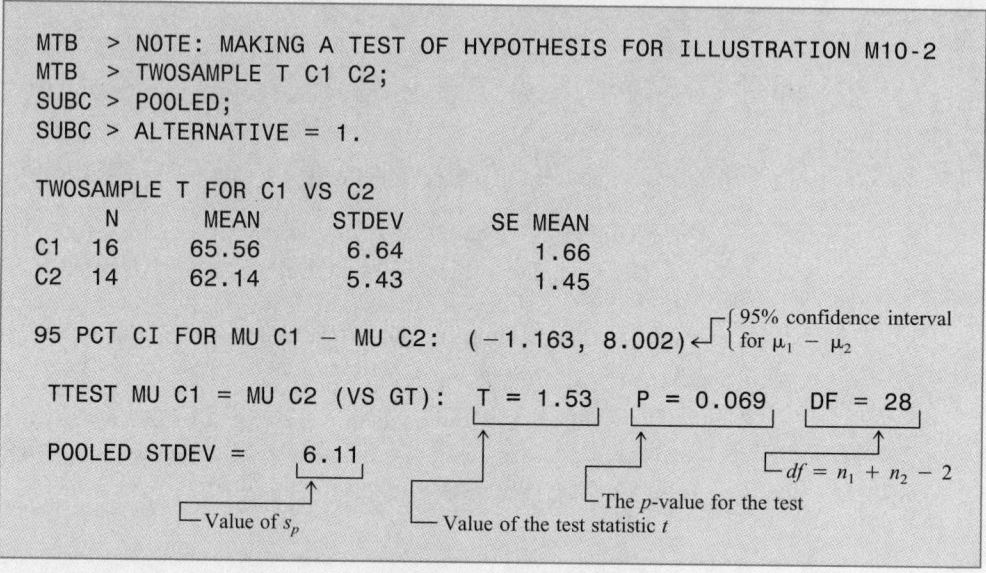

From the MINITAB solution of Figure 10.13, the pooled standard deviation is

$$s_p = 6.11$$

The value of the test statistic t for $\bar{x}_1 - \bar{x}_2$ is

$$t = 1.53$$

The test is right-tailed. The significance level is given to be 1%. Hence,

$$\alpha = .01 \quad \text{and} \quad df = n_1 + n_2 - 2 = 16 + 14 - 2 = 28$$

From the t distribution table, the critical value of t for $df = 28$ and .01 area in the right tail is 2.467. Because the value of the test statistic $t = 1.53$ is less than the critical value of $t = 2.467$, it falls in the nonrejection region. (The reader should draw a graph showing the rejection and nonrejection regions.) Consequently, we fail to reject the null hypothesis.

We can reach the same conclusion using the p-value printed in the MINITAB solution. From the MINITAB solution, the p-value is .069. Since the value of $\alpha = .01$ is less than the p-value of .069, we do not reject the null hypothesis.

Note that the MINITAB solution of Figure 10.13 also gives a 95% confidence interval for $\mu_1 - \mu_2$, which is −1.2 to 8.0.

INFERENCES ABOUT THE DIFFERENCE BETWEEN TWO POPULATION MEANS FOR PAIRED SAMPLES

To make a confidence interval and to test a hypothesis about the difference between two population means for paired samples, first we enter the given data in columns C1 and C2. Then we take the difference between the corresponding data values of columns C1 and C2 and put them in column C3 using the command given in Figure 10.14.

Figure 10.14

```
MTB > LET C3 = C1 - C2
```

The values recorded in column C3 are the paired differences denoted by d in Section 10.3.

To make a confidence interval for μ_d, we use the MINITAB command given in Figure 10.15. In this command, $(1 - \alpha)100\%$ is replaced by the given confidence level, e.g., 99%.

Figure 10.15

```
MTB > TINTERVAL (1 - α)100% CONFIDENCE LEVEL C3
```

To test a hypothesis about μ_d, we use the MINITAB commands given in Figure 10.16.

Figure 10.16

```
MTB > TTEST DIFFERENCE = 0 C3;
SUBC > ALTERNATIVE = k.
```
This command instructs MINITAB that the test is to be made using the t distribution and that the null hypothesis is $\mu_d = 0$.

This number will be -1, 0, or 1 depending on whether the alternative hypothesis is less than, not equal to, or greater than zero, respectively.

Illustration M10–3 describes how to use MINITAB to construct a confidence interval for μ_d using data from paired samples.

Illustration M10–3   Recall Example 10–10. Sham Corporation is considering installing a new assembly procedure. The company randomly selected seven employees and gathered information on the times taken by them to assemble one unit of the product by using the old procedure. Then, the same employees were asked to assemble one unit of the product by using the new procedure. The following table gives the assembly times (in minutes) for these seven employees for each procedure.

Employee	Old Procedure	New Procedure
1	64	60
2	71	66
3	68	66
4	66	69
5	73	63
6	62	57
7	70	62

Let μ_d be the mean of the population differences between the assembly times for the two procedures. Using MINITAB, construct a 95% confidence interval for μ_d. Assume that the population of paired differences is (approximately) normally distributed.

Solution   Figure 10.17 gives the MINITAB input and output for this illustration.

Figure 10.17

```
MTB  > NOTE: SOLUTION FOR ILLUSTRATION M10-3
MTB  > SET C1
DATA > 64 71 68 66 73 62 70
DATA > END
MTB  > SET C2
DATA > 60 66 66 69 63 57 62
DATA > END
MTB  > LET C3 = C1 - C2
MTB  > TINTERVAL 95% CONFIDENCE LEVEL C3


           N      MEAN     STDEV     SE MEAN      95.0 PERCENT C.I.
C3         7      4.43      4.20      1.59      (    0.55,    8.31)
                    ↑         ↑         ↑              ↑
                    d̄        s_d       s_d̄         95% confidence
                                                   interval for μ_d
```

From the MINITAB solution given in Figure 10.17, the 95% confidence interval for μ_d is 0.55 to 8.31.

Illustration M10–4 describes the procedure to make a test of hypothesis about μ_d using MINITAB.

Illustration M10–4   Refer to Illustration M10–3. Test at the 5% significance level if the mean μ_d of the population paired differences is different from zero.

Solution   We entered the given data in columns C1 and C2 in Figure 10.17 and also calculated the values of d and put them in column C3. We do not repeat those steps in this illustration. The MINITAB commands and MINITAB solution for the hypothesis test about the mean μ_d of the population paired differences are given in Figure 10.18.

Figure 10.18

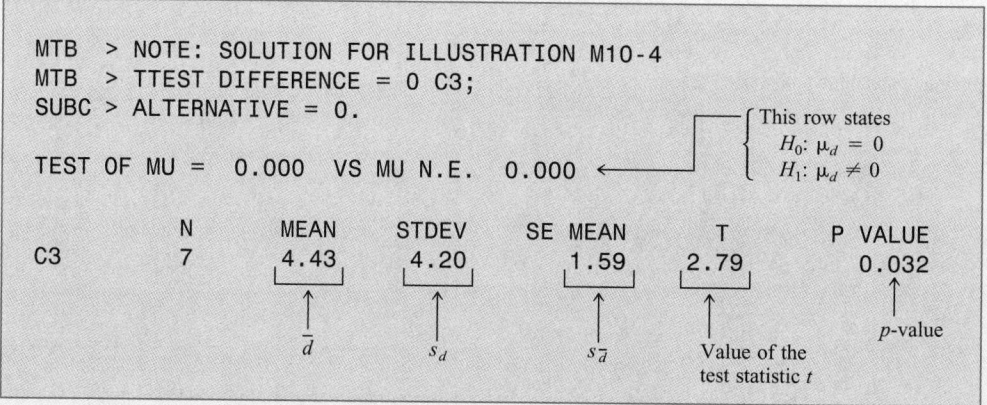

```
MTB  > NOTE: SOLUTION FOR ILLUSTRATION M10-4
MTB  > TTEST DIFFERENCE = 0 C3;
SUBC > ALTERNATIVE = 0.                        ⎰ This row states
                                               ⎱  H_0: μ_d = 0
TEST OF MU =   0.000   VS MU N.E.   0.000  ←      H_1: μ_d ≠ 0

           N     MEAN     STDEV     SE MEAN      T      P VALUE
C3         7     4.43      4.20      1.59      2.79      0.032
                   ↑         ↑         ↑         ↑          ↑
                   d̄        s_d       s_d̄    Value of    p-value
                                            the test
                                            statistic t
```

From the MINITAB output of Figure 10.18, the value of the test statistic t for $\bar{d}$ is 2.79. The two critical values of t for $df = 7 - 1 = 6$ and $\alpha/2 = .025$ area in each tail are -2.447

and 2.447. The value of the test statistic $t = 2.79$ falls in the rejection region (see Figure 10.7 in Example 10–12). As a result, we reject the null hypothesis.

We can reach the same conclusion using the p-value, which is .032 in the MINITAB solution. Since $\alpha = .05$ is greater than the p-value, we reject the null hypothesis. ▪

COMPUTER ASSIGNMENTS

M10.1 A random sample of 13 male college students who hold jobs gave the following data on their GPAs.

| 3.12 | 2.84 | 2.43 | 2.15 | 3.92 | 2.45 | 2.73 |
| 3.06 | 2.36 | 1.93 | 2.81 | 3.27 | 1.83 | |

Another random sample of 16 female college students who also hold jobs gave the following data on their GPAs.

| 2.76 | 3.84 | 2.24 | 2.81 | 1.79 | 3.89 | 2.96 | 3.77 |
| 2.36 | 2.81 | 3.29 | 2.08 | 3.11 | 1.69 | 2.84 | 3.02 |

    a. Using MINITAB, construct a 99% confidence interval for the difference between the mean GPAs of all male and all female college students who hold jobs.

    b. Using MINITAB, test at the 5% significance level if the mean GPAs of all male and all female college students who hold jobs are different.

Assume that the GPAs of all such male and female college students are both normally distributed with equal but unknown population standard deviations.

M10.2 A company recently opened two supermarkets in two different areas. The management wants to know if the mean sales per day for these two supermarkets are different. A sample of 10 days for Supermarket A produced the following data on daily sales.

| 47.56 | 57.66 | 51.23 | 58.29 | 43.71 |
| 49.33 | 52.35 | 50.13 | 47.45 | 53.86 |

A sample of 12 days for Supermarket B produced the following data on daily sales.

| 56.34 | 63.55 | 61.64 | 63.75 | 54.78 | 58.19 |
| 55.40 | 59.44 | 62.33 | 67.82 | 56.65 | 67.90 |

Assume that the daily sales of the two supermarkets are both normally distributed with equal standard deviations.

    a. Using MINITAB, construct a 99% confidence interval for the difference between the mean daily sales for these two supermarkets.

    b. Using MINITAB, test at the 1% significance level if the mean daily sales for these two supermarkets are different.

M10.3 Refer to Exercise 10.45. The manufacturer of a gasoline additive claims that the use of this additive increases gasoline mileage. A random sample of six cars was selected. These cars were driven for one week without the gasoline additive and then for one week with the gasoline additive. The following table gives the miles per gallon used by these cars without and with the gasoline additive.

Without	24.6	28.3	18.9	23.7	15.4	29.5
With	26.3	31.7	18.2	25.3	18.3	30.9

a. Using MINITAB, construct a 99% confidence interval for the mean μ_d of the population paired differences.

b. Using MINITAB, test at the 1% significance level if the use of the gasoline additive increases the gasoline mileage.

Assume that the population of paired differences is (approximately) normally distributed.

M10.4 Refer to Exercise 10.46. A company is considering installing new machines to assemble its products. The company is considering two types of machines, but it will buy only one type. The company selected eight assembly workers and asked them to use these two types of machines to assemble products. The following table gives the time taken (in minutes) to assemble one unit of the product on each type of machine for each of these eight workers.

Machine I	23	26	19	24	26	22	20	18
Machine II	21	24	23	25	24	25	24	23

a. Using MINITAB, construct a 98% confidence interval for the mean μ_d of the population paired differences where a paired difference is equal to the time taken to assemble a unit of the product on Machine I minus the time taken to assemble a unit of the product on Machine II.

b. Using MINITAB, test at the 5% significance level if the mean time taken to assemble a unit of the product is different for the two types of machines.

Assume that the population of paired differences is (approximately) normally distributed.

11 CHI-SQUARE TESTS

The tests of hypotheses about the mean, the difference between two means, the proportion, and the difference between two proportions were discussed in Chapters 9 and 10. The tests about proportions dealt with countable or categorical data. In the case of a proportion and the difference between two proportions, the tests concerned experiments with only two categories. Recall from Chapter 5 that such experiments are called binomial experiments.

This chapter describes three types of tests:

1. Tests of hypotheses for experiments with more than two categories, called goodness-of-fit tests

2. Tests of hypotheses about contingency tables, called independence and homogeneity tests

3. Tests of hypotheses about the variance and standard deviation of a single population

All of these tests are performed by using the **chi-square distribution**. The chi-square distribution is sometimes written as the χ^2 *distribution*, and read as ''chi-square distribution.'' The symbol χ is the Greek letter **chi**, pronounced ''ki.'' The values of a chi-square distribution are denoted by the symbol χ^2 (read as chi-square), just as the values of the standard normal distribution and the t distribution are denoted by z and t, respectively. Section 11.1 describes the chi-square distribution.

11.1   THE CHI-SQUARE DISTRIBUTION

Like the t distribution, the chi-square distribution has only one parameter called the degrees of freedom (df). The shape of a specific chi-square distribution depends on the number of degrees of freedom.[1] (The degrees of freedom for a chi-square distribution are calculated by using different formulas for different tests. This will be explained when we discuss those tests.) The random variable χ^2 assumes nonnegative values only. Hence, a chi-square curve starts at the origin (zero point) and lies entirely to the right of the vertical axis. Figure 11.1 shows three chi-square distribution curves. They are for 2, 7, and 12 degrees of freedom, respectively.

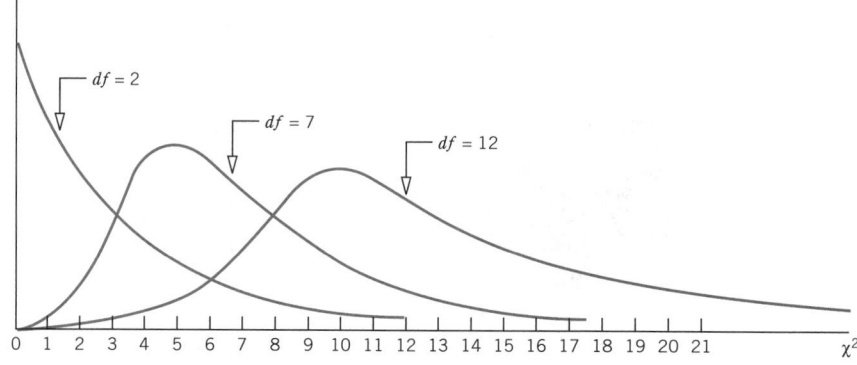

Figure 11.1   Three chi-square distribution curves.

As we can see from Figure 11.1, the shape of a chi-square distribution curve is skewed for very small degrees of freedom and it changes drastically as the degrees of freedom increase. Eventually, for large degrees of freedom, the chi-square distribution curve looks like a normal distribution curve. The peak (or mode) of a chi-square distribution curve with 1 or 2 degrees of freedom occurs at zero and for a curve with 3 or more degrees of freedom at $df - 2$. For instance, the peak of the chi-square distribution curve with $df = 2$ in Figure 11.1 occurs at zero. The peak for the curve with $df = 7$ occurs at $7 - 2 = 5$. Finally, the peak for the curve with $df = 12$ occurs at $12 - 2 = 10$. Like all other continuous distribution curves, the total area under a chi-square distribution curve is 1.0.

THE CHI-SQUARE DISTRIBUTION

The chi-square distribution has only one parameter called the degrees of freedom. The shape of a chi-square distribution curve is skewed to the right for small df and becomes symmetric for large df. The entire chi-square distribution curve lies to the right of the vertical axis. The chi-square distribution assumes nonnegative values only, and these are denoted by the symbol χ^2 (read as chi-square).

If we know the degrees of freedom and the area in the right tail of a chi-square distribution, we can find the value of χ^2 from Table IX of Appendix B. Examples 11−1 and 11−2 show how to read that table.

---

[1]The mean of a chi-square distribution is equal to its df, and the standard deviation is equal to $\sqrt{2\ df}$.

Reading the chi-square distribution table: area in the right tail known.

EXAMPLE 11–1 Find the value of χ^2 for 7 degrees of freedom and an area of .10 in the right tail of the chi-square distribution curve.

Solution To find the required value of χ^2, we locate 7 in the column for *df* and .10 in the top row in Table IX of Appendix B. The required χ^2 value is given by the entry at the intersection of the row for 7 and the column for .10. This value is 12.017. The relevant portion of Table IX is presented as Table 11.1 below.

Table 11.1 χ^2 for 7 *df* and .10 Area in the Right Tail

| *df* | Area in the Right Tail | | | | |
	.995	...	**.100**	...	**.005**
1	0.000	...	2.706	...	7.879
2	0.010	...	4.605	...	10.597
.	...	...	...	...	...
.	...	...	...	...	...
7	0.989	...	12.017←	...	20.278
.	...	...	...	...	...
.	...	...	...	...	...
100	67.328	...	118.498	...	140.169

Required value of χ^2

As shown in Figure 11.2, the χ^2 value for 7 *df* and an area of .10 in the right tail of the chi-square distribution curve is 12.017.

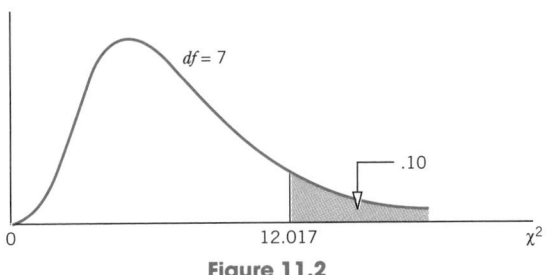

Figure 11.2

Reading the chi-square distribution table: area in the left tail known.

EXAMPLE 11–2 Find the value of χ^2 for 12 degrees of freedom and an area of .05 in the left tail of the chi-square distribution curve.

Solution We can read Table IX only when an area in the right tail of the chi-square distribution curve is known. When the given area is in the left tail, as in this example, the first step is to find the area in the right tail of the chi-square distribution curve as follows.

$$\text{Area in the right tail} = 1 - \text{area in the left tail}$$

Therefore, for our example,

$$\text{Area in the right tail} = 1 - .05 = .95$$

Next, we locate 12 in the column for *df* and .95 in the top row in Table IX. The required value of χ^2, given by the entry at the intersection of the row for 12 and the column for .95, is 5.226. The relevant portion of Table IX is presented here as Table 11.2.

Table 11.2 χ^2 for 12 *df* and .95 Area in the Right Tail

	Area in the Right Tail				
df	**.995**	**. . .**	**.950**	**. . .**	**.005**
1	0.000	. . .	0.004	. . .	7.879
2	0.010	. . .	0.103	. . .	10.597
.	. . .	. . .	. . .	. . .	. . .
.	. . .	. . .	. . .	. . .	. . .
.	. . .	. . .	. . .	. . .	. . .
12	3.074	. . .	5.226 ←	. . .	28.300
.	. . .	. . .	. . .	. . .	. . .
.	. . .	. . .	. . .	. . .	. . .
.	. . .	. . .	. . .	. . .	. . .
100	67.328	. . .	77.930	. . .	140.169

Required value of χ^2

As shown in Figure 11.3, the χ^2 value for 12 *df* and .05 area in the left tail is 5.226.

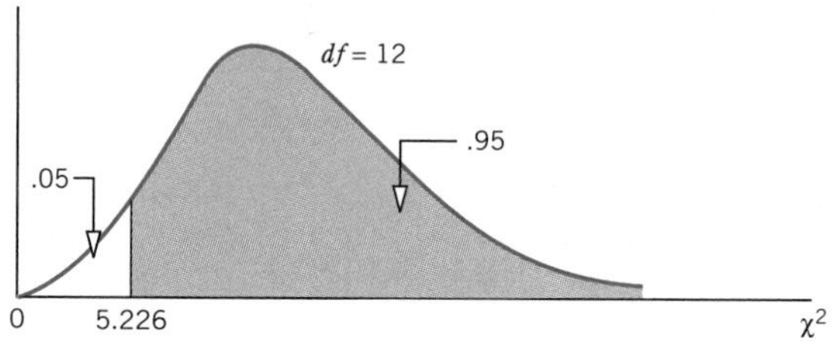

Figure 11.3

EXERCISES

Concepts and Procedures

11.1 Describe the chi-square distribution. What is the parameter (parameters) of such a distribution?

11.2 Find the value of χ^2 for 10 degrees of freedom and an area of .025 in the right tail of the chi-square distribution curve.

11.3 Find the value of χ^2 for 30 degrees of freedom and an area of .05 in the right tail of the chi-square distribution curve.

11.4 Determine the value of χ^2 for 15 degrees of freedom and an area of .10 in the left tail of the chi-square distribution curve.

11.5 Determine the value of χ^2 for 19 degrees of freedom and an area of .990 in the left tail of the chi-square distribution curve.

11.6 Find the value of χ^2 for 4 degrees of freedom and

- **a.** .005 area in the right tail of the chi-square distribution curve
- **b.** .05 area in the left tail of the chi-square distribution curve

11.7 Determine the value of χ^2 for 13 degrees of freedom and

- **a.** .025 area in the left tail of the chi-square distribution curve
- **b.** .995 area in the right tail of the chi-square distribution curve

11.2 A GOODNESS-OF-FIT TEST

This section explains how to make tests of hypotheses about experiments with more than two possible outcomes (or categories). Such experiments, called **multinomial experiments**, possess four characteristics. Note that a binomial experiment is a special case of a multinomial experiment.

A MULTINOMIAL EXPERIMENT

An experiment with the following characteristics is called a multinomial experiment.

1. It consists of n identical trials (repetitions).
2. Each trial results in one of the k possible outcomes (or categories) where $k > 2$.
3. The trials are independent.
4. The probabilities of the various outcomes remain constant for each trial.

An experiment of many rolls of a die is an example of a multinomial experiment. It consists of many identical rolls (trials); each roll (trial) results in one of the six possible outcomes; each roll is independent of the other rolls; and the probabilities of the six outcomes remain constant for each roll.

As a second example of a multinomial experiment, suppose we select a random sample of people and ask them whether or not the quality of American cars is better than that of Japanese cars. The response of a person can be *yes*, *no*, or *do not know*. Each person included in the sample can be considered as one trial (repetition) of the experiment. There will be as many trials for this experiment as the number of persons selected. Each person can belong to any of the three categories—*yes*, *no*, or *do not know*. The response of each selected person is independent of the responses of other persons. Given that the population is large, the probabilities of a person belonging to the three categories remain the same. Consequently, this is an example of a multinomial experiment.

The frequencies obtained from the actual performance of an experiment are called the **observed frequencies**. In a **goodness-of-fit test**, we test the null hypothesis that the observed frequencies for an experiment follow a certain pattern or theoretical distribution. The test is called a goodness-of-fit test because the hypothesis tested is how *good* the observed frequencies *fit* a given pattern.

For our first example involving the experiment of many rolls of a die, we may test the null hypothesis that the given die is fair. The die will be fair if the observed frequency for each outcome is close to one-sixth of the total number of rolls.

For our second example involving opinions of people on the quality of American cars, suppose such a survey was conducted in 1985 and in that survey 41% of the people said *yes*, 48% said *no*, and 11% said *do not know*. We want to test if these percentages still hold true. Suppose we take a random sample of 1000 adults and observe that 536 of them think that the quality of American cars is better than that of Japanese cars, 362 say it is worse, and 102 have no opinion. The frequencies 536, 362, and 102 are the observed frequencies. These frequencies are obtained by actually performing the survey. Now, assuming that the 1985 percentages are still true (which will be our null hypothesis), in a sample of 1000 adults we will expect 410 to say *yes*, 480 to say *no*, and 110 to say *do not know*. These frequencies are obtained by multiplying the sample size (1000) by the 1985 proportions. These frequencies are called the **expected frequencies**. Then, we will make a decision to reject or not to reject the null hypothesis based on how large the difference between the observed frequencies and the expected frequencies is. To perform this test, we will use the chi-square distribution. Note that in this case, we are testing the null hypothesis that all three percentages (or proportions) are unchanged. However, if we want to make a test for only one of the three proportions, we use the procedure learned in Section 9.6 of Chapter 9. For example, if we are testing the hypothesis that the percentage of people who think the quality of American cars is better than that of the Japanese cars is different from 41%, then we will test the null hypothesis $H_0: p = .41$ against the alternative hypothesis $H_1: p \neq .41$. This test will be conducted using the procedure discussed in Section 9.6 of Chapter 9.

As mentioned earlier, the frequencies obtained from the performance of an experiment are called the observed frequencies. They are denoted by "*O*." To make a goodness-of-fit test, we calculate the expected frequencies for all categories of the experiment. The expected frequency for a category, denoted by "*E*," is given by the product of n and p where n is the total number of trials and p is the probability for that category.

OBSERVED AND EXPECTED FREQUENCIES

The frequencies obtained from the performance of an experiment are called the observed frequencies and are denoted by O. The expected frequencies, denoted by E, are the frequencies that we will expect to obtain if the null hypothesis is true. The expected frequency for a category is obtained as

$$E = np$$

where n is the sample size and p is the probability that an element belongs to that category if the null hypothesis is true.

DEGREES OF FREEDOM FOR A GOODNESS-OF-FIT TEST

In a goodness-of-fit test, the degrees of freedom are

$$df = k - 1$$

where k denotes the number of possible outcomes (or categories) for the experiment.

The procedure to make a goodness-of-fit test involves the same five steps that were used in the preceding chapters. *The chi-square goodness-of-fit test is always a right-tailed test.*

TEST STATISTIC FOR A GOODNESS-OF-FIT TEST

The test statistic for a goodness-of-fit test is χ^2 and its value is calculated as

$$\chi^2 = \Sigma \frac{(O - E)^2}{E}$$

where O = observed frequency for a category

              E = expected frequency for a category = np

Remember that a chi-square goodness-of-fit test is always a right-tailed test.

Whether or not the null hypothesis is rejected depends on how much the observed and expected frequencies differ from each other. To find how large the difference between the observed frequencies and the expected frequencies is, we do not just look at $\Sigma(O - E)$ because some of the $O - E$ values will be positive and others will be negative. The net result of the sum of these differences will always be zero. Therefore, we square each of the $O - E$ values to obtain $(O - E)^2$ and then weight them according to the reciprocals of their expected frequencies. The sum of the resulting numbers gives the computed value of the test statistic χ^2.

To make a goodness-of-fit test, the sample size should be large enough so that the expected frequency for each category is at least 5. If there is a category with an expected frequency of less than 5, either increase the sample size or combine two or more categories to make each expected frequency at least 5.

Examples 11–3 and 11–4 describe the procedure for performing goodness-of-fit tests using the chi-square distribution.

Goodness-of-fit test: equal proportions for all categories.

EXAMPLE 11–3 A large shopping mall in New York City has five department stores. A business consultant to one of these stores was asked by the management to investigate if the proportion of visitors to this shopping mall who prefer each of the five department stores is the same. The consultant took a random sample of 1000 visitors to the shopping mall and obtained the frequencies listed in the following table. In this table, the five stores are listed as A, B, C, D, and E. The frequency for any store gives the number of persons in the sample who expressed a preference for that store.

Store	A	B	C	D	E
Frequency	214	231	182	154	219

Using the 1% significance level, test the null hypothesis that the proportion of visitors to this shopping mall who prefer each of the five department stores is the same.

Solution   To make this test, we proceed as follows.

Step 1.   *State the null and alternative hypotheses*

Because there are five categories (stores) listed in the table, the proportion of visitors who prefer a given store will be the same if one-fifth of all visitors to the shopping mall prefer each of the five stores. The null and alternative hypotheses are

H_0: The proportion of visitors to this shopping mall who prefer
each of the five department stores is the same.

H_1: The proportion of visitors to this shopping mall who prefer
each of the five department stores is not the same.

If the proportion of visitors who prefer a given store is the same for all five stores, then the probability for any randomly selected visitor to prefer any of the five stores will be $1/5 = .20$. Let p_1, p_2, p_3, p_4, and p_5 be the probabilities of any randomly selected person preferring stores A, B, C, D, and E, respectively. Then, the null hypothesis can also be written as

$$H_0: p_1 = p_2 = p_3 = p_4 = p_5 = .20$$

and the alternative hypothesis can be stated as

H_1: At least two of the five probabilities are not equal to .20

Step 2.   *Select the distribution to use*

Because there are five categories (i.e., five stores listed in the table), this is a multinomial experiment. Consequently, we use the chi-square distribution to make the test.

Step 3.   *Determine the rejection and nonrejection regions*

The significance level is given to be .01 and the goodness-of-fit test is always right-tailed. Therefore,

$$\text{Area in the right tail} = \alpha = .01$$

$$k = \text{the number of categories} = 5$$

$$df = k - 1 = 5 - 1 = 4$$

From the chi-square distribution table (Table IX of Appendix B), the critical value of χ^2 for 4 *df* and .01 area in the right tail of the chi-square distribution curve is 13.277, as shown in Figure 11.4.

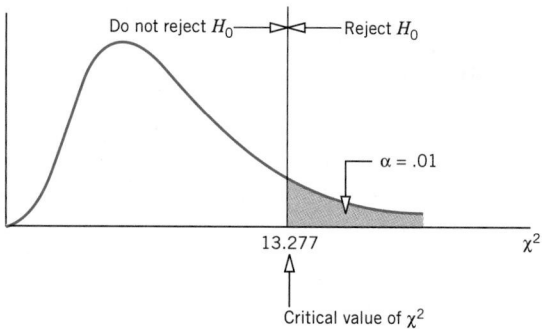

Figure 11.4

Step 4.   *Calculate the value of the test statistic*

All the required calculations to find the value of the test statistic χ^2 are shown in Table 11.3.

Table 11.3

Store	Observed Frequency O	p	Expected Frequency $E = np$	$(O - E)$	$(O - E)^2$	$\dfrac{(O - E)^2}{E}$
A	214	.20	$1000(.20) = 200$	14	196	.980
B	231	.20	$1000(.20) = 200$	31	961	4.805
C	182	.20	$1000(.20) = 200$	-18	324	1.620
D	154	.20	$1000(.20) = 200$	-46	2116	10.580
E	219	.20	$1000(.20) = 200$	19	361	1.805
	$n = 1000$					Sum $= 19.790$

The calculations made in Table 11.3 are explained below.

1. The first two columns in Table 11.3 list the five categories (stores) and the observed frequencies for the sample of 1000 visitors to the shopping mall, respectively. The third column contains the probabilities for the five categories assuming that the null hypothesis is true.

2. The fourth column contains the expected frequencies. These frequencies are obtained by multiplying the sample size ($n = 1000$) by the probabilities listed in the third column. If the null hypothesis is true (that is, if the proportion of visitors to this shopping mall who prefer each of the five department stores is the same), then we will expect 200 out of 1000 persons to show a preference for each store. Consequently, each category in the fourth column has the same expected frequency.

3. The fifth column lists the differences between the observed and expected frequencies, that is, $O - E$. These values are squared and recorded in the sixth column.

4. Finally, we divide the squared differences (of the sixth column) by the corresponding expected frequencies (listed in the fourth column) and write the resulting numbers in the seventh column.

5. The sum of the seventh column gives the value of the test statistic χ^2. Thus,

$$\chi^2 = \Sigma \frac{(O - E)^2}{E} = 19.790$$

Step 5. *Make a decision*

The value of the test statistic $\chi^2 = 19.790$ is greater than the critical value of $\chi^2 = 13.277$ and it falls in the rejection region. Hence, we reject the null hypothesis and state that the proportion of all visitors to this shopping mall who prefer each of the five department stores is not the same. In other words, more visitors prefer one or more of the stores over others. However, based on this test we cannot say which store (stores) is (are) preferred over others. ▬

Goodness-of-fit test: testing if results of a survey fit a given distribution.

EXAMPLE 11–4 The following table gives the 1992 U.S. car market shares held by various auto companies (*The Wall Street Journal*, January 8, 1993). Market shares are usually defined as percentages of the total revenue earned by various companies. Suppose for convenience that in this case the market shares are synonymous with percentages of the total new cars sold by various auto companies in 1992.

Company	GM	Ford	Honda	Toyota	Chrysler	Others
Market share	34.5%	21.6%	9.3%	9.2%	8.3%	17.1%

A business organization wanted to investigate if the current U.S. car market shares of these companies are the same as for 1992. A sample of 2000 recently sold new cars showed that 715 of them were manufactured by General Motors, 446 by Ford, 175 by Honda, 187 by Toyota, 178 by Chrysler, and 299 were manufactured by other companies. Testing at the 2.5% significance level, will you reject the null hypothesis that the current U.S. car market shares held by these companies are the same as for 1992?

Solution

Step 1.   *State the null and alternative hypotheses*

The null and alternative hypotheses are

H_0: The current U.S. car market shares of various auto
companies are the same as those for 1992

H_1: The current U.S. car market shares of various auto
companies are not the same as those for 1992

Note that the current U.S. car market shares held by various auto companies will be the same as for 1992 if the percentages given in the table above are still true. Let p_1, p_2, p_3, p_4, p_5, and p_6 be the current U.S. car market shares held by General Motors, Ford, Honda, Toyota, Chrysler, and other auto companies, respectively. Then, the null hypothesis can also be written as

H_0: $p_1 = .345$, $p_2 = .216$, $p_3 = .093$, $p_4 = .092$, $p_5 = .083$, and $p_6 = .171$

and the alternative hypothesis can be stated as

H_1: At least two of the six probabilities are different
from those mentioned in the null hypothesis

Step 2.   *Select the distribution to use*

Because there are six categories (companies), this is a multinomial experiment. Consequently, we use the chi-square distribution to make the test.

Step 3.   *Determine the rejection and nonrejection regions*

The significance level is .025. Because a goodness-of-fit test is right-tailed,

$$\text{Area in the right tail} = \alpha = .025$$

$$k = \text{number of categories} = 6$$

$$df = k - 1 = 6 - 1 = 5$$

From the chi-square table (Table IX of Appendix B), the critical value of χ^2 for 5 df and .025 area in the right tail of the chi-square distribution curve is 12.833. This value is shown in Figure 11.5.

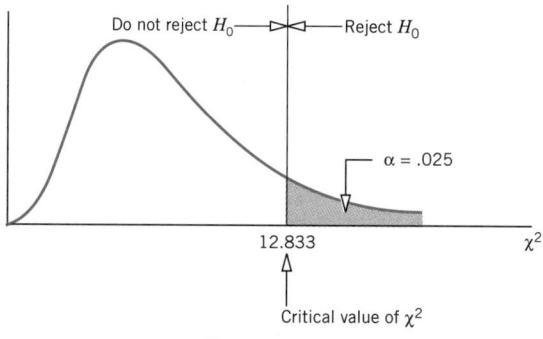

Do not reject H_0 ——▷◁—— Reject H_0

$\alpha = .025$

12.833

χ^2

Critical value of χ^2

Figure 11.5

Step 4. *Calculate the value of the test statistic*

All the required calculations to find the value of the test statistic χ^2 are shown in Table 11.4.

Note that the given percentages have been converted to probabilities and recorded in the third column of Table 11.4. The value of the test statistic χ^2 is given by the sum of the seventh column. Thus,

$$\chi^2 = \Sigma \frac{(O - E)^2}{E} = 8.333$$

Table 11.4

Company	Observed Frequency O	p	Expected Frequency $E = np$	$(O - E)$	$(O - E)^2$	$\dfrac{(O - E)^2}{E}$
GM	715	.345	2000(.345) = 690	25	625	.906
Ford	446	.216	2000(.216) = 432	14	196	.454
Honda	175	.093	2000(.093) = 186	−11	121	.651
Toyota	187	.092	2000(.092) = 184	3	9	.049
Chrysler	178	.083	2000(.083) = 166	12	144	.867
Others	299	.171	2000(.171) = 342	−43	1849	5.406
	$n = 2000$					Sum = 8.333

Step 5. *Make a decision*

The value of the test statistic $\chi^2 = 8.333$ is smaller than the critical value of $\chi^2 = 12.833$ and it falls in the nonrejection region. Consequently, we fail to reject the null hypothesis and state that the current U.S. car market shares held by various auto companies are the same as those for 1992. The difference between the observed frequencies and the expected frequencies has occurred only because of sampling error. ■

Case Study 11–1 shows the application of the goodness-of-fit test to determine whether or not leadership styles are evenly distributed.

CASE STUDY 11-1   ARE LEADERSHIP STYLES EVENLY DISTRIBUTED?

Professor Donald C. Lueder conducted a study on the distribution of leadership styles based on a sample of 95 doctoral students enrolled in a basic administration course. These students were experienced teachers and administrators. The students were given the Leader Effectiveness and Description Instrument of Self-perception Test (called LEAD-Self for short) developed by Paul Hersey and Kenneth Blanchard to measure leadership style and leader adaptability.

The LEAD-Self instrument contains 12 situations. Each respondent was required to select one of the four alternative action responses for each situation. Each of the four responses corresponds to a different leadership style. The four leadership styles are ''high task/low relationship (S1), or 'telling'; high task/high relationship (S2), or 'selling'; low task/high relationship (S3), or 'participating'; and low task/low relationship (S4), or 'delegating.' '' The researcher hypothesized that a significantly high number of respondents would choose S2. To test this hypothesis, he applied three statistical tests, one of which was the chi-square test.

The category having the largest number of a student's responses was considered to be the leadership style of that student. For 10 of the 95 respondents, at least two categories had the same number of responses. Hence, the researcher dropped these 10 respondents from the chi-square test. Of the 85 remaining respondents, 4 were observed to belong to the S1 leadership style, 66 to the S2, 14 to the S3, and 1 to the S4. Table 11.5 gives all the calculations required to make the chi-square test. The expected frequencies are based on the following null and alternative hypotheses.

H_0: The respondents are evenly distributed over different leadership styles

H_1: The respondents are not evenly distributed over different leadership styles

If the null hypothesis is true, then we will expect the 85 respondents to be equally distributed over the four categories. Hence, the expected frequency for each category is $85/4 = 21.25$.

Table 11.5

Leadership Style	Observed Frequency (O)	Expected Frequency (E)	O − E	$\dfrac{(O-E)^2}{E}$
S1	4	21.25	−17.25	14.00
S2	66	21.25	44.75	94.24
S3	14	21.25	−7.25	2.47
S4	1	21.25	−20.25	19.30
Total	85	85		130.01

The value of the test statistic χ^2 is

$$\chi^2 = \Sigma \frac{(O - E)^2}{E} = 130.01$$

The researcher made the test at $\alpha = .001$. The critical value of χ^2 for 3 df and .001 area in the right tail of the chi-square distribution curve is 16.27. (Note that Table IX of Appendix B does not contain $\alpha = .001$ area.) Because the value of the test statistic $\chi^2 = 130.01$ is larger than the critical value of $\chi^2 = 16.27$, the null hypothesis is rejected. The decision is that the respondents are not evenly distributed over the different leadership styles.

Source: Donald C. Lueder: ''Don't Be Misled by LEAD.'' *The Journal of Applied Behavioral Science*, 21(2), 1985, pp. 143–151. Copyright © 1985 by NTL Institute. Data and excerpts reprinted with permission.

EXERCISES

Concepts and Procedures

11.8 Describe the four characteristics of a multinomial experiment.

11.9 What is a goodness-of-fit test and when is it applied? Explain.

11.10 Explain the difference between the observed and expected frequencies for a goodness-of-fit test.

11.11 How is the expected frequency of a category calculated for a goodness-of-fit test? What are the degrees of freedom for such a test?

11.12 To make a goodness-of-fit test, what should be the minimum expected frequency for each category? What are the alternatives if this condition is not satisfied?

11.13 The following table lists the frequency distribution for 60 rolls of a die.

Outcome	1-spot	2-spot	3-spot	4-spot	5-spot	6-spot
Frequency	7	12	8	15	11	7

Test at the 5% significance level if the null hypothesis that the given die is fair is true.

Applications

11.14 The following table gives the percentage distribution of the 1992 U.S. labor force by educational attainment. (Source: U.S. Bureau of Labor Statistics.)

Educational Attainment	Percentage of Workers
Less than a high school diploma	12.4
High school graduate, no college	35.7
Some college, no degree	18.0
Associate degree	7.4
Bachelor's degree	17.2
Master's degree	6.5
Professional or doctoral degree	2.8

A recent random sample of 1000 workers showed that 116 have less than a high school diploma, 363 are high school graduates with no college, 164 have some college but no degree, 71 possess an associate degree, 187 hold a bachelor's degree, 61 have a master's degree, and 38 possess a professional or doctoral degree. Test at the 1% significance level if the percentage distribution of the U.S. labor force by educational attainment has changed since 1992.

11.15 DOM Corporation owns department stores in almost all states. Forty-eight percent of all persons who possess credit cards for these department stores usually make payments on their credit card bills within 15 days of the issue of the statements, 44% pay within 16 to 30 days, and 8% take more than 30 days to pay. The management of DOM Corporation wants to know if the credit card holders in a specific state are different from credit card holders in other states. The research department at the company took a random sample of 800 credit card holders from this state and found that 397 of them made the payments on their credit card bills within 15 days of the issue of the statement last month, 342 paid within 16 to 30 days, and 61 took more than 30 days to pay. Test at the 2.5% significance level if the credit card holders in this state are different from all credit card holders with respect to making the payments on their credit card bills.

11.16 Home Mail Corporation sells products by mail. The company's management wants to find out if the number of orders received at the company's office on each of the five days of the week is the same. The company took a sample of 400 orders received during a four-week period. The following table lists the frequency distribution for these orders by the day of the week.

Day of the week	Mon	Tue	Wed	Thu	Fri
Number of orders received	92	71	65	83	89

Test at the 5% significance level if the null hypothesis that the orders are evenly distributed over all days of the week is true.

11.17   The following table gives the average number of computers assembled by an employee of a small electronics company on different days of the week for a sample of five weeks.

Day	Mon	Tue	Wed	Thu	Fri
Computers assembled	16	22	26	21	15

Using the 1% significance level, can you reject the null hypothesis that the number of items assembled by this employee on different days of the week is the same?

11.18   The following table lists the frequency distribution of cars sold at an auto dealership during the past 12 months.

Month	Jan	Feb	Mar	Apr	May	Jun	Jul	Aug	Sep	Oct	Nov	Dec
Cars sold	21	17	15	12	14	12	13	19	23	26	24	28

Using the 10% significance level, will you reject the null hypothesis that the number of cars sold at this dealership is the same for each month?

11.19   A company that owns department stores advertised for one month on television and radio and in newspapers and spent the same amount of money on each of these three media. A random sample of 1500 customers who visited this company's department stores during that month as a result of these promotional advertisements was asked which advertising medium was most effective in getting each of them to visit the stores. The responses of these customers are recorded in the following table.

Advertising medium	Television	Radio	Newspapers
Medium that was most effective	704	368	428

Using the 5% significance level, test the null hypothesis that these three media sources have the same promotional impact on customers.

11.20   Chance Corporation produces beauty products. Two years ago the quality control department at the company conducted a survey of users of one of the company's products. The survey revealed that 53% of the users said the product was excellent, 31% said it was satisfactory, 7% said it was unsatisfactory, and 9% had no opinion. Assume that these percentages are true for the population of all users of this product at that time. After this survey was conducted, the company redesigned this product. A recent survey of 800 users of the redesigned product conducted by the quality control department at the company showed that 495 of the users think the product is excellent, 259 think that it is satisfactory, 31 think it is unsatisfactory, and 15 have no opinion. Do you think the percentage distribution of the opinions of the users of the redesigned product is different from the percentage distribution of the users of this product before it was redesigned? Use $\alpha = .025$.

11.21   Henderson Corporation makes metal sheets among other products. When the process that is used to make metal sheets works properly, 92% of the metal sheets contain no defects, 5% have one defect each, and 3% have two or more defects each. The quality control inspectors at the company take samples of metal sheets quite often and check them for defects. If the distribution of defects for a sample is significantly different from the above mentioned percentage distribution, the process is stopped and adjusted. A recent sample of 300 sheets produced the frequency distribution of defects listed in the following table.

Number of defects	None	One	Two or More
Number of metal sheets	269	18	13

Does the evidence from this sample suggest that the process needs an adjustment? Use $\alpha = .01$.

11.3 CONTINGENCY TABLES

We often may have information on more than one variable for each element. Such information can be summarized and presented using a two-way classification table, which is also called a *contingency table* or *cross-tabulation*. Table 11.6 is an example of a contingency table. It gives information on a sample of men and women workers selected from a large company who are in favor of approving a new labor–management contract, are against it, or have no opinion. Table 11.6 has two rows (one for men workers and the other for women workers) and three columns (corresponding to workers who are in favor of approving the contract, are against it, or have no opinion, respectively). Hence, it is also called a 2×3 (read as ''two by three'') contingency table.

Table 11.6 Opinions of Workers on a Labor–Management Contract

	In Favor	**Against**	**No Opinion**	
Men	93	70	12 ←	⎤ Men workers who have no opinion
Women	87	32	6	

A contingency table can be of any size. For example, it can be 2×3, 3×2, 3×3, or 4×2. Note that in these notations, the first digit refers to the number of rows in the table and the second digit refers to the number of columns. For example, a 3×2 table will contain 3 rows and 2 columns. In general, an $R \times C$ table contains R rows and C columns.

Each of the six boxes that contain numbers in Table 11.6 is called a *cell*. The number of cells for a contingency table is obtained by multiplying the number of rows by the number of columns. Thus, Table 11.6 contains $2 \times 3 = 6$ cells. The subjects that belong to a cell of a contingency table possess two characteristics. For example, 12 workers listed in the third cell of the first row in Table 11.6 are *men* and *have no opinion*. The numbers entered in the cells are usually called the *joint frequencies*. For example, 12 workers belong to the joint category of *men* and *have no opinion*. Hence, it is the joint frequency of this category.

11.4 A TEST OF INDEPENDENCE OR HOMOGENEITY

This section is concerned with tests of independence and homogeneity, which are performed using the contingency tables. Except for a few modifications, the procedure used to make such tests is almost the same as the one applied in Section 11.2 for a goodness-of-fit test.

11.4.1 A TEST OF INDEPENDENCE

In a **test of independence** for a contingency table, we test the null hypothesis that the two attributes (characteristics) of the elements of a given population are not related (that is, they are independent) against the alternative hypothesis that the two characteristics are related (that is, they are dependent). For example, we may want to test if the gender and opinions of workers about the labor–management contract mentioned in Table 11.6 are dependent. We perform such a test by using the chi-square distribution. As another example, we may want to test whether or not an association exists between the job satisfaction index and the absentee rate of employees.

DEGREES OF FREEDOM FOR A TEST OF INDEPENDENCE

A test of independence involves a test of the null hypothesis that two attributes of a population are not related. The degrees of freedom for a test of independence are

$$df = (R - 1)(C - 1)$$

where R and C are the number of rows and the number of columns, respectively, in the given contingency table.

The value of the test statistic χ^2 in the case of a test of independence is obtained using the same formula as in the goodness-of-fit test described in Section 11.2.

TEST STATISTIC FOR A TEST OF INDEPENDENCE

The value of the test statistic χ^2 for a test of independence is calculated as

$$\chi^2 = \Sigma \frac{(O - E)^2}{E}$$

where O and E are the observed and expected frequencies, respectively, for a cell.

The null hypothesis in a test of independence is always that the two attributes are not related. The alternative hypothesis is that the two attributes are related.

The frequencies obtained from the performance of an experiment for a contingency table are called the **observed frequencies**. The procedure to calculate the **expected frequencies** for a contingency table for a test of independence is different from the one for a goodness-of-fit test. Example 11–5 describes this procedure.

Calculating expected frequencies for a test of independence.

EXAMPLE 11–5 The labor union at a large auto company has drafted a new labor–management contract. Before this contract is presented to and negotiated with management, the labor union leaders want to seek opinions of workers on this proposed contract. The union leaders selected a random sample of 300 workers and asked their opinions. The two-way classification of the responses of workers was presented earlier in Table 11.6. Calculate the expected frequencies for this table assuming that the gender and opinions of workers are independent.

Solution   Table 11.6 is reproduced below as Table 11.7. Note that we have added the row and column totals to Table 11.7.

Table 11.7

	In Favor (F)	Against (A)	No Opinion (N)	Row Totals
Men (M)	93	70	12	175
Women (W)	87	32	6	125
Column totals	180	102	18	300

The numbers 93, 70, 12, 87, 32, and 6 in Table 11.7 are called the *observed frequencies* for the six cells.

As mentioned earlier, the null hypothesis in a test of independence is that the two attributes (or classifications) are independent. In an independence test of hypothesis, first we assume that the null hypothesis is true and that the two attributes are independent. Assuming that the null hypothesis is true and that the gender and opinions of workers are not related, the expected frequency for the cell corresponding to *men* and *in favor* is calculated as follows. From Table 11.7,

$$P(\text{a worker is a man}) = P(M) = 175/300$$

$$P(\text{a worker is in favor of the proposed contract}) = P(F) = 180/300$$

Because we are assuming that M and F are independent (by assuming that the null hypothesis is true), using the formula learned in Chapter 4 the joint probability of these two events is

$$P(M \text{ and } F) = P(M) \cdot P(F) = (175/300) \cdot (180/300)$$

Then, assuming M and F are independent, the number of workers expected (E) to be *men* and *in favor* of the proposed contract in a sample of 300 is

$$E \text{ for } men \text{ and } in\ favor = 300 \times P(M \text{ and } F)$$

$$= 300 \times \frac{175}{300} \times \frac{180}{300} = \frac{175 \times 180}{300}$$

$$= \frac{(\text{Row total})(\text{Column total})}{\text{Sample size}}$$

Thus, the rule to obtain the expected frequency for a cell is to divide the product of the corresponding row and column totals by the sample size.

EXPECTED FREQUENCIES FOR A TEST OF INDEPENDENCE

The expected frequency E for a cell is calculated as

$$E = \frac{(\text{Row total})(\text{Column total})}{n}$$

Using this rule, we calculate the expected frequencies of the six cells as follows.

$$E \text{ for } men \text{ and } in \text{ } favor \text{ cell} = (175)(180)/300 = 105.00$$

$$E \text{ for } men \text{ and } against \text{ cell} = (175)(102)/300 = 59.50$$

$$E \text{ for } men \text{ and } no \text{ } opinion \text{ cell} = (175)(18)/300 = 10.50$$

$$E \text{ for } women \text{ and } in \text{ } favor \text{ cell} = (125)(180)/300 = 75.00$$

$$E \text{ for } women \text{ and } against \text{ cell} = (125)(102)/300 = 42.50$$

$$E \text{ for } women \text{ and } no \text{ } opinion \text{ cell} = (125)(18)/300 = 7.50$$

The expected frequencies are usually written in parentheses below the observed frequencies within the corresponding cells, as shown in Table 11.8.

Table 11.8

	In Favor (F)	Against (A)	No Opinion (N)	Row Totals
Men (M)	93 (105.00)	70 (59.50)	12 (10.50)	175
Women (W)	87 (75.00)	32 (42.50)	6 (7.50)	125
Column totals	180	102	18	300

Like a goodness-of-fit test, *a test of independence is always right-tailed*. To apply a chi-square test of independence, *the sample size should be large enough so that the expected frequency for each cell is at least 5*. If the expected frequency for a cell is not at least 5, we either increase the sample size or combine some categories. Examples 11–6 and 11–7 describe the procedure to make tests of independence using the chi-square distribution.

A test of independence: 2 × 3 table.

EXAMPLE 11–6 Reconsider the two-way classification of 300 workers based on gender and opinions about the proposed labor–management contract given in Table 11.6. Using the 5% significance level, can you conclude that the gender and opinions of workers are dependent?

Solution The test involves the following five steps.

Step 1. *State the null and alternative hypotheses*

As mentioned earlier, in an independence test of hypothesis, the null hypothesis must be that the two attributes are independent. Consequently, the alternative hypothesis is that these attributes are dependent.

H_0: Gender and opinions of workers are independent

H_1: Gender and opinions of workers are dependent

Step 2. *Select the distribution to use*

We use the chi-square distribution to make a test of independence for a contingency table.

Step 3. *Determine the rejection and nonrejection regions*

The significance level is 5%. Because a test of independence is always right-tailed, the area of the rejection region is .05 and it falls in the right tail of the chi-square distribution curve. The contingency table contains two rows (*men workers* and *women workers*) and three columns (*in favor*, *against*, and *no opinion*). Note that we do not count the row and column containing the totals. The degrees of freedom are

$$df = (R - 1)(C - 1) = (2 - 1)(3 - 1) = 2$$

From Table IX of Appendix B, the critical value of χ^2 for $df = 2$ and $\alpha = .05$ is 5.991. This value is shown in Figure 11.6.

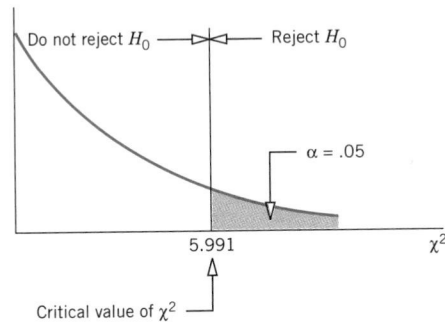

Figure 11.6

Step 4. *Calculate the value of the test statistic*

Table 11.8, with the observed and expected frequencies constructed in Example 11–5, is reproduced as Table 11.9 below.

Table 11.9

	In Favor (F)	Against (A)	No Opinion (N)	Row Totals
Men (M)	93 (105.00)	70 (59.50)	12 (10.50)	175
Women (W)	87 (75.00)	32 (42.50)	6 (7.50)	125
Column totals	180	102	18	300

To compute the value of the test statistic χ^2, we take the difference between the observed and expected frequencies listed in Table 11.9, square those differences, and then divide each of the squared differences by the respective expected frequencies. The sum of the resulting numbers gives the value of the test statistic χ^2. We make these calculations as follows.

$$\chi^2 = \Sigma \frac{(O - E)^2}{E}$$

$$= \frac{(93 - 105.00)^2}{105.00} + \frac{(70 - 59.50)^2}{59.50} + \frac{(12 - 10.50)^2}{10.50}$$

$$+ \frac{(87 - 75.00)^2}{75.00} + \frac{(32 - 42.50)^2}{42.50} + \frac{(6 - 7.50)^2}{7.50}$$

$$= 1.371 + 1.853 + .214 + 1.920 + 2.594 + .300 = 8.252$$

Step 5. *Make a decision*

The value of the test statistic $\chi^2 = 8.252$ is greater than the critical value of $\chi^2 = 5.991$ and it falls in the rejection region. Hence, we reject the null hypothesis and state that there

598

is strong evidence from the sample that the two characteristics, gender and worker's opinions about the proposed contract, are dependent. ▬

A test of independence:
2 × 2 table.

EXAMPLE 11–7   Windham Electronics Company manufactures cassettes. The company has two machines that are used to manufacture cassettes. From time to time the quality control inspector at the company takes a sample of cassettes and checks them for being good or defective. A recent such sample of 200 cassettes produced the two-way classification presented in the following table.

	Good (G)	Defective (D)
Machine I (A)	109	11
Machine II (B)	66	14

Does the sample provide sufficient evidence to conclude that the two attributes, the machine type and cassettes being good or defective, are dependent? Use $\alpha = .01$.

Solution

Step 1.   *State the null and alternative hypotheses*

The null and alternative hypotheses are

H_0: Machine type and cassettes being good or defective are independent

H_1: Machine type and cassettes being good or defective are dependent

Step 2.   *Select the distribution to use*

Because we are making a test of independence, we use the chi-square distribution to make the test.

Step 3.   *Determine the rejection and nonrejection regions*

With a significance level of 1%, the area of the rejection region is .01, and it lies in the right tail of the chi-square distribution curve. The contingency table contains two rows (of Machine I and Machine II) and two columns (of cassettes being good or defective). Hence, the degrees of freedom are

$$df = (R - 1)(C - 1) = (2 - 1)(2 - 1) = 1$$

From Table IX of Appendix B, the critical value of χ^2 for $df = 1$ and $\alpha = .01$ is 6.635, as shown in Figure 11.7.

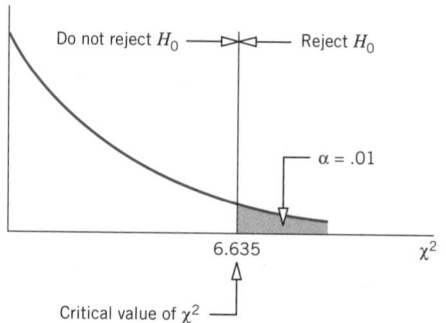

Figure 11.7

Step 4. *Calculate the value of the test statistic*

The expected frequencies for the various cells are calculated as follows, and they are listed within parentheses in Table 11.10.

$$E \text{ for } Machine\ I \text{ and } good \text{ cell} = (120)(175)/200 = 105$$

$$E \text{ for } Machine\ I \text{ and } defective \text{ cell} = (120)(25)/200 = 15$$

$$E \text{ for } Machine\ II \text{ and } good \text{ cell} = (80)(175)/200 = 70$$

$$E \text{ for } Machine\ II \text{ and } defective \text{ cell} = (80)(25)/200 = 10$$

Table 11.10

	Good (G)	Defective (D)	Row Total
Machine I (A)	109 (105)	11 (15)	120
Machine II (B)	66 (70)	14 (10)	80
Column totals	175	25	200

The value of the test statistic χ^2 is computed as follows.

$$\chi^2 = \Sigma \frac{(O - E)^2}{E}$$

$$= \frac{(109 - 105)^2}{105} + \frac{(11 - 15)^2}{15} + \frac{(66 - 70)^2}{70} + \frac{(14 - 10)^2}{10}$$

$$= .152 + 1.067 + .229 + 1.600 = 3.048$$

Step 5. *Make a decision*

The value of the test statistic $\chi^2 = 3.048$ is smaller than the critical value of $\chi^2 = 6.635$ and it falls in the nonrejection region. Consequently, we fail to reject the null hypothesis and conclude that the two characteristics, machine type and cassettes being good or defective, are not related. In other words, the sample does not provide sufficient evidence to conclude that the two machines are producing different percentages of good and defective cassettes. ▪

11.4.2 A TEST OF HOMOGENEITY

In a **test of homogeneity**, we test if two (or more) populations are homogeneous (similar) with regard to the distribution of a certain characteristic. For example, we might be interested in testing the null hypothesis that the proportions of households that belong to different income groups are the same in California and Wisconsin. Or we may want to test whether or not the preferences of people in Florida, Arizona, and Vermont are similar with regard to Coke, Pepsi, and 7-Up.

A TEST OF HOMOGENEITY

A test of homogeneity involves testing the null hypothesis that the proportions of elements with certain characteristics in two or more different populations are the same against the alternative hypothesis that these proportions are not the same.

The economic recession and budget deficits in the early 1990s led many states to cut benefits and freeze salaries of state employees. Combined with an increase in taxes, this may have affected the morale of many state employees. A citizens group wanted to test the null hypothesis that the distribution of the job satisfaction index of state employees in California and New York is the same. (Note that in a test of homogeneity the null hypothesis will always be that the proportions of elements with certain characteristics are the same in two or more populations. The alternative hypothesis will be that these proportions are not the same.) Suppose the researcher, who was assigned this job by the citizens group, took one sample of 500 state employees from California and another sample of 400 state employees from New York State. The information she collected from these employees is tabulated and recorded in Table 11.11.

Table 11.11

	California	New York	Row Totals
Very satisfied	60	75	135
Somewhat satisfied	100	125	225
Somewhat dissatisfied	184	140	324
Very dissatisfied	156	60	216
Column totals	500	400	900

Note that in this example the column totals are fixed. That is, we decide in advance to take samples of 500 state employees from California and 400 from New York. However, the row totals of 135, 225, 324, and 216 are determined randomly by the outcomes of the two samples. If we compare this example to Example 11–6 about the sample of men and women workers and their opinions on the proposed contract, we will note that neither the column nor the row totals were fixed in that example. Instead, the union took just one sample of 300 workers, collected the information on their gender and opinions, and prepared the contingency table. Consequently, in that example, the row and column totals were all determined randomly by the outcome of the survey. Thus, when both the row and column totals are determined randomly, we make a test of independence. However, when either the column totals or the row totals are fixed, we make a test of homogeneity. In the case of state employees in California and New York, we will make a test of homogeneity to test for the similarity of job satisfaction levels for workers in the two states.

The procedure to make a test of homogeneity is similar to that used to make a test of independence discussed earlier. Like a test of independence, a test of homogeneity is also right-tailed. Example 11–8 illustrates the procedure to make a homogeneity test.

A test of homogeneity.

EXAMPLE 11–8 Consider the data on the job satisfaction of state employees in California and New York given in Table 11.11. At the 2.5% significance level, test the null hypothesis that the distribution of state employees with regard to job satisfaction is similar (homogeneous) for the two states.

Solution To conduct this test of homogeneity, we perform the following five steps.

Step 1. *State the null and alternative hypotheses*

The two hypotheses are[2]

H_0: The proportions of state employees who belong to the four job satisfaction categories are the same in both states

H_1: The proportions of state employees who belong to the four job satisfaction categories are not the same in both states

Step 2. *Select the distribution to use*

We use the chi-square distribution to make a homogeneity test.

Step 3. *Determine the rejection and nonrejection regions*

The significance level is 2.5%. Because the homogeneity test is right-tailed, the area of the rejection region is .025 and it lies in the right tail of the chi-square distribution curve. Table 11.11 contains four rows and two columns. Hence, the degrees of freedom are

$$df = (R - 1)(C - 1) = (4 - 1)(2 - 1) = 3$$

From Table IX of Appendix B, the value of χ^2 for 3 *df* and a .025 area in the right tail of the chi-square distribution curve is 9.348. This value is shown in Figure 11.8.

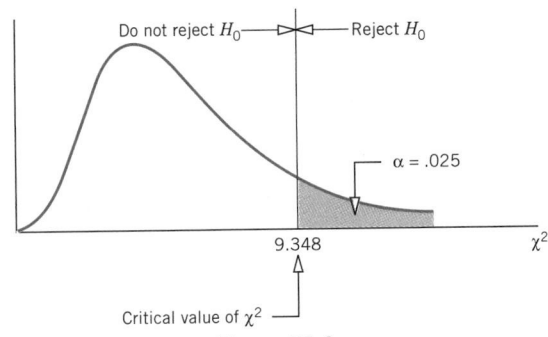

Figure 11.8

Step 4. *Calculate the value of the test statistic*

To compute the value of the test statistic χ^2, we need to calculate the expected frequencies first. Table 11.12 lists the observed as well as expected frequencies for Table 11.11.

The numbers in parentheses in Table 11.12 are expected frequencies, which are calculated using the formula

$$E = \frac{(\text{Row total})(\text{Column total})}{\text{Total of both samples}}$$

For instance,

$$E \text{ for } \textit{very satisfied} \text{ and } \textit{California} \text{ cell} = \frac{(135)(500)}{(900)} = 75$$

<hr>

[2]Let p_{AC}, p_{BC}, p_{CC}, and p_{DC} be the proportions of state employees in California who belong to very satisfied, somewhat satisfied, somewhat dissatisfied, and very dissatisfied categories, respectively. Let p_{AN}, p_{BN}, p_{CN}, and p_{DN} be the corresponding proportions for New York. Then, we can also write the null hypothesis as

$$H_0: p_{AC} = p_{AN}, p_{BC} = p_{BN}, p_{CC} = p_{CN}, \text{ and } p_{DC} = p_{DN}$$

and the alternative hypothesis as

$$H_1: \text{At least two of the equalities mentioned in } H_0 \text{ are not true}$$

Table 11.12

	California	New York	Row Totals
Very satisfied	60 (75)	75 (60)	135
Somewhat satisfied	100 (125)	125 (100)	225
Somewhat dissatisfied	184 (180)	140 (144)	324
Very dissatisfied	156 (120)	60 (96)	216
Column totals	500	400	900

The remaining expected frequencies are calculated in the same way. Note that the expected frequencies in a test of homogeneity are calculated in the same way as in a test of independence.

The value of the test statistic χ^2 is computed as follows.

$$\chi^2 = \Sigma \frac{(O - E)^2}{E}$$

$$= \frac{(60 - 75)^2}{75} + \frac{(75 - 60)^2}{60} + \frac{(100 - 125)^2}{125} + \frac{(125 - 100)^2}{100} + \frac{(184 - 180)^2}{180}$$

$$+ \frac{(140 - 144)^2}{144} + \frac{(156 - 120)^2}{120} + \frac{(60 - 96)^2}{96}$$

$$= 3.000 + 3.750 + 5.000 + 6.250 + .089 + .111 + 10.800 + 13.500$$

$$= 42.500$$

Step 5. *Make a decision*

The value of the test statistic $\chi^2 = 42.500$ is greater than the critical value of $\chi^2 = 9.348$, and it falls in the rejection region. As a result, we reject the null hypothesis and state that the distributions of job satisfaction for state employees in California and New York are not homogeneous. ▬

EXERCISES

Concepts and Procedures

11.22 Describe in your own words a test of independence and a test of homogeneity. Give one example of each.

11.23 Explain how the expected frequencies for cells of a contingency table are calculated in a test of independence or homogeneity. How do you find the degrees of freedom for such tests?

11.24 To make a test of independence or homogeneity, what should be the minimum expected frequency for each cell? What are the alternatives if this condition is not satisfied?

11.25 Consider the following contingency table.

	Column 1	Column 2	Column 3
Row 1	137	67	102
Row 2	98	71	65
Row 3	110	83	118

  a. Write the null and alternative hypotheses for a test of independence for this table.
  b. Calculate the expected frequencies for all cells assuming that the null hypothesis is true.
  c. For $\alpha = .01$, find the critical value of χ^2. Show the rejection and nonrejection regions on the chi-square distribution curve.
  d. Find the value of the test statistic χ^2.
  e. Using $\alpha = .01$, would you reject the null hypothesis?

11.26 Consider the following contingency table that records the results obtained for four samples of fixed sizes selected from four populations.

	Sample Selected from			
	Population 1	**Population 2**	**Population 3**	**Population 4**
Row 1	27	81	55	123
Row 2	46	64	91	72
Row 3	18	39	105	93

  a. Write the null and alternative hypotheses for a test of homogeneity for this table.
  b. Calculate the expected frequencies for all cells assuming that the null hypothesis is true.
  c. For $\alpha = .025$, find the critical value of χ^2. Show the rejection and nonrejection regions on the chi-square distribution curve.
  d. Find the value of the test statistic χ^2.
  e. Using $\alpha = .025$, would you reject the null hypothesis?

Applications

11.27 During the recession in the early 1990s, many families faced hard times financially. Some studies observed that more and more people stopped buying national brand products and purchased less expensive store brand products instead. Data produced by a sample of 700 adults on whether they usually buy store brand or name brand products are recorded in the following table.

	More Often Buy	
	National Brand	**Store Brand**
Men	172	143
Women	178	207

Using the 1% significance level, can you reject the null hypothesis that the two attributes, gender and buying national or store brand products, are independent?

11.28 A management consultant thinks that job satisfaction affects the absentee rate of employees. He took a sample of 400 employees of companies and collected information on job satisfaction (on a scale of 1 to 10, 1 being the lowest and 10 the highest) and the number of days they were absent for last year. The following table gives the tabulated results.

		Job Satisfaction Index		
		Less than 4	**4 to 7**	**Higher than 7**
Number of	Less than 6	12	61	107
Days Absent	6 to 12	22	80	50
Last Year	More than 12	41	18	9

Do you think the sample information provides sufficient evidence to reject the null hypothesis that job satisfaction and absentee rate of employees are not related? Use $\alpha = .025$.

11.29 Two hundred adults were asked whether they prefer to watch sports or soap operas on television. The following table lists the preferences of these men and women.

	Sports	Soap Operas
Men	51	39
Women	68	42

Test at the 10% significance level if gender and preference for watching sports or soap operas are related.

11.30 A sample of 150 chief executive officers (CEOs) were tested for personality type. The following table gives the results of this survey.

	Type A	Type B
Men	78	42
Women	19	11

Test at the 5% significance level if gender and personality type are related for all CEOs.

11.31 National Electronics Company buys parts from two subsidiaries. The quality control department at this company wanted to check if the distribution of good and defective parts is the same for the supplies of parts received from both subsidiaries. The quality control inspector selected a sample of 300 parts received from Subsidiary A and a sample of 400 parts received from Subsidiary B. These parts were checked for being good or defective. The following table records the results of this investigation.

	Subsidiary A	Subsidiary B
Good	284	381
Defective	16	19

Using the 5% significance level, test the null hypothesis that the distribution of good and defective parts is the same for both subsidiaries.

11.32 Two drugs were administered to two groups of randomly assigned patients to cure the same disease. One group of 60 patients and another group of 40 patients were selected. The following table gives information about the number of patients who were cured and not cured by each of the two drugs.

	Cured	Not Cured
Drug I	46	14
Drug II	18	22

Test at the 1% significance level whether or not the two drugs are similar in curing and not curing the patients.

11.33 Two random samples, one of 95 blue-collar workers and a second of 50 white-collar workers, were taken from a large company. These workers were asked about their views on a certain company issue. The following table gives the results of the survey.

	Opinion		
	Favor	Oppose	Uncertain
Blue-collar worker	47	39	9
White-collar worker	21	26	3

Using the 2.5% significance level, test the null hypothesis that the distributions of opinions are homogeneous for the two groups of workers.

11.34 A company introduced a new product in the market a few months ago. The management wants to determine the reaction of customers in different regions to this product. The research department at the company selected four different samples of 400 users of this product from four regions—the East, South, Midwest, and West. The users of the product were asked whether or not they like the product. The responses of these people are recorded in the following table.

	East	South	Midwest	West
Like	274	203	291	257
Do not like	126	197	109	143

Based on evidence from these samples, can you conclude that the distribution of opinions of users of this product are not homogeneous for all four regions with regard to liking and not liking this product? Use $\alpha = .05$.

11.5 INFERENCES ABOUT THE POPULATION VARIANCE

Earlier chapters explained how to make inferences (confidence intervals and hypothesis tests) about the population mean and population proportion. However, we may often need to control the variance (or standard deviation). Consequently, there may be a need to estimate and to test a hypothesis about the population variance σ^2. Section 11.5.1 describes how to make a confidence interval for the population variance (or standard deviation). Section 11.5.2 explains how to test a hypothesis about the population variance.

As an example, suppose a machine is set up to fill packages of cookies so that the net weight of cookies per package is 32 ounces. Note that the machine will not put exactly 32 ounces of cookies in each package. Some of the packages will contain less and some will contain more than 32 ounces. However, if the variance (and, hence, the standard deviation) is too large, some of the packages will contain quite a bit less than 32 ounces of cookies and some others will contain quite a bit more than 32 ounces. The manufacturer will not want a large variation in the amounts of cookies put in different packages. To keep this variation within some specified acceptable limit, the machine will be adjusted from time to time. Before the manager decides to adjust the machine at any time, he must estimate the variance or test a hypothesis or do both to find out if the variance exceeds the maximum acceptable value.

Like every sample statistic, the sample variance is a random variable and it possesses a sampling distribution. If all the possible samples of a given size are taken from a population and their variances are calculated, the probability distribution of these variances is called the *sampling distribution of the sample variance.*

SAMPLING DISTRIBUTION OF $(n - 1)s^2/\sigma^2$

If the population from which the sample is selected is (approximately) normally distributed, then

$$\frac{(n - 1)\, s^2}{\sigma^2}$$

has a chi-square distribution with $n - 1$ degrees of freedom.

Thus, the chi-square distribution is used to construct a confidence interval and test a hypothesis about the population variance σ^2.

11.5.1 ESTIMATION OF THE POPULATION VARIANCE

The value of the sample variance s^2 is a point estimate of the population variance σ^2. The $(1 - \alpha)100\%$ confidence interval for σ^2 is given by the following formula.

CONFIDENCE INTERVAL FOR THE POPULATION VARIANCE σ^2

Assuming that the population from which the sample is selected is (approximately) normally distributed, the $(1 - \alpha)100\%$ confidence interval for the population variance σ^2 is

$$\frac{(n - 1)\, s^2}{\chi^2_{\alpha/2}} \quad \text{to} \quad \frac{(n - 1)\, s^2}{\chi^2_{1-\alpha/2}}$$

where $\chi^2_{\alpha/2}$ and $\chi^2_{1-\alpha/2}$ are obtained from the chi-square distribution table for $\alpha/2$ and $1 - \alpha/2$ areas in the right tail of the chi-square distribution curve, respectively, and for $n - 1$ degrees of freedom.

The confidence interval for the population standard deviation can be obtained by simply taking the positive square root of the two limits of the confidence interval for the population variance.

The procedure for making a confidence interval for σ^2 involves the following three steps.

1. Take a sample of size n and compute s^2 using the formula learned in Chapter 3. However, if n and s^2 are given, then perform only steps 2 and 3.
2. Calculate $\alpha/2$ and $1 - \alpha/2$. Find two values of χ^2 from the chi-square distribution table (Table IX of Appendix B): one for $\alpha/2$ area in the right tail of the chi-square distribution curve and $df = n - 1$, and the second for $1 - \alpha/2$ area in the right tail and $df = n - 1$.
3. Substitute all the values in the formula for the confidence interval for σ^2 and simplify.

Example 11–9 illustrates the estimation of the population variance and population standard deviation.

Constructing confidence intervals for σ^2 and σ.

EXAMPLE 11–9 One type of cookie manufactured by Haddad Food Company is Cocoa Cookies. The machine that fills packages of these cookies is set up in such a way that the average net weight of these packages is 32 ounces with a variance of .015 square ounces. From time to time the quality control inspector at the company selects a sample of a few such packages, calculates the variance of the net weights of these packages, and constructs a 95% confidence interval for the population variance. If either both or one of the two limits of this confidence interval is not in the interval .008 to .030, the machine is stopped and adjusted. A recently taken random sample of 25 packages from the production line gave a sample variance of .029 square ounces. Based on this sample information, do you think the machine needs an adjustment? Assume that the net weights of cookies in all packages are normally distributed.

Solution The following three steps are performed to estimate the population variance and to make a decision.

Step 1. From the given information,

$$n = 25 \quad \text{and} \quad s^2 = .029$$

Step 2. The confidence level is $1 - \alpha = .95$. Hence,

$$\alpha = 1 - .95 = .05$$
$$\alpha/2 = .05/2 = .025$$
$$1 - \alpha/2 = 1 - .025 = .975$$
$$df = n - 1 = 25 - 1 = 24$$

From Table IX of Appendix B,

$$\chi^2 \text{ for 24 } df \text{ and .025 area in the right tail } = 39.364$$
$$\chi^2 \text{ for 24 } df \text{ and .975 area in the right tail } = 12.401$$

These values are shown in Figure 11.9.

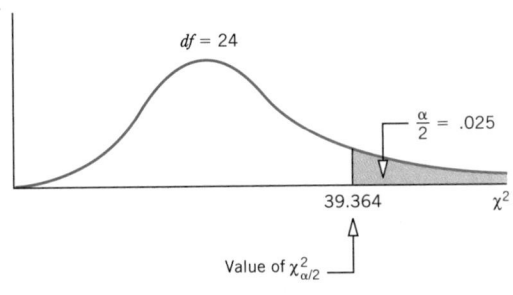

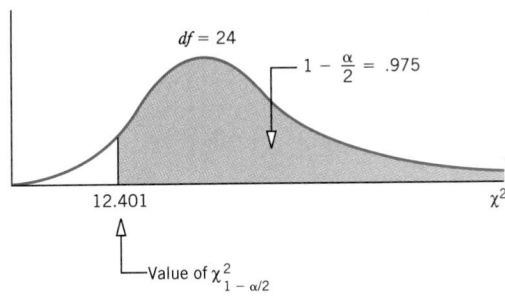

Figure 11.9

Step 3. The 95% confidence interval for σ^2 is

$$\frac{(n - 1) s^2}{\chi^2_{\alpha/2}} \quad \text{to} \quad \frac{(n - 1) s^2}{\chi^2_{1 - \alpha/2}}$$

or

$$\frac{(25 - 1) (.029)}{39.364} \quad \text{to} \quad \frac{(25 - 1) (.029)}{12.401}$$

or

$$\textbf{.0177} \quad \text{to} \quad \textbf{.0561}$$

Thus, with 95% confidence, we can state that the variance for all packages of Cocoa Cookies lies between .0177 and .0561 square ounces. Note that the lower limit (.0177) of this confidence interval is between .008 and .030 but the upper limit (.0561) is larger than .030 and falls outside the interval .008 to .030. Since the upper limit is larger than .030, we can state that the machine needs to be stopped and adjusted.

We can obtain the confidence interval for the population standard deviation σ by taking the positive square root of the two limits of the above confidence interval for the population variance. Thus, a 95% confidence interval for the population standard deviation is

$$\sqrt{.0177} \quad \text{to} \quad \sqrt{.0561} \quad \text{or} \quad \textbf{.133 to .237}$$

Hence, the standard deviation of all packages of Cocoa Cookies is between .133 and .237 ounces at a 95% confidence level. ◼

11.5.2 HYPOTHESIS TESTS ABOUT THE POPULATION VARIANCE

A test of hypothesis about the population variance can be one-tailed or two-tailed. To make a test of hypothesis about σ^2, we perform the same five steps we have used earlier in hypothesis-testing examples. The procedure to test a hypothesis about σ^2 discussed in this section is applied only when the population from which a sample is taken is (approximately) normally distributed.

TEST STATISTIC FOR A TEST OF HYPOTHESIS ABOUT σ^2

The value of the test statistic χ^2 is calculated as

$$\chi^2 = \frac{(n-1)\, s^2}{\sigma^2}$$

where s^2 is the sample variance, σ^2 is the hypothesized value of the population variance, and $n-1$ represents the degrees of freedom. The population from which the sample is taken is assumed to be (approximately) normally distributed.

Examples 11–10 and 11–11 illustrate the procedure to make tests of hypothesis about σ^2.

Making a right-tailed test of hypothesis about σ^2.

EXAMPLE 11–10 One type of cookie manufactured by Haddad Food Company is Cocoa Cookies. The machine that fills packages of these cookies is set up in such a way that the average net weight of these packages is 32 ounces with a variance of .015 square ounces. From time to time the quality control inspector at the company selects a sample of a few such packages, calculates the variance of the net weights of these packages, and makes a test of hypothesis about the population variance. She always uses $\alpha = .01$. The acceptable value of the population variance is .015 square ounces or less. If the conclusion from the test of hypothesis is that the population variance is not within the acceptable limit, the machine is stopped and adjusted. A recently taken random sample of 25 packages from the production line gave a sample variance of .029 square ounces. Based on this sample information, do you think the machine needs an adjustment? Assume that the net weights of cookies in all packages are normally distributed.

Solution From the given information,

$$n = 25, \qquad \alpha = .01, \qquad \text{and} \qquad s^2 = .029$$

The population variance should not exceed .015.

Step 1. *State the null and alternative hypotheses*

We are to test whether or not the population variance is within the acceptable limit. The population variance is within the acceptable limit if it is less than or equal to .015; otherwise, it is not. Thus, the two hypotheses are

$H_0: \sigma^2 \leq .015$ (the population variance is within the acceptable limit)

$H_1: \sigma^2 > .015$ (the population variance exceeds the acceptable limit)

Step 2. *Select the distribution to use*

We use the chi-square distribution to test a hypothesis about σ^2.

Step 3. *Determine the rejection and nonrejection regions*

The significance level is 1% and, because of the $>$ sign in H_1, the test is right-tailed. The rejection region lies in the right tail of the chi-square distribution curve with its area equal to .01. The degrees of freedom for a chi-square test about σ^2 are $n - 1$, that is,

$$df = n - 1 = 25 - 1 = 24$$

From Table IX of Appendix B, the critical value of χ^2 for 24 degrees of freedom and .01 area in the right tail is 42.980. This value is shown in Figure 11.10.

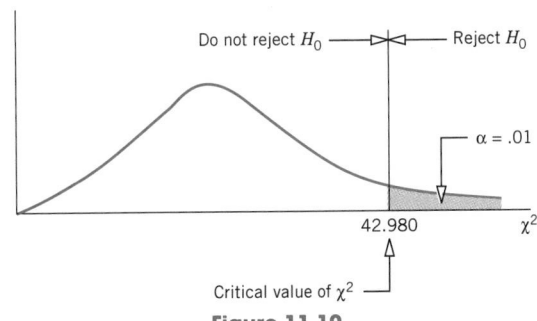

Figure 11.10

Step 4. *Calculate the value of the test statistic*

The value of the test statistic χ^2 for the sample variance is calculated as follows.

$$\chi^2 = \frac{(n - 1) s^2}{\sigma^2} = \frac{(25 - 1) (.029)}{(.015)} = 46.400$$

From H_0

Step 5. *Make a decision*

The value of the test statistic $\chi^2 = 46.400$ is greater than the critical value of $\chi^2 = 42.980$ and it falls in the rejection region. Consequently, we reject H_0 and conclude that the population variance is not within the acceptable limit. The machine should be stopped and adjusted. ■

Conducting a two-tailed test of hypothesis about σ^2.

EXAMPLE 11–11   It is known that the variance of monthly earnings of all state employees for all 50 states in the United States is 490,000 square dollars. A sample of 29 state employees selected from New Jersey produced a variance of their monthly earnings equal to 600,000 square dollars. Test at the 5% significance level if the variance of monthly earnings of state employees in New Jersey is different from 490,000 square dollars. Assume that the monthly earnings of all state employees in New Jersey have an (approximate) normal distribution.

Solution From the given information,

$$n = 29, \quad \alpha = .05, \quad \text{and} \quad s^2 = 600,000$$

Step 1. *State the null and alternative hypotheses*

The null and alternative hypotheses are

$H_0: \sigma^2 = 490,000$ (the population variance is not different from 490,000)

$H_1: \sigma^2 \neq 490,000$ (the population variance is different from 490,000)

Step 2. *Select the distribution to use*

We use the chi-square distribution to test a hypothesis about σ^2.

Step 3. *Determine the rejection and nonrejection regions*

The significance level is 5%. The $\neq$ sign in H_1 indicates that the test is two-tailed. The rejection region lies in both tails of the chi-square distribution curve with its total area equal to .05. Consequently, the area in each tail of the chi-square distribution curve is .025. The values of $\alpha/2$ and $1 - \alpha/2$ are

$$\alpha/2 = .05/2 = .025 \quad \text{and} \quad 1 - \alpha/2 = 1 - .025 = .975$$

The degrees of freedom are

$$df = n - 1 = 29 - 1 = 28$$

From Table IX, the critical values of χ^2 for 28 degrees of freedom and for $\alpha/2$ and $1 - \alpha/2$ areas in the right tail are

χ^2 for 28 *df* and .025 area in the right tail = 44.461

χ^2 for 28 *df* and .025 area in the left tail

     = χ^2 for 28 *df* and .975 area in the right tail = 15.308

These two values are shown in Figure 11.11.

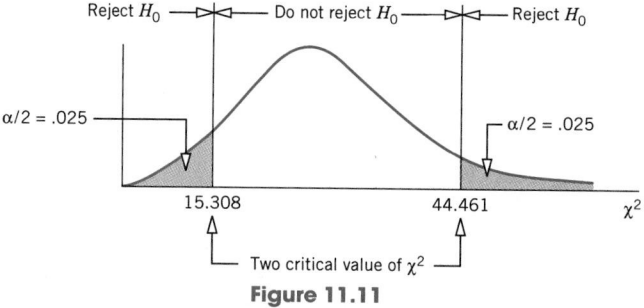

Figure 11.11

Step 4. *Calculate the value of the test statistic*

The value of the test statistic χ^2 for the sample variance is calculated as follows.

$$\chi^2 = \frac{(n-1)\,s^2}{\sigma^2} = \frac{(29-1)\,(600,000)}{(490,000)} = 34.286$$

From H_0

Step 5. *Make a decision*

The value of the test statistic $\chi^2 = 34.286$ is between the two critical values of χ^2, 15.308 and 44.461, and it falls in the nonrejection region. Consequently, we fail to reject H_0 and conclude that the population variance of monthly earnings of all state employees in New Jersey is not different from 490,000 square dollars.   ▬

Note that we can make a test of hypothesis about the population standard deviation σ using the same procedure as that for the population variance σ^2.

EXERCISES

Concepts and Procedures

11.35   A sample of certain observations selected from a normally distributed population produced a sample variance of 38. Construct a 95% confidence interval for σ^2 for each of the following cases and comment on what happens to the confidence interval of σ^2 when the sample size increases.

    a. $n = 12$ **b.** $n = 16$ **c.** $n = 25$

11.36   A sample of 25 observations selected from a normally distributed population produced a sample variance of 47. Construct a confidence interval for σ^2 for each of the following confidence levels and comment on what happens to the confidence interval of σ^2 when the confidence level decreases.

    a. $1 - \alpha = .99$ **b.** $1 - \alpha = .95$ **c.** $1 - \alpha = .90$

11.37   A sample of 20 observations selected from a normally distributed population produced a sample variance of 14.

    a. Write the null and alternative hypotheses to test if the population variance is different from 11.

    b. Using $\alpha = .05$, find the critical values of χ^2. Show the rejection and nonrejection regions on a chi-square distribution curve.

    c. Find the value of the test statistic χ^2.

    d. Using the 5% significance level, will you reject the null hypothesis stated in part a?

11.38   A sample of 16 observations selected from a normally distributed population produced a sample variance of .9.

    a. Write the null and alternative hypotheses to test if the population variance is greater than .50.

    b. Using $\alpha = .01$, find the critical value of χ^2. Show the rejection and nonrejection regions on a chi-square distribution curve.

    c. Find the value of the test statistic χ^2.

    d. Using the 1% significance level, will you reject the null hypothesis stated in part a?

11.39   A sample of 25 observations selected from a normally distributed population produced a sample variance of 1.8.

    a. Write the null and alternative hypotheses to test if the population variance is less than 2.5.

    b. Using $\alpha = .025$, find the critical value of χ^2. Show the rejection and nonrejection regions on a chi-square distribution graph.

    c. Find the value of the test statistic χ^2.

    d. Using the 2.5% significance level, will you reject the null hypothesis stated in part a?

11.40 A sample of 14 observations selected from a normally distributed population produced a sample variance of 3.6.

    a. Write the null and alternative hypotheses to test if the population variance is different from 1.7.

    b. Using $\alpha = .05$, find the critical values of χ^2. Show the rejection and nonrejection regions on a chi-square distribution graph.

    c. Find the value of the test statistic χ^2.

    d. Using the 5% significance level, will you reject the null hypothesis stated in part a?

Applications

11.41 The management of a soft-drink company does not want the variance of the amount of soda in 12-ounce cans to be more than .01 square ounces. (Recall from Chapter 3 that the variance is always in square units.) The company manager takes a sample of certain cans and estimates the population variance quite often. A random sample of twenty 12-ounce cans taken from the production line of this company showed that the variance for this sample was .014 square ounces.

    a. Construct a 99% confidence interval for the population variance. Assume that the amount of soda in all 12-ounce cans has a normal distribution.

    b. Test at the 2.5% significance level if the variance of the amounts of soda in all 12-ounce cans for this company is greater than .01 square ounces.

11.42 The 2-inch-long bolts manufactured by a company must have a variance of .003 square inches or less for acceptance by a buyer. A random sample of 29 such bolts gave a variance of .0058 square inches.

    a. Test at the 1% significance level if the variance of all such bolts is greater than .003 square inches. Assume that the lengths of all 2-inch-long bolts manufactured by this company are (approximately) normally distributed.

    b. Make the 98% confidence intervals for the population variance and standard deviation.

11.43 An auto manufacturing company wants to estimate the variance of miles per gallon for its auto model AST727. A random sample of 24 cars of this model showed that the variance of miles per gallon is .58.

    a. Construct the 95% confidence intervals for the population variance and standard deviation. Assume that the miles per gallon for all such cars are (approximately) normally distributed.

    b. Test at the 1% significance level if the sample result indicates that the population variance is different from .30.

11.44 The manufacturer of a certain brand of light bulbs claims that the variance of the lives of these bulbs is 4000 square hours. A consumer agency took a random sample of 25 such bulbs and tested them. The variance of the lives of these bulbs was found to be 4970 square hours. Assume that the lives of all such bulbs are (approximately) normally distributed.

    a. Make the 99% confidence intervals for the variance and standard deviation of the lives of all such bulbs.

    b. Test at the 5% significance level if the variance of such bulbs is different from 4000 square hours.

GLOSSARY

Chi-square distribution A distribution, with degrees of freedom as the only parameter, that is skewed to the right for small df and looks like a normal curve for large df.

Expected frequencies The frequencies for different categories of a multinomial experiment or for different cells of a contingency table that are expected to occur when a given null hypothesis is true.

Goodness-of-fit test A test of the null hypothesis that the observed frequencies for an experiment follow a certain pattern or theoretical distribution.

Multinomial experiment An experiment with n trials for which (1) the trials are identical, (2) there are more than two possible outcomes per trial, (3) the trials are independent, and (4) the probabilities of various outcomes remain constant for each trial.

Observed frequencies The frequencies actually obtained from the performance of an experiment.

Test of homogeneity A test of the null hypothesis that the proportions of elements that belong to different groups in two (or more) populations are similar.

Test of independence A test of the null hypothesis that two attributes of a population are not related.

KEY FORMULAS

1. **Expected frequency for a category for a goodness-of-fit test**

$$E = np$$

where n is the sample size and p is the probability that an element belongs to this category.

2. **Value of the test statistic χ^2 for a goodness-of-fit test and a test of independence or homogeneity**

$$\chi^2 = \Sigma \frac{(O - E)^2}{E}$$

where O and E are the observed and expected frequencies, respectively, for a category or cell.

3. **Degrees of freedom for a goodness-of-fit test**

$$df = k - 1$$

where k denotes the number of categories for the experiment.

4. **Degrees of freedom for a test of independence or homogeneity**

$$df = (R - 1)(C - 1)$$

where R and C are, respectively, the number of rows and columns for the contingency table.

5. **Expected frequency for a cell for an independence or homogeneity test**

$$E = \frac{(\text{Row total}) (\text{Column total})}{\text{Sample size}}$$

6. **The $(1 - \alpha)100\%$ confidence interval for population variance σ^2**

$$\frac{(n - 1) s^2}{\chi^2_{\alpha/2}} \quad \text{to} \quad \frac{(n - 1) s^2}{\chi^2_{1 - \alpha/2}}$$

where $\chi^2_{\alpha/2}$ and $\chi^2_{1-\alpha/2}$ are obtained from the chi-square table for $\alpha/2$ and $1 - \alpha/2$ areas in the right tail of the chi-square distribution curve and $n - 1$ degrees of freedom.

7. **Value of the test statistic χ^2 in a hypothesis test about σ^2**

$$\chi^2 = \frac{(n - 1) s^2}{\sigma^2}$$

where s^2 is the sample variance, σ^2 is the hypothesized value of the population variance, and $n - 1$ are the degrees of freedom.

SUPPLEMENTARY EXERCISES

11.45   In a 1987 Roper poll, schoolchildren aged 8 to 17 were asked about the overall effect on children aged 12 or younger of both parents working outside the home. Twenty-five percent of the children surveyed said that the effect is good, 39% said it is bad, 29% said there is no effect, and 7% said they did not know (*The American Chicle Youth Poll*, commissioned by The American Chicle Group, Warner-Lambert Company, March 1987). A random sample of 250 schoolchildren aged 8 to 17 taken recently produced the results listed in the following table in response to the same question.

Category	Good	Bad	No Effect	Did Not Know
Number of children	56	111	61	22

Test at the 1% significance level if the current percentage distribution of schoolchildren aged 8 to 17 who belong to the four categories is different from the one for 1987.

11.46   In 1980, 28% of the CEOs (chief executive officers) of major U.S. corporations had a degree beyond the bachelor's, 59% possessed a bachelor's degree, and 13% had a high school diploma. A recently taken sample of 500 CEOs showed that 208 of them have a degree beyond the bachelor's, 252 have a bachelor's degree, and 40 possess a high school diploma. Using the 1% significance level, can you reject the null hypothesis that the current percentage distribution by degree of all CEOs is the same as for 1980?

11.47   The following table lists the number of persons in a random sample of 210 according to the day of the week on which they prefer to do their grocery shopping.

Day	Mon	Tue	Wed	Thu	Fri	Sat	Sun
Number of persons	9	15	10	28	38	69	41

Using the 2.5% significance level, test the null hypothesis that the proportion of persons who prefer to do their grocery shopping on a particular day is the same for all days of the week.

11.48   One of the products produced by Branco Food Company is All-Bran Cereal, which competes with three other brands of similar all-bran cereals. The company's research office wants to investigate if the percentage of people who consume all-bran cereal is the same for each of these four brands. Let us denote the four brands of cereal by A, B, C, and D. A sample of 1000 persons who consume all-bran cereal was taken and they were asked which brand they most often consume. Of the respondents, 212 said they usually consume Brand A, 284 consume Brand B, 259 consume Brand C, and 245 consume Brand D. Does the sample provide enough evidence to reject the null hypothesis that the percentage of people who consume all-bran cereal is the same for all four brands? Use $\alpha = .05$.

11.49   The information on the age distribution of people who shop by mail is very important for mail-order companies. By knowing such a distribution, they can stock and sell the products that appeal to specific age groups of shoppers. The following table, based on a survey conducted by Deloitte and Touche, gives the 1992 percentage distribution by age of people who shop by mail (*USA TODAY*, October 1992). Assume that this percentage distribution is true for all people who shopped by mail in 1992.

Age (in years)	18–24	25–34	35–44	45–54	55–64	65 and Older
Percentage of shoppers	11	26	25	16	11	11

A mail-order company wants to investigate if the 1992 percentage distribution by age of mail-order shoppers is still true. A random sample of 2000 such shoppers taken recently by this company produced the frequency distribution listed in the following table.

Age (in years)	18–24	25–34	35–44	45–54	55–64	65 and Older
Number of shoppers	213	542	486	329	227	203

Does the sample information provide sufficient evidence to reject the null hypothesis that the 1992 percentage distribution by age of all mail-order shoppers is still true? Use $\alpha = .05$.

11.50 The following table gives the U.S. dentifrice market shares for 52 weeks from October 1991 to September 1992 held by various dentifrice brands. (*Source:* Information Resources Inc., *The Wall Street Journal*, November 10, 1992.) Market shares are usually defined as percentages of total sales revenue for different brands. Suppose for convenience that the market shares are synonymous with percentages of the total number of persons using different dentifrice brands.

Dentifrice Brand	Market Share (percentage)
Crest	31.2
Colgate	22.1
Arm & Hammer	10.1
Aquafresh	9.2
Close-up	5.1
Sensodyne	3.5
Ultra Brite	2.5
Others	16.3

A business organization wanted to investigate if the current market shares of these dentifrice brands are the same as for 1991–1992. A random sample of 2000 persons was selected and each was asked what dentifrice brand he or she uses. The following table gives the frequency distribution of these 2000 persons by dentifrice brand.

Dentifrice Brand	Frequency
Crest	618
Colgate	430
Arm & Hammer	193
Aquafresh	190
Close-up	107
Sensodyne	72
Ultra Brite	53
Others	337

Testing at the 2.5% significance level, will you reject the null hypothesis that the current dentifrice market shares held by various brands are the same as for 1991–1992?

11.51 The president of a bank selected a sample of 200 loan applications to check if the approval or rejection of an application depends on which one of the two loan officers, Thurow or Webber, handles that application. The information obtained from the sample is summarized in the following table.

	Approved	Rejected
Thurow	57	38
Webber	69	36

Test at the 2.5% significance level if the approval or rejection of a loan application depends on which loan officer handles the application.

11.52 The following table gives the two-way classification of a sample of 1000 students selected from certain colleges and universities.

	Business Major	Nonbusiness Major
Male	77	389
Female	52	482

Test at the 1% significance level if being a male or a female and being a business or nonbusiness major are related.

11.53 A random sample of 100 jurors was selected and asked whether or not each of them had ever been a victim of crime. The jurors were also asked whether they are strict, fair, or lenient regarding punishment for crime. The following table gives the results of the survey.

	Strict	Fair	Lenient
Have been a victim	20	8	3
Have never been a victim	22	38	9

Test at the 5% significance level if the two attributes for all jurors are dependent.

11.54 The recession and bad economic conditions have forced many people to hold more than one job to make ends meet. A sample of 500 persons who held more than one job produced the following two-way table.

	Single	Married	Other
Male	69	212	39
Female	33	102	45

Test at the 10% significance level if gender and marital status are related for all people who hold more than one job.

11.55 A random sample of 100 persons was selected from each of four regions in the United States. These people were asked whether or not they support a certain farm subsidy program. The results of the survey are summarized in the following table.

	Favor	Oppose	Uncertain
Northeast	56	33	11
Midwest	73	23	4
South	67	28	5
West	59	35	6

Using the 1% significance level, test the null hypothesis that the percentages of people with different opinions are similar for all four regions.

11.56 A company owns four restaurants in four different parts of New York City. The management wanted to determine the opinions of its patrons on the quality of food and service at each of the four restaurants. A random sample of 100 patrons was selected for each restaurant and asked to rate the quality of food and service at that restaurant. The results of the survey are recorded in the following table.

	Opinions of Patrons		
	Excellent	Average	Below Average
Restaurant 1	59	32	9
Restaurant 2	48	44	8
Restaurant 3	64	26	10
Restaurant 4	54	42	4

Based on these sample surveys, can you conclude that the distributions of opinions are homogeneous for all four restaurants? Use $\alpha = .025$.

11.57   Usually people do not like waiting in line for service for a long time. A bank manager does not want the variance of the waiting times for her customers to be higher than 4.0 square minutes. A random sample of 25 customers taken from this bank gave the variance of the waiting times equal to 7.9 square minutes.

    a.   Test at the 1% significance level if the variance of the waiting times for all customers at this bank is higher than 4.0 square minutes. Assume that the waiting times for all customers are normally distributed.

    b.   Construct a 99% confidence interval for the population variance.

11.58   The variance of the SAT scores for all students who took that test this year is 5000. The variance of the SAT scores for a random sample of 20 students from one school is equal to 3200.

    a.   Test at the 2.5% significance level if the variance of the SAT scores for students from this school is lower than 5000. Assume that SAT scores for all students at this school are (approximately) normally distributed.

    b.   Construct the 98% confidence intervals for the variance and the standard deviation of SAT scores for all students at this school.

11.59   A company manufactures ball bearings that it supplies to other companies. The machine that is used to manufacture these ball bearings produces them with a variance of diameters of .025 square millimeters or less. The quality control officer takes a sample of such ball bearings quite often and checks, using confidence intervals and tests of hypotheses, whether or not the variance of these bearings is within .025 square millimeters. If it is not, the machine is stopped and adjusted. A recently taken random sample of 23 ball bearings gave a variance of the diameters equal to .031 square millimeters.

    a.   Using the 5% significance level, can you conclude that the machine needs an adjustment? Assume that the diameters of all ball bearings have a normal distribution.

    b.   Construct a 95% confidence interval for the population variance.

11.60   The manager at the assembly department of a company that makes toys claims that the variance of the assembly times for all workers is 28 square minutes. A sample of 10 workers was taken and the assembly time for one toy for each worker under normal conditions was recorded. This information produced a sample variance of 34 square minutes. Assume that the assembly times for all workers have a normal distribution.

    a.   Test at the 1% significance level if the variance of the assembly times for all workers is different from 28 square minutes.

    b.   Construct the 99% confidence intervals for the variance and the standard deviation of assembly times for all workers.

11.61   A sample of seven observations taken from a population produced the following data.

$$10 \quad 8 \quad 13 \quad 15 \quad 6 \quad 8 \quad 13$$

    a.   Using the formula learned in Chapter 3, find the sample variance s^2 for these data.

    b.   Make the 98% confidence intervals for the population variance and standard deviation. Assume that the population from which this sample is selected is normally distributed.

    c.   Test at the 1% significance level if the population variance is different from 10.

11.62   The following are the prices of the same brand of camcorder found at eight stores in Los Angeles.

$$\$749 \quad 810 \quad 789 \quad 799 \quad 732 \quad 825 \quad 799 \quad 774$$

    a.   Using the formula learned in Chapter 3, find the sample variance s^2 for these data.

    b.   Make the 95% confidence intervals for the population variance and standard deviation. Assume that the prices of this camcorder at all stores in Los Angeles follow a normal distribution.

    c.   Test at the 5% significance level if the population variance is different from 500 square dollars.

SELF-REVIEW TEST

1. The random variable χ^2 assumes only
   a. positive   b. nonnegative   c. nonpositive values

2. The parameter(s) of the chi-square distribution is(are)
   a. degrees of freedom   b. df and n c. χ^2

3. Which of the following is *not* a characteristic of a multinomial experiment?
   a. It consists of n identical trials.
   b. There are k possible outcomes for each trial and $k > 2$.
   c. The occurrences are random.
   d. The trials are independent.
   e. The probabilities of outcomes remain constant for each trial.

4. The observed frequencies for a goodness-of-fit test are
   a. the frequencies obtained from the performance of an experiment
   b. the frequencies given by the product of n and p
   c. the frequencies obtained by adding the results of a and b

5. The expected frequencies for a goodness-of-fit test are
   a. the frequencies obtained from the performance of an experiment
   b. the frequencies given by the product of n and p
   c. the frequencies obtained by adding the results of a and b

6. The degrees of freedom for a goodness-of-fit test are
   a. $n - 1$ b. $k - 1$ c. $n + k - 1$

7. The chi-square goodness-of-fit test is always
   a. two-tailed   b. left-tailed   c. right-tailed

8. To apply a goodness-of-fit test, the expected frequency of each category must be at least
   a. 10   b. 5   c. 8

9. The degrees of freedom for a test of independence are
   a. $(R - 1)(C - 1)$ b. $n - 2$ c. $(n - 1)(k - 1)$

10. According to an estimate, more than 4 million Americans were fired from their jobs in 1992. A study done by Robert Half International Inc. investigated the reasons why employees are fired by their employers. The following table lists the various reasons and the percentages of employees who are fired due to those reasons. (*Source: U.S. News & World Report*, August 2, 1993.)

Reason for Firing	Percentage
Incompetence	39
Not getting along with others	17
Dishonesty or lying	12
Negative attitude	10
Lack of motivation	7
Other	15

A recent sample of 1000 employees who were fired by their employers produced the following frequency distribution.

Reason for Firing	Number of Employees Fired
Incompetence	416
Not getting along with others	181
Dishonesty or lying	114
Negative attitude	108
Lack of motivation	65
Other	116

Test at the 1% significance level if the current distribution of reasons why employees are fired is different from the one obtained by Robert Half International Inc.

11. A management agency wants to investigate whether or not the job satisfaction and ages of employees of large companies are associated. The agency took a random sample of 1000 employees from large companies. The following table gives the two-way classification of these employees by job satisfaction and age.

		Job Satisfaction			
		Very Satisfied	Somewhat Satisfied	Somewhat Dissatisfied	Very Dissatisfied
Age	<30	35	39	62	44
	30 to 39	48	54	72	46
	40 to 49	71	59	61	39
	50 to 59	74	52	47	37
	≥60	62	46	38	14

Test at the 1% significance level if the job satisfaction and ages of employees of large companies are related.

12. An economist wanted to investigate if people belonging to different income groups are homogeneous with regard to playing lotteries. She took a sample of 600 people from the low-income group, another sample of 500 people from the middle-income group, and a third sample of 400 people from the high-income group. All these people were asked whether they play the lottery often, sometimes, or never. The results of the survey are summarized in the following table.

	Income Group		
	Low	Middle	High
Play often	170	160	90
Play sometimes	290	220	120
Never play	140	120	190

Using the 5% significance level, can you reject the null hypothesis that the percentages of people who play the lottery often, sometimes, and never are the same for each income group?

13. A cough syrup drug manufacturer requires that the variance for a chemical contained in the bottles of this drug should not exceed .03 square ounces. A sample of 25 such bottles gave the variance for this chemical as .06 square ounces.

   a. Construct the 99% confidence intervals for the population variance and the population standard deviation. Assume that the amount of this chemical in all such bottles is (approximately) normally distributed.

   b. Test at the 1% significance level if the variance of this chemical in all such bottles exceeds .03 square ounces.

USING MINITAB

In this section we first discuss how MINITAB is used to make a goodness-of-fit test and then how to make a test of independence.

A GOODNESS-OF-FIT TEST

MINITAB does not have a direct command to make a goodness-of-fit test. However, by combining a few commands, we can make such a test. Illustration M11–1 describes the procedure to perform a goodness-of-fit test using MINITAB.

Illustration M11–1 Refer to Example 11–4. The following table gives the 1992 U.S. car market shares held by various auto companies (*The Wall Street Journal*, January 8, 1993). Market shares are usually defined as percentages of the total revenue earned by various companies. Suppose for convenience that in this case the market shares are synonymous with percentages of the total new cars sold by various auto companies in 1992.

Company	GM	Ford	Honda	Toyota	Chrysler	Others
Market share	34.5%	21.6%	9.3%	9.2%	8.3%	17.1%

A business organization wanted to investigate if the current U.S. car market shares of these companies are the same as for 1992. A sample of 2000 recently sold new cars showed that 715 of them were manufactured by General Motors, 446 by Ford, 175 by Honda, 187 by Toyota, 178 by Chrysler, and 299 by other companies. Testing at the 2.5% significance level, will you reject the null hypothesis that the current U.S. car market shares held by these companies are the same as for 1992?

Solution   The null and alternative hypotheses are

H_0: The current U.S. car market shares of various auto
companies are the same as those for 1992

H_1: The current U.S. car market shares of various auto
companies are not the same as those for 1992

To make a goodness-of-fit test for this illustration, perform the following steps.

1. Enter the data on observed frequencies and probabilities in columns C1 and C2, respectively, using the READ command.

2. Find the sum of column C1 (the column of observed frequencies) and put it in K1. This gives the sample size.

3. Create column C3 by multiplying K1 by the probabilities listed in column C2. Column C3 lists the expected frequencies.

4. Create column C4 using the formula $(O - E)^2/E$. In MINITAB, this formula will be written as LET C4 = (C1 − C3)**2/C3.

5. Find the sum of the values listed in column C4, which gives the value of the test statistic χ^2.

6. Find the value of χ^2 for the given α and df from the chi-square table and make a decision by comparing it with the value of the test statistic χ^2.

All these steps are shown in Figure 11.12, which presents the MINITAB input and output display.

Figure 11.12   MINITAB input and output for goodness-of-fit test.

```
MTB  > NOTE: MAKING A GOODNESS-OF-FIT TEST FOR ILLUSTRATION M11-1
MTB  > READ C1 C2 ←——⎧ This command enters the data on
DATA > 715    .345    ⎨ observed frequencies and probabilities
DATA > 446    .216    ⎩ in columns C1 and C2, respectively.
DATA > 175    .093
DATA > 187    .092
DATA > 178    .083
DATA > 299    .171
DATA > END
        6 ROWS READ


MTB  > SUM C1 PUT IN K1 ←——⎧ This command calculates the sum of the
                            ⎨ values entered in column C1 and puts
                            ⎩ it in K1.


    SUM       =        2000.0 ←——This is the sample size.


MTB  > LET C3 = K1 * C2 ←——⎧ This command creates column C3
                            ⎨ of expected frequencies by
                            ⎩ multiplying K1 and column C2 values.


MTB  > LET C4 = (C1 − C3)**2/C3 ←——⎧ This command calculates
                                    ⎨ (O − E)²
                                    ⎩ ──────── for each category.
                                       E


MTB  > PRINT C1 C2 C3 C4

    ROW       C1        C2       C3         C4  ⎫
     1       715     0.345     690    0.90579   ⎪
     2       446     0.216     432    0.45370   ⎪ Compare this table
     3       175     0.093     186    0.65054   ⎬ with Table 11.4 of
     4       187     0.092     184    0.04891   ⎪ Example 11–4.
     5       178     0.083     166    0.86747   ⎪
     6       299     0.171     342    5.40644   ⎭


MTB  > SUM C4 ←————————— ⎧ This command prints the sum of column C4,
    SUM    =    8.3329    ⎩ which is the value of the test statistic χ².
```

From the MINITAB solution given in Figure 11.12, the value of the test statistic χ^2 is 8.3329. For this illustration, $\alpha = .025$ and $df = 6 - 1 = 5$. The critical value of χ^2 from the chi-square distribution table for $\alpha = .025$ and $df = 5$ is 12.833. (See Figure 11.5 of Example 11–4.) The value of the test statistic $\chi^2 = 8.3329$ is smaller than the critical value of $\chi^2 = 12.833$ and it falls in the nonrejection region. Consequently, we fail to reject the null hypothesis and state that the current U.S. car market shares held by various auto companies are the same as for 1992. The difference between the observed frequencies and the expected frequencies seems to have occurred only because of sampling error. ■

A TEST OF INDEPENDENCE OR HOMOGENEITY

To make a chi-square test of independence or homogeneity, we first enter the data for all columns of a given table by using the READ command. Remember, we do not enter the data listed in the row and column of totals. MINITAB command CHISQUARE followed by the names of the data columns gives us the value of the test statistic. Finally, we compare this value of the test statistic with the value of χ^2 obtained from the chi-square distribution table and make a decision. Illustration M11–2 describes the procedure used to make a chi-square test of independence. The procedure for a chi-square test of homogeneity is similar.

Illustration M11–2   According to Example 11–7, Windham Electronics Company has two machines that are used to manufacture cassettes. From time to time the quality control inspector at the company takes a sample of cassettes and checks them for being good or defective. A recent such sample of 200 cassettes produced the two-way classification presented in the following table.

	Good (G)	Defective (D)
Machine I (A)	109	11
Machine II (B)	66	14

Does the sample provide sufficient evidence to conclude that the two attributes, the machine type and cassettes being good or defective, are dependent? Use $\alpha = .01$.

Solution   The null and alternative hypotheses are

H_0: Machine type and cassettes being good or defective are independent

H_1: Machine type and cassettes being good or defective are dependent

First we enter the data given in the contingency table into MINITAB. We use the MINITAB command READ C1 C2 to enter the data given in the two columns of the table. Note that the number of columns in the READ command is equal to the number of columns in the contingency table. The table contains two columns, one for good cassettes and the second for defective cassettes. The MINITAB commands and the MINITAB output for a test of independence for these data are shown in Figure 11.13.

Figure 11.13 MINITAB input and output for a test of independence.

```
MTB  > NOTE: MAKING A TEST OF INDEPENDENCE FOR ILLUSTRATION M11-2
MTB  > READ C1 C2  ←──── ⎰This command is used to enter the values
DATA > 109 11            ⎱given in the contingency table.
DATA > 66   14
DATA > END
        2 ROWS READ


MTB  > CHISQUARE C1 C2  ←──── ⎧This command instructs MINITAB to
                                ⎨make a chi-square test using the
                                ⎩data of columns C1 and C2.


Expected counts are printed below observed counts

              C1           C2      Total
    1         109          11       120
            105.00       15.00  ←──────────────────┐
                                                    ⎱These rows give
                                                     the expected
    2          66          14        80             frequencies.
             70.00       10.00  ←──────────────────┘

  Total       175          25       200

  CHISQ =   0.152 +   1.067 +            ⎧This is the value
            0.229 +   1.600 = 3.048  ←── ⎨of the test
                                          ⎩statistic.
  DF  =  1
```

From the MINITAB output presented in Figure 11.13, the value of the test statistic χ^2 is 3.048. The critical value of χ^2 from Table IX of Appendix B for $\alpha = .01$ and $df = (2 - 1)(2 - 1) = 1$ is 6.635. Because the value of the test statistic $\chi^2 = 3.048$ is smaller than the critical value of $\chi^2 = 6.635$ and it falls in the nonrejection region, we fail to reject the null hypothesis (see Figure 11.7 of Example 11–7). Consequently, we conclude that the two characteristics, machine type and the cassettes being good or defective, are not related. In other words, the sample does not provide sufficient evidence to conclude that the two machines are producing different percentages of good and defective cassettes. ■

COMPUTER ASSIGNMENTS

M11.1 The following table gives the percentage distribution of the U.S. labor force by educational attainment for 1992. (Source: U.S. Bureau of Labor Statistics.)

Educational Attainment	Percentage of Workers
Less than a high school diploma	12.4
High school graduates, no college	35.7
Some college, no degree	18.0
Associate degree	7.4
Bachelor's degree	17.2
Master's degree	6.5
Professional or doctoral degree	2.8

A recent random sample of 1000 workers showed that 116 of them have less than a high school diploma, 363 are high school graduates with no college, 164 have some college but no degree, 71 possess an associate degree, 187 hold a bachelor's degree, 61 have a master's degree, and 38 possess a professional or doctoral degree. Using MINITAB, test at the 1% significance level if the percentage distribution of the U.S. labor force by educational attainment has changed since 1992.

M11.2　A management consultant thinks that job satisfaction affects the absentee rate of employees. He took a sample of 400 employees of companies and collected information on job satisfaction (on a scale of 1 to 10, 1 being the lowest and 10 the highest) and the number of days they were absent for last year. The following table gives the tabulated results.

		Job Satisfaction Index		
		Less than 4	4 to 7	Higher than 7
Number of	Less than 6	12	61	107
Days Absent	6 to 12	22	80	50
Last Year	More than 12	41	18	9

Testing at the 2.5% significance level, do you think the sample provides sufficient evidence to reject the null hypothesis that job satisfaction and absentee rate of employees are not related? Use MINITAB to make this test.

M11.3　A company owns four restaurants in four different parts of New York City. The management wanted to seek opinions of its patrons on the quality of food and service at each of these four restaurants. A random sample of 100 patrons for each restaurant was selected and asked to rate the quality of food and service at that restaurant. The results of the survey are recorded in the following table.

	Opinions of Patrons		
	Excellent	Average	Below Average
Restaurant 1	59	32	9
Restaurant 2	48	44	8
Restaurant 3	64	26	10
Restaurant 4	54	42	4

Based on these sample surveys and using $\alpha = .025$, can you conclude that the distributions of opinions are homogeneous for all four restaurants? Use MINITAB.

12 ANALYSIS OF VARIANCE

Chapter 10 described the procedures that are used to test hypotheses about the difference between two population means using the normal and t distributions. Also described in that chapter were the hypothesis-testing procedures for the difference between two population proportions using the normal distribution. Then, Chapter 11 explained the procedures to test hypotheses about the equality of more than two population proportions using the chi-square distribution.

This chapter explains how to test the null hypothesis that the means of more than two populations are equal. For example, suppose a company that produces detergents is planning to buy new machines to fill 64-ounce detergent jugs. There are three types of such machines that the company is considering. Before the company decides which type of machine to buy, it puts all three machines to a test. One of each type of machine is randomly selected and used to fill the 64-ounce jugs for a few hours. Let μ_1, μ_2, and μ_3 be the mean number of 64-ounce jugs filled per hour by each of these three machines, respectively. To test if the three types of machines produce equal means, we test the null hypothesis

$$H_0: \mu_1 = \mu_2 = \mu_3 \qquad \text{(All three population means are equal)}$$

against the alternative hypothesis

$$H_1: \text{All three population means are not equal}$$

We use the analysis of variance procedure to perform this test of hypothesis.

Note that the analysis of variance procedure can be used to compare two population means. However, the procedures learned in Chapter 10 are more efficient for performing tests of hypotheses about the difference between two population means, and the analysis of variance procedure, to be discussed in this chapter, is used to compare three or more population means.

The *analysis of variance* tests are made using the F distribution. First, the F distribution is described in Section 12.1. Then, Section 12.2 discusses the application of the one-way analysis of variance procedure to perform tests of hypotheses.

12.1   THE *F* DISTRIBUTION

Like the *t* and chi-square distributions, the shape of a particular **F distribution**[1] curve depends on the number of degrees of freedom. However, the *F* distribution has *two* numbers of degrees of freedom: *degrees of freedom for the numerator* and *degrees of freedom for the denominator*. These two numbers of degrees of freedom are the *parameters of the F distribution*. Each set of degrees of freedom for the numerator and for the denominator gives a different *F* distribution curve. The units of an *F* distribution are denoted by *F*, which assumes only nonnegative values. Like the normal, *t*, and chi-square distributions, the *F* distribution is also a continuous distribution. The shape of an *F* distribution curve is skewed to the right, but the skewness decreases as the number of degrees of freedom increases.

THE *F* DISTRIBUTION

1. The *F* distribution is continuous and skewed to the right.
2. The *F* distribution has two numbers of degrees of freedom: *df* for the numerator and *df* for the denominator.
3. The units of an *F* distribution, denoted by *F*, are nonnegative.

For an *F* distribution, degrees of freedom for the numerator and degrees of freedom for the denominator are usually written as follows.

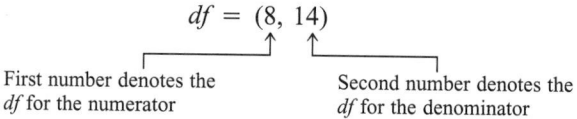

$$df = (8, 14)$$

First number denotes the
df for the numerator

Second number denotes the
df for the denominator

Figure 12.1 gives three *F* distribution curves for three sets of degrees of freedom for the numerator and for the denominator. In the figure, the first number gives the degrees of freedom associated with the numerator and the second number gives the degrees of freedom associated with the denominator. We can observe from this figure that as the degrees of freedom increase, the peak of the curve moves to the right, that is, the skewness decreases.

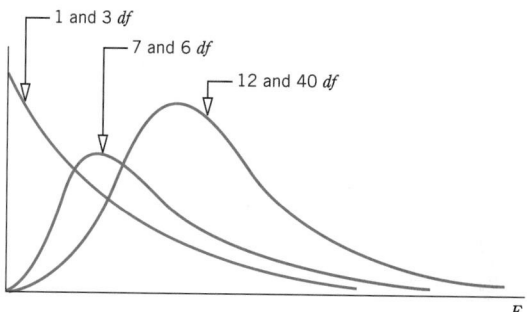

Figure 12.1   Three *F* distribution curves.

---

[1]The *F* distribution is named after Sir Ronald Fisher.

Table X in Appendix B lists the values of *F* for the *F* distribution. To read Table X, we need to know three quantities: the degrees of freedom for the numerator, the degrees of freedom for the denominator, and an area in the right tail of an *F* distribution curve. Note that the *F* distribution table (Table X) is read only for an area in the right tail of the *F* distribution curve. Also note that Table X has four parts. These four parts give the *F* values for an area of .01, .025, .05, and .10, respectively, in the right tail of the *F* distribution curve. Example 12–1 illustrates how to read Table X.

Reading the F distribution table.

EXAMPLE 12–1   Find the *F* value for 8 degrees of freedom for the numerator, 14 degrees of freedom for the denominator, and .05 area in the right tail of the *F* curve.

Solution   To find the required value of *F*, we consult the portion of Table X of Appendix B that corresponds to .05 area in the right tail of the *F* distribution curve. The relevant portion of that table is shown as Table 12.1 below. To find the required *F* value, we locate 8 in the row for degrees of freedom for the numerator (at the top of Table X) and 14 in the column for degrees of freedom in the denominator (the first column on the left side in Table X). The entry where the column for 8 and the row for 14 intersect gives the required *F* value. This value of *F* is 2.70, as shown in Figure 12.2. The *F* value taken from this table for a test of hypothesis is called the critical value of *F*.

Table 12.1

		Degrees of Freedom for the Numerator				
		1	**2**	**. . .**	**8**	**. . .**	**100**
	1	161.5	199.5		238.9	. . .	253.0
Degrees of	2	18.51	19.00	. . .	19.37	. . .	19.49
Freedom for	.	. . .	. . .	. . .	. . .	. . .	. . .
the Denominator	.	. . .	. . .	. . .	. . .	. . .	. . .
	14	4.60	3.74	. . .	2.70←	. . .	2.19
	.	. . .	. . .	. . .	. . .	. . .	. . .
	100	3.94	3.09	. . .	2.03	. . .	1.39

The *F* value for 8 *df* for the numerator, 14 *df* for the denominator, and .05 area in the right tail.

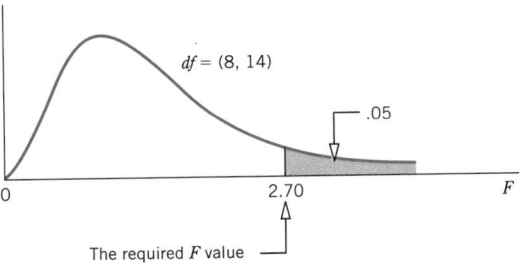

df = (8, 14)

.05

0

2.70

F

The required *F* value

Figure 12.2   The critical value of *F* for 8 *df* for the numerator, 14 *df* for the denominator, and .05 area in the right tail.

EXERCISES

Concepts and Procedures

12.1 Describe the main characteristics of an F distribution.

12.2 Find the critical value of F for the following.
    a. $df = (5, 12)$ and area in the right tail $= .05$
    b. $df = (4, 18)$ and area in the right tail $= .05$
    c. $df = (12, 7)$ and area in the right tail $= .05$

12.3 Find the critical value of F for the following.
    a. $df = (5, 12)$ and area in the right tail $= .025$
    b. $df = (4, 18)$ and area in the right tail $= .025$
    c. $df = (12, 7)$ and area in the right tail $= .025$

12.4 Determine the critical value of F for the following.
    a. $df = (8, 11)$ and area in the right tail $= .01$
    b. $df = (6, 12)$ and area in the right tail $= .01$
    c. $df = (15, 5)$ and area in the right tail $= .01$

12.5 Determine the critical value of F for the following.
    a. $df = (3, 14)$ and area in the right tail $= .10$
    b. $df = (9, 10)$ and area in the right tail $= .10$
    c. $df = (11, 4)$ and area in the right tail $= .10$

12.6 Find the critical value of F for an F distribution with $df = (4, 12)$ and
    a. area in the right tail $= .05$
    b. area in the right tail $= .10$

12.7 Find the critical value of F for an F distribution with $df = (11, 6)$ and
    a. area in the right tail $= .01$
    b. area in the right tail $= .025$

12.8 Find the critical value of F for an F distribution with .025 area in the right tail and
    a. $df = (5, 11)$ **b.** $df = (15, 5)$

12.9 Find the critical value of F for an F distribution with .01 area in the right tail and
    a. $df = (10, 10)$ **b.** $df = (9, 25)$

12.2 ONE-WAY ANALYSIS OF VARIANCE

As mentioned in the beginning of this chapter, the analysis of variance procedure is used to test the null hypothesis that the means of three or more populations are the same against the alternative hypothesis that all population means are not the same. The analysis of variance procedure can be used to compare two population means. However, the procedures learned in Chapter 10 are more efficient for performing tests of hypotheses about the difference between two population means, and the analysis of variance procedure is used to compare three or more population means.

Reconsider the example of the company that produces detergents and is planning to buy a new machine to fill 64-ounce detergent jugs. There are three types of such machines that the company is considering. Before it decides which type of machine to buy, the company puts all three machines to a test. One of each type of machine is randomly selected and used to fill the 64-ounce jugs for a few hours. Let μ_1, μ_2, and μ_3 be the mean number of 64-ounce jugs filled per hour by each of these three machines, respectively. To test if the three types of machines produce different means, we test the null hypothesis

$$H_0: \mu_1 = \mu_2 = \mu_3 \qquad \text{(All three population means are equal)}$$

against the alternative hypothesis

$$H_1: \text{All three population means are not equal}$$

One method to test such a hypothesis is to test the three hypotheses $H_0: \mu_1 = \mu_2$, $H_0: \mu_1 = \mu_3$, and $H_0: \mu_2 = \mu_3$ separately using the procedure discussed in Chapter 10. Besides being time consuming, such a procedure has other disadvantages. First, if we reject even one of these three hypotheses, then we must reject the null hypothesis $H_0: \mu_1 = \mu_2 = \mu_3$. Second, combining the Type I errors for the three tests (one for each test) will give a very large Type I error for the test $H_0: \mu_1 = \mu_2 = \mu_3$. Hence, we should prefer a procedure that can test the equality of three means in one test. The **ANOVA**, short for **analysis of variance**, provides such a procedure. It is used to compare three or more population means in a single test.

ANOVA

ANOVA is a procedure used to test the null hypothesis that the means of three or more populations are equal.

This section discusses the **one-way ANOVA** procedure to make tests comparing the means of several populations. By using a one-way ANOVA test, we analyze only one factor or variable. For instance, in the example of testing for the equality of means of three filling machines, we are considering only one factor, which is the effect of different machines on the number of jugs filled. Sometimes we may analyze the effects of two factors. For example, if the three filling machines are operated by different persons, we can analyze the effects of machines and operators on the number of jugs filled. This is done using a two-way ANOVA. The procedure under discussion is called the analysis of variance because the test is based on the analysis of variation in the data obtained from samples. The application of one-way ANOVA requires that the following assumptions hold true.

ASSUMPTIONS OF ONE-WAY ANOVA

The following assumptions must hold true to use one-way ANOVA.

1. The populations from which the samples are drawn are (approximately) normally distributed.
2. The populations from which the samples are drawn have the same variance (or standard deviation).
3. The samples drawn from different populations are random and independent.

For instance, in the example about three types of machines being considered by a company to fill 64-ounce jugs, we first assume that the number of jugs filled per hour by each machine is (approximately) normally distributed. Second, the means of the distributions of the number of jugs filled per hour for three machines may or may not be the same, but all three distributions have the same variance σ^2. Third, when we take samples to perform an

ANOVA test, these samples are drawn independently and randomly from three different populations.

The ANOVA test is applied by calculating two estimates of the variance σ^2 of population distributions: the **variance between samples** and the **variance within samples**. The variance between samples is also called the **mean square between samples** or **MSB**. The variance within samples is also called the **mean square within samples** or **MSW**.

The variance between samples MSB gives an estimate of σ^2 based on the variation among the means of samples taken from different populations. For the example of three filling machines, MSB will be based on the values of the means for three samples, which represent the number of jugs filled per hour by three machines, respectively. If the means of all populations under consideration are equal, the means of respective samples will still be different but the variation among them is expected to be small and, consequently, the value of MSB is expected to be small. However, if the means of populations under consideration are not all equal, the variation among the means of respective samples is expected to be large and, consequently, the value of MSB is expected to be large.

The variance within samples MSW gives an estimate of σ^2 based on the variation within the data of samples. For the example of three filling machines, MSW will be based on the measurements (the number of jugs filled per hour) included in the three samples taken from three populations. The concept of MSW is similar to the concept of the pooled standard deviation s_p for two samples discussed in Section 10.2 of Chapter 10.

The one-way ANOVA test is always right-tailed with the rejection region in the right tail of the F distribution curve. The hypothesis-testing procedure using ANOVA involves the same five steps that were used in earlier chapters. The next subsection explains how to calculate the value of the test statistic F for an ANOVA test.

12.2.1   CALCULATING THE VALUE OF THE TEST STATISTIC

The value of the test statistic F for a test of hypothesis using ANOVA is given by the ratio of two variances, the variance between samples (MSB) and the variance within samples (MSW).

> **TEST STATISTIC F FOR A ONE-WAY ANOVA TEST**
>
> The value of the test statistic F for an ANOVA test is calculated as
>
> $$F = \frac{\text{Variance between samples}}{\text{Variance within samples}} \quad \text{or} \quad \frac{\text{MSB}}{\text{MSW}}$$
>
> The calculation of MSB and MSW is explained in Example 12–2.

Example 12–2 describes the calculation of MSB, MSW, and the value of the test statistic F. Since the basic formulas are laborious to use, they are not presented here. We have used only the short-cut formulas to make calculations in this chapter.

Calculating the value of the test statistic F.

EXAMPLE 12–2   One of the products made at Abe Chemicals Company is detergents. Due to increased sales, the company is planning to buy a few new machines that will be used to fill the 64-ounce detergent jugs. The company is considering three types of such machines but will eventually buy only one type of machine. Before making such a decision, the company wanted to test the three machines for the number of jugs filled per hour. To do so, the company used each of the three types of machines for five hours and recorded the number of jugs filled during each hour by these machines. The following table gives the number of jugs filled by these machines during each of the five hours.

Machine I	Machine II	Machine III
54	53	49
49	56	53
52	57	47
55	51	50
48	59	54

Calculate the value of the test statistic F. Assume that all the required assumptions mentioned in Section 12.2 hold true.

Solution In ANOVA terminology, the three machines that are used to fill the jugs are called **treatments**. The table contains data on the number of jugs filled during each of the five hours. Note that the five hours for which each machine is used give the sample size. Let

x = the number of jugs filled by a machine during a given hour

k = the number of different machines (or treatments)

n_i = the size of sample i (the number of hours for which each machine is used)

T_i = the sum of the values in sample i

n = the number of values in all samples = $n_1 + n_2 + n_3 + \cdots$

Σx = the sum of the values in all samples = $T_1 + T_2 + T_3 + \cdots$

Σx^2 = the sum of the squares of the values in all samples

To calculate MSB and MSW, we first compute the **between-samples sum of squares** denoted by **SSB** and the **within-samples sum of squares** denoted by **SSW**. The sum of SSB and SSW is called the **total sum of squares** and it is denoted by **SST**, that is,

$$\text{SST} = \text{SSB} + \text{SSW}$$

The SSB and SSW are calculated using the following formulas.

BETWEEN- AND WITHIN-SAMPLES SUM OF SQUARES

The between-samples sum of squares, denoted by SSB, is calculated as

$$\text{SSB} = \left(\frac{T_1^2}{n_1} + \frac{T_2^2}{n_2} + \frac{T_3^2}{n_3} + \cdots \right) - \frac{(\Sigma x)^2}{n}$$

The within-sample sum of squares, denoted by SSW, is calculated as

$$\text{SSW} = \Sigma x^2 - \left(\frac{T_1^2}{n_1} + \frac{T_2^2}{n_2} + \frac{T_3^2}{n_3} + \cdots \right)$$

Table 12.2 lists the number of jugs filled by each of the three machines during the selected five hours, the values of T_1, T_2, and T_3, and the values of n_1, n_2, and n_3.

Table 12.2

Machine I	Machine II	Machine III
54	53	49
49	56	53
52	57	47
55	51	50
48	59	54
$T_1 = 258$	$T_2 = 276$	$T_3 = 253$
$n_1 = 5$	$n_2 = 5$	$n_3 = 5$

In Table 12.2, T_1 is obtained by adding the five numbers in the first column, which makes the first sample. Thus, $T_1 = 54 + 49 + 52 + 55 + 48 = 258$. Similarly, the sums of the values in the second and third columns (or samples) give $T_2 = 276$ and $T_3 = 253$. Because there are five observations in each sample, $n_1 = n_2 = n_3 = 5$. The values of Σx and n are

$$\Sigma x = T_1 + T_2 + T_3 = 258 + 276 + 253 = 787$$

$$n = n_1 + n_2 + n_3 = 5 + 5 + 5 = 15$$

To calculate Σx^2, we square all numbers included in all three samples and then add them. Thus,

$$\Sigma x^2 = (54)^2 + (49)^2 + (52)^2 + (55)^2 + (48)^2 + (53)^2 + (56)^2 + (57)^2$$
$$+ (51)^2 + (59)^2 + (49)^2 + (53)^2 + (47)^2 + (50)^2 + (54)^2$$
$$= 41,461$$

Substituting all the values in the formulas for SSB and SSW, we obtain the following values of SSB and SSW.

$$\text{SSB} = \left(\frac{(258)^2}{5} + \frac{(276)^2}{5} + \frac{(253)^2}{5} \right) - \frac{(787)^2}{15} = 58.5333$$

$$\text{SSW} = 41,461 - \left(\frac{(258)^2}{5} + \frac{(276)^2}{5} + \frac{(253)^2}{5} \right) = 111.2000$$

The value of SST is obtained by adding the values of SSB and SSW. Thus,

$$\text{SST} = 58.5333 + 111.2000 = 169.7333$$

The variance between samples MSB and the variance within samples MSW are calculated using the following formulas.

CALCULATING THE VALUES OF MSB AND MSW

The MSB and MSW are calculated as

$$\text{MSB} = \frac{\text{SSB}}{k-1} \quad \text{and} \quad \text{MSW} = \frac{\text{SSW}}{n-k}$$

where $k - 1$ and $n - k$ are respectively the df for the numerator and the df for the denominator for the F distribution.

Consequently, the variance between samples is

$$\text{MSB} = \frac{\text{SSB}}{k-1} = \frac{58.5333}{3-1} = 29.2667$$

The variance within samples is

$$\text{MSW} = \frac{\text{SSW}}{n-k} = \frac{111.2000}{15-3} = 9.2667$$

The value of the test statistic F is given by the ratio of MSB and MSW. Therefore,

$$F = \frac{\text{MSB}}{\text{MSW}} = \frac{29.2667}{9.2667} = \textbf{3.16}$$

For convenience, these calculations are often recorded in a table called the *ANOVA table*. Table 12.3 gives the general form of an ANOVA table.

Table 12.3   ANOVA Table

Source of Variation	Degrees of Freedom	Sum of Squares	Mean Square	Value of the Test Statistic
Between	$k-1$	SSB	MSB	$F = \dfrac{\text{MSB}}{\text{MSW}}$
Within	$n-k$	SSW	MSW	
Total	$n-1$	SST		

Substituting the values of various quantities in Table 12.3, we write an ANOVA table for our example as Table 12.4.

Table 12.4   ANOVA Table

Source of Variation	Degrees of Freedom	Sum of Squares	Mean Square	Value of the Test Statistic
Between	2	58.5333	29.2667	$F = \dfrac{29.2667}{9.2667} = 3.16$
Within	12	111.2000	9.2667	
Total	14	169.7333		

12.2.2   ONE-WAY ANOVA TEST

Now suppose we want to test the null hypothesis that the mean number of 64-ounce jugs filled per hour by each of the three machines mentioned in Example 12–2 is the same against the alternative hypothesis that these three population means are not all equal. Note that in a one-way ANOVA test, the null hypothesis is that the means for all populations are equal. The alternative hypothesis is that all population means are not equal. In other words, the alternative hypothesis states that at least one of the population means is different from the others. Example 12–3 demonstrates how we use the one-way ANOVA procedure to make such a test.

Performing a one-way ANOVA test: all samples of the same size.

EXAMPLE 12–3   Reconsider Example 12–2 about the number of jugs filled during each of the five hours by three different machines. At the 1% significance level, can we reject the null hypothesis that the mean number of jugs filled per hour by each of these three machines is the same? Assume that all the assumptions required to apply the one-way ANOVA procedure hold true.

Solution   To make a test about the equality of means of three populations, we follow our standard five-step procedure.

Step 1.   *State the null and alternative hypotheses*

Let μ_1, μ_2, and μ_3 be the mean number of jugs filled per hour by each of the three machines under consideration. The null and alternative hypotheses are

$$H_0\colon \mu_1 = \mu_2 = \mu_3 \quad \text{(the means for three machines are equal)}$$

$$H_1\colon \text{All three means are not equal}$$

Note that the alternative hypothesis states that the mean number of jugs filled per hour by at least one of the three machines is different from the mean number of jugs filled per hour by the other two machines.

Step 2.   *Select the distribution to use*

Because we are comparing three means for three normally distributed populations, we use the F distribution to make the test.

Step 3.   *Determine the rejection and nonrejection regions*

The significance level is .01. Because a one-way ANOVA test is always right-tailed, the area in the right tail of the F distribution curve is .01, which is the rejection region in Figure 12.3.

Next we need to know the df for the numerator and the denominator. In our example, the jugs were filled on three different machines. As mentioned earlier, these machines are called treatments. The number of treatments is denoted by k. The total number of observations in all samples taken together is denoted by n. Then the number of degrees of freedom for the numerator is equal to $k - 1$ and the number of degrees of freedom for the denominator is equal to $n - k$. In our example, there are 3 treatments (machines) and 15 total observations (total number of hours for which the three machines are used) in all three samples. Thus,

$$\text{Degrees of freedom for the numerator} = k - 1 = 3 - 1 = 2$$

$$\text{Degrees of freedom for the denominator} = n - k = 15 - 3 = 12$$

From Table X of Appendix B, we find the critical value of F for 2 df for the numerator, 12 df for the denominator, and .01 area in the right tail of the F distribution curve. The required value of F is 6.93, which is shown in Figure 12.3.

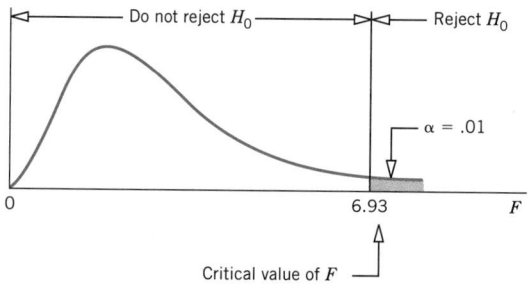

Figure 12.3   Critical value of F for $df = (2, 12)$ and $\alpha = .01$

Thus, we will fail to reject H_0 if the calculated value of the test statistic F is less than 6.93 and we will reject H_0 if it is greater than 6.93.

Step 4.   *Calculate the value of the test statistic*

We computed the value of the test statistic F for these data in Example 12–2. This value is

$$F = 3.16$$

Step 5.   *Make a decision*

Because the value of the test statistic $F = 3.16$ is less than the critical value of $F = 6.93$, it falls in the nonrejection region. Hence, we fail to reject the null hypothesis and conclude that the means of the three populations are equal. In other words, the three machines fill the same number of jugs per hour on average. The difference in the three means obtained from the three samples occurred only because of sampling error.   ▬

In Example 12–3, the sample sizes were the same for all treatments. Example 12–4 describes a case in which the sample sizes are not the same for all treatments.

Performing a one-way ANOVA test: all samples not of the same size.

EXAMPLE 12–4   From time to time, unknown to its employees, the research department at Post Bank observes various employees for work productivity. Recently this department wanted to check whether the four tellers at a branch of this bank serve, on average, the same number of customers per hour. The research manager observed each of the four tellers for a certain number of hours. The following table gives the number of customers served by the four tellers during each of the observed hours.

Teller A	Teller B	Teller C	Teller D
19	14	11	24
21	16	14	19
26	14	21	21
24	13	13	26
18	17	16	20
	13	18	

At the 5% significance level, test the null hypothesis that the mean number of customers served per hour by each of these four tellers is the same. Assume that all the assumptions required to apply the one-way ANOVA procedure hold true.

Solution

Step 1.   *State the null and alternative hypotheses*

Let μ_1, μ_2, μ_3, and μ_4 be the mean number of customers served per hour by tellers A, B, C, and D, respectively. The null and alternative hypotheses are

$H_0: \mu_1 = \mu_2 = \mu_3 = \mu_4$ (the mean number of customers served per hour by each of the four tellers is the same)

$H_1:$ All four population means are not equal

Step 2.   *Select the distribution to use*

Because we are testing for the equality of four means for four normally distributed populations, we use the F distribution to make the test.

Step 3.   *Determine the rejection and nonrejection regions*

The significance level is .05, which means the area in the right tail of the F distribution curve is .05.

In this example, there are four treatments (tellers) and 22 total observations in all four samples. Thus,

$$\text{Degrees of freedom for the numerator} = k - 1 = 4 - 1 = 3$$

$$\text{Degrees of freedom for the denominator} = n - k = 22 - 4 = 18$$

The critical value of F from Table X for 3 df for the numerator, 18 df for the denominator, and .05 area in the right tail of the F distribution curve is 3.16. This value is shown in Figure 12.4.

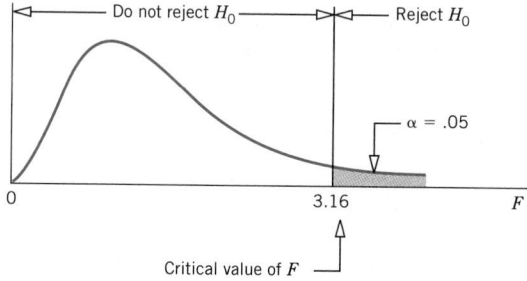

Figure 12.4   Critical value of F for $df = (3, 18)$ and $\alpha = .05$.

Step 4.   *Calculate the value of the test statistic*

First we calculate SSB and SSW. Table 12.5 lists the number of customers served by the four tellers during the selected hours, the values of T_1, T_2, T_3, and T_4, and the values of n_1, n_2, n_3, and n_4.

Table 12.5

Teller A	Teller B	Teller C	Teller D
19	14	11	24
21	16	14	19
26	14	21	21
24	13	13	26
18	17	16	20
	13	18	
$T_1 = 108$	$T_2 = 87$	$T_3 = 93$	$T_4 = 110$
$n_1 = 5$	$n_2 = 6$	$n_3 = 6$	$n_4 = 5$

The values of Σx and n are

$$\Sigma x = T_1 + T_2 + T_3 + T_4 = 108 + 87 + 93 + 110 = 398$$

$$n = n_1 + n_2 + n_3 + n_4 = 5 + 6 + 6 + 5 = 22$$

The value of Σx^2 is calculated as follows.

$$\Sigma x^2 = (19)^2 + (21)^2 + (26)^2 + (24)^2 + (18)^2 + (14)^2 + (16)^2 + (14)^2$$
$$+ (13)^2 + (17)^2 + (13)^2 + (11)^2 + (14)^2 + (21)^2 + (13)^2$$
$$+ (16)^2 + (18)^2 + (24)^2 + (19)^2 + (21)^2 + (26)^2 + (20)^2$$
$$= 7614$$

Substituting all the values in the formulas for SSB and SSW, we obtain the following values of SSB and SSW.

$$SSB = \left(\frac{T_1^2}{n_1} + \frac{T_2^2}{n_2} + \frac{T_3^2}{n_3} + \frac{T_4^2}{n_4}\right) - \frac{(\Sigma x)^2}{n}$$

$$= \left(\frac{(108)^2}{5} + \frac{(87)^2}{6} + \frac{(93)^2}{6} + \frac{(110)^2}{5}\right) - \frac{(398)^2}{22} = 255.6182$$

$$SSW = \Sigma x^2 - \left(\frac{T_1^2}{n_1} + \frac{T_2^2}{n_2} + \frac{T_3^2}{n_3} + \frac{T_4^2}{n_4}\right)$$

$$= 7614 - \left(\frac{(108)^2}{5} + \frac{(87)^2}{6} + \frac{(93)^2}{6} + \frac{(110)^2}{5}\right) = 158.2000$$

Hence, the variance between samples MSB and the variance within samples MSW are

$$MSB = \frac{SSB}{k - 1} = \frac{255.6182}{4 - 1} = 85.2061$$

$$MSW = \frac{SSW}{n - k} = \frac{158.2000}{22 - 4} = 8.7889$$

The value of the test statistic F is given by the ratio of MSB and MSW, which is

$$F = \frac{MSB}{MSW} = \frac{85.2061}{8.7889} = 9.69$$

Writing the values of various quantities in the ANOVA table, we obtain Table 12.6.

Table 12.6 ANOVA Table

Source of Variation	Degrees of Freedom	Sum of Squares	Mean Square	Value of the Test Statistic
Between	3	255.6182	85.2061	
Within	18	158.2000	8.7889	$F = \dfrac{85.2061}{8.7889} = 9.69$
Total	21	413.8182		

Step 5. *Make a decision*

Because the value of the test statistic $F = 9.69$ is greater than the critical value of $F = 3.16$, it falls in the rejection region. Consequently, we reject the null hypothesis and conclude that the mean number of customers served per hour by each of the four tellers is not the same. In other words, at least one of the four means is different from the other three. ∎

EXERCISES

Concepts and Procedures

12.10 Briefly explain when a one-way ANOVA is used to make a test of hypothesis.

12.11 Describe the assumptions that must hold true to apply the one-way analysis of variance procedure to test hypotheses.

12.12 Three samples randomly selected from three independent populations that are normally distributed with equal variances produced the following data.

Sample I	Sample II	Sample III
41	44	46
32	42	33
28	37	42
37	38	36
39	45	39
34	34	36
39	36	40

a. We are to test if the means of these three populations are all equal. Write the null and alternative hypotheses.
b. Show the rejection and nonrejection regions on the F distribution curve for $\alpha = .01$.
c. Calculate SSB, SSW, and SST.
d. What are the degrees of freedom for the numerator and denominator, respectively?
e. Calculate the between-samples and within-samples variances.
f. What is the critical value of F for $\alpha = .01$?
g. What is the calculated value of the test statistic F?
h. Write the ANOVA table for this exercise.
i. Will you reject the null hypothesis stated in part a at a significance level of 1%?

12.13 Four samples randomly selected from four independent populations that are normally distributed with equal variances produced the following data.

Sample I	Sample II	Sample III	Sample IV
18	15	12	17
11	21	18	14
14	16	10	22
17	13	15	16
13	12	17	24
16	19	20	20
	16	11	

a. We are to test if the means of these four populations are all equal. Write the null and alternative hypotheses.
b. Show the rejection and nonrejection regions on the F distribution curve for $\alpha = .025$.
c. Calculate SSB, SSW, and SST.
d. What are the degrees of freedom for the numerator and denominator, respectively?
e. Calculate the between-samples and within-samples variances.
f. What is the critical value of F for $\alpha = .025$?
g. What is the calculated value of the test statistic F?
h. Write the ANOVA table for this exercise.
i. Will you reject the null hypothesis stated in part a at a significance level of 2.5%?

12.14 Consider the following data obtained for two samples selected at random from two populations that are independent and normally distributed with equal variances.

Sample I	Sample II
32	27
26	37
31	33
29	36
27	38
34	31

a. Calculate the means and standard deviations for these samples using the formulas learned in Chapter 3.
b. Using the procedure learned in Section 10.2 of Chapter 10, test at the 1% significance level whether the means of the populations from which these samples are drawn are equal.
c. Using the one-way ANOVA procedure, test at the 1% significance level whether the means of the populations from which these samples are drawn are equal.
d. Are the conclusions reached in parts b and c the same?

12.15 Consider the following data obtained for two samples selected at random from two populations that are independent and normally distributed with equal variances.

Sample I	Sample II
14	11
16	8
11	12
9	13
13	15
15	7
17	9

a. Calculate the means and standard deviations for these samples using the formulas learned in Chapter 3.
b. Using the procedure learned in Section 10.2 of Chapter 10, test at the 5% significance level whether the means of the populations from which these samples are drawn are equal.
c. Using the one-way ANOVA procedure, test at the 5% significance level whether the means of the populations from which these samples are drawn are equal.
d. Are the conclusions reached in parts b and c the same?

12.16 The following ANOVA table, based on information obtained for three samples selected from three independent populations that are normally distributed with equal variances, has a few missing values.

Source of Variation	Degrees of Freedom	Sum of Squares	Mean Square	Value of the Test Statistic
Between	2		19.2813	
Within		89.3677		$F = \underline{\quad} = \underline{\quad}$
Total	12			

a. Find the missing values and complete the ANOVA table.
b. Using $\alpha = .01$, what is your conclusion for the test with the null hypothesis that the means of the three populations are all equal against the alternative hypothesis that the means of the three populations are not all equal?

12.17   The following ANOVA table, based on information obtained for four samples selected from four independent populations that are normally distributed with equal variances, has a few missing values.

Source of Variation	Degrees of Freedom	Sum of Squares	Mean Square	Value of the Test Statistic
Between				
Within	15		9.2154	$F = \dfrac{\quad}{\quad} = 4.07$
Total	18			

  a. Find the missing values and complete the ANOVA table.
  b. Using $\alpha = .05$, what is your conclusion for the test with the null hypothesis that the means of the four populations are all equal against the alternative hypothesis that the means of the four populations are not all equal?

Applications

For the following exercises assume that all the assumptions required to apply the one-way ANOVA procedure hold true.

12.18   A large manufacturer of copying machines recently hired three new salespersons with degrees in marketing, mathematics, and sociology. The company wants to check if the fields of study have any effect on the mean number of sales made by salespersons. The following table lists the number of sales made by these three salespersons during certain randomly selected days.

Salesperson with Marketing Degree	Salesperson with Mathematics Degree	Salesperson with Sociology Degree
9	2	4
10	1	1
3	3	1
7	2	3
4	5	6
12	3	8
8	1	1

Using the 5% significance level, can you reject the null hypothesis that the mean number of copying machines sold per day by all salespersons with degrees in each of these three areas is the same?

12.19   A consumer agency wanted to investigate if four insurance companies differed with regard to the premiums charged for auto insurance. The agency randomly selected a few auto drivers who were insured by each of these four companies and had almost similar driving records, autos, and insurance policies. The following table gives premiums paid per month by these drivers insured with these four insurance companies.

Company A	Company B	Company C	Company D
65	48	57	62
73	69	61	53
54	88	89	45
43		77	51
		69	

Using the 1% significance level, test the null hypothesis that the mean auto insurance premium paid per month by all drivers insured by each of these four companies is the same.

12.20 Three new brands of fertilizer that a farmer can use to grow crops just came on the market. Before deciding which brand he should use permanently for all crops, a farmer decided to experiment for one season. To do so, he randomly assigned each fertilizer to eight one-acre tracts of land that he used to grow wheat. The following table gives the production of wheat (in bushels) for each acre for three brands of fertilizer.

Fertilizer I	Fertilizer II	Fertilizer III
72	58	61
69	42	58
75	53	63
59	47	68
64	45	55
68	52	65
71	47	59
67	57	63

At the 5% significance level, can you conclude that the mean yield of wheat for each of these three brands of fertilizer is the same?

12.21 Simsbury Inc. opened five supermarkets in Connecticut four years ago. Among other things, the quality control department at the company has always given a high priority to customer satisfaction. From time to time this department surveys the shoppers at the five stores to find out how satisfied they are. A recent such survey produced the following data on customer satisfaction, which is measured on a scale of 1 to 10, 1 being the lowest level of satisfaction and 10 being the highest.

Store I	Store II	Store III	Store IV	Store V
8	7	9	5	8
6	9	7	8	9
9	7	8	6	4
7	5	8	7	7
8	8	9	6	9
6	6	7	9	8
9	4		4	

Using the 1% significance level, test the null hypothesis that the mean consumer satisfaction index is the same for each of these five stores.

12.22 COPE Inc. owns a chain of department stores, including three in three different sections of New York City. The management wants to find out if the mean gross sales per day are the same for these three stores. The research department at the company collected data on the gross sales for randomly selected days for each of these three stores. The following table lists the gross sales (rounded to thousands of dollars) for these stores for the selected days.

Store I	Store II	Store III
43	52	38
37	49	45
57	59	40
39	57	49
46	61	36
51	58	42
47	63	39

At the 1% significance level, will you reject the null hypothesis that the mean gross sales per day for each of these three stores are the same?

12.23 A large company buys thousands of lightbulbs every year. The company is currently considering four brands of lightbulbs to choose from. Before the company decides which lightbulbs to buy, it wants to investigate if the mean life of the four types of lightbulbs is the same. The company's research department randomly selected a few bulbs of each type and tested them. The following table lists the number of hours (in thousands) that each of the bulbs in each brand survived before being burned out.

Brand I	Brand II	Brand III	Brand IV
23	19	23	26
24	23	27	24
21	18	25	21
26	24	26	29
22	21	23	28
23	22	21	27
25	19	27	28

At the 2.5% significance level, test the null hypothesis that the mean life of bulbs for each of these four brands is the same.

GLOSSARY

Analysis of variance (ANOVA)   A statistical technique used to test whether the means of three or more populations are equal.

F distribution   A continuous distribution that has two parameters: df for the numerator and df for the denominator.

Mean square between samples or **MSB**   A measure of the variation among means of samples taken from different populations.

Mean square within samples or **MSW**   A measure of the variation within data of all samples taken from different populations.

One-way ANOVA   The analysis of variance technique that analyzes one variable only.

SSB   The sum of squares between samples. Also called the sum of squares of the factor or treatment.

SST   The total sum of squares given by the sum of SSB and SSW.

SSW   The sum of squares within samples. Also called the sum of squares of errors.

KEY FORMULAS

Let

$$k = \text{the number of different samples (or treatments)}$$
$$n_i = \text{the size of sample } i$$
$$T_i = \text{the sum of the values in sample } i$$
$$n = \text{the number of values in all samples} = n_1 + n_2 + n_3 + \cdots$$
$$\Sigma x = \text{the sum of the values in all samples} = T_1 + T_2 + T_3 + \cdots$$
$$\Sigma x^2 = \text{the sum of the squares of values in all samples}$$

1. **Degrees of freedom for the F distribution**

$$\text{Degrees of freedom for the numerator} = k - 1$$
$$\text{Degrees of freedom for the denominator} = n - k$$

2. **Between-samples sum of squares**

$$\text{SSB} = \left(\frac{T_1^2}{n_1} + \frac{T_2^2}{n_2} + \frac{T_3^2}{n_3} + \cdots \right) - \frac{(\Sigma x)^2}{n}$$

3. **Within-samples sum of squares**

$$\text{SSW} = \Sigma x^2 - \left(\frac{T_1^2}{n_1} + \frac{T_2^2}{n_2} + \frac{T_3^2}{n_3} + \cdots \right)$$

4. **Total sum of squares**

$$\text{SST} = \text{SSB} + \text{SSW}$$

5. **Variance between samples**

$$\text{MSB} = \frac{\text{SSB}}{k - 1}$$

6. **Variance within samples**

$$\text{MSW} = \frac{\text{SSW}}{n - k}$$

7. **Value of the test statistic F**

$$F = \frac{\text{Variance between samples}}{\text{Variance within samples}} \quad \text{or} \quad \frac{\text{MSB}}{\text{MSW}}$$

SUPPLEMENTARY EXERCISES

For the following exercises, assume that all the assumptions required to apply the one-way ANOVA procedure hold true.

12.24 A consumer agency wants to check if the mean lives of four brands of auto batteries, which sell for nearly the same price, are the same. The agency randomly selected a few batteries of each brand and tested them. The following table gives the lives of these batteries in thousands of hours.

Brand A	Brand B	Brand C	Brand D
74	53	57	56
68	51	71	51
51	47	81	45
59	55	72	43
65		68	

a. At the 5% significance level, will you reject the null hypothesis that the mean life of each of these four brands of batteries is the same?
b. What is the Type I error in this case and what is the probability of committing such an error? Explain.

12.25 An economist wanted to investigate if recent college graduates with different majors who got jobs in San Francisco are commanding the same average salary. She selected a random sample of recent graduates in four areas—engineering, business, mathematics, and sociology. The following table gives the starting salaries (in thousands of dollars) for these samples.

Engineering	Business	Mathematics	Sociology
29.2	23.3	23.3	18.6
24.5	29.8	21.4	19.2
34.3	25.6	28.3	23.9
27.4	27.5	23.6	
32.1	24.8		

a. At the 1% significance level, test the null hypothesis that the mean starting salaries of all recent college graduates with these four majors who got jobs in San Francisco are equal.
b. What is the Type I error in this case and what is the probability of commiting such an error? Explain.

12.26 The following table gives the response time (in minutes) of three fire companies in a city for certain randomly selected incidents after a fire was reported.

Company A	Company B	Company C
1.6	1.4	.8
.8	2.6	1.3
2.7	.9	1.7
1.2	3.5	.9
3.4	1.2	1.1
1.9	1.5	.7
4.3		2.1

a. At the 2.5% significance level, can you conclude that the mean response time for each of these three fire companies for all fire incidents is the same?
b. If you did not reject the null hypothesis in part a, explain the Type II error that you may have made in this case. Note that you cannot calculate the probability of committing a Type II error without additional information.

12.27 The following table lists the prices of certain randomly selected college textbooks in statistics, psychology, economics, and business.

Statistics	Psychology	Economics	Business
57	51	67	64
53	59	54	61
57	54	51	56
62	61	59	60
51		62	53

a. Using the 1% significance level, test the null hypothesis that the mean prices of college textbooks in statistics, psychology, economics, and business are all equal.
b. If you did not reject the null hypothesis in part a, explain the Type II error that you may have made in this case. Note that you cannot calculate the probability of committing a Type II error without additional information.

12.28 A consumer agency that wanted to compare drying times for paints made by three companies, tested a few samples of paints from each of these three companies. The following table records the drying times (in minutes) for these samples of paints.

Company A	Company B	Company C
42	57	45
53	63	49
43	61	51
47	54	58
42	51	44
51	60	41
56		47

a. Using the 5% significance level, test the null hypothesis that the mean drying times for paints of these three companies are equal.

b. What will your decision be if the probability of making a Type I error is zero? Explain.

12.29 An auto company wanted to find out if a specific auto model gave the same average miles per gallon for three different brands of gas. The company took a random sample of 20 cars of this model and randomly divided them into three groups to test for miles per gallon traveled on each of three different brands of gasoline. The following table lists the miles per gallon obtained by the cars from the use of these three brands of gasoline.

Brand A	Brand B	Brand C
26.2	24.6	27.3
28.5	23.6	29.4
25.0	22.7	25.7
26.7	25.1	27.3
25.8	26.0	24.8
27.3	22.2	26.1
28.4		24.6

a. At the 2.5% significance level, test the null hypothesis that the mean miles per gallon given for this model are the same for each of these three types of gasoline.

b. What will your decision be if the probability of making a Type I error is zero? Explain.

SELF-REVIEW TEST

1. The F distribution is
   a. continuous   **b.** discrete

2. The F distribution is always
   a. symmetric   **b.** skewed to the right   **c.** skewed to the left

3. The units of the F distribution, denoted by F, are always
   a. nonpositive   **b.** positive   **c.** nonnegative

4. The one-way ANOVA test analyzes only one
   a. variable   **b.** population   **c.** sample

5. The one-way ANOVA test is always
   a. right-tailed   **b.** left-tailed   **c.** two-tailed

6. For a one-way ANOVA with k treatments and n observations in all samples taken together, the number of degrees of freedom for the numerator are

       a. $k - 1$ **b.** $n - k$ **c.** $n - 1$

7. For a one-way ANOVA with k treatments and n observations in all samples taken together, the number of degrees of freedom for the denominator are

       a. $k - 1$ **b.** $n - k$ **c.** $n - 1$

8. The ANOVA test can be applied to compare

   a. three or more population means
   b. more than four population means
   c. more than three population means

9. Briefly describe the assumptions that must hold true to apply the one-way ANOVA procedure as mentioned in this chapter.

10. The following table gives the hourly wage of computer programmers for samples taken from three cities.

New York	Boston	Los Angeles
$15.45	$23.50	$17.50
28.80	18.60	11.40
26.45	14.75	29.40
22.10	30.00	22.30
31.50	35.40	16.35
39.30	26.40	19.50
28.75		21.30

    a. Using the 1% significance level, test the null hypothesis that the mean hourly wage for all computer programmers in each of these three cities is the same.

    b. Is it a Type I error or a Type II error that may have been committed in part a? Explain.

USING MINITAB

The first step in using MINITAB to solve a problem using the one-way analysis of variance procedure is to enter the given data in different columns. If all columns contain the same number of values (i.e., all samples are of the same size), we can use the READ command. However, if different columns contain a different number of values (as in Illustration M12–1 below), we will use the SET command to enter data for each column one at a time. Suppose our example contains data on four samples. After entering data for these four samples in four columns, C1, C2, C3, and C4, we will use the MINITAB command given in Figure 12.5 to perform the one-way analysis of variance.

Figure 12.5

```
MTB > AOVONEWAY C1 C2 C3 C4
```

In the MINITAB command AOVONEWAY, AOV stands for analysis of variance and ONEWAY stands for one-way. Illustration M12–1 describes the use of MINITAB to perform a test of hypothesis using the one-way analysis of variance procedure.

Illustration M12–1   According to Example 12–4, the research department at Post Bank wants to know if the mean number of customers served per hour by each of the four tellers at a branch of this bank is the same. The research manager observed each of the four tellers for a certain number of hours. The following table gives the number of customers served by each of the four tellers during each of the observed hours.

Teller A	Teller B	Teller C	Teller D
19	14	11	24
21	16	14	19
26	14	21	21
24	13	13	26
18	17	16	20
	13	18	

Using MINITAB, test the null hypothesis that the mean number of customers served per hour by each of these four tellers is the same. Use $\alpha = .05$ and assume that all the required assumptions to apply the one-way analysis of variance hold true.

Solution   Let μ_1, μ_2, μ_3, and μ_4 be the mean number of customers served per hour by each of the four tellers, respectively. Then the null and alternative hypotheses are

$$H_0: \mu_1 = \mu_2 = \mu_3 = \mu_4 \qquad \text{(All four population means are equal)}$$

$$H_1: \text{All four population means are not equal}$$

The number of values in each of the four samples is not the same. Hence, we use the SET command to enter data on the four samples in the four columns.

Figure 12.6   Minitab output for Illustration M12–1.

```
MTB  > NOTE: APPLICATION OF ONE-WAY ANOVA TO ILLUSTRATION M12-1
MTB  > SET C1
DATA > 19 21 26 24 18
DATA > END
MTB  > SET C2
DATA > 14 16 14 13 17 13
DATA > END
MTB  > SET C3
DATA > 11 14 21 13 16 18
DATA > END
MTB  > SET C4
DATA > 24 19 21 26 20
DATA > END
MTB  > AOVONEWAY C1-C4  ←──
```
{This command instructs MINITAB to perform a one-way analysis of variance test for the data of columns C1, C2, C3, and C4.

```
ANALYSIS OF VARIANCE
SOURCE     DF        SS        MS        F        P
FACTOR      3     255.62     85.21     9.69     0.000
ERROR      18     158.20      8.79
TOTAL      21     413.82
```
←Compare this table to the ANOVA Table 12.6 of Example 12–4.

```
                                   INDIVIDUAL 95 PCT CI'S FOR MEAN
                                   BASED ON POOLED STDEV
LEVEL      N      MEAN     STDEV   ------+---------+---------+---------+
C1         5    21.600     3.362                       (-------*-------)
C2         6    14.500     1.643   (------*-------)
C3         6    15.500     3.619      (------*-------)
C4         5    22.000     2.915                        (-------*-------)
                                   ------+---------+---------+---------+
POOLED STDEV =    2.965            14.0      17.5      21.0      24.5
```

Compare the analysis of variance table in the MINITAB solution of Figure 12.6 with Table 12.6 of Example 12–4. In Table 12.6, the two sources of variation were called the variations between- and within-samples in the column labeled Source of Variation. In the MINITAB solution of Figure 12.6 these two sources are called the *factor* and *error*, respectively. Also, MINITAB prints the *p*-value for the test.

The MINITAB solution also gives the following information.

1. The mean and standard deviation for data contained in each of the columns C1, C2, C3, and C4. Thus, for example, the mean and standard deviation for the data of column C1 (for teller A) are 21.6 and 3.362, respectively.

2. The 95% confidence interval for the mean of the population corresponding to each of the four samples, that is, the 95% confidence interval for the mean number of customers served per hour by each of the four tellers.

3. The pooled standard deviation, which is 2.965. This pooled standard deviation is nothing but the square root of what we called MSW in this chapter. The MSW calculated in Example 12–4 was 8.79. The square root of 8.79 is 2.965, which is printed as the pooled standard deviation in the MINITAB solution.

From the MINITAB printout of Figure 12.6, the value of the test statistic F is 9.69. The critical value of F from the F distribution table for $\alpha = .05$, df for the numerator $= 3$, and df for the denominator $= 18$ is 3.16 (see Figure 12.4 of Example 12–4). The value of the test statistic $F = 9.69$ is larger than the critical value of $F = 3.16$ and it falls in the rejection region. Consequently, we reject the null hypothesis and conclude that the mean number of customers served per hour by each of the four tellers is not the same.

We can reach the same conclusion by considering the p-value. The p-value from the MINITAB solution is 0.000. Because this p-value is less than $\alpha = .05$, we reject the null hypothesis. ▬

COMPUTER ASSIGNMENTS

M12.1 Refer to Exercise 12.18. Solve that exercise using MINITAB. (*Note:* Because the number of data values in each of the three columns in Exercise 12.18 is the same, the READ command can be used to enter data into MINITAB.)

M12.2 Refer to Exercise 12.25. Solve that exercise using MINITAB.

MORE CHALLENGING EXERCISES (Optional)
CHAPTERS 10 TO 12

1. Does the use of cellular telephones increase the risk of brain tumors? Suppose a manufacturer of cellular telephones hires you to answer this question because of concern about product liability suits. How would you conduct an experiment to address this question? Be specific. Explain who you would observe, what you would observe, how many observations you would take, and how you would analyze the data once you collect it. What are your null and alternative hypotheses? Would you want to use a high or low alpha level for the test? Explain.

2. Do rock music CDs and country music CDs give the consumer the same amount of music listening time? A sample of 12 randomly selected single rock music CDs and a sample of 14 randomly selected single country music CDs have the following total lengths (in minutes).

Rock Music	Country Music
43.0	45.3
44.3	40.2
63.8	42.8
32.8	33.0
54.2	33.5
51.3	37.7
64.8	36.8
36.1	34.6
33.9	33.4
51.7	36.5
36.5	43.3
59.7	31.7
	44.0
	42.7

  a. Compute the value of the test statistic t for testing the null hypothesis that the mean lengths of the rock and country music single CDs are the same against the alternative hypothesis that these mean lengths are not the same. Use the value of this t statistic to compute the (approximate) p-value.

  b. Compute the value of the (one-way ANOVA) test statistic F for performing the test of equality of the mean lengths of the rock and country music single CDs and use it to find the (approximate) p-value.

  c. How do the test statistics in parts a and b compare? How do the p-values computed in parts a and b compare? Do you think this is a coincidence or will this always happen?

3. A student who needs to pass Elementary Business Statistics wonders if it makes a difference with which of two possible instructors she takes the class. Observing the final grades given by each instructor in a recent Elementary Business Statistics course, she finds that Instructor I gave 48 passing grades in a class of 52 students and Instructor II gave 44 passing grades in a class of 54 students.

  a. Compute the value of the standard normal test statistic z, of Section 10.4.3, for the data and use it to find the p-value when testing for a difference between the proportions of passing grades given by these instructors.

  b. Construct a 2×2 contingency table for these data. Compute the value of the χ^2 test statistic for the test of homogeneity and use it to find the p-value.

    c. How do the test statistics in parts *a* and *b* compare? How do the *p*-values for the tests in parts *a* and *b* compare? Do you think this is a coincidence or do you think this will always happen?

4. Economic growth used to be considered necessary for a reduction in birth rates in developing countries. The following chart reports the results of studies measuring the effectiveness of family planning programs in a number of developing countries. (*Source: The New York Times*, January 2, 1994. Copyright © 1994/The New York Times Company. Chart reproduced with permission.)

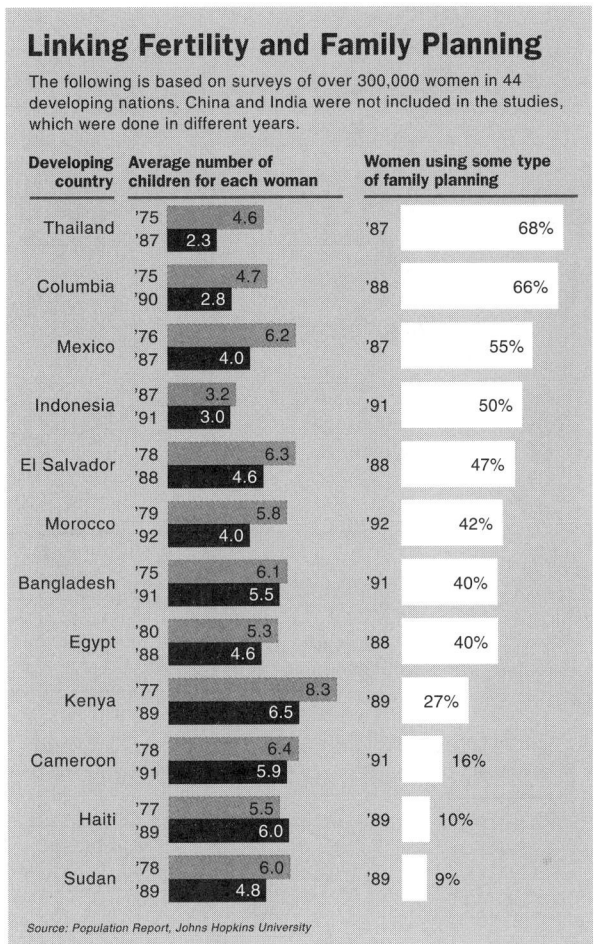

Linking Fertility and Family Planning

The following is based on surveys of over 300,000 women in 44 developing nations. China and India were not included in the studies, which were done in different years.

Developing country	Average number of children for each woman		Women using some type of family planning	
Thailand	'75	4.6	'87	68%
	'87	2.3		
Columbia	'75	4.7	'88	66%
	'90	2.8		
Mexico	'76	6.2	'87	55%
	'87	4.0		
Indonesia	'87	3.2	'91	50%
	'91	3.0		
El Salvador	'78	6.3	'88	47%
	'88	4.6		
Morocco	'79	5.8	'92	42%
	'92	4.0		
Bangladesh	'75	6.1	'91	40%
	'91	5.5		
Egypt	'80	5.3	'88	40%
	'88	4.6		
Kenya	'77	8.3	'89	27%
	'89	6.5		
Cameroon	'78	6.4	'91	16%
	'91	5.9		
Haiti	'77	5.5	'89	10%
	'89	6.0		
Sudan	'78	6.0	'89	9%
	'89	4.8		

Source: Population Report, Johns Hopkins University

The New York Times

Assume that the countries presented here were randomly selected from the 44 nations studied.

    a. Using methods you have learned, how would you test for evidence that modern contraceptive methods significantly reduce birth rates?

    b. Compute the *p*-value for the test in part *a* and use it to draw your own conclusion for this issue.

5. Jill Poultry Farm supplies four brands of eggs to the PSM Company. The eggs are shipped in cartons containing 144 eggs each. The company's quality control department inspects these eggs quite often for grade specifications. A recently taken sample of one carton of each brand produced 10, 3, 12, and 20 eggs, respectively, that failed to meet grade specifications. Would you conclude that the four brands are not comparable in meeting grade specifications? In other words, are the brands of eggs homogeneous with regard to the number of eggs that fail to meet grade specifications?

6. A dietician wanted to investigate if the mean weight loss for each of three diet plans is the same. She took a random sample of 12 persons who wanted to lose weight. Then she randomly assigned four of these persons to Diet A, four to Diet B, and four to Diet C. The following table gives the MINITAB output for a one-way ANOVA test based on the mean weight lost by persons on each diet plan at the end of a 3-week period. Find the missing entries.

```
ANALYSIS OF VARIANCE

SOURCE          DF          SS          MS          F          P
FACTOR        _____       _____       _____       .65        .545
ERROR         _____       _____       56.92
TOTAL         _____       _____

LEVEL           N          MEAN        STDEV

C1            _____       12.25        9.323
C2            _____       11.75        5.560
C3            _____        6.75        7.274

POOLED STDEV = _____
```

7. Case Study 3–1 considers the average annual compensation (including pay and benefits) for CEOs in the United States, Japan, and Germany. Suppose the averages given there for companies with revenues of $1 billion or more are based on the sample sizes listed in the following table, with SST equal to 52.88.

Country	Number of CEOs sampled	Average Compensation (millions of dollars)
United States	15	3.0
Japan	10	.6
Germany	8	1.1

Further assume that the averages given there for companies with revenues of approximately $250 million are based on sample sizes listed in the following table with SST equal to 8.05.

Country	Number of CEOs sampled	Average Compensation (millions of dollars)
United States	10	.748
Japan	12	.370
Germany	11	.3645

What conclusions would you make about the compensations for CEOs of the United States, Japan, and Germany?

13 SIMPLE LINEAR REGRESSION

This chapter considers the relationship between two variables in two ways: (1) by using the regression analysis and (2) by computing the correlation coefficient. By using the regression model, we can evaluate the magnitude of change in one variable due to a certain change in another variable. For example, an economist can estimate the amount of change in food expenditure due to a certain change in the income of a household by using the regression model. A company's research department can estimate the increase in gross sales due to a particular increase in its advertising expenditure by using the regression model. Besides answering these questions, a regression model also helps to predict the value of one variable for a given value of another variable. For example, by using the regression line, we can predict the (approximate) food expenditure of a household with a given income.

The correlation coefficient, on the other hand, simply tells us how strongly two variables are related. It does not provide any information about the size of change in one variable as a result of a certain change in the other variable. For example, the correlation coefficient tells us how strongly income and food expenditure or an advertising expenditure and gross sales are related.

13.1 THE SIMPLE LINEAR REGRESSION MODEL

Only simple linear regression will be discussed in this chapter.[1] In the next two subsections the meaning of the words *simple* and *linear* as used in *simple linear regression* is explained.

13.1.1 SIMPLE REGRESSION

Let us return to the example of an economist investigating the relationship between food expenditure and income. What factors or variables does a household consider when deciding how much money should be spent on food every week or every month? Certainly, income of the household is one factor. However, many other variables also affect food expenditure. For instance, the assets owned by the household, the size of the household, the preferences and tastes of household members, and any special dietary needs of household members are some of the variables that will influence a household's decision about food expenditure. These variables are called **independent** or **explanatory variables** because they all vary independently and they explain the variation in food expenditure among different households. In other words, these variables explain why different households spend different amounts of money on food. Food expenditure is called the **dependent variable** because it depends on the independent variables. Studying the effect of two or more independent variables on a dependent variable using regression analysis is called **multiple regression**. However, if we choose only one (usually the most important) independent variable and study the effect of that single variable on a dependent variable, it is called a **simple regression**. Thus, a simple regression includes only two variables: one independent and one dependent. Note that whether it is a simple or a multiple regression analysis, it always includes one and only one dependent variable. It is the number of independent variables that changes in simple and multiple regressions.

> **SIMPLE REGRESSION**
>
> A regression model is a mathematical equation that describes the relationship between two or more variables. A simple regression model includes only two variables: one independent and one dependent. The dependent variable is the one being explained and the independent variable is the one used to explain the variation in the dependent variable.

13.1.2 LINEAR REGRESSION

The relationship between two variables in a regression analysis is expressed by a mathematical equation called a **regression equation** or **model**. A regression equation, when plotted, may assume one of many possible shapes, including that of a straight line. A regression equation that gives a straight line relationship between two variables is called a **linear regression model**; otherwise, it is called a **nonlinear regression model**. In this chapter, only linear regression models are studied.

[1]The term *regression* was first used by Sir Francis Galton (1822–1911), who studied the relationship between the heights of children and the heights of their parents.

LINEAR REGRESSION

A (simple) regression model that gives a straight line relationship between two variables is called a linear regression model.

The two diagrams in Figure 13.1 show a linear and a nonlinear relationship between the dependent variable food expenditure and the independent variable income.

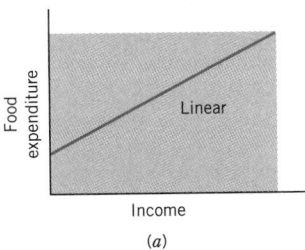

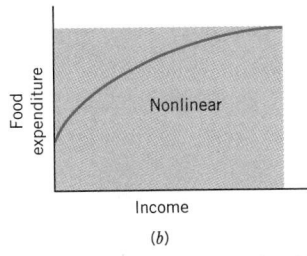

Figure 13.1 Relationship between food expenditure and income. (a) Linear relationship. (b) Nonlinear relationship.

A linear relationship between income and food expenditure, which is shown in Figure 13.1a, indicates that as income increases the food expenditure always increases at the same rate. However, a nonlinear relationship between income and food expenditure, as depicted in Figure 13.1b, shows that as income increases the food expenditure increases, although, after a point, the rate of increase in food expenditure is lower for every subsequent increase in income.

The **equation of a linear relationship** between two variables x and y is written as

$$y = a + bx$$

Each set of values of a and b gives a different straight line. For instance, when $a = 50$ and $b = 5$, then the above equation becomes

$$y = 50 + 5x$$

To plot a straight line, we need to know two points that lie on that line. We can find two points on a line by assigning any two values to x and then calculating the corresponding values of y. For the equation $y = 50 + 5x$,

1. When $x = 0$, then $y = 50 + 5(0) = 50$

2. When $x = 10$, then $y = 50 + 5(10) = 100$

These two points are plotted in Figure 13.2. By joining these two points we obtain the line representing the equation $y = 50 + 5x$.

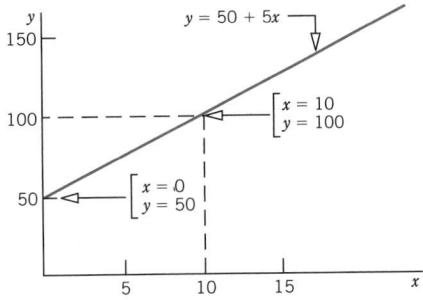

Figure 13.2

Note that in Figure 13.2 the line intersects the y (vertical) axis at 50. Consequently, 50 is called the **y-intercept**. The y-intercept is given by the constant term in the equation. It is the value of y when x is zero.

In the equation $y = 50 + 5x$, 5 is called the **coefficient of x** or the **slope** of the line. It gives the amount of change in y due to a change of one unit in x. For example,

$$\text{If } x = 10, \text{ then } \quad y = 50 + 5(10) = 100$$

$$\text{If } x = 11, \text{ then } \quad y = 50 + 5(11) = 105$$

Hence, as x increases by 1 unit (from 10 to 11), y increases by 5 units (from 100 to 105). This is true for any value of x. Such changes in x and y are shown in Figure 13.3.

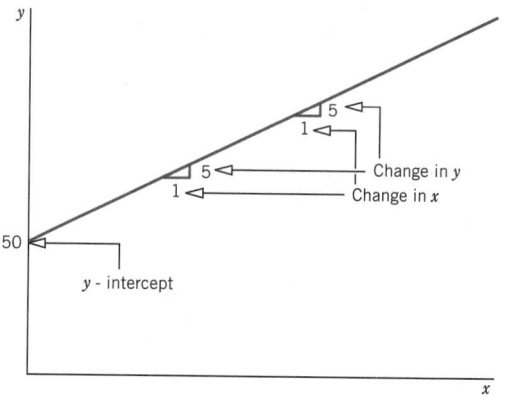

Figure 13.3

In general, when an equation is written in the form

$$y = a + bx$$

a gives the y-intercept and b represents the slope of the line. In other words, a represents the point where the line intersects the y-axis and b gives the amount of change in y due to a change of one unit in x. Note that b is also called the coefficient of x.

13.2   THE SIMPLE LINEAR REGRESSION ANALYSIS

In a regression model, the independent variable is usually denoted by x and the dependent variable is usually denoted by y. The x variable, with its coefficient, is written on the right side of the "$=$" sign, whereas the y variable is written on the left side of the "$=$" sign. The y-intercept and the slope, which we earlier denoted by a and b, respectively, can be represented by any of the many commonly used symbols. Let us denote the y-intercept (which is also called the *constant term*) by A, and the slope (or the coefficient of x variable) by B. Then, our simple linear regression model is written as

Constant term or y-intercept ———┐   ┌——Slope

$$y = A + Bx \tag{1}$$

Dependent variable   Independent variable

In model (1), A gives the value of y for $x = 0$, and B gives the change in y due to a change of one unit in x.

Model (1) is called a **deterministic model**. It gives an **exact relationship** between x and y. This model simply states that y is determined exactly by x and for a given value of x there is one and only one (unique) value of y.

However, in many cases the relationship between variables is not exact. For instance, if y is food expenditure and x is income, then model (1) would state that food expenditure is determined by income only and that all households with the same income will spend the same amount on food. But as mentioned earlier, food expenditure is determined by many variables, only one of which is included in model (1). In reality, different households with the same income spend different amounts of money on food because of the differences in the size of the household, the assets they own, and their preferences and tastes. Hence, to take these variables into consideration and to make our model complete, we add another term to the right side of model (1). This term is called the **random error term**. It is denoted by ϵ (Greek letter *epsilon*). The complete regression model is written as

$$y = A + Bx + \epsilon \qquad (2)$$

$$\uparrow$$

Random error term

The regression model (2) is called a **probabilistic model** (or a **statistical relationship.**)

EQUATION OF A REGRESSION MODEL

In the regression model $y = A + Bx + \epsilon$, A is called the y-intercept or constant term, B is the slope, and ϵ is the random error term. The dependent and independent variables are y and x, respectively.

The random error term ϵ is included in the model to represent the following two phenomena.

1. *Missing or omitted variables.* As mentioned earlier, food expenditure is affected by many variables other than income. The random error term ϵ is included to capture the effect of all those missing or omitted variables that have not been included in the model.

2. *Random variation.* Human behavior is unpredictable. For example, a household may have many parties during one month and may spend more than usual on food during that month. The same household may spend less than usual during another month because it spent quite a bit of money to buy furniture. The variation in food expenditure for such reasons may be called random variation.

In model (2), A and B are the **population parameters**. The regression line obtained for model (2) by using the population data is called the **population regression line**. The values of A and B in the population regression line are called the **true values of the y-intercept and slope**.

However, population data are difficult to obtain. As a result, we almost always use sample data to estimate model (2). The values of the y-intercept and slope calculated from sample data on x and y are called the **estimated values of A and B and are denoted by a and b**. Using a and b we write the estimated model as

$$\hat{y} = a + bx \qquad (3)$$

where $\hat{y}$ (read as *y hat*) is the **estimated or predicted value of** y for a given value of x. Equation (3) is called the **estimated model**; it gives the **regression of** y **on** x.

ESTIMATES OF *A* AND *B*

In the model $\hat{y} = a + bx$, a and b, which are calculated using sample data, are called the estimates of A and B.

13.2.1 SCATTER DIAGRAM

Suppose we take a sample of seven households and collect information on their incomes and food expenditures for the past month. The information obtained (in hundreds of dollars) is given in Table 13.1.

Table 13.1 Incomes and Food Expenditures of Seven Households

Income (hundreds of dollars)	Food Expenditure (hundreds of dollars)
35	9
49	15
21	7
39	11
15	5
28	8
25	9

In Table 13.1, we have a pair of observations for each of the seven households. Each pair consists of one observation on income and a second on food expenditure. For example, the first household's income for the past month was \$3500 and its food expenditure was \$900. By plotting all seven pairs of values, we obtain a **scatter diagram** or **scattergram**. Figure 13.4 gives the scatter diagram for the data of Table 13.1. Each dot in this diagram represents one household. A scatter diagram is helpful in detecting a relationship between two variables. For example, by looking at the scatter diagram of Figure 13.4, we can observe that there exists a strong linear relationship between food expenditure and income. If a straight line is drawn through the points, the points will be scattered closely around the line.

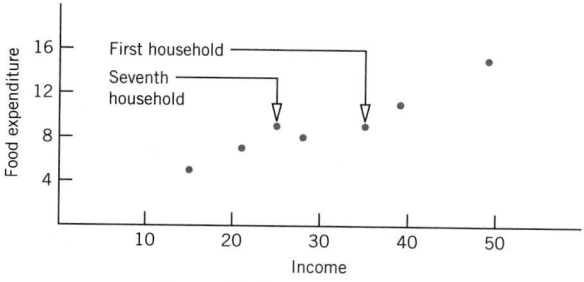

Figure 13.4 Scatter diagram.

Emphasize the importance of scatter diagrams.

> **SCATTER DIAGRAM**
>
> A plot of paired observations is called a scatter diagram.

As shown in Figure 13.5, a large number of straight lines can be drawn through the scatter diagram of Figure 13.4. Each of these lines will give different values for a and b of model (3).

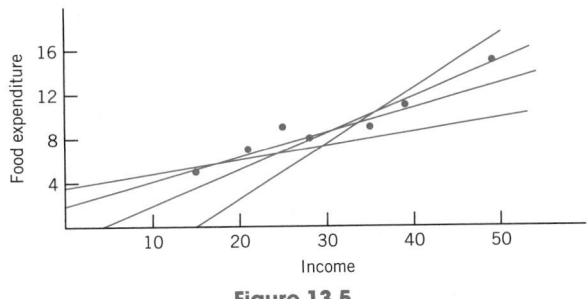

Figure 13.5

In regression analysis, we try to find a line that best fits the points in the scatter diagram. Such a line provides the best possible description of the relationship between the dependent and independent variables. The **least squares method**, discussed in the next section, gives such a line. The line obtained by using the least squares method is called the **least squares regression line**.

13.2.2 LEAST SQUARES LINE

The value of y obtained for a member from the survey is called the **observed or actual value of y**. As mentioned in Section 13.2, the value of y, denoted by $\hat{y}$, obtained for a given x by using the regression line is called the **predicted value of y**. The random error ϵ denotes the difference between the actual value of y and the predicted value of y for population data. For example, for a given household, ϵ is the difference between what this household actually spent on food during the past month and what is predicted using the population regression line. The ϵ is also called the *residual*, as it measures the surplus (positive or negative) of actual food expenditure over what is predicted by using the regression model. If we estimate model (2) by using sample data, the difference between the actual y and predicted y based on this estimation cannot be denoted by ϵ. *The random error for the sample regression model is denoted by e.* Thus, e is an estimator of ϵ. If we estimate model (2) using sample data, then the value of e is given by

$$e = \text{Actual food expenditure} - \text{Predicted food expenditure} = y - \hat{y}$$

In Figure 13.6, e is the vertical distance between the actual position of a household and the point on the regression line. Note that in such a diagram, we always measure the dependent variable on the vertical axis and independent variable on the horizontal axis.

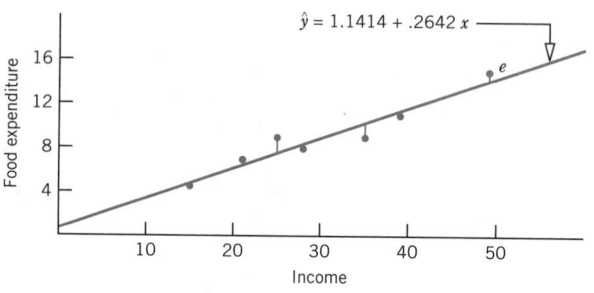

Figure 13.6

The value of an error is positive if the point that gives the actual food expenditure is above the regression line and negative if it is below the regression line. *The sum of these errors is always zero.* In other words, the sum of the actual food expenditures for seven households included in the sample will be the same as the sum of the food expenditures predicted from the regression model. Thus,

$$\Sigma e = \Sigma(y - \hat{y}) = 0$$

Hence, to find the line that best fits the scatter of points, we cannot minimize the sum of errors. Instead, we minimize the **error sum of squares**, denoted by **SSE**, which is obtained by adding the squares of errors. Thus,

$$\text{SSE} = \Sigma e^2 = \Sigma(y - \hat{y})^2$$

The least squares method gives the values of a and b for model (3) such that the sum of squared errors (SSE) is minimum.

ERROR SUM OF SQUARES (SSE)

The error sum of squares, denoted by SSE, is

$$\text{SSE} = \Sigma e^2 = \Sigma(y - \hat{y})^2$$

The values of a and b which give the minimum SSE are called the **least squares estimates** of A and B and the regression line obtained with these estimates is called the least squares line.

The least squares values of a and b are computed using the following formulas.

THE LEAST SQUARES LINE

For the least squares regression line $\hat{y} = a + bx$

$$b = \frac{\text{SS}_{xy}}{\text{SS}_{xx}} \quad \text{and} \quad a = \bar{y} - b\bar{x}$$

where $\text{SS}_{xy} = \Sigma xy - \dfrac{(\Sigma x)(\Sigma y)}{n}$ and $\text{SS}_{xx} = \Sigma x^2 - \dfrac{(\Sigma x)^2}{n}$

and "SS" stands for "sum of squares."[2]
The least squares regression line $\hat{y} = a + bx$ is also called the regression of y on x.

---

[2]The values of SS_{xy} and SS_{xx} can also be obtained by using the following basic formulas.

$$\text{SS}_{xy} = \Sigma(x - \bar{x})(y - \bar{y}) \quad \text{and} \quad \text{SS}_{xx} = \Sigma(x - \bar{x})^2$$

However, these formulas usually take longer to make calculations.

The formulas given above are for estimating a sample regression line. Suppose we have access to a population data set. We can find the population regression line by using the same formulas with a little adaptation. For a population data set, in the above formulas we replace a by A, b by B, n by N, and use the values of Σx, Σy, Σxy, and Σx^2 calculated for population data. The population regression line is written as

$$\mu_{y|x} = A + Bx$$

where $\mu_{y|x}$ is read as *the mean value of y for a given x*. When plotted on a graph, the points on this population regression line will give the average values of y for the corresponding values of x. These average values of y are denoted by $\mu_{y|x}$.

Example 13–1 illustrates how to estimate a regression line for sample data.

Estimating the least squares regression line.

EXAMPLE 13–1   Find the least squares regression line for the data on incomes and food expenditures of seven households given in Table 13.1. Use income as an independent variable and food expenditure as a dependent variable.

Solution   We are to find the values of a and b for the regression model $\hat{y} = a + bx$. Table 13.2 shows the calculations required for the computation of a and b. We denote the independent variable (income) by x and the dependent variable (food expenditure) by y.

Table 13.2

Income	Food Expenditure		
x	y	xy	x^2
35	9	315	1225
49	15	735	2401
21	7	147	441
39	11	429	1521
15	5	75	225
28	8	224	784
25	9	225	625
$\Sigma x = 212$	$\Sigma y = 64$	$\Sigma xy = 2150$	$\Sigma x^2 = 7222$

The following steps are performed to compute a and b.

Step 1.   Compute Σx, Σy, $\bar{x}$, and $\bar{y}$.

$$\Sigma x = 212, \quad \Sigma y = 64$$

$$\bar{x} = \Sigma x/n = 212/7 = 30.2857$$

$$\bar{y} = \Sigma y/n = 64/7 = 9.1429$$

Step 2.   Compute Σxy and Σx^2.

To calculate Σxy, we multiply the corresponding values of x and y. Then, we sum all the products. The products of x and y are recorded in the third column of Table 13.2. To compute Σx^2, we square each of the x values and then add them. The squared values of x are listed in the fourth column of Table 13.2. From these calculations,

$$\Sigma xy = 2150 \quad \text{and} \quad \Sigma x^2 = 7222$$

Step 3.   Compute SS_{xy} and SS_{xx}.

$$SS_{xy} = \Sigma xy - \frac{(\Sigma x)(\Sigma y)}{n} = 2150 - \frac{(212)(64)}{7} = 211.7143$$

$$SS_{xx} = \Sigma x^2 - \frac{(\Sigma x)^2}{n} = 7222 - \frac{(212)^2}{7} = 801.4286$$

Step 4. Compute a and b.

$$b = \frac{SS_{xy}}{SS_{xx}} = \frac{211.7143}{801.4286} = .2642$$

$$a = \bar{y} - b\bar{x} = 9.1429 - (.2642)(30.2857) = 1.1414$$

Thus, our estimated regression model $\hat{y} = a + bx$ is

$$\hat{y} = 1.1414 + .2642 \, x$$

This regression line is called the least squares regression line. It gives the *regression of food expenditure on income.*

Note that we have rounded all calculations to four decimal places. We can round the values of a and b in the regression line to two decimal places, but it is not done here because we will use this regression line for prediction and estimation purposes later on. ■

Using this estimated model, we can find the predicted value of y for a specific value of x. For instance, suppose we randomly select a household whose monthly income is \$3500 so that $x = 35$ (recall that x denotes income in hundreds of dollars). The predicted value of food expenditure for this household is

$$\hat{y} = 1.1414 + (.2642)(35) = \$10.3884 \text{ hundred}$$

In other words, based on our regression line, we predict that a household with a monthly income of \$3500 is expected to spend \$1038.84 per month on food. This value of $\hat{y}$ can also be interpreted as the mean value of y for $x = 35$. Thus, we can state that, on average, all households with a monthly income of \$3500 spend \$1038.84 per month on food.

In our data on seven households, there is one household whose income is \$3500. The actual food expenditure for that household is \$900 (see Table 13.1). The difference between the actual and predicted values gives the error of prediction. Thus, the error of prediction for this household, which is shown in Figure 13.7, is

$$e = y - \hat{y} = 9.00 - 10.3884 = -\$1.3884 \text{ hundreds}$$

Therefore, the error of prediction is $-\$138.84$. The negative error indicates that the predicted value of y is greater than the actual value of y. Thus, if we use the regression model, this household's food expenditure is overestimated by \$138.84.

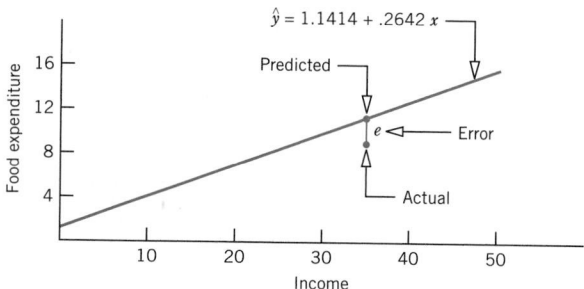

Figure 13.7 Errors of prediction.

13.2.3 INTERPRETATION OF *a* AND *b*

How do we interpret $a = 1.1414$ and $b = .2642$ obtained in Example 13–1 for the regression of food expenditure on income? A brief explanation of the y-intercept and slope of a regression line was given in Section 13.1.2. The next two parts of this subsection explain the meaning of *a* and *b* in more detail.

Interpretation of *a*

Consider a household with zero income. Using the estimated regression line obtained in Example 13–1, the predicted value of *y* for $x = 0$ is

$$\hat{y} = 1.1414 + .2642\,(0) = \$1.1414 \text{ hundred}$$

Thus, we can state that a household with no income is expected to spend $114.14 per month on food. Alternatively, we can also state that the average monthly food expenditure for all households with zero income is $114.14. Note that here we have used $\hat{y}$ as a point estimate of $\mu_{y|x}$. Thus, $a = 1.1414$ gives the predicted or mean value of *y* for $x = 0$ based on the regression model estimated for the sample data.

However, we should be very careful while making this interpretation of *a*. In our sample of seven households, the incomes vary from a minimum of $1500 to a maximum of $4900. (Note that in Table 13.1, the minimum value of *x* is 15 and the maximum value is 49.) Hence, our regression line is valid only for the values of *x* between 15 and 49. If we predict *y* for a value of *x* outside this range, the prediction usually will not hold true. Thus, since $x = 0$ is outside the range of household incomes that we have in the sample data, the prediction that a household with zero income spends $114.14 per month on food does not carry much credibility. The same is true if we try to predict *y* for an income greater than $4900, which is the maximum value of *x* in Table 13.1.

Interpretation of *b*

The value of *b* in a regression model gives the change in *y* (dependent variable) due to a change of one unit in *x* (independent variable). For example, by using the regression line obtained in Example 13–1,

$$\text{when } x = 30, \quad \hat{y} = 1.1414 + .2642\,(30) = 9.0674$$
$$\text{when } x = 31, \quad \hat{y} = 1.1414 + .2642\,(31) = 9.3316$$

Hence, when *x* increased by one unit, from 30 to 31, $\hat{y}$ increased by $9.3316 - 9.0674 = .2642$, which is the value of *b*. Because our unit of measurement is hundreds of dollars, we can state that, on average, a $100 increase in income will cause a $26.42 increase in food expenditure. We can also state that, on average, a $1 increase in income of a household will increase the food expenditure by $.2642. Note the phrase "on average" in these statements. The regression line is seen as a measure of the mean value of *y* for a given value of *x*. If one household's income is increased by $100, that household's food expenditure may or may not increase by $26.42. But if the incomes of all households are increased by $100 each, the average increase in their food expenditures will be very close to $26.42.

Note that when *b* is positive, an increase in *x* will lead to an increase in *y* and a decrease in *x* will lead to a decrease in *y*. In other words, when *b* is positive, the movements in *x* and *y* are in the same direction. Such a relationship between *x* and *y* is called a **positive linear relationship**. The regression line in this case slopes upward from left to right. On the other hand, if the value of *b* is negative, an increase in *x* will cause a decrease in *y* and a decrease in *x* will cause an increase in *y*. The changes in *x* and *y* in this case are in opposite directions. Such a relationship between *x* and *y* is called a **negative linear relationship**. The regression line in this case slopes downward from left to right. The two diagrams in Figure 13.8 show these two cases.

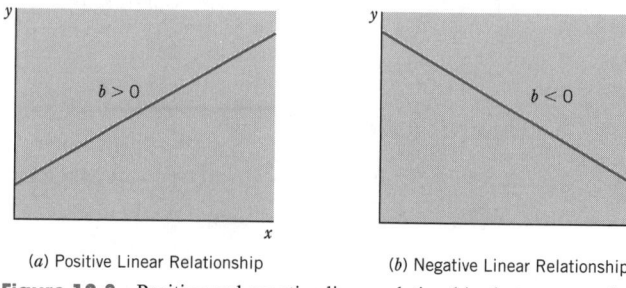

(a) Positive Linear Relationship   (b) Negative Linear Relationship

Figure 13.8   Positive and negative linear relationships between x and y.

☞ *Remember*   For a regression model, b is computed as $b = SS_{xy}/SS_{xx}$. The value of SS_{xx} is always positive and that of SS_{xy} can be positive or negative. Hence, the sign of b depends on the sign of SS_{xy}. If SS_{xy} is positive (as in our example on incomes and food expenditures of seven households) then b will be positive, and if SS_{xy} is negative then b will be negative.

13.2.4   ASSUMPTIONS OF THE REGRESSION MODEL

Like any other theory, the linear regression analysis is also based on certain assumptions. Consider the population regression model

$$y = A + Bx + \epsilon \tag{4}$$

There are four assumptions made about this model, which are described next. These assumptions are explained with reference to the example regarding incomes and food expenditures of households. Note that these assumptions are made about the population regression model and not about the sample regression model.

Assumption 1:   The random error term ϵ has a mean equal to zero for each x. In other words, among all households with the same income, some spend more than the predicted food expenditure (and, hence, have positive errors) and others spend less than the predicted food expenditure (and, consequently, have negative errors). This assumption simply states that the sum of the positive errors is equal to the sum of the negative errors so that the mean of errors for all households with the same income is zero. Thus, when the mean value of ϵ is zero, the mean value of y for a given x is equal to $A + Bx$ and it is written as

$$\mu_{y|x} = A + Bx$$

As mentioned earlier in this chapter, $\mu_{y|x}$ is read as *the mean value of y for a given value of x*. When we find the values of A and B for model (4) using the population data, the points on the regression line give the average values of y, denoted by $\mu_{y|x}$, for the corresponding values of x for the population.

Assumption 2:   The errors associated with different observations are independent. According to this assumption, the errors for any two households in our example are independent. In other words, all households decide independently how much to spend on food.

Assumption 3:   For any given x, the distribution of errors is normal. The corollary of this assumption is that the food expenditures for all households with the same income are normally distributed.

Assumption 4: The distribution of population errors for each x has the same (constant) standard deviation, which is denoted by σ_ϵ. This assumption indicates that the spread of points around the regression line is similar for all x values.

Figure 13.9 illustrates the meaning of the first, third, and fourth assumptions for households with incomes of $2000 and $3500 per month. The same assumptions hold true for any other income level. In the population of all households, there will be many households with a monthly income of $2000. Using the population regression line, if we calculate the errors for all these households and prepare the distribution of these errors, it will look like the distribution given in Figure 13.9a. Its standard deviation will be σ_ϵ. Similarly, Figure 13.9b gives the distribution of errors for all those households in the population whose monthly income is $3500. Its standard deviation is also σ_ϵ. Both these distributions are identical. Note that the mean of both of these distributions is $E(\epsilon) = 0$.

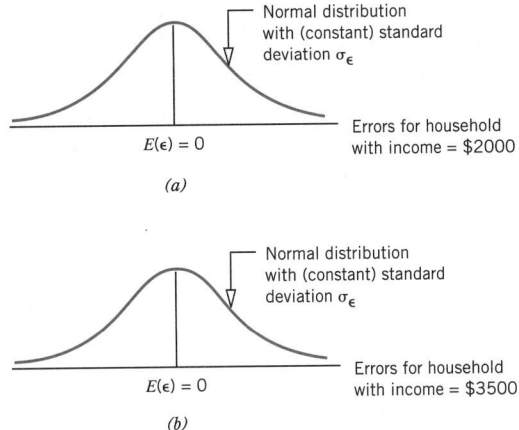

Figure 13.9 (a) Errors for households with an income of $2000 per month.
(b) Errors for households with an income of $3500 per month.

Figure 13.10 shows how these distributions look when they are imposed on the same diagram with the population regression line. The points on the vertical line through $x = 20$ give the food expenditures for various households in the population, each of whom has the same monthly income of $2000. The same is true about the vertical line through $x = 35$ or any other vertical line for some other value of x.

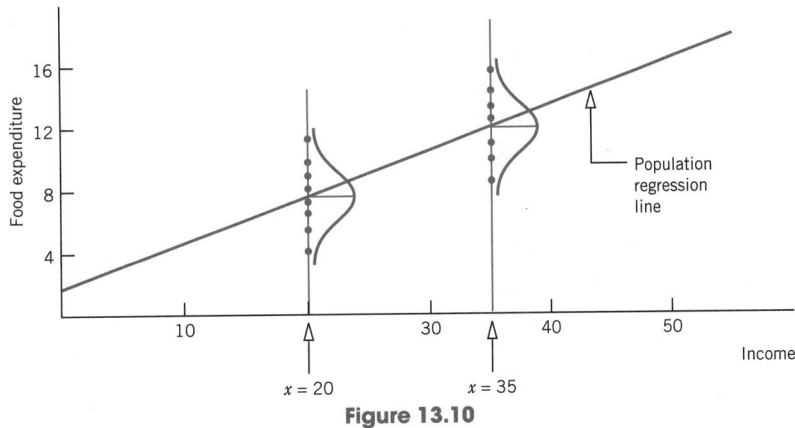

Figure 13.10

13.2.5   A NOTE ON THE USE OF SIMPLE LINEAR REGRESSION

We should apply linear regression with caution. When we use simple linear regression, we assume that the relationship between two variables is described by a straight line. In the real world, the relationship between variables may not be linear. Hence, before we use a simple linear regression, it is better to construct a scatter diagram and look at the plot of the data points. We should estimate a linear regression model only if the scatter diagram indicates such a relationship. The scatter diagrams of Figure 13.11 give two examples where the relationship between x and y is not linear. Consequently, fitting linear regression in such cases would be wrong.

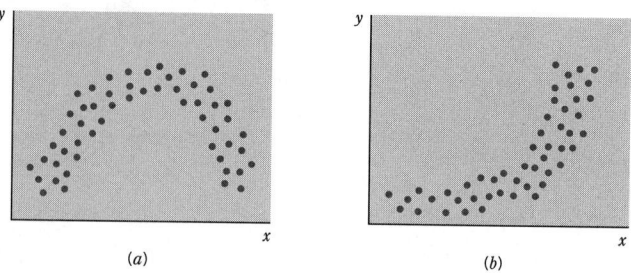

(a)   (b)

Figure 13.11   Nonlinear relationship between x and y.

EXERCISES

Concepts and Procedures

13.1   Explain the meaning of the words *simple* and *linear* as used in *simple linear regression.*

13.2   Explain the meaning of independent and dependent variables for a regression model.

13.3   Explain the difference between exact and nonexact relationships between two variables.

13.4   Explain the difference between linear and nonlinear relationships between two variables.

13.5   Explain the difference between a simple and a multiple regression model.

13.6   Briefly explain the difference between a deterministic and a probabilistic regression model.

13.7   Why is the random error term included in a regression model?

13.8   Explain the least squares method and least squares regression line. Why are they called by these names?

13.9   Explain the meaning and concept of SSE. You may use a graph for illustration purposes.

13.10   Explain the difference between y and $\hat{y}$.

13.11   Two variables x and y have a positive linear relationship. Explain what happens to the value of y when x increases.

13.12   Two variables x and y have a negative linear relationship. Explain what happens to the value of y when x increases.

13.13   Explain the following.

    a.   Population regression line   **b.**   Sample regression line
    c.   True values of A and B
    d.   Estimated values of A and B that are denoted by a and b, respectively

13.14   Briefly explain the assumptions of the population regression model.

13.15   Plot the following straight lines. Give the values of the y-intercept and slope for each of these lines and interpret them. Indicate whether each of the lines gives a positive or a negative relationship between x and y.

a. $y = 100 + 5x$ b. $y = 400 - 4x$

13.16 Plot the following straight lines. Give the values of the y-intercept and slope for each of these lines and interpret them. Indicate whether each of the lines gives a positive or a negative relationship between x and y.

a. $y = -60 + 8x$ b. $y = 300 - 6x$

13.17 A population data set produced the following information.

$$N = 250, \quad \Sigma x = 9880, \quad \Sigma y = 1456, \quad \Sigma xy = 85,080, \quad \Sigma x^2 = 485,870$$

Find the population regression line.

13.18 A population data set produced the following information.

$$N = 460, \quad \Sigma x = 3920, \quad \Sigma y = 2650, \quad \Sigma xy = 26,570, \quad \Sigma x^2 = 48,530$$

Find the population regression line.

13.19 The following information is obtained from a sample data set.

$$n = 10, \quad \Sigma x = 100, \quad \Sigma y = 220, \quad \Sigma xy = 3680, \quad \Sigma x^2 = 1140$$

Find the estimated regression line.

13.20 The following information is obtained from a sample data set.

$$n = 12, \quad \Sigma x = 66, \quad \Sigma y = 588, \quad \Sigma xy = 2244, \quad \Sigma x^2 = 396$$

Find the estimated regression line.

Applications

13.21 A car rental company charges $30 a day and 10 cents per mile for renting a car. Let y be the total rental charges (in dollars) for a car for one day and x be the miles driven. The equation for the relationship between x and y is

$$y = 30 + .10x$$

a. How much will a person pay who rents a car for one day and drives it 100 miles?
b. Suppose each of 20 persons rents a car from this agency for one day and drives it 100 miles. Will each of them pay the same amount for renting a car for a day or is each person expected to pay a different amount? Explain.
c. Is the relationship between x and y exact or nonexact?

13.22 Ben is an electrician who makes house calls for electrical repairs. He charges $25 to go to a house plus $22 per hour. Let y be the total amount (in dollars) paid by a household who uses Ben's services and x be the number of hours Ben spends doing repairs in that household's home. The equation for the relationship between x and y is

$$y = 25 + 22x$$

a. Ben spent six hours doing repairs in Kristine's home. How much will he be paid?
b. Suppose seven persons called Ben for repairs during a week. Surprisingly, each of these jobs took six hours. Would each of these home owners pay the same amount for repairs or do you expect each to pay a different amount? Explain.
c. Is the relationship between x and y exact or nonexact?

13.23 A researcher took a sample of 25 electronic companies and found the following relationship between x and y where x is the amount of money (in millions of dollars) spent on advertising by a company in 1993 and y represents the total gross sales (in millions of dollars) of that company for 1993.

$$\hat{y} = 3.4 + 11.55x$$

a. An electronic company spent $2 million on advertising in 1993. What are its expected gross sales for 1993?
b. Suppose four electronic companies spent $2 million each on advertising in 1993. Do you expect these four companies to have the same actual gross sales for 1993? Explain.
c. Is the relationship between x and y exact or nonexact?

13.24   A researcher took a sample of 10 years and found the following relationship between x and y where x is the number of major natural calamities (such as tornadoes, hurricanes, earthquakes, floods, etc.) that occurred during a year and y represents the average total profits (in millions of dollars) of all insurance companies in the United States.

$$\hat{y} = 212.6 - 1.80x$$

    a. A randomly selected year had 24 major calamities. What are the expected average profits of all U.S. insurance companies for that year?

    b. Suppose the number of major calamities was the same for each of three years. Do you expect the average profits for all U.S. insurance companies to be the same for these three years? Explain.

    c. Is the relationship between x and y exact or nonexact?

13.25   An economist wanted to determine whether or not the amount of phone bills and incomes of households are related. The following table gives information on the monthly incomes (in hundreds of dollars) and monthly telephone bills (in dollars) for a random sample of 10 households.

Income	16	45	36	32	30	13	41	15	36	40
Phone bill	35	78	102	56	75	26	130	42	59	85

    a. Find the regression line with income as an independent variable and the amount of the phone bill as a dependent variable.

    b. Give a brief interpretation of the values of a and b calculated in part a.

    c. Estimate the amount of the monthly phone bill for a household with a monthly income of $2500.

13.26   An auto manufacturing company wanted to investigate how the price of one of its car models depreciates with age. The research department at the company took a sample of eight cars of this model and collected the following information on the ages (in years) and prices (in hundreds of dollars) of these cars.

Age	8	3	6	9	2	5	6	3
Price	16	74	38	19	102	36	33	69

    a. Construct a scatter diagram for these data. Does the scatter diagram exhibit a linear relationship between ages and prices of cars?

    b. Find the regression line with price as a dependent variable and age as an independent variable.

    c. Give a brief interpretation of the values of a and b calculated in part b.

    d. Plot the regression line on the scatter diagram of part a and show the errors by drawing vertical lines between scatter points and the regression line.

    e. Predict the price of a 7-year-old car of this model.

    f. Estimate the price of an 18-year-old car of this model. Comment on this finding.

13.27   The management at Picaso Electronics wants to investigate the relationship between the years of experience and the number of CD players assembled by its employees working in the assembly department. The management took a sample of seven employees from the assembly department and observed them for a week. The following table gives data on the years of experience for these employees and the average number of CD players each of them assembled per day.

Experience	5	11	15	7	2	10	9
CD players assembled	14	21	20	18	13	16	18

    a. Construct a scatter diagram for these data. Does the scatter diagram exhibit a linear relationship between experience and the number of units assembled?

b. Find the regression line with experience as an independent variable and units assembled as a dependent variable.
c. Give a brief interpretation of the values of a and b calculated in part b.
d. Plot the regression line on the scatter diagram of part a and show the errors by drawing vertical lines between scatter points and the regression line.
e. Predict the number of CD players assembled by an employee with 12 years of experience.
f. Estimate the average number of CD players assembled per day by a worker with 25 years of experience. Comment on this finding.

13.28 A consumer welfare agency wants to investigate the relationship between the sizes of houses and rents paid by tenants in a small city. The agency collected the following information on the sizes (in hundreds of square feet) of houses and the monthly rents (in dollars) paid by tenants for six houses.

Size of the house	21	16	19	27	34	23
Monthly rent	700	580	720	850	1050	800

a. Construct a scatter diagram for these data. Does the scatter diagram show a linear relationship between the sizes of houses and monthly rents?
b. Find the regression line $\hat{y} = a + bx$ with the size of a house as an independent variable and monthly rent as a dependent variable.
c. Give a brief interpretation of the values of a and b calculated in part b.
d. Plot the regression line on the scatter diagram of part a and show the errors by drawing vertical lines between the scatter points and the regression line.
e. Predict the monthly rent for a house with 2500 square feet.
f. One of the houses in our sample is 2700 square feet and its rent is $850. What is the predicted rent for this house? Find the error for this observation.

13.29 The following table gives the total payroll (rounded to millions of dollars) as of March 1989 and the percentage of games won during the 1988 season by each of the National League baseball teams.

Team	Total Payroll	Percentage of Games Won
Atlanta Braves	9	34
Chicago Cubs	12	48
Cincinnati Reds	11	54
Houston Astros	16	51
Los Angeles Dodgers	22	58
Montreal Expos	12	50
New York Mets	20	63
Philadelphia Phillies	10	40
Pittsburgh Pirates	12	53
St. Louis Cardinals	15	47
San Diego Padres	14	52
San Francisco Giants	14	51

a. Find the least squares regression line with total payroll as an independent variable and percentage of games won as a dependent variable.
b. Is the regression line obtained in part a the population regression line? Why or why not? Do the values of the y-intercept and the slope in the regression line give A and B or a and b?
c. Give a brief interpretation of the values of the y-intercept and the slope.
d. Predict the percentage of games won for a team with a total payroll of $13.40 million.

13.30 The following table gives the percentage of games won and the average attendance (rounded to nearest thousand) per home game for the 1988 season for each of the American League baseball teams.

Team	Percentage of Games Won	Average Attendance per Home Game (in thousands)
Baltimore Orioles	34	22
Boston Red Sox	55	31
California Angels	46	29
Chicago White Sox	44	14
Cleveland Indians	48	18
Detroit Tigers	54	26
Kansas City Royals	52	29
Milwaukee Brewers	54	24
Minnesota Twins	56	37
New York Yankees	53	34
Oakland Athletics	64	29
Seattle Mariners	42	13
Texas Rangers	44	20
Toronto Blue Jays	54	32

a. Find the least squares regression line with percentage of games won as an independent variable and average attendance as a dependent variable.
b. Is the regression line obtained in part a the population regression line? Why or why not? Do the values of the y-intercept and the slope give A and B or a and b?
c. Give a brief interpretation of the values of the y-intercept and the slope.
d. Predict the average attendance per home game for a team with 49.6 percent of games won.

13.3 THE STANDARD DEVIATION OF RANDOM ERRORS

When we consider income and food expenditures, all households with the same income are expected to spend different amounts on food. Consequently, the random error ϵ will assume different values for these households. The standard deviation σ_ϵ measures the spread of these errors around the population regression line. The **standard deviation of errors** tells us how widely the errors and, hence, the values of y are spread for a given x. In Figure 13.10, which is reproduced below as Figure 13.12, the points on the vertical line through $x = 20$ give the

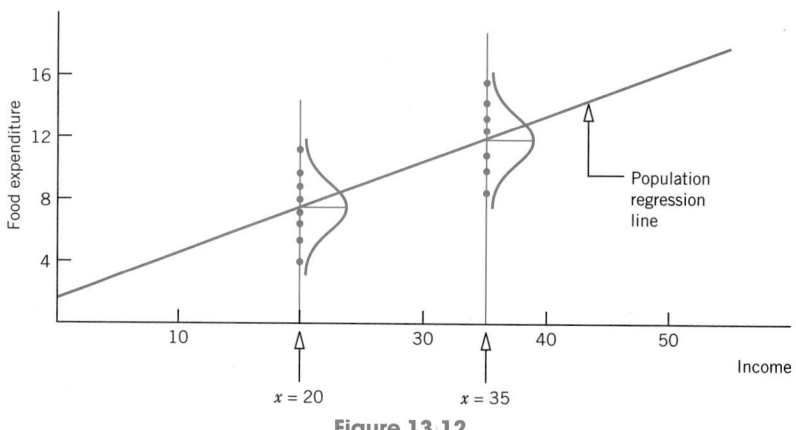

Figure 13.12

monthly food expenditures for all households with a monthly income of $2000. The distance of each dot from the point on the regression line gives the value of the corresponding error. The standard deviation of errors σ_ϵ measures the spread of such points around the population regression line. The same is true for $x = 35$ or any other value of x.

Note that σ_ϵ denotes the standard deviation of errors for the population. However, usually σ_ϵ is unknown. In such cases, it is estimated by s_e, which is the standard deviation of errors for the sample data. The following is the basic formula to calculate the standard deviation s_e.

$$s_e = \sqrt{\frac{SSE}{n-2}} \quad \text{where} \quad SSE = \Sigma(y - \hat{y})^2$$

In the above formula, $n - 2$ represents the **degrees of freedom** for the regression model. The reason that $df = n - 2$ is that we lose one degree of freedom to calculate $\bar{x}$ and one for $\bar{y}$.

DEGREES OF FREEDOM FOR A SIMPLE LINEAR REGRESSION MODEL

The degrees of freedom for a simple linear regression model are

$$df = n - 2$$

For computational purposes, it is more convenient to use the following formula to calculate the standard deviation of errors s_e.

STANDARD DEVIATION OF ERRORS

The standard deviation s_e of errors is calculated as

$$s_e = \sqrt{\frac{SS_{yy} - b\,SS_{xy}}{n-2}}$$

where

$$SS_{yy} = \Sigma y^2 - \frac{(\Sigma y)^2}{n}$$

The calculation of SS_{xy} was discussed earlier in this chapter.[3]

Like the value of SS_{xx}, the value of SS_{yy} is always positive.

Example 13–2 illustrates the calculation of the standard deviation of errors for the data of Table 13.1.

Calculating the standard deviation of errors.

EXAMPLE 13–2 Compute the standard deviation of errors s_e for the data on monthly incomes and food expenditures of seven households given in Table 13.1.

Solution To compute s_e, we need to know the values of SS_{yy}, SS_{xy}, and b. Earlier in Example 13–1 on page 665, we computed SS_{xy} and b. These values are

$$SS_{xy} = 211.7143 \quad \text{and} \quad b = .2642$$

To compute SS_{yy}, we calculate Σy^2 as in Table 13.3.

---

[3]The basic formula to calculate SS_{yy} is $SS_{yy} = \Sigma(y - \bar{y})^2$.

Table 13.3

Income	Food Expenditure	
x	y	y^2
35	9	81
49	15	225
21	7	49
39	11	121
15	5	25
28	8	64
25	9	81
$\Sigma x = 212$	$\Sigma y = 64$	$\Sigma y^2 = 646$

The value of SS_{yy} is

$$SS_{yy} = \Sigma y^2 - \frac{(\Sigma y)^2}{n} = 646 - \frac{(64)^2}{7} = 60.8571$$

Hence, the standard deviation of errors s_e is

$$s_e = \sqrt{\frac{SS_{yy} - b\, SS_{xy}}{n - 2}} = \sqrt{\frac{60.8571 - .2642\,(211.7143)}{7 - 2}} = .9922$$

13.4 THE COEFFICIENT OF DETERMINATION

We may ask the question, "How good is the regression model?" In other words, "How well does the independent variable explain the dependent variable in the regression model?" The *coefficient of determination* is one concept that answers this question.

For a moment, assume that we possess information only on food expenditures of households and not on their incomes. Hence, in this case, we cannot use the regression line to predict the food expenditure for any household. As we did in earlier chapters, in the absence of a regression model, we use $\bar{y}$ to estimate or predict every household's food expenditure. Consequently, the error of prediction for each household is now given by $y - \bar{y}$, which is the difference between the actual food expenditure of a household and the mean food expenditure. If we calculate such errors for all households, then square and add them, the resulting sum is called the **total sum of squares** and is denoted by **SST**. Actually SST is the same as SS_{yy} and is defined as

$$SST = SS_{yy} = \Sigma(y - \bar{y})^2$$

However, for computational purposes, SST is calculated using the following formula.

TOTAL SUM OF SQUARES (SST)

The total sum of squares, denoted by SST, is

$$SST = \Sigma y^2 - \frac{(\Sigma y)^2}{n}$$

Note that this is the same formula that we used to calculate SS_{yy}.

The value of SS$_{yy}$, which is 60.8571, was calculated in Example 13–2. Consequently, the value of SST is

$$SST = 60.8571$$

From Example 13–1, $\bar{y} = 9.1429$. Figure 13.13 shows the total errors for each of the seven households in our sample.

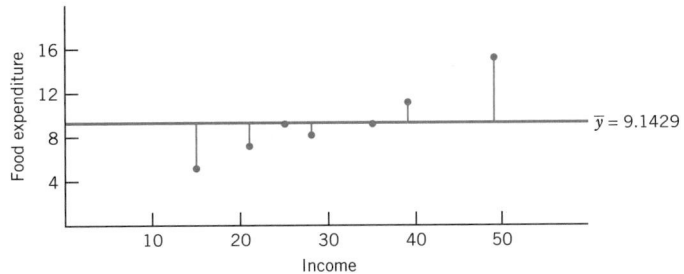

Figure 13.13 Total errors.

Now suppose we use the simple linear regression model to predict the food expenditure of each of the seven households in our sample. In this case, we predict each household's food expenditure by using the regression line we estimated earlier in Example 13–1, which is

$$\hat{y} = 1.1414 + .2642x$$

The predicted food expenditures, denoted by $\hat{y}$, for all households are shown in Table 13.4. Also shown are the errors and error squares.

Table 13.4

x	y	$\hat{y} = 1.1414 + .2642x$	$e = y - \hat{y}$	$e^2 = (y - \hat{y})^2$
35	9	10.3884	− 1.3884	1.9277
49	15	14.0872	.9128	.8332
21	7	6.6896	.3104	.0963
39	11	11.4452	− .4452	.1982
15	5	5.1044	− .1044	.0109
28	8	8.5390	− .5390	.2905
25	9	7.7464	1.2536	1.5715
				$\Sigma e^2 = \Sigma(y - \hat{y})^2 = 4.9283$

We calculate the values of $\hat{y}$ (given in the third column of Table 13.4) by substituting the values of x in the estimated regression model. For example, the value of x for the first household is 35. Substituting this value of x in the regression line, we obtain

$$\hat{y} = 1.1414 + .2642 (35) = 10.3884$$

Similarly we find the other values of $\hat{y}$.

The error sum of squares SSE is given by the sum of the fifth column in Table 13.4. Thus,

$$SSE = \Sigma(y - \hat{y})^2 = 4.9283$$

The errors of prediction for the regression model for seven households are shown in Figure 13.14.

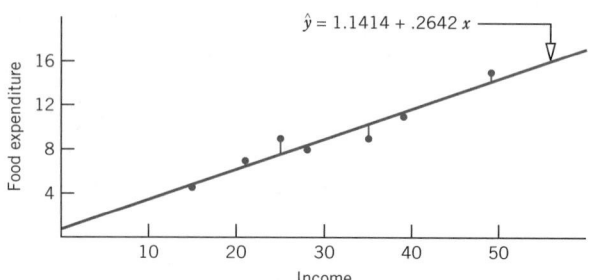

Figure 13.14   Errors of prediction when regression model is used.

Thus, from the foregoing calculations,

$$\text{SST} = 60.8571 \quad \text{and} \quad \text{SSE} = 4.9283$$

These values indicate that the sum of squared errors decreased from 60.8571 to 4.9283 when we used $\hat{y}$ to predict food expenditures in place of $\bar{y}$. This reduction in squared errors is called the **regression sum of squares** and is denoted by **SSR**. Consequently,

$$\text{SSR} = \text{SST} - \text{SSE} = 60.8571 - 4.9283 = 55.9288$$

The value of SSR can also be computed by using the formula

$$\text{SSR} = \Sigma(\hat{y} - \bar{y})^2$$

REGRESSION SUM OF SQUARES (SSR)

The regression sum of squares, denoted by SSR, is

$$\text{SSR} = \text{SST} - \text{SSE}$$

Thus, SSR is that portion of SST that is explained by the use of the regression model and SSE is that portion of SST that is not explained by the use of the regression model. The sum of SSR and SSE is always equal to SST. Thus,

$$\text{SST} = \text{SSR} + \text{SSE}$$

The ratio of SSR to SST gives the **coefficient of determination**. The coefficient of determination calculated for population data is denoted by ρ^2 (ρ is the Greek letter *rho*) and the one calculated for sample data is denoted by r^2. The coefficient of determination gives the proportion of SST that is explained by the use of the regression model. The value of the coefficient of determination always lies in the range zero to 1. The coefficient of determination can be calculated by using the formula

$$r^2 = \frac{\text{SSR}}{\text{SST}} \quad \text{or} \quad \frac{\text{SST} - \text{SSE}}{\text{SST}}$$

However, for computational purposes, the following formula is more efficient to use to calculate coefficient of determination.

> **COEFFICIENT OF DETERMINATION**
>
> The coefficient of determination, denoted by r^2, represents the proportion of SST that is explained by the use of the regression model. The computational formula for r^2 is
>
> $$r^2 = \frac{b\ SS_{xy}}{SS_{yy}}$$
>
> and $$0 \le r^2 \le 1$$

To calculate the coefficient of determination ρ^2 for a population data set, we replace b by B in the above formula and use the values of SS_{xy} and SS_{yy} calculated for the population data set.

Example 13–3 illustrates the calculation of the coefficient of determination for a sample data set.

Calculating the coefficient of determination.

EXAMPLE 13–3 For the data of Table 13.1 on monthly incomes and food expenditures of seven households, calculate the coefficient of determination.

Solution From earlier calculations made in Examples 13–1 and 13–2,

$$b = .2642, \quad SS_{xy} = 211.7143, \quad \text{and} \quad SS_{yy} = 60.8571$$

Hence,

$$r^2 = \frac{b\ SS_{xy}}{SS_{yy}} = \frac{(.2642)\ (211.7143)}{60.8571} = .92$$

Thus, we can state that SST is reduced by approximately 92% (from 60.8571 to 4.9283) when we use $\hat{y}$, instead of $\bar{y}$, to predict the food expenditures of households. Note that r^2 is usually rounded to two decimal places. ▄

The total sum of squares SST is a measure of the total variation in food expenditures, SSR is the portion of total variation explained by the regression model (or by income), and the error sum of squares SSE is the portion of total variation not explained by the regression model. Hence, for Example 13–3 we can state that 92% of the total variation in food expenditures of households occurs because of the variation in their incomes, and the remaining 8% is due to randomness and other variables.

Usually, the higher the value of r^2, the better the regression model. This is so because if r^2 is larger, a greater portion of the total errors is explained by the included independent variable and a smaller portion of errors is attributed to other variables and randomness.

EXERCISES

Concepts and Procedures

13.31 What are the degrees of freedom for a simple linear regression model?

13.32 Explain the meaning of coefficient of determination.

13.33 Explain the meaning of SST and SSR. You may use graphs for illustration purposes.

13.34 A population data set produced the following information.

$$N = 250, \quad \Sigma x = 9880, \quad \Sigma y = 1456, \quad \Sigma xy = 85,080,$$

$$\Sigma x^2 = 485,870, \quad \text{and} \quad \Sigma y^2 = 135,675$$

Find the values of σ_ϵ and ρ^2.

13.35 A population data set produced the following information.

$$N = 460, \quad \Sigma x = 3920, \quad \Sigma y = 2650, \quad \Sigma xy = 26,570,$$

$$\Sigma x^2 = 48,530, \quad \text{and} \quad \Sigma y^2 = 39,347$$

Find the values of σ_ϵ and ρ^2.

13.36 The following information is obtained from a sample data set.

$$n = 10, \quad \Sigma x = 100, \quad \Sigma y = 220, \quad \Sigma xy = 3680,$$

$$\Sigma x^2 = 1140, \quad \text{and} \quad \Sigma y^2 = 5272$$

Find the values of s_e and r^2.

13.37 The following information is obtained from a sample data set.

$$n = 12, \quad \Sigma x = 66, \quad \Sigma y = 588, \quad \Sigma xy = 2244,$$

$$\Sigma x^2 = 396, \quad \text{and} \quad \Sigma y^2 = 58,734$$

Find the values of s_e and r^2.

Applications

13.38 The following table, reproduced from Exercise 13.25, gives information on the monthly incomes (in hundreds of dollars) and monthly telephone bills (in dollars) for a random sample of 10 households.

Income	16	45	36	32	30	13	41	15	36	40
Phone bill	35	78	102	56	75	26	130	42	59	85

Find the following.

a. SS_{xx}, SS_{yy}, and SS_{xy} b. Standard deviation of errors
c. SST, SSE, and SSR   d. Coefficient of determination

13.39 Refer to Exercise 13.26. The following table, which gives the ages (in years) and prices (in hundreds of dollars) of eight cars of a specific model, is reproduced from that exercise.

Age	8	3	6	9	2	5	6	3
Price	16	74	38	19	102	36	33	69

a. Calculate the standard deviation of errors.
b. Compute the coefficient of determination and give a brief interpretation of it.

13.40 The following data on the years of experience and the number of CD players assembled by employees working in the assembly department at Picaso Electronics are reproduced from Exercise 13.27.

Experience	5	11	15	7	2	10	9
CD players assembled	14	21	20	18	13	16	18

a. Determine the standard deviation of errors.
b. Find the coefficient of determination and give a brief interpretation of it.

13.41 The following table, reproduced from Exercise 13.28, lists the sizes of six houses (in hundreds of square feet) and the monthly rents (in dollars) paid by tenants for those houses.

Size of the house	21	16	19	27	34	23
Monthly rent	700	580	720	850	1050	800

a. Compute the standard deviation of errors.
b. Calculate the coefficient of determination. What percentage of the variation in monthly rents is explained by the sizes of the houses? What percentage of this variation is not explained?

13.42 The following table, reproduced from Exercise 13.29, gives the total payroll (rounded to millions of dollars) as of March 1989 and the percentage of games won during the 1988 season by each of the National League baseball teams.

Team	Total Payroll	Percentage of Games Won
Atlanta Braves	9	34
Chicago Cubs	12	48
Cincinnati Reds	11	54
Houston Astros	16	51
Los Angeles Dodgers	22	58
Montreal Expos	12	50
New York Mets	20	63
Philadelphia Phillies	10	40
Pittsburgh Pirates	12	53
St. Louis Cardinals	15	47
San Diego Padres	14	52
San Francisco Giants	14	51

a. Find the standard deviation of errors σ_ϵ. (Note that this data set belongs to a population.)
b. Compute the coefficient of determination ρ^2.

13.43 The following table, reproduced from Exercise 13.30, gives the percentage of games won and the average attendance (rounded to nearest thousand) per home game for the 1988 season for each of the American League baseball teams.

Team	Percentage of Games Won	Average Attendance per Home Game (in thousands)
Baltimore Orioles	34	22
Boston Red Sox	55	31
California Angels	46	29
Chicago White Sox	44	14
Cleveland Indians	48	18
Detroit Tigers	54	26
Kansas City Royals	52	29
Milwaukee Brewers	54	24
Minnesota Twins	56	37
New York Yankees	53	34
Oakland Athletics	64	29
Seattle Mariners	42	13
Texas Rangers	44	20
Toronto Blue Jays	54	32

a. Find the standard deviation of errors σ_ϵ. (Note that this data set belongs to a population.)
b. Compute the coefficient of determination ρ^2.

13.5   INFERENCES ABOUT *B*

This section is concerned with estimation and tests of hypotheses about the population regression slope *B*. We can also make confidence intervals and test hypotheses about the *y*-intercept *A* of the population regression line. However, making inferences about *A* is beyond the scope of this text.

13.5.1   SAMPLING DISTRIBUTION OF *b*

One of the main purposes for determining a regression line is to find the true value of the slope *B* of the population regression line. However, in almost all cases, the regression line is estimated using sample data. Then, based on the sample regression line, inferences are made about the population regression line. The slope *b* of a sample regression line is a point estimator of the slope *B* of the population regression line. The different sample regression lines estimated for different samples taken from the same population will give different values of *b*. If only one sample is taken and the regression line for that sample is estimated, the value of *b* will depend on which elements are included in the sample. Thus, *b* is a random variable and it possesses a probability distribution that is more commonly called its sampling distribution. The shape of the sampling distribution of *b*, its mean, and standard deviation are as follows.

MEAN, STANDARD DEVIATION, AND SAMPLING DISTRIBUTION OF *b*

Because of the assumption of normally distributed random errors, the sampling distribution of *b* is normal. The mean and standard deviation of *b*, denoted by μ_b and σ_b respectively, are

$$\mu_b = B \quad \text{and} \quad \sigma_b = \frac{\sigma_\epsilon}{\sqrt{SS_{xx}}}$$

However, usually the standard deviation of population errors σ_ϵ is not known. Hence, the sample standard deviation of errors s_e is used to estimate σ_ϵ. In such a case, when σ_ϵ is unknown, the standard deviation of *b* is estimated by s_b, which is calculated as

$$s_b = s_e / \sqrt{SS_{xx}}$$

If σ_ϵ is not known and the sample size is large ($n \geq 30$), the normal distribution can be used to make inferences about *B*. However, if σ_ϵ is not known and the sample size is small ($n < 30$), the normal distribution is replaced by the *t* distribution to make inferences about *B*.

13.5.2   ESTIMATION OF *B*

The value of *b* obtained from the sample regression line is a point estimate of slope *B* of the population regression line. As mentioned in Section 13.5.1, if σ_ϵ is not known and the sample size is small, the *t* distribution is used to make a confidence interval for *B*.

CONFIDENCE INTERVAL FOR *B*

The $(1 - \alpha)100\%$ confidence interval for B is given by

$$b \pm ts_b$$

where

$$s_b = \frac{s_e}{\sqrt{SS_{xx}}}$$

and the value of t is obtained from the t distribution table for $\alpha/2$ area in the right tail of the t distribution and $n - 2$ degrees of freedom.

Example 13–4 describes the procedure for making a confidence interval for B.

Constructing a confidence interval for B.

EXAMPLE 13-4 Construct a 95% confidence interval for B for the data on incomes and food expenditures of seven households given in Table 13.1.

Solution From the given information and earlier calculations,

$$n = 7, \quad b = .2642, \quad SS_{xx} = 801.4286, \quad \text{and} \quad s_e = .9922$$

The confidence level is 95%.

$$s_b = \frac{s_e}{\sqrt{SS_{xx}}} = \frac{.9922}{\sqrt{801.4286}} = .0350$$

$$df = n - 2 = 7 - 2 = 5$$

$$\alpha/2 = .5 - (.95/2) = .025$$

From the t distribution table, the value of t for 5 df and .025 area in the right tail of the t distribution curve is 2.571. The 95% confidence interval for B is

$$b \pm ts_b = .2642 \pm 2.571 (.0350) = .2642 \pm .0900 = \mathbf{.17 \text{ to } .35}$$

Thus, we are 95% confident that slope B of the population regression line is between .17 and .35. ■

13.5.3 HYPOTHESIS TESTING ABOUT *B*

Testing a hypothesis about B when the null hypothesis is $B = 0$ (that is, the slope of the regression line is zero) is equivalent to testing that x does not determine y and that the regression line is of no use in predicting y for a given x. However, we should remember that we are testing for a linear relationship between x and y. It is possible that x may determine y nonlinearly. Hence, a nonlinear relationship may exist between x and y.

To test the hypothesis that x does not determine y linearly, we will test the null hypothesis that the slope of the regression line is zero, that is, $B = 0$. The alternative hypothesis can be: (1) x determines y, that is, $B \neq 0$, (2) x determines y positively, that is, $B > 0$, or (3) x determines y negatively, that is, $B < 0$.

The procedure used to make a hypothesis test about B is similar to the one used in earlier chapters. It involves the same five steps.

TEST STATISTIC FOR b

The value of the test statistic t for b is calculated as

$$t = \frac{b - B}{s_b}$$

The value of B is substituted from the null hypothesis.

Example 13–5 illustrates the procedure for testing a hypothesis about B.

Conducting a test of
hypothesis about B.

EXAMPLE 13-5   Test at the 1% significance level if the slope of the regression line for the example on incomes and food expenditures of seven households is positive.

Solution   From the given information and earlier calculations in Examples 13–1 and 13–4,

$$n = 7, \quad b = .2642, \quad \text{and} \quad s_b = .0350$$

Step 1.   *State the null and alternative hypotheses*

We are to test whether or not slope B of the population regression line is positive. Hence, the two hypotheses are

$$H_0\colon B = 0 \quad \text{(slope is zero)}$$

$$H_1\colon B > 0 \quad \text{(slope is positive)}$$

Note that we can also write the null hypothesis as $H_0\colon B \leq 0$, which states that the slope is either zero or negative.

Step 2.   *Select the distribution to use*

The sample size is small ($n < 30$) and σ_ϵ is not known. Hence, we use the t distribution to make the test about B.

Step 3.   *Determine the rejection and nonrejection regions*

The significance level is .01. The $>$ sign in the alternative hypothesis indicates that the test is right-tailed. Therefore,

$$\text{Area in the right tail of the } t \text{ distribution} = \alpha = .01$$

$$df = n - 2 = 7 - 2 = 5$$

From the t distribution table, the critical value of t for 5 df and .01 area in the right tail of the t distribution is 3.365, as shown in Figure 13.15.

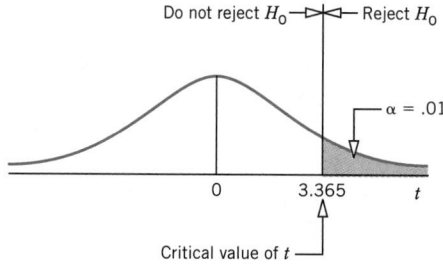

Figure 13.15

Step 4. *Calculate the value of the test statistic*

The value of the test statistic t for b is calculated as follows.

$$t = \frac{b - B}{s_b} = \frac{.2642 - 0}{.0350} = 7.549$$

where the arrow points to $-$From H_0 above the B.

Step 5. *Make a decision*

The value of the test statistic $t = 7.549$ is greater than the critical value of $t = 3.365$ and it falls in the rejection region. Hence, we reject the null hypothesis and conclude that x (income) determines y (food expenditure) positively. That is, food expenditure increases with an increase in income and it decreases with a decrease in income. ▬

Note that the null hypothesis does not always have to be $B = 0$. We may test the null hypothesis that B is equal to a certain value. See Exercises 13.45 to 13.48, and 13.54 for such cases.

EXERCISES

Concepts and Procedures

13.44 Describe the mean, standard deviation, and shape of the sampling distribution of the slope b of the simple linear regression model.

13.45 The following information is obtained for a sample of 16 observations taken from a population.

$$SS_{xx} = 340.700, \quad s_e = 1.951, \quad \text{and} \quad \hat{y} = 12.45 + 6.32x$$

    a. Make a 99% confidence interval for B.
    b. Using a significance level of .025, can you conclude that B is positive?
    c. Using a significance level of .01, can you conclude that B is different from zero?
    d. Using a significance level of .02, test whether B is different from 4.50. (*Hint:* The null hypothesis here will be H_0: $B = 4.50$, and the alternative hypothesis will be H_1: $B \neq 4.50$. Notice that the value of $B = 4.50$ will be used to calculate the value of the test statistic t.)

13.46 The following information is obtained for a sample of 25 observations taken from a population.

$$SS_{xx} = 274.600, \quad s_e = .932, \quad \text{and} \quad \hat{y} = 280.56 - 3.77x$$

    a. Make a 95% confidence interval for B.
    b. Using a significance level of .01, test whether B is negative.
    c. Testing at the 5% significance level, can you conclude that B is different from zero?
    d. Test if B is different from 5.20. Use $\alpha = .01$.

13.47 The following information is obtained for a sample of 100 observations taken from a population. (Note that because $n > 30$, we can use the normal distribution to make a confidence interval and test a hypothesis about B.)

$$SS_{xx} = 524.884, \quad s_e = 1.464, \quad \text{and} \quad \hat{y} = 5.48 + 2.50x$$

    a. Make a 98% confidence interval for B.
    b. Test at the 2% significance level whether B is positive.
    c. Can you conclude that B is different from zero? Use $\alpha = .01$.
    d. Using a significance level of .01, test whether B is greater than 1.75.

13.48 The following information is obtained for a sample of 80 observations taken from a population.

$$SS_{xx} = 380.592, \qquad s_e = .961, \qquad \text{and} \qquad \hat{y} = 160.24 - 2.70x$$

    a. Make a 97% confidence interval for B.
    b. Test at the 1% significance level whether B is negative.
    c. Can you conclude that B is different from zero? Use $\alpha = .01$.
    d. Using a significance level of .02, test whether B is less than 4.25.

Applications

13.49 Refer to Exercise 13.26. The data on ages (in years) and prices (in hundreds of dollars) for eight cars of a specific model are reproduced below from that exercise.

Age	8	3	6	9	2	5	6	3
Price	16	74	38	19	102	36	33	69

    a. Construct a 95% confidence interval for B. You can use results obtained in Exercises 13.26 and 13.39 here.
    b. Test at the 5% significance level if B is negative.

13.50 A city with higher rents for office buildings will usually experience a lower office occupancy rate. Many businesses will relocate to cities with lower rents. The following data give the annual rent per square foot for office buildings and the occupancy rate (percentage of the office buildings that are occupied) for seven cities.

Rent per square foot	12	17	9	15	11	7	19
Occupancy rate	83	77	92	82	88	90	75

    a. Find the regression line with occupancy rate as a dependent variable and rent per square foot as an independent variable.
    b. Construct a 99% confidence interval for B.
    c. Test at the 2.5% significance level whether B is less than zero.

13.51 The following data give the experience (in years) and monthly salaries (in hundreds of dollars) of nine randomly selected secretaries.

Experience	14	3	5	6	4	9	18	5	16
Monthly salary	22	12	15	17	15	19	24	13	27

    a. Find the least squares regression line with experience as an independent variable and monthly salary as a dependent variable.
    b. Construct a 98% confidence interval for B.
    c. Test at the 2.5% significance level whether B is greater than zero.

13.52 Refer to Exercise 13.27. The following table gives the data on the years of experience for seven employees and the average number of CD players each of them assembled per day.

Experience	5	11	15	7	2	10	9
CD players assembled	14	21	20	18	13	16	18

    a. Make a 95% confidence interval for B. You can use the results obtained in Exercises 13.27 and 13.40 here.
    b. Using the 1% significance level, can you conclude that B is positive?

13.53 The data on the sizes of six houses (in hundreds of square feet) and the monthly rents (in dollars) paid by tenants for those houses are reproduced below from Exercise 13.28.

Size of the house	21	16	19	27	34	23
Monthly rent	700	580	720	850	1050	800

a. Construct a 99% confidence interval for *B*. You can use the calculations made in Exercises 13.28 and 13.41 here.

b. Testing at the 5% significance level, can you conclude that *B* is different from zero?

13.54 Refer to Exercise 13.25. The following table gives information on the monthly incomes (in hundreds of dollars) and monthly telephone bills (in dollars) for a random sample of 10 households.

Income	16	45	36	32	30	13	41	15	36	40
Phone bill	35	78	102	56	75	26	130	42	59	85

a. Make a 98% confidence interval for *B*. Use the results obtained in Exercises 13.25 and 13.38.

b. An earlier study claims that for the relationship between incomes and amounts of phone bills *B* = 1.50. Test at the 1% significance level if *B* is greater than 1.50.

13.6 LINEAR CORRELATION

Another measure of the relationship between two variables is the correlation coefficient. This section describes the simple linear correlation, for short **linear correlation**, which measures the strength of the linear association between two variables. In other words, the linear correlation coefficient measures how closely the points in a scatter diagram are spread around the regression line. The correlation coefficient calculated for the population data is denoted by **ρ** (Greek letter *rho*) and the one calculated for sample data is denoted by **r**. (Note that the square of the correlation coefficient is equal to the coefficient of determination.)

VALUE OF THE CORRELATION COEFFICIENT

The value of the correlation coefficient always lies in the range −1 to 1, that is,

$$-1 \le \rho \le 1 \quad \text{and} \quad -1 \le r \le 1$$

Although we can explain the linear correlation using the population correlation coefficient ρ, we will do so using the sample correlation coefficient *r*.

If *r* = 1, it is said to be a case of *perfect positive linear correlation*. In such a case, all points in the scatter diagram lie on a straight line that slopes upward from left to right, as shown in Figure 13.16*a*. If *r* = −1, the correlation is said to be a *perfect negative linear correlation*. In this case, all points in the scatter diagram fall on a straight line that slopes downward from left to right, as shown in Figure 13.16*b*. If the points are scattered all over the diagram, as shown in Figure 13.16*c*, then there is *no linear correlation* between the two variables and consequently *r* = 0.

We do not usually encounter an example with perfect positive or perfect negative correlation. What we observe in real-world problems is either a positive linear correlation with

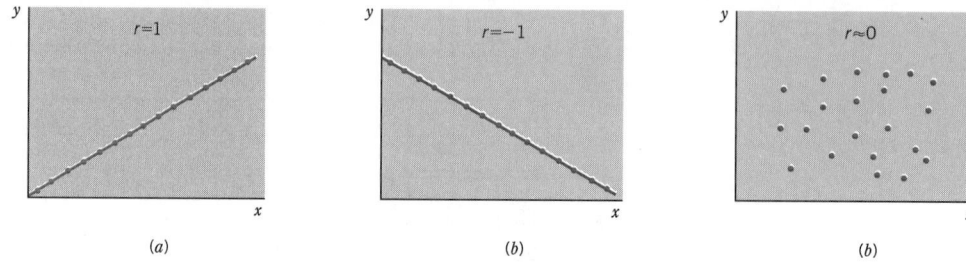

Figure 13.16 Linear correlation between variables. (*a*) Perfect positive linear correlation, $r = 1$. (*b*) Perfect negative linear correlation, $r = -1$. (*c*) No linear correlation, $r = 0$.

$0 < r < 1$ (that is, the correlation coefficient is greater than zero but less than 1) or a negative linear correlation with $-1 < r < 0$ (that is, the correlation coefficient is greater than -1 but less than zero).

If the correlation between two variables is positive and close to 1, we say that the variables have a *strong positive linear correlation*. If the correlation between two variables is positive but close to zero, then the variables have a *weak positive linear correlation*. On the other hand, if the correlation between two variables is negative and close to -1, then the variables are said to have a *strong negative linear correlation*. Also, if the correlation between variables is negative but close to zero, there exists a *weak negative linear correlation* between the variables. Graphically, a strong correlation indicates that the points in the scatter diagram are very close to the regression line and a weak correlation indicates that the points in the scatter diagram are widely spread around the regression line. These four cases are shown in Figure 13.17*a–d*.

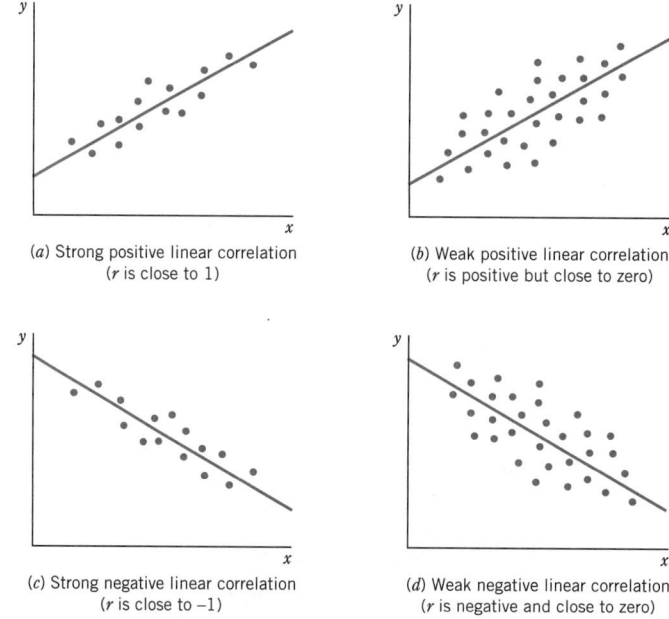

(*a*) Strong positive linear correlation
(*r* is close to 1)

(*b*) Weak positive linear correlation
(*r* is positive but close to zero)

(*c*) Strong negative linear correlation
(*r* is close to –1)

(*d*) Weak negative linear correlation
(*r* is negative and close to zero)

Figure 13.17 Linear correlation between variables.

The linear correlation coefficient is calculated by using the following formula. (This correlation coefficient is also called the *Pearson product moment correlation coefficient*.)

LINEAR CORRELATION COEFFICIENT

The simple linear correlation, denoted by r, measures the strength of the linear relationship between two variables for a sample and is calculated as

$$r = \frac{SS_{xy}}{\sqrt{SS_{xx}\, SS_{yy}}}$$

If we have access to population data, we can calculate the population correlation coefficient ρ by using the same formula as the one used to calculate r. But now the values of SS_{xx}, SS_{xy}, and SS_{yy} used in the formula will be based on the population data.

As both SS_{xx} and SS_{yy} are always positive, the sign of the correlation coefficient r depends on the sign of SS_{xy}. If SS_{xy} is positive then r will be positive and if SS_{xy} is negative then r will be negative. Another important observation to remember is that r and b, *calculated for the same sample*, *will always have the same sign. That is, both r and b are either positive or negative.* This is so because both r and b provide information about the relationship between x and y. Likewise, the corresponding population parameters ρ and B will always have the same sign.

Example 13–6 illustrates the calculation of the linear correlation coefficient r.

Calculating the linear correlation coefficient.

EXAMPLE 13–6 Calculate the correlation coefficient for the example on incomes and food expenditures of seven households.

Solution From earlier calculations made in Examples 13–1 and 13–2,

$$SS_{xy} = 211.7143, \qquad SS_{xx} = 801.4286, \qquad \text{and} \qquad SS_{yy} = 60.8571$$

Substituting these values in the formula for r, we obtain

$$r = \frac{SS_{xy}}{\sqrt{SS_{xx}\, SS_{yy}}} = \frac{211.7143}{\sqrt{(801.4286)\,(60.8571)}} = .96$$

Thus, the linear correlation coefficient is .96. The correlation coefficient is usually rounded to two decimal places. ∎

The linear correlation coefficient simply tells us how strongly the two variables are (linearly) related. The correlation coefficient of .96 for incomes and food expenditures of seven households indicates that income and food expenditure are very strongly and positively correlated. This correlation coefficient does not, however, provide us with any more information.

The square of the correlation coefficient gives the coefficient of determination, which was explained in Section 13.4. Thus, $(.96)^2$ is .92, which is the value of r^2 calculated in Example 13–3.

Sometimes the calculated value of r may indicate that the two variables are very strongly linearly correlated but in reality they are not. For example, if we calculate the correlation coefficient between the price of Coke and the size of families in the United States using data for the last 30 years, we will find a strong negative linear correlation. Over time, the price of Coke has increased and the size of families has decreased. This finding does not mean that family size and price of Coke are related. As a result, before we calculate the correlation coefficient, we must seek help from a theory or from common sense to postulate whether or not the two variables have a causal relationship.

Another point to note is that in a simple regression model one of the two variables is categorized as an independent variable and the other is classified as a dependent variable. However, no such distinction is made between the two variables when the correlation coefficient is calculated.

EXERCISES

Concepts and Procedures

13.55   What does a linear correlation coefficient tell you about the relationship between two variables? Within what range can a correlation coefficient assume a value?

13.56   What is the difference between ρ and r? Explain.

13.57   Explain each of the following concepts. You may use graphs to illustrate each concept.
  a.   Perfect positive linear correlation
  b.   Perfect negative linear correlation
  c.   Strong positive linear correlation
  d.   Strong negative linear correlation
  e.   Weak positive linear correlation
  f.   Weak negative linear correlation
  g.   No linear correlation

13.58   Can the values of B and ρ calculated for the same population data have different signs? Explain.

13.59   For a sample data set, the linear correlation coefficient r has a positive value. Which of the following is true about the slope b of the regression line estimated for the same sample data?
  a.   The value of b will be positive
  b.   The value of b will be negative
  c.   The value of b can be positive or negative

13.60   For a sample data set, the slope b of the regression line has a positive value. Which of the following is true about the linear correlation coefficient r calculated for the same sample data?
  a.   The value of r will be positive
  b.   The value of r will be negative
  c.   The value of r can be positive or negative

13.61   For a sample data set on two variables, the value of the linear correlation coefficient is zero. Does this mean that these variables are not related? Explain.

13.62   Will you expect a positive, zero, or negative correlation between the two variables for each of the following examples?
  a.   Income of a household and the number of newspaper/magazine subscriptions
  b.   Income and entertainment expenditure of a household
  c.   Ages of husbands and wives
  d.   Price of a computer and consumption of Coke
  e.   Price and consumption of wine
  f.   Unemployment rate and the number of houses sold in a city

13.63   Will you expect a positive, zero, or negative correlation between the two variables for each of the following examples?
  a.   Number of miles certain cars have been driven and their prices
  b.   Stress level and blood pressure of individuals
  c.   Amount of fertilizer used and yield of corn per acre
  d.   Ages and prices of houses

e. Heights of husbands and incomes of their wives

f. Index of job satisfaction and employee turnover

13.64 A population data set produced the following information.

$$N = 250, \quad \Sigma x = 9880, \quad \Sigma y = 1456, \quad \Sigma xy = 85,080,$$

$$\Sigma x^2 = 485,870, \quad \text{and} \quad \Sigma y^2 = 135,675$$

Find the linear correlation coefficient ρ.

13.65 A population data set produced the following information.

$$N = 460, \quad \Sigma x = 3920, \quad \Sigma y = 2650, \quad \Sigma xy = 26,570,$$

$$\Sigma x^2 = 48,530, \quad \text{and} \quad \Sigma y^2 = 39,347$$

Find the linear correlation coefficient ρ.

13.66 A sample data set produced the following information.

$$n = 10, \quad \Sigma x = 100, \quad \Sigma y = 220, \quad \Sigma xy = 3680,$$

$$\Sigma x^2 = 1140, \quad \text{and} \quad \Sigma y^2 = 52,272$$

Calculate the linear correlation coefficient *r*.

13.67 A sample data set produced the following information.

$$n = 12, \quad \Sigma x = 66, \quad \Sigma y = 588, \quad \Sigma xy = 2244,$$

$$\Sigma x^2 = 396, \quad \text{and} \quad \Sigma y^2 = 58,734$$

Calculate the linear correlation coefficient *r*.

Applications

13.68 The following data, reproduced from Exercise 13.50, give the annual rent per square foot for office buildings and the occupancy rates (percentages of the office buildings that are occupied) for seven cities.

Rent per square foot	12	17	9	15	11	7	19
Occupancy rate	83	77	92	82	88	90	75

a. Do you expect these two variables to be positively or negatively related? Explain.

b. Calculate the correlation coefficient.

13.69 The following table, reproduced from Exercise 13.51, gives the experience (in years) and monthly salaries (in hundreds of dollars) of nine randomly selected secretaries.

Experience	14	3	5	6	4	9	18	5	16
Monthly salary	22	12	15	17	15	19	24	13	27

a. Do you expect the experience and monthly salaries to be positively or negatively related? Explain.

b. Compute the correlation coefficient.

13.70 Refer to Exercise 13.26. The data on ages (in years) and prices (in hundreds of dollars) for eight cars of a specific model are reproduced below from that exercise.

Age	8	3	6	9	2	5	6	3
Price	16	74	38	19	102	36	33	69

Find the correlation coefficient. Is the sign of the correlation coefficient the same as that of *b* calculated in Exercise 13.26?

13.71   The following data on years of experience and the number of CD players assembled by employees at Picaso Electronics are reproduced from Exercise 13.27.

Experience	5	11	15	7	2	10	9
CD players assembled	14	21	20	18	13	16	18

Find the correlation coefficient. Is the sign of the correlation coefficient the same as that of b calculated in Exercise 13.27?

13.72   The following table, reproduced from Exercise 13.29, gives the total payroll (rounded to millions of dollars) as of March 1989 and the percentage of games won during the 1988 season by each of the National League baseball teams.

Team	Total Payroll	Percentage of Games Won
Atlanta Braves	9	34
Chicago Cubs	12	48
Cincinnati Reds	11	54
Houston Astros	16	51
Los Angeles Dodgers	22	58
Montreal Expos	12	50
New York Mets	20	63
Philadelphia Phillies	10	40
Pittsburgh Pirates	12	53
St. Louis Cardinals	15	47
San Diego Padres	14	52
San Francisco Giants	14	51

Compute the coefficient of linear correlation ρ.

13.73   The following table, reproduced from Exercise 13.30, gives the percentage of games won and the average attendance (rounded to nearest thousand) per home game for the 1988 season for each of the American League baseball teams.

Team	Percentage of Games Won	Average Attendance per Home Game (in thousands)
Baltimore Orioles	34	22
Boston Red Sox	55	31
California Angels	46	29
Chicago White Sox	44	14
Cleveland Indians	48	18
Detroit Tigers	54	26
Kansas City Royals	52	29
Milwaukee Brewers	54	24
Minnesota Twins	56	37
New York Yankees	53	34
Oakland Athletics	64	29
Seattle Mariners	42	13
Texas Rangers	44	20
Toronto Blue Jays	54	32

Compute the coefficient of linear correlation ρ.

13.7 REGRESSION ANALYSIS: A COMPLETE EXAMPLE

This section works out an example that includes all the topics we have discussed so far in this chapter.

A complete example of regression analysis.

EXAMPLE 13-7   A random sample of eight auto drivers insured with a company and having similar auto insurance policies was selected. The following table lists their driving experience (in years) and the monthly auto insurance premium (in dollars) paid by them.

Driving Experience (years)	Monthly Auto Insurance Premium (dollars)
5	64
2	87
12	50
9	71
15	44
6	56
25	42
16	60

(a) Does the insurance premium depend on driving experience or does the driving experience depend on insurance premium? Do you expect a positive or a negative relationship between these two variables?

(b) Compute SS_{xx}, SS_{yy}, and SS_{xy}.

(c) Find the least squares regression line by choosing appropriate dependent and independent variables based on your answer in part (a).

(d) Interpret the meaning of the values of a and b calculated in part (c).

(e) Plot the scatter diagram and the regression line.

(f) Calculate r and r^2 and explain what they mean.

(g) Predict the monthly auto insurance premium for a driver with 10 years of driving experience.

(h) Compute the standard deviation of errors.

(i) Construct a 90% confidence interval for B.

(j) Test at the 5% significance level if B is negative.

Solution

(a) Based on theory and intuition, we expect the insurance premium to depend on driving experience. Consequently, the insurance premium would be a dependent variable and driving experience an independent variable in the regression model. A new driver is considered a high risk by the insurance companies and he or she has to pay a higher premium for auto insurance. On average, the insurance premium is expected to decrease with an increase in the years of driving experience. Therefore, we expect a negative relationship between these two variables. In other words, both the population correlation coefficient ρ and the population regression slope B are expected to be negative.

(b) Table 13.5 shows the calculation of Σx, Σy, Σxy, Σx^2, and Σy^2.
The values of $\bar{x}$ and $\bar{y}$ are

$$\bar{x} = \Sigma x/n = 90/8 = 11.25$$

$$\bar{y} = \Sigma y/n = 474/8 = 59.25$$

Table 13.5

Experience x	Premium y	xy	x^2	y^2
5	64	320	25	4096
2	87	174	4	7569
12	50	600	144	2500
9	71	639	81	5041
15	44	660	225	1936
6	56	336	36	3136
25	42	1050	625	1764
16	60	960	256	3600
$\Sigma x = 90$	$\Sigma y = 474$	$\Sigma xy = 4739$	$\Sigma x^2 = 1396$	$\Sigma y^2 = 29{,}642$

The values of SS_{xy}, SS_{xx}, and SS_{yy} are computed as follows.

$$SS_{xy} = \Sigma xy - \frac{(\Sigma x)(\Sigma y)}{n} = 4739 - \frac{(90)(474)}{8} = -593.5000$$

$$SS_{xx} = \Sigma x^2 - \frac{(\Sigma x)^2}{n} = 1396 - \frac{(90)^2}{8} = 383.5000$$

$$SS_{yy} = \Sigma y^2 - \frac{(\Sigma y)^2}{n} = 29642 - \frac{(474)^2}{8} = 1557.5000$$

(c)   To find the regression line, we calculate a and b as follows.

$$b = \frac{SS_{xy}}{SS_{xx}} = \frac{-593.5000}{383.5000} = -1.5476$$

$$a = \bar{y} - b\bar{x} = 59.25 - (-1.5476)(11.25) = 76.6605$$

Thus, our estimated regression line $\hat{y} = a + bx$ is

$$\hat{y} = 76.6605 - 1.5476\,x$$

(d)   The value of $a = 76.6605$ gives the value of $\hat{y}$ for $x = 0$, that is, it gives the monthly auto insurance premium for a driver with no driving experience. However, as mentioned earlier in this chapter, we should not attach much importance to this statement because the sample contains drivers with only 2 or more years of experience.

The value of b gives the change in $\hat{y}$ due to a change of one unit in x. Thus, $b = -1.5476$ indicates that, on average, for every extra year of driving experience the monthly auto insurance premium decreases by \$1.55. Note that when b is negative, y decreases as x increases.

(e)   Figure 13.18 shows the scatter diagram and the regression line for the data on eight

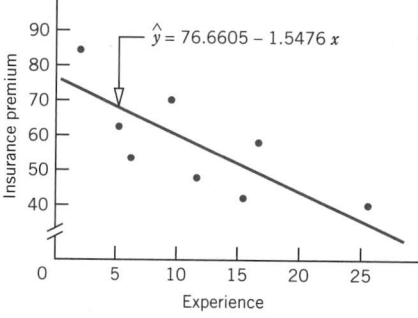

Figure 13.18

auto drivers. Note that the regression line slopes downward from left to right. This result is consistent with the negative relationship we anticipated between driving experience and insurance premium.

(f) The values of r and r^2 are computed as follows.

$$r = \frac{SS_{xy}}{\sqrt{SS_{xx}\ SS_{yy}}} = \frac{-593.5000}{\sqrt{(383.5000)\ (1557.5000)}} = -.77$$

$$r^2 = \frac{b\ SS_{xy}}{SS_{yy}} = \frac{(-1.5476)\ (-593.5000)}{1557.5000} = .59$$

The value of $r = -.77$ indicates that the driving experience and the monthly auto insurance premium are negatively related. The (linear) relationship is strong but not very strong.

The value of $r^2 = .59$ states that 59% of the total variation in insurance premiums is explained by years of driving experience and 41% is not. The low value of r^2 indicates that there may be many other important variables that contribute to the determination of auto insurance premiums. For example, the premium is expected to depend on the driving record of a driver and the type and age of the car.

(g) Using the estimated regression line, the predicted value of y for $x = 10$ is

$$\hat{y} = 76.6605 - 1.5476\ x = 76.6605 - 1.5476\ (10) = \$61.18$$

Thus, we expect the monthly auto insurance premium of a driver with 10 years of driving experience to be \$61.18.

(h) The standard deviation of errors is

$$s_e = \sqrt{\frac{SS_{yy} - b\ SS_{xy}}{n - 2}} = \sqrt{\frac{1557.5000 - (-1.5476)\ (-593.5000)}{8 - 2}} = 10.3199$$

(i) To construct a 90% confidence interval for B, first we calculate the standard deviation of b.

$$s_b = \frac{s_e}{\sqrt{SS_{xx}}} = \frac{10.3199}{\sqrt{383.5000}} = .5270$$

For a 90% confidence level, the area in each tail of the t distribution is

$$\alpha/2 = .5 - (.90/2) = .05$$

The degrees of freedom are

$$df = n - 2 = 8 - 2 = 6$$

From the t distribution table, the t value for .05 area in the right tail of the t distribution and 6 df is 1.943. The 90% confidence interval for B is

$$b \pm ts_b = -1.5476 \pm 1.943\ (.5270)$$

$$= -1.5476 \pm 1.0240 = -2.57\ \text{to}\ -.52$$

Thus, we can state with 90% confidence that B lies in the interval -2.57 to $-.52$. That is, on average, the monthly auto insurance premium of a driver decreases by an amount between \$.52 and \$2.57 for every extra year of driving experience.

(j) We perform the following five steps to test the hypothesis about B.

Step 1. *State the null and alternative hypotheses*

The null and alternative hypotheses are written as follows.

$$H_0: B = 0 \qquad (B \text{ is not negative})$$

$$H_1: B < 0 \qquad (B \text{ is negative})$$

Note that the null hypothesis can also be written as $H_0: B \geq 0$.

Step 2.   *Select the distribution to use*

As the sample size is small ($n < 30$) and σ_ϵ is not known, we use the t distribution to make the hypothesis test.

Step 3.   *Determine the rejection and nonrejection regions*

The significance level is .05. The $<$ sign in the alternative hypothesis indicates that it is a left-tailed test.

$$\text{Area in the left tail of the } t \text{ distribution} = \alpha = .05$$

$$df = n - 2 = 8 - 2 = 6$$

From the t distribution table, the critical value of t for .05 area in the left tail of the t distribution and 6 df is -1.943, as shown in Figure 13.19.

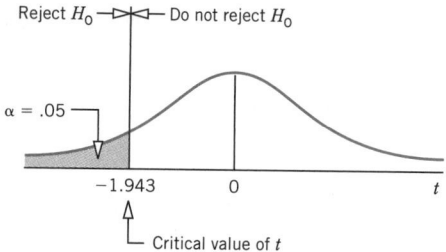

Figure 13.19

Step 4.   *Calculate the value of the test statistic*

The value of the test statistic t for b is calculated as follows.

$$t = \frac{b - B}{s_b} = \frac{-1.5476 - 0}{.5270} = -2.937$$

From H_0

Step 5.   *Make a decision*

The value of the test statistic $t = -2.937$ falls in the rejection region. Hence, we reject the null hypothesis and conclude that B is negative. That is, the monthly auto insurance premium decreases with an increase in years of driving experience.   ■

EXERCISES

Applications

13.74   The American Manufacturing Company makes baseball bats. The management of the company wants to know the relationship between the number of bats produced and the total cost of production. The manager of the company gathered data on the number of baseball bats (in thousands) manufactured and the total cost (in thousands of dollars) of production for eight selected months. These data are given in the following table.

Baseball bats	44	56	48	54	51	46	59	40
Total cost	664	795	682	770	747	692	806	658

a. Determine which of the two variables is the independent variable and which is the dependent variable. Do you expect a positive or a negative relationship between these two variables?

b. Taking the number of baseball bats manufactured as an independent variable and the total cost of production as a dependent variable, compute SS_{xx}, SS_{yy}, and SS_{xy}.

c. Find the least squares regression line.

d. Interpret the meaning of the values of a and b calculated in part c.

e. Plot the scatter diagram and the regression line.

f. Calculate r and r^2 and briefly explain what they mean.

g. Predict the total cost of production for 49 thousand baseball bats.

h. Compute the standard deviation of errors.

i. Construct a 95% confidence interval for B.

j. Test at the 1% significance level if B is positive.

13.75 A farmer wanted to find the relationship between the amount of fertilizer used and the yield of corn. He selected seven acres of his land on which he used different amounts of fertilizer to grow corn. The following table gives the amount (in pounds) of fertilizer used and the yield (in bushels) of corn for each of the seven acres.

Fertilizer Used	Yield of Corn
120	138
80	112
100	129
70	96
88	119
75	104
110	134

a. With the amount of fertilizer used as an independent variable and yield of corn as a dependent variable, compute SS_{xx}, SS_{yy}, and SS_{xy}.

b. Find the least squares regression line.

c. Interpret the meaning of the values of a and b calculated in part b.

d. Calculate r and r^2 and explain what they mean.

e. Compute the standard deviation of errors.

f. Predict the yield of corn per acre for $x = 105$.

g. Construct a 98% confidence interval for B.

h. Test at the 5% significance level if B is different from zero.

13.76 The following table gives information on the incomes (in thousands of dollars) and charitable contributions (in hundreds of dollars) for the past year for a random sample of 10 households.

Income (thousands of dollars)	Charitable Contributions (hundreds of dollars)
33	10
23	4
82	29
47	23
26	3
71	28
28	8
39	16
58	18
17	1

a. With income as an independent variable and charitable contributions as a dependent variable, compute SS_{xx}, SS_{yy}, and SS_{xy}.
b. Find the regression of charitable contributions on income.
c. Briefly explain the meaning of the values of a and b.
d. Calculate r and r^2 and briefly explain what they mean.
e. Compute the standard deviation of errors.
f. Construct a 99% confidence interval for B.
g. Test at the 1% significance level if B is positive.

13.77 The following table gives information on GPAs (grade point averages) and starting salaries (rounded to the nearest thousand dollars) of seven recent college graduates.

GPA	2.90	3.81	3.20	2.42	3.94	2.05	2.25
Starting salary	23	28	23	21	32	19	22

a. With GPA as an independent variable and starting salary as a dependent variable, compute SS_{xx}, SS_{yy}, and SS_{xy}.
b. Find the least squares regression line.
c. Interpret the meaning of the values of a and b calculated in part b.
d. Calculate r and r^2 and briefly explain what they mean.
e. Compute the standard deviation of errors.
f. Construct a 98% confidence interval for B.
g. Test at the 5% significance level if B is different from zero.

13.78 The following data give information on the ages (in years) and the monthly maintenance costs (in dollars) for a sample of eight cars.

Age	6	4	11	8	13	9	6	3
Maintenance cost	64	35	116	92	125	78	72	33

a. Taking age as an independent variable and maintenance cost as a dependent variable, compute SS_{xx}, SS_{yy}, and SS_{xy}.
b. Find the regression line $\hat{y} = a + bx$.
c. Briefly explain the meaning of the values of a and b calculated in part b.
d. Calculate r and r^2 and briefly explain what they mean.
e. Compute the standard deviation of errors.
f. Construct a 95% confidence interval for B.
g. Test at the 1% significance level if B is different from zero.

13.79 The following data give information on the lowest cost ticket price (in dollars) and the average attendance (rounded to the nearest thousand) for the past year for six football teams.

Ticket price	12.50	9.50	10.00	14.50	16.00	12.00
Attendance	56	65	71	69	55	42

a. Taking ticket price as an independent variable and attendance as a dependent variable, compute SS_{xx}, SS_{yy}, and SS_{xy}.
b. Find the least squares regression line.
c. Briefly explain the meaning of the values of a and b calculated in part b.
d. Calculate r and r^2 and briefly explain what they mean.
e. Compute the standard deviation of errors.
f. Construct a 90% confidence interval for B.
g. Test at the 2.5% significance level if B is negative.

13.8 USING THE REGRESSION MODEL

Let us return to the example on incomes and food expenditures to discuss two major uses of a regression model. These two uses are

1. Estimating the mean value of y for a given value of x. For instance, we can use our food expenditure regression model to estimate the mean food expenditure of all households with a specific income (say $3500 per month).

2. Predicting a particular value of y for a given value of x. For instance, we can determine the expected food expenditure of a randomly selected household with a particular monthly income (say $3500) using our food expenditure regression model.

13.8.1 USING THE REGRESSION MODEL FOR ESTIMATING THE MEAN VALUE OF y

Our population regression model is

$$y = A + Bx + \epsilon$$

As mentioned earlier in this chapter, the mean value of y for a given x is denoted by $\mu_{y|x}$, read as "the mean value of y for a given value of x." Because of the assumption that the mean value of ϵ is zero, the mean value of y is given by

$$\mu_{y|x} = A + Bx$$

Our objective is to estimate this mean value. The value of $\hat{y}$, obtained from the sample regression line by substituting the value of x, is the *point estimate of* $\mu_{y|x}$.

For our example on incomes and food expenditures, the estimated sample regression line is (see Example 13–1)

$$\hat{y} = 1.1414 + .2642x$$

Suppose we want to estimate the mean food expenditure for all households with a monthly income of $3500. We will denote this population mean by $\mu_{y|x=35}$ or $\mu_{y|35}$. Note that we have written $x = 35$ and not $x = 3500$ in $\mu_{y|35}$ because the units of measurement for the data used to estimate the above regression line in Example 13–1 were hundreds of dollars. Using the regression line, we find that the point estimate of $\mu_{y|35}$ is

$$\hat{y} = 1.1414 + .2642 (35) = 10.3884$$

Thus, based on the sample regression line, the point estimate for the mean food expenditure $\mu_{y|35}$ for all households with a monthly income of $3500 is $1038.84 per month.

However, suppose we take a second sample of seven households from the same population and estimate the regression line for this sample. The point estimate of $\mu_{y|35}$ obtained from the regression line for the second sample is expected to be different. All possible samples of the same size taken from the same population will give different regression lines as shown in Figure 13.20, and, consequently, a different point estimate of $\mu_{y|x}$. Therefore, a confidence interval constructed for $\mu_{y|x}$ based on one sample will give a more reliable estimate of $\mu_{y|x}$ than will a point estimate.

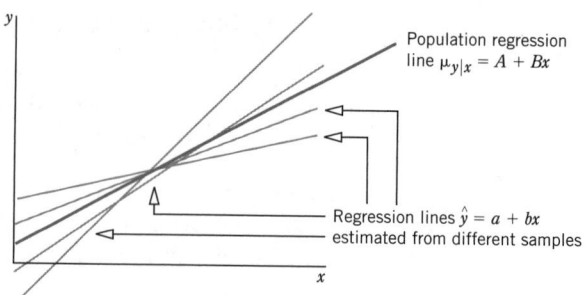

Figure 13.20 Population and sample regression lines.

To construct a confidence interval for $\mu_{y|x}$, we must know the mean, the standard deviation, and the shape of the sampling distribution of its point estimator $\hat{y}$.

The point estimator $\hat{y}$ of $\mu_{y|x}$ is normally distributed with a mean of $A + Bx$ and a standard deviation of

$$\sigma_{\hat{y}_m} = \sigma_\epsilon \sqrt{\frac{1}{n} + \frac{(x_0 - \bar{x})^2}{SS_{xx}}}$$

where $\sigma_{\hat{y}_m}$ is the standard deviation of $\hat{y}$ when it is used to estimate $\mu_{y|x}$, x_0 is the value of x for which we are estimating $\mu_{y|x}$, and σ_ϵ is the population standard deviation of ϵ.

However, usually σ_ϵ is not known. Rather, it is estimated by the standard deviation of sample errors s_e. In this case, we replace σ_ϵ by s_e and $\sigma_{\hat{y}_m}$ by $s_{\hat{y}_m}$ in the foregoing expression. To make a confidence interval for $\mu_{y|x}$, we use the normal distribution when the sample size is large ($n \geq 30$) and the t distribution when the sample size is small ($n < 30$).

CONFIDENCE INTERVAL FOR $\mu_{y|x}$

The $(1 - \alpha)100\%$ confidence interval for $\mu_{y|x}$ for $x = x_0$ is

$$\hat{y} \pm t s_{\hat{y}_m}$$

where the value of t is obtained from the t distribution table for $\alpha/2$ area in the right tail of the t distribution curve and $df = n - 2$. The value of $s_{\hat{y}_m}$ is calculated as follows.

$$s_{\hat{y}_m} = s_e \sqrt{\frac{1}{n} + \frac{(x_0 - \bar{x})^2}{SS_{xx}}}$$

Example 13–8 illustrates how to make a confidence interval for the mean value of y, $\mu_{y|x}$.

Constructing a confidence interval for the mean value of y.

EXAMPLE 13–8 Refer to Example 13–1 on incomes and food expenditures. Find a 99% confidence interval for the mean food expenditure for all households with a monthly income of $3500.

Solution Using the regression line estimated in Example 13–1, the point estimate of the mean food expenditure for $x = 35$ is

$$\hat{y} = 1.1414 + .2642 (35) = 10.3884$$

The confidence level is 99%. Hence, the area in each tail of the t distribution is

$$\alpha/2 = .5 - (.99/2) = .005$$

The degrees of freedom are

$$df = n - 2 = 7 - 2 = 5$$

From the t distribution table, the t value for .005 area in the right tail of the t distribution and 5 df is 4.032. From calculations in Examples 13–1 and 13–2 we know that

$$s_e = .9922, \qquad \bar{x} = 30.2857, \qquad \text{and} \qquad SS_{xx} = 801.4286$$

The standard deviation of $\hat{y}$ as an estimate of $\mu_{y|x}$ for $x = 35$ is calculated as follows.

$$s_{\hat{y}_m} = s_e \sqrt{\frac{1}{n} + \frac{(x_0 - \bar{x})^2}{SS_{xx}}} = (.9922) \sqrt{\frac{1}{7} + \frac{(35 - 30.2857)^2}{801.4286}} = .4098$$

Hence, the 99% confidence interval for $\mu_{y|35}$ is

$$\hat{y} \pm ts_{\hat{y}_m} = 10.3884 \pm 4.032 \, (.4098)$$

$$= 10.3884 \pm 1.6523 = 8.7361 \text{ to } 12.0407$$

Thus, with 99% confidence we can state that the mean food expenditure for all households with a monthly income of $3500 is between $873.61 and $1204.07. ▬

13.8.2   USING THE REGRESSION MODEL FOR PREDICTING A PARTICULAR VALUE OF *y*

The second major use of a regression model is to predict a particular value of y for a given value of x, say x_0. For example, we may want to predict the food expenditure of a randomly selected household with a monthly income of $3500. In this case, we are not interested in the mean food expenditure of all households with a monthly income of $3500 but in the food expenditure of one particular household with a monthly income of $3500. This predicted value of y is denoted by y_p. Again, to predict a single value of y for $x = x_0$ from the estimated sample regression line, we use the value of $\hat{y}$ *as a point estimate of* y_p. Using the estimated regression line, we find that $\hat{y}$ for $x = 35$ is

$$\hat{y} = 1.1414 + .2642(35) = 10.3884$$

Thus, based on our regression line, the point estimate for the food expenditure of a given household with a monthly income of $3500 is $1038.84 per month. Note that $\hat{y} = 1038.84$ is the point estimate for the mean food expenditure for all households with $x = 35$ as well as for the predicted value of food expenditure of one household with $x = 35$.

Different regression lines estimated by using different samples of seven households each taken from the same population will give different values of the point estimator for the predicted value of y for $x = 35$. Hence, a confidence interval constructed for y_p based on one sample will give a more reliable estimate of y_p than will a point estimate. The confidence interval constructed for y_p is more commonly called a **prediction interval**.

The procedure to construct a prediction interval for y_p is similar to that for constructing a confidence interval for $\mu_{y|x}$ except that the standard deviation of $\hat{y}$ is larger when we predict a single value of y than when we estimate $\mu_{y|x}$.

The point estimator $\hat{y}$ of y_p is normally distributed with a mean of $A + Bx$ and a standard deviation of

$$\sigma_{\hat{y}_p} = \sigma_\epsilon \sqrt{1 + \frac{1}{n} + \frac{(x_0 - \bar{x})^2}{SS_{xx}}}$$

where $\sigma_{\hat{y}_p}$ is the standard deviation of the predicted value of y, x_0 is the value of x for which we are predicting y, and σ_ϵ is the population standard deviation of ϵ.

However, usually σ_ϵ is not known. In this case, we replace σ_ϵ by s_e and $\sigma_{\hat{y}_p}$ by $s_{\hat{y}_p}$ in the foregoing expression. To make a prediction interval for y_p, we use the normal distribution when the sample size is large ($n \geq 30$) and the t distribution when the sample size is small ($n < 30$).

PREDICTION INTERVAL FOR y_p

The $(1 - \alpha)100\%$ prediction interval for the predicted value of y, denoted by y_p, for $x = x_0$ is

$$\hat{y} \pm t s_{\hat{y}_p}$$

where the value of t is obtained from the t distribution table for $\alpha/2$ area in the right tail of the t distribution curve and $df = n - 2$. The value of $s_{\hat{y}_p}$ is calculated as follows

$$s_{\hat{y}_p} = s_e \sqrt{1 + \frac{1}{n} + \frac{(x_0 - \bar{x})^2}{SS_{xx}}}$$

Example 13–9 illustrates the procedure to make a prediction interval for a particular value of y.

Making a prediction interval for a particular value of y.

EXAMPLE 13–9   Refer to Example 13–1 on incomes and food expenditures. Find a 99% prediction interval for the predicted food expenditure for a randomly selected household with a monthly income of \$3500.

Solution   Using the regression line estimated in Example 13–1, the point estimate of the predicted food expenditure for $x = 35$ is given by

$$\hat{y} = 1.1414 + .2642\,(35) = 10.3884$$

The area in each tail of the t distribution for a 99% prediction level is

$$\alpha/2 = .5 - (.99/2) = .005$$

The degrees of freedom are

$$df = n - 2 = 7 - 2 = 5$$

From the t distribution table, the t value for .005 area in the right tail of the t distribution and 5 df is 4.032. From calculations in Examples 13–1 and 13–2,

$$s_e = .9922, \quad \bar{x} = 30.2857, \quad \text{and} \quad SS_{xx} = 801.4286$$

The standard deviation of $\hat{y}$ as an estimator of y_p for $x = 35$ is calculated as follows.

$$s_{\hat{y}_p} = s_e \sqrt{1 + \frac{1}{n} + \frac{(x_0 - \bar{x})^2}{SS_{xx}}}$$

$$= (.9922) \sqrt{1 + \frac{1}{7} + \frac{(35 - 30.2857)^2}{801.4286}} = 1.0735$$

Hence, the 99% prediction interval for y_p for $x = 35$ is

$$\hat{y} \pm ts_{\hat{y}_p} = 10.3884 \pm 4.032\,(1.0735)$$

$$= 10.3884 \pm 4.3284 = 6.0600 \text{ to } 14.7168$$

Thus, with 99% confidence we can state that the predicted food expenditure of a household with a monthly income of $3500 is between $606.00 and $1471.68. ▬

As we can observe, this interval is much wider than the one for the mean value of y for $x = 35$ calculated in Example 13–8, which was $873.61 to $1204.07. This is always true. The prediction interval for predicting a single value of y is always larger than the confidence interval for estimating the mean value of y for a certain value of x.

13.9 CAUTIONS IN USING REGRESSION

When carefully applied, regression is a very helpful technique for making predictions and estimations about one variable for a certain value of another variable. However, we need to be cautious while using the regression analysis, for it can give us misleading results and predictions. Following are the two most important points to remember while using regression.

Extrapolation

The regression line estimated for the sample data is true only for the range of x values observed in the sample. For example, the values of x in our example on incomes and food expenditures vary from a minimum of 15 to a maximum of 49. Hence, our estimated regression line is only applicable for values of x between 15 and 49, that is, we should use this regression line to estimate the mean food expenditure or to predict the food expenditure of a single household only for income levels between $1500 and $4900. If we estimate or predict y for a value of x either less than 15 or greater than 49, it is called *extrapolation*. This does not mean that we should never use the regression line for extrapolation. Instead, we should interpret such predictions cautiously and not attach much value to them.

Similarly, if the data used for the regression estimation are time series data, the predicted values of y for periods outside the time interval used for the estimation of the regression line should be interpreted very cautiously. When using the estimated regression line for extrapolation, we are assuming that the same linear relationship between the two variables holds true for the values of x outside the given range. It is possible that the relationship between the two variables may not be linear outside that range. Nonetheless, even if it is linear, adding a few more observations at either end will probably give a new estimation of the regression line.

Causality

The regression line does not prove causality between two variables. That is, it does not predict that a change in y is caused by a change in x. The information about causality is based on theory or common sense. A regression line only describes whether or not a significant quantitative relationship between x and y exists. Significant relationship means that we reject the null hypothesis H_0: $B = 0$ at a given significance level. The estimated regression line gives the change in y due to a change of one unit in x. Note that it does not indicate that the reason y has changed is that x has changed. In our example on incomes and food expenditures, it is economic theory and common sense, not the regression line, that tells us that food expenditure depends on income. The regression analysis simply helps to determine whether or not this dependence is significant.

EXERCISES

Concepts and Procedures

13.80　Briefly explain the difference between estimating the mean value of y and predicting a particular value of y using a regression model.

13.81　Construct a 99% confidence interval for the mean value of y and a 99% prediction interval for the predicted value of y for the following.

　　a.　$\hat{y} = 3.25 + .80x$ for $x = 15$ given $s_e = .954$, $\bar{x} = 18.52$, $SS_{xx} = 144.65$, and $n = 10$
　　b.　$\hat{y} = -27 + 7.67x$ for $x = 12$ given $s_e = 2.46$, $\bar{x} = 13.43$, $SS_{xx} = 369.77$, and $n = 10$

13.82　Construct a 95% confidence interval for the mean value of y and a 95% prediction interval for the predicted value of y for the following.

　　a.　$\hat{y} = 13.40 + 2.58x$ for $x = 8$ given $s_e = 1.29$, $\bar{x} = 11.30$, $SS_{xx} = 210.45$, and $n = 12$
　　b.　$\hat{y} = -8.6 + 3.72x$ for $x = 24$ given $s_e = 1.89$, $\bar{x} = 19.70$, $SS_{xx} = 315.40$, and $n = 10$

Applications

13.83　Refer to Exercise 13.51. Construct a 90% confidence interval for the mean monthly salary of secretaries with 10 years of experience. Construct a 90% prediction interval for the monthly salary of a randomly selected secretary with 10 years of experience.

13.84　Refer to the data on the years of experience and the number of CD players assembled by workers given in Exercise 13.52. Construct a 99% confidence interval for $\mu_{y|x}$ for $x = 12$ and a 99% prediction interval for y_p for $x = 12$.

13.85　Refer to Exercise 13.75. Construct a 99% confidence interval for the mean yield of corn per acre for all acres on which 90 pounds of fertilizer are used. Determine a 99% prediction interval for the yield of corn for a randomly selected acre on which 90 pounds of fertilizer are used.

13.86　Refer to Exercise 13.76. Construct a 95% confidence interval for the mean charitable contributions made by all households with an income of $64,000. Make a 95% prediction interval for the charitable contributions made by a randomly selected household with an income of $64,000.

13.87　Refer to Exercise 13.77. Construct a 98% confidence interval for the mean starting salary of recent college graduates with a GPA of 3.15. Construct a 98% prediction interval for the starting salary of a randomly selected recent college graduate with a GPA of 3.15.

13.88　Using the data on ages and maintenance costs of eight cars given in Exercise 13.78, find a 90% confidence interval for the mean maintenance cost per month for all 10-year-old cars. Make a 90% prediction interval for the maintenance cost per month for a randomly selected 10-year-old car.

GLOSSARY

Coefficient of determination A measure that gives the proportion (or percentage) of the total variation in a dependent variable that is explained by a given independent variable.

Degrees of freedom for a simple linear regression model Sample size minus 2, that is, $n - 2$.

Dependent variable The variable to be predicted or explained.

Deterministic model A model in which the independent variable determines the dependent variable exactly. Such a model gives an **exact relationship** between two variables.

Estimated or predicted value of y The value of the dependent variable, denoted by $\hat{y}$, that is calculated for a given value of x using the estimated regression model.

Independent or **explanatory variable** The variable included in a model to explain the variation in the dependent variable.

Least squares estimates of A **and** B The values of a and b that are calculated by using the sample data.

Least squares method The method used to fit a regression line through a scatter diagram such that the error sum of squares is minimum.

Least squares regression line A regression line obtained by using the least squares method.

Linear correlation coefficient A measure of the strength of the linear relationship between two variables.

Linear regression model A regression model that gives a straight line relationship between two variables.

Multiple regression model A regression model that contains two or more independent variables.

Negative relationship between two variables The value of the slope in the regression line and the correlation coefficient between two variables are both negative.

Nonlinear (simple) regression model A regression model that does not give a straight line relationship between two variables.

Population parameters for a simple regression model The values of A and B for the regression model $y = A + Bx + \epsilon$ that are obtained by using population data.

Positive relationship between two variables The value of the slope in the regression line and the correlation coefficient between two variables are both positive.

Prediction interval The confidence interval for a particular value of y for a given value of x.

Probabilistic or **statistical model** A model in which the independent variable does not determine the dependent variable exactly.

Random error term (ϵ) The difference between the actual and predicted values of y.

Scatter diagram or **scattergram** A plot of the paired observations of x and y.

Simple linear regression A regression model with one dependent and one independent variable that assumes a straight line relationship.

Slope The coefficient of x in a regression model that gives the change in y for a change of one unit in x.

SSE (error sum of squares) The sum of the squared differences between the actual and predicted values of y. It is that portion of the SST that is not explained by the regression model.

SSR (regression sum of squares) That portion of the SST that is explained by the regression model.

SST (total sum of squares) The sum of the squared differences between actual y values and $\bar{y}$.

Standard deviation of errors A measure of spread for the random errors.

y-intercept The point at which the regression line intersects the vertical axis on which the dependent variable is marked. It is the value of y when x is zero.

KEY FORMULAS

1. **Simple linear regression model**

$$y = A + Bx + \epsilon$$

2. **Estimated simple linear regression model**

$$\hat{y} = a + bx$$

3. **Least squares estimates of A and B**

$$b = \frac{SS_{xy}}{SS_{xx}} \quad \text{and} \quad a = \bar{y} - b\bar{x}$$

where

$$SS_{xy} = \Sigma xy - \frac{(\Sigma x)(\Sigma y)}{n}$$

$$SS_{xx} = \Sigma x^2 - \frac{(\Sigma x)^2}{n}$$

4. **Standard deviation of the sample errors**

$$s_e = \sqrt{\frac{SS_{yy} - b\,SS_{xy}}{n - 2}}$$

where

$$SS_{yy} = \Sigma y^2 - \frac{(\Sigma y)^2}{n}$$

5. **Error sum of squares**

$$SSE = \Sigma e^2 = \Sigma(y - \hat{y})^2$$

6. **Total sum of squares**

$$SST = \Sigma y^2 - \frac{(\Sigma y)^2}{n}$$

7. **Regression sum of squares**

$$SSR = SST - SSE$$

8. **Coefficient of determination**

$$r^2 = \frac{b\,SS_{xy}}{SS_{yy}}$$

9. **Standard deviation of b**

$$s_b = \frac{s_e}{\sqrt{SS_{xx}}}$$

10. **The $(1 - \alpha)100\%$ confidence interval for B**

$$b \pm ts_b$$

11. **Value of the test statistic t for b**

$$t = \frac{b - B}{s_b}$$

12. **Linear correlation coefficient**

$$r = \frac{SS_{xy}}{\sqrt{SS_{xx}\,SS_{yy}}}$$

13. The $(1 - \alpha)100\%$ confidence interval for $\mu_{y|x}$

$$\hat{y} \pm ts_{\hat{y}_m}$$

where
$$s_{\hat{y}_m} = s_e \sqrt{\frac{1}{n} + \frac{(x_0 - \bar{x})^2}{SS_{xx}}}$$

14. The $(1 - \alpha)100\%$ prediction interval for y_p

$$\hat{y} \pm ts_{\hat{y}_p}$$

where
$$s_{\hat{y}_p} = s_e \sqrt{1 + \frac{1}{n} + \frac{(x_0 - \bar{x})^2}{SS_{xx}}}$$

SUPPLEMENTARY EXERCISES

13.89 The following table gives the total payroll (in millions of dollars) as of March 1989 and the percentage of games won during the 1988 season by six of the American League baseball teams.

Team	Total Payroll (x)	Percentage of Games Won (y)
Baltimore Orioles	9	34
California Angels	14	46
Kansas City Royals	18	52
Oakland Athletics	15	64
Seattle Mariners	10	42
Texas Rangers	11	44

    a. Construct a scatter diagram for these data. Take total payroll as an independent variable and percentage of games won as a dependent variable. Does the scatter diagram exhibit a linear relationship between the two variables?

    b. Find the least squares regression line $\hat{y} = a + bx$.

    c. Give a brief interpretation of the values of a and b calculated in part b.

    d. Plot the regression line on the scatter diagram and show the errors by drawing vertical lines between scatter points and the regression line.

    e. Predict the percentage of games won for a team with a total payroll of $13 million.

    f. Compute r and r^2.

    g. Calculate the standard deviation of errors.

    h. Make a 99% confidence interval for B.

    i. Test at the 1% significance level if B is different from zero.

13.90 The following table gives the percentage of games won and the average attendance (in thousands) per home game for the 1988 season for seven of the National League baseball teams.

Team	Percentage of Games Won (x)	Average Attendance per Home Game (y)
Atlanta Braves	34	11
Chicago Cubs	48	27
Los Angeles Dodgers	58	38
New York Mets	63	40
Philadelphia Phillies	40	26
St. Louis Cardinals	47	36
San Diego Padres	52	19

a. Construct a scatter diagram for these data. Take the percentage of games won as an independent variable and average attendance as a dependent variable. Does the scatter diagram exhibit a linear relationship between the two variables?

b. Find the least squares regression line $\hat{y} = a + bx$.

c. Give a brief interpretation of the values of the y-intercept and the slope calculated in part b.

d. Plot the regression line on the scatter diagram and show the errors by drawing vertical lines between the scatter points and the regression line.

e. Predict the average attendance per home game for a team with 49 percent of games won.

f. Compute r and r^2.

g. Calculate the standard deviation of errors.

h. Make a 95% confidence interval for B.

i. Test at the 5% significance level if B is different from zero.

13.91 The following data give information on the ages (in years) and the number of breakdowns during the past month for a sample of seven machines at a large company.

Age	12	7	2	8	13	9	4
Number of breakdowns	9	5	1	4	11	7	2

a. Taking age as an independent variable and number of breakdowns as a dependent variable, what is your hypothesis about the sign of B in the regression line? (In other words, do you expect B to be positive or negative?)

b. Find the least squares regression line. Is the sign of b the same as you hypothesized for B in part a?

c. Give a brief interpretation of the values of a and b calculated in part b.

d. Compute r and r^2 and explain what they mean.

e. Compute the standard deviation of errors.

f. Construct a 98% confidence interval for B.

g. Test at the 2.5% significance level if B is positive.

13.92 The following table gives information on the number of hours that eight bank loan officers slept the previous night and the number of loan applications that they processed the next day.

Number of hours slept	8	5	7	6	4	8	6	5
Number of applications processed	14	10	16	11	8	15	10	8

a. Taking the number of hours slept as an independent variable and the number of applications processed as a dependent variable, do you expect B to be positive or negative in the regression model $y = A + Bx + \epsilon$?

b. Find the least squares regression line. Is the sign of b the same as you hypothesized for B in part a?

c. Compute r and r^2 and explain what they mean.

d. Compute the standard deviation of errors.

e. Construct a 90% confidence interval for B.

f. Test at the 5% significance level if B is positive.

13.93 The management of a supermarket wanted to check the effect of the number of commercials broadcast on local radio on the gross sales at the store. The management experimented for eight weeks by broadcasting a different number of commercials each week on local radio. The following table gives the number of commercials broadcast during each of the eight weeks and the gross sales (in millions of dollars) for those weeks.

Number of Commercials Broadcast per Week	Gross Sales per Week (millions of dollars)
22	3.64
16	3.12
28	4.08
12	2.84
30	3.98
19	3.55
24	4.02
32	4.38

   a. Find the least squares regression line $\hat{y} = a + bx$. Take the number of commercials broadcast as an independent variable and the gross sales as a dependent variable.

   b. Predict the gross sales for a week with 25 commercials.

   c. Compute r and r^2 and explain what they mean.

   d. Compute the standard deviation of errors.

   e. Construct a 95% confidence interval for B.

   f. Test at the 1% significance level if B is positive.

13.94 The management of a supermarket wants to find if there is a relationship between the number of times a specific product is promoted on the intercom system in the store and the number of units of that product sold. To experiment, the management selected a product and promoted it on the intercom system for seven days. The following table gives the number of times this product was promoted each day and the number of units sold.

Number of Promotions per Day	Number of Units Sold per Day (hundreds)
15	11
22	18
42	26
30	24
18	17
12	15
38	21

   a. With the number of promotions as an independent variable and the number of units sold as a dependent variable, what do you expect the sign of B in the regression line $y = A + Bx + \epsilon$ will be?

   b. Find the least squares regression line $\hat{y} = a + bx$. Is the sign of b the same as you hypothesized for B in part a?

   c. Give a brief interpretation of the values of a and b calculated in part b.

   d. Compute r and r^2 and explain what they mean.

   e. Predict the number of units of this product sold on a day with 35 promotions.

   f. Compute the standard deviation of errors.

   g. Construct a 98% confidence interval for B.

   h. Testing at the 1% significance level, can you conclude that B is positive?

13.95 The following table gives information on the temperature in a city and the volume of ice cream (in pounds) sold at an ice cream parlor for a random sample of eight days during the summer of 1993.

Temperature	93	86	77	89	98	102	87	79
Ice cream sold	187	169	123	198	232	267	158	117

a. Find the least squares regression line $\hat{y} = a + bx$. Take temperature as an independent variable and volume of ice cream sold as a dependent variable.
b. Give a brief interpretation of the values of a and b.
c. Compute r and r^2 and explain what they mean.
d. Predict the amount of ice cream sold on a day with a temperature of 95°.
e. Compute the standard deviation of errors.
f. Construct a 99% confidence interval for B.
g. Testing at the 1% significance level, can you conclude that B is different from zero?

13.96 The following table gives the average mortgage rate (in percentages) and the median prices of homes (in thousands of dollars) in the United States for the period 1981–1987. (Source: National Association of Realtors.)

Average mortgage rate	15.1	15.4	12.9	12.5	11.8	9.7	9.3
Median price	66.4	67.8	70.3	72.4	75.5	80.8	85.0

a. Find the least squares regression line with average mortgage rate as an independent variable and median price of homes as a dependent variable.
b. What is the predicted median price of a home if the average mortgage rate is 13.5%?
c. Compute r and r^2 and explain what they mean.
d. Compute the standard deviation of errors.
e. Construct a 99% confidence interval for B.
f. Testing at the 1% significance level, can you conclude that B is negative?

13.97 The following table gives the milk production (rounded to billions of pounds) in the United States for the years 1980 to 1991. (Source: U.S. Department of Agriculture.)

Year	Milk Production
1980	128
1981	133
1982	136
1983	140
1984	135
1985	143
1986	144
1987	143
1988	145
1989	144
1990	148
1991	149

a. Assign a value of 0 to 1980, 1 to 1981, 2 to 1982, and so on. Call this new variable *time*. Write a new table with the variables *time* and *milk production*.
b. With time as an independent variable and milk production as a dependent variable, compute SS_{xx}, SS_{yy}, and SS_{xy}.
c. Construct a scatter diagram for these data. Does the scatter diagram exhibit a linear positive relationship between time and milk production?
d. Find the least squares regression line $\hat{y} = a + bx$.
e. Give a brief interpretation of the values of a and b calculated in part d. (*Hint:* The value of b will give an increase in the milk production [in billions of pounds] per year for the period 1980 to 1991 based on a linear relationship between the two variables.)
f. Compute the correlation coefficient r.
g. Predict the milk production for $x = 16$. Comment on this prediction.

13.98 The following table gives the gross national product (in hundreds of billions of dollars) for the United States for the years 1980 to 1990.

Year	Gross National Product
1980	27.32
1981	30.53
1982	31.66
1983	34.06
1984	38.02
1985	40.54
1986	42.78
1987	45.45
1988	49.08
1989	52.48
1990	55.25

a. Assign a value of 0 to 1980, 1 to 1981, 2 to 1982, and so on. Call this new variable *time*. Write a new table with the variables *time* and *gross national product*.
b. With time as an independent variable and gross national product as a dependent variable, compute SS_{xx}, SS_{yy}, and SS_{xy}.
c. Construct a scatter diagram for these data. Does the scatter diagram exhibit a linear positive relationship between time and gross national product?
d. Find the least squares regression line $\hat{y} = a + bx$.
e. Give a brief interpretation of the values of a and b calculated in part d. (See the hint in Exercise 13.97 part e.)
f. Compute the correlation coefficient r.
g. Predict the gross national product for $x = 15$. Comment on this prediction.

13.99 Refer to the data on ages and number of breakdowns for seven machines given in Exercise 13.91. Construct a 99% confidence interval for the mean number of breakdowns per month for all machines with an age of eight years. Find a 99% prediction interval for the number of breakdowns per month for a randomly selected machine with an age of eight years.

13.100 Refer to the data on the number of commercials broadcast on local radio per week by a supermarket and the gross sales per week at this supermarket given in Exercise 13.93. Determine a 95% confidence interval for the mean gross sales for a week with 25 commercials. Make a 95% prediction interval for the gross sales for a randomly selected week with 25 commercials.

13.101 Refer to the data given in Exercise 13.94 on the number of times a specific product is promoted on the intercom system in a supermarket and the number of units of that product sold. Make a 90% confidence interval for the mean number of units of that product sold on days with 35 promotions. Construct a 90% prediction interval for the number of units of that product sold on a randomly selected day with 35 promotions.

13.102 Refer to the data given in Exercise 13.95 on temperatures and the volumes of ice cream sold at an ice cream parlor for a sample of eight days. Construct a 98% confidence interval for the mean volume of ice cream sold at this parlor on all days with a temperature of 95°. Determine a 98% prediction interval for the volume of ice cream sold at this parlor on a randomly selected day with a temperature of 95°.

SELF-REVIEW TEST

1. A simple regression is a regression model that contains
   a. only one independent variable
   b. only one dependent variable

   c.   more than one independent variable
   d.   both a and b

2.   The relationship between independent and dependent variables represented by the (simple) linear regression is that of

   a.   a straight line   b.   a curve   c.   both a and b

3.   A deterministic regression model is a model that

   a.   contains the random error term
   b.   does not contain the random error term
   c.   gives a nonlinear relationship

4.   A probabilistic regression model is a model that

   a.   contains the random error term
   b.   does not contain the random error term
   c.   shows an exact relationship

5.   The least squares regression line minimizes the sum of

   a.   errors   b.   squared errors   c.   predictions

6.   The degrees of freedom for a simple regression model are

   a. $n - 1$ b. $n - 2$ c. $n - 5$

7.   Indicate if the following statement is true or false.

      The coefficient of determination gives the proportion of total squared errors (SST) that is explained by the use of the regression model.

8.   Indicate if the following statement is true or false.

      The simple linear correlation measures the strength of the linear association between two variables.

9.   The value of the coefficient of determination is always in the range

   a.   0 to 1   b. -1 to 1   c. -1 to 0

10.   The value of the correlation coefficient is always in the range

   a.   0 to 1   b. -1 to 1   c. -1 to 0

11.   Explain why the random error term ϵ is added to the regression model.

12.   Explain the difference between A and a and between B and b for a regression model.

13.   Briefly explain the assumptions of a regression model.

14.   Briefly explain the difference between the population regression line and a sample regression line.

15.   The following table gives the experience (in years) and the number of computers sold during the previous three months by seven salespersons.

Experience	4	12	9	6	10	16	7
Computers sold	19	42	28	31	39	35	21

   a.   Do you think experience depends on the number of computers sold or the number of computers sold depends on experience?
   b.   With experience as an independent variable and the number of computers sold as a dependent variable, what is your hypothesis about the sign of B in the regression model?
   c.   Construct a scatter diagram for these data. Does the scatter diagram exhibit a linear relationship between the two variables?

d. Find the least squares regression line. Is the sign of b the same as the one you hypothesized for B in part b?

e. Give a brief interpretation of the values of the y-intercept and slope calculated in part d.

f. Compute r and r^2 and explain what they mean.

g. Predict the number of computers sold during the past three months by a salesperson with 11 years of experience.

h. Compute the standard deviation of errors.

i. Construct a 99% confidence interval for B.

j. Testing at the 1% significance level, can you conclude that B is positive?

k. Construct a 95% confidence interval for the mean number of computers sold by all salespersons with 8 years of experience.

l. Make a 95% prediction interval for the number of computers sold by a randomly selected salesperson with 8 years of experience.

USING MINITAB

The MINITAB commands given in Figure 13.21 can be used to do the regression analysis discussed in this chapter.

Figure 13.21   MINITAB commands for regression analysis.

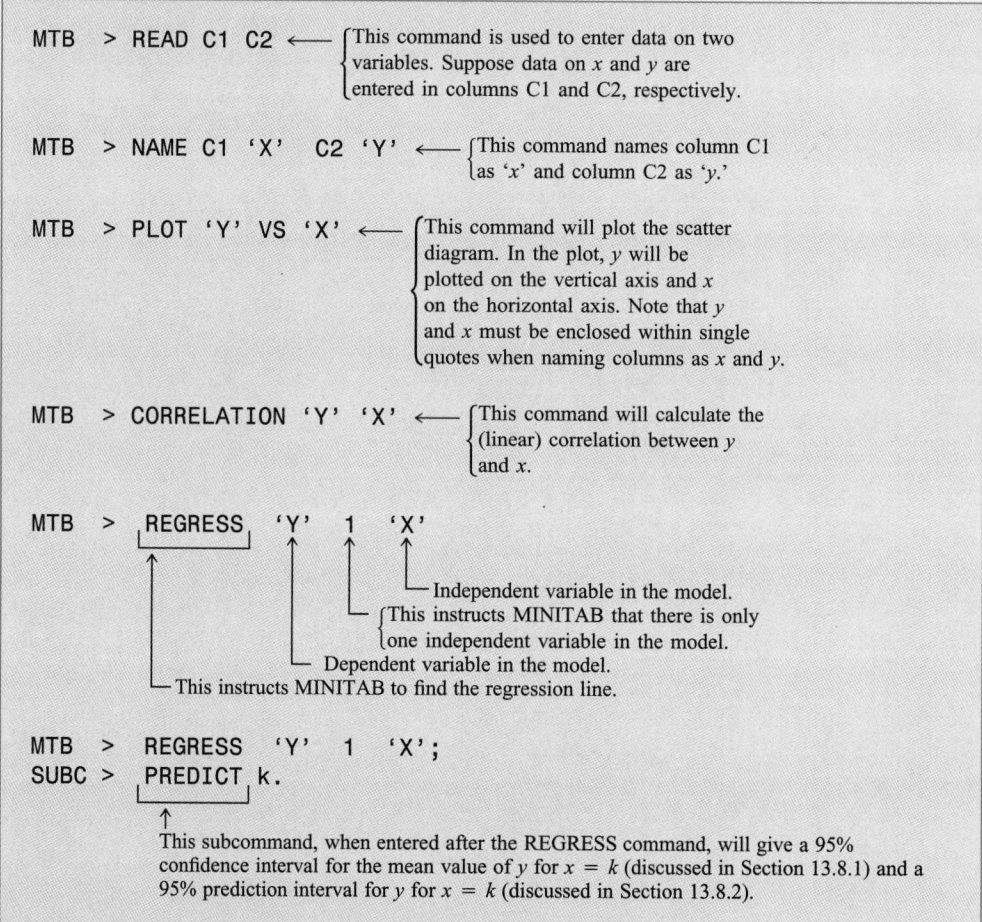

Illustration M13−1 describes the use of these MINITAB commands.

Illustration M13-1 Refer to the data of Example 13–7 on driving experience (in years) and monthly auto insurance premiums (in dollars) for eight auto drivers. That data set is reproduced below.

Driving Experience (years)	Monthly Auto Insurance Premium (dollars)
5	64
2	87
12	50
9	71
15	44
6	56
25	42
16	60

Use MINITAB to answer the following.

(a) Construct a scatter diagram for these data.

(b) Find the correlation between these two variables.

(c) Find the regression line with experience as an independent variable and premium as a dependent variable.

(d) Make a 90% confidence interval for B.

(e) Test at the 5% significance level if B is negative.

(f) Make a 95% confidence interval for the mean monthly auto insurance premium for all drivers with 10 years of driving experience. Construct a 95% prediction interval for the monthly auto insurance premium for a randomly selected driver with 10 years of driving experience.

Solution First we enter the data into MINITAB using the READ command, as shown in Figure 13.22.

Figure 13.22 Entering data into MINITAB.

```
MTB  > READ C1 C2
DATA >   5    64
DATA >   2    87
DATA >  12    50
DATA >   9    71
DATA >  15    44
DATA >   6    56
DATA >  25    42
DATA >  16    60
DATA > END
```

In our example, experience is the independent variable and premium is the dependent variable. Hence, the data entered in column C1 represent the 'x' variable and those entered in column C2 represent the 'y' variable. In Figure 13.23, we name column C1 as 'x' and column C2 as 'y'.

Figure 13.23   Naming columns.

```
MTB  > NAME C1 'X' C2 'Y'
```

(a)   The scatter diagram is constructed using the PLOT command, as shown in Figure 13.24.

Figure 13.24   Scatter diagram.

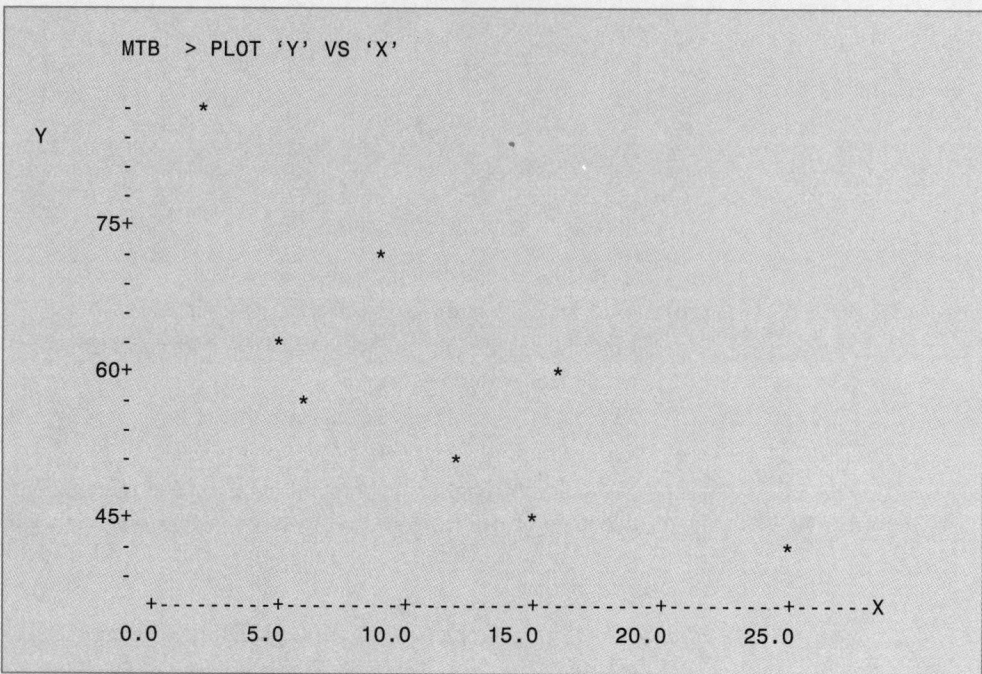

(b)   The CORRELATION command gives the linear correlation coefficient between x and y, as shown in Figure 13.25.

Figure 13.25   Correlation coefficient.

```
MTB > CORRELATION 'Y' 'X'

CORRELATION OF Y AND X = -0.768  ⟵  This is the linear
                                     correlation coefficient r.
```

Hence, the linear correlation coefficient between x and y is

$$r = -.768$$

(c)   The regression line is obtained using the REGRESS command. Figure 13.26 shows the MINITAB output obtained in response to the REGRESS command.

Figure 13.26 MINITAB regression output.

```
MTB  > REGRESS 'Y' 1 'X'

THE REGRESSION EQUATION IS
Y = 76.7 - 1.55 X

PREDICTOR        COEF      STDEV    T-RATIO        P
CONSTANT       76.660      6.961      11.01    0.000
X             -1.5476      0.5270     -2.94    0.026


S = 10.32      R-SQ = 59.0%     R-SQ(ADJ) = 52.1%

ANALYSIS OF VARIANCE

SOURCE          DF          SS         MS        F          P
REGRESSION       1       918.5      918.5     8.62      0.026
ERROR            6       639.0      106.5
TOTAL            7      1557.5
```

The MINITAB output given in Figure 13.26 has four main parts. The first part gives the regression equation, which is

$$\hat{Y} = 76.7 - 1.55X$$

Thus, $a = 76.7$ and $b = -1.55$

The second part of the MINITAB output gives the values of a and b, their standard deviations, the values of the test statistic t for the hypothesis tests about A and B, and the p-values for these two tests. Figure 13.27 explains this part of the output.

Figure 13.27

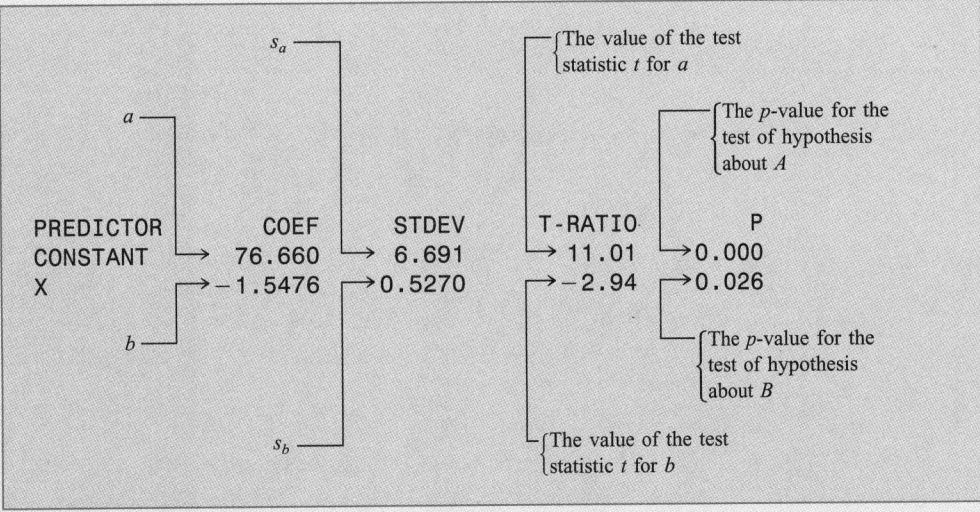

Note that we have not discussed the standard deviation s_a of a and the test of hypothesis about a (and, hence, the value of the test statistic t for a and the p-value for the test of hypothesis about a) in this chapter. Therefore, in the row for CONSTANT in Figure 13.27, all the values except the coefficient value are irrelevant for this chapter. If we compare the various values in this figure with the ones calculated in Example 13–7, we notice a slight difference due to rounding.

The third part of the MINITAB output in Figure 13.26 gives the standard deviation s_e of errors, the coefficient of determination r^2, and the adjusted r^2 (the value of r^2 adjusted for the degrees of freedom). However, we have not discussed the concept of adjusted r^2 in this chapter. The explanation of this part of the output appears in Figure 13.28.

Figure 13.28

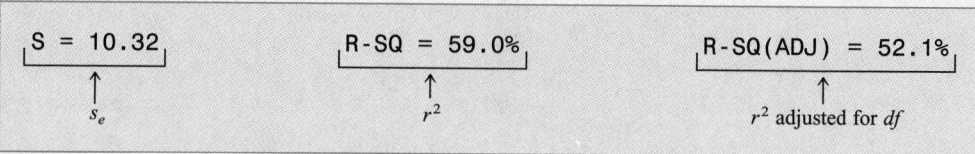

The fourth part of the MINITAB output gives the analysis of variance table. We have not discussed such a table in this chapter except for the SS column. However, this table is similar to the analysis of variance table discussed in Chapter 12. The column of SS gives the values of SSR, SSE, and SST as shown in Figure 13.29.

Figure 13.29

SOURCE	SS	
REGRESSION	918.5	◄——SSR
ERROR	639.0	◄——SSE
TOTAL	1557.5	◄——SST

(d)   The 90% confidence interval for B is given by the formula

$$b \pm ts_b$$

From the row corresponding to x in Figure 13.27,

$$b = -1.5476, \quad \text{and} \quad s_b = 0.5270$$

For a 90% confidence interval, $\alpha = .10$ and $\alpha/2 = .05$. The degrees of freedom are $n - 2 = 8 - 2 = 6$. From the t distribution table, the t value for .05 area in the right tail of the t distribution curve and 6 df is 1.943. The 90% confidence interval for B is

$$b \pm ts_b = -1.5476 \pm 1.943\,(.5270) = -1.5476 \pm 1.0240 = -2.57 \text{ to } -.52$$

(e)   The null and alternative hypotheses are

$$H_0: B = 0 \quad (B \text{ is not negative})$$

$$H_1: B < 0 \quad (B \text{ is negative})$$

The significance level is .05. The degrees of freedom are 6. The test is left-tailed. From the t distribution table, the critical value of t for $\alpha = .05$ and $df = 6$ is -1.943.

We again use the information given in the row corresponding to x in Figure 13.27. From that row of information,

$$\text{Value of the test statistic } t \text{ for } b = -2.94$$

Because the value of the test statistic $t = -2.94$ is less than the critical value of $t = -1.943$, it falls in the rejection region, which is in the left tail of the t distribution curve (see Figure 13.19 of Example 13–7). Consequently, we reject the null hypothesis and state that our sample information supports the alternative hypothesis that B is negative.

(f) To make a 95% confidence interval for $\mu_{y|x}$ and a 95% prediction interval for y_p for $x = 10$, we use the subcommand PREDICT with the REGRESS command. This is illustrated in Figure 13.30. Note that when we use the command and subcommand given in Figure 13.30, MINITAB will give all the output that we listed in Figure 13.26 and the output given in Figure 13.30. We have omitted the output of Figure 13.26 from Figure 13.30.

Figure 13.30

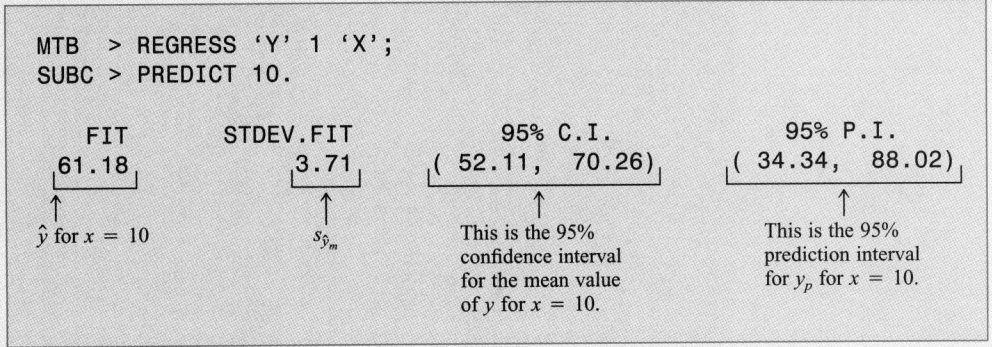

Thus, from Figure 13.30, the value of $\hat{y}$ for $x = 10$ is 61.18. This can be considered as the point estimate of the mean value of y as well as the point estimate of the predicted value of y for $x = 10$. The interval 52.11 to 70.26 in the printout gives a 95% confidence interval for the mean value $\mu_{y|10}$ of y for $x = 10$ as was discussed in Section 13.8.1. The interval 34.34 to 88.02 gives a 95% prediction interval for y for $x = 10$ as was discussed in Section 13.8.2. ◼

COMPUTER ASSIGNMENTS

M13.1 Professor Hamid Zangenehzadeh studied the relationship between student evaluations of a teacher and the expected grades of students in a course. The following table lists the (average of the) ratings of teachers by students and (the average of the) students' expected grades for 39 faculty members of the three departments of the School of Management at Widener University as reported in this study. (*Note:* The ratings of teachers reported in this table are what the author called the *unadjusted ratings of teachers.*)

Ratings of Teachers	Students' Expected Grades	Ratings of Teachers	Students' Expected Grades
3.833	3.500	2.739	3.000
3.769	3.769	2.543	2.829
3.642	3.214	2.286	3.143
3.625	3.250	2.278	2.833
3.529	3.529	2.133	2.800
3.500	3.300	2.103	2.620
3.500	3.500	2.053	2.368
3.409	3.864	2.043	2.696
3.380	3.048	1.944	2.944
3.333	3.200	1.923	2.846
3.294	3.059	1.800	2.800
3.267	3.000	1.800	3.000
3.263	3.368	1.692	2.769
3.120	3.440	1.692	3.462
3.045	2.909	1.688	3.125
3.000	3.500	1.667	4.000
3.000	3.500	1.625	2.375
2.923	2.538	1.333	3.555
2.826	3.086	0.521	2.652
2.778	3.111		

Source: Hamid Zangenehzadeh: "Grade Inflation: A Way Out." *Journal of Economic Education*, Summer 1988, 217–226. Reprinted with permission of the Helen Dwight Reid Educational Foundation. Published by Heldref Publications, Washington, D.C. Copyright © 1988.

Use MINITAB to answer the following.

 a. Construct a scatter diagram for these data.
 b. Find the correlation between the two variables.
 c. Find the regression line with ratings of teachers as a dependent variable and students' expected grades as an independent variable.
 d. Make a 95% confidence interval for B.
 e. Test at the 1% significance level if B is positive.

(*Note:* As the sample size is large, $n = 39$, the normal distribution can be applied to construct a confidence interval and to make a test of hypothesis about B in parts d and e if we so desire.)

M13.2 Refer to Data Set I of Appendix A. Using MINITAB, find the regression line with sales as an independent variable and profits as a dependent variable. Explain your results.

M13.3 Refer to the data on the average mortgage rates (in percentages) and the median prices of homes (in thousands of dollars) given in Exercise 13.96. Using MINITAB,

 a. Construct a scatter diagram for these data.
 b. Find the least squares regression line with average mortgage rate as an independent variable and median price of homes as a dependent variable.
 c. Compute the correlation coefficient.
 d. Construct a 99% confidence interval for B.
 e. Test at the 5% significance level if B is negative.

14 MULTIPLE REGRESSION

In Chapter 13 we discussed simple linear regression and linear correlation. A simple regression model includes one independent and one dependent variable, and it presents a very simplified scenario of real-world situations. In the real world, a dependent variable is usually influenced by a number of independent variables. For example, the sales of a company's product may be determined by the price of its product, the quality of the product, and advertising expenditure incurred by the company to promote that product. Therefore it makes more sense to use a regression model that includes more than one independent variable. Such a model is called a multiple regression model. In this chapter we will discuss multiple regression models.

14.1 MULTIPLE REGRESSION ANALYSIS[1]

The simple linear regression model discussed in Chapter 13 was written as

$$y = A + Bx + \epsilon$$

This model includes one independent variable, which is denoted by x, and one dependent variable, which is denoted by y.

Usually a dependent variable is affected by more than one independent variable. When we include two or more independent variables in a regression model, it is called a **multiple regression model**. Remember, whether it is a simple or a multiple regression model, it always includes one dependent variable.

A multiple regression model with y as a dependent variable and $x_1, x_2, x_3, \ldots,$ and x_k as independent variables is written as

$$y = A + B_1x_1 + B_2x_2 + B_3x_3 + \cdots + B_kx_k + \epsilon \tag{1}$$

where A represents the constant term and $B_1, B_2, B_3, \ldots,$ and B_k are the regression coefficients of independent variables $x_1, x_2, x_3, \ldots,$ and x_k, respectively. The ϵ represents the random error term. Note that this model contains k independent variables, $x_1, x_2, x_3, \ldots,$ and x_k.

In regression model (1), A represents the constant term, which gives the value of y when all independent variables assume zero values. The coefficients $B_1, B_2, B_3, \ldots,$ and B_k are called the **partial regression coefficients**. For example, B_1 is a partial regression coefficient of x_1. It gives the change in y due to a one-unit change in x_1 when all other independent variables included in the model are held constant. In other words, if we change x_1 by one unit but keep $x_2, x_3, \ldots,$ and x_k unchanged, then the resulting change in y is measured by B_1. Similarly the value of B_2 gives a change in y due to a one-unit change in x_2 when all other independent variables are held constant. In model (1) above, $A, B_1, B_2, B_3, \ldots,$ and B_k are called the *true regression coefficients* or *population parameters*.

A positive value for a particular B_i in model (1) will indicate a positive relationship between y and the corresponding x_i variable. A negative value for a particular B_i in that model will indicate a negative relationship between y and the corresponding x_i variable.

Note that regression model (1) is linear. The relationship between each x_i and y is a straight line relationship. In this model, $A + B_1x_1 + B_2x_2 + B_3x_3 + \cdots + B_kx_k$ is called the *deterministic portion* and ϵ is called the *stochastic portion* of the model.

When we use the t distribution to make inferences about a single parameter of a multiple regression model, the **degrees of freedom** are calculated as

$$df = n - k - 1$$

where n represents the sample size and k is the number of independent variables in the model.

However, when we use the F distribution to make inferences about parameters of a multiple regression model, the degrees of freedom for the numerator and denominator are calculated as follows.

$$\text{degrees of freedom for the numerator} = k$$

$$\text{degrees of freedom for the denominator} = n - k - 1$$

---

[1]We will discuss only multiple linear regression models in this chapter. Multiple nonlinear regression models will not be discussed here.

> **MULTIPLE REGRESSION MODEL**
>
> A regression model that includes two or more independent variables is called a multiple regression model. It is written as
>
> $$y = A + B_1 x_1 + B_2 x_2 + B_3 x_3 + \cdots + B_k x_k + \epsilon$$
>
> where y is the dependent variable and x_1, x_2, x_3, ..., and x_k are the k independent variables. The ϵ is the random error term.

In a situation where a multiple regression model includes only two independent variables, model (1) reduces to

$$y = A + B_1 x_1 + B_2 x_2 + \epsilon$$

and in this case $k = 2$.

A multiple regression model that includes three independent variables is written as

$$y = A + B_1 x_1 + B_2 x_2 + B_3 x_3 + \epsilon$$

and in this case $k = 3$.

If model (1) is estimated using sample data, which is usually the case, the estimated regression model is written as

$$\hat{y} = a + b_1 x_1 + b_2 x_2 + b_3 x_3 + \cdots + b_k x_k \qquad (2)$$

In model (2), a, b_1, b_2, b_3, ..., and b_k are the sample statistics, which are the point estimators of A, B_1, B_2, B_3, ..., and B_k, respectively.

In model (1), y denotes the actual values of the dependent variable. In model (2), $\hat{y}$ denotes the predicted or estimated values of the dependent variable. The difference between y and $\hat{y}$ gives the error of prediction. For a multiple regression model,

$$\text{SSE} = \Sigma(y - \hat{y})^2$$

where SSE stands for the error sum of squares.

As in Chapter 13, the estimated regression model (2) is obtained by minimizing the sum of squared errors, that is,

$$\text{Minimize } \Sigma(y - \hat{y})^2$$

The estimated model (2) obtained by minimizing the sum of squared errors is called the **least squares regression model**.

Usually the calculations for a multiple regression model are made by using statistical software packages for computers, such as MINITAB, instead of using the formulas manually. Even for a multiple regression model with two independent variables, the formulas are complex and manual calculations are time consuming. In this chapter we will analyze the multiple regression models using MINITAB statistical software. The solutions obtained by using other statistical software packages such as SAS, SPSS, BMDP, or MYSTAT can be interpreted the same way.

14.2 ASSUMPTIONS OF THE MULTIPLE REGRESSION MODEL

Like the simple linear regression model, the multiple (linear) regression analysis is also based on certain assumptions. Consider the multiple regression model (1), which is

$$y = A + B_1x_1 + B_2x_2 + B_3x_3 + \cdots + B_kx_k + \epsilon$$

There are six assumptions made about this model that are described next. Note that these assumptions are made about the population regression model and not about the sample regression model.

Assumption 1: The mean of the probability distribution of ϵ is zero, that is,

$$E(\epsilon) = 0$$

In other words, if we calculate errors for all measurements for a given set of values of independent variables for a population data set, the mean of these errors will be zero. When this assumption holds true, the mean value of y is given by the deterministic part of regression model (1). Thus,

$$E(y) = A + B_1x_1 + B_2x_2 + B_3x_3 + \cdots + B_kx_k$$

where $E(y)$ is the expected or mean value of y. The mean value of y is also denoted by $\mu_{y|x_1,x_2,\ldots,x_k}$.

Assumption 2: The errors associated with different sets of values of independent variables are independent.

Assumption 3: The probability distribution of errors is normal.

Assumption 4: The probability distribution of population errors for each set of values of independent variables has the same (constant) standard deviation, which is denoted by σ_ϵ.

Assumption 5: The independent variables are not linearly related. However, they can have a nonlinear relationship. When independent variables are highly linearly correlated, it is referred to as **multicollinearity**. This assumption is about the nonexistence of the multicollinearity problem. For example, consider the following multiple regression model.

$$y = A + B_1x_1 + B_2x_2 + B_3x_3 + \epsilon$$

All of the following relationships (and other similar relationships) between x_1, x_2, and x_3 are linear and, consequently, should be invalid for this model.

$$x_1 = x_2 + 4x_3$$
$$x_2 = 5x_1 - 2x_3$$
$$x_1 = 3.5x_2$$

If any of these linear relationships is true, by making a substitution we can eliminate one variable from our model and reduce the number of independent variables to two.

However, the following type of nonlinear relationships between x_1, x_2, and x_3 are permissible.

$$x_1 = 4x_2^2$$

$$x_2 = 2x_1 + 6x_3^2$$

$$x_1 = 5x_2^3 - x_3$$

$$x_3 = 4x_1^2$$

Assumption 6: There exists a zero correlation between the random error term ϵ and each independent variable x_i.

14.3 STANDARD DEVIATION OF ERRORS

The **standard deviation of errors** (also called the standard deviation of estimate) for the multiple regression model (1) is denoted by σ_ϵ and it is a measure of the variation among errors. However, when sample data are used to estimate multiple regression model (1), the standard deviation of errors is denoted by s_e, which is also called the standard error of estimate. The formula for calculating s_e is as follows.

$$s_e = \sqrt{\frac{SSE}{n - k - 1}} \quad \text{where} \quad SSE = \Sigma(y - \hat{y})^2$$

We will not use this formula to calculate s_e manually. Rather we will obtain it from the computer solution.

14.4 COEFFICIENT OF MULTIPLE DETERMINATION

In Chapter 13, we denoted the coefficient of determination for a simple linear regression model by r^2 and defined it as the proportion of the total sum of squares SST that is explained by the regression model. The coefficient of determination for a multiple regression model, usually called the **coefficient of multiple determination**, is denoted by R^2 and is defined as the proportion of the total sum of squares SST that is explained by the multiple regression model. It tells us how good the multiple regression model is and how well the independent variables included in the model explain the dependent variable.

Like r^2, the value of the coefficient of multiple determination R^2 always lies in the range 0 to 1, that is,

$$0 \leq R^2 \leq 1$$

Just as in the case of the simple linear regression model, **SST** is the total sum of squares, **SSR** is the regression sum of squares, and **SSE** is the error sum of squares. SST is always equal to the sum of SSE and SSR. They are calculated as follows.

$$SSE = \Sigma e^2 = \Sigma(y - \hat{y})^2$$

$$SST = SS_{yy} = \Sigma(y - \bar{y})^2$$

$$SSR = \Sigma(\hat{y} - \bar{y})^2$$

SSR is the portion of SST that is explained by the use of the regression model and SSE is the portion of SST that is not explained by the use of the regression model. The coefficient of multiple determination is given by the ratio of SSR and SST as follows.

$$R^2 = \frac{SSR}{SST}$$

The coefficient of multiple determination R^2 has one major shortcoming. The value of R^2 generally increases as we add more and more explanatory variables to the regression model. Therefore, by adding a large number of explanatory variables to our regression model (even if they do not belong in the model) we can make the value of R^2 very close to 1. Such a value of R^2 will be misleading, and it will not represent the true explanatory power of the regression model. To eliminate this shortcoming of R^2, it is preferable to use the **adjusted coefficient of multiple determination**, which is denoted by $\overline{R}^2$. Note that $\overline{R}^2$ is the coefficient of multiple determination adjusted for degrees of freedom. The value of $\overline{R}^2$ may increase, decrease, or stay the same as we add more explanatory variables to our regression model. If a new variable added to the regression model contributes significantly to explain the variation in y, then $\overline{R}^2$ increases; otherwise it decreases. The value of $\overline{R}^2$ is calculated as follows.

$$\overline{R}^2 = 1 - (1 - R^2)\frac{n-1}{n-k-1} \quad or \quad 1 - \frac{SSE/(n-k-1)}{SST/(n-1)}$$

Thus, if we know R^2, we can find the value of $\overline{R}^2$. Almost all statistical software packages give the values of both R^2 and $\overline{R}^2$ for a regression model.

Another property of $\overline{R}^2$ to remember is that whereas R^2 can never be negative, $\overline{R}^2$ can be negative.

14.5   COMPUTER SOLUTION OF MULTIPLE REGRESSION

In this section we take an example of a multiple regression model, solve it using MINITAB, interpret the solution, and make inferences about the population parameters of the regression model.

Figure 14.1 gives the MINITAB command that is used to estimate a multiple regression model. The data are entered in columns C1, C2, C3, C4, etc. Column C1 contains data on the dependent variable and columns C2, C3, C4, . . . contain data on the independent variables. In this MINITAB command, k is replaced by the number of independent variables in the multiple regression model.

Figure 14.1   MINITAB command for multiple regression analysis.

```
MTB  > REGRESS  C1  k  C2  C3  C4 ...
                    ↑
             In this command, k refers to the number of independent
             variables in the regression model.
```

Obtaining MINITAB solution for a multiple regression model.

EXAMPLE 14–1   A researcher wanted to find the effect of driving experience and the number of driving violations on auto insurance premiums. A random sample of 12 drivers insured with a company and having similar auto insurance policies was selected. Table 14.1 lists the monthly auto insurance premiums (in dollars) paid by these drivers, their driving experience (in years), and the number of driving violations that each of them has committed during the past three years.

Table 14.1

Monthly Premium (dollars)	Driving Experience (years)	Number of Driving Violations (past 3 years)
74	5	2
38	14	0
50	6	1
63	10	3
97	4	6
55	8	2
57	11	3
43	16	1
99	3	5
46	9	1
35	19	0
60	13	3

Using the MINITAB REGRESS command, find the regression of monthly premium paid by drivers on the driving experience and the number of driving violations.

Solution   Let

y = the monthly auto insurance premium (in dollars) paid by a driver

x_1 = the driving experience (in years) of a driver

x_2 = the number of driving violations committed by a driver during the past three years

We are to estimate the regression model

$$y = A + B_1 x_1 + B_2 x_2 + \epsilon \tag{3}$$

Figure 14.2 lists the MINITAB solution for this regression model for the data of Table 14.1. First we enter the given data on three variables in columns C1, C2, and C3 using the READ command. Then we name column C1 as 'Y,' C2 as 'X1,' and C3 as 'X2.' Note that the names given to columns are enclosed within single quotes. To obtain the estimated regression equation, we could use either the command REGRESS C1 2 C2 C3 or the command RE-GRESS 'Y' 2 'X1' 'X2.' If we use the latter command, we will enclose the names Y, X1, and X2 within single quotes.

Figure 14.2   Estimation of the multiple regression model.

```
MTB  >  READ  C1  C2  C3
DATA >  74     5   2
DATA >  38    14   0
DATA >  50     6   1
DATA >  63    10   3
DATA >  97     4   6
DATA >  55     8   2
DATA >  57    11   3
DATA >  43    16   1
DATA >  99     3   5
DATA >  46     9   1
DATA >  35    19   0
DATA >  60    13   3
DATA >  END
MTB  >  NAME  C1  'Y'   C2   'X1'   C3   'X2'
MTB  >  REGRESS  C1   2   C2   C3
```
{ This number indicates that the regression model has two independent variables.

```
THE REGRESSION EQUATION IS
Y = 55.1 − 1.37 X1 + 8.05 X2        }  (I)


PREDICTOR         COEF          STDEV      T-RATIO         P
CONSTANT         55.138         7.309        7.54      0.000
X1               −1.3736        0.4885      −2.81      0.020       (II)
X2                8.053         1.307        6.16      0.000


S = 6.073      R-SQ = 93.1%     R-SQ(ADJ) = 91.6%   }  (III)


ANALYSIS OF VARIANCE

SOURCE         DF         SS          MS          F          P
REGRESSION      2       4490.3      2245.2      60.88      0.000    (IV)
ERROR           9        331.9        36.9
TOTAL          11       4822.2


SOURCE         DF       SEQ SS
X1              1       3089.5                              (V)
X2              1       1400.9
```

The MINITAB solution presented in Figure 14.2 contains five parts that are explained in the next few subsections.

14.5.1   ESTIMATED MULTIPLE REGRESSION MODEL

Example 14–2 describes how the coefficients of the multiple regression model are interpreted.

Interpreting an estimated multiple regression model.

EXAMPLE 14–2   Refer to Example 14–1 and the MINITAB solution given in Figure 14.2.

(a)   Write the estimated regression equation.

(b)   Explain the meaning of the estimated regression coefficients.

(c)   What are the values of the standard deviation of errors, the coefficient of multiple determination, and the adjusted coefficient of multiple determination?

(d)   What is the predicted auto insurance premium paid per month by a driver with seven years of driving experience and three driving violations?

(e)   What is the point estimate of the expected (or mean) auto insurance premium paid per month by all drivers with 12 years of driving experience and 4 driving violations?

Solution

(a)   From the portion of the MINITAB solution that is marked I in Figure 14.2, the estimated regression equation is

$$\hat{y} = 55.1 - 1.37x_1 + 8.05x_2$$

From this equation,

$$a = 55.1, \qquad b_1 = -1.37, \qquad \text{and} \qquad b_2 = 8.05$$

We can also read the values of these coefficients from the column labeled COEF in the portion marked II in the MINITAB solution of Figure 14.2. From that column we obtain

$$a = 55.138, \qquad b_1 = -1.3736, \qquad \text{and} \qquad b_2 = 8.053$$

Notice that in this column the coefficients appear with more digits after the decimal point. With these coefficient values, we can write the estimated regression equation as

$$\hat{y} = 55.138 - 1.3736x_1 + 8.053x_2 \tag{4}$$

(b)   The value of $a = 55.138$ in the estimated regression equation (4) gives the value of $\hat{y}$ for $x_1 = 0$ and $x_2 = 0$. Thus, a driver with no experience and no driving violations is expected to pay an auto insurance premium of $55.138 (or $55.14) per month. Again, this is the technical interpretation of a. In reality that may not be true because none of the drivers in our sample has both zero experience *and* zero driving violations.

The value of $b_1 = -1.3736$ in the estimated regression model gives the change in $\hat{y}$ for a one-unit change in x_1 when x_2 is held constant. Thus, we can state that a driver with one extra year of experience but the same number of driving violations is expected to pay $1.3736 (or $1.37) less for the auto insurance premium per month. Note that because b_1 is negative, an increase in experience decreases the premium paid. In other words, y and x_1 have a negative relationship.

The value of $b_2 = 8.053$ in the estimated regression model gives the change in $\hat{y}$ for a one-unit change in x_2 when x_1 is held constant. Thus, a driver with one extra driving violation but with the same years of driving experience is expected to pay $8.053 (or $8.05) more per month for the auto insurance premium.

(c)   The values of the standard deviation of errors, the coefficient of multiple determination, and the adjusted coefficient of multiple determination are given in part III of the MINITAB solution in Figure 14.2. That portion of the MINITAB solution is reproduced below as Figure 14.3.

Figure 14.3

$$S = 6.073 \qquad\qquad R\text{-}SQ = 93.1\% \qquad\qquad R\text{-}SQ(ADJ) = 91.6\%$$
$$\uparrow \qquad\qquad\qquad\qquad \uparrow \qquad\qquad\qquad\qquad\qquad \uparrow$$
$$s_e \qquad\qquad\qquad\qquad R^2 \qquad\qquad\qquad\qquad\qquad \overline{R}^2$$

From Figure 14.3,

$$s_e = \textbf{6.073}, \qquad R^2 = \textbf{93.1\%}, \qquad \text{and} \qquad \overline{R}^2 = \textbf{91.6\%}$$

Thus, the standard deviation of errors is 6.073. The value of $R^2 = 93.1\%$ tells us that the two independent variables, years of driving experience and number of driving violations, included in our model explain 93.1% of the variation in the dependent variable. The value of $\overline{R}^2 = 91.6\%$ is the value of the coefficient of multiple determination adjusted for degrees of freedom. It states that when adjusted for degrees of freedom, the two independent variables explain 91.6% of the variation in the dependent variable.

(d) To find the predicted auto insurance premium paid per month by a driver with seven years of driving experience and three driving violations, we substitute $x_1 = 7$ and $x_2 = 3$ in the estimated regression model (4).

$$\hat{y} = 55.138 - 1.3736x_1 + 8.053x_2$$
$$= 55.138 - 1.3736\,(7) + 8.053\,(3) = 69.6818 = \textbf{\$69.68}$$

Note that this value of $\hat{y}$ is a point estimate of the predicted value of y, which is denoted by y_p. The concept of the predicted value of y is the same as that for a simple linear regression model discussed in Section 13.8.2 of Chapter 13.

(e) To obtain the point estimate of the expected (mean) auto insurance premium paid per month by all drivers with 12 years of driving experience and four driving violations, we substitute $x_1 = 12$ and $x_2 = 4$ in the estimated regression model (4).

$$\hat{y} = 55.138 - 1.3736x_1 + 8.053x_2$$
$$= 55.138 - 1.3736\,(12) + 8.053\,(4) = 70.8668 = \textbf{\$70.87}$$

This value of $\hat{y}$ is a point estimate of the mean value of y, which is denoted by $E(y)$ or $\mu_{y|x_1,x_2}$. The concept of the mean value of y is the same as that for a simple linear regression model discussed in Section 13.8.1 of Chapter 13. ▬

14.5.2 CONFIDENCE INTERVAL FOR INDIVIDUAL COEFFICIENTS

The values of a, b_1, b_2, ..., and b_k obtained by estimating model (1) using sample data give the point estimates of A, B_1, B_2, ..., and B_k, respectively, which are the population parameters. Using the values of sample statistics a, b_1, b_2, ..., and b_k, we can make confidence intervals for the corresponding population parameters A, B_1, B_2, ..., and B_k, respectively.

Because of the assumption that the errors are normally distributed, the sampling distribution of each b_i is normal with its mean equal to B_i and standard deviation equal to σ_{b_i}. For example, the sampling distribution of b_1 is normal with its mean equal to B_1 and standard deviation equal to σ_{b_1}. However, if σ_ϵ is not known, then σ_{b_i} cannot be calculated. In such a case we use s_{b_i} as an estimator of σ_{b_i}. Then, if the sample size is large ($n \geq 30$), we use the normal distribution to make a confidence interval for B_i. If the sample size is small ($n < 30$), we replace the normal distribution by the t distribution.

The formula for a confidence interval for a parameter B_i is given below. This is the same formula we used to make a confidence interval for B in Section 13.5.2 of Chapter 13. The only difference is that to make a confidence interval for a particular B_i for a multiple regression model, the degrees of freedom are $n - k - 1$.

CONFIDENCE INTERVAL FOR B_i

The $(1 - \alpha)100\%$ confidence interval for B_i is given by

$$b_i \pm ts_{b_i}$$

The value of t that is used in this formula is obtained from the t distribution table for $\alpha/2$ area in the right tail of the t distribution curve and $(n - k - 1)$ degrees of freedom.

However, if the sample size is large ($n \geq 30$), we can use the normal distribution to make a confidence interval for B_i.

Example 14–3 describes the procedure for making a confidence interval for an individual B_i.

Constructing a confidence interval for an individual B_i.

EXAMPLE 14–3 Determine a 95% confidence interval for B_1 (the coefficient of experience) for the multiple regression of auto insurance premium on driving experience and the number of driving violations. Use the MINITAB solution of Figure 14.2.

Solution To make a confidence interval for B_1, we use the portion marked ⓘ in the MINITAB solution of Figure 14.2. That portion of the solution is shown below in Figure 14.4.

Figure 14.4

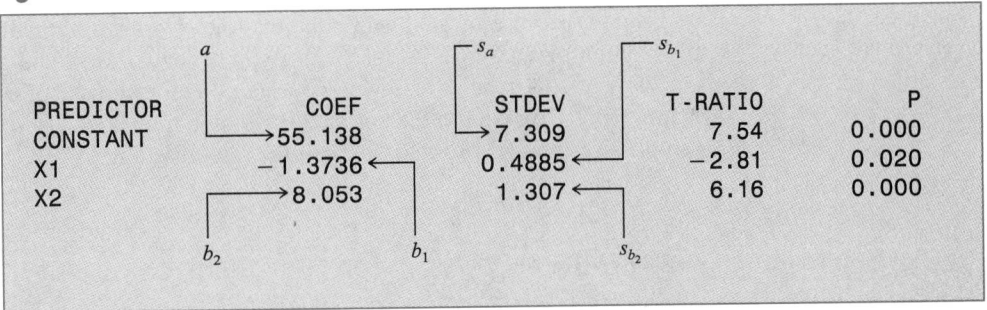

From the given information and from Figure 14.4,

$$n = 12, \quad b_1 = -1.3736, \quad \text{and} \quad s_{b_1} = .4885$$

The confidence level is 95%. The area in each tail of the t distribution curve is obtained as follows.

$$\text{Area in each tail of the } t \text{ distribution} = .5 - (.95/2) = .025$$

Because there are two independent variables, $k = 2$. Therefore,

$$\text{degrees of freedom} = n - k - 1 = 12 - 2 - 1 = 9$$

From the t distribution table, the value of t for .025 area in the right tail of the t distribution curve and 9 degrees of freedom is 2.262. The 95% confidence interval for B_1 is

$$b_1 \pm ts_{b_1} = -1.3736 \pm 2.262 \,(.4885)$$
$$= -1.3736 \pm 1.1050 = \mathbf{-2.48 \text{ to } -.27}$$

Thus, the 95% confidence interval for b_1 is -2.48 to $-.27$. That is, we can state with 95% confidence that for one extra year of driving experience, the monthly auto insurance premium changes by an amount between $-\$2.48$ and $-\$.27$. Note that since both limits of the confidence interval have a negative sign, we can also state that for each extra year of driving experience, the monthly auto insurance premium decreases by an amount between \$.27 and \$2.48.

Using the procedure shown in Example 14–3, we can make a confidence interval for any of the coefficients of a multiple regression model, such as A and B_2 in model (3). For example, the 95% confidence intervals for A and B_2, respectively, are

$$a \pm ts_a = 55.138 \pm 2.262 \,(7.309) = 38.61 \text{ to } 71.67$$
$$b_2 \pm ts_{b_2} = 8.053 \pm 2.262 \,(1.307) = 5.10 \text{ to } 11.01$$

14.5.3   TEST OF HYPOTHESIS ABOUT INDIVIDUAL COEFFICIENTS

We can make a test of hypothesis about any of the B_i coefficients of model (1) using the same procedure that we used to make a test of hypothesis about B for a simple regression model in Section 13.5.3 of Chapter 13. The only difference is the degrees of freedom, which are equal to $n - k - 1$ for a multiple regression model.

Again, because of the assumption that the errors are normally distributed, the sampling distribution of each b_i is normal with its mean equal to B_i and standard deviation equal to σ_{b_i}. However, if σ_ϵ is not known, we use s_{b_i} as an estimator of σ_{b_i}. Then, if the sample size is large ($n \geq 30$), we use the normal distribution to make a test of hypothesis about B_i. If the sample size is small ($n < 30$), we use the t distribution.

TEST STATISTIC FOR b_i

The value of the test statistic t for b_i is calculated as

$$t = \frac{b_i - B_i}{s_{b_i}}$$

The value of B_i is substituted from the null hypothesis.

Example 14–4 illustrates the procedure for testing a hypothesis about a single coefficient.

Conducting a test of hypothesis about B_2 of model (3).

EXAMPLE 14-4 Using the 1% significance level, can you conclude that the slope of the number of driving violations in regression model (3) is positive? Use the MINITAB output obtained in Example 14–1 and shown in Figure 14.2 to make this test.

Solution From Example 14–1, our multiple regression model (3) is

$$y = A + B_1x_1 + B_2x_2 + \epsilon$$

where y is the monthly auto insurance premium (in dollars) paid by a driver, x_1 is the driving experience (in years), and x_2 is the number of driving violations committed during the past three years.

From Example 14–2, the estimated regression model is

$$\hat{y} = 55.138 - 1.3736x_1 + 8.053x_2$$

To conduct a test of hypothesis about B_2, we use the portion marked II in the MINITAB solution of Figure 14.2, which was reproduced in Figure 14.4. That portion of the solution is again reproduced below in Figure 14.5.

Figure 14.5

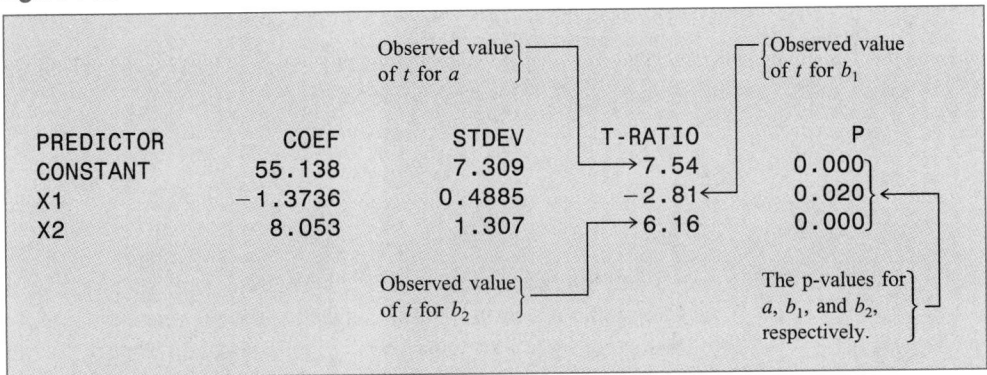

To make a test of hypothesis about B_2, we perform the following five steps.

Step 1. *State the null and alternative hypotheses*

We are to test whether or not the slope of the number of driving violations in the regression model (3) is positive, that is, whether or not B_2 is positive. The two hypotheses are

$$H_0: B_2 = 0 \quad \text{(slope is zero)}$$

$$H_1: B_2 > 0 \quad \text{(slope is positive)}$$

Note that we can also write the null hypothesis as $H_0: B_2 \le 0$, which states that the slope of the number of driving violations in the regression model (3) is either zero or negative.

Step 2. *Select the distribution to use*

The sample size is small ($n < 30$) and σ_ϵ is not known. Hence, we use the t distribution to make a test of hypothesis about B_2.

Step 3. *Determine the rejection and nonrejection regions*

The significance level is .01. The $>$ sign in the alternative hypothesis indicates that the test is right-tailed. Therefore,

Area in the right tail of the t distribution curve $= \alpha = .01$

$$df = n - k - 1 = 12 - 2 - 1 = 9$$

From the t distribution table, the critical value of t for 9 degrees of freedom and .01 area in the right tail of the t distribution curve is 2.821, as shown in Figure 14.6.

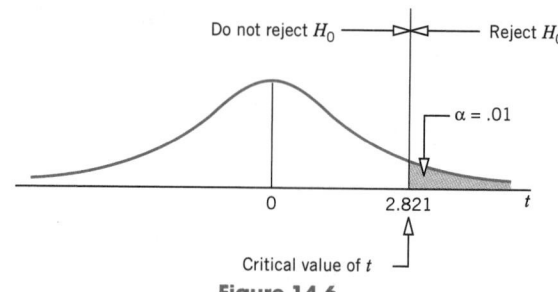

Figure 14.6

Step 4.   *Calculate the value of the test statistic*

The value of the test statistic t for b_2 can be obtained from the MINITAB solution given in Figure 14.5. From that solution,

$$t = \frac{b_2 - B_2}{s_{b_2}} = 6.16$$

Step 5.   *Make a decision*

The value of the test statistic $t = 6.16$ is greater than the critical value of $t = 2.821$ and it falls in the rejection region. Consequently, we reject the null hypothesis and conclude that the slope of x_2 in regression model (3) is positive. That is, the number of driving violations committed by drivers is significant and an increase in the number of driving violations increases the auto insurance premium.

We can also use the p-value to make this decision. From Figure 14.5, the p-value for x_2 is .000 (see the last entry in the row for x_2). This p-value is for a two-tailed test. The p-value for a one-tailed test is obtained by dividing this value by 2. Thus, the required p-value is

$$p\text{-value} = .000/2 = .000$$

The significance level α is .01. Since the p-value is smaller than α, we reject the null hypothesis. ■

Note that the observed value of t in Step 4 of Example 14–4 is obtained from the MINITAB solution only if the null hypothesis is $H_0: B_2 = 0$. However, if the null hypothesis is that B_2 is equal to a number other than zero, then the t value obtained from the MINITAB solution is no longer valid. For example, suppose the null hypothesis in Example 14–4 is

$$H_0: B_2 = 5$$

and the alternative hypothesis is

$$H_1: B_2 > 5$$

In this case the observed value of t will be calculated as

$$t = \frac{b_2 - B_2}{s_{b_2}} = \frac{8.053 - 5}{1.307} = 2.336$$

To calculate this value of t, the values of b_2 and s_{b_2} are obtained from the MINITAB solution of Figure 14.5. The value of B_2 is substituted from H_0.

☞   *Remember*   Note that the p-value given in a MINITAB solution for a coefficient is for a two-tailed test of H_0: $B_i = 0$. If we are performing a one-tailed test, we must divide the given p-value by 2 before making a decision. In this example, the p-value for x_2 is .000. Dividing it by 2 gives the same p-value.

14.5.4   TESTING FOR THE OVERALL SIGNIFICANCE OF THE MULTIPLE REGRESSION MODEL

In regression analysis, we can also test for the overall significance of the model. In other words, we perform a test of hypothesis with the null hypothesis that the coefficients of all independent variables in the regression model are equal to zero and the alternative hypothesis that the coefficients of all independent variables are not zero. For the multiple regression model

$$y = A + B_1 x_1 + B_2 x_2 + B_3 x_3 + \cdots + B_k x_k + \epsilon$$

the two hypotheses for such a test are written as

$$H_0: B_1 = B_2 = B_3 = \cdots = B_k = 0$$

$$H_1: \text{Not all } B_i\text{'s are zero}$$

Note that the alternative hypothesis states that there is at least one coefficient among B_1, B_2, B_3, ..., and B_k that is different from zero. There can be more than one of the B_i's that is different from zero.

A test of hypothesis for the overall significance of a multiple regression model is performed by using the F distribution. (The reader is advised to review at this point the F distribution in Section 12.1 of Chapter 12.)

Although the value of the test statistic F is obtained from the computer solution, it can also be calculated by using the following formula.

TEST STATISTIC FOR THE TEST OF OVERALL SIGNIFICANCE OF THE REGRESSION MODEL

The value of the test statistic F for the test of overall significance of the multiple regression model is given by the formula

$$F = \frac{\text{SSR}/k}{\text{SSE}/(n - k - 1)} \quad \text{or} \quad \frac{\text{MSR}}{\text{MSE}}$$

where **MSR** stands for the mean square regression and **MSE** for the mean square error. The MSR and MSE are calculated as follows.

$$\text{MSR} = \text{SSR}/k$$

$$\text{MSE} = \text{SSE}/(n - k - 1)$$

A test of hypothesis for the overall significance of a multiple regression model is always a right-tailed test. The value of F for the given significance level is obtained from Table X of Appendix B for k degrees of freedom for the numerator and $n - k - 1$ degrees of freedom for the denominator.

To conduct a test of overall significance for a multiple regression model, we will use the portion marked ④ and labeled *Analysis of Variance* in Figure 14.2. Example 14–5 shows how such a test is performed.

Conducting a test of overall significance for the multiple regression model.

EXAMPLE 14–5   Using the 2.5% significance level, can you conclude that the coefficients of all independent variables in the regression model (3) of Example 14–1 are equal to zero? Use the MINITAB output obtained in Example 14–1 and shown in Figure 14.2 to make this test.

Solution   From Example 14–1, our multiple regression model (3) is

$$y = A + B_1 x_1 + B_2 x_2 + \epsilon$$

where y is the monthly auto insurance premium (in dollars) paid by a driver, x_1 is the driving experience (in years), and x_2 is the number of driving violations committed during the past three years.

From Example 14–2, the estimated regression model is

$$\hat{y} = 55.138 - 1.3736 x_1 + 8.053 x_2$$

We are to test whether or not the coefficients of all independent variables are equal to zero. This is a test for the overall significance of the regression model. To make this test, we use the portion marked ④ in the MINITAB solution of Figure 14.2. That portion of the solution is reproduced below in Figure 14.7.

Figure 14.7

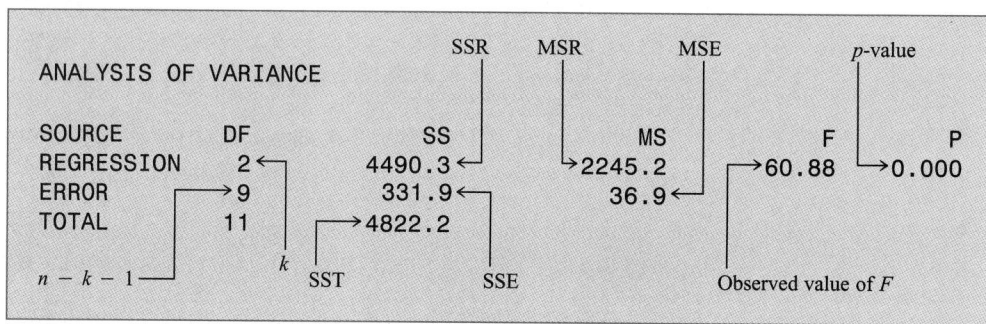

To make the required test, we perform the following five steps.

Step 1.   *State the null and alternative hypotheses*

We are to test whether or not coefficients of all independent variables are equal to zero in the regression model (3). The two hypotheses are

$$H_0: B_1 = B_2 = 0$$

$$H_1: \text{Both } B_i\text{'s are not equal to zero}$$

Step 2. *Select the distribution to use*

Because we are performing a test of overall significance, such a test is always performed using the F distribution. Consequently we use the F distribution to make this test.

Step 3. *Determine the rejection and nonrejection regions*

The significance level is .025. The F distribution test is always a right-tailed test. Therefore,

Area in the right tail of the F distribution curve $= \alpha = .025$

degrees of freedom for the numerator $= k = 2$

degrees of freedom for the denominator $= n - k - 1 = 12 - 2 - 1 = 9$

From the F distribution table, the critical value of F for 2 df for the numerator, 9 df for the denominator, and .025 area in the right tail of the F distribution curve is 5.72, as shown in Figure 14.8.

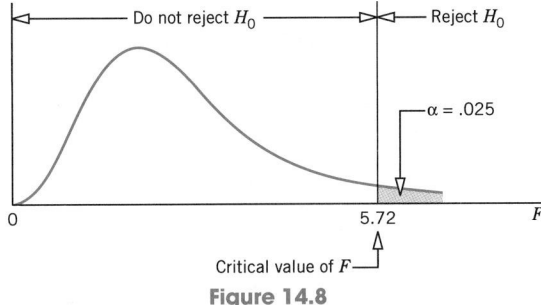

Figure 14.8

Thus, we will reject H_0 if the observed value of the test statistic F is greater than 5.72 and we will fail to reject H_0 otherwise.

Step 4. *Calculate the value of the test statistic*

The observed value of the test statistic F from the MINITAB solution given in Figure 14.7 is

$$F = 60.88$$

Step 5. *Make a decision*

Because the value of the test statistic $F = 60.88$ is greater than the critical value of $F = 5.72$, it falls in the rejection region. Consequently, we reject the null hypothesis and conclude that both B_i's are not zero. At least one of the two B_i's (B_1 or B_2) is different from zero.

We can also use the p-value to make this decision. From Figure 14.7, the p-value is .000. The significance level α is .025. Since the p-value is smaller than α, we reject the null hypothesis.

Note that by rejecting the null hypothesis we are stating that at least one of the two B_i's is significant. However it is possible that both B_i's may be significant. ∎

In Figure 14.2 we marked a portion of the MINITAB solution as $\text{\small(v)}$, which we have not used in any of the examples. That portion of the solution is reproduced in Figure 14.9.

Figure 14.9

```
SOURCE       DF       SEQ SS
X1           1        3089.5
X2           1        1400.9
```

From the ANALYSIS OF VARIANCE section of the MINITAB solution of Figure 14.2 (which was also reproduced in Figure 14.7), the values of SST, SSR, and SSE are

$$SST = 4822.2, \quad SSR = 4490.3, \quad \text{and} \quad SSE = 331.9$$

We can state that when we use x_1 and x_2 to explain y, the value of SST is reduced by 4490.3, which is the value of SSR.

Now, if we estimate the simple linear regression of y on x_1, the value of SSR will be 3089.5, which is the value in the row of x_1 and the column labeled SEQ SS in Figure 14.9. In other words, if we estimate the simple linear regression model

$$y = A + B_1x_1 + \epsilon \tag{5}$$

then the value of SSR will be 3089.5. That is, x_1 alone will reduce SST by 3089.5.

Then if we add x_2 to model (5), the SST will further be reduced by 1400.9, which is the value in the row of x_2 and the column labeled SEQ SS in Figure 14.9.

The sum of the two numbers in the column of SEQ SS of Figure 14.9 is

$$3089.5 + 1400.9 = 4490.4$$

which is the value of SSR in Figure 14.2. The difference is due to rounding off error.

Thus, the SEQ SS column in Figure 14.9 (or Figure 14.2) gives the sequential reduction in SST as we add independent variables to our regression model.

14.6   CORRELATION COEFFICIENTS

When we are considering more than two variables (as in a multiple regression model), we can calculate the **simple linear correlation coefficient** for each pair of variables just as we calculated the linear correlation coefficient between y and x in Section 13.6 of Chapter 13. We will use the same formula that we used in that section to calculate the simple linear correlation coefficient between any pair of variables. Suppose we have three variables y, x_1, and x_2. The notation used to denote simple linear correlation coefficients between different pairs of these variables is as follows.

r_{y1} = the simple linear correlation coefficient between y and x_1

r_{y2} = the simple linear correlation coefficient between y and x_2

r_{12} = the simple linear correlation coefficient between x_1 and x_2

These correlation coefficients measure the strength of the linear correlation between the corresponding variables. Each of these correlation coefficients can be calculated using a formula similar to the one used in Section 13.6 of Chapter 13. For example, the simple linear correlation coefficient between x_1 and x_2 will be calculated using the formula

$$r_{12} = \frac{SS_{x_1x_2}}{\sqrt{SS_{x_1x_1} SS_{x_2x_2}}}$$

Alternatively, we can obtain these correlation coefficients by using a statistical software package.

Besides these simple linear correlation coefficients, we can find the **partial linear correlation coefficients** between variables. For example, we can find the correlation coefficient between y and x_1 while all other variables are held constant. This correlation coefficient is called the partial correlation coefficient between y and x_1. However, we will not discuss partial correlation coefficients in this text.

Example 14–6 shows how we can find correlation coefficients between variables using MINITAB.

Finding simple linear correlation coefficients.

EXAMPLE 14–6   Using the data of Table 14.1 and MINITAB, find the simple linear correlation coefficients between each pair of the three variables.

Solution   Figure 14.10 shows the MINITAB commands to find the required correlation coefficients and the MINITAB solution.

Figure 14.10   Finding simple linear correlation coefficients.

```
MTB  > READ C1 C2 C3
DATA > 74     5    2
DATA > 38    14    0
DATA > 50     6    1
DATA > 63    10    3
DATA > 97     4    6
DATA > 55     8    2
DATA > 57    11    3
DATA > 43    16    1
DATA > 99     3    5
DATA > 46     9    1
DATA > 35    19    0
DATA > 60    13    3
DATA > END

MTB  > NAME  C1  'Y'  C2  'X1'  C3  'X2'

MTB  > CORRELATION  C1  C2

CORRELATION OF Y AND X1 = -0.800

MTB  > CORRELATION  C1  C3

CORRELATION OF Y AND X2 = 0.933

MTB  > CORRELATION  C2  C3

CORRELATION OF X1 AND X2 = -0.660
```

From the MINITAB output shown in Figure 14.10, the simple linear correlation coefficient between each pair of variables is as follows.

$$r_{y1} = -.800$$

$$r_{y2} = .933$$

$$r_{12} = -.660$$

where r_{y1} is the simple linear correlation coefficient between y and x_1, r_{y2} is the simple linear correlation coefficient between y and x_2, and r_{12} is the simple linear correlation coefficient between x_1 and x_2.

As in Chapter 13, these correlation coefficients indicate the strength of the linear relationship between the respective pairs of variables.   ■

EXERCISES

Concepts and Procedures

14.1 How are the coefficients of independent variables in a multiple regression model interpreted? Explain.

14.2 What are the degrees of freedom for a multiple regression model to make inferences about individual parameters?

14.3 What kinds of relationships among independent variables are permissible and which ones are not permissible in a linear multiple regression model?

14.4 Explain the meaning of the coefficient of multiple determination and the adjusted coefficient of multiple determination for a multiple regression model. What is the difference between the two?

14.5 What are the assumptions of a multiple regression analysis?

14.6 Following is the MINITAB solution for a regression of y on x_1 and x_2.

```
THE REGRESSION EQUATION IS
Y = 13.2 + 0.244 X1 - 0.0431 X2

PREDICTOR           COEF        STDEV      T-RATIO          P
CONSTANT          13.230        6.234         2.12      0.060
X1                0.24367      0.03262         7.47      0.000
X2               -0.04314      0.02690        -1.60      0.140

S = 1.807         R-SQ = 96.3%      R-SQ(ADJ) = 95.6%

ANALYSIS OF VARIANCE

SOURCE          DF           SS          MS           F          P
REGRESSION       2       849.65      424.83      130.10      0.000
ERROR           10        32.65        3.27
TOTAL           12       882.30

SOURCE          DF       SEQ SS
X1               1       841.25
X2               1         8.40
```

Using this MINITAB solution, answer the following questions for the population regression model

$$y = A + B_1 x_1 + B_2 x_2 + \epsilon$$

a. Write the estimated regression equation.
b. Write the values of a, b_1, and b_2 and explain the meaning of these estimated regression coefficients.
c. What are the values of the standard deviation of errors, the coefficient of multiple determination, the adjusted coefficient of multiple determination, SST, SSR, SSE, MSR, and MSE?
d. Write the values of the standard deviation, the observed value of t, and the p-value for each of the coefficients a, b_1, and b_2.
e. What is the value of SSR if we estimate the regression of y on x_1 alone? What is the increase in this value of SSR if we add x_2 as an independent variable to this regression model?
f. What is the predicted value of y for $x_1 = 74$ and $x_2 = 150$?
g. What is the point estimate of the expected (mean) value of y for all elements given that $x_1 = 58$ and $x_2 = 122$?
h. Construct a 99% confidence interval for the coefficient of x_1 in the population regression model.
i. Make a 95% confidence interval for the coefficient of x_2 in the population regression model.
j. Determine a 90% confidence interval for A, the constant term in the population regression model.
k. Using the 2.5% significance level, test whether or not the coefficient of x_1 in the population regression model is positive.
l. Using the 1% significance level, can you conclude that the coefficient of x_2 in the population regression model is negative?
m. At the 5% significance level, test if the constant term A in the population regression model is different from zero. (*Hint:* The procedure to make a test of hypothesis about A is similar to the one used to test a hypothesis about any of the coefficients of independent variables. Follow the same five steps. The observed value of t for a can be obtained from the computer solution.)
n. Using the 1% significance level, test whether or not the coefficients of all independent variables in the population regression model are equal to zero.

14.7 Following is the MINITAB solution for a regression of y on x_1, x_2, and x_3.

```
THE REGRESSION EQUATION IS
Y = 51.6 - 0.0599 X1 + 0.0850 X2 - 0.00477 X3

PREDICTOR          COEF          STDEV      T-RATIO          P
CONSTANT          51.61          11.25         4.59      0.000
X1             -0.05993        0.01526        -3.93      0.003
X2              0.08497        0.02875         2.96      0.014
X3            -0.004773       0.008707        -0.55      0.596

S = 0.8995        R-SQ = 99.7%        R-SQ(ADJ) = 99.6%

ANALYSIS OF VARIANCE

SOURCE        DF            SS            MS          F          P
REGRESSION     3       2909.91        969.97    1198.86      0.000
ERROR         10          8.09          0.81
TOTAL         13       2918.00

SOURCE        DF        SEQ SS
X1             1       2901.82
X2             1          7.85
X3             1          0.24
```

Using the MINITAB solution, answer the following questions for the population regression model

$$y = A + B_1x_1 + B_2x_2 + B_3x_3 + \epsilon$$

a. Write the estimated regression equation.

b. Write the values of a, b_1, b_2, and b_3 and explain the meaning of these estimated regression coefficients.

c. What are the values of the standard deviation of errors, the coefficient of multiple determination, the adjusted coefficient of multiple determination, SST, SSR, SSE, MSR, and MSE?

d. Write the values of the standard deviation, the observed value of t, and the p-value for each of the estimated coefficients a, b_1, b_2, and b_3.

e. What is the value of SSR if we estimate the regression of y on x_1 alone? What is the increase in this value of SSR if we add x_2 as an independent variable to this regression model? What is the increase in this value of SSR if we add x_3 as an independent variable to this regression model after adding x_1 and x_2?

f. What is the predicted value of y for $x_1 = 310$, $x_2 = 260$, and $x_3 = 180$?

g. What is the point estimate of the expected (mean) value of y for all elements given that $x_1 = 360$, $x_2 = 220$, and $x_3 = 165$?

h. Construct a 95% confidence interval for the coefficient of x_1 in the population regression model.

i. Determine a 99% confidence interval for the coefficient of x_2 in the population regression model.

j. Make a 98% confidence interval for the coefficient of x_3 in the population regression model.

k. Construct a 95% confidence interval for A, the constant term in the population regression model.

l. Using the 5% significance level, test whether or not the coefficient of x_1 in the population regression model is negative.

m. Using the 1% significance level, can you conclude that the coefficient of x_2 in the population regression model is positive?

n. At the 2.5% significance level, test if the coefficient of x_3 in the population regression model is negative.

o. At the 1% significance level, test if the constant term A in the population regression model is different from zero.

p. Using the 2.5% significance level, can you conclude that the coefficients of all independent variables in the population regression model are equal to zero?

14.8 The following table gives data on variables y, x_1, x_2, and x_3.

y	x_1	x_2	x_3
8	18	38	74
11	26	25	64
19	34	24	47
21	38	44	31
7	13	12	79
23	49	48	35
16	28	38	42
27	59	52	18
9	14	17	71
13	21	39	57

Using MINITAB (or any other statistical software package) estimate the regression model

$$y = A + B_1x_1 + B_2x_2 + B_3x_3 + \epsilon$$

Using the solution obtained, answer the following questions.

a. Write the estimated regression equation.
b. Explain the meaning of a, b_1, b_2, and b_3 obtained by estimating the given regression model.
c. What are the values of the standard deviation of errors, the coefficient of multiple determination, and the adjusted coefficient of multiple determination?
d. What is the predicted value of y for $x_1 = 35$, $x_2 = 40$, and $x_3 = 65$?
e. What is the point estimate of the expected (mean) value of y for all elements given that $x_1 = 40$, $x_2 = 30$, and $x_3 = 55$?
f. Construct a 95% confidence interval for the coefficient of x_3.
g. Using the 2.5% significance level, test whether or not the coefficient of x_1 is positive.
h. Using the 2.5% significance level, test whether or not the coefficients of all independent variables in the population regression model are equal to zero.
i. Find the simple linear correlation coefficient between each pair of the four variables.

14.9 The following table gives data on variables y, x_1, and x_2.

y	x_1	x_2
24	98	52
14	51	69
18	74	63
31	108	35
10	33	88
29	119	54
26	99	51
33	141	31
13	47	67
27	103	41
26	111	46

Using MINITAB (or any other statistical software package), find the regression of y on x_1 and x_2. Using the solution obtained, answer the following questions.

a. Write the estimated regression equation.
b. Explain the meaning of the estimated regression coefficients of the independent variables.
c. What are the values of the standard deviation of errors, the coefficient of multiple determination, and the adjusted coefficient of multiple determination?
d. What is the predicted value of y for $x_1 = 87$ and $x_2 = 54$?
e. What is the point estimate of the expected (mean) value of y for all elements given that $x_1 = 95$ and $x_2 = 49$?
f. Construct a 99% confidence interval for the coefficient of x_1.
g. Using the 1% significance level, test if the coefficient of x_2 in the population regression model is negative.
h. Using the 1% significance level, can you conclude that the coefficients of both independent variables in the population regression model are equal to zero?
i. Find the simple linear correlation coefficient between each pair of the three variables.

Applications

14.10 The salaries of workers are expected to be dependent, among other factors, on the number of years they have spent in school and their work experience. The following table gives information on the annual salaries (in thousands of dollars) for 12 persons, the number of years each of them spent in school, and the total number of years of work experience.

Salary	32	24	28	57	48	28	39	43	18	41	17	49
Schooling	16	12	13	20	18	16	14	18	12	16	12	16
Experience	6	10	15	8	11	2	12	4	6	9	2	18

Using MINITAB (or any other statistical software package), find the regression of salary on schooling and experience. Using the solution obtained, answer the following questions.

a.   Write the estimated regression equation.
b.   Explain the meaning of the estimates of the constant term and the regression coefficients of independent variables.
c.   What are the values of the standard deviation of errors, the coefficient of multiple determination, and the adjusted coefficient of multiple determination?
d.   What is the value of the total sum of squares? What portion of SST is explained by our regression model? What portion of SST is not explained by our regression model?
e.   How much salary is a person with 18 years of schooling and 7 years of work experience expected to earn?
f.   What is the point estimate of the expected (mean) salary for all people with 16 years of schooling and 10 years of work experience?
g.   Determine a 99% confidence interval for the coefficient of schooling.
h.   Using the 1% significance level, test whether or not the coefficient of experience is positive.
i.   Using the 2.5% significance level, can you conclude that the coefficients of both independent variables in the population regression model are equal to zero?
j.   Determine the simple linear correlation coefficient between each pair of the three variables.

14.11   The CTO Corporation has a large number of chain restaurants throughout the United States. The research department at the company wanted to find if the sales of restaurants depend on the size of the population within a certain area surrounding the restaurants and the mean income of households in those areas. The company collected information on these variables for 11 restaurants. The following table gives information on the weekly sales (in thousands of dollars) of these restaurants, the population (in thousands) within five miles of the restaurants, and the mean annual income (in thousands of dollars) of the households for those areas.

Sales	19	29	17	21	14	30	33	22	18	27	24
Population	21	15	32	18	47	69	29	43	75	39	53
Income	38	49	29	32	27	36	41	26	19	34	28

Using MINITAB (or any other statistical software package), find the regression of sales on population and income. Using the solution obtained, answer the following questions.

a.   Write the estimated regression equation.
b.   Explain the meaning of the estimates of the constant term and the regression coefficients of population and income.
c.   What are the values of the standard deviation of errors, the coefficient of multiple determination, and the adjusted coefficient of multiple determination?
d.   What is the value of the total sum of squares? What portion of SST is explained by our regression model? What portion of SST is not explained by our regression model?
e.   What are the predicted sales for a restaurant with 50 thousand people living within a five-mile area surrounding it and $35 thousand mean annual income of households in that area?
f.   What is the point estimate of the expected (mean) sales for all restaurants with 45 thousand people living within a five-mile area surrounding them and $40 thousand mean annual income of households living in those areas?
g.   Determine a 95% confidence interval for the coefficient of *income*.
h.   Using the 1% significance level, test whether or not the coefficient of *population* is different from zero.
i.   Using the 1% significance level, can you conclude that the coefficients of both independent variables in the population regression model are equal to zero?
j.   Determine the simple linear correlation coefficient between each pair of the three variables.

14.7 DUMMY VARIABLES IN THE REGRESSION MODEL

The multiple regression model considered in Examples 14–1 through 14–5 of this chapter contained independent variables that are quantitative. Many times we may include in a regression model a variable that is qualitative. Such a variable contains different categories instead of numerical values. To include such variables in regression models, we use **dummy variables** that are usually denoted by $D, D_1, D_2, \ldots$, etc. A dummy variable indicates whether or not an element in the sample possesses a characteristic. It is also called a *binary*, *dichotomous*, or *0–1 variable*.

The number of dummy variables for a qualitative variable included in a regression model is equal to the number of categories for that variable minus 1. For example, if a variable contains two categories, we will introduce one dummy variable in the regression model for this variable. However, if a qualitative variable contains three categories, we will include two dummy variables in our regression model to represent this qualitative variable.

If a qualitative variable has two categories, one of the categories is assigned a value of zero and the other category is assigned a value of 1. The category that is assigned a value of zero is called the **reference group**.

DUMMY VARIABLE

Dummy variables are used to represent qualitative variables in a regression model. A dummy variable indicates whether or not an element possesses an attribute. The number of dummy variables introduced in a regression model for a variable is equal to the number of categories for that variable minus 1. A dummy variable is usually denoted by $D, D_1, D_2, D_3, \ldots$, etc.

Example 14–7 shows how a dummy variable is used in a regression model.

A regression model with a dummy variable.

EXAMPLE 14–7 Refer to Example 14–1. Table 14.2 reproduces the data from Table 14.1 of that example with an additional column that contains information on gender for each of the 12 drivers.

Table 14.2

Monthly Premium (dollars)	Driving Experience (years)	Number of Driving Violations (past 3 years)	Gender
74	5	2	Male
38	14	0	Female
50	6	1	Female
63	10	3	Male
97	4	6	Male
55	8	2	Male
57	11	3	Female
43	16	1	Male
99	3	5	Female
46	9	1	Female
35	19	0	Male
60	13	3	Male

Using MINITAB, find the regression of monthly auto insurance premium on the years of driving experience, the number of driving violations, and the gender of drivers. Using the solution obtained, answer the following questions.

(a) Write the estimated regression equation.

(b) Explain the meaning of the estimated regression coefficient of the independent variable *gender*.

(c) What are the values of the standard deviation of errors, the coefficient of multiple determination, and the adjusted coefficient of multiple determination?

(d) What is the predicted auto insurance premium paid per month by a male driver with 10 years of driving experience and 2 driving violations?

(e) What is the predicted auto insurance premium paid per month by a female driver with 10 years of driving experience and 2 driving violations?

(f) Construct a 99% confidence interval for the coefficient of *gender*.

(g) Using the 1% significance level, can you conclude that the coefficient of *gender* is different from zero?

Solution   Let

$$y = \text{the monthly auto insurance premium (in dollars) paid by drivers}$$

$$x_1 = \text{driving experience (in years)}$$

$$x_2 = \text{the number of driving violations committed during the past three years}$$

As we notice, *gender* is not a quantitative variable but it is a qualitative variable. Consequently, we will use a dummy variable for it in the regression model. Although we can denote this dummy variable by x_3, in order to differentiate it from quantitative (independent) variables (which are denoted by x_1 and x_2), we prefer to denote it by D, which indicates that *gender* is a dummy variable. This variable has two categories, *male* and *female* drivers. We assign a value of zero to one of these categories and a value of 1 to the other. Thus,

$$D = \text{the dummy variable representing the } \textit{gender} \text{ variable}$$

Suppose

$$D = 0 \qquad \text{if a driver is a male}$$

$$D = 1 \qquad \text{if a driver is a female}$$

As mentioned earlier, the category for which the dummy variable assumes a zero value is called the *reference group*. In our example *male* drivers make up the reference group. Note that there is no reason to assign a zero value to male drivers. We could have assigned a zero value to female drivers and 1 to male drivers. In that case female drivers would have belonged to the reference group.

Our population regression model is

$$y = A + B_1 x_1 + B_2 x_2 + B_3 D + \epsilon \tag{6}$$

Assigning values of 0 and 1 to male and female drivers, respectively, we rewrite the data of Table 14.2 as in Table 14.3.

Table 14.3

Monthly Premium (dollars)	Driving Experience (years)	Number of Driving Violations (past 3 years)	Gender D
74	5	2	0
38	14	0	1
50	6	1	1
63	10	3	0
97	4	6	0
55	8	2	0
57	11	3	1
43	16	1	0
99	3	5	1
46	9	1	1
35	19	0	0
60	13	3	0

Figure 14.11 gives the MINITAB solution for regression model (6) using the data of Table 14.3.

Figure 14.11   Estimation of the multiple regression model.

```
MTB  > READ  C1  C2  C3  C4
DATA > 74     5    2    0
DATA > 38    14    0    1
DATA > 50     6    1    1
DATA > 63    10    3    0
DATA > 97     4    6    0
DATA > 55     8    2    0
DATA > 57    11    3    1
DATA > 43    16    1    0
DATA > 99     3    5    1
DATA > 46     9    1    1
DATA > 35    19    0    0
DATA > 60    13    3    0
DATA > END
MTB  > NAME  C1  'Y'  C2  'X1'  C3  'X2'  C4  'D'
MTB  > REGRESS  C1  3  C2  C3  C4

THE REGRESSION EQUATION IS
Y = 58.8 - 1.53 X1 + 7.67 X2 - 2.95 D

PREDICTOR          COEF        STDEV      T-RATIO          P
CONSTANT         58.795        8.992         6.54      0.000
X1              -1.5336       0.5463        -2.81      0.023
X2                7.674        1.436         5.34      0.000
D                -2.954        4.005        -0.74      0.482

S = 6.233        R-SQ = 93.6%     R-SQ(ADJ) = 91.1%
```

Figure 14.11 *(continued)*

```
ANALYSIS OF VARIANCE

SOURCE        DF        SS          MS        F         P
REGRESSION    3      4511.5      1503.8     38.71     0.000
ERROR         8       310.8        38.8
TOTAL        11      4822.2

SOURCE        DF      SEQ SS
X1            1      3089.5
X2            1      1400.9
D             1        21.1
```

(a) From the MINITAB solution given in Figure 14.11, the estimated regression equation is

$$\hat{y} = 58.8 - 1.53x_1 + 7.67x_2 - 2.95D$$

Or by using the column labeled COEF in Figure 14.11, we can write the estimated model as follows.

$$\hat{y} = 58.795 - 1.5336x_1 + 7.674x_2 - 2.954D \qquad (7)$$

Notice that this estimated regression model contains coefficients with three or four digits after the decimal point.

(b) The coefficient of the variable *gender* is $b_3 = -2.954$. This coefficient is interpreted differently from b_1 and b_2. The value $b_3 = -2.954$ indicates that the female drivers pay, on average, $2.95 less than the male drivers with similar driving experiences and the same number of driving violations. Note that the female drivers pay less than the male drivers because the sign of b_3 is negative. Also, female drivers pay less than the male drivers because $D = 1$ for female drivers. Actually the dummy variable D belongs to the female drivers, the category for which $D = 1$.

In fact, by using the dummy variable D in our regression model, we have estimated two regression models, one for the male drivers and another for the female drivers. For the male drivers, $D = 0$. Substituting $D = 0$ in the estimated regression model (7), we obtain the estimated regression model for the male drivers as

$$\hat{y} = 58.795 - 1.5336x_1 + 7.674x_2 \qquad (8)$$

For the female drivers, $D = 1$. Substituting $D = 1$ in the estimated regression model (7), we obtain the estimated regression model for the female drivers as

$$\hat{y} = 58.795 - 1.5336x_1 + 7.674x_2 - 2.954(1)$$

$$= (58.795 - 2.954) - 1.5336x_1 + 7.674x_2$$

or $\qquad$ $$\hat{y} = 55.841 - 1.5336x_1 + 7.674x_2 \qquad (9)$$

Note that the constant term for model (9) is 2.954 smaller than that for model (8). This means that the value of $\hat{y}$ is 2.954 smaller for model (9) than for model (8) for the same values of x_1 and x_2. Thus, on average, female drivers pay a monthly auto insurance premium that is $2.95 less than the monthly auto insurance premium paid by male drivers with similar driving experiences and the same number of driving violations.

(c)   From Figure 14.11, the values of the standard deviation of errors, the coefficient of multiple determination, and the adjusted coefficient of multiple determination are as follows.

$$s_e = 6.233, \quad R^2 = 93.6\%, \quad \text{and} \quad \overline{R}^2 = 91.1\%$$

Thus, the standard deviation of errors is 6.233. The value of $R^2 = 93.6\%$ tells us that the three independent variables (the years of driving experience, the number of driving violations, and the gender) taken together explain 93.6% of the variation in the dependent variable. The value of $\overline{R}^2 = 91.1\%$ is the value of the coefficient of multiple determination adjusted for degrees of freedom. It states that when adjusted for degrees of freedom, the three independent variables explain 91.1% of the variation in the dependent variable.

Comparing these results to the ones given in Figure 14.2 for model (3), we observe that while the value of R^2 increased from 93.1% to 93.6% when we added the *gender* variable to model (3), the value of $\overline{R}^2$ decreased from 91.6% to 91.1%. As we will see later on in part (g) of this example, the value of $\overline{R}^2$ decreased because the variable *gender* is not significant in determining the auto insurance premium.

(d)   To find the predicted auto insurance premium for a male driver with 10 years of driving experience and 2 driving violations, we substitute $x_1 = 10$, $x_2 = 2$, and $D = 0$ in the estimated regression model (7).

$$\hat{y} = 58.795 - 1.5336\,(10) + 7.674\,(2) - 2.954\,(0) = 58.807 = \textbf{\$58.81}$$

Thus, a male driver with 10 years of driving experience and 2 driving violations is expected to pay a monthly auto insurance premium of $58.81.

(e)   To find the predicted auto insurance premium for a female driver with 10 years of driving experience and 2 violations, we substitute $x_1 = 10$, $x_2 = 2$, and $D = 1$ in the estimated regression model (7).

$$\hat{y} = 58.795 - 1.5336\,(10) + 7.674\,(2) - 2.954\,(1) = 55.853 = \textbf{\$55.85}$$

Thus, a female driver with 10 years of driving experience and 2 driving violations is expected to pay a monthly auto insurance premium of $55.85.

(f)   We are to make a 99% confidence interval for B_3. From the given information and from the MINITAB solution given in Figure 14.11,

$$n = 12, \quad b_3 = -2.954, \quad \text{and} \quad s_{b_3} = 4.005$$

For a 99% confidence level,

$$\text{Area in each tail of the } t \text{ distribution curve} = .5 - (.99/2) = .005$$

Because there are three independent variables in our regression model (6), $k = 3$. Therefore,

$$\text{degrees of freedom} = n - k - 1 = 12 - 3 - 1 = 8$$

From the t distribution table, the t value for .005 area in the right tail of the t distribution curve and 8 degrees of freedom is 3.355. The 99% confidence interval for B_3 is

$$b_3 \pm t s_{b_3} = -2.954 \pm 3.355\,(4.005)$$

$$= -2.954 \pm 13.437 = -\textbf{\$16.39 to \$10.48}$$

Thus, the 99% confidence interval for b_3 is $-$\$16.39 to \$10.48. We can state with 99% confidence that female drivers pay somewhere between \$16.39 less than to \$10.48 more than male drivers with similar values for the x_1 and x_2 variables.

(g) We are to test whether or not the coefficient B_3 of *gender* in model (6) is different from zero. We perform the following five steps to make this test of hypothesis.

Step 1. *State the null and alternative hypotheses*

We are to test whether or not B_3 in the regression model (6) is different from zero. The two hypotheses are

$$H_0: B_3 = 0 \quad (B_3 \text{ is zero})$$

$$H_1: B_3 \neq 0 \quad (B_3 \text{ is different from zero})$$

Step 2. *Select the distribution to use*

The sample size is small ($n < 30$) and σ_ϵ is not known. Hence, we use the t distribution to make a test of hypothesis about B_3.

Step 3. *Determine the rejection and nonrejection regions*

The significance level is .01. The $\neq$ sign in the alternative hypothesis indicates that the test is two-tailed. Therefore,

Area in each tail of the t distribution curve $= \alpha/2 = .01/2 = .005$

degrees of freedom $= n - k - 1 = 12 - 3 - 1 = 8$

From the t distribution table, the critical values of t for 8 degrees of freedom and .005 area in each tail of the t distribution curve are -3.355 and 3.355, as shown in Figure 14.12.

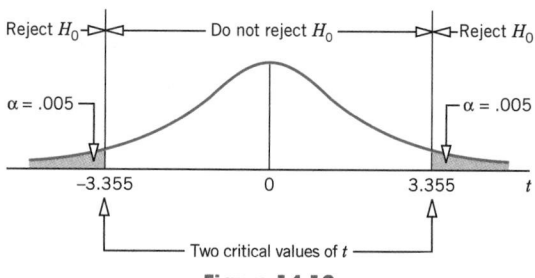

Figure 14.12

Step 4. *Calculate the value of the test statistic*

The value of the test statistic t for b_3 can be obtained from the MINITAB solution given in Figure 14.11. From that solution,

$$t = -.74$$

Step 5. *Make a decision*

The value of the test statistic $t = -.74$ falls between the two critical values of t, -3.355 and 3.355, which is in the nonrejection region in Figure 14.12. Consequently, we fail to reject the null hypothesis and conclude that B_3 in regression model (6) is not different from zero. That is, the variable *gender* has no effect on the auto insurance premiums paid by drivers.

We can also use the p-value to make this decision. From Figure 14.11, the p-value for D is .482 (see the last entry in the row for D in Figure 14.11). The significance level α is .01. Since the p-value is greater than α, we fail to reject the null hypothesis.

Note that this p-value is for a two-tailed test. If we were making a one-tailed test, we would divide .482 by 2 to obtain the p-value before we compared it to α. ∎

As mentioned earlier, the number of dummy variables used for a qualitative variable in a regression model is one less than the number of categories for that variable. In Example 14–7, the variable *gender* had only two categories—male and female. Consequently, we used only one dummy variable.

Suppose a company wants to find whether or not its sales depend on advertising expenditures and time of year. Table 14.4 gives data on sales (in millions of dollars) and advertising expenditures (in millions of dollars) for this company for the past 14 quarters.

Table 14.4

Sales	Advertising Expenditure	Quarter
21	1.2	Fall
29	1.6	Winter
23	1.4	Spring
26	1.7	Summer
28	1.1	Fall
31	.9	Winter
24	1.3	Spring
27	1.7	Summer
26	1.9	Fall
30	1.1	Winter
24	1.5	Spring
27	1.3	Summer
28	1.7	Fall
33	1.8	Winter

Now we want to estimate the regression of sales on the advertising expenditure and quarters. Because the variable *quarter* is a qualitative variable, we will use dummy variable(s) to represent it in our regression model. Since there are four quarters—fall, winter, spring, and summer—we will use three dummy variables in the model. One of the four quarters will be the reference group and each of the remaining three quarters will be represented by a dummy variable. Suppose we decide to take the fall quarter as the reference group and introduce one dummy variable for each of the other three quarters. Let D_1 be the dummy variable that represents the winter quarter, D_2 be the dummy variable that represents the spring quarter, and D_3 be the dummy variable that represents the summer quarter. Then,

$$D_1 = 1 \text{ for the winter quarter, and zero for other quarters}$$

$$D_2 = 1 \text{ for the spring quarter, and zero for other quarters}$$

$$D_3 = 1 \text{ for the summer quarter, and zero for other quarters}$$

Using these dummy variables, we rewrite the data of Table 14.4 as in Table 14.5. In this table, y represents the sales of the company and x_1 refers to the advertising expenditure.

Table 14.5

y	x_1	D_1	D_2	D_3
21	1.2	0	0	0
29	1.6	1	0	0
23	1.4	0	1	0
26	1.7	0	0	1
28	1.1	0	0	0
31	.9	1	0	0
24	1.3	0	1	0
27	1.7	0	0	1
26	1.9	0	0	0
30	1.1	1	0	0
24	1.5	0	1	0
27	1.3	0	0	1
28	1.7	0	0	0
33	1.8	1	0	0

Using the data of Table 14.5, we will estimate the regression model

$$y = A + B_1x_1 + B_2D_1 + B_3D_2 + B_4D_3 + \epsilon \tag{10}$$

By estimating this regression model, we will actually estimate the following four models.

1. The regression model for the fall quarter is obtained by substituting $D_1 = 0$, $D_2 = 0$, and $D_3 = 0$ in model (10). This regression model is

$$y = A + B_1x_1 + \epsilon$$

2. The regression model for the winter quarter is obtained by substituting $D_1 = 1$, $D_2 = 0$, and $D_3 = 0$ in model (10). This regression model is

$$y = (A + B_2) + B_1x_1 + \epsilon$$

3. The regression model for the spring quarter is obtained by substituting $D_1 = 0$, $D_2 = 1$, and $D_3 = 0$ in model (10). This regression model is

$$y = (A + B_3) + B_1x_1 + \epsilon$$

4. The regression model for the summer quarter is obtained by substituting $D_1 = 0$, $D_2 = 0$, and $D_3 = 1$ in model (10). This regression model is

$$y = (A + B_4) + B_1x_1 + \epsilon$$

Thus, by estimating regression model (10), we will estimate these four regression models.

EXERCISES

Concepts and Procedures

14.12 Explain the concept of dummy variables. When is such a variable included in a regression model?

14.13 A qualitative (independent) variable has two categories. How many dummy variables will you include in the regression model to represent this variable?

14.14 A qualitative (independent) variable has four categories. How many dummy variables will you include in the regression model to represent this variable?

14.15 The following MINITAB solution was obtained for the regression model

$$y = A + B_1x_1 + B_2x_2 + B_3D + \epsilon$$

for a sample data set.

```
THE REGRESSION EQUATION IS
Y = 12.1 + 0.250 X1 - 0.0359 X2 - 0.79 D

PREDICTOR            COEF         STDEV      T-RATIO        P
CONSTANT           12.138        6.564         1.85    0.097
X1                0.24972      0.03446         7.25    0.000
X2               -0.03595      0.02930        -1.23    0.251
D                  -0.795        1.100        -0.72    0.488


S = 1.852       R-SQ = 96.5%       R-SQ(ADJ) = 95.3%


ANALYSIS OF VARIANCE

SOURCE           DF           SS          MS         F         P
REGRESSION        3       851.45      283.82     82.77     0.000
ERROR             9        30.86        3.43
TOTAL            12       882.31


SOURCE           DF       SEQ SS
X1                1       841.25
X2                1         8.40
D                 1         1.79
```

Using the MINITAB solution, answer the following questions.

    a.   Write the estimated regression equation.
    b.   Explain the meaning of b_3 obtained by estimating the given regression model.
    c.   What are the values of the standard deviation of errors, the coefficient of multiple determination, the adjusted coefficient of multiple determination, SST, SSR, SSE, MSR, and MSE?
    d.   Write the values of the standard deviation, the observed value of t, and the p-value for the estimated coefficient b_3.
    e.   What is the predicted value of y for $x_1 = 98$, $x_2 = 145$, and $D = 0$?
    f.   What is the point estimate of the expected (mean) value of y for all elements given that $x_1 = 87$, $x_2 = 165$, and $D = 1$?
    g.   Construct a 95% confidence interval for the coefficient of D.
    h.   Using the 1% significance level, test whether or not the coefficient of D is negative.
    i.   At the 2.5% significance level, can you conclude that the coefficients of all independent variables in the population regression model are equal to zero?

14.16 The following MINITAB solution was obtained for the regression model

$$y = A + B_1x_1 + B_2x_2 + B_3x_3 + B_4D + \epsilon$$

for a sample data set.

```
THE REGRESSION EQUATION IS
Y = 46.4 - 0.0520 X1 + 0.0921 X2 - 0.00170 X3 + 1.03 D

PREDICTOR            COEF        STDEV    T-RATIO        P
CONSTANT            46.36        12.34       3.76    0.005
X1               -0.05201      0.01709      -3.04    0.014
X2                0.09208      0.02951       3.12    0.012
X3              -0.001704      0.009193     -0.19    0.857
D                   1.032        1.010       1.02    0.334

S = 0.8975        R-SQ = 99.8%    R-SQ(ADJ) = 99.6%

ANALYSIS OF VARIANCE

SOURCE        DF          SS          MS         F        P
REGRESSION     4     2910.75      727.69    903.35    0.000
ERROR          9        7.25        0.81
TOTAL         13     2918.00

SOURCE        DF      SEQ SS
X1             1     2901.82
X2             1        7.85
X3             1        0.24
D              1        0.84
```

Using the MINITAB solution, answer the following questions.

    a. Write the estimated regression equation.
    b. Explain the meaning of b_4 obtained by estimating the given regression model.
    c. What are the values of the standard deviation of errors, the coefficient of multiple determination, the adjusted coefficient of multiple determination, SST, SSR, SSE, MSR, and MSE?
    d. Write the values of the standard deviation, the observed value of t, and the p-value for the estimated coefficient b_4.
    e. What is the predicted value of y for $x_1 = 210$, $x_2 = 195$, $x_3 = 260$, and $D = 1$?
    f. What is the point estimate of the expected (mean) value of y for all elements given that $x_1 = 290$, $x_2 = 230$, $x_3 = 195$, and $D = 0$?
    g. Determine a 99% confidence interval for the coefficient of D.
    h. Using the 2.5% significance level, can you conclude that the coefficient of D is positive?
    i. Using the 1% significance level, test whether or not the coefficients of all independent variables in the population regression model are equal to zero.

14.17 The following table gives data on variables y, x_1, x_2, x_3, and a dummy variable D.

y	x_1	x_2	x_3	D
8	18	38	74	1
11	26	25	64	1
19	34	24	47	0
21	38	44	31	0
7	13	12	79	1
23	49	48	35	0
16	28	38	42	1
27	59	52	18	0
9	14	17	71	1
13	21	39	57	0

Using MINITAB (or any other statistical software package), estimate the regression model

$$y = A + B_1x_1 + B_2x_2 + B_3x_3 + B_4D + \epsilon$$

Using the solution obtained, answer the following questions.

    a.   Write the estimated regression equation.

    b.   Explain the meaning of b_4 obtained by estimating the given regression model.

    c.   What are the values of the standard deviation of errors, the coefficient of multiple determination, and the adjusted coefficient of multiple determination?

    d.   What is the predicted value of y for $x_1 = 45$, $x_2 = 40$, $x_3 = 60$, and $D = 0$?

    e.   What is the point estimate of the expected (mean) value of y for all elements given that $x_1 = 40$, $x_2 = 30$, $x_3 = 55$, and $D = 1$?

    f.   Construct a 99% confidence interval for the coefficient of D.

    g.   Using the 5% significance level, test whether or not the coefficient of D is different from zero.

    h.   Using the 1% significance level, test whether or not the coefficients of all independent variables in the population regression model are equal to zero.

14.18   The following table gives data on variables y, x_1, x_2, and a dummy variable D.

y	x_1	x_2	D
24	98	52	1
14	51	69	0
18	74	63	0
31	108	35	1
10	33	88	0
29	119	54	1
26	99	51	1
33	141	31	1
13	47	67	0
27	103	41	0
26	111	46	1

Using MINITAB (or any other statistical software package), estimate the regression model

$$y = A + B_1x_1 + B_2x_2 + B_3D + \epsilon$$

Using the solution obtained, answer the following questions.

    a.   Write the estimated regression equation.

    b.   Explain the meaning of the estimated regression coefficient of the dummy variable.

    c.   What are the values of the standard deviation of errors, the coefficient of multiple determination, and the adjusted coefficient of multiple determination?

    d.   What is the predicted value of y for $x_1 = 87$, $x_2 = 54$, and $D = 1$?

    e.   What is the point estimate of the expected (mean) value of y for all elements given that $x_1 = 95$, $x_2 = 49$, and $D = 0$?

    f.   Make a 95% confidence interval for the coefficient of D.

    g.   Using the 1% significance level, test if the coefficient of D is different from zero.

    h.   Using the 2.5% significance level, can you conclude that the coefficients of all independent variables in the population regression model are equal to zero?

Applications

14.19   Refer to Exercise 14.10. The following table includes information on gender in addition to the information on annual salaries (in thousands of dollars), number of years of schooling, and years of work experience for 12 persons. In the table, M represents males and F refers to females.

Salary	32	24	28	57	48	28	39	43	18	41	17	49
Schooling	16	12	13	20	18	16	14	18	12	16	12	16
Experience	6	10	15	8	11	2	12	4	6	9	2	18
Gender	F	F	M	M	F	F	M	M	F	M	M	F

Using MINITAB (or any other statistical software package), find the regression of salary on schooling, experience, and gender. Using the solution obtained, answer the following questions.

a. Write the estimated regression equation.
b. Which of the two groups, males or females, is the reference group in your regression model?
c. Explain the meaning of the estimated regression coefficient of the dummy variable.
d. By estimating the regression model with *gender* as a dummy variable, you have actually estimated two regression models—one for males and the other for females. Write these two regression models.
e. What are the values of the standard deviation of errors, the coefficient of multiple determination, and the adjusted coefficient of multiple determination?
f. What is the value of the total sum of squares? What portion of SST is explained by our regression model? What portion of SST is not explained by our regression model?
g. How much salary is a male worker with 18 years of schooling and 7 years of work experience expected to earn?
h. How much salary is a female worker with 18 years of schooling and 7 years of work experience expected to earn?
i. What is the point estimate of the expected (mean) salary for all female workers with 16 years of schooling and 10 years of work experience?
j. What is the point estimate of the expected (mean) salary for all male workers with 16 years of schooling and 10 years of work experience?
k. Determine a 99% confidence interval for the coefficient of the dummy variable.
l. Using the 2.5% significance level, can you conclude that female workers are paid lower salaries than male workers?
m. Using the 1% significance level, can you conclude that the coefficients of all independent variables in the population regression model are equal to zero?

14.20 Refer to Exercise 14.11. Some of the restaurants owned by the CTO corporation have takeout facilities. The research department at the company wanted to determine if takeout facilities at restaurants affect sales. The following table gives information on the weekly sales (in thousands of dollars) of 11 restaurants, the population (in thousands) within five miles of the restaurants, the mean annual income (in thousands of dollars) of the households for those areas, and whether or not each of these restaurants has a takeout facility.

Sales	19	29	17	21	14	30	33	22	18	27	24
Population	21	15	32	18	47	69	29	43	75	39	53
Income	38	49	29	32	27	36	41	26	19	34	28
Takeout facility	No	Yes	No	No	Yes	Yes	No	Yes	Yes	No	No

Using MINITAB (or any other statistical software package), find the regression of sales on population, income, and takeout facility. Using the solution obtained, answer the following questions.

a. Write the estimated regression equation.
b. Explain the meaning of the estimated regression coefficient of the dummy variable.
c. What are the values of the standard deviation of errors, the coefficient of multiple determination, and the adjusted coefficient of multiple determination?

d. By estimating the regression model with *takeout facility* as a dummy variable, you have actually estimated two regression models—one for restaurants without takeout facilities and the other for restaurants with takeout facilities. Write these two regression models.
e. What is the value of the total sum of squares? What portion of SST is explained by our regression model? What portion of SST is not explained by our regression model?
f. What are the expected sales for a restaurant with 50 thousand people living within a five-mile area surrounding it, $35 thousand mean annual income of households in that area, and a takeout facility?
g. What is the point estimate of the expected (mean) sales for all restaurants with 55 thousand people living within a five-mile area surrounding them, $38 thousand mean annual income of households in those areas, and no takeout facility?
h. Determine a 95% confidence interval for the coefficient of the dummy variable.
i. Using the 1% significance level, test whether or not the coefficient of the dummy variable is different from zero.
j. Using the 2.5% significance level, can you conclude that the coefficients of all independent variables in the population regression model are equal to zero?

GLOSSARY

Adjusted coefficient of multiple determination Denoted by $\overline{R}^2$, it gives the proportion of SST that is explained by the multiple regression model and is adjusted for the degrees of freedom.

Coefficient of multiple determination Denoted by R^2, it gives the proportion of SST that is explained by the multiple regression model.

Dummy variable Dummy variables are used to represent qualitative variables in a regression model. A dummy variable indicates whether or not an element possesses an attribute. The number of dummy variables introduced in a regression model for a variable is equal to the number of categories for that variable minus 1.

Least squares regression model The estimated regression model obtained by minimizing the sum of squared errors.

MSE The mean square error given by $SSE/(n - k - 1)$.

MSR The mean square regression given by SSR/k.

Multiple regression model A regression model that contains two or more independent variables.

Partial linear correlation coefficient The linear correlation coefficient between two variables when all other variables are held constant.

Partial regression coefficients The coefficients of independent variables in a multiple regression model are called the partial regression coefficients because each of them gives the effect of the corresponding independent variable on the dependent variable when all other independent variables are held constant.

Simple linear correlation coefficient The linear correlation coefficient between two variables without controlling any other variable.

Standard deviation of errors Also called the *standard deviation of estimate*, it is a measure of the variation among errors.

SSE (error sum of squares) The sum of the squared differences between the actual and predicted values of y. It is the portion of SST that is not explained by the regression model.

SSR (regression sum of squares) The portion of SST that is explained by the regression model.

SST (total sum of squares) The sum of squared differences between actual y values and $\overline{y}$.

KEY FORMULAS

1. **Multiple linear regression model**

$$y = A + B_1 x_1 + B_2 x_2 + B_3 x_3 + \cdots + B_k x_k + \epsilon$$

2. **Estimated multiple regression model**

$$\hat{y} = a + b_1 x_1 + b_2 x_2 + b_3 x_3 + \cdots + b_k x_k$$

3. **Degrees of freedom for a multiple regression model**

When we use the t distribution to make inferences about a single parameter of a multiple regression model, the degrees of freedom are calculated as

$$df = n - k - 1$$

where n is the sample size and k is the number of independent variables.

However, when we use the F distribution to make inferences about parameters of a multiple regression model, the degrees of freedom for the numerator and denominator are calculated as follows.

$$\text{degrees of freedom for the numerator} = k$$
$$\text{degrees of freedom for the denominator} = n - k - 1$$

4. **Standard deviation of errors for a multiple regression model**

$$s_e = \sqrt{\frac{\text{SSE}}{n - k - 1}} \qquad \text{where} \qquad \text{SSE} = \Sigma(y - \hat{y})^2$$

5. **Coefficient of multiple determination**

$$R^2 = \frac{\text{SSR}}{\text{SST}}$$

where
$$\text{SST} = \text{SS}_{yy} = \Sigma(y - \bar{y})^2$$
$$\text{SSR} = \Sigma(\hat{y} - \bar{y})^2$$

6. **Adjusted coefficient of multiple determination**

$$\bar{R}^2 = 1 - (1 - R^2)\frac{n - 1}{n - k - 1} \qquad or \qquad 1 - \frac{\text{SSE}/(n - k - 1)}{\text{SST}/(n - 1)}$$

7. **The $(1 - \alpha)100\%$ confidence interval for B_i**

$$b_i \pm t s_{b_i}$$

8. **Value of the test statistic t for b_i for the test of hypothesis about a single coefficient**

$$t = \frac{b_i - B_i}{s_{b_i}}$$

9. **Value of the test statistic F for the test of overall significance of the multiple regression model**

$$F = \frac{\text{SSR}/k}{\text{SSE}/(n - k - 1)} \qquad or \qquad \frac{\text{MSR}}{\text{MSE}}$$

where MSR stands for the mean square regression and MSE for the mean square error. The MSR and MSE are calculated as

$$\text{MSR} = \text{SSR}/k$$
$$\text{MSE} = \text{SSE}/(n - k - 1)$$

10. **Simple linear correlation coefficient between x_1 and x_2**

$$r_{12} = \frac{SS_{x_1 x_2}}{\sqrt{SS_{x_1 x_1} \, SS_{x_2 x_2}}}$$

where r_{12} is the simple linear correlation coefficient between x_1 and x_2. We can calculate r_{y1} (the simple linear correlation coefficient between y and x_1) and r_{y2} (the simple linear correlation coefficient between y and x_2) using similar formulas.

SUPPLEMENTARY EXERCISES

14.21 Brown Corporation produces many consumer products that are sold all over the country. The company employs a large number of salespersons who visit dealers and stores to promote these products and obtain orders. Each salesperson is allocated a specified market region in which to sell these products. The company's research department wanted to find how sales depend on the experience, gender, and major of the salespersons. The company only hires salespersons with a bachelor's degree in finance, management, or marketing. The research department took a sample of 15 salespersons and collected information on them that is given in the following table. In the table, sales represent the amount (in thousands of dollars) of orders obtained per week by each salesperson. This amount is the average for many weeks. The experience is the number of years each of these employees has worked as a salesperson.

Sales	Experience	Gender	Major
48	6	Female	Marketing
41	10	Male	Marketing
29	2	Male	Management
38	5	Female	Finance
43	12	Male	Marketing
33	7	Female	Management
43	8	Female	Finance
42	4	Female	Marketing
37	10	Male	Finance
46	9	Male	Marketing
30	6	Male	Management
46	13	Female	Finance
35	8	Male	Finance
37	11	Male	Management
51	12	Female	Marketing

Define the following dummy variables.

$$D_1 = 1 \text{ if gender is male, } 0 \text{ if female}$$

$$D_2 = 1 \text{ if major is management, } 0 \text{ otherwise}$$

$$D_3 = 1 \text{ if major is marketing, } 0 \text{ otherwise}$$

Note that for the variable *gender*, *females* are the reference group. For the variable *major*, *finance* is the reference group. Also note that we have included two dummy variables for the variable *major* because it has three categories: finance, management, and marketing. Recall that the number of dummy variables included in a regression model for a variable is one less than the number of categories for that variable.

Now write a new table with variables sales, experience, D_1, D_2, and D_3. Using these data and MINITAB (or any other statistical software package), estimate the following regression model.

$$y = A + B_1x_1 + B_2D_1 + B_3D_2 + B_4D_3 + \epsilon$$

where y represents sales and x_1 represents experience. Using the solution obtained, answer the following questions.

a.   What is your expectation of the sign of B_1 in the above regression model? In other words, do you expect B_1 to be positive or negative? Explain why.
b.   Write the estimated regression equation. Is the sign of b_1 in the solution consistent with your expectation of part a?
c.   Explain the meaning of the estimated regression coefficients of all independent variables.
d.   What are the values of the standard deviation of errors, the coefficient of multiple determination, and the adjusted coefficient of multiple determination?
e.   What is the value of the total sum of squares? What portion of SST is explained by our regression model? What portion of SST is not explained by our regression model?
f.   By estimating the above regression model, you have actually estimated six regression models, one for each pair of the values of *gender* and *major*. Write these six estimated regression equations.
g.   Find the expected sales per week for the salespersons employed by this company with the following characteristics.
    1.   Experience = 9 years,   gender = female,   and   major = finance
    2.   Experience = 3 years,   gender = male,   and   major = management
    3.   Experience = 13 years,   gender = female,   and   major = marketing
h.   Determine a 95% confidence interval for the coefficient of experience.
i.   Make a 99% confidence interval for each of the coefficients of the dummy variables.
j.   Construct a 98% confidence interval for the constant term in the population regression model.
k.   Using the 1% significance level, can you conclude that B_1 is positive?
l.   At the 2% significance level, test if the coefficient of D_1 is different from zero.
m.   Using the 2.5% significance level, test whether or not the coefficient of D_2 is negative.
n.   At the 1% significance level, test if the coefficient of D_3 is positive.
o.   Using the 5% significance level, can you conclude that the coefficients of all independent variables in the population regression model are equal to zero?
p.   Determine the simple linear correlation coefficient between each pair of variables.

14.22   A real estate expert wanted to find the relationship between the sale price of houses and various characteristics of the houses. She collected data on six variables, recorded in the table, for 13 houses that were sold recently. The six variables are

Price = Sale price of a house in thousands of dollars

Lot size = Size of the lot in acres

Living area = Living area in square feet

Age = Age of a house in years

Corner = Whether or not a house is on a corner lot

Garage = Whether or not a house has a garage.

Price	Lot Size	Living Area	Age	Corner Lot	Garage
255	1.4	2500	8	Yes	Yes
178	.9	2250	12	No	Yes
263	1.8	2900	5	No	Yes
127	.7	1800	24	No	No
305	2.6	3200	10	Yes	Yes
164	1.2	2400	18	No	No
245	2.1	2700	9	No	Yes
146	1.1	2050	28	No	No
287	2.8	2850	13	Yes	Yes
189	1.6	2400	4	No	No
234	3.2	2600	9	No	Yes
211	1.7	2300	8	No	Yes
123	.5	1700	11	No	No

Using MINITAB (or any other statistical software package), find the regression of price on lot size, living area, age, corner lot, and garage. For "corner lot," take "Yes" as the reference group. For "garage," take "No" as the reference group. Using the solution obtained, answer the following questions.

    a.   Indicate whether you expect a positive or a negative relationship between the dependent variable and each of the independent variables.

    b.   Write the estimated regression equation. Are the signs of the coefficients of independent variables obtained in the solution consistent with your expectations of part a?

    c.   Explain the meaning of the estimated regression coefficients of all independent variables.

    d.   What are the values of the standard deviation of errors, the coefficient of multiple determination, and the adjusted coefficient of multiple determination?

    e.   What is the value of the total sum of squares? What portion of SST is explained by our regression model? What portion of SST is not explained by our regression model?

    f.   By estimating the said regression model, you have actually estimated four regression models, one for each pair of the values of *corner lot* and *garage*. Write these four estimated regression equations.

    g.   What is the predicted sale price of a house that has a lot size of 2.5 acres and a living area of 3000 square feet, is 14 years old, is built on a corner lot, and has a garage?

    h.   What is the point estimate of the mean sale price of all houses that have a lot size of 2.2 acres and a living area of 2500 square feet, are 7 years old, are built on a lot that is not a corner lot, and have a garage?

    i.   Determine a 99% confidence interval for each of the coefficients of the quantitative independent variables.

    j.   Make a 95% confidence interval for each of the coefficients of the dummy variables.

    k.   Construct a 98% confidence interval for the constant term in the population regression model.

    l.   Using the 1% significance level, test whether or not the coefficient of *lot size* is positive.

    m.   At the 2.5% significance level, test if the coefficient of *living area* is positive.

    n.   At the 5% significance level, test if the coefficient of *age* is negative.

    o.   At the 1% significance level, can you conclude that the coefficient of *corner lot* is different from zero?

    p.   Using the 5% significance level, test if the coefficient of *garage* is positive.

    q.   Using the 1% significance level, can you conclude that the coefficients of all independent variables in the population regression model are equal to zero?

    r.   Determine the simple linear correlation coefficient between each pair of variables.

14.23   The total output depends on the amounts of labor and capital employed. The relationship between these variables is called the production function. The following table gives information on GNP, labor, and capital for the United States for the years 1972 through 1989. The GNP represents the gross national product in billions of current dollars; labor gives the thousands of persons (16 years of age and over) employed; and capital represents gross investment in billions of current dollars. (Note that the appropriate measure of labor is the number of standardized hours worked by all persons. However, this information is not available, so we have to use the number of persons employed to represent this variable. Such a variable is called a *proxy variable*.)

Year	GNP	Labor	Capital
1972	1212.8	82153	199.1
1973	1359.3	85064	247.6
1974	1472.8	86794	246.2
1975	1598.4	85846	241.2
1976	1782.8	88752	286.6
1977	1990.5	92017	335.3
1978	2249.7	96048	406.7
1979	2508.2	98824	457.4
1980	2732.0	99303	450.0
1981	3052.6	100397	526.1
1982	3166.0	99526	446.3
1983	3405.7	100834	468.8
1984	3772.2	105005	573.9
1985	4014.9	107150	528.7
1986	4231.6	109597	523.6
1987	4524.3	112440	549.0
1988	4880.6	114968	632.8
1989	5233.2	117342	677.4

Using MINITAB (or any other statistical software package), estimate the regression model

$$y = A + B_1 x_1 + B_2 x_2 + \epsilon$$

where y = GNP, x_1 = labor, and x_2 = capital. Using the solution obtained, answer the following questions.

a. Write the estimated regression equation.
b. Explain the meaning of the estimated regression coefficients of both independent variables.
c. What are the values of the standard deviation of errors, the coefficient of multiple determination, and the adjusted coefficient of multiple determination?
d. What is the value of the total sum of squares? What portion of SST is explained by our regression model? What portion of SST is not explained by our regression model?
e. Determine a 99% confidence interval for the coefficient of *labor*.
f. Using the 2.5% significance level, can you conclude that B_1 is positive?
g. We expect B_2 to be positive. In other words, we expect GNP to increase if we increase the amount of capital. In your estimated model, is b_2 positive? If not, can you explain what could be the reason?
h. Using the 1% significance level, test if the coefficients of both independent variables in the population regression model are equal to zero.

14.24 Refer to Data Set III of Appendix A. Using those data and MINITAB (or any other statistical software package), estimate the regression model

$$y = A + B_1 x_1 + B_2 x_2 + \epsilon$$

where y = annual food expenditure, x_1 = annual income before taxes, and x_2 = members in the household. Using the solution obtained, answer the following questions.

a. Explain whether you expect a positive or a negative relationship between the dependent variable and each of the two independent variables.
b. Write the estimated regression equation. Are the signs of the coefficients of independent variables obtained in the solution consistent with your expectations of part a?

c. Explain the meaning of the estimated regression coefficients of both independent variables.

d. What are the values of the standard deviation of errors, the coefficient of multiple determination, and the adjusted coefficient of multiple determination?

e. What is the value of the total sum of squares? What portion of SST is explained by our regression model? What portion of SST is not explained by our regression model?

f. What is the predicted annual food expenditure for a household with $45,000 income before taxes and 4 members?

g. What is the point estimate of the mean annual food expenditure for all households with $55,000 income before taxes and 3 members?

h. Determine a 98% confidence interval for each of the coefficients of independent variables.

i. Using the 1% significance level, test whether or not the coefficient of *income before taxes* is positive.

j. At the 5% significance level, test if the coefficient of *members in the household* is different from zero.

k. Using the 1% significance level, can you conclude that the coefficients of both independent variables in the population regression model are equal to zero?

l. Determine the simple linear correlation coefficient between each pair of variables.

SELF-REVIEW TEST

1. When using the t distribution to make inferences about a single parameter, the degrees of freedom for a multiple regression model with k independent variables and a sample size of n are equal to

    a. $n + k - 1$ **b.** $n - k + 1$ **c.** $n - k - 1$

2. The value of R^2 is always in the range

    a. zero to 1   **b.** -1 to 1   **c.** -1 to zero

3. The value of $\overline{R}^2$ is

    a. always positive   **b.** always nonnegative   **c.** can be positive, zero, or negative

4. What is the difference between the population multiple regression model and the estimated multiple regression model?

5. Why are the regression coefficients in a multiple regression model called the partial regression coefficients?

6. What is the difference between R^2 and $\overline{R}^2$? Explain.

7. When is a dummy variable used as an independent variable in a regression model? Give an example.

8. Following is the MINITAB solution obtained for a regression of y on x_1, x_2, D_1, and D_2.

```
THE REGRESSION EQUATION IS
Y = 4.09 + 0.332 X1 - 0.0077 X2 + 3.97 D1 - 0.679 D2

PREDICTOR          COEF        STDEV      T-RATIO        P
CONSTANT          4.089        5.460         0.75    0.468
X1              0.33157      0.03115        10.64    0.000
X2             -0.00775      0.01240        -0.62    0.544
D1                3.969        1.501         2.64    0.021
D2              -0.6788       0.9362        -0.73    0.482

S = 1.727      R-SQ = 97.5%     R-SQ(ADJ) = 96.7%
```

```
ANALYSIS OF VARIANCE

SOURCE        DF           SS          MS          F           P
REGRESSION     4       1402.10      350.53      117.57      0.000
ERROR         12         35.78        2.98
TOTAL         16       1437.88


SOURCE        DF        SEQ SS
X1             1       1362.72
X2             1         18.065
D1             1         19.764
D2             1          1.57
```

Using the MINITAB solution, answer the following questions for the population regression model

$$y = A + B_1 x_1 + B_2 x_2 + B_3 D_1 + B_4 D_2 + \epsilon$$

a. Write the estimated regression equation.

b. Write the values of a, b_1, b_2, b_3, and b_4 and explain the meaning of these estimated regression coefficients.

c. What are the values of the standard deviation of errors, the coefficient of multiple determination, the adjusted coefficient of multiple determination, SST, SSR, SSE, MSR, and MSE?

d. Write the values of the standard deviation, the observed value of t, and the p-value for each of the estimated coefficients a, b_1, b_2, b_3, and b_4.

e. What is the value of SSR if we estimate the regression of y on x_1 alone? What is the increase in this value of SSR as we add x_2, D_1, and D_2 to our regression model in that order?

f. What is the predicted value of y for $x_1 = 86$, $x_2 = 312$, $D_1 = 0$, and $D_2 = 1$?

g. What is the point estimate of the expected (mean) value of y for all elements given that $x_1 = 94$, $x_2 = 295$, $D_1 = 1$, and $D_2 = 1$?

h. Construct a 99% confidence interval for each of the coefficients of x_1 and x_2 in the population regression model.

i. Determine a 95% confidence interval for each of the coefficients of D_1 and D_2 in the population regression model.

j. Make a 98% confidence interval for A, the constant term in the population regression model.

k. Using the 5% significance level, test whether or not the coefficient of x_1 is positive.

l. Using the 1% significance level, can you conclude that the coefficient of x_2 in the population regression model is negative?

m. At the 1% significance level, test if the coefficient of D_1 in the population regression model is different from zero.

n. At the 2.5% significance level, test if the coefficient of D_2 in the population regression model is negative.

o. At the 5% significance level, test if the constant term A in the population regression model is different from zero.

p. Using the 1% significance level, can you conclude that the coefficients of all independent variables in the population regression model are equal to zero?

9. A researcher wanted to find if the starting salaries of recent college graduates depend on GPA (grade point average) and major. She considered only three majors: business, economics, and computers. She took a sample of 16 recent college graduates and collected the required information. The following table records that information. The salaries are in thousands of dollars per year.

Salary	GPA	Major
28	3.6	Economics
27	3.1	Business
24	2.9	Computers
22	2.7	Business
31	3.3	Business
33	3.0	Computers
23	2.8	Economics
29	3.2	Economics
24	2.6	Business
30	2.9	Business
35	3.8	Computers
28	3.5	Computers
26	2.5	Economics
29	2.8	Business
26	2.4	Business
23	2.2	Computers

Using MINITAB (or any other statistical software package), estimate the regression model of salaries on the GPA and major. For the qualitative variable *major*, take business as the reference group. Using the solution obtained, answer the following questions.

a. What is your expectation of the sign of the coefficient of GPA in the population regression model? In other words, do you expect this coefficient to be positive or negative? Explain why.

b. Write the estimated regression equation. Is the sign of the coefficient of GPA in the solution consistent with your expectation of part a?

c. Explain the meaning of the estimated regression coefficients of all independent variables.

d. What are the values of the standard deviation of errors, the coefficient of multiple determination, and the adjusted coefficient of multiple determination?

e. What is the value of the total sum of squares? What portion of SST is explained by our regression model? What portion of SST is not explained by our regression model?

f. By estimating the given regression model, you have actually estimated three regression models, one for each of the three majors. Write these three estimated regression equations.

g. Find the expected starting salaries per annum for college graduates with the following characteristics.

　　1.　GPA = 3.4　and　major = economics
　　2.　GPA = 2.9　and　major = business
　　3.　GPA = 3.1　and　major = computers

h. Determine a 95% confidence interval for the coefficient of GPA.

i. Make a 99% confidence interval for each of the coefficients of the dummy variables.

j. Construct a 98% confidence interval for the constant term in the population regression model.

k. Using the 1% significance level, can you conclude that the coefficient of GPA is positive?

l. At the 2.5% significance level, test if the coefficient of the dummy variable representing economics major is negative.

m. Using the 2% significance level, test whether or not the coefficient of the dummy variable representing computer major is different from zero.

n. At the 1% significance level, test if the constant term A is different from zero.

o. Using the 2.5% significance level, can you conclude that the coefficients of all independent variables in the population regression model are equal to zero?

15 | TIME SERIES ANALYSIS

As mentioned in Chapter 2, data collected on the same element for the same variable at different points in time or for different periods of time are called *time series data* or a **time series**. Such data may be daily, weekly, biweekly, monthly, bimonthly, quarterly, or yearly. The following are a few examples of time series data.

1. The sales of a company for each of the past 10 years
2. The money supply in circulation at the end of each of the past 20 quarters
3. Unemployment rates for the past 20 months
4. Price of a company's stock at the end of each of the past 40 working days

Making short-run and long-run decisions is critical for the success of any business entity. Companies often make decisions and plans for the future about production, sales, inventories, investments, raw materials, employee needs, development of new products, etc. For example, a company may want to forecast the demand for its product for next year and the year after. As another example, a department store may want to predict the inventories needed for the coming holiday season. Such decisions and forecasts are usually made using time series data.

In this chapter we discuss methods used to analyze time series data and to forecast future values of the variables. We first discuss the components of a time series and then learn how to isolate these components using the regression method and the moving average method. Then we learn how to obtain seasonally adjusted data—data that do not contain seasonal variations.

15.1   COMPONENTS OF A TIME SERIES

A time series may contain one or more of the following four components: trend, cyclical, seasonal, and irregular. These components, explained next, describe how a time series behaves over time.

TREND COMPONENT

Trend is defined as the long-term movement in a time series. In other words, an increase or decrease in the values of a variable occurring over a period of several years gives a trend. If the values of a time series do not change over time, then that time series has no trend.

> **TREND COMPONENT**
>
> Trend is the movement (increase or decrease) in a time series over a long period of time.

A trend curve shows the long-run pattern in the growth or decline in a time series that may represent a variable such as sales, inventories, revenues, profits, employment, population, housing starts, gross national product, consumer expenditure, investment, or interest rates, etc. The long-term movement in a time series may occur due to many factors such as changes in tastes and preferences of consumers, technological changes, demographic changes (changes in population and structure of population), and social and cultural changes.

The trend analysis of historical data helps evaluate past policies. For example, the trend patterns of variables such as sales, revenues, and profits of a company for the past decade can tell us how successful the managerial policies of this period have been.

The trend in a time series is shown by a smooth (linear or nonlinear) curve that slopes upward or downward. Figure 15.1 shows the actual sales (in millions of dollars) of a company over time and the trend for that period.

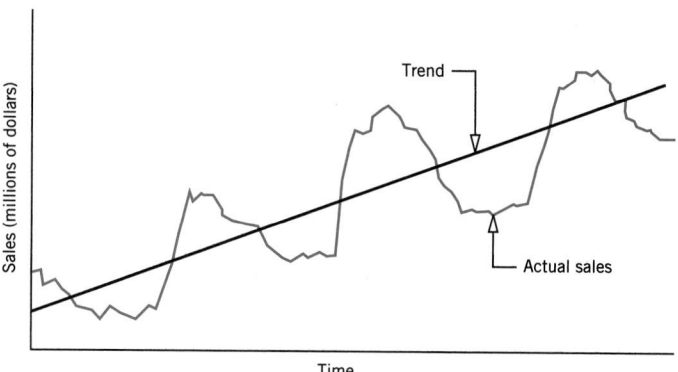

Figure 15.1   Trend of a time series.

Figure 15.2 shows a few possible shapes of linear and nonlinear trend patterns. Figure 15.2a shows the linear trend for a time series that grows continuously, and Figure 15.2b shows the linear trend for a time series that declines continuously. Figure 15.2c shows the (nonlinear) trend curve for a time series for which the increase in the value of the variable occurs at an increasing rate for each successive period. Figure 15.2d gives the (nonlinear) trend curve for a time series that grows at an increasing rate in the beginning but eventually reaches a maximum point so that there is very little increase in the value of the variable.

Figure 15.2e shows the nonlinear trend pattern for a time series that has a slow decline in the initial periods but very rapid decline in later periods. Figure 15.2f exhibits the nonlinear trend pattern for a time series that has a rapid decline in the initial periods but a very slow decline in later periods. Knowing the trend of a time series is useful for making decisions about the future values of business and economic activities.

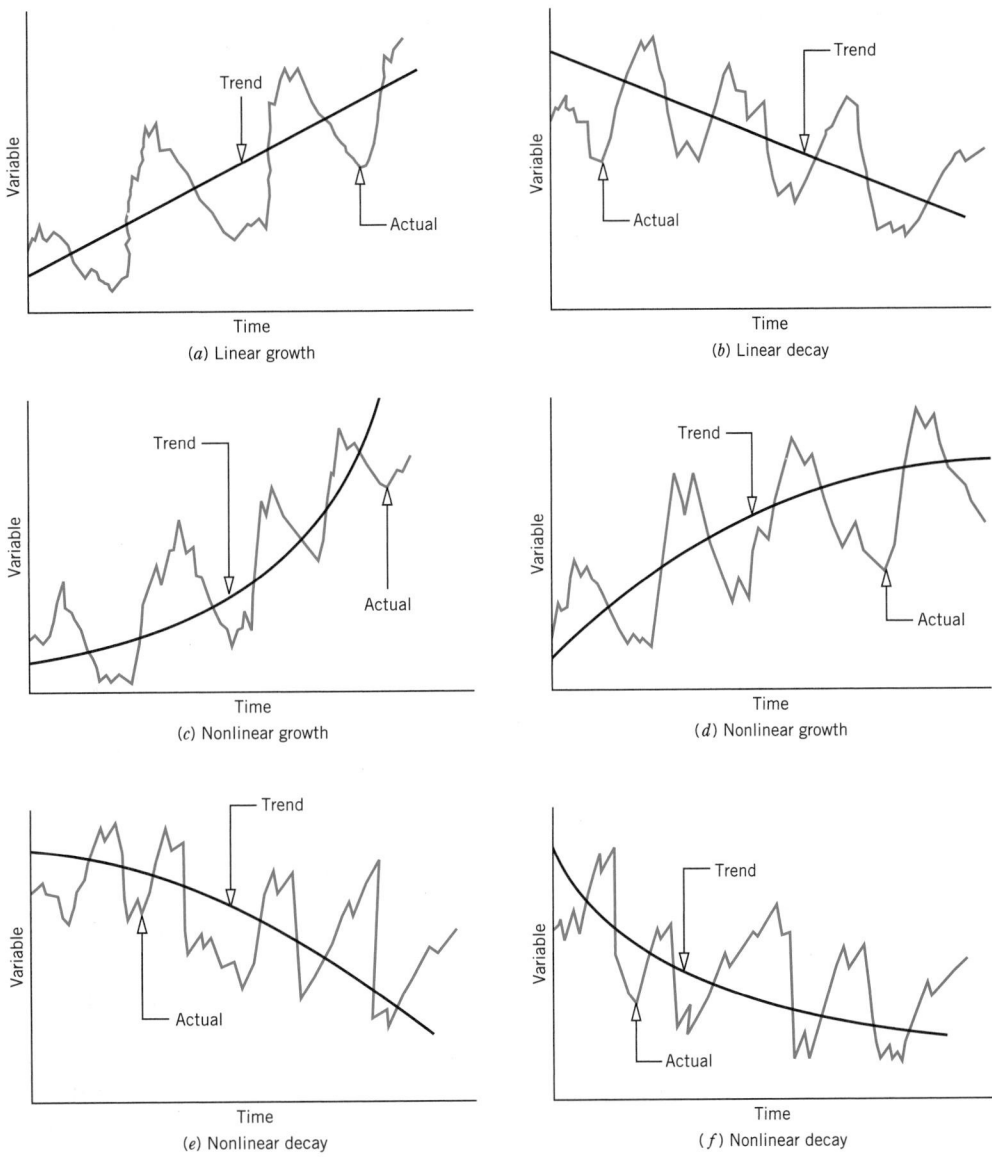

Figure 15.2 A few possible trend patterns.

CYCLICAL COMPONENT

Usually a time series exhibits fluctuations around the trend line. That is, for some periods the actual time series curve is above the trend line and for other periods it is below the trend line. The nonperiodic fluctuations in time series data around the trend line are referred to as the **cyclical component** or **cyclical fluctuations**. Note that the cyclical fluctuations do not last for any specific amount of time. They may last from more than one year to a period of 20 years.

> **CYCLICAL COMPONENT**
>
> The nonperiodic fluctuations in a time series around the trend line are referred to as the cyclical component. The cyclical fluctuations may last for any period more than one year.

Recession, depression, recovery, and boom cause variations in time series data, and these variations make up business cycles. These business cycle fluctuations are called the cyclical component or fluctuations. Remember that the cyclical fluctuations last for a period of more than one year. Figure 15.3 shows cyclical fluctuations for a time series.

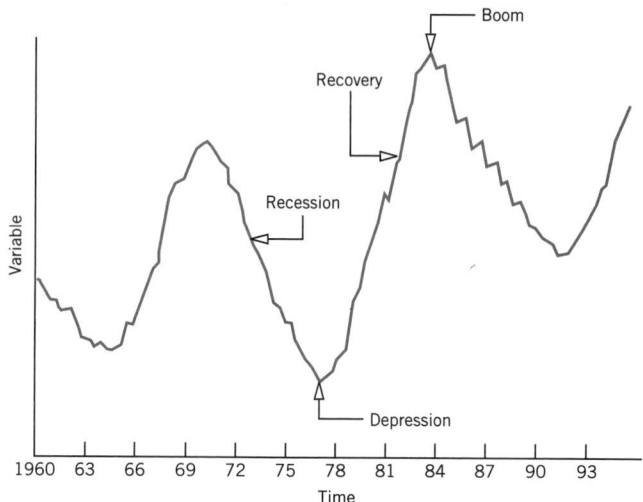

Figure 15.3   Cyclical fluctuations.

SEASONAL COMPONENT

Seasonal variations in a time series are defined as the movements that occur in a time series within a one-year period. Many business activities, such as production and sales, exhibit seasonal patterns over different time periods (e.g., months or quarters) of a year. These seasonal variations may occur due to weather, holidays, or institutional factors. For example, sales of ski equipment are very high during winter and almost nonexistent during other seasons. Sales at department stores around Christmas holidays are much higher than during other times of the year. Utility bills of families are expected to be higher during winter than in other seasons. Such fluctuations are repeated every year, although not in an identical fashion. Nevertheless, with good data and appropriate statistical analysis, many seasonal variations in business and economic activities can be predicted.

> **SEASONAL COMPONENT**
>
> The movements that occur in a time series within a one-year period are referred to as the seasonal component. Such fluctuations are repeated every year, although not in an identical fashion.

Figure 15.4 exhibits a time series with seasonal variations. Note that the cyclical fluctuations last for more than one year, while seasonal variations last for a period that is less than one year. Thus, the major difference between cyclical and seasonal variations is one of duration.

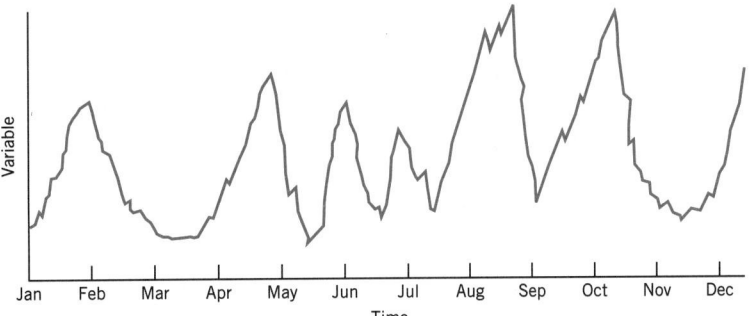

Figure 15.4 Seasonal fluctuations.

IRREGULAR COMPONENT

Irregular fluctuations are random or chance variations in a time series. They are the residuals after trend, cyclical, and seasonal components have been removed from the time series data. Such movements occur due to unpredictable events such as strikes, wars, earthquakes, hurricanes, floods, etc. These events can significantly affect the production of goods and services and other economic and business activities. For example, floods usually cause a reduction in agricultural and industrial production. Due to the nature of irregular variations, they are impossible to predict.

IRREGULAR COMPONENT

The random or chance variations in a time series are referred to as the irregular component. They are the variations in a time series that are left over after trend, cyclical, and seasonal components have been removed.

15.2 ADDITIVE AND MULTIPLICATIVE TIME SERIES MODELS

The value of a time series depends on the four components discussed in Section 15.1. Let

$$Y = \text{observed or actual value of the time series}$$
$$T = \text{value of the trend component}$$
$$C = \text{value of the cyclical component}$$
$$S = \text{value of the seasonal component}$$
$$I = \text{value of the irregular component}$$

The value of Y is determined by the values of T, C, S, and I. Two classical time series models that present Y as a function of the four components are the additive and multiplicative time series models. These models are explained next.

ADDITIVE TIME SERIES MODEL

According to this model, the actual value of a time series is given by the sum of the four components—trend, cyclical, seasonal, and irregular. This model is written as

$$Y = T + C + S + I \qquad (1)$$

ADDITIVE TIME SERIES MODEL

According to the additive model, the value of the time series is equal to the sum of the four components, that is,

$$Y = T + C + S + I$$

Suppose we have data on quarterly output (in millions of dollars) of an auto company. Then, for this time series, Y will be the actual output for a given quarter. The values of T, C, S, and I will give the four components that when added together equal the value of Y. For example, suppose for the fourth quarter of 1993 for this company,

$$T = \$171 \text{ million}$$

$$C = -\$18 \text{ million}$$

$$S = \$6 \text{ million}$$

$$I = \$1 \text{ million}$$

Then, model (1) gives

$$Y = T + C + S + I = 171 - 18 + 6 + 1 = \$160 \text{ million}$$

These values tell us that based on past historical data, the trend indicates that this company's output should have been $171 million for this quarter. However, it is $18 million less than this value because of poor economic conditions (recession or depression); it is $6 million more than this trend (average) value because the production of autos in the fourth quarter for this company usually are higher than the production in other quarters; and it is $1 million above the average output because of chance.

Note that in an additive time series model, the values of T, C, S, and I are all measured in the same units as the value of Y. In the above example, Y is measured in millions of dollars and so are T, C, S, and I.

MULTIPLICATIVE TIME SERIES MODEL

According to the multiplicative time series model, the actual value of a time series is obtained by multiplying the values of the four components. This model is written as

$$Y = T \cdot C \cdot S \cdot I \qquad (2)$$

MULTIPLICATIVE TIME SERIES MODEL

According to the multiplicative model, the value of the time series is equal to the product of the four components, that is,

$$Y = T \cdot C \cdot S \cdot I$$

In this model, whereas the value of T is measured in the same units as that of Y, the values of C, S, and I are measured in relative terms. For example, suppose that for the output of the auto company in the first quarter of 1994,

$$T = \$100 \text{ million}$$
$$C = .80$$
$$S = 1.10$$
$$I = 1.50$$

Then, model (2) gives

$$Y = T \cdot C \cdot S \cdot I = 100(.80)(1.10)(1.50) = \$132 \text{ million}$$

Note that a value of more than 1.0 for any of the components C, S, or I indicates that the relative effect of that component is above the trend, and a value of less than 1.0 for any of these three components means that the relative effect of that component is below the trend.

EXERCISES

Concepts and Procedures

15.1 Briefly explain the four components of a time series.

15.2 What kind of fluctuations in a time series last for more than one year? Explain.

15.3 What kind of fluctuations in a time series last for less than one year? Explain.

15.4 Describe the additive and multiplicative time series models.

15.5 For an additive time series model, for a given period, $T = \$100$ million, $C = \$21$ million, $S = -\$5$ million, and $I = \$3$ million. Find the value of Y for this period.

15.6 For an additive time series model, for a given period, $Y = \$74,000$, $T = \$75,000$, $C = -\$10,000$, and $S = \$12,000$. Find the value of I for this period.

15.7 For a multiplicative time series model, the values of Y and T for a given period are \$200 million and \$150 million, respectively. The relative values of C and I for the same period are 1.2 and 1.1, respectively. Find the relative value of S.

15.8 For a multiplicative time series model, the value of T for a period is \$100 thousand, and the relative values of C, S, and I are .80, 1.2, and .90, respectively. Find the value of Y for this period.

Applications

15.9 For the fourth quarter of 1993, the revenue from the sale of products produced by Mills Corporation was \$25 million. The values of the trend, seasonal, and irregular components for this quarter were \$30 million, $-\$8$ million, and \$2 million, respectively. Find the value of the cyclical component for this quarter using the additive model.

15.10 R. B. Chabot Corporation manufactures many kinds of furniture. For the second quarter of 1993, the values of the trend, cyclical, seasonal, and irregular components for the production of the company were \$12 million, \$4 million, $-\$3.5$ million, and \$.5 million, respectively. Using the additive model, find the value of the company's production for this quarter.

15.11 Seaport Corporation owns a chain of department stores. For the fourth quarter of 1993, the value of the trend component for the inventory held by the company was \$410 million. The relative values of the cyclical, seasonal, and irregular components for the same period were 1.2, 1.4, and .90, respectively. Find the value of inventory for this quarter using the multiplicative model.

15.12   Mandel Corporation owns a chain of 85 restaurants. For the month of June 1993, the total sales of these restaurants were $9.4 million. The relative values of the cyclical, seasonal, and irregular components for this month were .90, 1.3, and 1.05, respectively. Using the multiplicative model, find the value of the trend component for this month.

15.3   MEASURING LINEAR TREND: REGRESSION METHOD

A number of methods can be used to find the long-term trend in a time series. One method to measure the linear trend is using the simple linear regression model. Let

Y = the dependent variable that represents the actual values of the time series

t = the independent variable that represents time

The simple linear regression model can be written as

$$Y = A + Bt + \epsilon \tag{3}$$

In this model, A is the constant term or Y-intercept, B is the coefficient of t or the slope of the linear trend line, and ϵ is the random error term. This trend line is estimated using the least squares method that was studied in Chapter 13. The estimated trend line is

$$\hat{Y} = a + bt \tag{4}$$

The values of a and b, which give estimates of A and B, are calculated as follows.

THE LINEAR TREND LINE USING THE LEAST SQUARES METHOD

For the linear trend line $\hat{Y} = a + bt$, the values of b and a are calculated as

$$b = \frac{SS_{Yt}}{SS_{tt}} \quad \text{and} \quad a = \overline{Y} - b\overline{t}$$

where

$$\overline{Y} = \frac{\Sigma Y}{n} \qquad \overline{t} = \frac{\Sigma t}{n}$$

$$SS_{Yt} = \Sigma Yt - \frac{(\Sigma Y)(\Sigma t)}{n} \quad \text{and} \quad SS_{tt} = \Sigma t^2 - \frac{(\Sigma t)^2}{n}$$

Here "SS" stands for "sum of squares."

Note that the procedure used to estimate model (3) is the same as the one used in Chapter 13 to estimate the simple linear regression model

$$y = A + Bx + \epsilon$$

The only difference is that we have replaced x by t in model (3) above.

Example 15–1 illustrates how to estimate a linear trend line using the least squares method.

Estimating the linear trend line.

EXAMPLE 15–1   Table 15.1 gives information on the gross private domestic investment (in billions of dollars) for the United States for the years 1980 through 1992.

Table 15.1

Year	Gross Private Domestic Investment (billions of dollars)
1980	467.6
1981	558.0
1982	503.4
1983	546.7
1984	718.9
1985	714.5
1986	717.6
1987	749.3
1988	793.6
1989	832.3
1990	799.5
1991	721.1
1992	770.4

(a)   Find the linear trend line using the least squares method. Show the actual time series curve and the trend line for the gross private domestic investment on a graph.

(b)   Using the estimated trend line, predict the gross private domestic investment for the year 1999.

Solution

(a)   Let

$$Y = \text{gross private domestic investment in billions of dollars}$$

$$t = \text{time representing years}$$

We are to find the values of a and b for the linear trend line

$$\hat{Y} = a + bt$$

Although we can use years as the independent variable t, it is usually preferable to assign code numbers to the various years in a time series. To obtain the variable t, we assign a value of zero to 1980 (the first year in our data), 1 to 1981, 2 to 1982, and so on. Then using the values of the Y and t variables, we construct columns for Yt and t^2. This is done in Table 15.2.

Table 15.2

Year	Y	t	Yt	t^2
1980	467.6	0	0.0	0
1981	558.0	1	558.0	1
1982	503.4	2	1006.8	4
1983	546.7	3	1640.1	9
1984	718.9	4	2875.6	16
1985	714.5	5	3572.5	25
1986	717.6	6	4305.6	36
1987	749.3	7	5245.1	49
1988	793.6	8	6348.8	64
1989	832.3	9	7490.7	81
1990	799.5	10	7995.0	100
1991	721.1	11	7932.1	121
1992	770.4	12	9244.8	144
	$\Sigma Y = 8892.9$	$\Sigma t = 78$	$\Sigma Yt = 58{,}215.1$	$\Sigma t^2 = 650$

From the calculations made in Table 15.2,

$$\Sigma Y = 8892.9, \quad \Sigma t = 78, \quad \Sigma Yt = 58{,}215.1, \quad \text{and} \quad \Sigma t^2 = 650$$

There are 13 observations in the data set. Hence, $n = 13$. The values of $\overline{Y}, \overline{t}, SS_{Yt}$, and SS_{tt} are calculated as follows.

$$\overline{Y} = \Sigma Y/n = 8892.9/13 = 684.0692$$

$$\overline{t} = \Sigma t/n = 78/13 = 6.0$$

$$SS_{Yt} = \Sigma Yt - \frac{(\Sigma Y)(\Sigma t)}{n} = 58{,}215.1 - \frac{(8892.9)(78)}{13} = 4857.7$$

$$SS_{tt} = \Sigma t^2 - \frac{(\Sigma t)^2}{n} = 650 - \frac{(78)^2}{13} = 182.0$$

Using the above calculations, the values of b and a are

$$b = \frac{SS_{Yt}}{SS_{tt}} = \frac{4857.7}{182.0} = 26.6907$$

$$a = \overline{Y} - b\overline{t} = 684.0692 - (26.6907)(6.0) = 523.9250$$

Thus, our estimated linear trend line $\hat{Y} = a + bt$ is

$$\hat{Y} = 523.9250 + 26.6907t \tag{5}$$

Note that $\hat{Y}$ gives the trend value for a particular year. Earlier in Section 15.2, we denoted the trend value by T. Thus,

$$\text{Trend value} = T = \hat{Y}$$

Using the estimated regression model (5), we can calculate the trend values for various years. Table 15.3 lists the trend values for 1980 through 1992 for our example.

Table 15.3 Calculating Trend Values

Year	t	Trend Values $\hat{Y} = 523.9250 + 26.6907t$	
1980	0	$523.9250 + 26.6907\,(0)$	$= 523.9250$
1981	1	$523.9250 + 26.6907\,(1)$	$= 550.6157$
1982	2	$523.9250 + 26.6907\,(2)$	$= 577.3064$
1983	3	$523.9250 + 26.6907\,(3)$	$= 603.9971$
1984	4	$523.9250 + 26.6907\,(4)$	$= 630.6878$
1985	5	$523.9250 + 26.6907\,(5)$	$= 657.3785$
1986	6	$523.9250 + 26.6907\,(6)$	$= 684.0692$
1987	7	$523.9250 + 26.6907\,(7)$	$= 710.7599$
1988	8	$523.9250 + 26.6907\,(8)$	$= 737.4506$
1989	9	$523.9250 + 26.6907\,(9)$	$= 764.1413$
1990	10	$523.9250 + 26.6907\,(10)$	$= 790.8320$
1991	11	$523.9250 + 26.6907\,(11)$	$= 817.5227$
1992	12	$523.9250 + 26.6907\,(12)$	$= 844.2134$

Figure 15.5 shows the plot of the actual time series (the gross private domestic investment given in Table 15.1) and the trend line obtained above. The trend line is drawn using the $\hat{Y}$ values obtained in Table 15.3.

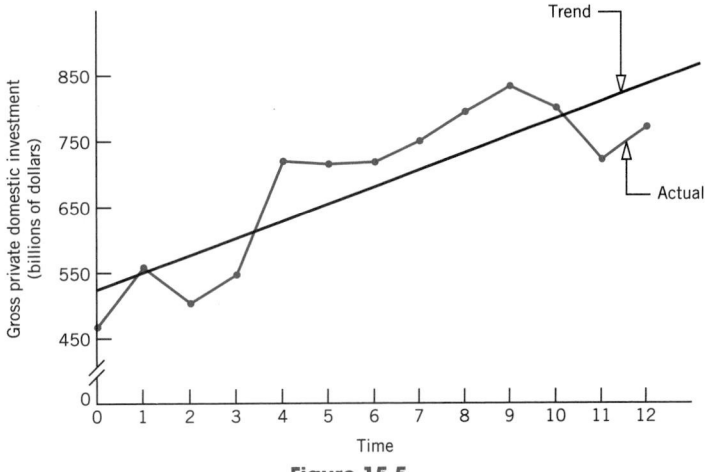

Figure 15.5

(b) In Table 15.2, $t = 0$ for 1980 and $t = 12$ for 1992. We are to predict the gross private domestic investment for the year 1999. The value of t for 1999 is 19. Substituting $t = 19$ in the estimated model (5), we obtain

$$\hat{Y} = 523.9250 + 26.6907\,(19) = \textbf{\$1031.0483 billion}$$

Thus, the gross private domestic investment for 1999 is expected to be \$1031.0483 billion.

Note that here we have used the trend line to forecast the value of the dependent variable for a future period. By using the trend line in this way, we have assumed that the same linear trend holds true for future years, a useful though not necessarily accurate assumption. ∎

15.4 MEASURING THE CYCLICAL EFFECTS: ANNUAL DATA

A time series based on annual data contains only three components: trend, cyclical, and irregular. Such a time series contains no seasonal component. Only the time series containing data collected for a period less than a year can have seasonal variations. For example, daily, weekly, monthly, or quarterly data will contain seasonal variations. Remember that the seasonal fluctuations are the movements in a time series that occur within a one-year period. For a time series with no seasonal variation, the multiplicative model is

$$Y = T \cdot C \cdot I \tag{6}$$

By removing the trend component from such a time series, we can obtain the cyclical and irregular effects. Then, we can show how the cyclical and irregular effects are measured using the multiplicative model. Note that in this chapter we will analyze the cyclical and irregular effects for the multiplicative model only.

When we use the linear regression method to find the trend line, the trend component T is equal to $\hat{Y}$. Substituting $\hat{Y}$ for T in model (6), we obtain

$$Y = \hat{Y} \cdot C \cdot I$$

or

$$C \cdot I = \frac{Y}{\hat{Y}} \tag{7}$$

Thus, the product of cyclical and irregular effects is given by the ratio of the actual value of the time series to the trend value. The product of C and I, which gives the cyclical and irregular components, is called the **cyclical-irregular relative**. Note that the cyclical-irregular component measured in this case is in relative terms and not in original units.

CYCLICAL-IRREGULAR RELATIVES

Using the multiplicative model, the cyclical-irregular relative is given by the ratio of the actual value of the time series to the trend value. That is,

$$C \cdot I = \frac{Y}{\hat{Y}}$$

Example 15–2 illustrates the calculation of cyclical and irregular effects using the multiplicative model.

Calculating the cyclical-irregular relatives using the multiplicative model.

EXAMPLE 15-2 Refer to the data on gross private domestic investment for the United States for 1980 through 1992 given in Table 15.1 of Example 15–1. Using the linear trend line estimated in that example, find the cyclical-irregular relatives for the multiplicative model.

Solution In Table 15.4, the second column lists the actual values of Y from Table 15.1 and the third column contains the trend values, denoted by $\hat{Y}$. The trend values are recorded from Table 15.3. The fourth column shows the ratio of Y to $\hat{Y}$. This column gives the cyclical-irregular relatives. Note that these relatives give the cyclical-irregular component.

Table 15.4 Calculating Cyclical-Irregular Relatives

Year	Y	Trend Values $\hat{Y}$	Cyclical-Irregular Relatives $C \cdot I = Y/\hat{Y}$
1980	467.6	523.9250	.8925
1981	558.0	550.6157	1.0134
1982	503.4	577.3064	.8720
1983	546.7	603.9971	.9051
1984	718.9	630.6878	1.1399
1985	714.5	657.3785	1.0869
1986	717.6	684.0692	1.0490
1987	749.3	710.7599	1.0542
1988	793.6	737.4506	1.0761
1989	832.2	764.1413	1.0891
1990	799.5	790.8320	1.0110
1991	721.1	817.5227	.8821
1992	770.4	844.2134	.9126

A value of less than 1.0 for a cyclical-irregular relative shows that the actual value of the time series for that period is below the trend value due to the negative effect of cyclical and irregular components. For example, a value of .8925 for the cyclical-irregular relative for 1980 in Table 15.4 indicates that the actual gross private domestic investment for 1980 is $(1 - .8925) \times 100 = 10.75\%$ below the trend value for that year. On the other hand, a value of greater than 1.0 for a cyclical-irregular relative indicates that the actual value of the time series for that period is above the trend value due to the positive effect of cyclical and irregular components. For example, a value of 1.0134 for the cyclical-irregular relative for 1981 in Table 15.4 shows that the actual gross private domestic investment for 1981 is $(1.0134 - 1) \times 100 = 1.34\%$ above the trend value for that year.

Figure 15.6 shows the plot of cyclical-irregular relatives recorded in the fourth column of Table 15.4.

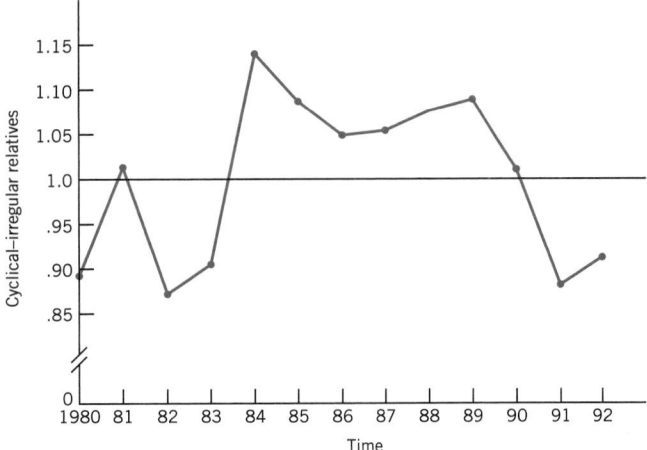

Figure 15.6   Plot of cyclical-irregular relatives.

EXERCISES

Concepts and Procedures

15.13   Explain how the linear regression method is used to measure the trend component for a time series.

15.14   Explain how the cyclical and irregular effects are measured for annual data using the linear regression model and the multiplicative time series model.

15.15   The following table gives the values of a variable Y for the years 1984 through 1993.

Year	1984	1985	1986	1987	1988	1989	1990	1991	1992	1993
Y	43	46	56	53	58	52	61	53	58	67

    a.   Find the linear trend line using the least squares method.
    b.   Using the trend line estimated in part a, predict the trend value for 1997.
    c.   Using the trend line estimated in part a, find the cyclical-irregular relatives.

15.16   The following table gives the values of a variable Y for the years 1983 through 1993.

Year	1983	1984	1985	1986	1987	1988	1989	1990	1991	1992	1993
Y	85	80	83	76	72	75	70	72	67	69	63

    a.   Estimate the linear trend line $Y = A + Bt + \epsilon$.
    b.   Predict the value of Y for 1998. Use the estimated trend line of part a.
    c.   Using the estimated trend line of part a, find the cyclical-irregular relatives.

Applications

15.17 The following table gives the U.S. exports (in billions of dollars) for the years 1977 through 1991.

Year	Exports (billions of dollars)
1977	158.8
1978	186.1
1979	228.9
1980	279.2
1981	303.0
1982	282.6
1983	276.7
1984	302.4
1985	302.1
1986	319.2
1987	364.0
1988	444.2
1989	508.0
1990	557.0
1991	598.2

a. Estimate the linear trend line $Y = A + Bt + \epsilon$, where Y represents the exports and t is the time variable.
b. Assuming the linear trend is valid, predict the value of exports for 1996.
c. Using the estimated trend line of part a, find the cyclical-irregular relatives.

15.18 The following table gives the U.S. population (in millions) for the years 1975 through 1991.

Year	U.S. Population (millions)
1975	215.981
1976	218.086
1977	220.289
1978	222.629
1979	225.106
1980	227.715
1981	229.989
1982	232.201
1983	234.326
1984	236.393
1985	238.510
1986	240.691
1987	242.860
1988	245.093
1989	247.397
1990	249.961
1991	252.711

a. Find the linear trend line with population as the dependent variable and time as the independent variable.
b. Assuming that the U.S. population will grow linearly in the future, predict the U.S. population for the year 2000.
c. Using the estimated trend line of part a, find the cyclical-irregular relatives.

15.19 The following table gives the value of the total output of autos (in billions of dollars) for the United States for the years 1976 through 1991.

Year	U.S. Auto Output (billions of dollars)
1976	55.1
1977	64.2
1978	67.9
1979	66.2
1980	59.2
1981	68.3
1982	65.3
1983	88.3
1984	104.2
1985	115.8
1986	120.4
1987	118.9
1988	129.1
1989	135.1
1990	129.7
1991	119.7

a. Estimate the linear trend line $Y = A + Bt + \epsilon$, where Y is the value of the output of autos and t is the time variable.

b. Using the trend line estimated in part a, predict the value of the output of autos for the year 1998. What assumption have you made to make this prediction?

c. Using the estimated trend line of part a, find the cyclical and irregular effects in relative terms.

15.20 The following table gives the gross private personal savings (in billions of dollars) for the United States for the years 1980 through 1991.

Year	Gross Private Personal Savings (billions of dollars)
1980	153.8
1981	191.8
1982	199.5
1983	168.7
1984	222.0
1985	189.3
1986	187.5
1987	142.0
1988	155.7
1989	152.1
1990	175.6
1991	199.6

a. Find the linear trend line using the least squares method.

b. What is the predicted value of the gross private personal savings for the year 1999? What assumption must hold true for your prediction to be valid?

c. Using the estimated trend line of part a, find the cyclical and irregular effects in relative terms.

15.21   The following table gives the number of civilian people (in millions) employed in the United States during the years 1976 through 1992.

Year	Civilian Population Employed (millions of persons)
1976	88.75
1977	92.02
1978	96.05
1979	98.82
1980	99.30
1981	100.40
1982	99.53
1983	100.83
1984	105.01
1985	107.15
1986	109.60
1987	112.44
1988	114.97
1989	117.34
1990	117.91
1991	116.88
1992	117.60

a.   Estimate the linear trend line $Y = A + Bt + \epsilon$, where Y represents the number of civilian people employed and t is the time variable.
b.   Plot the actual time series data and the estimated trend line.
c.   Predict the number of civilian people who will be employed in the United States in 2000. What assumption have you made?
d.   Using the estimated trend line of part a, find the cyclical-irregular relatives.

15.22   The following table gives the U.S. corporate profits (in billions of dollars) with inventory valuation and capital consumption adjustment for the years 1977 through 1991.

Year	Corporate Profits (billions of dollars)
1977	175.7
1978	199.7
1979	202.5
1980	177.7
1981	182.0
1982	151.5
1983	212.7
1984	264.2
1985	280.8
1986	271.6
1987	319.8
1988	365.0
1989	362.8
1990	361.7
1991	346.3

    a. Estimate the linear trend line using the least squares method.
    b. Plot the actual time series data and the estimated trend line.
    c. Predict the value of corporate profits for the year 1999. What assumption must hold true for your prediction to be valid?
    d. Using the estimated trend line of part a, find the cyclical-irregular relatives.

15.5 MEASURING THE SEASONAL EFFECTS

In this section we discuss how to measure seasonal effects. The method we use to find the seasonal pattern in a time series is called the **ratio-to-moving-average method**. This method eliminates the trend, cyclical, and irregular components from a time series and isolates the seasonal variations. We learn, in this section, how to calculate the moving averages, obtain the seasonal-irregular relatives, construct seasonal indexes, and obtain deseasonalized data.

Most of the business and economic time series with variables such as production, sales, inventories, revenue, employment, housing starts, and consumption expenditures contain seasonal variations. In other words, the values of each of these variables fluctuate during different seasons or periods of a year. Therefore, before we use such data to forecast future values of these variables, we need to eliminate the seasonal component from the respective time series.

In this section we measure seasonal variations using the multiplicative model, which is

$$Y = T \cdot C \cdot S \cdot I$$

A **moving average** for k time periods can be defined as the average of the measurements for the most recent k time periods (e.g., weeks, months, quarters, years, etc.). To calculate each successive moving average, we drop the first observation from the previous k observations and add the next observation to the remaining $k - 1$ observations. The average of the observations for any k time periods is placed at the center of those periods. For example, if we are to calculate 3-period moving averages for a time series, we will add the values for the first three periods, divide this sum by 3, and place the resulting average next to the second period, which is the period in the center. To calculate the next moving average, we will drop the value for the first period but include the value for the fourth period. Now we add the values for the second, third, and fourth periods, divide this sum by 3, and place the resulting average next to the third period. We continue this process until we calculate all 3-period moving averages. Note that when we calculate the 3-period moving averages, there are no moving averages for the first and the last periods.

If we are to calculate 5-period moving averages, we add the values for the first five periods, divide this sum by 5, and place the resulting average next to the third period. Then we drop the observation for the first period but include the observation for the sixth period. The average for the second to sixth periods will be placed next to the fourth period. In the case of 5-period moving averages, there will be no moving averages for the first two and the last two periods.

THE k-PERIOD MOVING AVERAGES

The k-period moving averages are the means of each set of k consecutive observations. To calculate each successive k-period moving average, we drop the first observation from the k observations used to calculate the previous k-period moving average and include the next available observation.

If for k-period moving averages k is an odd number (such as 3-period or 5-period moving averages), each moving average is placed next to the middle period. However, if k is an even number (such as 4-period moving averages), each moving average is placed in the center of the two middle periods. For example, to calculate a 4-period moving average, we add the values for the first four periods and divide this sum by 4. This average is placed in the middle of the second and third periods. The next moving average is placed in the middle of the third and fourth periods, and so on. Then, after calculating the 4-period moving averages, we calculate the *centered moving averages*. To do so, we add the first two moving averages, divide this sum by 2, and place the resulting centered moving average next to the third period, which is in the middle of the first two moving averages.

CENTERED MOVING AVERAGES

The centered moving averages are calculated only when k is an even number for k-period moving averages. In such a case, we find the mean of each pair of consecutive k-period moving averages and place this mean at the center of the corresponding pair of the k-period moving averages. The resulting series gives the centered moving averages.

By taking the ratio of the original time series data to the centered moving averages we obtain the **seasonal-irregular relatives**. These relatives are also called the *original series as proportion of the centered moving averages* or **ratio-to-moving-averages**. The seasonal-irregular relatives give the seasonal and irregular effects in relative terms.

SEASONAL-IRREGULAR RELATIVES

The seasonal-irregular relatives give the seasonal and irregular effects in relative terms. They are obtained by dividing each value of the original time series by the corresponding centered moving average. In other words, the seasonal-irregular relative for a period is

$$\frac{\text{The value of the original time series for that period}}{\text{Centered moving average for that period}}$$

This method of calculating the seasonal-irregular relatives is called the *ratio-to-moving-average method*. Note that if k is an odd number, we will not need to calculate the centered moving averages. In this case we will use the k-period moving averages to obtain the seasonal-irregular relatives.

Example 15–3 illustrates how to calculate the moving averages, the centered moving averages, and the seasonal-irregular relatives. In Example 15–4, we construct the seasonal indexes by using the seasonal-irregular relatives of Example 15–3. These seasonal indexes are used to find the deseasonalized time series in Example 15–5.

Calculating the moving averages, centered

EXAMPLE 15–3   Table 15.5 gives the quarterly gasoline sales (in millions of dollars) for the years 1988 through 1993 for J.B. Oil Corporation.

Table 15.5

Year	Quarter			
	1	**2**	**3**	**4**
1988	8.2	10.5	12.8	9.3
1989	9.1	11.4	13.2	9.8
1990	9.3	11.7	13.4	9.7
1991	9.5	12.1	13.9	9.9
1992	9.4	12.5	13.7	10.2
1993	9.9	12.8	14.2	10.7

moving averages, and seasonal-irregular relatives.

Calculate the four-quarter moving averages, the centered moving averages, and the seasonal-irregular relatives.

Solution   Table 15.6 lists the years in the first column, quarters in the second column, and quarterly gasoline sales in the third column. These three columns are reproduced from Table 15.5. The fourth column in Table 15.6 lists the four-quarter totals, the fifth column records the four-quarter moving averages, the sixth column lists the centered moving averages, and the seventh column gives the seasonal-irregular relatives.

Note that the centered moving averages, given in the sixth column of Table 15.6, are determined only when the moving averages are calculated for an even number of periods, which is the case (four-quarter moving averages) in this example. However, if we are calculating the moving averages for an odd number of periods (such as 3-period or 5-period moving averages), we do not need to calculate the centered moving averages because 3-period or 5-period moving averages are already centered.

The following steps explain how we construct columns four through seven in Table 15.6.

Step 1.   *Calculate the four-quarter moving totals*

To find the four-quarter moving averages, first we find the four-quarter moving totals. We take the gasoline sales for the first four quarters and add them together, which gives

$$8.2 + 10.5 + 12.8 + 9.3 = 40.8$$

We place this sum for the first four quarters at the center of these four quarters, which is in the middle of the second and third quarters of 1988.

To calculate the next four-quarter moving total, we drop the first quarter of 1988 but include the first quarter of 1989. By adding these four values, we obtain the second four-quarter moving total, which is

$$10.5 + 12.8 + 9.3 + 9.1 = 41.7$$

We place this sum at the center of these four quarters, which is in the middle of the third and fourth quarters of 1988.

To find the next four-quarter total, we drop the second quarter of 1988 but include the second quarter of 1989 and add these four values together. We continue this procedure until we calculate the sum of the gasoline sales for the last four quarters, which is

$$9.9 + 12.8 + 14.2 + 10.7 = 47.6$$

Table 15.6   Calculating Moving Averages and Seasonal-Irregular Relatives

Year	Quarter	Gasoline Sales (millions of dollars)	Four-quarter Moving Total	Four-quarter Moving Average	Centered Moving Average	Seasonal-Irregular Relative
1988	1	8.2				
	2	10.5				
			40.8	10.2000		
	3	12.8			10.3125	1.2412
			41.7	10.4250		
	4	9.3			10.5375	.8826
			42.6	10.6500		
1989	1	9.1			10.7000	.8505
			43.0	10.7500		
	2	11.4			10.8125	1.0543
			43.5	10.8750		
	3	13.2			10.9000	1.2110
			43.7	10.9250		
	4	9.8			10.9625	.8940
			44.0	11.0000		
1990	1	9.3			11.0250	.8435
			44.2	11.0500		
	2	11.7			11.0375	1.0600
			44.1	11.0250		
	3	13.4			11.0500	1.2127
			44.3	11.0750		
	4	9.7			11.1250	.8719
			44.7	11.1750		
1991	1	9.5			11.2375	.8454
			45.2	11.3000		
	2	12.1			11.3250	1.0684
			45.4	11.3500		
	3	13.9			11.3375	1.2260
			45.3	11.3250		
	4	9.9			11.3750	.8703
			45.7	11.4250		
1992	1	9.4			11.4000	.8246
			45.5	11.3750		
	2	12.5			11.4125	1.0953
			45.8	11.4500		
	3	13.7			11.5125	1.1900
			46.3	11.5750		
	4	10.2			11.6125	.8784
			46.6	11.6500		
1993	1	9.9			11.7125	.8453
			47.1	11.7750		
	2	12.8			11.8375	1.0813
			47.6	11.9000		
	3	14.2				
	4	10.7				

This four-quarter moving total is placed in the middle of the second and third quarters of 1993.

Step 2. *Calculate the four-quarter moving averages*

By dividing each of the four-quarter moving totals by 4 we obtain the four-quarter moving averages. For example,

$$\text{First four-quarter moving average} = 40.8/4 = 10.2000$$

$$\text{Second four-quarter moving average} = 41.7/4 = 10.4250$$

$$\text{Third four-quarter moving average} = 42.6/4 = 10.6500$$

$$\vdots$$

$$\text{Last four-quarter moving average} = 47.6/4 = 11.9000$$

These moving averages are placed in the fifth column in Table 15.6 next to the respective four-quarter moving totals.

Step 3. *Compute the centered moving averages*

Because the moving averages calculated in Step 2 are for an even number of periods (four quarters), we need to compute the centered moving averages so that they can be placed next to the respective quarters. To compute the first centered moving average, we add the first two four-quarter moving averages, divide this sum by 2, and place this centered moving average in the middle of the first two four-year moving averages. Thus, this centered moving average belongs to the third quarter of 1988. This centered moving average is

$$\text{First centered moving average} = \frac{10.2000 + 10.4250}{2} = 10.3125$$

To calculate the second centered moving average, we add the second and third moving averages and divide this sum by 2. Thus,

$$\text{Second centered moving average} = \frac{10.4250 + 10.6500}{2} = 10.5375$$

This centered moving average is placed in the sixth column in Table 15.6 next to the fourth quarter of 1988. We calculate the remaining centered moving averages the same way.

By calculating the centered moving averages, we have removed the seasonal and irregular effects from the given time series. The time series given by the centered moving averages contains the trend and cyclical effects only.

Step 4. *Calculate the seasonal-irregular relatives*

The column labeled *seasonal-irregular relative* is also called the *ratio-to-moving-average* or the original series as a proportion of the centered moving averages. We obtain these relatives by dividing each value of the original time series by the corresponding centered moving average. Because the first two quarters in Table 15.6 do not contain a centered moving average, there is no seasonal-irregular relative for these quarters. The third quarter of 1988 has actual gasoline sales of $12.8 million and a centered moving average of $10.3125 million. Therefore, the seasonal-irregular relative for this quarter is

$$\frac{12.8}{10.3125} = 1.2412$$

We calculate the remaining seasonal-irregular relatives the same way.

As mentioned earlier, by calculating the centered moving averages in Step 3, we eliminated the seasonal and irregular components from the given time series. Thus, the column of centered moving averages in Table 15.6 contains only trend and cyclical components. As we can observe from Table 15.6, the centered moving averages do not fluctuate from quarter to quarter as much as the original gasoline sales do. This is because of the elimination of seasonal and irregular effects from the original time series.

The column of seasonal-irregular relatives gives the seasonal and irregular components in relative terms.

Next we want to eliminate the irregular component from the seasonal-irregular relatives calculated in Example 15–3. We do so by constructing **seasonal indexes**. These seasonal indexes measure the seasonal effects only. The seasonal indexes are the adjusted averages of the seasonal-irregular relatives (given in the last column of Table 15.6) for the same quarter for all years. By averaging these seasonal-irregular relatives for the same quarter for all years, we will have the same index for the same quarter of each year. By calculating these seasonal indexes, we remove the irregular effects from seasonal-irregular relatives.

SEASONAL INDEXES

The seasonal indexes measure the effects (in relative terms) of different seasons on the time series.

Example 15–4 illustrates how the seasonal indexes are constructed.

Calculating the seasonal indexes.

EXAMPLE 15–4   Refer to the seasonal-irregular relatives for the quarterly gasoline sales of J.B. Oil Corporation calculated in Table 15.6 of Example 15–3. Using those relatives, calculate the seasonal indexes.

Solution   The seasonal indexes are computed from the seasonal-irregular relatives using the following steps.

Step 1.   *Arrange the seasonal-irregular relatives in a table*

In the first step we arrange the seasonal-irregular relatives of Table 15.6 for the same quarter of all years in the same column. This is done in Table 15.7.

Table 15.7   Arranged Seasonal-Irregular Relatives

Year	Quarter			
	1	**2**	**3**	**4**
1988	—	—	1.2412	.8826
1989	.8505	1.0543	1.2110	.8940
1990	.8435	1.0600	1.2127	.8719
1991	.8454	1.0684	1.2260	.8703
1992	.8246	1.0953	1.1900	.8784
1993	.8453	1.0813	—	—

Note that the seasonal-irregular relatives for the first quarter of all years are arranged in the column for the first quarter in Table 15.7. The same is done for the relatives belonging to other quarters.

Step 2. *Find the mean of the relatives for the same quarter of all years*

In the second step we find the mean for the seasonal-irregular relatives for the same quarter of all years. This is done in Table 15.8.

Table 15.8 Calculating Means of Seasonal-Irregular Relatives

	Quarter			
Year	1	2	3	4
1988	—	—	1.2412	.8826
1989	.8505	1.0543	1.2110	.8940
1990	.8435	1.0600	1.2127	.8719
1991	.8454	1.0684	1.2260	.8703
1992	.8246	1.0953	1.1900	.8784
1993	.8453	1.0813	—	—
Total	4.2093	5.3593	6.0809	4.3972
Mean	.84186	1.07186	1.21618	.87944

To calculate these means, first we add all the seasonal-irregular relatives for the same quarter of all years and list those sums in the row labeled *Total*. Then we divide each of these totals by the number of relatives in that column. For example, the sum of the seasonal-irregular relatives for the first quarter of all years is

$$.8505 + .8435 + .8454 + .8246 + .8453 = 4.2093$$

This column contains five seasonal-irregular relatives. Therefore, the mean for these relatives is

$$\text{Mean of relatives for the first quarters} = \frac{4.2093}{5} = .84186$$

Similarly we calculate the means of seasonal-irregular relatives for the other three quarters.

Step 3. *Compute the seasonal indexes*

In the third step we calculate the seasonal indexes. These seasonal indexes are simply the means of the four quarters calculated in Table 15.8 adjusted for rounding-off errors. The sum of the four means calculated in Table 15.8 should be 4.0. However, due to the rounding-off errors, this sum may not actually be equal to 4.0. From Table 15.8,

$$\text{Sum of the four means} = .84186 + 1.07186 + 1.21618 + .87944 = 4.00934$$

This sum of the four means is shown in the row of means in Table 15.9.

The seasonal indexes for the four quarters are calculated using the following formula.

SEASONAL INDEX FOR A PERIOD

The seasonal index for a period is calculated as follows.

$$\text{Seasonal index for a period} = \frac{\text{Mean for that period}}{\text{Sum of means for all periods}} \times (\text{Number of periods})$$

In our example, the sum of the means of seasonal-irregular relatives for all periods (four quarters) is 4.00934 and the number of periods (quarters) is four. Therefore, to calculate the seasonal indexes for our example, we divide the mean of each quarter by 4.00934 and multiply the resulting number by 4. The seasonal indexes for the four quarters are as follows.

$$\text{Seasonal index for the first quarter} = \frac{.84186}{4.00934} \times 4 = .83990$$

$$\text{Seasonal index for the second quarter} = \frac{1.07186}{4.00934} \times 4 = 1.06936$$

$$\text{Seasonal index for the third quarter} = \frac{1.21618}{4.00934} \times 4 = 1.21335$$

$$\text{Seasonal index for the fourth quarter} = \frac{.87944}{4.00934} \times 4 = .87739$$

These seasonal indexes are recorded in the bottom row of Table 15.9.

Table 15.9   Calculating Seasonal Indexes

Year	Quarter				
	1	2	3	4	
1988	—	—	1.2412	.8826	
1989	.8505	1.0543	1.2110	.8940	
1990	.8435	1.0600	1.2127	.8719	
1991	.8454	1.0684	1.2260	.8703	
1992	.8246	1.0953	1.1900	.8784	
1993	.8453	1.0813	—	—	
Total	4.2093	5.3593	6.0809	4.3972	
Mean	.84186	1.07186	1.21618	.87944	Sum = 4.00934
Seasonal Index	.83990	1.06936	1.21335	.87739	Sum = 4.00

If the seasonal index for a quarter is greater than 1.0, this quarter has an above average seasonal effect on the original time series. In Table 15.9, the seasonal indexes for the second and third quarters are 1.06936 and 1.21335, respectively. These indexes indicate that the gasoline sales for this company during the second quarter are 6.936% [obtained as $(1.06936 - 1.0) \times 100 = 6.936\%$] above the average for all four quarters, and the sales during the third quarter are 21.335% above the average sales for all four quarters.

A seasonal index less than 1.0 for a quarter indicates that the seasonal effect for that quarter is below the average seasonal effect on the original time series. Thus, the seasonal index of .83990 for the first quarter shows that the gasoline sales during this quarter are 16.01% [$(1 - .83990) \times 100 = 16.01\%$] below the average sales for all four quarters. Similarly the seasonal index of .87739 for the fourth quarter indicates that the gasoline sales during this quarter are 12.261% below the average sales for all four quarters. ▬

Using the seasonal indexes calculated in Table 15.9 of Example 15–4, we can adjust the original time series on gasoline sales and obtain a new time series that is free of seasonal fluctuations. Such a time series is called the **deseasonalized times series** or **seasonally adjusted data**, and this series contains only trend, cyclical, and irregular components. The deseasonalized value of a time series for any period is obtained by dividing the actual value for that period by the seasonal index for that period.

DESEASONALIZED DATA

The time series obtained from the original time series by removing the seasonal component is called the deseasonalized time series. Such data are obtained by using the following formula.

$$\text{Deseasonalized value for a period} = \frac{\text{Actual value for that period}}{\text{Seasonal index for that period}}$$

Example 15–5 illustrates the calculation of deseasonalized time series.

Obtaining deseasonalized data.

EXAMPLE 15–5 Compute the deseasonalized gasoline sales for the J.B. Oil Corporation using the seasonal indexes obtained in Example 15–4.

Solution In Table 15.10, the third column lists the actual gasoline sales, which are reproduced from Table 15.5. The fourth column records the seasonal indexes, which are reproduced from Table 15.9. Note that the seasonal index is .83990 for the first quarter of each year, 1.06936 for the second quarter of each year, and so on.

Table 15.10 Obtaining the Deseasonalized Data

Year	Quarter	Actual Sales (millions of dollars)	Seasonal Index	Deseasonalized Sales (millions of dollars)
1988	1	8.2	.83990	9.76
	2	10.5	1.06936	9.82
	3	12.8	1.21335	10.55
	4	9.3	.87739	10.60
1989	1	9.1	.83990	10.83
	2	11.4	1.06936	10.66
	3	13.2	1.21335	10.88
	4	9.8	.87739	11.17
1990	1	9.3	.83990	11.07
	2	11.7	1.06936	10.94
	3	13.4	1.21335	11.04
	4	9.7	.87739	11.06
1991	1	9.5	.83990	11.31
	2	12.1	1.06936	11.32
	3	13.9	1.21335	11.46
	4	9.9	.87739	11.28
1992	1	9.4	.83990	11.19
	2	12.5	1.06936	11.69
	3	13.7	1.21335	11.29
	4	10.2	.87739	11.63
1993	1	9.9	.83990	11.79
	2	12.8	1.06936	11.97
	3	14.2	1.21335	11.70
	4	10.7	.87739	12.20

The deseasonalized gasoline sales for each quarter are obtained as follows.

$$\text{Deseasonalized sales for a quarter} = \frac{\text{Actual sales for that quarter}}{\text{Seasonal index for that quarter}}$$

Thus, the deseasonalized sales for the first quarter of 1988 are

$$\frac{8.2}{.83990} = \$9.76 \text{ million}$$

The deseasonalized sales for the second quarter of 1988 are

$$\frac{10.5}{1.06936} = \$9.82 \text{ million}$$

The deseasonalized sales for the remaining periods are calculated the same way.

Note that these deseasonalized sales do not contain seasonal fluctuations. They contain trend, cyclical, and irregular effects only. Now using the regression method (as was done in Examples 15–1 and 15–2), we can remove the trend component from these deseasonalized sales and obtain the cyclical and irregular effects. ■

Most of the data obtained by federal government agencies (such as the Census Bureau, Labor Bureau, etc.) and published in government publications are deseasonalized data. The predictions and forecasts made based on deseasonalized data are more reliable.

EXERCISES

Concepts and Procedures

15.23   Can a time series containing annual data have a seasonal component? Explain why or why not.

15.24   Explain how the seasonal-irregular relatives are calculated for a time series using the ratio-to-moving-average method.

15.25   Explain what seasonal indexes are and how they are calculated.

15.26   Describe briefly how a deseasonalized time series is obtained.

15.27   The following table gives the quarterly values of a variable for the periods from the third quarter of 1989 to the second quarter of 1994.

Year	Quarter			
	1	2	3	4
1989	—	—	12	10
1990	9	15	11	9
1991	10	16	13	11
1992	11	17	12	13
1993	13	20	15	14
1994	12	19	—	—

a.   Calculate the four-quarter moving averages, the centered moving averages, and the seasonal-irregular relatives.
b.   Using the seasonal-irregular relatives of part a, compute the seasonal indexes.
c.   Obtain the deseasonalized data.

15.28 The following table gives the quarterly values of a variable for the periods from the second quarter of 1990 to the first quarter of 1994.

	Quarter			
Year	**1**	**2**	**3**	**4**
1990	—	27	33	30
1991	26	28	35	32
1992	27	29	34	31
1993	29	32	38	33
1994	28	—	—	—

a. Calculate the four-quarter moving averages, the centered moving averages, and the seasonal-irregular relatives.
b. Using the seasonal-irregular relatives of part a, compute the seasonal indexes.
c. Obtain the seasonally adjusted data.

Applications

15.29 The table below gives the value of inventory (in millions of dollars) for a company for the periods from the first quarter of 1988 to the fourth quarter of 1993.

	Quarter			
Year	**1**	**2**	**3**	**4**
1988	3.5	4.2	4.6	5.9
1989	3.7	4.3	4.9	6.3
1990	3.4	4.5	4.8	6.1
1991	4.2	4.9	5.3	6.9
1992	3.9	4.4	5.1	6.6
1993	4.0	4.3	5.2	6.8

a. Calculate the four-quarter moving averages, the centered moving averages, and the seasonal-irregular relatives for these data.
b. Explain the seasonal-irregular relatives calculated in part a.
c. Using the seasonal-irregular relatives of part a, compute the seasonal indexes. Explain what these seasonal indexes represent.
d. Obtain the seasonally adjusted data. Explain how this seasonally adjusted time series is different from the original time series.

15.30 The following table gives the total production (in millions of dollars) of a company for the periods from the second quarter of 1989 to the second quarter of 1994.

	Quarter			
Year	**1**	**2**	**3**	**4**
1989	—	14	20	25
1990	17	18	21	29
1991	16	19	23	28
1992	14	17	24	26
1993	17	20	27	30
1994	18	22	—	—

a. Calculate the four-quarter moving averages, the centered moving averages, and the seasonal-irregular relatives for these data.
b. Explain the seasonal-irregular relatives calculated in part a.
c. Using the seasonal-irregular relatives of part a, compute the seasonal indexes. Explain what these seasonal indexes represent.
d. Obtain the deseasonalized data. Explain how this deseasonalized time series is different from the original time series.

GLOSSARY

Cyclical component or **cyclical fluctuations**   Fluctuations around the trend line for a time series that last for more than one year.

Deseasonalized time series or **seasonally adjusted data**   A time series that does not contain seasonal effects. This time series is obtained by dividing the original time series by the seasonal indexes.

Irregular component   Variations in a time series that occur due to chance.

Moving averages   The k-period moving averages are the means of each set of k consecutive observations. To calculate each successive k-period moving average, we drop the first observation from the k observations used to calculate the previous k-period moving average and include the next available observation.

Seasonal component   Movements in a time series that occur during different seasons or periods of less than one year.

Seasonal indexes   Measures of the seasonal effects that are obtained by averaging the seasonal-irregular relatives for the same season/period of all the years.

Seasonal-irregular relatives   The ratio of the original time series values to the centered moving averages. Also called the *ratio-to-moving averages*.

Trend component   Long-run movement (increase or decrease) in a time series.

KEY FORMULAS

1. **Additive time series model**

$$Y = T + C + S + I$$

where T, C, S, and I are the trend, cyclical, seasonal, and irregular components, respectively, and Y is the value of the time series at a given time.

2. **Multiplicative time series model**

$$Y = T \cdot C \cdot S \cdot I$$

where T, C, S, and I are the trend, cyclical, seasonal, and irregular components, respectively, and Y is the value of the time series at a given time.

3. **Regression model for linear trend**

$$Y = A + Bt + \epsilon$$

where Y is the value of the time series, t is the time period, A is the constant term, B is the slope of the trend line, and ϵ is the random error term.

4. **Estimated linear trend line**

$$\hat{Y} = a + bt$$

where a and b are estimates of A and B, respectively.

5. **Multiplicative time series model for annual data (no seasonal variations)**

$$Y = T \cdot C \cdot I$$

6. **Seasonal index for a period**

$$\text{Seasonal index for a period} = \frac{\text{Mean for that period}}{\text{Sum of means for all periods}} \times (\text{Number of periods})$$

7. **Deseasonalized value for a period**

$$\text{Deseasonalized value for a period} = \frac{\text{Actual value for that period}}{\text{Seasonal index for that period}}$$

SUPPLEMENTARY EXERCISES

15.31 The following table gives the gross domestic product (in billions of dollars) of the United States for the years 1975 through 1991.

Year	Gross Domestic Product (billions of dollars)
1975	1585.9
1976	1768.4
1977	1974.1
1978	2232.7
1979	2488.6
1980	2708.0
1981	3030.6
1982	3149.6
1983	3405.0
1984	3777.2
1985	4038.7
1986	4268.6
1987	4539.9
1988	4900.4
1989	5250.8
1990	5522.2
1991	5677.5

a. Estimate the linear trend line $Y = A + Bt + \epsilon$, where Y represents the gross domestic product and t is the time variable.

b. Assuming that the linear trend is valid, predict the value of the gross domestic product of the United States for the year 2000.

c. Using the estimated trend line of part a, find the cyclical-irregular relatives.

15.32 The following table gives the U.S. imports (in billions of dollars) for the years 1977 through 1991.

Year	U.S. Imports (billions of dollars)
1977	182.4
1978	212.3
1979	252.7
1980	293.9
1981	317.7
1982	303.2
1983	328.1
1984	405.1
1985	417.6
1986	451.7
1987	507.1
1988	552.2
1989	587.7
1990	625.9
1991	620.0

a. Estimate the linear trend line $Y = A + Bt + \epsilon$, where Y represents the U.S. imports and t is the time variable.

b. Assuming that the linear trend holds true, predict the value of the U.S. imports for the year 1999.

c. Using the estimated trend line of part a, find the cyclical-irregular relatives.

15.33 The following table gives the net private domestic investment (in billions of dollars) for the United States for the years 1978 through 1991.

Year	Net Private Domestic Investment (billions of dollars)
1978	199.3
1979	207.8
1980	155.7
1981	195.6
1982	104.3
1983	128.2
1984	285.6
1985	260.0
1986	239.1
1987	247.1
1988	259.6
1989	251.9
1990	196.7
1991	95.0

a. Using the least squares method, find the linear trend line with net private domestic investment as the dependent variable and time as the independent variable.

b. Assuming that the linear trend holds true, predict the value of the net private domestic investment for the year 2002.

c. Using the estimated trend line of part a, find the cyclical-irregular relatives.

15.34 The following table gives the value of the sales (in billions of dollars) of the manufacturing sector of the United States for the years 1980 through 1991.

Year	Manufacturing Sales (billions of dollars)
1980	154.39
1981	168.13
1982	163.35
1983	172.55
1984	190.68
1985	194.54
1986	194.66
1987	206.33
1988	223.54
1989	232.72
1990	239.46
1991	235.14

a. Find the linear trend line using the least squares method.
b. Using the linear trend line estimated in part a, predict the value of the manufacturing sector sales for the year 2003. What assumption must hold true for your prediction to be valid?
c. Using the estimated trend line of part a, find the cyclical-irregular relatives.

15.35 The following table gives the total U.S. consumer credit (in billions of dollars) for the years 1975 through 1991.

Year	Total Consumer Credit (billions of dollars)
1975	204.96
1976	228.16
1977	263.81
1978	308.27
1979	347.51
1980	350.27
1981	366.87
1982	383.13
1983	431.17
1984	511.32
1985	591.29
1986	647.98
1987	680.04
1988	729.12
1989	777.61
1990	793.98
1991	780.57

a. Estimate the linear trend line with total consumer credit as the dependent variable and time as the independent variable.
b. Plot the actual time series data and the estimated trend line.
c. Using the estimated linear trend line of part a, predict the value of the total consumer credit for the year 1999. What assumption must hold true for this prediction to be valid?
d. Using the estimated trend line of part a, find the cyclical-irregular relatives.

15.36 The table below gives the total U.S. corporate profits tax liability (in billions of dollars) for the years 1977 through 1991.

Year	Corporate Profits Tax Liability (billions of dollars)
1977	73.0
1978	83.5
1979	88.0
1980	84.8
1981	81.1
1982	63.1
1983	77.2
1984	94.0
1985	96.5
1986	106.5
1987	127.1
1988	137.0
1989	141.3
1990	136.7
1991	124.0

a. Estimate the linear trend line $Y = A + Bt + \epsilon$, where Y represents the corporate profits tax liability and t is the time variable.

b. Plot the actual time series data and the estimated trend line.

c. Assuming the linear trend is valid, predict the value of U.S. corporate profits tax liability for the year 2002.

d. Using the estimated trend line of part a, find the cyclical-irregular relatives.

15.37 The following table gives the total revenue (in millions of dollars) of a company for the periods from the first quarter of 1988 through the fourth quarter of 1993.

Year	Quarter			
	1	2	3	4
1988	20	25	24	31
1989	22	29	26	34
1990	19	26	23	29
1991	21	29	27	35
1992	25	31	30	38
1993	27	30	33	40

a. Calculate the four-quarter moving averages, the centered moving averages, and the seasonal-irregular relatives for these data.

b. Explain the seasonal-irregular relatives calculated in part a.

c. Using the seasonal-irregular relatives of part a, compute the seasonal indexes. Explain what these seasonal indexes represent.

d. Obtain the deseasonalized data. Explain how this deseasonalized time series is different from the original time series.

e. Let Y be the deseasonalized quarterly revenue. Let t be the time variable that assumes a value of zero for the first quarter of 1988, 1 for the second quarter of 1988, and so on. This will give you 24 pairs of values on Y and t. Using these values of Y and t, find the linear trend line.

f. Using the estimated trend line of part e, find the cyclical-irregular relatives.

15.38 The following table gives the number of housing starts (in thousands) in a large city for the period from the second quarter of 1988 through the second quarter of 1994.

	Quarter			
Year	1	2	3	4
1988	—	22	25	10
1989	11	20	21	8
1990	10	18	18	7
1991	11	18	19	8
1992	12	23	26	13
1993	14	24	28	15
1994	16	27	—	—

a. Calculate the four-quarter moving averages, the centered moving averages, and the seasonal-irregular relatives for these data.

b. Explain the seasonal-irregular relatives calculated in part a.

c. Using the seasonal-irregular relatives of part a, compute the seasonal indexes. Explain what these seasonal indexes represent.

d. Obtain the deseasonalized data. Explain how this deseasonalized time series is different from the original time series.

e. Let Y be the deseasonalized quarterly data on the number of housing starts. Let t be the time variable that assumes a value of zero for the second quarter of 1988, 1 for the third quarter of 1988, and so on. This will give you 25 pairs of values on Y and t. Using these values of Y and t, find the linear trend line.

f. Using the estimated trend line of part e, find the cyclical-irregular relatives.

SELF-REVIEW TEST

1. Explain the four components of a time series.

2. Cyclical fluctuations in a time series last for

   a. exactly one year   b. more than one year   c. less than one year

3. Trend in a time series

   a. is always linear   b. is always nonlinear   c. can be linear or nonlinear

4. Seasonal fluctuations in a time series occur

   a. within a one-year period   b. within a five-year period   c. over a long period

5. Irregular fluctuations in a time series occur because of

   a. irregularity in the cyclical fluctuations
   b. irregularity in the cyclical and seasonal fluctuations
   c. random factors

6. For an additive time series model, the values for Y, C, S, and I for the second quarter 1994 sales for a company are $90 million, $11 million, $-$3 million, and $-$2 million, respectively. The trend value for this quarter's sales is

   a. $84 million   b. $96 million   c. $106 million

7. For a multiplicative time series model, the values for Y, T, S, and I are $50 million, $40 million, 1.20, and 1.10, respectively. The value of C is approximately

   a. .947   b. 7.576   c. 1.359

8. Briefly explain the additive and multiplicative time series models.

9. Explain the meaning of the cyclical-irregular relatives and describe how they are obtained using the least squares regression model.

10. Describe how the k-period moving averages are obtained. What are the centered moving averages and when are they calculated?

11. Explain the meaning of the seasonal-irregular relatives and describe how they are obtained using the moving averages or centered moving averages.

12. Explain the meaning of seasonal indexes and describe how they are calculated.

13. What is a seasonally adjusted time series? How is such a time series obtained?

14. The following table gives the total U.S. personal consumption expenditure (in billions of dollars) for the years 1975 through 1991.

Year	Personal Consumption Expenditure (billions of dollars)
1975	1024.9
1976	1143.1
1977	1271.5
1978	1421.2
1979	1583.7
1980	1748.1
1981	1926.2
1982	2059.2
1983	2257.5
1984	2460.3
1985	2667.4
1986	2850.6
1987	3052.2
1988	3296.1
1989	3523.1
1990	3748.4
1991	3887.7

a. Find the linear trend line using the least squares method.
b. Plot the actual time series data and the estimated trend line.
c. Using the linear trend line estimated in part a, predict the total U.S. personal consumption expenditure for the year 1999. What assumption must hold true for your prediction to be valid?
d. Using the estimated trend line of part a, find the cyclical-irregular relatives.

15. The following table gives the total sales (in millions of dollars) of a company for the period from the second quarter of 1988 through the second quarter of 1994.

Year	Quarter 1	2	3	4
1988	—	18	20	25
1989	17	20	24	29
1990	15	17	20	24
1991	16	19	23	30
1992	18	23	28	33
1993	17	21	25	31
1994	19	24	—	—

a. Calculate the four-quarter moving averages, the centered moving averages, and the seasonal-irregular relatives for these data.

b. Explain the seasonal-irregular relatives calculated in part a.

c. Using the seasonal-irregular relatives of part a, compute the seasonal indexes. Explain what these seasonal indexes represent.

d. Obtain the deseasonalized data. Explain how this deseasonalized time series is different from the original time series.

e. Let Y be the deseasonalized quarterly sales. Let t be the time variable that assumes a value of zero for the second quarter of 1988, 1 for the third quarter of 1988, and so on. This will give you 25 pairs of values on Y and t. Using these values of Y and t, find the linear trend line.

f. Using the estimated trend line of part e, find the cyclical-irregular relatives.

USING MINITAB

The MINITAB REGRESS command can be used to find the trend line for time series data. The procedure to use this command and the interpretation of the MINITAB output are similar to those given in the "Using MINITAB" section of Chapter 13. The following illustration shows how we can obtain the trend line using the REGRESS command.

Illustration M15–1   Refer to the data on gross private domestic investment for the United States given in Example 15–1. Find the trend line for that data.

Solution   First we enter the data into MINITAB using the READ command, then we find the trend line using the REGRESS command. Figure 15.7 shows this procedure.

Figure 15.7   Finding the trend line using the REGRESS command.

```
MTB  > READ C1 C2
DATA > 467.6    0
DATA > 558.0    1
DATA > 503.4    2
DATA > 546.7    3
DATA > 718.9    4
DATA > 714.5    5
DATA > 717.6    6
DATA > 749.3    7
DATA > 793.6    8
DATA > 832.3    9
DATA > 799.5    10
DATA > 721.1    11
DATA > 770.4    12
DATA > END

MTB  > NAME C1 'Y' C2 't'

MTB  > REGRESS 'Y' 1 't'

THE REGRESSION EQUATION IS
Y = 524 + 26.7 t
```

```
PREDICTOR       COEF      STDEV     T-RATIO        P
CONSTANT      523.93      34.71       15.09    0.000
t             26.691      4.909        5.44    0.000

S = 66.23     R-SQ = 72.9%      R-SQ(ADJ) = 70.4%

ANALYSIS OF VARIANCE

SOURCE          DF          SS          MS        F        P
REGRESSION       1      129655      129655    29.56    0.000
ERROR           11       48245        4386
TOTAL           12      177901
```

In our example, C1 or Y is the dependent variable and C2 or t is the independent variable. In Figure 15.7, we name column C1 as 'Y' and column C2 as 't' before using the REGRESS command. The regression equation given in the MINITAB output of Figure 15.7 gives the trend line, which is

$$\hat{Y} = 524 + 26.7t$$

Thus, $\qquad\qquad\qquad\qquad a = 524 \qquad$ and $\qquad b = 26.7$

Using the information given in the MINITAB output of Figure 15.7, we can make the confidence interval and test hypotheses about B using the procedures explained in the "Using MINITAB" section of Chapter 13.

COMPUTER ASSIGNMENTS

M15.1. Using the MINITAB REGRESS command, find the trend line for the data of Exercise 15.19.

M15.2 Refer to the data given in Exercise 15.20. Find the trend line for those data using MINITAB.

1. A bowling establishment owner interested in the relation between the price she charges for a game of bowling and the number of games bowled per day, collects data on the number of games bowled per day at 15 different prices. Fill in the missing entries in the following MINITAB output that was obtained for these data. In this output, X represents the price of a game of bowling and Y is the games bowled per day.

```
THE REGRESSION EQUATION IS

Y=____ ____ X


PREDICTOR    COEF    STDEV    T-RATIO    P
CONSTANT     691.02  21.70    ____       .000
X           -141.30  ____     -16.83     .000

S=____       R-SQ=____        R-SQ(ADJ)=95.3%


ANALYSIS OF VARIANCE

SOURCE      DF      SS          MS       F         P
REGRESSION  ____    148484.0    ____     283.13    .000
ERROR       ____    ____        ____
TOTAL       ____    ____
```

2. A researcher observed measurements on intoxication and reaction times for a sample of 25 intoxicated drivers. According to her calculations, the intoxication measures have a mean of .5 and a standard deviation of .2, and the reaction times have a mean of 20 seconds and a standard deviation of 6 seconds. She also found a correlation coefficient of .8 between the intoxication measures and reaction times.

   a. Find the appropriate regression line. Define the variables you use and interpret all the parameter estimates you find.
   b. What proportion of variation in reaction time is explained by the intoxication measure?
   c. Find the standard deviation of errors.
   d. Test if B is positive using a .025 level of significance.
   e. What would you predict the reaction time to be of a person with a .3 intoxication measure? Based on your conclusion in part d, how valid do you think this predicted value is? Find a 95% confidence interval for this reaction time.

3. The following chart summarizes some of the results of studies relating birth rates and family planning. (*The New York Times*, January 2, 1994. Copyright © 1994 by The New York Times Company. Reprinted with permission.) These studies, largely financed by the United States Agency for International Development, began 20 years ago.

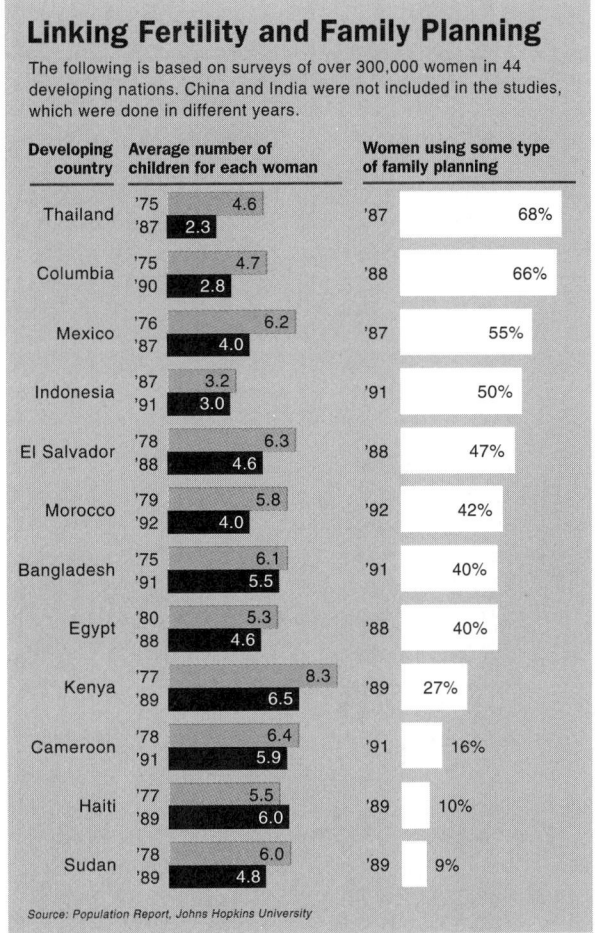

Linking Fertility and Family Planning

The following is based on surveys of over 300,000 women in 44 developing nations. China and India were not included in the studies, which were done in different years.

Developing country	Average number of children for each woman		Women using some type of family planning	
Thailand	'75	4.6		
	'87	2.3	'87	68%
Columbia	'75	4.7		
	'90	2.8	'88	66%
Mexico	'76	6.2		
	'87	4.0	'87	55%
Indonesia	'87	3.2		
	'91	3.0	'91	50%
El Salvador	'78	6.3		
	'88	4.6	'88	47%
Morocco	'79	5.8		
	'92	4.0	'92	42%
Bangladesh	'75	6.1		
	'91	5.5	'91	40%
Egypt	'80	5.3		
	'88	4.6	'88	40%
Kenya	'77	8.3		
	'89	6.5	'89	27%
Cameroon	'78	6.4		
	'91	5.9	'91	16%
Haiti	'77	5.5		
	'89	6.0	'89	10%
Sudan	'78	6.0		
	'89	4.8	'89	9%

Source: Population Report, Johns Hopkins University

The New York Times

Assume that the countries included here were randomly selected from the 44 nations studied. Using methods you have learned, what conclusions would you make about the relation between birth rates and family planning? That is, how is birth rate related to the degree of participation in family planning? How does the prior birth rate relate to the more recent birth rate? Is there any relation between the past birth rate and the level of participation in family planning?

4. The manufacturer of a portable CD player wishes to determine the effects that the amount of a rebate, length of the warranty, and inclusion of batteries have on the number of CD players sold in a test market area. The base net profit to the manufacturer on each portable CD player sold is $50. Providing batteries with a CD player costs the manufacturer $1, and each month of the warranty is figured to cost the manufacturer $2 for each CD player sold. Based on the information given in the table on the next page, what recommendations would you make to the manufacturer about rebates, warranties, and the inclusion of batteries? In the column labeled "Batteries," 0 represents "batteries not included" and 1 represents "batteries included."

Sales	Rebate (dollars)	Warranty (months)	Batteries
20	0	12	0
16	0	6	1
23	5	12	0
38	10	12	1
32	8	9	1
35	15	3	0
27	10	6	0
36	12	6	1
45	15	12	1
28	5	3	1
16	0	3	0
23	0	12	1

5.   The health-care expenditures in the United States have steadily risen since 1981 as displayed in the following chart. (*The New York Times*, December 29, 1993. Copyright © 1993 by The New York Times Company. Reprinted with permission.)

The Steady Growth in Health Spending

Spending on health care in the United States, in billions of dollars.

Year	Percent of gross domestic product
1981	9.6%
1982	10.4
1983	10.5
1984	10.3
1985	10.5
1986	10.7
1987	10.9
1988	11.1
1989	11.5
1990	12.2
1991	13.2
1992	13.9
1993	Over 14.0
1994	About 15.0

Figures for 1992-94 are Commerce Department estimates.

Source: Department of Health and Human Services, Department of Commerce

BREAKDOWN OF ESTIMATED COSTS	1993	1994
Total	$942.5	$1,060.5
HEALTH SERVICES AND SUPPLIES	914.0	1,029.6
Personal health care	830.2	934.8
Hospitals	363.4	408.8
Physicians' services	175.2	194.9
Dentists' services	44.2	47.5
Other professional services	47.4	54.0
Home health	16.5	22.2
Nondurable medical products	72.6	80.2
Durable medical equipment	14.2	15.5
Nursing home care	76.0	85.5
Other personal health care	20.7	26.2
Administration	54.3	61.9
Government public health activity	29.4	32.9
RESEARCH	14.1	15.0
CONSTRUCTION	14.4	15.8

The New York Times

The figures for health-care spending as a percentage of gross domestic product for the years 1981 through 1991 are actual, but for the years 1992 through 1994 are estimates.

   a.   Based on a linear trend model, how do your predictions for 1992 through 1994 compare with those of the Department of Commerce?

   b.   Do you think a linear trend line is appropriate here? Assuming a multiplicative time series model, compute the cyclical-irregular relatives. Do they support your answer about a linear trend?

   c.   Based on a quadratic trend curve, how do your predictions for 1992 through 1994 compare with those of the Department of Commerce?

APPENDIX A

DATA SETS[1]

[1]These data sets are available on a floppy diskette in MINITAB and ASCII format. To obtain this diskette, contact either John Wiley's College Division or your area representative for John Wiley. The diskette contains the following files.
1. SAMDATA (This file contains Data Set I)
2. CITYDATA (This file contains Data Set II)
3. SURVDATA (This file contains Data Set III)
4. COMDATA (This file contains data on all companies taken from *The Business Week 1000* that contained information on all variables included in Data Set I).

DATA SET I: INFORMATION ON A SAMPLE OF 50 U.S. COMPANIES SELECTED FROM THE *BUSINESS WEEK 1000*[2]

Explanation of Columns

C1 Name of the company

C2 Market value: Share price (in millions of dollars) on March 5, 1993 multiplied by latest available common shares outstanding

C3 Sales: 1992 sales (in millions of dollars)

C4 Profits: Net income (in millions of dollars) for 1992 from continuing operations before extraordinary items

C5 Assets: Total assets (in millions of dollars) as reported at the end of the company's latest available 1992 quarter

C6 Recent share price: Price (in dollars) of a single share of a company's most widely traded issue of common stock as of the close of trading March 5, 1993

C7 P–E Ratio: Price–earnings ratio based on 1992 earnings and March 5, 1993 stock price

C8 Yield: Annual dividend rate as a percentage of March 5, 1993 stock price

C1	C2	C3	C4	C5	C6	C7	C8
1. MERCK	42926	9663	2446.6	11086	38	18	2.67
2. DU PONT	31557	37799	975.0	38870	47	33	3.76
3. PFIZER	19143	7230	1093.5	9590	59	18	2.85
4. TEXAS UTILITIES	9915	4908	737.6	19429	46	16	6.66
5. NOVERLL	9094	989	267.7	1203	30	37	0.00
6. MERRILL LYNCH	7206	13428	952.4	111896	70	8	1.72
7. PPG INDUSTRIES	7164	5814	319.4	5662	68	22	2.96
8. PHILADELPHIA ELECTRIC	6340	3962	478.9	12578	29	15	4.87
9. MASCO	4996	3525	183.1	3987	33	27	1.92
10. TIMES MIRROR	4211	3702	56.8	4070	33	74	3.30
11. FIRST INTERSTATE BANCORP	4041	4102	282.3	50863	54	17	2.23
12. NUCOR	3789	1619	79.2	1490	87	48	0.32
13. TRW	3744	8311	194.0	5458	60	19	3.16
14. SUN MICROSYSTEMS	3618	3832	128.0	2769	35	20	0.00
15. GENERAL PUBLIC UTILITIES	3200	3434	190.0	7483	29	21	5.54
16. KNIGHT-RIDDER	3167	2330	146.1	2458	58	22	2.43
17. CBS	2759	3472	138.6	3175	205	19	0.49
18. MBNA	2636	1172	172.7	6455	27	15	3.31
19. DRESSER INDUSTRIES	2567	3858	91.4	3188	19	36	3.20
20. COSTCO WHOLESALE	2416	6895	115.6	1812	21	22	0.00
21. CINCINNATI GAS & ELECTRIC	2300	1553	202.3	4802	27	13	6.23
22. PROGRESSIVE	2206	1739	139.6	3005	33	16	0.61
23. DOLE FOOD	2012	3376	65.2	2994	34	31	1.18
24. GULF STATES UTILITIES	2010	1773	133.8	6859	18	24	0.00
25. DIAL	1994	3389	28.0	3266	43	66	2.64
26. GAYLORD ENTERTAINMENT	1707	644	29.4	845	41	58	0.98
27. NORTHROP	1612	5550	120.7	3162	34	13	4.71
28. MONTANA POWER	1404	505	107.1	2273	27	14	5.77
29. POLAROID	1318	2152	99.0	2008	28	14	2.12
30. WENDY'S INTERNATIONAL	1285	1220	64.7	920	13	20	1.85
31. PROVIDENT LIFE AND ACCIDENT	1281	2867	112.6	15925	28	11	3.67

[2]Reproduced from the 1993 special bonus issue of *Business Week*. The companies are ranked by market value. Data on all companies from *The Business Week 1000* that contained information on all variables included in this data set are available on diskette.

DATA SET I

C1	C2	C3	C4	C5	C6	C7	C8
32. MUTIMEDIA	1261	577	60.5	628	34	21	0.00
33. SUNAMERICA	1253	840	84.9	13626	38	21	0.74
34. UNION BANK	1216	1386	102.6	16844	38	13	3.73
35. SPIEGEL	1066	2219	39.1	1662	21	27	1.76
36. SMITH'S FOOD AND DRUG CENTERS	1060	2650	53.7	1479	35	20	1.47
37. COMPUWARE	1034	219	10.6	250	30	57	0.00
38. MIRAGE RESORTS	1021	819	34.3	1566	37	28	0.00
39. STOP & SHOP	1011	3352	54.4	1319	21	20	0.00
40. KEYSTONE INTERNATIONAL	956	528	42.5	438	27	22	2.48
41. IMC FERTILIZER GROUP	832	966	69.7	1844	38	9	2.86
42. READ-RITE	824	461	55.4	502	24	15	0.00
43. STAPLES	805	883	16.1	495	35	47	0.00
44. WESTERN GAS RESOURCES	788	600	39.7	584	31	22	0.65
45. R.P. SCHERER	692	389	26.6	497	30	25	0.00
46. UGI	633	708	35.8	1020	24	20	5.45
47. WALLACE COMPUTER SERVICES	616	526	40.9	475	27	16	2.12
48. FIRST BANCORPORATION OF OHIO	602	351	50.7	3916	48	12	3.60
49. JUSTIN INDUSTRIES	585	453	27.1	331	44	23	0.63
50. STANDARD REGISTER	581	705	39.4	472	20	15	3.16

DATA SET II: CITY DATA[3]

Data on prices of selected products for selected cities across the country

Explanation of Columns

C1   Name of the city

C2   Price of T-bone steak per pound

C3   Price of half-gallon carton of whole milk

C4   Price of 2-liter Coca-Cola bottle, excluding any deposit

C5   Monthly rent of an unfurnished two-bedroom apartment (excluding all utilities except water), 1-1/2 or 2 baths, approximately 950 square feet

C6   Purchase price of 1800-square-foot living area new house, on 8000-square-foot lot in urban area with all utilities

C7   Monthly telephone charges for a private residential line (customer owns instruments)

C8   Price of one gallon regular unleaded gas, national brand, including all taxes; cash price at self-service pump if available

C9   Average cost per day of a semiprivate room in a hospital

C10   Price of 100-tablet bottle of aspirin, Bayer brand, 325-mg tablets

C11   Price for woman's shampoo, trim, and blow dry

C12   Price of Gallo Chablis blanc wine, 1.5-liter bottle

C1	C2	C3	C4	C5	C6	C7	C8	C9	C10	C11	C12
ALABAMA											
1. Birmingham	5.69	1.38	1.32	471	116,400	25.98	1.069	363.00	3.89	18.67	5.99
2. Decatur-Hartselle	5.11	1.43	1.28	385	89,500	24.79	1.145	233.33	4.28	15.00	6.31
3. Dothan	5.97	1.40	1.07	350	96,000	20.25	1.033	223.00	4.21	18.20	6.37
4. Florence	5.18	1.39	1.50	482	103,417	24.94	0.999	251.25	3.63	18.30	6.66
5. Huntsville	5.55	1.27	1.00	441	105,111	23.69	1.093	276.67	4.12	13.00	5.95
6. Mobile	5.87	1.42	1.19	352	92,260	24.65	1.013	196.20	4.55	20.10	6.39
7. Montgomery	5.87	1.41	1.06	416	91,897	22.16	0.999	316.60	3.82	22.60	6.29
8. Tuscaloosa	5.15	1.36	1.12	425	93,760	22.10	1.051	308.00	4.23	19.38	5.89
ALASKA											
9. Anchorage	5.46	2.05	2.35	720	155,913	15.91	1.250	532.50	5.39	26.93	6.13
10. Fairbanks	5.02	1.97	2.25	695	142,600	13.67	1.196	426.00	5.16	23.20	6.34
11. Juneau	6.49	1.97	2.04	800	140,138	14.27	1.634	380.00	7.39	23.08	7.40
ARIZONA											
12. Flagstaff	4.64	1.46	1.18	583	125,400	18.14	1.123	366.00	5.70	18.30	4.14
13. Lake Havasu	4.54	1.43	1.13	555	108,333	21.80	1.176	320.00	4.92	17.17	4.46
14. Phoenix	4.74	1.44	1.17	531	100,748	17.79	1.167	358.14	5.67	19.25	4.40
15. Prescott	4.65	1.40	1.18	520	144,375	19.34	1.181	405.00	5.29	22.00	4.14
16. Scottsdale	4.79	1.30	1.17	496	128,343	17.05	1.193	377.50	5.39	24.99	4.02
17. Tucson	4.63	1.37	1.06	477	112,000	18.00	1.151	485.20	4.93	24.40	4.73
ARKANSAS											
18. Fayetteville	4.91	1.33	1.09	400	88,675	22.74	1.049	185.50	6.18	20.50	5.01
19. Hot Springs	5.03	1.48	1.27	350	91,033	19.61	1.039	210.00	6.39	20.75	5.61

[3]Data Set II is excerpted from *ACCRA Cost of Living Index*, 26(1): First Quarter 1993. Copyright © 1993 ACCRA, P.O. Box 6749, Louisville, KY 40206-0749. Reproduced with permission of ACCRA.

C1	C2	C3	C4	C5	C6	C7	C8	C9	C10	C11	C12
20. Jonesboro	5.21	1.34	1.05	408	94,667	21.85	0.952	192.50	4.67	13.33	5.50
21. Little Rock	4.81	1.49	1.01	412	92,750	23.63	1.013	221.80	5.31	24.20	4.95
CALIFORNIA											
22. Bakersfield	5.35	1.41	1.33	573	129,148	12.45	1.193	469.25	6.31	22.20	3.89
23. Blythe	5.24	1.46	1.29	438	122,400	19.21	1.379	385.00	5.81	20.40	4.16
24. Indio	4.25	1.32	1.20	552	131,453	16.38	1.321	530.00	5.72	24.80	3.83
25. Los Angeles–Long Beach	4.02	1.44	1.31	741	206,960	14.07	1.282	562.40	6.60	15.20	4.09
26. Palm Springs	4.99	1.32	1.28	660	152,102	19.88	1.313	575.00	5.46	34.80	3.99
27. Riverside City	4.98	1.31	1.33	619	149,450	12.85	1.301	539.00	5.70	25.50	3.84
28. San Diego	4.72	1.45	1.20	900	223,600	12.65	1.286	527.88	6.02	23.95	5.41
COLORADO											
29. Colorado Springs	4.69	1.40	1.00	384	111,000	16.77	1.065	482.50	4.91	18.20	4.35
30. Denver	4.75	1.45	1.16	604	125,200	19.72	1.100	433.00	5.09	20.44	4.39
31. Fort Collins	4.79	1.46	1.15	507	137,190	20.21	1.143	395.00	5.18	15.70	4.41
32. Pueblo	4.10	1.38	1.10	353	82,033	17.35	1.067	295.00	4.83	12.74	4.52
CONNECTICUT											
33. Hamden	6.36	1.46	1.09	750	182,500	19.26	1.246	427.71	5.75	30.00	5.49
34. Hartford	5.59	1.41	1.09	735	176,000	17.76	1.301	555.17	5.81	26.37	5.35
DELAWARE											
35. Dover	5.22	1.20	0.91	545	125,333	15.28	1.061	395.00	4.95	18.00	6.23
36. Wilmington	5.09	1.39	1.29	525	143,680	15.02	1.084	472.67	5.39	18.71	5.63
FLORIDA											
37. Boca Raton	5.85	1.49	1.07	676	137,140	15.90	1.123	280.00	3.33	27.70	4.90
38. Jacksonville	5.26	1.43	1.01	467	87,957	15.83	1.090	273.80	4.04	21.90	4.87
39. Miami–Dade County	5.91	1.50	1.04	656	114,939	18.00	1.099	409.00	3.90	22.00	4.92
40. Orlando	4.82	1.60	1.14	556	104,112	20.41	1.083	333.56	3.97	22.00	5.13
41. Tallahassee	5.34	1.43	1.06	437	108,177	16.60	1.139	363.00	3.83	17.38	4.11
42. Tampa	4.92	1.50	1.15	554	101,400	20.72	1.102	311.80	3.95	16.59	4.97
GEORGIA											
43. Atlanta	5.79	1.44	1.18	549	106,220	20.84	0.935	279.20	3.50	18.60	5.35
44. Augusta	4.22	1.47	1.29	409	115,750	21.71	0.905	268.50	5.86	19.39	5.31
45. Dalton	5.95	1.42	1.08	386	95,000	22.41	0.936	285.00	4.39	19.80	5.69
46. Macon	5.92	1.64	0.95	492	107,917	19.34	0.996	226.67	4.55	16.54	5.43
47. Rome	4.98	1.40	1.20	403	95,460	20.63	0.941	270.00	5.19	14.40	5.49
IDAHO											
48. Boise	4.04	1.43	1.15	676	132,675	16.74	1.083	350.50	4.65	21.40	4.77
ILLINOIS											
49. Bloomington-Normal	4.81	1.50	1.09	602	105,400	21.09	1.099	297.00	5.96	22.00	4.35
50. Champaign-Urbana	5.83	1.57	1.37	477	111,077	21.69	1.089	415.50	5.03	20.24	3.99
51. Decatur	5.29	1.54	0.99	453	86,500	17.32	1.126	320.50	4.36	15.00	4.50
52. DeKalb	4.68	1.52	1.12	512	126,002	34.65	1.103	330.00	3.88	18.25	4.82
53. Peoria	5.61	1.69	1.10	484	137,700	19.98	1.093	278.33	5.13	24.70	4.63
54. Quincy	4.97	1.41	1.05	384	109,800	19.12	1.046	290.00	5.36	14.60	4.19
55. Rockford	4.99	1.25	1.33	525	120,650	21.60	1.084	345.00	5.88	19.90	3.79
56. Springfield	5.41	1.61	1.06	414	97,850	21.83	1.133	297.00	4.97	18.57	4.03

C1	C2	C3	C4	C5	C6	C7	C8	C9	C10	C11	C12
INDIANA											
57. Anderson	5.25	1.40	1.19	435	111,600	19.39	1.013	309.50	3.94	17.80	4.05
58. Bloomington	5.27	1.55	1.23	518	105,613	21.14	1.039	433.00	5.76	16.60	4.67
59. Evansville	5.39	1.55	1.13	412	100,440	23.08	1.001	347.33	3.27	19.30	5.13
60. Fort Wayne	5.23	1.21	1.15	455	97,000	20.21	0.993	310.33	3.77	19.10	4.26
61. Indianapolis	5.29	1.52	1.03	518	104,800	21.98	1.003	327.56	4.95	14.95	4.18
62. Muncie	4.93	1.35	1.29	540	112,378	20.61	0.999	265.00	5.40	20.80	4.42
63. South Bend	5.24	1.27	1.26	532	96,228	22.04	0.967	433.50	3.72	22.90	3.97
64. Warsaw	5.61	1.36	1.02	422	117,667	20.83	1.019	345.00	5.56	15.20	4.61
IOWA											
65. Ames	4.92	1.13	1.07	490	110,325	18.03	1.054	295.00	3.88	21.50	5.36
66. Cedar Rapids	5.15	1.20	1.37	443	120,620	24.25	1.055	305.50	5.46	18.70	5.06
67. Des Moines	4.53	1.33	0.98	506	131,400	19.74	0.992	303.00	6.14	17.95	5.25
68. Dubuque	4.39	1.29	1.07	575	138,000	20.29	1.129	240.00	5.28	13.79	4.67
69. Fort Dodge	4.62	1.22	1.13	435	115,000	13.97	1.066	325.00	3.77	16.67	5.05
70. Mason City	5.07	1.18	1.24	360	102,416	20.38	1.042	302.50	4.83	13.67	5.22
71. Sioux City	4.99	1.16	1.11	507	134,950	20.08	1.079	347.50	4.67	16.50	4.72
KANSAS											
72. Garden City	5.10	1.29	0.99	400	92,000	16.64	1.089	255.00	4.27	18.70	5.48
73. Lawrence	5.56	1.38	1.18	488	108,527	19.21	1.021	270.00	4.09	17.00	4.92
74. Manhattan	5.25	1.38	1.32	455	105,600	17.34	0.999	355.00	3.85	14.30	5.74
75. Salina	5.32	1.26	1.04	352	115,000	17.40	1.076	318.50	5.01	16.67	5.66
KENTUCKY											
76. Bowling Green	5.13	1.27	1.12	373	96,333	19.83	0.986	297.00	3.72	23.00	5.16
77. Lexington	5.26	1.57	1.08	550	110,427	26.22	1.022	318.13	4.22	22.56	6.21
78. Louisville	4.33	1.56	0.97	450	96,920	25.31	1.005	334.57	3.81	14.90	4.77
LOUISIANA											
79. Alexandria	5.43	1.36	1.15	430	93,700	20.61	1.138	225.00	4.16	24.30	4.39
80. Baton Rouge	5.64	1.28	1.15	476	105,000	22.46	1.097	324.00	4.41	19.60	4.55
81. Lake Charles	4.30	1.34	1.22	425	90,250	22.42	1.108	276.75	5.69	18.00	5.44
82. Monroe	5.17	1.34	1.10	442	87,040	19.65	1.101	209.50	4.18	14.56	4.71
83. New Orleans	6.15	1.42	0.96	508	88,000	22.90	1.149	318.80	3.89	22.80	4.52
MARYLAND											
84. Cumberland	5.63	1.15	1.36	426	123,250	26.83	1.067	265.00	5.36	14.40	5.53
85. Hagerstown	4.91	1.00	1.04	477	138,887	20.75	1.116	320.00	4.94	17.30	5.43
86. Worcester County	4.81	1.35	0.95	560	125,000	16.95	1.112	345.33	5.78	19.33	6.48
MASSACHUSETTS											
87. Boston	5.93	1.60	1.39	729	220,738	26.35	1.143	581.40	5.99	20.50	5.57
MICHIGAN											
88. Ann Arbor	5.31	1.43	1.51	749	150,538	18.34	1.049	530.00	4.15	25.00	5.55
89. Holland	4.87	1.38	1.44	460	128,478	15.71	1.025	238.00	4.84	15.70	5.42
90. Lansing	4.71	1.45	1.23	514	136,030	17.55	1.046	424.50	4.63	15.50	5.53
MINNESOTA											
91. Minneapolis	4.98	1.41	1.09	503	128,112	20.74	1.121	476.60	4.45	18.50	5.18
92. Rochester	4.93	1.36	1.39	628	109,960	23.88	1.122	403.33	4.80	15.90	5.18

C1	C2	C3	C4	C5	C6	C7	C8	C9	C10	C11	C12
93. St. Cloud	4.11	1.41	0.97	448	89,269	22.35	1.114	349.00	4.91	16.18	4.19
94. St. Paul	4.86	1.41	1.09	654	124,715	23.34	1.115	442.20	4.54	18.00	5.79
MISSISSIPPI											
95. Laurel	4.84	1.53	1.00	385	75,000	28.86	1.012	175.00	3.79	16.67	4.95
MISSOURI											
96. Columbia	5.39	1.32	1.16	406	97,000	12.63	0.953	374.33	4.07	21.60	4.93
97. Joplin	4.83	1.31	0.99	382	95,669	16.60	0.937	308.33	4.31	14.80	4.99
98. Kansas City	5.30	1.26	1.29	570	104,500	21.01	1.013	391.92	5.39	15.55	5.38
99. Kirksville	5.29	1.23	1.04	420	114,333	18.62	0.955	302.00	3.72	14.83	4.98
100. St. Charles	4.59	1.73	1.01	460	109,333	17.72	0.869	326.25	3.70	16.25	4.85
101. St. Louis	4.58	1.38	1.01	554	103,525	17.97	1.016	353.50	4.41	21.50	4.65
102. Springfield	5.24	1.27	1.02	424	93,267	17.92	0.934	309.33	5.47	17.75	5.22
MONTANA											
103. Bozeman	4.19	1.43	1.44	500	126,467	18.22	1.199	310.00	5.87	17.06	4.74
104. Great Falls	4.38	1.43	1.01	351	98,900	20.78	1.139	315.00	4.95	14.12	5.09
105. Missoula	4.57	1.43	1.51	545	100,225	18.22	1.199	343.87	6.11	19.30	5.59
NEBRASKA											
106. Hastings	4.97	1.34	1.09	382	93,000	17.48	1.131	255.00	4.17	17.95	5.67
107. Lincoln	4.56	1.33	1.20	432	92,925	16.46	1.076	275.00	5.23	21.20	5.18
108. Omaha	4.33	1.36	1.01	446	93,941	22.60	1.123	285.20	3.37	18.29	4.45
NEVADA											
109. Carson City	3.95	1.28	1.28	532	138,650	14.69	1.211	405.00	5.12	18.00	3.94
110. Reno-Sparks	4.51	1.31	1.40	606	142,300	13.50	1.205	445.72	5.12	15.59	3.97
NEW HAMPSHIRE											
111. Manchester	5.46	1.03	1.26	667	125,750	21.04	1.107	379.50	5.41	16.75	4.24
NEW MEXICO											
112. Albuquerque	4.49	1.55	1.61	547	130,550	20.86	1.091	327.00	5.35	21.39	4.93
113. Carlsbad	4.76	1.32	1.56	354	104,139	20.74	1.122	315.00	4.22	18.33	4.99
114. Farmington	4.46	1.43	1.27	445	106,150	19.94	1.156	265.00	4.79	17.30	4.85
115. Las Cruces	4.40	1.39	1.23	462	117,180	19.15	1.168	273.00	4.99	20.49	5.09
116. Sante Fe	4.99	1.56	1.72	692	159,260	20.46	1.253	270.00	4.73	22.79	4.83
NEW YORK											
117. Albany	6.23	1.25	1.57	622	116,363	36.47	1.127	287.40	5.36	16.00	5.91
118. Glens Falls	6.03	1.19	1.57	532	109,848	25.80	1.153	228.00	5.52	20.80	5.99
119. New York	7.33	1.69	1.33	1877	391,250	47.95	1.379	868.75	7.57	41.97	7.49
120. Rochester	5.81	1.28	1.77	539	138,000	20.68	1.138	496.00	4.74	17.28	6.14
121. Syracuse	5.25	1.29	1.65	500	104,200	32.90	1.149	347.50	4.85	15.00	5.29
NORTH CAROLINA											
122. Burlington	4.32	1.39	1.19	447	115,450	16.64	1.061	336.00	4.06	17.60	4.71
123. Charlotte	5.37	1.41	1.11	412	115,800	18.12	1.023	293.67	4.13	19.80	4.69
124. Fayetteville	4.51	1.60	1.35	434	97,200	18.16	0.984	350.00	5.23	19.00	4.70
125. Greenville	5.39	1.41	1.01	426	102,000	17.55	1.001	295.00	5.36	16.00	4.69
126. Hickory	5.06	1.42	0.93	435	125,375	17.45	1.019	270.00	4.23	17.50	4.76
127. Raleigh-Durham	5.12	1.41	1.18	480	113,760	18.93	1.092	274.00	5.31	19.28	4.68
128. Winston-Salem	5.01	1.40	1.08	416	114,780	16.76	1.087	227.00	4.14	19.60	4.72

C1	C2	C3	C4	C5	C6	C7	C8	C9	C10	C11	C12
NORTH DAKOTA											
129. Fargo	5.33	1.47	1.23	466	114,931	20.30	1.161	309.00	5.13	18.40	5.19
130. Minot	4.66	1.46	1.19	370	118,750	9.45	1.199	310.68	5.53	14.00	5.21
OHIO											
131. Akron	5.51	1.20	1.01	446	101,000	22.45	1.075	385.80	4.72	14.09	5.51
132. Canton	5.57	1.14	0.94	397	90,000	22.17	0.992	213.20	5.11	17.10	5.48
133. Cincinnati	5.07	1.31	1.03	570	121,329	20.30	1.079	351.25	3.92	19.69	5.49
134. Cleveland	6.12	1.24	1.06	628	127,926	24.90	1.099	506.00	5.57	24.42	5.62
135. Toledo	6.03	1.38	1.09	525	117,625	21.96	1.025	289.00	4.74	18.09	5.73
136. Youngstown	6.05	1.27	1.00	410	98,500	21.29	1.019	323.50	4.55	21.00	5.59
OKLAHOMA											
137. McAlester	5.53	1.32	1.14	362	87,822	18.03	1.022	225.00	5.05	17.17	5.44
138. Oklahoma City	4.97	1.30	1.09	424	85,374	20.27	0.984	246.30	5.59	18.67	5.04
139. Stillwater	5.31	1.18	1.12	442	102,600	18.56	0.939	250.00	5.33	18.17	5.22
140. Tulsa	5.04	1.27	1.06	417	88,350	19.64	0.979	261.80	5.79	17.20	4.80
OREGON											
141. Portland	4.95	1.34	1.27	655	142,500	22.43	1.257	456.67	4.90	20.39	4.65
142. Salem	4.48	1.24	1.18	490	108,656	17.91	1.235	325.00	4.40	17.99	4.51
PENNSYLVANIA											
143. Erie	6.73	1.26	0.95	550	138,000	21.88	1.261	437.75	5.36	16.00	5.99
144. Hanover	4.83	1.19	1.28	424	130,625	14.11	1.082	309.00	6.04	14.64	5.99
145. Harrisburg	4.87	1.17	1.07	537	112,340	19.67	1.061	456.25	5.42	18.90	5.99
146. Lancaster	4.81	1.18	1.09	543	136,540	16.32	1.087	257.75	5.69	19.00	5.99
147. Mercer County	5.84	1.26	1.29	444	131,430	14.26	1.059	319.50	5.37	13.21	5.99
148. Philadelphia	5.66	1.33	1.34	721	169,000	14.92	1.209	443.75	4.90	25.62	5.99
149. Pittsburgh	4.43	1.24	1.10	530	109,875	22.80	1.039	477.00	4.79	21.14	5.99
150. York County	4.81	1.18	1.01	465	121,750	20.21	1.055	314.00	5.69	12.35	5.99
SOUTH CAROLINA											
151. Charleston	5.43	1.54	0.98	547	112,683	25.45	0.989	282.00	5.99	22.00	5.59
152. Columbia	4.61	1.41	1.04	545	112,400	23.05	0.967	337.67	5.39	14.79	5.47
153. Florence	4.96	1.39	1.04	401	95,575	21.85	0.993	267.67	4.75	18.40	6.87
154. Greenville	5.66	1.44	1.01	502	113,000	25.61	0.923	220.00	4.16	22.40	4.79
155. Myrtle Beach	4.98	1.38	0.91	442	107,400	22.50	1.047	351.25	4.24	18.80	5.49
156. Spartanburg	5.83	1.45	1.11	456	93,900	22.35	0.953	219.67	3.93	17.00	4.91
157. Sumter	5.16	1.64	1.07	418	89,120	21.80	0.983	194.00	5.88	17.40	5.49
SOUTH DAKOTA											
158. Sioux Falls	4.95	1.21	1.05	541	99,890	23.11	0.969	310.00	5.59	14.90	5.07
159. Vermillion	4.49	1.44	1.09	435	112,000	19.96	0.989	287.00	5.43	16.33	5.99
TENNESSEE											
160. Chattanooga	5.43	1.29	1.19	537	91,380	20.38	0.979	271.00	5.29	16.99	5.77
161. Dyersburg	5.22	1.48	0.92	413	88,776	15.63	1.059	195.00	5.18	15.83	7.62
162. Johnson City	4.98	1.33	1.02	490	116,932	21.60	1.079	231.67	5.36	18.00	6.36
163. Knoxville	5.35	1.31	0.84	470	98,500	19.40	0.983	278.60	4.48	24.40	5.59
164. Memphis	5.27	1.45	1.00	425	91,297	20.04	1.061	246.20	6.22	14.20	5.91
165. Morristown	5.52	1.29	0.93	410	84,500	16.93	0.999	210.00	5.30	16.00	6.84
166. Nashville-Franklin	4.70	1.41	1.23	452	96,433	18.99	1.027	218.00	4.10	17.20	5.75

C1	C2	C3	C4	C5	C6	C7	C8	C9	C10	C11	C12
TEXAS											
167. Amarillo	5.23	1.29	1.16	438	88,833	15.13	1.081	282.38	4.51	21.60	5.11
168. Corpus Christi	5.05	1.14	1.89	473	86,350	17.82	0.991	279.00	4.79	20.35	4.93
169. Dallas	5.23	1.40	1.37	594	101,204	17.21	1.054	358.56	5.75	27.70	4.74
170. El Paso	4.73	1.57	1.15	522	98,855	16.53	1.263	288.00	4.67	26.20	4.95
171. Fort Worth	6.28	1.42	1.47	528	88,167	16.62	1.107	323.20	4.24	13.54	5.39
172. Georgetown	4.34	1.16	1.54	455	99,663	24.67	1.066	285.00	5.09	18.50	5.44
173. Houston	5.26	1.44	1.30	579	96,900	18.35	1.025	326.36	5.77	22.66	5.64
174. San Antonio	4.79	1.21	1.50	473	97,700	17.33	1.047	303.40	5.32	20.19	4.85
175. Waco	4.24	1.20	1.15	474	89,480	17.56	1.017	293.50	4.98	21.50	4.90
176. Wichita Falls	4.87	1.37	1.06	383	91,075	15.08	0.990	307.50	3.93	15.20	4.73
UTAH											
177. Cedar City	4.26	1.36	1.19	362	91,125	13.52	1.139	293.80	5.62	14.33	4.95
178. St. George	3.86	1.49	1.16	470	118,500	13.67	1.155	357.00	5.32	17.00	4.95
179. Salt Lake City	3.99	1.59	1.01	505	92,281	17.95	1.023	375.00	4.50	15.99	4.95
VERMONT											
180. Montpelier-Barre	5.65	1.38	1.51	492	137,333	27.48	1.066	486.00	5.51	16.83	5.99
VIRGINIA											
181. Lynchburg	5.39	1.39	0.95	413	96,833	17.31	1.009	280.00	4.80	19.87	5.25
182. Prince William	5.77	1.36	1.24	615	177,800	21.89	1.145	238.00	5.30	18.00	5.61
183. Richmond	5.94	1.38	1.00	568	113,406	21.08	1.057	307.00	5.53	22.20	4.85
WASHINGTON											
184. Bellingham	5.04	1.41	1.09	575	131,000	14.96	1.219	485.00	5.05	19.79	4.20
185. Olympia	4.72	1.42	1.22	476	139,558	14.58	1.187	529.00	5.03	18.60	4.02
186. Seattle	5.06	1.48	1.33	564	180,228	18.50	1.165	452.25	5.29	18.99	4.25
187. Spokane	4.69	1.51	1.16	563	134,768	16.01	1.143	441.25	6.31	17.60	4.07
188. Tacoma	5.18	1.52	1.27	515	116,000	16.57	1.123	448.33	5.30	16.20	4.01
189. Yakima	4.47	1.44	1.45	521	114,000	14.58	1.243	325.00	5.29	18.50	5.23
WEST VIRGINIA											
190. Charleston	5.19	1.31	1.11	551	126,000	29.93	1.159	289.60	5.71	20.40	6.05
191. Martinsburg/Berkeley County	4.28	1.09	1.04	430	107,978	22.17	1.067	231.00	5.62	10.68	5.55
WISCONSIN											
192. Eau Claire	4.63	1.46	1.01	548	130,000	16.31	1.109	260.00	4.83	14.38	4.65
193. Fond Du Lac	4.29	1.52	0.99	505	113,818	16.31	1.076	300.00	4.24	16.12	4.41
194. Green Bay	4.61	1.29	0.99	519	114,500	16.54	1.099	283.33	4.45	17.50	3.67
195. Madison	4.97	1.17	1.20	663	160,750	16.63	1.059	270.75	5.55	23.70	4.53
196. Marshfield	4.02	1.46	1.09	504	120,250	22.75	1.099	228.00	4.50	14.88	4.92
197. Milwaukee-Waukesha	5.00	1.33	1.01	648	132,680	16.63	1.087	342.00	4.39	14.69	4.09
198. Wausau	4.95	1.25	1.23	530	144,625	24.33	1.067	233.00	6.02	16.50	4.39
WYOMING											
199. Casper	4.89	1.73	1.84	317	130,000	18.76	0.989	400.00	4.55	19.00	5.46
200. Gillette	4.29	1.65	1.16	360	121,667	21.88	1.055	306.00	3.01	13.10	5.89

DATA SET III: **SAMPLE OF 200 OBSERVATIONS SELECTED FROM THE 1990–1991 INTERVIEW SURVEY CONDUCTED BY THE U.S. BUREAU OF LABOR[4]**

Explanation of Columns

C1   Observation number

C2   Members in the household

C3   Annual income (in dollars) before taxes

C4   Social security payments (in dollars) per year

C5   Number of earners in the household

C6   Number of vehicles owned by the household

C7   Annual food expenditure (in dollars)

C8   Annual entertainment expenditure (in dollars)

C1	C2	C3	C4	C5	C6	C7	C8
1	3	52930	3343	1	2	2580.00	472.53
2	1	20200	1530	1	1	2109.30	1434.67
3	3	72000	5508	2	4	4272.00	364.00
4	4	33110	1154	2	1	2020.00	324.00
5	1	21060	1607	1	1	1480.00	507.76
6	4	36000	2755	2	2	2973.33	376.00
7	5	81159	8950	4	2	2869.33	600.00
8	4	54080	4132	2	2	2088.00	863.39
9	2	83400	6350	2	2	1640.00	182.67
10	2	49355	3672	2	2	2760.00	230.80
11	3	8828	612	2	2	2217.33	114.00
12	5	19900	1522	1	2	1032.00	218.92
13	1	25060	1913	1	1	1006.67	111.67
14	2	35301	2678	2	1	2553.33	979.73
15	2	29045	1683	1	4	2193.33	197.33
16	5	55611	4107	1	2	4629.33	2819.23
17	2	40000	3060	2	2	5146.67	890.67
18	6	49170	3749	4	4	2145.33	173.00
19	7	96900	7413	5	3	8962.67	512.00
20	2	68990	5118	2	2	2533.33	3493.33
21	3	33410	1567	1	1	2154.67	720.93
22	3	97190	6747	2	3	1870.67	381.13
23	3	35989	2396	3	4	4653.33	956.93
24	2	10639	765	1	2	2280.00	185.33
25	1	23140	1759	1	2	4106.67	639.60
26	3	72000	4181	1	1	4440.00	389.33
27	2	56000	4108	1	2	3146.67	270.03
28	1	29400	2219	1	1	1880.00	432.00
29	4	20415	1317	3	3	2060.00	149.80
30	2	115000	7392	2	2	2896.00	581.91
31	3	37500	3710	1	2	1380.00	281.33
32	2	48764	1437	2	2	2013.33	2234.68
33	3	36527	2795	3	4	1073.33	357.29
34	4	27000	2066	3	3	1340.00	273.33
35	1	30966	1677	1	1	1405.33	388.88
36	5	51152	4070	3	3	1665.33	772.00

[4]This data set is extracted from the file for the first quarter 1991 from the public-use tape prepared by the Bureau of Labor based on the 1990–1991 Interview Survey that is conducted by the U.S. Bureau of Labor Statistics.

C1	C2	C3	C4	C5	C6	C7	C8
37	5	36705	2265	2	3	3944.00	1464.00
38	2	68000	5203	2	2	3626.67	1180.00
39	2	90114	6857	2	1	2300.00	882.60
40	3	59150	4514	2	1	2826.67	380.00
41	3	28754	2146	2	2	2893.33	444.93
42	2	21100	1492	2	1	1506.67	215.20
43	2	129100	12466	2	1	2926.67	2542.67
44	4	54750	4070	1	2	3496.00	1450.27
45	2	78000	1377	1	2	2160.00	1146.67
46	2	54900	7956	2	1	3176.00	677.33
47	4	55812	3902	2	5	1300.00	1172.00
48	4	64000	4897	2	3	1593.33	168.00
49	1	21000	1607	1	1	1146.67	269.33
50	3	14513	963	1	2	1121.33	213.33
51	4	63040	4820	2	3	2133.33	290.67
52	4	47095	3450	3	4	2312.00	2305.60
53	4	28000	2143	2	3	2786.67	182.67
54	4	102000	5123	1	3	5533.33	2172.00
55	1	33600	2570	1	0	1308.00	191.60
56	4	35000	2678	1	3	5325.33	160.00
57	4	51000	2166	4	7	5312.00	1098.80
58	3	34000	2601	2	1	3173.33	1066.67
59	3	4000	307	2	2	1985.33	1110.67
60	1	33607	2448	1	2	1613.33	208.99
61	5	19160	1454	2	2	1700.00	1364.67
62	4	93300	7115	2	3	3546.67	297.33
63	4	50050	5356	3	5	5040.00	727.60
64	4	58000	4438	2	2	1586.67	148.00
65	3	48008	3673	2	4	3400.00	2430.27
66	1	29650	2261	1	1	1853.33	728.00
67	2	12500	574	1	4	2666.67	160.00
68	5	72310	5394	3	3	8569.33	3144.00
69	6	23000	1760	2	2	2413.33	191.60
70	2	32304	2471	2	2	1840.00	913.33
71	3	60341	4616	2	4	2986.67	1892.61
72	4	39000	2984	3	4	1649.33	225.27
73	4	48028	3672	2	3	3653.33	801.33
74	2	32775	2502	2	2	1546.67	121.83
75	3	31055	2678	2	3	2957.33	1822.67
76	4	27600	1813	2	2	2473.33	321.33
77	2	35100	1286	2	3	3402.67	892.00
78	2	63643	2884	1	2	1786.67	111.80
79	6	25500	3863	2	4	2064.00	405.33
80	2	31253	1866	2	2	3213.33	160.80
81	1	31641	2419	1	1	1644.00	3647.40
82	4	64000	4896	2	3	3192.00	385.33
83	1	21900	1224	1	0	1380.00	394.27
84	2	10000	1071	2	4	1693.33	549.33
85	5	26921	3546	3	3	1708.00	193.33
86	2	108000	7109	2	2	2677.33	456.00
87	2	27050	1760	2	2	1606.67	221.80
88	4	68500	4297	1	2	5866.67	2301.33
89	1	26505	1728	1	1	892.00	132.00
90	2	43520	3283	2	3	2773.33	251.12
91	1	36000	2754	1	1	926.67	345.33

C1	C2	C3	C4	C5	C6	C7	C8
92	6	46000	3443	2	1	2277.33	576.00
93	1	29124	2009	1	1	1141.33	275.96
94	5	36140	2765	4	1	4560.00	970.67
95	3	36802	2755	2	2	2746.67	500.00
96	2	47455	3187	1	4	1993.33	146.43
97	3	52000	3749	2	1	1537.33	326.67
98	2	29594	765	1	3	1540.00	544.15
99	3	14600	1117	3	0	1853.33	426.67
100	2	25960	1913	2	1	3560.00	432.00
101	3	49000	3749	3	2	4536.00	1194.67
102	4	48000	3672	2	2	4830.67	528.00
103	4	60000	4590	2	2	2733.33	1277.33
104	1	26200	1989	1	1	2077.33	299.60
105	4	34850	2666	3	3	2120.00	178.00
106	1	40446	3060	1	1	2040.00	377.33
107	4	51580	6441	2	3	4698.67	4976.00
108	2	21000	1607	2	2	1840.00	213.33
109	3	45000	3443	2	1	5346.67	1806.67
110	4	29407	1709	2	1	1273.33	292.80
111	2	56005	4284	2	2	2253.33	234.61
112	4	64648	4650	4	4	1653.33	610.27
113	2	54140	4094	1	2	4466.67	4042.67
114	2	33000	689	2	5	1186.67	405.80
115	2	27000	2387	2	2	3000.00	585.33
116	3	56940	4361	3	4	2680.00	1846.80
117	1	62400	4210	1	1	1806.67	557.80
118	2	10000	765	2	1	1406.67	117.33
119	3	19000	1454	2	2	2426.67	334.67
120	1	25000	3825	1	1	2380.00	156.00
121	4	102000	5534	2	3	3181.33	752.00
122	6	131356	9962	5	5	6840.00	785.44
123	2	85000	6279	2	1	3184.00	241.00
124	3	74200	5676	3	2	2972.00	2197.12
125	1	23510	1798	1	1	2006.67	387.39
126	2	75000	5738	2	2	1680.00	357.33
127	3	29960	1913	1	3	1336.00	1066.99
128	1	17600	1301	1	1	2162.67	370.67
129	2	27300	2089	2	1	5320.00	448.00
130	4	54500	725	1	2	2053.33	540.00
131	1	42607	618	1	1	1176.00	178.67
132	2	31543	1989	1	1	1400.00	261.33
133	2	40300	3083	2	2	4106.67	215.04
134	2	10000	765	2	3	2846.67	380.00
135	5	25068	1629	3	3	3546.67	1360.00
136	3	39000	2984	3	1	2586.67	293.33
137	4	23000	1760	1	1	1720.00	200.00
138	3	37215	2792	3	3	1616.00	329.33
139	3	50700	3825	1	1	2760.00	240.00
140	4	59500	4452	2	2	5113.33	1064.00
141	1	30050	2295	1	1	2837.33	2538.67
142	2	36600	1760	1	2	2781.33	731.12
143	4	35000	2678	1	3	1541.33	256.00
144	7	35900	2830	2	2	3813.33	217.33
145	2	13208	899	1	4	2533.33	660.16
146	2	73239	5581	2	2	2934.67	764.00

C1	C2	C3	C4	C5	C6	C7	C8
147	1	36000	1836	1	1	3026.67	772.93
148	3	39384	2876	2	4	3560.00	719.25
149	3	27581	1530	2	3	2808.00	202.67
150	3	66400	4743	2	2	2344.00	1416.03
151	2	30000	2295	1	3	2386.67	440.00
152	4	52200	3990	3	3	3080.00	1041.56
153	4	57300	4142	2	2	2033.33	999.28
154	3	115000	1233	2	3	3461.33	1717.33
155	1	33600	2570	1	0	2810.67	2432.00
156	5	22000	1683	1	1	2901.33	2400.00
157	4	53300	2934	2	2	2416.00	202.67
158	3	94700	5759	2	2	3332.00	1069.33
159	2	68000	5202	2	2	5480.00	580.00
160	1	25257	1913	1	1	1301.33	233.33
161	2	66600	5049	2	2	2538.67	280.00
162	2	112138	8069	2	2	1368.00	846.67
163	4	54970	3826	4	5	1600.00	504.67
164	4	34000	2602	2	2	3418.67	782.00
165	4	44558	2766	2	4	4056.00	1601.33
166	3	48000	3672	1	0	3893.33	74.67
167	1	22300	1706	1	1	1066.67	168.00
168	2	33000	2525	2	3	2677.33	197.33
169	3	34250	2616	1	4	5744.00	816.00
170	1	24041	1836	1	1	1240.00	385.71
171	3	59100	8492	2	2	4953.33	1473.33
172	5	43651	2437	3	3	2708.00	472.27
173	5	24595	1408	1	1	3197.33	266.67
174	3	14640	746	1	2	1158.67	128.00
175	2	102040	6464	2	2	2186.67	512.00
176	3	14359	1071	1	1	2146.67	740.00
177	6	54082	4138	2	2	5677.33	4856.00
178	4	31790	2410	1	1	1588.00	614.67
179	3	40060	3060	2	2	2136.00	1014.27
180	1	22675	1607	1	1	1764.00	1016.80
181	2	168158	10247	1	2	4046.67	220.44
182	5	60000	4590	2	3	4173.33	475.13
183	3	42715	3265	2	1	3797.33	1554.67
184	1	16700	1278	1	1	1962.67	372.00
185	1	120830	5921	1	1	2764.00	5290.67
186	4	81383	946	2	3	2272.00	307.80
187	2	28000	2142	2	2	1146.67	160.00
188	1	34000	2601	1	0	1073.33	241.33
189	6	53094	4001	3	4	2045.33	189.13
190	4	34720	2647	2	2	3666.67	649.33
191	2	33809	2349	2	2	2013.33	733.33
192	4	24000	3672	1	1	1633.33	161.80
193	2	62830	4743	2	1	4425.33	486.67
194	2	37100	2831	1	2	3906.67	418.67
195	2	84538	6315	2	2	1926.67	116.80
196	2	42000	6350	2	1	5258.67	3334.67
197	4	33127	2534	2	1	1560.00	744.00
198	3	42456	3061	2	4	1713.33	476.80
199	4	39000	4514	3	4	3073.33	283.33
200	3	61500	4590	2	1	2181.33	1472.00

APPENDIX B

STATISTICAL TABLES[1]

[1]All tables were prepared by the author. Tables I, IV, VI, and VII were constructed using MINITAB, and Tables VIII through X were made using SAS statistical software. Tables II, III, and V were prepared by using a computer program.

TABLE I   RANDOM NUMBERS

TABLE I   RANDOM NUMBERS

57728	16308	27337	53884	60742	61693	39887	81779	36354
63962	45765	75060	46767	28844	32354	91463	25057	91907
51041	22252	38447	71567	95103	11124	34960	35710	91098
84048	53578	67379	42605	59122	39415	82869	86971	64817
17736	34458	67227	97041	77846	20338	52372	34645	56563
82238	83763	45464	18493	98489	72138	38942	97661	95788
28853	61793	44664	69427	68144	71949	57192	25592	49835
22251	73098	68108	87626	76724	56495	87357	83065	95316
66236	46591	69225	29867	60815	51931	40507	52568	47097
50006	91666	86406	92778	51232	38761	21861	98596	42673
68328	12840	61206	64298	27378	61452	13349	27223	79637
83039	25015	95983	82835	67268	23355	44647	25542	10536
53158	82329	81756	81429	54366	97530	51447	11324	49939
46802	61720	97508	73339	29277	17964	35421	39880	38180
25162	78468	44303	14425	42587	37212	58866	39008	91938
65957	15171	22417	95571	90679	54774	43979	71017	49647
10876	36062	91375	90128	14906	81447	49158	14703	89517
35354	66633	62311	58185	67310	95474	21878	89101	38299
70822	69983	23726	97422	46713	20340	42807	10859	26897
64299	12987	60370	70165	43306	14417	79261	53891	72816
74007	61658	86698	31571	75098	11676	35867	39764	47504
70909	68300	55074	42093	55745	80364	18488	47981	18702
67898	98830	97705	10723	82370	45586	19013	60915	84961
59386	25440	92441	14265	26123	85453	57326	72790	55243
71469	49833	95737	84195	78444	32104	89917	88361	35344
34064	12993	23818	28197	33755	96438	84223	10400	36797
86492	25367	65712	81581	89579	31759	56108	24476	47696
86914	87565	20344	39027	98338	95171	75562	54283	35342
88418	58064	13624	32978	90704	56218	84064	69990	45354
87948	83451	96217	40534	40775	74376	43157	74856	13950
13049	85293	32747	17728	50495	34617	73707	33976	86177
86544	52703	74990	98288	61833	48803	75258	83382	79099
77295	70694	97326	35430	53881	94007	70471	66815	73042
54637	32831	59063	72353	87365	15322	33156	40331	93942
50938	12004	18585	23896	62559	44470	27701	66780	56157
80999	49724	76745	25232	74291	74184	91055	58903	18172
71303	36255	77310	95847	30282	77207	34439	47763	99697
79264	16901	55814	89734	30255	87209	31629	19328	42532
30235	69368	38685	32790	58980	42159	88577	18427	73504
59110	69783	93713	29151	34933	95745	72271	38684	15426
28094	19560	27259	82736	49700	37876	52322	69562	75837
40341	20666	26662	16422	76351	70520	36890	86559	89160
30117	68850	28319	44992	68110	47007	22243	72813	60934
62287	44957	47690	79484	69449	27981	34770	34228	81686
96976	77830	61746	67846	15584	28070	79200	12663	63273
82584	34789	33494	55533	25040	84187	14479	26286	10665
35728	87881	70271	13115	35745	99145	92717	74357	16716

TABLE I (*Continued*)

88458	63625	59577	92037	99012	40836	58817	30757	37934
49789	20873	53858	91356	11387	75208	33643	88210	42440
49131	34078	45396	56884	81416	46292	36012	30806	65220
96256	82566	34796	88012	43066	35786	93715	15550	16690
43742	97487	68089	69887	23737	71136	21108	85204	60726
34527	87490	81183	95864	59430	19473	57978	39853	47877
58906	37390	88924	80917	58840	29907	99098	33761	50335
30438	12056	12104	61012	44674	49815	85298	94129	59542
79149	98261	48599	54336	71894	82889	51219	70291	60922
48703	25290	13835	35695	15440	52533	82849	25504	81623
96050	74505	18706	10572	66015	53509	48115	87578	86099
98859	48791	15048	73300	48045	33559	98939	98003	39453
32758	55597	11686	18385	31103	87621	39659	81413	68625
17238	73653	15557	79374	60965	75564	15872	34611	86497
79748	21687	94964	43348	26957	27085	81760	29099	23553
89199	75213	37815	99891	60990	37062	80331	54009	26812
87491	62544	51229	13028	81370	16309	28493	98555	24278
87338	17647	40018	48386	49992	44304	23330	38730	21601
48635	73063	37450	65403	65134	83119	16341	95766	83949
32197	94930	50586	88559	48025	97023	15372	18847	97168
83421	68819	69623	45088	54839	70855	86714	38202	98163
66167	84791	40631	33428	78200	41145	57816	86795	31646
26555	70521	69140	93495	38179	43253	78172	44239	60701
47786	37539	17452	88719	24423	59201	24979	51019	35458
47775	42564	15665	92454	98345	87963	81142	34356	41518
49414	83761	74309	82620	53677	34575	81871	76615	27653
75918	39825	60958	96584	26872	15379	84080	40371	35019
13440	85096	85668	24896	65261	83757	68388	29797	48376
39614	53926	97122	85279	15622	29329	59579	60250	73895
40067	48944	98882	39023	31677	41118	52818	29586	43848
65350	11148	63012	59418	54688	83692	95840	91627	84057
98902	62170	49281	29406	63143	43722	35838	98979	67024
10529	81048	29639	50740	93253	77339	80328	88580	30970
99123	48497	35247	33488	63781	19388	98534	27479	44269
65147	42913	50654	64220	13950	74293	53489	39014	86040
86886	15231	43834	88205	87159	30789	10959	81631	15575
17264	57846	52347	96649	69212	28053	62290	93328	98520
36110	87509	95913	66687	67149	81500	44107	27546	94868
58288	91109	66433	75388	80441	95720	64891	63049	68237
67834	18606	88840	39705	17329	15690	90382	35725	21362
67746	23016	87357	89427	98266	39452	58011	86665	70716
32196	36633	63350	73154	47699	15479	63905	81186	67181
33646	35175	41141	75793	58908	80681	88974	84611	46634
66973	98812	21094	45209	52503	51038	96306	75653	32482
74819	41419	91296	31736	99727	68791	93588	99566	98413
76495	37282	43051	17275	30370	76105	55926	98910	84767
39350	85262	59225	47343	63449	47004	94970	77067	16857
29657	66820	47420	37404	80296	94070	54249	60378	54670

TABLE I   RANDOM NUMBERS

825

TABLE I (*Continued*)

98059	31868	86468	80389	66521	23304	99582	48791	74154
78131	47852	62735	79575	48757	25712	22468	66035	29237
36071	71312	33098	38558	57088	26162	32752	24827	95562
23742	21969	82378	28923	19944	91024	63237	36022	76979
48418	36759	92342	93571	86923	26627	46138	86343	21083
74678	64188	39402	65189	30854	65086	43052	54042	79127
45693	86700	39667	53646	11663	86785	84727	83728	21758
68539	65113	68955	71627	38626	57160	63171	41707	51634
56807	61373	21941	71481	88523	72157	92088	41244	75735
29320	38387	89881	59789	50099	64811	31131	74334	65674
63399	73318	61578	28141	21655	65378	56261	69795	67096
33316	23627	55609	48463	92502	64287	99853	54497	85985
15403	81891	20190	72235	85636	16745	99483	43583	89137
76057	62447	71848	60035	76280	38017	26998	82690	16512
93885	29489	26222	77121	83244	74614	62527	36019	31265
66312	17182	96913	10736	52184	57082	58901	87749	34684
88960	52088	92432	18463	25562	20674	48988	41829	98681
38903	23457	87215	47089	38395	21929	94929	59489	79066
88281	90912	22965	16428	32289	99354	87068	55884	69518
98627	47123	32667	69196	70158	19828	42793	54593	53682
13183	37010	53184	53434	53631	96983	21201	10236	90134
69177	18284	72840	68433	88300	85396	10298	15680	13859
79501	74784	50483	37213	67077	59481	43976	35404	90683
52734	93419	47519	85203	27665	97179	47002	41258	39219
70256	58003	11565	67432	56505	18468	10293	46490	98191
47603	62142	37636	43374	19773	10538	27243	20800	50383
41507	43884	18253	81908	22803	84840	38968	70176	59393
56464	84865	65387	97484	95349	55548	54214	86814	20654
54523	35676	93542	20744	23942	22935	19794	53413	12979
86965	67669	42284	68532	24766	41411	97597	34998	70248
80707	35351	29958	12270	76227	31529	26105	94145	30469
57351	44045	67826	18191	75712	86420	36234	93377	20205
56860	89252	53362	82306	24114	91538	49114	67506	73489
87844	42168	83234	59134	17403	20418	65647	14702	59080
47157	27318	46686	59507	31598	16152	41184	95641	58835
63530	51286	46562	43739	51259	39836	72962	96998	89257
74595	35004	61728	28879	60412	81320	99003	20824	47086
26649	55512	58180	87954	91885	22660	31132	15752	11807
63596	44068	12648	91827	35448	59307	64466	68502	36292
66621	89136	46721	43322	78706	60249	90841	79917	18000
98128	99125	86432	87068	88376	65121	64402	55931	45748
85968	99264	64582	85694	29027	62883	53615	26692	73490
93011	71694	78514	63842	33754	84577	78698	38667	54673
36994	29619	36095	44782	85794	28498	25870	83655	97905
58857	32343	61392	65331	66939	51145	77060	85743	85278
97430	82854	28720	52153	37246	87152	95563	51769	79320
64642	69774	67582	95955	91433	95515	35211	39734	82631
56789	90056	28697	88922	39250	66008	55324	39129	63408

TABLE I (*Continued*)

78707	36317	69939	63529	88044	66897	16846	67664	99997
20752	71605	38186	18221	79499	14660	86115	80339	34321
20794	82021	37432	97568	85812	97016	15655	40601	39475
84832	45347	60186	66673	62148	37683	79034	46572	69243
39960	63046	99657	28301	19953	84261	30215	52274	91374
51835	19676	40685	45677	57150	73208	59526	76240	24209
88213	83367	21935	72494	87548	43000	72275	81974	54718
39746	38989	28721	38803	89668	57496	97127	59364	83335
95915	51291	14163	40972	33163	85169	66522	72010	53429
43937	78760	47672	69700	87058	19072	89435	13390	72315
28633	29330	20463	89033	16968	62815	65802	53006	70674
27415	32278	61924	61670	38880	13911	85037	93738	94913
10157	74513	43054	44601	35689	54559	91660	60035	83733
18041	40798	39274	72760	83644	48960	52193	95674	22516
22679	12792	60046	80515	12962	57351	36431	52277	50567
81468	99534	30455	17430	92600	85813	90223	15335	97102
50636	87932	25489	29395	87683	84579	10396	38276	33729
60635	13409	81824	77150	51472	65915	62520	33839	52209
68336	29892	94343	37822	55260	97321	20488	50172	45199
79273	96036	89979	78654	38959	36250	91126	90337	91381
71942	89335	75664	75278	40445	12818	24033	11809	44129
87842	60665	73523	55824	61257	22080	74425	54851	84786

TABLE II   FACTORIALS

827

TABLE II   FACTORIALS

n	$n!$
0	1
1	1
2	2
3	6
4	24
5	120
6	720
7	5,040
8	40,320
9	362,880
10	3,628,800
11	39,916,800
12	479,001,600
13	6,227,020,800
14	87,178,291,200
15	1,307,674,368,000
16	20,922,789,888,000
17	355,687,428,096,000
18	6,402,373,705,728,000
19	121,645,100,408,832,000
20	2,432,902,008,176,640,000
21	51,090,942,171,709,440,000
22	1,124,000,727,777,607,680,000
23	25,852,016,738,884,976,640,000
24	620,448,401,733,239,439,360,000
25	15,511,210,043,330,985,984,000,000

TABLE III   VALUES OF $\binom{n}{x}$ (COMBINATION)

n \ x	0	1	2	3	4	5	6	7	8	9	10	11	12	13	14	15	16	17	18	19	20
1	1	1																			
2	1	2	1																		
3	1	3	3	1																	
4	1	4	6	4	1																
5	1	5	10	10	5	1															
6	1	6	15	20	15	6	1														
7	1	7	21	35	35	21	7	1													
8	1	8	28	56	70	56	28	8	1												
9	1	9	36	84	126	126	84	36	9	1											
10	1	10	45	120	210	252	210	120	45	10	1										
11	1	11	55	165	330	462	462	330	165	55	11	1									
12	1	12	66	220	495	792	924	792	495	220	66	12	1								
13	1	13	78	286	715	1,287	1,716	1,716	1,287	715	286	78	13	1							
14	1	14	91	364	1,001	2,002	3,003	3,432	3,003	2,002	1,001	364	91	14	1						
15	1	15	105	455	1,365	3,003	5,005	6,435	6,435	5,005	3,003	1,365	455	105	15	1					
16	1	16	120	560	1,820	4,368	8,008	11,440	12,870	11,440	8,008	4,368	1,820	560	120	16	1				
17	1	17	136	680	2,380	6,188	12,376	19,448	24,310	24,310	19,448	12,376	6,188	2,380	680	136	17	1			
18	1	18	153	816	3,060	8,568	18,564	31,824	43,758	48,620	43,758	31,824	18,564	8,568	3,060	816	153	18	1		
19	1	19	171	969	3,876	11,628	27,132	50,388	75,582	92,378	92,378	75,582	50,388	27,132	11,628	3,876	969	171	19	1	
20	1	20	190	1,140	4,845	15,504	38,760	77,520	125,970	167,960	184,756	167,960	125,970	77,520	38,760	15,504	4,845	1,140	190	20	1
21	1	21	210	1,330	5,985	20,349	54,264	116,280	203,490	293,930	352,716	352,716	293,930	203,490	116,280	54,264	20,349	5,985	1,330	210	21
22	1	22	231	1,540	7,315	26,334	74,613	170,544	319,770	497,420	646,646	705,432	646,646	497,420	319,770	170,544	74,613	26,334	7,315	1,540	231
23	1	23	253	1,771	8,855	33,649	100,947	245,157	490,314	817,190	1,144,066	1,352,078	1,352,078	1,144,066	817,190	490,314	245,157	100,947	33,649	8,855	1,771
24	1	24	276	2,024	10,626	42,504	134,596	346,104	735,471	1,307,504	1,961,256	2,496,144	2,704,156	2,496,144	1,961,256	1,307,504	735,471	346,104	134,596	42,504	10,626
25	1	25	300	2,300	12,650	53,130	177,100	480,700	1,081,575	2,042,975	3,268,760	4,457,400	5,200,300	5,200,300	4,457,400	3,268,760	2,042,975	1,081,575	480,700	177,100	53,130

TABLE IV   TABLE OF BINOMIAL PROBABILITIES

829

TABLE IV   TABLE OF BINOMIAL PROBABILITIES

n	x	.05	.10	.20	.30	.40	.50	.60	.70	.80	.90	.95
1	0	.9500	.9000	.8000	.7000	.6000	.5000	.4000	.3000	.2000	.1000	.0500
	1	.0500	.1000	.2000	.3000	.4000	.5000	.6000	.7000	.8000	.9000	.9500
2	0	.9025	.8100	.6400	.4900	.3600	.2500	.1600	.0900	.0400	.0100	.0025
	1	.0950	.1800	.3200	.4200	.4800	.5000	.4800	.4200	.3200	.1800	.0950
	2	.0025	.0100	.0400	.0900	.1600	.2500	.3600	.4900	.6400	.8100	.9025
3	0	.8574	.7290	.5120	.3430	.2160	.1250	.0640	.0270	.0080	.0010	.0001
	1	.1354	.2430	.3840	.4410	.4320	.3750	.2880	.1890	.0960	.0270	.0071
	2	.0071	.0270	.0960	.1890	.2880	.3750	.4320	.4410	.3840	.2430	.1354
	3	.0001	.0010	.0080	.0270	.0640	.1250	.2160	.3430	.5120	.7290	.8574
4	0	.8145	.6561	.4096	.2401	.1296	.0625	.0256	.0081	.0016	.0001	.0000
	1	.1715	.2916	.4096	.4116	.3456	.2500	.1536	.0756	.0256	.0036	.0005
	2	.0135	.0486	.1536	.2646	.3456	.3750	.3456	.2646	.1536	.0486	.0135
	3	.0005	.0036	.0256	.0756	.1536	.2500	.3456	.4116	.4096	.2916	.1715
	4	.0000	.0001	.0016	.0081	.0256	.0625	.1296	.2401	.4096	.6561	.8145
5	0	.7738	.5905	.3277	.1681	.0778	.0312	.0102	.0024	.0003	.0000	.0000
	1	.2036	.3280	.4096	.3602	.2592	.1562	.0768	.0284	.0064	.0005	.0000
	2	.0214	.0729	.2048	.3087	.3456	.3125	.2304	.1323	.0512	.0081	.0011
	3	.0011	.0081	.0512	.1323	.2304	.3125	.3456	.3087	.2048	.0729	.0214
	4	.0000	.0004	.0064	.0283	.0768	.1562	.2592	.3601	.4096	.3281	.2036
	5	.0000	.0000	.0003	.0024	.0102	.0312	.0778	.1681	.3277	.5905	.7738
6	0	.7351	.5314	.2621	.1176	.0467	.0156	.0041	.0007	.0001	.0000	.0000
	1	.2321	.3543	.3932	.3025	.1866	.0937	.0369	.0102	.0015	.0001	.0000
	2	.0305	.0984	.2458	.3241	.3110	.2344	.1382	.0595	.0154	.0012	.0001
	3	.0021	.0146	.0819	.1852	.2765	.3125	.2765	.1852	.0819	.0146	.0021
	4	.0001	.0012	.0154	.0595	.1382	.2344	.3110	.3241	.2458	.0984	.0305
	5	.0000	.0001	.0015	.0102	.0369	.0937	.1866	.3025	.3932	.3543	.2321
	6	.0000	.0000	.0001	.0007	.0041	.0156	.0467	.1176	.2621	.5314	.7351
7	0	.6983	.4783	.2097	.0824	.0280	.0078	.0016	.0002	.0000	.0000	.0000
	1	.2573	.3720	.3670	.2471	.1306	.0547	.0172	.0036	.0004	.0000	.0000
	2	.0406	.1240	.2753	.3177	.2613	.1641	.0774	.0250	.0043	.0002	.0000
	3	.0036	.0230	.1147	.2269	.2903	.2734	.1935	.0972	.0287	.0026	.0002
	4	.0002	.0026	.0287	.0972	.1935	.2734	.2903	.2269	.1147	.0230	.0036
	5	.0000	.0002	.0043	.0250	.0774	.1641	.2613	.3177	.2753	.1240	.0406
	6	.0000	.0000	.0004	.0036	.0172	.0547	.1306	.2471	.3670	.3720	.2573
	7	.0000	.0000	.0000	.0002	.0016	.0078	.0280	.0824	.2097	.4783	.6983
8	0	.6634	.4305	.1678	.0576	.0168	.0039	.0007	.0001	.0000	.0000	.0000
	1	.2793	.3826	.3355	.1977	.0896	.0312	.0079	.0012	.0001	.0000	.0000
	2	.0515	.1488	.2936	.2965	.2090	.1094	.0413	.0100	.0011	.0000	.0000
	3	.0054	.0331	.1468	.2541	.2787	.2187	.1239	.0467	.0092	.0004	.0000
	4	.0004	.0046	.0459	.1361	.2322	.2734	.2322	.1361	.0459	.0046	.0004

TABLE IV   (*Continued*)

n	x	.05	.10	.20	.30	.40	.50	.60	.70	.80	.90	.95
	5	.0000	.0004	.0092	.0467	.1239	.2187	.2787	.2541	.1468	.0331	.0054
	6	.0000	.0000	.0011	.0100	.0413	.1094	.2090	.2965	.2936	.1488	.0515
	7	.0000	.0000	.0001	.0012	.0079	.0312	.0896	.1977	.3355	.3826	.2793
	8	.0000	.0000	.0000	.0001	.0007	.0039	.0168	.0576	.1678	.4305	.6634
9	0	.6302	.3874	.1342	.0404	.0101	.0020	.0003	.0000	.0000	.0000	.0000
	1	.2985	.3874	.3020	.1556	.0605	.0176	.0035	.0004	.0000	.0000	.0000
	2	.0629	.1722	.3020	.2668	.1612	.0703	.0212	.0039	.0003	.0000	.0000
	3	.0077	.0446	.1762	.2668	.2508	.1641	.0743	.0210	.0028	.0001	.0000
	4	.0006	.0074	.0661	.1715	.2508	.2461	.1672	.0735	.0165	.0008	.0000
	5	.0000	.0008	.0165	.0735	.1672	.2461	.2508	.1715	.0661	.0074	.0006
	6	.0000	.0001	.0028	.0210	.0743	.1641	.2508	.2668	.1762	.0446	.0077
	7	.0000	.0000	.0003	.0039	.0212	.0703	.1612	.2668	.3020	.1722	.0629
	8	.0000	.0000	.0000	.0004	.0035	.0176	.0605	.1556	.3020	.3874	.2985
	9	.0000	.0000	.0000	.0000	.0003	.0020	.0101	.0404	.1342	.3874	.6302
10	0	.5987	.3487	.1074	.0282	.0060	.0010	.0001	.0000	.0000	.0000	.0000
	1	.3151	.3874	.2684	.1211	.0403	.0098	.0016	.0001	.0000	.0000	.0000
	2	.0746	.1937	.3020	.2335	.1209	.0439	.0106	.0014	.0001	.0000	.0000
	3	.0105	.0574	.2013	.2668	.2150	.1172	.0425	.0090	.0008	.0000	.0000
	4	.0010	.0112	.0881	.2001	.2508	.2051	.1115	.0368	.0055	.0001	.0000
	5	.0001	.0015	.0264	.1029	.2007	.2461	.2007	.1029	.0264	.0015	.0001
	6	.0000	.0001	.0055	.0368	.1115	.2051	.2508	.2001	.0881	.0112	.0010
	7	.0000	.0000	.0008	.0090	.0425	.1172	.2150	.2668	.2013	.0574	.0105
	8	.0000	.0000	.0001	.0014	.0106	.0439	.1209	.2335	.3020	.1937	.0746
	9	.0000	.0000	.0000	.0001	.0016	.0098	.0403	.1211	.2684	.3874	.3151
	10	.0000	.0000	.0000	.0000	.0001	.0010	.0060	.0282	.1074	.3487	.5987
11	0	.5688	.3138	.0859	.0198	.0036	.0005	.0000	.0000	.0000	.0000	.0000
	1	.3293	.3835	.2362	.0932	.0266	.0054	.0007	.0000	.0000	.0000	.0000
	2	.0867	.2131	.2953	.1998	.0887	.0269	.0052	.0005	.0000	.0000	.0000
	3	.0137	.0710	.2215	.2568	.1774	.0806	.0234	.0037	.0002	.0000	.0000
	4	.0014	.0158	.1107	.2201	.2365	.1611	.0701	.0173	.0017	.0000	.0000
	5	.0001	.0025	.0388	.1321	.2207	.2256	.1471	.0566	.0097	.0003	.0000
	6	.0000	.0003	.0097	.0566	.1471	.2256	.2207	.1321	.0388	.0025	.0001
	7	.0000	.0000	.0017	.0173	.0701	.1611	.2365	.2201	.1107	.0158	.0014
	8	.0000	.0000	.0002	.0037	.0234	.0806	.1774	.2568	.2215	.0710	.0137
	9	.0000	.0000	.0000	.0005	.0052	.0269	.0887	.1998	.2953	.2131	.0867
	10	.0000	.0000	.0000	.0000	.0007	.0054	.0266	.0932	.2362	.3835	.3293
	11	.0000	.0000	.0000	.0000	.0000	.0005	.0036	.0198	.0859	.3138	.5688
12	0	.5404	.2824	.0687	.0138	.0022	.0002	.0000	.0000	.0000	.0000	.0000
	1	.3413	.3766	.2062	.0712	.0174	.0029	.0003	.0000	.0000	.0000	.0000
	2	.0988	.2301	.2835	.1678	.0639	.0161	.0025	.0002	.0000	.0000	.0000
	3	.0173	.0852	.2362	.2397	.1419	.0537	.0125	.0015	.0001	.0000	.0000
	4	.0021	.0213	.1329	.2311	.2128	.1208	.0420	.0078	.0005	.0000	.0000
	5	.0002	.0038	.0532	.1585	.2270	.1934	.1009	.0291	.0033	.0000	.0000
	6	.0000	.0005	.0155	.0792	.1766	.2256	.1766	.0792	.0155	.0005	.0000
	7	.0000	.0000	.0033	.0291	.1009	.1934	.2270	.1585	.0532	.0038	.0002
	8	.0000	.0000	.0005	.0078	.0420	.1208	.2128	.2311	.1329	.0213	.0021
	9	.0000	.0000	.0001	.0015	.0125	.0537	.1419	.2397	.2362	.0852	.0173

TABLE IV   (*Continued*)

n	x	.05	.10	.20	.30	.40	.50	.60	.70	.80	.90	.95
	10	.0000	.0000	.0000	.0002	.0025	.0161	.0639	.1678	.2835	.2301	.0988
	11	.0000	.0000	.0000	.0000	.0003	.0029	.0174	.0712	.2062	.3766	.3413
	12	.0000	.0000	.0000	.0000	.0000	.0002	.0022	.0138	.0687	.2824	.5404
13	0	.5133	.2542	.0550	.0097	.0013	.0001	.0000	.0000	.0000	.0000	.0000
	1	.3512	.3672	.1787	.0540	.0113	.0016	.0001	.0000	.0000	.0000	.0000
	2	.1109	.2448	.2680	.1388	.0453	.0095	.0012	.0001	.0000	.0000	.0000
	3	.0214	.0997	.2457	.2181	.1107	.0349	.0065	.0006	.0000	.0000	.0000
	4	.0028	.0277	.1535	.2337	.1845	.0873	.0243	.0034	.0001	.0000	.0000
	5	.0003	.0055	.0691	.1803	.2214	.1571	.0656	.0142	.0011	.0000	.0000
	6	.0000	.0008	.0230	.1030	.1968	.2095	.1312	.0442	.0058	.0001	.0000
	7	.0000	.0001	.0058	.0442	.1312	.2095	.1968	.1030	.0230	.0008	.0000
	8	.0000	.0000	.0011	.0142	.0656	.1571	.2214	.1803	.0691	.0055	.0003
	9	.0000	.0000	.0001	.0034	.0243	.0873	.1845	.2337	.1535	.0277	.0028
	10	.0000	.0000	.0000	.0006	.0065	.0349	.1107	.2181	.2457	.0997	.0214
	11	.0000	.0000	.0000	.0001	.0012	.0095	.0453	.1388	.2680	.2448	.1109
	12	.0000	.0000	.0000	.0000	.0001	.0016	.0113	.0540	.1787	.3672	.3512
	13	.0000	.0000	.0000	.0000	.0000	.0001	.0013	.0097	.0550	.2542	.5133
14	0	.4877	.2288	.0440	.0068	.0008	.0001	.0000	.0000	.0000	.0000	.0000
	1	.3593	.3559	.1539	.0407	.0073	.0009	.0001	.0000	.0000	.0000	.0000
	2	.1229	.2570	.2501	.1134	.0317	.0056	.0005	.0000	.0000	.0000	.0000
	3	.0259	.1142	.2501	.1943	.0845	.0222	.0033	.0002	.0000	.0000	.0000
	4	.0037	.0349	.1720	.2290	.1549	.0611	.0136	.0014	.0000	.0000	.0000
	5	.0004	.0078	.0860	.1963	.2066	.1222	.0408	.0066	.0003	.0000	.0000
	6	.0000	.0013	.0322	.1262	.2066	.1833	.0918	.0232	.0020	.0000	.0000
	7	.0000	.0002	.0092	.0618	.1574	.2095	.1574	.0618	.0092	.0002	.0000
	8	.0000	.0000	.0020	.0232	.0918	.1833	.2066	.1262	.0322	.0013	.0000
	9	.0000	.0000	.0003	.0066	.0408	.1222	.2066	.1963	.0860	.0078	.0004
	10	.0000	.0000	.0000	.0014	.0136	.0611	.1549	.2290	.1720	.0349	.0037
	11	.0000	.0000	.0000	.0002	.0033	.0222	.0845	.1943	.2501	.1142	.0259
	12	.0000	.0000	.0000	.0000	.0005	.0056	.0317	.1134	.2501	.2570	.1229
	13	.0000	.0000	.0000	.0000	.0001	.0009	.0073	.0407	.1539	.3559	.3593
	14	.0000	.0000	.0000	.0000	.0000	.0001	.0008	.0068	.0440	.2288	.4877
15	0	.4633	.2059	.0352	.0047	.0005	.0000	.0000	.0000	.0000	.0000	.0000
	1	.3658	.3432	.1319	.0305	.0047	.0005	.0000	.0000	.0000	.0000	.0000
	2	.1348	.2669	.2309	.0916	.0219	.0032	.0003	.0000	.0000	.0000	.0000
	3	.0307	.1285	.2501	.1700	.0634	.0139	.0016	.0001	.0000	.0000	.0000
	4	.0049	.0428	.1876	.2186	.1268	.0417	.0074	.0006	.0000	.0000	.0000
	5	.0006	.0105	.1032	.2061	.1859	.0916	.0245	.0030	.0001	.0000	.0000
	6	.0000	.0019	.0430	.1472	.2066	.1527	.0612	.0116	.0007	.0000	.0000
	7	.0000	.0003	.0138	.0811	.1771	.1964	.1181	.0348	.0035	.0000	.0000
	8	.0000	.0000	.0035	.0348	.1181	.1964	.1771	.0811	.0138	.0003	.0000
	9	.0000	.0000	.0007	.0116	.0612	.1527	.2066	.1472	.0430	.0019	.0000
	10	.0000	.0000	.0001	.0030	.0245	.0916	.1859	.2061	.1032	.0105	.0006
	11	.0000	.0000	.0000	.0006	.0074	.0417	.1268	.2186	.1876	.0428	.0049
	12	.0000	.0000	.0000	.0001	.0016	.0139	.0634	.1700	.2501	.1285	.0307
	13	.0000	.0000	.0000	.0000	.0003	.0032	.0219	.0916	.2309	.2669	.1348
	14	.0000	.0000	.0000	.0000	.0000	.0005	.0047	.0305	.1319	.3432	.3658
	15	.0000	.0000	.0000	.0000	.0000	.0000	.0005	.0047	.0352	.2059	.4633

TABLE IV   (*Continued*)

n	x	.05	.10	.20	.30	.40	.50	.60	.70	.80	.90	.95
16	0	.4401	.1853	.0281	.0033	.0003	.0000	.0000	.0000	.0000	.0000	.0000
	1	.3706	.3294	.1126	.0228	.0030	.0002	.0000	.0000	.0000	.0000	.0000
	2	.1463	.2745	.2111	.0732	.0150	.0018	.0001	.0000	.0000	.0000	.0000
	3	.0359	.1423	.2463	.1465	.0468	.0085	.0008	.0000	.0000	.0000	.0000
	4	.0061	.0514	.2001	.2040	.1014	.0278	.0040	.0002	.0000	.0000	.0000
	5	.0008	.0137	.1201	.2099	.1623	.0667	.0142	.0013	.0000	.0000	.0000
	6	.0001	.0028	.0550	.1649	.1983	.1222	.0392	.0056	.0002	.0000	.0000
	7	.0000	.0004	.0197	.1010	.1889	.1746	.0840	.0185	.0012	.0000	.0000
	8	.0000	.0001	.0055	.0487	.1417	.1964	.1417	.0487	.0055	.0001	.0000
	9	.0000	.0000	.0012	.0185	.0840	.1746	.1889	.1010	.0197	.0004	.0000
	10	.0000	.0000	.0002	.0056	.0392	.1222	.1983	.1649	.0550	.0028	.0001
	11	.0000	.0000	.0000	.0013	.0142	.0666	.1623	.2099	.1201	.0137	.0008
	12	.0000	.0000	.0000	.0002	.0040	.0278	.1014	.2040	.2001	.0514	.0061
	13	.0000	.0000	.0000	.0000	.0008	.0085	.0468	.1465	.2463	.1423	.0359
	14	.0000	.0000	.0000	.0000	.0001	.0018	.0150	.0732	.2111	.2745	.1463
	15	.0000	.0000	.0000	.0000	.0000	.0002	.0030	.0228	.1126	.3294	.3706
	16	.0000	.0000	.0000	.0000	.0000	.0000	.0003	.0033	.0281	.1853	.4401
17	0	.4181	.1668	.0225	.0023	.0002	.0000	.0000	.0000	.0000	.0000	.0000
	1	.3741	.3150	.0957	.0169	.0019	.0001	.0000	.0000	.0000	.0000	.0000
	2	.1575	.2800	.1914	.0581	.0102	.0010	.0001	.0000	.0000	.0000	.0000
	3	.0415	.1556	.2393	.1245	.0341	.0052	.0004	.0000	.0000	.0000	.0000
	4	.0076	.0605	.2093	.1868	.0796	.0182	.0021	.0001	.0000	.0000	.0000
	5	.0010	.0175	.1361	.2081	.1379	.0472	.0081	.0006	.0000	.0000	.0000
	6	.0001	.0039	.0680	.1784	.1839	.0944	.0242	.0026	.0001	.0000	.0000
	7	.0000	.0007	.0267	.1201	.1927	.1484	.0571	.0095	.0004	.0000	.0000
	8	.0000	.0001	.0084	.0644	.1606	.1855	.1070	.0276	.0021	.0000	.0000
	9	.0000	.0000	.0021	.0276	.1070	.1855	.1606	.0644	.0084	.0001	.0000
	10	.0000	.0000	.0004	.0095	.0571	.1484	.1927	.1201	.0267	.0007	.0000
	11	.0000	.0000	.0001	.0026	.0242	.0944	.1839	.1784	.0680	.0039	.0001
	12	.0000	.0000	.0000	.0006	.0081	.0472	.1379	.2081	.1361	.0175	.0010
	13	.0000	.0000	.0000	.0001	.0021	.0182	.0796	.1868	.2093	.0605	.0076
	14	.0000	.0000	.0000	.0000	.0004	.0052	.0341	.1245	.2393	.1556	.0415
	15	.0000	.0000	.0000	.0000	.0001	.0010	.0102	.0581	.1914	.2800	.1575
	16	.0000	.0000	.0000	.0000	.0000	.0001	.0019	.0169	.0957	.3150	.3741
	17	.0000	.0000	.0000	.0000	.0000	.0000	.0002	.0023	.0225	.1668	.4181
18	0	.3972	.1501	.0180	.0016	.0001	.0000	.0000	.0000	.0000	.0000	.0000
	1	.3763	.3002	.0811	.0126	.0012	.0001	.0000	.0000	.0000	.0000	.0000
	2	.1683	.2835	.1723	.0458	.0069	.0006	.0000	.0000	.0000	.0000	.0000
	3	.0473	.1680	.2297	.1046	.0246	.0031	.0002	.0000	.0000	.0000	.0000
	4	.0093	.0700	.2153	.1681	.0614	.0117	.0011	.0000	.0000	.0000	.0000
	5	.0014	.0218	.1507	.2017	.1146	.0327	.0045	.0002	.0000	.0000	.0000
	6	.0002	.0052	.0816	.1873	.1655	.0708	.0145	.0012	.0000	.0000	.0000
	7	.0000	.0010	.0350	.1376	.1892	.1214	.0374	.0046	.0001	.0000	.0000
	8	.0000	.0002	.0120	.0811	.1734	.1669	.0771	.0149	.0008	.0000	.0000
	9	.0000	.0000	.0033	.0386	.1284	.1855	.1284	.0386	.0033	.0000	.0000
	10	.0000	.0000	.0008	.0149	.0771	.1669	.1734	.0811	.0120	.0002	.0000
	11	.0000	.0000	.0001	.0046	.0374	.1214	.1892	.1376	.0350	.0010	.0000
	12	.0000	.0000	.0000	.0012	.0145	.0708	.1655	.1873	.0816	.0052	.0002

TABLE IV   TABLE OF BINOMIAL PROBABILITIES

833

TABLE IV   (*Continued*)

n	x	.05	.10	.20	.30	.40	.50	.60	.70	.80	.90	.95
	13	.0000	.0000	.0000	.0002	.0045	.0327	.1146	.2017	.1507	.0218	.0014
	14	.0000	.0000	.0000	.0000	.0011	.0117	.0614	.1681	.2153	.0700	.0093
	15	.0000	.0000	.0000	.0000	.0002	.0031	.0246	.1046	.2297	.1680	.0473
	16	.0000	.0000	.0000	.0000	.0000	.0006	.0069	.0458	.1723	.2835	.1683
	17	.0000	.0000	.0000	.0000	.0000	.0001	.0012	.0126	.0811	.3002	.3763
	18	.0000	.0000	.0000	.0000	.0000	.0000	.0001	.0016	.0180	.1501	.3972
19	0	.3774	.1351	.0144	.0011	.0001	.0000	.0000	.0000	.0000	.0000	.0000
	1	.3774	.2852	.0685	.0093	.0008	.0000	.0000	.0000	.0000	.0000	.0000
	2	.1787	.2852	.1540	.0358	.0046	.0003	.0000	.0000	.0000	.0000	.0000
	3	.0533	.1796	.2182	.0869	.0175	.0018	.0001	.0000	.0000	.0000	.0000
	4	.0112	.0798	.2182	.1491	.0467	.0074	.0005	.0000	.0000	.0000	.0000
	5	.0018	.0266	.1636	.1916	.0933	.0222	.0024	.0001	.0000	.0000	.0000
	6	.0002	.0069	.0955	.1916	.1451	.0518	.0085	.0005	.0000	.0000	.0000
	7	.0000	.0014	.0443	.1525	.1797	.0961	.0237	.0022	.0000	.0000	.0000
	8	.0000	.0002	.0166	.0981	.1797	.1442	.0532	.0077	.0003	.0000	.0000
	9	.0000	.0000	.0051	.0514	.1464	.1762	.0976	.0220	.0013	.0000	.0000
	10	.0000	.0000	.0013	.0220	.0976	.1762	.1464	.0514	.0051	.0000	.0000
	11	.0000	.0000	.0003	.0077	.0532	.1442	.1797	.0981	.0166	.0002	.0000
	12	.0000	.0000	.0000	.0022	.0237	.0961	.1797	.1525	.0443	.0014	.0000
	13	.0000	.0000	.0000	.0005	.0085	.0518	.1451	.1916	.0955	.0069	.0002
	14	.0000	.0000	.0000	.0001	.0024	.0222	.0933	.1916	.1636	.0266	.0018
	15	.0000	.0000	.0000	.0000	.0005	.0074	.0467	.1491	.2182	.0798	.0112
	16	.0000	.0000	.0000	.0000	.0001	.0018	.0175	.0869	.2182	.1796	.0533
	17	.0000	.0000	.0000	.0000	.0000	.0003	.0046	.0358	.1540	.2852	.1787
	18	.0000	.0000	.0000	.0000	.0000	.0000	.0008	.0093	.0685	.2852	.3774
	19	.0000	.0000	.0000	.0000	.0000	.0000	.0001	.0011	.0144	.1351	.3774
20	0	.3585	.1216	.0115	.0008	.0000	.0000	.0000	.0000	.0000	.0000	.0000
	1	.3774	.2702	.0576	.0068	.0005	.0000	.0000	.0000	.0000	.0000	.0000
	2	.1887	.2852	.1369	.0278	.0031	.0002	.0000	.0000	.0000	.0000	.0000
	3	.0596	.1901	.2054	.0716	.0123	.0011	.0000	.0000	.0000	.0000	.0000
	4	.0133	.0898	.2182	.1304	.0350	.0046	.0003	.0000	.0000	.0000	.0000
	5	.0022	.0319	.1746	.1789	.0746	.0148	.0013	.0000	.0000	.0000	.0000
	6	.0003	.0089	.1091	.1916	.1244	.0370	.0049	.0002	.0000	.0000	.0000
	7	.0000	.0020	.0545	.1643	.1659	.0739	.0146	.0010	.0000	.0000	.0000
	8	.0000	.0004	.0222	.1144	.1797	.1201	.0355	.0039	.0001	.0000	.0000
	9	.0000	.0001	.0074	.0654	.1597	.1602	.0710	.0120	.0005	.0000	.0000
	10	.0000	.0000	.0020	.0308	.1171	.1762	.1171	.0308	.0020	.0000	.0000
	11	.0000	.0000	.0005	.0120	.0710	.1602	.1597	.0654	.0074	.0001	.0000
	12	.0000	.0000	.0001	.0039	.0355	.1201	.1797	.1144	.0222	.0004	.0000
	13	.0000	.0000	.0000	.0010	.0146	.0739	.1659	.1643	.0545	.0020	.0000
	14	.0000	.0000	.0000	.0002	.0049	.0370	.1244	.1916	.1091	.0089	.0003
	15	.0000	.0000	.0000	.0000	.0013	.0148	.0746	.1789	.1746	.0319	.0022
	16	.0000	.0000	.0000	.0000	.0003	.0046	.0350	.1304	.2182	.0898	.0133
	17	.0000	.0000	.0000	.0000	.0000	.0011	.0123	.0716	.2054	.1901	.0596
	18	.0000	.0000	.0000	.0000	.0000	.0002	.0031	.0278	.1369	.2852	.1887
	19	.0000	.0000	.0000	.0000	.0000	.0000	.0005	.0068	.0576	.2702	.3774
	20	.0000	.0000	.0000	.0000	.0000	.0000	.0000	.0008	.0115	.1216	.3585

TABLE IV   (*Continued*)

							p					
n	*x*	.05	.10	.20	.30	.40	.50	.60	.70	.80	.90	.95
21	0	.3406	.1094	.0092	.0006	.0000	.0000	.0000	.0000	.0000	.0000	.0000
	1	.3764	.2553	.0484	.0050	.0003	.0000	.0000	.0000	.0000	.0000	.0000
	2	.1981	.2837	.1211	.0215	.0020	.0001	.0000	.0000	.0000	.0000	.0000
	3	.0660	.1996	.1917	.0585	.0086	.0006	.0000	.0000	.0000	.0000	.0000
	4	.0156	.0998	.2156	.1128	.0259	.0029	.0001	.0000	.0000	.0000	.0000
	5	.0028	.0377	.1833	.1643	.0588	.0097	.0007	.0000	.0000	.0000	.0000
	6	.0004	.0112	.1222	.1878	.1045	.0259	.0027	.0001	.0000	.0000	.0000
	7	.0000	.0027	.0655	.1725	.1493	.0554	.0087	.0005	.0000	.0000	.0000
	8	.0000	.0005	.0286	.1294	.1742	.0970	.0229	.0019	.0000	.0000	.0000
	9	.0000	.0001	.0103	.0801	.1677	.1402	.0497	.0063	.0002	.0000	.0000
	10	.0000	.0000	.0031	.0412	.1342	.1682	.0895	.0176	.0008	.0000	.0000
	11	.0000	.0000	.0008	.0176	.0895	.1682	.1342	.0412	.0031	.0000	.0000
	12	.0000	.0000	.0002	.0063	.0497	.1402	.1677	.0801	.0103	.0001	.0000
	13	.0000	.0000	.0000	.0019	.0229	.0970	.1742	.1294	.0286	.0005	.0000
	14	.0000	.0000	.0000	.0005	.0087	.0554	.1493	.1725	.0655	.0027	.0000
	15	.0000	.0000	.0000	.0001	.0027	.0259	.1045	.1878	.1222	.0112	.0004
	16	.0000	.0000	.0000	.0000	.0007	.0097	.0588	.1643	.1833	.0377	.0028
	17	.0000	.0000	.0000	.0000	.0001	.0029	.0259	.1128	.2156	.0998	.0156
	18	.0000	.0000	.0000	.0000	.0000	.0006	.0086	.0585	.1917	.1996	.0660
	19	.0000	.0000	.0000	.0000	.0000	.0001	.0020	.0215	.1211	.2837	.1981
	20	.0000	.0000	.0000	.0000	.0000	.0000	.0003	.0050	.0484	.2553	.3764
	21	.0000	.0000	.0000	.0000	.0000	.0000	.0000	.0006	.0092	.1094	.3406
22	0	.3235	.0985	.0074	.0004	.0000	.0000	.0000	.0000	.0000	.0000	.0000
	1	.3746	.2407	.0406	.0037	.0002	.0000	.0000	.0000	.0000	.0000	.0000
	2	.2070	.2808	.1065	.0166	.0014	.0001	.0000	.0000	.0000	.0000	.0000
	3	.0726	.2080	.1775	.0474	.0060	.0004	.0000	.0000	.0000	.0000	.0000
	4	.0182	.1098	.2108	.0965	.0190	.0017	.0001	.0000	.0000	.0000	.0000
	5	.0034	.0439	.1898	.1489	.0456	.0063	.0004	.0000	.0000	.0000	.0000
	6	.0005	.0138	.1344	.1808	.0862	.0178	.0015	.0000	.0000	.0000	.0000
	7	.0001	.0035	.0768	.1771	.1314	.0407	.0051	.0002	.0000	.0000	.0000
	8	.0000	.0007	.0360	.1423	.1642	.0762	.0144	.0009	.0000	.0000	.0000
	9	.0000	.0001	.0140	.0949	.1703	.1186	.0336	.0032	.0001	.0000	.0000
	10	.0000	.0000	.0046	.0529	.1476	.1542	.0656	.0097	.0003	.0000	.0000
	11	.0000	.0000	.0012	.0247	.1073	.1682	.1073	.0247	.0012	.0000	.0000
	12	.0000	.0000	.0003	.0097	.0656	.1542	.1476	.0529	.0046	.0000	.0000
	13	.0000	.0000	.0001	.0032	.0336	.1186	.1703	.0949	.0140	.0001	.0000
	14	.0000	.0000	.0000	.0009	.0144	.0762	.1642	.1423	.0360	.0007	.0000
	15	.0000	.0000	.0000	.0002	.0051	.0407	.1314	.1771	.0768	.0035	.0001
	16	.0000	.0000	.0000	.0000	.0015	.0178	.0862	.1808	.1344	.0138	.0005
	17	.0000	.0000	.0000	.0000	.0004	.0063	.0456	.1489	.1898	.0439	.0034
	18	.0000	.0000	.0000	.0000	.0001	.0017	.0190	.0965	.2108	.1098	.0182
	19	.0000	.0000	.0000	.0000	.0000	.0004	.0060	.0474	.1775	.2080	.0726
	20	.0000	.0000	.0000	.0000	.0000	.0001	.0014	.0166	.1065	.2808	.2070
	21	.0000	.0000	.0000	.0000	.0000	.0000	.0002	.0037	.0406	.2407	.3746
	22	.0000	.0000	.0000	.0000	.0000	.0000	.0000	.0004	.0074	.0985	.3235
23	0	.3074	.0886	.0059	.0003	.0000	.0000	.0000	.0000	.0000	.0000	.0000
	1	.3721	.2265	.0339	.0027	.0001	.0000	.0000	.0000	.0000	.0000	.0000
	2	.2154	.2768	.0933	.0127	.0009	.0000	.0000	.0000	.0000	.0000	.0000
	3	.0794	.2153	.1633	.0382	.0041	.0002	.0000	.0000	.0000	.0000	.0000

TABLE IV   TABLE OF BINOMIAL PROBABILITIES

835

TABLE IV   (*Continued*)

n	x	.05	.10	.20	.30	.40	.50	.60	.70	.80	.90	.95
	4	.0209	.1196	.2042	.0818	.0138	.0011	.0000	.0000	.0000	.0000	.0000
	5	.0042	.0505	.1940	.1332	.0350	.0040	.0002	.0000	.0000	.0000	.0000
	6	.0007	.0168	.1455	.1712	.0700	.0120	.0008	.0000	.0000	.0000	.0000
	7	.0001	.0045	.0883	.1782	.1133	.0292	.0029	.0001	.0000	.0000	.0000
	8	.0000	.0010	.0442	.1527	.1511	.0584	.0088	.0004	.0000	.0000	.0000
	9	.0000	.0002	.0184	.1091	.1679	.0974	.0221	.0016	.0000	.0000	.0000
	10	.0000	.0000	.0064	.0655	.1567	.1364	.0464	.0052	.0001	.0000	.0000
	11	.0000	.0000	.0019	.0332	.1234	.1612	.0823	.0142	.0005	.0000	.0000
	12	.0000	.0000	.0005	.0142	.0823	.1612	.1234	.0332	.0019	.0000	.0000
	13	.0000	.0000	.0001	.0052	.0464	.1364	.1567	.0655	.0064	.0000	.0000
	14	.0000	.0000	.0000	.0016	.0221	.0974	.1679	.1091	.0184	.0002	.0000
	15	.0000	.0000	.0000	.0004	.0088	.0584	.1511	.1527	.0442	.0010	.0000
	16	.0000	.0000	.0000	.0001	.0029	.0292	.1133	.1782	.0883	.0045	.0001
	17	.0000	.0000	.0000	.0000	.0008	.0120	.0700	.1712	.1455	.0168	.0007
	18	.0000	.0000	.0000	.0000	.0002	.0040	.0350	.1332	.1940	.0505	.0042
	19	.0000	.0000	.0000	.0000	.0000	.0011	.0138	.0818	.2042	.1196	.0209
	20	.0000	.0000	.0000	.0000	.0000	.0002	.0041	.0382	.1633	.2153	.0794
	21	.0000	.0000	.0000	.0000	.0000	.0000	.0009	.0127	.0933	.2768	.2154
	22	.0000	.0000	.0000	.0000	.0000	.0000	.0001	.0027	.0339	.2265	.3721
	23	.0000	.0000	.0000	.0000	.0000	.0000	.0000	.0003	.0059	.0886	.3074
24	0	.2920	.0798	.0047	.0002	.0000	.0000	.0000	.0000	.0000	.0000	.0000
	1	.3688	.2127	.0283	.0020	.0001	.0000	.0000	.0000	.0000	.0000	.0000
	2	.2232	.2718	.0815	.0097	.0006	.0000	.0000	.0000	.0000	.0000	.0000
	3	.0862	.2215	.1493	.0305	.0028	.0001	.0000	.0000	.0000	.0000	.0000
	4	.0238	.1292	.1960	.0687	.0099	.0006	.0000	.0000	.0000	.0000	.0000
	5	.0050	.0574	.1960	.1177	.0265	.0025	.0001	.0000	.0000	.0000	.0000
	6	.0008	.0202	.1552	.1598	.0560	.0080	.0004	.0000	.0000	.0000	.0000
	7	.0001	.0058	.0998	.1761	.0960	.0206	.0017	.0000	.0000	.0000	.0000
	8	.0000	.0014	.0530	.1604	.1360	.0438	.0053	.0002	.0000	.0000	.0000
	9	.0000	.0003	.0236	.1222	.1612	.0779	.0141	.0008	.0000	.0000	.0000
	10	.0000	.0000	.0088	.0785	.1612	.1169	.0318	.0026	.0000	.0000	.0000
	11	.0000	.0000	.0028	.0428	.1367	.1488	.0608	.0079	.0002	.0000	.0000
	12	.0000	.0000	.0008	.0199	.0988	.1612	.0988	.0199	.0008	.0000	.0000
	13	.0000	.0000	.0002	.0079	.0608	.1488	.1367	.0428	.0028	.0000	.0000
	14	.0000	.0000	.0000	.0026	.0318	.1169	.1612	.0785	.0088	.0000	.0000
	15	.0000	.0000	.0000	.0008	.0141	.0779	.1612	.1222	.0236	.0003	.0000
	16	.0000	.0000	.0000	.0002	.0053	.0438	.1360	.1604	.0530	.0014	.0000
	17	.0000	.0000	.0000	.0000	.0017	.0206	.0960	.1761	.0998	.0058	.0001
	18	.0000	.0000	.0000	.0000	.0004	.0080	.0560	.1598	.1552	.0202	.0008
	19	.0000	.0000	.0000	.0000	.0001	.0025	.0265	.1177	.1960	.0574	.0050
	20	.0000	.0000	.0000	.0000	.0000	.0006	.0099	.0687	.1960	.1292	.0238
	21	.0000	.0000	.0000	.0000	.0000	.0001	.0028	.0305	.1493	.2215	.0862
	22	.0000	.0000	.0000	.0000	.0000	.0000	.0006	.0097	.0815	.2718	.2232
	23	.0000	.0000	.0000	.0000	.0000	.0000	.0001	.0020	.0283	.2127	.3688
	24	.0000	.0000	.0000	.0000	.0000	.0000	.0000	.0002	.0047	.0798	.2920
25	0	.2774	.0718	.0038	.0001	.0000	.0000	.0000	.0000	.0000	.0000	.0000
	1	.3650	.1994	.0236	.0014	.0000	.0000	.0000	.0000	.0000	.0000	.0000
	2	.2305	.2659	.0708	.0074	.0004	.0000	.0000	.0000	.0000	.0000	.0000
	3	.0930	.2265	.1358	.0243	.0019	.0001	.0000	.0000	.0000	.0000	.0000

TABLE IV   (*Continued*)

n	x	.05	.10	.20	.30	.40	.50	.60	.70	.80	.90	.95
	4	.0269	.1384	.1867	.0572	.0071	.0004	.0000	.0000	.0000	.0000	.0000
	5	.0060	.0646	.1960	.1030	.0199	.0016	.0000	.0000	.0000	.0000	.0000
	6	.0010	.0239	.1633	.1472	.0442	.0053	.0002	.0000	.0000	.0000	.0000
	7	.0001	.0072	.1108	.1712	.0800	.0143	.0009	.0000	.0000	.0000	.0000
	8	.0000	.0018	.0623	.1651	.1200	.0322	.0031	.0001	.0000	.0000	.0000
	9	.0000	.0004	.0294	.1336	.1511	.0609	.0088	.0004	.0000	.0000	.0000
	10	.0000	.0001	.0118	.0916	.1612	.0974	.0212	.0013	.0000	.0000	.0000
	11	.0000	.0000	.0040	.0536	.1465	.1328	.0434	.0042	.0001	.0000	.0000
	12	.0000	.0000	.0012	.0268	.1140	.1550	.0760	.0115	.0003	.0000	.0000
	13	.0000	.0000	.0003	.0115	.0760	.1550	.1140	.0268	.0012	.0000	.0000
	14	.0000	.0000	.0001	.0042	.0434	.1328	.1465	.0536	.0040	.0000	.0000
	15	.0000	.0000	.0000	.0013	.0212	.0974	.1612	.0916	.0118	.0001	.0000
	16	.0000	.0000	.0000	.0004	.0088	.0609	.1511	.1336	.0294	.0004	.0000
	17	.0000	.0000	.0000	.0001	.0031	.0322	.1200	.1651	.0623	.0018	.0000
	18	.0000	.0000	.0000	.0000	.0009	.0143	.0800	.1712	.1108	.0072	.0001
	19	.0000	.0000	.0000	.0000	.0002	.0053	.0442	.1472	.1633	.0239	.0010
	20	.0000	.0000	.0000	.0000	.0000	.0016	.0199	.1030	.1960	.0646	.0060
	21	.0000	.0000	.0000	.0000	.0000	.0004	.0071	.0572	.1867	.1384	.0269
	22	.0000	.0000	.0000	.0000	.0000	.0001	.0019	.0243	.1358	.2265	.0930
	23	.0000	.0000	.0000	.0000	.0000	.0000	.0004	.0074	.0708	.2659	.2305
	24	.0000	.0000	.0000	.0000	.0000	.0000	.0000	.0014	.0236	.1994	.3650
	25	.0000	.0000	.0000	.0000	.0000	.0000	.0000	.0001	.0038	.0718	.2774

TABLE V   VALUES OF $e^{-\lambda}$

837

TABLE V   VALUES OF $e^{-\lambda}$

λ	$e^{-\lambda}$	λ	$e^{-\lambda}$
0.0	1.000000	5.5	.004087
0.1	.904837	5.6	.003698
0.2	.818731	5.7	.003346
0.3	.740818	5.8	.003028
0.4	.670320	5.9	.002739
0.5	.606531	6.0	.002479
0.6	.548812	6.1	.002243
0.7	.496585	6.2	.002029
0.8	.449329	6.3	.001836
0.9	.406570	6.4	.001662
1.0	.367879	6.5	.001503
1.1	.332871	6.6	.001360
1.2	.301194	6.7	.001231
1.3	.272532	6.8	.001114
1.4	.246597	6.9	.001008
1.5	.223130	7.0	.000912
1.6	.201897	7.1	.000825
1.7	.182684	7.2	.000747
1.8	.165299	7.3	.000676
1.9	.149569	7.4	.000611
2.0	.135335	7.5	.000553
2.1	.122456	7.6	.000500
2.2	.110803	7.7	.000453
2.3	.100259	7.8	.000410
2.4	.090718	7.9	.000371
2.5	.082085	8.0	.000335
2.6	.074274	8.1	.000304
2.7	.067206	8.2	.000275
2.8	.060810	8.3	.000249
2.9	.055023	8.4	.000225
3.0	.049787	8.5	.000203
3.1	.045049	8.6	.000184
3.2	.040762	8.7	.000167
3.3	.036883	8.8	.000151
3.4	.033373	8.9	.000136
3.5	.030197	9.0	.000123
3.6	.027324	9.1	.000112
3.7	.024724	9.2	.000101
3.8	.022371	9.3	.000091
3.9	.020242	9.4	.000083
4.0	.018316	9.5	.000075
4.1	.016573	9.6	.000068
4.2	.014996	9.7	.000061
4.3	.013569	9.8	.000055
4.4	.012277	9.9	.000050
4.5	.011109	10.0	.0000454
4.6	.010052	11.0	.0000167
4.7	.009095	12.0	.00000614
4.8	.008230	13.0	.00000226
4.9	.007447	14.0	.00000083
5.0	.006738	15.0	.00000031
5.1	.006097	16.0	.00000011
5.2	.005517	17.0	.00000004
5.3	.004992	18.0	.00000001
5.4	.004517	19.0	.000000006
		20.0	.000000002

TABLE VI   **TABLE OF POISSON PROBABILITIES**

x	0.1	0.2	0.3	0.4	λ 0.5	0.6	0.7	0.8	0.9	1.0
0	.9048	.8187	.7408	.6703	.6065	.5488	.4966	.4493	.4066	.3679
1	.0905	.1637	.2222	.2681	.3033	.3293	.3476	.3595	.3659	.3679
2	.0045	.0164	.0333	.0536	.0758	.0988	.1217	.1438	.1647	.1839
3	.0002	.0011	.0033	.0072	.0126	.0198	.0284	.0383	.0494	.0613
4	.0000	.0001	.0003	.0007	.0016	.0030	.0050	.0077	.0111	.0153
5	.0000	.0000	.0000	.0001	.0002	.0004	.0007	.0012	.0020	.0031
6	.0000	.0000	.0000	.0000	.0000	.0000	.0001	.0002	.0003	.0005
7	.0000	.0000	.0000	.0000	.0000	.0000	.0000	.0000	.0000	.0001

x	1.1	1.2	1.3	1.4	λ 1.5	1.6	1.7	1.8	1.9	2.0
0	.3329	.3012	.2725	.2466	.2231	.2019	.1827	.1653	.1496	.1353
1	.3662	.3614	.3543	.3452	.3347	.3230	.3106	.2975	.2842	.2707
2	.2014	.2169	.2303	.2417	.2510	.2584	.2640	.2678	.2700	.2707
3	.0738	.0867	.0998	.1128	.1255	.1378	.1496	.1607	.1710	.1804
4	.0203	.0260	.0324	.0395	.0471	.0551	.0636	.0723	.0812	.0902
5	.0045	.0062	.0084	.0111	.0141	.0176	.0216	.0260	.0309	.0361
6	.0008	.0012	.0018	.0026	.0035	.0047	.0061	.0078	.0098	.0120
7	.0001	.0002	.0003	.0005	.0008	.0011	.0015	.0020	.0027	.0034
8	.0000	.0000	.0001	.0001	.0001	.0002	.0003	.0005	.0006	.0009
9	.0000	.0000	.0000	.0000	.0000	.0000	.0001	.0001	.0001	.0002

x	2.1	2.2	2.3	2.4	λ 2.5	2.6	2.7	2.8	2.9	3.0
0	.1225	.1108	.1003	.0907	.0821	.0743	.0672	.0608	.0550	.0498
1	.2572	.2438	.2306	.2177	.2052	.1931	.1815	.1703	.1596	.1494
2	.2700	.2681	.2652	.2613	.2565	.2510	.2450	.2384	.2314	.2240
3	.1890	.1966	.2033	.2090	.2138	.2176	.2205	.2225	.2237	.2240
4	.0992	.1082	.1169	.1254	.1336	.1414	.1488	.1557	.1622	.1680
5	.0417	.0476	.0538	.0602	.0668	.0735	.0804	.0872	.0940	.1008
6	.0146	.0174	.0206	.0241	.0278	.0319	.0362	.0407	.0455	.0504
7	.0044	.0055	.0068	.0083	.0099	.0118	.0139	.0163	.0188	.0216
8	.0011	.0015	.0019	.0025	.0031	.0038	.0047	.0057	.0068	.0081
9	.0003	.0004	.0005	.0007	.0009	.0011	.0014	.0018	.0022	.0027
10	.0001	.0001	.0001	.0002	.0002	.0003	.0004	.0005	.0006	.0008
11	.0000	.0000	.0000	.0000	.0000	.0001	.0001	.0001	.0002	.0002
12	.0000	.0000	.0000	.0000	.0000	.0000	.0000	.0000	.0000	.0001

x	3.1	3.2	3.3	3.4	λ 3.5	3.6	3.7	3.8	3.9	4.0
0	.0450	.0408	.0369	.0334	.0302	.0273	.0247	.0224	.0202	.0183
1	.1397	.1304	.1217	.1135	.1057	.0984	.0915	.0850	.0789	.0733
2	.2165	.2087	.2008	.1929	.1850	.1771	.1692	.1615	.1539	.1465

TABLE VI   TABLE OF POISSON PROBABILITIES

839

TABLE VI   (*Continued*)

					λ					
x	3.1	3.2	3.3	3.4	3.5	3.6	3.7	3.8	3.9	4.0
3	.2237	.2226	.2209	.2186	.2158	.2125	.2087	.2046	.2001	.1954
4	.1733	.1781	.1823	.1858	.1888	.1912	.1931	.1944	.1951	.1954
5	.1075	.1140	.1203	.1264	.1322	.1377	.1429	.1477	.1522	.1563
6	.0555	.0608	.0662	.0716	.0771	.0826	.0881	.0936	.0989	.1042
7	.0246	.0278	.0312	.0348	.0385	.0425	.0466	.0508	.0551	.0595
8	.0095	.0111	.0129	.0148	.0169	.0191	.0215	.0241	.0269	.0298
9	.0033	.0040	.0047	.0056	.0066	.0076	.0089	.0102	.0116	.0132
10	.0010	.0013	.0016	.0019	.0023	.0028	.0033	.0039	.0045	.0053
11	.0003	.0004	.0005	.0006	.0007	.0009	.0011	.0013	.0016	.0019
12	.0001	.0001	.0001	.0002	.0002	.0003	.0003	.0004	.0005	.0006
13	.0000	.0000	.0000	.0000	.0001	.0001	.0001	.0001	.0002	.0002
14	.0000	.0000	.0000	.0000	.0000	.0000	.0000	.0000	.0000	.0001

					λ					
x	4.1	4.2	4.3	4.4	4.5	4.6	4.7	4.8	4.9	5.0
0	.0166	.0150	.0136	.0123	.0111	.0101	.0091	.0082	.0074	.0067
1	.0679	.0630	.0583	.0540	.0500	.0462	.0427	.0395	.0365	.0337
2	.1393	.1323	.1254	.1188	.1125	.1063	.1005	.0948	.0894	.0842
3	.1904	.1852	.1798	.1743	.1687	.1631	.1574	.1517	.1460	.1404
4	.1951	.1944	.1933	.1917	.1898	.1875	.1849	.1820	.1789	.1755
5	.1600	.1633	.1662	.1687	.1708	.1725	.1738	.1747	.1753	.1755
6	.1093	.1143	.1191	.1237	.1281	.1323	.1362	.1398	.1432	.1462
7	.0640	.0686	.0732	.0778	.0824	.0869	.0914	.0959	.1002	.1044
8	.0328	.0360	.0393	.0428	.0463	.0500	.0537	.0575	.0614	.0653
9	.0150	.0168	.0188	.0209	.0232	.0255	.0281	.0307	.0334	.0363
10	.0061	.0071	.0081	.0092	.0104	.0118	.0132	.0147	.0164	.0181
11	.0023	.0027	.0032	.0037	.0043	.0049	.0056	.0064	.0073	.0082
12	.0008	.0009	.0011	.0014	.0016	.0019	.0022	.0026	.0030	.0034
13	.0002	.0003	.0004	.0005	.0006	.0007	.0008	.0009	.0011	.0013
14	.0001	.0001	.0001	.0001	.0002	.0002	.0003	.0003	.0004	.0005
15	.0000	.0000	.0000	.0000	.0001	.0001	.0001	.0001	.0001	.0002

					λ					
x	5.1	5.2	5.3	5.4	5.5	5.6	5.7	5.8	5.9	6.0
0	.0061	.0055	.0050	.0045	.0041	.0037	.0033	.0030	.0027	.0025
1	.0311	.0287	.0265	.0244	.0225	.0207	.0191	.0176	.0162	.0149
2	.0793	.0746	.0701	.0659	.0618	.0580	.0544	.0509	.0477	.0446
3	.1348	.1293	.1239	.1185	.1133	.1082	.1033	.0985	.0938	.0892
4	.1719	.1681	.1641	.1600	.1558	.1515	.1472	.1428	.1383	.1339
5	.1753	.1748	.1740	.1728	.1714	.1697	.1678	.1656	.1632	.1606
6	.1490	.1515	.1537	.1555	.1571	.1584	.1594	.1601	.1605	.1606
7	.1086	.1125	.1163	.1200	.1234	.1267	.1298	.1326	.1353	.1377

TABLE VI   (*Continued*)

					λ					
x	5.1	5.2	5.3	5.4	5.5	5.6	5.7	5.8	5.9	6.0
8	.0692	.0731	.0771	.0810	.0849	.0887	.0925	.0962	.0998	.1033
9	.0392	.0423	.0454	.0486	.0519	.0552	.0586	.0620	.0654	.0688
10	.0200	.0220	.0241	.0262	.0285	.0309	.0334	.0359	.0386	.0413
11	.0093	.0104	.0116	.0129	.0143	.0157	.0173	.0190	.0207	.0225
12	.0039	.0045	.0051	.0058	.0065	.0073	.0082	.0092	.0102	.0113
13	.0015	.0018	.0021	.0024	.0028	.0032	.0036	.0041	.0046	.0052
14	.0006	.0007	.0008	.0009	.0011	.0013	.0015	.0017	.0019	.0022
15	.0002	.0002	.0003	.0003	.0004	.0005	.0006	.0007	.0008	.0009
16	.0001	.0001	.0001	.0001	.0001	.0002	.0002	.0002	.0003	.0003
17	.0000	.0000	.0000	.0000	.0000	.0001	.0001	.0001	.0001	.0001

					λ					
x	6.1	6.2	6.3	6.4	6.5	6.6	6.7	6.8	6.9	7.0
0	.0022	.0020	.0018	.0017	.0015	.0014	.0012	.0011	.0010	.0009
1	.0137	.0126	.0116	.0106	.0098	.0090	.0082	.0076	.0070	.0064
2	.0417	.0390	.0364	.0340	.0318	.0296	.0276	.0258	.0240	.0223
3	.0848	.0806	.0765	.0726	.0688	.0652	.0617	.0584	.0552	.0521
4	.1294	.1249	.1205	.1162	.1118	.1076	.1034	.0992	.0952	.0912
5	.1579	.1549	.1519	.1487	.1454	.1420	.1385	.1349	.1314	.1277
6	.1605	.1601	.1595	.1586	.1575	.1562	.1546	.1529	.1511	.1490
7	.1399	.1418	.1435	.1450	.1462	.1472	.1480	.1486	.1489	.1490
8	.1066	.1099	.1130	.1160	.1188	.1215	.1240	.1263	.1284	.1304
9	.0723	.0757	.0791	.0825	.0858	.0891	.0923	.0954	.0985	.1014
10	.0441	.0469	.0498	.0528	.0558	.0588	.0618	.0649	.0679	.0710
11	.0244	.0265	.0285	.0307	.0330	.0353	.0377	.0401	.0426	.0452
12	.0124	.0137	.0150	.0164	.0179	.0194	.0210	.0227	.0245	.0263
13	.0058	.0065	.0073	.0081	.0089	.0099	.0108	.0119	.0130	.0142
14	.0025	.0029	.0033	.0037	.0041	.0046	.0052	.0058	.0064	.0071
15	.0010	.0012	.0014	.0016	.0018	.0020	.0023	.0026	.0029	.0033
16	.0004	.0005	.0005	.0006	.0007	.0008	.0010	.0011	.0013	.0014
17	.0001	.0002	.0002	.0002	.0003	.0003	.0004	.0004	.0005	.0006
18	.0000	.0001	.0001	.0001	.0001	.0001	.0001	.0002	.0002	.0002
19	.0000	.0000	.0000	.0000	.0000	.0000	.0001	.0001	.0001	.0001

					λ					
x	7.1	7.2	7.3	7.4	7.5	7.6	7.7	7.8	7.9	8.0
0	.0008	.0007	.0007	.0006	.0006	.0005	.0005	.0004	.0004	.0003
1	.0059	.0054	.0049	.0045	.0041	.0038	.0035	.0032	.0029	.0027
2	.0208	.0194	.0180	.0167	.0156	.0145	.0134	.0125	.0116	.0107
3	.0492	.0464	.0438	.0413	.0389	.0366	.0345	.0324	.0305	.0286
4	.0874	.0836	.0799	.0764	.0729	.0696	.0663	.0632	.0602	.0573
5	.1241	.1204	.1167	.1130	.1094	.1057	.1021	.0986	.0951	.0916
6	.1468	.1445	.1420	.1394	.1367	.1339	.1311	.1282	.1252	.1221

TABLE VI   TABLE OF POISSON PROBABILITIES

TABLE VI   (*Continued*)

					λ					
x	7.1	7.2	7.3	7.4	7.5	7.6	7.7	7.8	7.9	8.0
7	.1489	.1486	.1481	.1474	.1465	.1454	.1442	.1428	.1413	.1396
8	.1321	.1337	.1351	.1363	.1373	.1381	.1388	.1392	.1395	.1396
9	.1042	.1070	.1096	.1121	.1144	.1167	.1187	.1207	.1224	.1241
10	.0740	.0770	.0800	.0829	.0858	.0887	.0914	.0941	.0967	.0993
11	.0478	.0504	.0531	.0558	.0585	.0613	.0640	.0667	.0695	.0722
12	.0283	.0303	.0323	.0344	.0366	.0388	.0411	.0434	.0457	.0481
13	.0154	.0168	.0181	.0196	.0211	.0227	.0243	.0260	.0278	.0296
14	.0078	.0086	.0095	.0104	.0113	.0123	.0134	.0145	.0157	.0169
15	.0037	.0041	.0046	.0051	.0057	.0062	.0069	.0075	.0083	.0090
16	.0016	.0019	.0021	.0024	.0026	.0030	.0033	.0037	.0041	.0045
17	.0007	.0008	.0009	.0010	.0012	.0013	.0015	.0017	.0019	.0021
18	.0003	.0003	.0004	.0004	.0005	.0006	.0006	.0007	.0008	.0009
19	.0001	.0001	.0001	.0002	.0002	.0002	.0003	.0003	.0003	.0004
20	.0000	.0000	.0001	.0001	.0001	.0001	.0001	.0001	.0001	.0002
21	.0000	.0000	.0000	.0000	.0000	.0000	.0000	.0000	.0001	.0001

					λ					
x	8.1	8.2	8.3	8.4	8.5	8.6	8.7	8.8	8.9	9.0
0	.0003	.0003	.0002	.0002	.0002	.0002	.0002	.0002	.0001	.0001
1	.0025	.0023	.0021	.0019	.0017	.0016	.0014	.0013	.0012	.0011
2	.0100	.0092	.0086	.0079	.0074	.0068	.0063	.0058	.0054	.0050
3	.0269	.0252	.0237	.0222	.0208	.0195	.0183	.0171	.0160	.0150
4	.0544	.0517	.0491	.0466	.0443	.0420	.0398	.0377	.0357	.0337
5	.0882	.0849	.0816	.0784	.0752	.0722	.0692	.0663	.0635	.0607
6	.1191	.1160	.1128	.1097	.1066	.1034	.1003	.0972	.0941	.0911
7	.1378	.1358	.1338	.1317	.1294	.1271	.1247	.1222	.1197	.1171
8	.1395	.1392	.1388	.1382	.1375	.1366	.1356	.1344	.1332	.1318
9	.1255	.1269	.1280	.1290	.1299	.1306	.1311	.1315	.1317	.1318
10	.1017	.1040	.1063	.1084	.1104	.1123	.1140	.1157	.1172	.1186
11	.0749	.0775	.0802	.0828	.0853	.0878	.0902	.0925	.0948	.0970
12	.0505	.0530	.0555	.0579	.0604	.0629	.0654	.0679	.0703	.0728
13	.0315	.0334	.0354	.0374	.0395	.0416	.0438	.0459	.0481	.0504
14	.0182	.0196	.0210	.0225	.0240	.0256	.0272	.0289	.0306	.0324
15	.0098	.0107	.0116	.0126	.0136	.0147	.0158	.0169	.0182	.0194
16	.0050	.0055	.0060	.0066	.0072	.0079	.0086	.0093	.0101	.0109
17	.0024	.0026	.0029	.0033	.0036	.0040	.0044	.0048	.0053	.0058
18	.0011	.0012	.0014	.0015	.0017	.0019	.0021	.0024	.0026	.0029
19	.0005	.0005	.0006	.0007	.0008	.0009	.0010	.0011	.0012	.0014
20	.0002	.0002	.0002	.0003	.0003	.0004	.0004	.0005	.0005	.0006
21	.0001	.0001	.0001	.0001	.0001	.0002	.0002	.0002	.0002	.0003
22	.0000	.0000	.0000	.0000	.0001	.0001	.0001	.0001	.0001	.0001

TABLE VI (*Continued*)

					λ					
x	9.1	9.2	9.3	9.4	9.5	9.6	9.7	9.8	9.9	10
0	.0001	.0001	.0001	.0001	.0001	.0001	.0001	.0001	.0001	.0000
1	.0010	.0009	.0009	.0008	.0007	.0007	.0006	.0005	.0005	.0005
2	.0046	.0043	.0040	.0037	.0034	.0031	.0029	.0027	.0025	.0023
3	.0140	.0131	.0123	.0115	.0107	.0100	.0093	.0087	.0081	.0076
4	.0319	.0302	.0285	.0269	.0254	.0240	.0226	.0213	.0201	.0189
5	.0581	.0555	.0530	.0506	.0483	.0460	.0439	.0418	.0398	.0378
6	.0881	.0851	.0822	.0793	.0764	.0736	.0709	.0682	.0656	.0631
7	.1145	.1118	.1091	.1064	.1037	.1010	.0982	.0955	.0928	.0901
8	.1302	.1286	.1269	.1251	.1232	.1212	.1191	.1170	.1148	.1126
9	.1317	.1315	.1311	.1306	.1300	.1293	.1284	.1274	.1263	.1251
10	.1198	.1209	.1219	.1228	.1235	.1241	.1245	.1249	.1250	.1251
11	.0991	.1012	.1031	.1049	.1067	.1083	.1098	.1112	.1125	.1137
12	.0752	.0776	.0799	.0822	.0844	.0866	.0888	.0908	.0928	.0948
13	.0526	.0549	.0572	.0594	.0617	.0640	.0662	.0685	.0707	.0729
14	.0342	.0361	.0380	.0399	.0419	.0439	.0459	.0479	.0500	.0521
15	.0208	.0221	.0235	.0250	.0265	.0281	.0297	.0313	.0330	.0347
16	.0118	.0127	.0137	.0147	.0157	.0168	.0180	.0192	.0204	.0217
17	.0063	.0069	.0075	.0081	.0088	.0095	.0103	.0111	.0119	.0128
18	.0032	.0035	.0039	.0042	.0046	.0051	.0055	.0060	.0065	.0071
19	.0015	.0017	.0019	.0021	.0023	.0026	.0028	.0031	.0034	.0037
20	.0007	.0008	.0009	.0010	.0011	.0012	.0014	.0015	.0017	.0019
21	.0003	.0003	.0004	.0004	.0005	.0006	.0006	.0007	.0008	.0009
22	.0001	.0001	.0002	.0002	.0002	.0002	.0003	.0003	.0004	.0004
23	.0000	.0001	.0001	.0001	.0001	.0001	.0001	.0001	.0002	.0002
24	.0000	.0000	.0000	.0000	.0000	.0000	.0000	.0001	.0001	.0001

					λ					
x	11	12	13	14	15	16	17	18	19	20
0	.0000	.0000	.0000	.0000	.0000	.0000	.0000	.0000	.0000	.0000
1	.0002	.0001	.0000	.0000	.0000	.0000	.0000	.0000	.0000	.0000
2	.0010	.0004	.0002	.0001	.0000	.0000	.0000	.0000	.0000	.0000
3	.0037	.0018	.0008	.0004	.0002	.0001	.0000	.0000	.0000	.0000
4	.0102	.0053	.0027	.0013	.0006	.0003	.0001	.0001	.0000	.0000
5	.0224	.0127	.0070	.0037	.0019	.0010	.0005	.0002	.0001	.0001
6	.0411	.0255	.0152	.0087	.0048	.0026	.0014	.0007	.0004	.0002
7	.0646	.0437	.0281	.0174	.0104	.0060	.0034	.0019	.0010	.0005
8	.0888	.0655	.0457	.0304	.0194	.0120	.0072	.0042	.0024	.0013
9	.1085	.0874	.0661	.0473	.0324	.0213	.0135	.0083	.0050	.0029
10	.1194	.1048	.0859	.0663	.0486	.0341	.0230	.0150	.0095	.0058
11	.1194	.1144	.1015	.0844	.0663	.0496	.0355	.0245	.0164	.0106
12	.1094	.1144	.1099	.0984	.0829	.0661	.0504	.0368	.0259	.0176
13	.0926	.1056	.1099	.1060	.0956	.0814	.0658	.0509	.0378	.0271
14	.0728	.0905	.1021	.1060	.1024	.0930	.0800	.0655	.0514	.0387

TABLE VI   TABLE OF POISSON PROBABILITIES   **843**

TABLE VI   (*Continued*)

x	11	12	13	14	λ 15	16	17	18	19	20
15	.0534	.0724	.0885	.0989	.1024	.0992	.0906	.0786	.0650	.0516
16	.0367	.0543	.0719	.0866	.0960	.0992	.0963	.0884	.0772	.0646
17	.0237	.0383	.0550	.0713	.0847	.0934	.0963	.0936	.0863	.0760
18	.0145	.0255	.0397	.0554	.0706	.0830	.0909	.0936	.0911	.0844
19	.0084	.0161	.0272	.0409	.0557	.0699	.0814	.0887	.0911	.0888
20	.0046	.0097	.0177	.0286	.0418	.0559	.0692	.0798	.0866	.0888
21	.0024	.0055	.0109	.0191	.0299	.0426	.0560	.0684	.0783	.0846
22	.0012	.0030	.0065	.0121	.0204	.0310	.0433	.0560	.0676	.0769
23	.0006	.0016	.0037	.0074	.0133	.0216	.0320	.0438	.0559	.0669
24	.0003	.0008	.0020	.0043	.0083	.0144	.0226	.0328	.0442	.0557
25	.0001	.0004	.0010	.0024	.0050	.0092	.0154	.0237	.0336	.0446
26	.0000	.0002	.0005	.0013	.0029	.0057	.0101	.0164	.0246	.0343
27	.0000	.0001	.0002	.0007	.0016	.0034	.0063	.0109	.0173	.0254
28	.0000	.0000	.0001	.0003	.0009	.0019	.0038	.0070	.0117	.0181
29	.0000	.0000	.0001	.0002	.0004	.0011	.0023	.0044	.0077	.0125
30	.0000	.0000	.0000	.0001	.0002	.0006	.0013	.0026	.0049	.0083
31	.0000	.0000	.0000	.0000	.0001	.0003	.0007	.0015	.0030	.0054
32	.0000	.0000	.0000	.0000	.0001	.0001	.0004	.0009	.0018	.0034
33	.0000	.0000	.0000	.0000	.0000	.0001	.0002	.0005	.0010	.0020
34	.0000	.0000	.0000	.0000	.0000	.0000	.0001	.0002	.0006	.0012
35	.0000	.0000	.0000	.0000	.0000	.0000	.0000	.0001	.0003	.0007
36	.0000	.0000	.0000	.0000	.0000	.0000	.0000	.0001	.0002	.0004
37	.0000	.0000	.0000	.0000	.0000	.0000	.0000	.0000	.0001	.0002
38	.0000	.0000	.0000	.0000	.0000	.0000	.0000	.0000	.0000	.0001
39	.0000	.0000	.0000	.0000	.0000	.0000	.0000	.0000	.0000	.0001

TABLE VII STANDARD NORMAL DISTRIBUTION TABLE

The entries in the table give the
areas under the standard normal
curve from 0 to z.

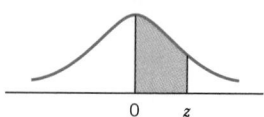

z	.00	.01	.02	.03	.04	.05	.06	.07	.08	.09
0.0	.0000	.0040	.0080	.0120	.0160	.0199	.0239	.0279	.0319	.0359
0.1	.0398	.0438	.0478	.0517	.0557	.0596	.0636	.0675	.0714	.0753
0.2	.0793	.0832	.0871	.0910	.0948	.0987	.1026	.1064	.1103	.1141
0.3	.1179	.1217	.1255	.1293	.1331	.1368	.1406	.1443	.1480	.1517
0.4	.1554	.1591	.1628	.1664	.1700	.1736	.1772	.1808	.1844	.1879
0.5	.1915	.1950	.1985	.2019	.2054	.2088	.2123	.2157	.2190	.2224
0.6	.2257	.2291	.2324	.2357	.2389	.2422	.2454	.2486	.2517	.2549
0.7	.2580	.2611	.2642	.2673	.2704	.2734	.2764	.2794	.2823	.2852
0.8	.2881	.2910	.2939	.2967	.2995	.3023	.3051	.3078	.3106	.3133
0.9	.3159	.3186	.3212	.3238	.3264	.3289	.3315	.3340	.3365	.3389
1.0	.3413	.3438	.3461	.3485	.3508	.3531	.3554	.3577	.3599	.3621
1.1	.3643	.3665	.3686	.3708	.3729	.3749	.3770	.3790	.3810	.3830
1.2	.3849	.3869	.3888	.3907	.3925	.3944	.3962	.3980	.3997	.4015
1.3	.4032	.4049	.4066	.4082	.4099	.4115	.4131	.4147	.4162	.4177
1.4	.4192	.4207	.4222	.4236	.4251	.4265	.4279	.4292	.4306	.4319
1.5	.4332	.4345	.4357	.4370	.4382	.4394	.4406	.4418	.4429	.4441
1.6	.4452	.4463	.4474	.4484	.4495	.4505	.4515	.4525	.4535	.4545
1.7	.4554	.4564	.4573	.4582	.4591	.4599	.4608	.4616	.4625	.4633
1.8	.4641	.4649	.4656	.4664	.4671	.4678	.4686	.4693	.4699	.4706
1.9	.4713	.4719	.4726	.4732	.4738	.4744	.4750	.4756	.4761	.4767
2.0	.4772	.4778	.4783	.4788	.4793	.4798	.4803	.4808	.4812	.4817
2.1	.4821	.4826	.4830	.4834	.4838	.4842	.4846	.4850	.4854	.4857
2.2	.4861	.4864	.4868	.4871	.4875	.4878	.4881	.4884	.4887	.4890
2.3	.4893	.4896	.4898	.4901	.4904	.4906	.4909	.4911	.4913	.4916
2.4	.4918	.4920	.4922	.4925	.4927	.4929	.4931	.4932	.4934	.4936
2.5	.4938	.4940	.4941	.4943	.4945	.4946	.4948	.4949	.4951	.4952
2.6	.4953	.4955	.4956	.4957	.4959	.4960	.4961	.4962	.4963	.4964
2.7	.4965	.4966	.4967	.4968	.4969	.4970	.4971	.4972	.4973	.4974
2.8	.4974	.4975	.4976	.4977	.4977	.4978	.4979	.4979	.4980	.4981
2.9	.4981	.4982	.4982	.4983	.4984	.4984	.4985	.4985	.4986	.4986
3.0	.4987	.4987	.4987	.4988	.4988	.4989	.4989	.4989	.4990	.4990

TABLE VIII   THE *t* DISTRIBUTION TABLE

TABLE VIII   THE *t* DISTRIBUTION TABLE

The entries in the table give the critical values
of *t* for the specified number of degrees
of freedom and areas in the right tail.

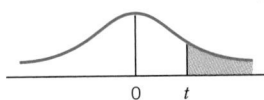

	Area in the Right Tail under the *t* Distribution Curve					
df	.10	.05	.025	.01	.005	.001
1	3.078	6.314	12.706	31.821	63.657	318.309
2	1.886	2.920	4.303	6.965	9.925	22.327
3	1.638	2.353	3.182	4.541	5.841	10.215
4	1.533	2.132	2.776	3.747	4.604	7.173
5	1.476	2.015	2.571	3.365	4.032	5.893
6	1.440	1.943	2.447	3.143	3.707	5.208
7	1.415	1.895	2.365	2.998	3.499	4.785
8	1.397	1.860	2.306	2.896	3.355	4.501
9	1.383	1.833	2.262	2.821	3.250	4.297
10	1.372	1.812	2.228	2.764	3.169	4.144
11	1.363	1.796	2.201	2.718	3.106	4.025
12	1.356	1.782	2.179	2.681	3.055	3.930
13	1.350	1.771	2.160	2.650	3.012	3.852
14	1.345	1.761	2.145	2.624	2.977	3.787
15	1.341	1.753	2.131	2.602	2.947	3.733
16	1.337	1.746	2.120	2.583	2.921	3.686
17	1.333	1.740	2.110	2.567	2.898	3.646
18	1.330	1.734	2.101	2.552	2.878	3.610
19	1.328	1.729	2.093	2.539	2.861	3.579
20	1.325	1.725	2.086	2.528	2.845	3.552
21	1.323	1.721	2.080	2.518	2.831	3.527
22	1.321	1.717	2.074	2.508	2.819	3.505
23	1.319	1.714	2.069	2.500	2.807	3.485
24	1.318	1.711	2.064	2.492	2.797	3.467
25	1.316	1.708	2.060	2.485	2.787	3.450
26	1.315	1.706	2.056	2.479	2.779	3.435
27	1.314	1.703	2.052	2.473	2.771	3.421
28	1.313	1.701	2.048	2.467	2.763	3.408
29	1.311	1.699	2.045	2.462	2.756	3.396
30	1.310	1.697	2.042	2.457	2.750	3.385
31	1.309	1.696	2.040	2.453	2.744	3.375
32	1.309	1.694	2.037	2.449	2.738	3.365
33	1.308	1.692	2.035	2.445	2.733	3.356
34	1.307	1.691	2.032	2.441	2.728	3.348
35	1.306	1.690	2.030	2.438	2.724	3.340
36	1.306	1.688	2.028	2.434	2.719	3.333
37	1.305	1.687	2.026	2.431	2.715	3.326

TABLE VIII   (*Continued*)

df	Area in the Right Tail under the *t* Distribution Curve					
	.10	**.05**	**.025**	**.01**	**.005**	**.001**
38	1.304	1.686	2.024	2.429	2.712	3.319
39	1.304	1.685	2.023	2.426	2.708	3.313
40	1.303	1.684	2.021	2.423	2.704	3.307
41	1.303	1.683	2.020	2.421	2.701	3.301
42	1.302	1.682	2.018	2.418	2.698	3.296
43	1.302	1.681	2.017	2.416	2.695	3.291
44	1.301	1.680	2.015	2.414	2.692	3.286
45	1.301	1.679	2.014	2.412	2.690	3.281
46	1.300	1.679	2.013	2.410	2.687	3.277
47	1.300	1.678	2.012	2.408	2.685	3.273
48	1.299	1.677	2.011	2.407	2.682	3.269
49	1.299	1.677	2.010	2.405	2.680	3.265
50	1.299	1.676	2.009	2.403	2.678	3.261
51	1.298	1.675	2.008	2.402	2.676	3.258
52	1.298	1.675	2.007	2.400	2.674	3.255
53	1.298	1.674	2.006	2.399	2.672	3.251
54	1.297	1.674	2.005	2.397	2.670	3.248
55	1.297	1.673	2.004	2.396	2.668	3.245
56	1.297	1.673	2.003	2.395	2.667	3.242
57	1.297	1.672	2.002	2.394	2.665	3.239
58	1.296	1.672	2.002	2.392	2.663	3.237
59	1.296	1.671	2.001	2.391	2.662	3.234
60	1.296	1.671	2.000	2.390	2.660	3.232
61	1.296	1.670	2.000	2.389	2.659	3.229
62	1.295	1.670	1.999	2.388	2.657	3.227
63	1.295	1.669	1.998	2.387	2.656	3.225
64	1.295	1.669	1.998	2.386	2.655	3.223
65	1.295	1.669	1.997	2.385	2.654	3.220
66	1.295	1.668	1.997	2.384	2.652	3.218
67	1.294	1.668	1.996	2.383	2.651	3.216
68	1.294	1.668	1.995	2.382	2.650	3.214
69	1.294	1.667	1.995	2.382	2.649	3.213
70	1.294	1.667	1.994	2.381	2.648	3.211
71	1.294	1.667	1.994	2.380	2.647	3.209
72	1.293	1.666	1.993	2.379	2.646	3.207
73	1.293	1.666	1.993	2.379	2.645	3.206
74	1.293	1.666	1.993	2.378	2.644	3.204
75	1.293	1.665	1.992	2.377	2.643	3.202
∞	1.282	1.645	1.960	2.326	2.576	3.090

TABLE IX   CHI-SQUARE DISTRIBUTION TABLE

847

TABLE IX   CHI-SQUARE DISTRIBUTION TABLE

The entries in the table give
the critical values of χ^2 for the
specified number of degrees of freedom
and areas in the right tail.

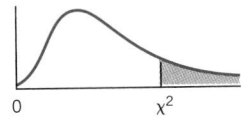

df	Area in the Right Tail under the Chi-square Distribution Curve									
	.995	.990	.975	.950	.900	.100	.050	.025	.010	.005
1	0.000	0.000	0.001	0.004	0.016	2.706	3.841	5.024	6.635	7.879
2	0.010	0.020	0.051	0.103	0.211	4.605	5.991	7.378	9.210	10.597
3	0.072	0.115	0.216	0.352	0.584	6.251	7.815	9.348	11.345	12.838
4	0.207	0.297	0.484	0.711	1.064	7.779	9.488	11.143	13.277	14.860
5	0.412	0.554	0.831	1.145	1.610	9.236	11.070	12.833	15.086	16.750
6	0.676	0.872	1.237	1.635	2.204	10.645	12.592	14.449	16.812	18.548
7	0.989	1.239	1.690	2.167	2.833	12.017	14.067	16.013	18.475	20.278
8	1.344	1.646	2.180	2.733	3.490	13.362	15.507	17.535	20.090	21.955
9	1.735	2.088	2.700	3.325	4.168	14.684	16.919	19.023	21.666	23.589
10	2.156	2.558	3.247	3.940	4.865	15.987	18.307	20.483	23.209	25.188
11	2.603	3.053	3.816	4.575	5.578	17.275	19.675	21.920	24.725	26.757
12	3.074	3.571	4.404	5.226	6.304	18.549	21.026	23.337	26.217	28.300
13	3.565	4.107	5.009	5.892	7.042	19.812	22.362	24.736	27.688	29.819
14	4.075	4.660	5.629	6.571	7.790	21.064	23.685	26.119	29.141	31.319
15	4.601	5.229	6.262	7.261	8.547	22.307	24.996	27.488	30.578	32.801
16	5.142	5.812	6.908	7.962	9.312	23.542	26.296	28.845	32.000	34.267
17	5.697	6.408	7.564	8.672	10.085	24.769	27.587	30.191	33.409	35.718
18	6.265	7.015	8.231	9.390	10.865	25.989	28.869	31.526	34.805	37.156
19	6.844	7.633	8.907	10.117	11.651	27.204	30.144	32.852	36.191	38.582
20	7.434	8.260	9.591	10.851	12.443	28.412	31.410	34.170	37.566	39.997
21	8.034	8.897	10.283	11.591	13.240	29.615	32.671	35.479	38.932	41.401
22	8.643	9.542	10.982	12.338	14.041	30.813	33.924	36.781	40.289	42.796
23	9.260	10.196	11.689	13.091	14.848	32.007	35.172	38.076	41.638	44.181
24	9.886	10.856	12.401	13.848	15.659	33.196	36.415	39.364	42.980	45.559
25	10.520	11.524	13.120	14.611	16.473	34.382	37.652	40.646	44.314	46.928
26	11.160	12.198	13.844	15.379	17.292	35.563	38.885	41.923	45.642	48.290
27	11.808	12.879	14.573	16.151	18.114	36.741	40.113	43.195	46.963	49.645
28	12.461	13.565	15.308	16.928	18.939	37.916	41.337	44.461	48.278	50.993
29	13.121	14.256	16.047	17.708	19.768	39.087	42.557	45.722	49.588	52.336
30	13.787	14.953	16.791	18.493	20.599	40.256	43.773	46.979	50.892	53.672
40	20.707	22.164	24.433	26.509	29.051	51.805	55.758	59.342	63.691	66.766
50	27.991	29.707	32.357	34.764	37.689	63.167	67.505	71.420	76.154	79.490
60	35.534	37.485	40.482	43.188	46.459	74.397	79.082	83.298	88.379	91.952
70	43.275	45.442	48.758	51.739	55.329	85.527	90.531	95.023	100.425	104.215
80	51.172	53.540	57.153	60.391	64.278	96.578	101.879	106.629	112.329	116.321
90	59.196	61.754	65.647	69.126	73.291	107.565	113.145	118.136	124.116	128.299
100	67.328	70.065	74.222	77.929	82.358	118.498	124.342	129.561	135.807	140.169

TABLE X   THE *F* DISTRIBUTION TABLE

Area in the Right Tail under the *F* Distribution Curve = .01

	Degrees of Freedom for the Numerator																		
	1	2	3	4	5	6	7	8	9	10	11	12	15	20	25	30	40	50	100
1	4052	5000	5403	5625	5764	5859	5928	5981	6022	6056	6083	6106	6157	6209	6240	6261	6287	6303	6334
2	98.50	99.00	99.17	99.25	99.30	99.33	99.36	99.37	99.39	99.40	99.41	99.42	99.43	99.45	99.46	99.47	99.47	99.48	99.49
3	34.12	30.82	29.46	28.71	28.24	27.91	27.67	27.49	27.35	27.23	27.13	27.05	26.87	26.69	26.58	26.50	26.41	26.35	26.24
4	21.20	18.00	16.69	15.98	15.52	15.21	14.98	14.80	14.66	14.55	14.45	14.37	14.20	14.02	13.91	13.84	13.75	13.69	13.58
5	16.26	13.27	12.06	11.39	10.97	10.67	10.46	10.29	10.16	10.05	9.96	9.89	9.72	9.55	9.45	9.38	9.29	9.24	9.13
6	13.75	10.92	9.78	9.15	8.75	8.47	8.26	8.10	7.98	7.87	7.79	7.72	7.56	7.40	7.30	7.23	7.14	7.09	6.99
7	12.25	9.55	8.45	7.85	7.46	7.19	6.99	6.84	6.72	6.62	6.54	6.47	6.31	6.16	6.06	5.99	5.91	5.86	5.75
8	11.26	8.65	7.59	7.01	6.63	6.37	6.18	6.03	5.91	5.81	5.73	5.67	5.52	5.36	5.26	5.20	5.12	5.07	4.96
9	10.56	8.02	6.99	6.42	6.06	5.80	5.61	5.47	5.35	5.26	5.18	5.11	4.96	4.81	4.71	4.65	4.57	4.52	4.41
10	10.04	7.56	6.55	5.99	5.64	5.39	5.20	5.06	4.94	4.85	4.77	4.71	4.56	4.41	4.31	4.25	4.17	4.12	4.01
11	9.65	7.21	6.22	5.67	5.32	5.07	4.89	4.74	4.63	4.54	4.46	4.40	4.25	4.10	4.01	3.94	3.86	3.81	3.71
12	9.33	6.93	5.95	5.41	5.06	4.82	4.64	4.50	4.39	4.30	4.22	4.16	4.01	3.86	3.76	3.70	3.62	3.57	3.47
13	9.07	6.70	5.74	5.21	4.86	4.62	4.44	4.30	4.19	4.10	4.02	3.96	3.82	3.66	3.57	3.51	3.43	3.38	3.27
14	8.86	6.51	5.56	5.04	4.69	4.46	4.28	4.14	4.03	3.94	3.86	3.80	3.66	3.51	3.41	3.35	3.27	3.22	3.11
15	8.68	6.36	5.42	4.89	4.56	4.32	4.14	4.00	3.89	3.80	3.73	3.67	3.52	3.37	3.28	3.21	3.13	3.08	2.98
16	8.53	6.23	5.29	4.77	4.44	4.20	4.03	3.89	3.78	3.69	3.62	3.55	3.41	3.26	3.16	3.10	3.02	2.97	2.86
17	8.40	6.11	5.18	4.67	4.34	4.10	3.93	3.79	3.68	3.59	3.52	3.46	3.31	3.16	3.07	3.00	2.92	2.87	2.76
18	8.29	6.01	5.09	4.58	4.25	4.01	3.84	3.71	3.60	3.51	3.43	3.37	3.23	3.08	2.98	2.92	2.84	2.78	2.68
19	8.18	5.93	5.01	4.50	4.17	3.94	3.77	3.63	3.52	3.43	3.36	3.30	3.15	3.00	2.91	2.84	2.76	2.71	2.60
20	8.10	5.85	4.94	4.43	4.10	3.87	3.70	3.56	3.46	3.37	3.29	3.23	3.09	2.94	2.84	2.78	2.69	2.64	2.54
21	8.02	5.78	4.87	4.37	4.04	3.81	3.64	3.51	3.40	3.31	3.24	3.17	3.03	2.88	2.79	2.72	2.64	2.58	2.48
22	7.95	5.72	4.82	4.31	3.99	3.76	3.59	3.45	3.35	3.26	3.18	3.12	2.98	2.83	2.73	2.67	2.58	2.53	2.42
23	7.88	5.66	4.76	4.26	3.94	3.71	3.54	3.41	3.30	3.21	3.14	3.07	2.93	2.78	2.69	2.62	2.54	2.48	2.37
24	7.82	5.61	4.72	4.22	3.90	3.67	3.50	3.36	3.26	3.17	3.09	3.03	2.89	2.74	2.64	2.58	2.49	2.44	2.33
25	7.77	5.57	4.68	4.18	3.85	3.63	3.46	3.32	3.22	3.13	3.06	2.99	2.85	2.70	2.60	2.54	2.45	2.40	2.29
30	7.56	5.39	4.51	4.02	3.70	3.47	3.30	3.17	3.07	2.98	2.91	2.84	2.70	2.55	2.45	2.39	2.30	2.25	2.13
40	7.31	5.18	4.31	3.83	3.51	3.29	3.12	2.99	2.89	2.80	2.73	2.66	2.52	2.37	2.27	2.20	2.11	2.06	1.94
50	7.17	5.06	4.20	3.72	3.41	3.19	3.02	2.89	2.78	2.70	2.63	2.56	2.42	2.27	2.17	2.10	2.01	1.95	1.82
100	6.90	4.82	3.98	3.51	3.21	2.99	2.82	2.69	2.59	2.50	2.43	2.37	2.22	2.07	1.97	1.89	1.80	1.74	1.60

Degrees of Freedom for the Denominator

TABLE X THE F DISTRIBUTION TABLE

TABLE X (*Continued*)

Area in the Right Tail under the F Distribution Curve = .025

df (den)	\	1	2	3	4	5	6	7	8	9	10	11	12	15	20	25	30	40	50	100
1		647.8	799.5	864.2	899.6	921.8	937.1	948.2	956.7	963.3	968.6	973.0	976.7	984.9	993.1	998.1	1001	1006	1008	1013
2		38.51	39.00	39.17	39.25	39.30	39.33	39.36	39.37	39.39	39.40	39.41	39.41	39.43	39.45	39.46	39.46	39.47	39.48	39.49
3		17.44	16.04	15.44	15.10	14.88	14.73	14.62	14.54	14.47	14.42	14.37	14.34	14.25	14.17	14.12	14.08	14.04	14.01	13.96
4		12.22	10.65	9.98	9.61	9.36	9.20	9.07	8.98	8.90	8.84	8.79	8.75	8.66	8.56	8.50	8.46	8.41	8.38	8.32
5		10.01	8.43	7.76	7.39	7.15	6.98	6.85	6.76	6.68	6.62	6.57	6.52	6.43	6.33	6.27	6.23	6.18	6.14	6.08
6		8.81	7.26	6.60	6.23	5.99	5.82	5.70	5.60	5.52	5.46	5.41	5.37	5.27	5.17	5.11	5.07	5.01	4.98	4.92
7		8.07	6.54	5.89	5.52	5.29	5.12	4.99	4.90	4.82	4.76	4.71	4.67	4.57	4.47	4.40	4.36	4.31	4.28	4.21
8		7.57	6.06	5.42	5.05	4.82	4.65	4.53	4.43	4.36	4.30	4.24	4.20	4.10	4.00	3.94	3.89	3.84	3.81	3.74
9		7.21	5.72	5.08	4.72	4.48	4.32	4.20	4.10	4.03	3.96	3.91	3.87	3.77	3.67	3.60	3.56	3.51	3.47	3.40
10		6.94	5.46	4.83	4.47	4.24	4.07	3.95	3.85	3.78	3.72	3.66	3.62	3.52	3.42	3.35	3.31	3.26	3.22	3.15
11		6.72	5.26	4.63	4.28	4.04	3.88	3.76	3.66	3.59	3.53	3.47	3.43	3.33	3.23	3.16	3.12	3.06	3.03	2.96
12		6.55	5.10	4.47	4.12	3.89	3.73	3.61	3.51	3.44	3.37	3.32	3.28	3.18	3.07	3.01	2.96	2.91	2.87	2.80
13		6.41	4.97	4.35	4.00	3.77	3.60	3.48	3.39	3.31	3.25	3.20	3.15	3.05	2.95	2.88	2.84	2.78	2.74	2.67
14		6.30	4.86	4.24	3.89	3.66	3.50	3.38	3.29	3.21	3.15	3.09	3.05	2.95	2.84	2.78	2.73	2.67	2.64	2.56
15		6.20	4.77	4.15	3.80	3.58	3.41	3.29	3.20	3.12	3.06	3.01	2.96	2.86	2.76	2.69	2.64	2.59	2.55	2.47
16		6.12	4.69	4.08	3.73	3.50	3.34	3.22	3.12	3.05	2.99	2.93	2.89	2.79	2.68	2.61	2.57	2.51	2.47	2.40
17		6.04	4.62	4.01	3.66	3.44	3.28	3.16	3.06	2.98	2.92	2.87	2.82	2.72	2.62	2.55	2.50	2.44	2.41	2.33
18		5.98	4.56	3.95	3.61	3.38	3.22	3.10	3.01	2.93	2.87	2.81	2.77	2.67	2.56	2.49	2.44	2.38	2.35	2.27
19		5.92	4.51	3.90	3.56	3.33	3.17	3.05	2.96	2.88	2.82	2.76	2.72	2.62	2.51	2.44	2.39	2.33	2.30	2.22
20		5.87	4.46	3.86	3.51	3.29	3.13	3.01	2.91	2.84	2.77	2.72	2.68	2.57	2.46	2.40	2.35	2.29	2.25	2.17
21		5.83	4.42	3.82	3.48	3.25	3.09	2.97	2.87	2.80	2.73	2.68	2.64	2.53	2.42	2.36	2.31	2.25	2.21	2.13
22		5.79	4.38	3.78	3.44	3.22	3.05	2.93	2.84	2.76	2.70	2.65	2.60	2.50	2.39	2.32	2.27	2.21	2.17	2.09
23		5.75	4.35	3.75	3.41	3.18	3.02	2.90	2.81	2.73	2.67	2.62	2.57	2.47	2.36	2.29	2.24	2.18	2.14	2.06
24		5.72	4.32	3.72	3.38	3.15	2.99	2.87	2.78	2.70	2.64	2.59	2.54	2.44	2.33	2.26	2.21	2.15	2.11	2.02
25		5.69	4.29	3.69	3.35	3.13	2.97	2.85	2.75	2.68	2.61	2.56	2.51	2.41	2.30	2.23	2.18	2.12	2.08	2.00
30		5.57	4.18	3.59	3.25	3.03	2.87	2.75	2.65	2.57	2.51	2.46	2.41	2.31	2.20	2.12	2.07	2.01	1.97	1.88
40		5.42	4.05	3.46	3.13	2.90	2.74	2.62	2.53	2.45	2.39	2.33	2.29	2.18	2.07	1.99	1.94	1.88	1.83	1.74
50		5.34	3.97	3.39	3.05	2.83	2.67	2.55	2.46	2.38	2.32	2.26	2.22	2.11	1.99	1.92	1.87	1.80	1.75	1.66
100		5.18	3.83	3.25	2.92	2.70	2.54	2.42	2.32	2.24	2.18	2.12	2.08	1.97	1.85	1.77	1.71	1.64	1.59	1.48

Degrees of Freedom for the Numerator

Degrees of Freedom for the Denominator

TABLE X (*Continued*)

Area in the Right Tail under the *F* Distribution Curve = .05

	1	2	3	4	5	6	7	8	9	10	11	12	15	20	25	30	40	50	100
1	161.5	199.5	215.7	224.6	230.2	234.0	236.8	238.9	240.5	241.9	243.0	243.9	246.0	248.0	249.3	250.1	251.1	251.8	253.0
2	18.51	19.00	19.16	19.25	19.30	19.33	19.35	19.37	19.38	19.40	19.40	19.41	19.43	19.45	19.46	19.46	19.47	19.48	19.49
3	10.13	9.55	9.28	9.12	9.01	8.94	8.89	8.85	8.81	8.79	8.76	8.74	8.70	8.66	8.63	8.62	8.59	8.58	8.55
4	7.71	6.94	6.59	6.39	6.26	6.16	6.09	6.04	6.00	5.96	5.94	5.91	5.86	5.80	5.77	5.75	5.72	5.70	5.66
5	6.61	5.79	5.41	5.19	5.05	4.95	4.88	4.82	4.77	4.74	4.70	4.68	4.62	4.56	4.52	4.50	4.46	4.44	4.41
6	5.99	5.14	4.76	4.53	4.39	4.28	4.21	4.15	4.10	4.06	4.03	4.00	3.94	3.87	3.83	3.81	3.77	3.75	3.71
7	5.59	4.74	4.35	4.12	3.97	3.87	3.79	3.73	3.68	3.64	3.60	3.57	3.51	3.44	3.40	3.38	3.34	3.32	3.27
8	5.32	4.46	4.07	3.84	3.69	3.58	3.50	3.44	3.39	3.35	3.31	3.28	3.22	3.15	3.11	3.08	3.04	3.02	2.97
9	5.12	4.26	3.86	3.63	3.48	3.37	3.29	3.23	3.18	3.14	3.10	3.07	3.01	2.94	2.89	2.86	2.83	2.80	2.76
10	4.96	4.10	3.71	3.48	3.33	3.22	3.14	3.07	3.02	2.98	2.94	2.91	2.85	2.77	2.73	2.70	2.66	2.64	2.59
11	4.84	3.98	3.59	3.36	3.20	3.09	3.01	2.95	2.90	2.85	2.82	2.79	2.72	2.65	2.60	2.57	2.53	2.51	2.46
12	4.75	3.89	3.49	3.26	3.11	3.00	2.91	2.85	2.80	2.75	2.72	2.69	2.62	2.54	2.50	2.47	2.43	2.40	2.35
13	4.67	3.81	3.41	3.18	3.03	2.92	2.83	2.77	2.71	2.67	2.63	2.60	2.53	2.46	2.41	2.38	2.34	2.31	2.26
14	4.60	3.74	3.34	3.11	2.96	2.85	2.76	2.70	2.65	2.60	2.57	2.53	2.46	2.39	2.34	2.31	2.27	2.24	2.19
15	4.54	3.68	3.29	3.06	2.90	2.79	2.71	2.64	2.59	2.54	2.51	2.48	2.40	2.33	2.28	2.25	2.20	2.18	2.12
16	4.49	3.63	3.24	3.01	2.85	2.74	2.66	2.59	2.54	2.49	2.46	2.42	2.35	2.28	2.23	2.19	2.15	2.12	2.07
17	4.45	3.59	3.20	2.96	2.81	2.70	2.61	2.55	2.49	2.45	2.41	2.38	2.31	2.23	2.18	2.15	2.10	2.08	2.02
18	4.41	3.55	3.16	2.93	2.77	2.66	2.58	2.51	2.46	2.41	2.37	2.34	2.27	2.19	2.14	2.11	2.06	2.04	1.98
19	4.38	3.52	3.13	2.90	2.74	2.63	2.54	2.48	2.42	2.38	2.34	2.31	2.23	2.16	2.11	2.07	2.03	2.00	1.94
20	4.35	3.49	3.10	2.87	2.71	2.60	2.51	2.45	2.39	2.35	2.31	2.28	2.20	2.12	2.07	2.04	1.99	1.97	1.91
21	4.32	3.47	3.07	2.84	2.68	2.57	2.49	2.42	2.37	2.32	2.28	2.25	2.18	2.10	2.05	2.01	1.96	1.94	1.88
22	4.30	3.44	3.05	2.82	2.66	2.55	2.46	2.40	2.34	2.30	2.26	2.23	2.15	2.07	2.02	1.97	1.94	1.91	1.85
23	4.28	3.42	3.03	2.80	2.64	2.53	2.44	2.37	2.32	2.27	2.24	2.20	2.13	2.05	2.00	1.96	1.91	1.88	1.82
24	4.26	3.40	3.01	2.78	2.62	2.51	2.42	2.36	2.30	2.25	2.22	2.18	2.11	2.03	1.97	1.94	1.89	1.86	1.80
25	4.24	3.39	2.99	2.76	2.60	2.49	2.40	2.34	2.28	2.24	2.20	2.16	2.09	2.01	1.96	1.92	1.87	1.84	1.78
30	4.17	3.32	2.92	2.69	2.53	2.42	2.33	2.27	2.21	2.16	2.13	2.09	2.01	1.93	1.88	1.84	1.79	1.76	1.70
40	4.08	3.23	2.84	2.61	2.45	2.34	2.25	2.18	2.12	2.08	2.04	2.00	1.92	1.84	1.78	1.74	1.69	1.66	1.59
50	4.03	3.18	2.79	2.56	2.40	2.29	2.20	2.13	2.07	2.03	1.99	1.95	1.87	1.78	1.73	1.69	1.63	1.60	1.52
100	3.94	3.09	2.70	2.46	2.31	2.19	2.10	2.03	1.97	1.93	1.89	1.85	1.77	1.68	1.62	1.57	1.52	1.48	1.39

Degrees of Freedom for the Numerator

Degrees of Freedom for the Denominator

TABLE X   THE F DISTRIBUTION TABLE

851

TABLE X (*Continued*)

Area in the Right Tail under the F Distribution Curve = .10

	Degrees of Freedom for the Numerator																		
	1	2	3	4	5	6	7	8	9	10	11	12	15	20	25	30	40	50	100
1	39.86	49.50	53.59	55.83	57.24	58.20	58.91	59.44	59.86	60.19	60.47	60.71	61.22	61.74	62.05	62.26	62.53	62.69	63.01
2	8.53	9.00	9.16	9.24	9.29	9.33	9.35	9.37	9.38	9.39	9.40	9.41	9.42	9.44	9.45	9.46	9.47	9.47	9.48
3	5.54	5.46	5.39	5.34	5.31	5.28	5.27	5.25	5.24	5.23	5.22	5.22	5.20	5.18	5.17	5.17	5.16	5.15	5.14
4	4.54	4.32	4.19	4.11	4.05	4.01	3.98	3.95	3.94	3.92	3.91	3.90	3.87	3.84	3.83	3.82	3.80	3.80	3.78
5	4.06	3.78	3.62	3.52	3.45	3.40	3.37	3.34	3.32	3.30	3.28	3.27	3.24	3.21	3.19	3.17	3.16	3.15	3.13
6	3.78	3.46	3.29	3.18	3.11	3.05	3.01	2.98	2.96	2.94	2.92	2.90	2.87	2.84	2.81	2.80	2.78	2.77	2.75
7	3.59	3.26	3.07	2.96	2.88	2.83	2.78	2.75	2.72	2.70	2.68	2.67	2.63	2.59	2.57	2.56	2.54	2.52	2.50
8	3.46	3.11	2.92	2.81	2.73	2.67	2.62	2.59	2.56	2.54	2.52	2.50	2.46	2.42	2.40	2.38	2.36	2.35	2.32
9	3.36	3.01	2.81	2.69	2.61	2.55	2.51	2.47	2.44	2.42	2.40	2.38	2.34	2.30	2.27	2.25	2.23	2.22	2.19
10	3.29	2.92	2.73	2.61	2.52	2.46	2.41	2.38	2.35	2.32	2.30	2.28	2.24	2.20	2.17	2.16	2.13	2.12	2.09
11	3.23	2.86	2.66	2.54	2.45	2.39	2.34	2.30	2.27	2.25	2.23	2.21	2.17	2.12	2.10	2.08	2.05	2.04	2.01
12	3.18	2.81	2.61	2.48	2.39	2.33	2.28	2.24	2.21	2.19	2.17	2.15	2.10	2.06	2.03	2.01	1.99	1.97	1.94
13	3.14	2.76	2.56	2.43	2.35	2.28	2.23	2.20	2.16	2.14	2.12	2.10	2.05	2.01	1.98	1.96	1.93	1.92	1.88
14	3.10	2.73	2.52	2.39	2.31	2.24	2.19	2.15	2.12	2.10	2.07	2.05	2.01	1.96	1.93	1.91	1.89	1.87	1.83
15	3.07	2.70	2.49	2.36	2.27	2.21	2.16	2.12	2.09	2.06	2.04	2.02	1.97	1.92	1.89	1.87	1.85	1.83	1.79
16	3.05	2.67	2.46	2.33	2.24	2.18	2.13	2.09	2.06	2.03	2.01	1.99	1.94	1.89	1.86	1.84	1.81	1.79	1.76
17	3.03	2.64	2.44	2.31	2.22	2.15	2.10	2.06	2.03	2.00	1.98	1.96	1.91	1.86	1.83	1.81	1.78	1.76	1.73
18	3.01	2.62	2.42	2.29	2.20	2.13	2.08	2.04	2.00	1.98	1.95	1.93	1.89	1.84	1.80	1.78	1.75	1.74	1.70
19	2.99	2.61	2.40	2.27	2.18	2.11	2.06	2.02	1.98	1.96	1.93	1.91	1.86	1.81	1.78	1.76	1.73	1.71	1.67
20	2.97	2.59	2.38	2.25	2.16	2.09	2.04	2.00	1.96	1.94	1.91	1.89	1.84	1.79	1.76	1.74	1.71	1.69	1.65
21	2.96	2.57	2.36	2.23	2.14	2.08	2.02	1.98	1.95	1.92	1.90	1.87	1.83	1.78	1.74	1.72	1.69	1.67	1.63
22	2.95	2.56	2.35	2.22	2.13	2.06	2.01	1.97	1.93	1.90	1.88	1.86	1.81	1.76	1.73	1.70	1.67	1.65	1.61
23	2.94	2.55	2.34	2.21	2.11	2.05	1.99	1.95	1.92	1.89	1.87	1.84	1.80	1.74	1.71	1.69	1.66	1.64	1.59
24	2.93	2.54	2.33	2.19	2.10	2.04	1.98	1.94	1.91	1.88	1.85	1.83	1.78	1.73	1.70	1.67	1.64	1.62	1.58
25	2.92	2.53	2.32	2.18	2.09	2.02	1.97	1.93	1.89	1.87	1.84	1.82	1.77	1.72	1.68	1.66	1.63	1.61	1.56
30	2.88	2.49	2.28	2.14	2.05	1.98	1.93	1.88	1.85	1.82	1.79	1.77	1.72	1.67	1.63	1.61	1.57	1.55	1.51
40	2.84	2.44	2.23	2.09	2.00	1.93	1.87	1.83	1.79	1.76	1.74	1.71	1.66	1.61	1.57	1.54	1.51	1.48	1.43
50	2.81	2.41	2.20	2.06	1.97	1.90	1.84	1.80	1.76	1.73	1.70	1.68	1.63	1.57	1.53	1.50	1.46	1.44	1.39
100	2.76	2.36	2.14	2.00	1.91	1.83	1.78	1.73	1.69	1.66	1.64	1.61	1.56	1.49	1.45	1.42	1.38	1.35	1.29

Degrees of Freedom for the Denominator

ANSWERS TO SELECTED ODD-NUMBERED EXERCISES AND SELF-REVIEW TESTS.

(*Note:* Due to differences in rounding, the answers obtained by readers may differ slightly from the ones given in this Appendix.)

CHAPTER 1

1.5   a. sample   b. population   c. population   d. sample
1.11 a. pay   b. six observations   c. six elements
1.15 a. quantitative   b. quantitative   c. qualitative   d. quantitative
1.17 a. discrete   b. continuous   d. continuous
1.19 a. ratio   b. ratio   c. nominal   d. ratio
1.21 a. 13   b. 7   c. 169   d. 35
1.23 a. 61   b. 33   c. 3721   d. 555
1.25 a. 59   b. 819   c. 576   d. 7614   e. 1994
1.27 a. 88   b. 58   c. 855   d. 1590   e. 622
1.31 a. population   b. sample   c. sample   d. population
1.33 a. 56   b. 20   c. 3136   d. 586
1.35 a. 59   b. 2662   c. 1508   d. 24,884   e. 12,354

Self-Review Test

1. b.   2. c
3. a. sampling without replacement   b. sampling with replacement
4. a. qualitative   b. quantitative (discrete)   c. quantitative (continuous)
6. a. 76   b. 1348   c. 5776   d. 52   e. 946
7. a. 45   b. 112   c. 495   d. 975   e. 9855

CHAPTER 2

2.3   c. 30%   d. 70%   2.5 c. 52%   2.7 c. 53.3%
2.15 b. Yes, each class has a width of 13.   d. 58%
2.17 a. class limits: $1–$4, $5–$8, $9–$12, $13–$16, $17–$20
      b. class boundaries: $0.5, $4.5, $8.5, $12.5, $16.5, $20.5; width = $4
      c. class midpoints: $2.5, $6.5, $10.5, $14.5, $18.5
2.19 c. skewed   d. 16.7%   2.21 d. 16%   2.29 c. 5.55%   d. skewed to the right
2.31 c. 77.5%   2.37 c. 42%   e. about 50%
2.51 218, 245, 256, 329, 367, 383, 397, 404, 427, 433, 471, 523, 537,
      551, 563, 581, 592, 622, 636, 647, 655, 678, 689, 810, 841
2.61 a. time-series data   b. cross-section data
      c. time-series data   d. cross-section data
2.67 d. 15%   2.69 d. 62.5%   2.71 d. nonsymmetric   2.73 c.   10%
2.75 c. class width = 10

Self-Review Test

2. a. 5   b. 5   c. 12   d. 4.5   e. 9   f. 100   g. .08
4. a. cross-section data   b. time-series data   c. cross-section data

5. **c.** 30%   6. **c.** 37.5%   7. about 60%

9. 30, 33, 37, 42, 44, 46, 47, 49, 51, 53, 53, 56, 60, 67, 67, 71, 79

CHAPTER 3

3.5 mode   **3.9** mean = 2.50; median = 3.50; no mode   **3.11** mean = 565.92; median = 539

3.13 mean = 21.85; median = 20.95; no mode   **3.15** mean = 4.97; median = 4.68

3.17 mean = 194.30; median = 142.50; no mode   **3.19** mean = 82,878.71; median = 42,100; no mode

3.21 mean = 47.33; median =47; mode = 48

3.23 **a.** mean = 9132.78; median = 2000
      b. outlier = 70,000. When the outlier is dropped: mean = 1524.38; median = 1600. The mean changes by a larger amount
           when we drop the outlier.   **c.** median

3.25 mean = 3.15. The machine needs to be adjusted.   **3.27** combined mean = \$41.89

3.29 age of the sixth person = 48 years

3.31 mean for data set I = 9.40; mean for data set II = 18.80.
      The mean of the second data set is twice that of the first data set.

3.37 range = 21; σ^2 = 50.250; σ = 7.09;   CV = 283.60%

3.39 CV for data set I = 20%;   CV for data set II = 12.5%. The relative variation for data set I is higher.

3.41 $\bar{x}$ = 86;   deviations from the mean: $-4, 30, -21, -11, 6$

3.43 range = 11; s^2 = 10.788; s = 3.28;   CV = 49.18%

3.45 range = 17.54; s^2 = 38.126; s = 6.17;   CV = 5141.67%

3.47 range = 58; s^2 = 613.071; s = 24.76;   CV = 83.23%

3.49 range = 11; s^2 = 11.720; s = 3.42;   CV = 39.86%

3.51 range = 725; s^2 = 41,784.424; s = 204.41;   CV = 105.55%

3.53 CV for assets is 24.67%;   CV for price-earnings ratios is 25.81%;   CV for assets is lower.

3.55 s = .14. The machine needs to be adjusted.

3.57 s = 14.64 for both data sets

3.61 $\bar{x}$ = 9.40; s^2 = 37.711; s = 6.14   **3.63** $\bar{x}$ = 27.20; s^2 = 176.694; s = 13.29

3.65 $\bar{x}$ = 74.40; s^2 = 506.092; s = 22.50   **3.67** $\bar{x}$ = 14.50; s^2 = 18.205; s = 4.27

3.69 $\bar{x}$ = 36.80; s^2 = 597.714; s = 24.45   **3.71** $\bar{x}$ = 1.74; s^2 = .930; s = .96

3.75 75%; 84%; 88.89%   **3.77** 68%; 95%; 99.7%

3.79 **a.** at least 75%   **b.** at least 84%   **c.** at least 88.89%

3.81 **a.** **i.** at least 75%   **ii.** at least 88.89%   **b.** \$765 to \$1965

3.83 **a.** 95%   **b.** 68%   **c.** 99.7%

3.85 **a.** **i.** 68%   **ii.** 99.7%   **b.** 3 to 11 years old

3.91 **a.** Q_1 = 45.50; Q_2 = 48; Q_3 = 52   **b.** 48   c56%

3.93 **a.** Q_1 = 4; Q_2 = 6.5; Q_3 = 8.5   **b.** 6.5   **c.** 66.67%

3.95 **a.** Q_1 = 24; Q_2 = 27.5; Q_3 = 31   **b.** 29   **c.** 23.33%

3.97 **a.** Q_1 = 5.9; Q_2 = 13.35; Q_3 = 23.1   **b.** 11.2   **c.** 60%

3.99 no outlier   **3.101** Data are slightly skewed to the right.

3.103 The data set is slightly skewed to the right.   **3.105** The data set seems to be close to symmetric.

3.107 **a.** mean = 5483.8; median = 1771.5; mode = 4000
      b. outlier = 34,899; mean = 2215.44; median = 1500; mean changes by a larger amount
      c. median

3.109 **a.** mean = 344.43; median = 311; no mode
b. range = 673; variance = 68,540.952; standard deviation = 261.80; CV = 76.01%

3.111 **a.** mean = 44.88; median = 43.5; no mode
b. range = 81; s^2 = 557.268; s = 23.61; CV = 52.61%

3.113 μ = 9.08; σ^2 = 7.760; σ = 2.79; These are population parameters.

3.115 $\bar{x}$ = 14.74; s^2 = 16.905; s = 4.11; These are sample statistics.

3.117 **a.** **i.** at least 75% **ii.** at least 88.89% **b.** 160 to 240 minutes

3.119 **a.** **i.** 68% **ii.** 95% **b.** 140 to 260 minutes

3.121 **a.** Q_1 = 2.6; Q_2 = 5.15; Q_3 = 10.0 **b.** 25th percentile is 2.05 **c.** 60%

3.123 The data set is slightly skewed to the rght.

Self-Review Test

1. b **2.** a and d **3.** c **4.** c **5.** b **6.** b **7.** a
8. a **9.** b **10.** a **11.** b **12.** c **13.** a **14.** a
15. mean = 9.8; median = 8; mode = 6; range = 20; s^2 = 42.178; s = 6.49; CV = 66.22%
19. **b.** $\bar{x}$ = 18.98; s^2 = 44.760; s = 6.69
20. **a.** **i.** at least 84% **ii.** at least 88.89% **b.** 2.9 to 11.7 years
21. **a.** **i.** 68% **ii.** 99.7% **c.** 2.9 to 11.7 years
22. **a.** Q_1 = 6; Q_2 = 10.5; Q_3 = 16.5 **b.** 15 **c.** 68.75%
23. Data are slightly skewed to the right. **24.** $466.43 **25.** $18,136
26. mean = 417.4; trimmed mean = 197.75. Trimmed mean is a better measure.
27. **a.** $\bar{x}_I$ = 19.75; $\bar{x}_{II}$ = 16.75. The mean of the second data set is equal to the mean of the first data set minus 3.
b. s = 11.32 for both data sets

CHAPTER 4

4.3 S = {AB, AC, BA, BC, CA, CB} **4.5** S = {RR, RG, GR, GG}; four possible outcomes

4.7 S = {DD, DG, GD, GG}; four possible outcomes **4.9** S = {$HHH, HHT, HTH, HTT, THH, THT, TTH, TTT$}

4.11 **a.** {RG, GR}: a compound event **b.** {RG, GR, RR}; a compound event
c. {RG, GR, RR}; a compound event **d.** {GR}; a simple event

4.13 **a.** {DG, GD, GG}; a compound event **b.** {DG, GD}; a compound event
c. {GD}; a simple event **d.** {DD, DG, GD}; a compound event

4.19 $-.35, 1.56, 5/3, -2/7$ **4.21** 1/8

4.23 not equally likely events; use relative frequency approach **4.25** subjective probability

4.27 .540 **4.29** .381; .619 **4.31** **a.** .115 **b.** .885 **4.33** .300; .700

4.35 **a.** .190 **b.** .266 **4.41** $P(A) + P(\bar{A})$ = 1.0 **4.43** 2^{10} = 1024

4.45 **a.** yes **b.** no
c. $\bar{A}$ = {1, 2, 5, 7, 8, 10}; $\bar{B}$ = {3, 4, 6, 7, 8, 9, 10}; $P(\bar{A})$ = 6/10; $P(\bar{B})$ = 7/10

4.47 288 **4.49** 960 **4.51** **a.** .65 **b.** .25 **c.** .733 **d.** .60

4.53 **a.** **i.** .40 **ii.** .75 **iii.** .75 **iv.** .40
b. "Men" and "yes" are not mutually exclusive events.
"Yes" and "no" are mutually exclusive events.
c. "Women" and "yes" are independent events.

4.55 **a.** **i.** .505 **ii.** .350 **iii.** .300 **iv.** .183

b. "Better off" and "high school education" are not mutually exclusive events.
"Less than high school" and "more than high school" are mutually exclusive events.

c. "Worse off" and "more than high school" are not independent events.

4.57 "Female" and "business major" are not independent events.
"Female" and "business major" are not mutually exclusive events.

4.59 $P(A) = .333$; $P(\overline{A}) = .667$ **4.61** .35   **4.65** zero

4.67 a. .454   b. .042   **4.69** a. .172   b. .182   **4.71** a. .196   b. .099

4.73 .692   **4.75** .725

4.77 a. i. .30   ii. .15   b. $P(M \text{ and } Y) = .45$; $P(W \text{ and } N) = .10$

4.79 a. i. .225   ii. .035   b. This probability is zero because these two events are mutually exclusive.

4.81 a. i. .200   ii. .550

b. This probability is zero because these two events are mutually exclusive.

c. $P(\text{incurred losses in 91 and earned profit in 92}) = .100$
$P(\text{incurred losses in 91 and incurred losses in 92}) = .150$

4.83 .357   **4.85** .253   **4.87** a. .09   b. .49   **4.89** .0025   **4.91** .000125

4.97 a. .54   b. .86   **4.99** a. .42   b. .47   **4.101** a. .85   b. .70

4.103 a. .78   b. .55   c. .79   **4.105** a. .90   b. .85   c. 1.00   **4.107** .28

4.109 .81   **4.111** .88   **4.113** b. .392

4.115 $P(A_1|B) = .645$; $P(A_2|B) = .355$; $P(A_1|C) = .682$; $P(A_2|C) = .318$

4.117 .286   **4.119** .586   **4.121** .104

4.123 a. .625   b. .375   c. .333   d. .200   e. .125   f. .875

4.127 $P(A \text{ and } B) = 0$; $P(A|B) = 0$; A and B are mutually exclusive but not independent events.

4.129 There are four possible outcomes.

4.131 a. {MS, SM, SS}; a compound event
b. {MM, MS, SM}; a compound event
c. {MS, SM}; a compound event

4.133 a. .35   b. .20

4.135 a. i. .290   ii. .350   iii. .690   iv. .371   v. .160   vi. .840
b. "Woman" and "not covered" are not mutually exclusive events.
"Covered" and "not covered" are mutually exclusive events.
c. "Covered" and "man" are not independent events.

4.137 a. i. .750   ii. .700   iii. .225   iv. .775   b. not independent; not mutually exclusive

4.139 a. .397   b. .603   **4.141** .159   **4.143** .027   **4.145** .311

Self-Review Test

1. a   2. b   3. c   4. a   5. a   6. b

7. c   8. b   9. b   10. c   11. b

12. a. .333   b. .667

13. a. The two events are neither independent nor mutually exclusive.
b. i. .400   ii. .679

14. .78   15. .313   16. .116

17. a. i. .460   ii. .250   iii. .060   iv. .680   v. .000   vi. .600
b. These events are neither independent nor mutually exclusive.

18. b. .534

CHAPTER 5

5.3   **a.** discrete   **b.** continuous   **c.** continuous   **d.** discrete   **e.** discrete   **f.** continuous

5.5   discrete random variable

5.7   **1.** $0 \le P(x) \le 1$ for each value of x **2.** $\Sigma P(x) = 1$

5.9   **a.** This is not a valid probablity distribution.
     b. This is not a valid probablity distribution.
     c. This is a valid probablity distribution.

5.11   **a.** .13   **b.** .16   **c.** .62   **d.** .38   **e.** .38   **f.** .31   **g.** .72

5.13   **b. i.** .10   **ii.** .45   **iii.** .75   **iv.** .45

5.15   **b. i.** .31   **ii.** .25   **iii.** .71   **iv.** .75

5.17   **a.**

x	2	3	4	5	6
$P(x)$	.12	.21	.34	.19	.14

     b. i. .21   **ii.** .88   **iii.** .67   **iv.** .33

5.19   **a.**

x	4	5	6	7	8	9	10
$P(x)$	.07	.18	.24	.21	.17	.09	.04

     b. i. .18   **ii.** .51   **iii.** .70   **iv.** .49

5.21   **a.** $\mu = 1.67$; $\sigma = .906$ **b.** $\mu = 7.19$; $\sigma = 1.102$

5.23 $\mu = 7.22$; $\sigma = 1.712$ **5.25** $\mu = 1.93$; $\sigma = 1.366$ **5.27** $\mu = .44$; $\sigma = .852$

5.29 $\mu = 1.896$; $\sigma = 1.079$ **5.31** $\mu = 2.376$; $\sigma = 1.464$ **5.33** $\mu = 38.25$; $\sigma = 18.660$

5.35 $3! = 6$; $(7 - 3)! = 24$; $9! = 362,880$; $(14 - 12)! = 2$; $\binom{5}{3} = 10$; $\binom{7}{4} = 35$; $\binom{9}{3} = 84$; $\binom{6}{0} = 1$; $\binom{3}{3} = 1$

5.37   15,504   **5.39**   220   **5.41**   125,970

5.45   **a.** not a binomial experiment   **b.** a binomial experiment   **c.** a binomial experiment

5.47   **a.** .2787   **b.** .0756   **c.** .2458   **5.49**   **b.** $\mu = 2.1$; $\sigma = 1.2124$

5.53   **a.** 0, 1, 2, 3, 4, 5, 6, 7, 8, 9, 10, 11, 12   **b.**   .0923

5.55   **a.** .0573   **b.** .6164   **c.** .4382   **5.57**   **a.** .0773   **b.** .3010   **c.** .1216   **d.** .0218

5.59   .0981   **5.61**   **a.** .0600   **b.** .3311   **c.** .278   **5.63**   **a.** $\mu = 5.6$; $\sigma = 1.296$ **b.** .0467

5.67   **a.** .0599   **b.** .0030   **c.** .2378   **5.69**   **a.** .0577   **b.** .0014   **c.** .0357

5.71   **a.** .3571   **b.** .1071   **c.** .6428   **5.73**   **a.** .5165   **b.** .4835   **5.77**   **a.** .1992;   **b.** .0771

5.79   **a.** $\mu = .6$; $\sigma = .7746$ **b.** $\mu = 1.8$; $\sigma = 1.3416$ **5.81**   .0978   **5.83**   .1008

5.85   **a.** .0998   **b. i.** .3301   **ii.** .0430   **iii.** .8571

5.87   **a.** .2681   **b. i.** .0615   **ii.** .0008   **iii.** .9384

5.89   **a.** .0050   **b. i.** .0062   **ii.** .4126   **5.91**   **a.** .5488   **c.** $\mu = .6$; $\sigma = .7746$

5.93   **a.** .26   **b.** .38   **c.** .28   **d.** .78   **5.95** $\mu = 1.88$; $\sigma = 1.1426$

5.97 $\mu = 2.04$; $\sigma = 1.1741$

5.99   **a.**

x	\$350	$-100,000$
$P(x)$	.998	.002

     b. $\mu = \$149.30$; $\sigma = \$4483.30$

5.101 $9! = 362,880$; $7! = 5040$; $(11 - 0)! = 39,916,800$
     $\binom{6}{3} = 20$; $\binom{11}{7} = 330$; $\binom{9}{0} = 1$

5.103   792 combinations   **5.105**   **a.** .0393   **b.** .3781   **c.** .4186

5.107   **a.** .5842   **b.** .0082   **c.** .0930   **5.109**   **a.** $\mu = 1.5$; $\sigma = 1.1619$ **b.** .0105

5.111   a.   .1174   b.   .0004   c.   .0139   **5.113**   a.   .4091   b.   .5455   c.   .0455

5.115   a.   .0915   b.   i.   .2833   ii.   .0996   **5.117**   a.   .0383

Self-Review Test

2.   probability distribution table   **3.**   a   **4.**   b   **5.**   a   **7.**   b   **8.**   a   **9.**   b

10.   a   **11.**   c   **13.**   a   **15.** $\mu = 1.98$; $\sigma = 1.3855$

16.   a.   i.   .0112   ii.   .0016   iii.   .9298   b. $\mu = 1$; $\sigma = .9487$

17.   a.   .3394   b.   .0020   c.   .5939   **18.**   a.   i.   .0361   ii.   .9473   iii.   .0526

CHAPTER 6

6.11   .8664   **6.13**   .9876

6.15   a.   .4713   b.   .4599   c.   .0967   d.   .0603   e.   .9429

6.17   a.   .0594   b.   .0244   c.   .9798   d.   .9686

6.19   a.   .5 approximately   b.   .5 approximately
          c.   .00 approximately   d.   .00 approximately

6.21   a.   .9626   b.   .4830   c.   .4706   d.   .0838

6.23   a.   .0207   b.   .2430   c.   .1841   d.   .9564

6.25   a.   .7823   b.   .8553   c.   .5 approximately   d.   .5 approximately   e.   .00 approximately
          f.   .00 approximately

6.27   a.   1.40   b. -2.20 c. -1.40 d.   2.80

6.29   a.   .4599   b.   .1210   c.   .2223

6.31   a.   .3336   b.   .9564   c.   .9564   d.   .00 approximately

6.33   a.   .2178   b.   .5997   **6.35**   a.   .7967   b.   .3372   c.   .0475   d.   .7734

6.37   a.   .2646   b.   .6368 or 63.68%   **6.39**   a.   .0618   b.   21.02%

6.41   a.   .6808   b.   12.83%   **6.43**   a.   1.04%   b.   1.04%

6.45   a.   .1711   b.   23.37%   c.   .44%   **6.47**   a.   .51%   b.   1.10%   **6.49**   2.64%

6.51   a.   2.00   b.   approximately -2.02 c.   approximately $-.37$ d.   approximately 1.02

6.53   a.   1.65   b. -1.96 c.   approximately -2.33 d.   approximately 2.58

6.55   a.   approximately 208.50   b.   approximately 241.25   c.   178.50   d.   145.75   e.   158.25
          f.   approximately 251.25

6.57   approximately 18.96 minutes   **6.59**   approximately 2060 kilowatt hours   **6.61**   approximately $62.02

6.63 $np > 5$ and $nq > 5$ **6.65**   a.   .6928   b.   .6922;   difference is .0006

6.67   a. $\mu = 72$; $\sigma = 5.367$ b.   .3897   c.   .3102

6.69   a.   .0764   b.   .6793   c.   .8413   d.   .8238

6.71   .1863   **6.73**   a.   .0451   b.   .3050   c.   .3911   **6.75**   a.   .0371   b.   .7517   c.   .3758

6.77   a.   .0692   b.   .4602   c.   .1630

6.81   b. $\mu = 50$; $\sigma = 11.547$ c.   .2500   d.   .1750   e.   .3000

6.83   b. $\mu = 17$; $\sigma = 1.732$ c.   .5000   d.   .3333   e.   .3333

6.85   b. $\mu = 12.50$; $\sigma = 4.330$ c.   .3333   d.   .2000   e.   .2667

6.89   a.   .0498   b.   .9817   c.   .1329   **6.91**   .2636

6.93   a.   .1353   b.   .2986   **6.95**   a.   .0357   b.   .7364

6.97   a.   .2142   b.   .5710   c.   .6915   d.   .6915

6.99 a. 3.59%   b. .1104   c. .7333

6.101 a. .0359   b. 2.35%   c. 29.02%   d. It is possible but its probability is close to zero.

6.103 1.24%

6.105 a. 2.81   b. approximately $-.82$ c. approximately -1.34 d. 2.41   e. .28   f. $-.55$

6.107 a. 663.35   b. 678.2   c. approximately 552.8   d. 585.8   e. approximately 546.2
         f. approximately 652.35

6.109 a. 848 hours   b. approximately 792 hours   **6.111** 16.23 ounces

6.113 a. .0151   b. .0465   c. .8340   d. .2540   **6.115** .1635

6.117 b. $\mu = 35$; $\sigma = 8.660$ c. .2000   d. .2667   e. .1333

6.119 a. .2865   b. .4866   c. .1710   **6.121** a. .0067   b. .9179   c. .0753

Self-Review Test

1. a   **2.** a   **3.** d   **4.** b   **5.** a   **6.** c   **7.** b   **8.** b   **9.** b

10. a. .1823   b. .9264   c. .1170   d. .7611

11. a. approximately -1.28 b. .61   c. approximately 1.65   d. approximately -1.07

12. a. 1.88%   b. .0475   c. .6911   d. 18.45%   e. almost impossible

13. a. approximately 17.604 ounces   b. approximately 18.307 ounces

14. 18.307 ounces

15. a. .0401   b. .2743   c. .6664   d. .2376   e. .7662

16. b. $\mu = 5.25$; $\sigma = .144$ c. .3000   d. .3000   e. .4000

17. a. .7135   b. .0067   c. .0586

CHAPTER 7

7.7 a. 15.1   b. sampling error $= -.43$ c. sampling error $= -.433$; nonsampling error $= 1.11$
       d. $\bar{x}_1 = 14.56$; $\bar{x}_2 = 14.89$; $\bar{x}_3 = 15.33$; $\bar{x}_4 = 14.67$; $\bar{x}_5 = 15.78$; $\bar{x}_6 = 15.11$; $\bar{x}_7 = 15.56$; $\bar{x}_8 = 16$;
       $\bar{x}_9 = 14.89$; $\bar{x}_{10} = 14.22$

7.9 b. $\bar{x}_1 = 36$; $\bar{x}_2 = 35.2$; $\bar{x}_3 = 34$; $\bar{x}_4 = 33.4$; $\bar{x}_5 = 28.4$; $\bar{x}_6 = 28$ c. $\mu = 32.5$

7.15 a. $\mu_{\bar{x}} = 60$; $\sigma_{\bar{x}} = 2.828$ b. $\mu_{\bar{x}} = 60$; $\sigma_{\bar{x}} = 1.265$

7.17 a. $\sigma_{\bar{x}} = 1.120$ b. $\sigma_{\bar{x}} = 1.414$ c. $\sigma_{\bar{x}} = .849$ d. $\sigma_{\bar{x}} = 2.000$

7.19 a. $n = 25$ b. $n = 64$ **7.21** $\mu_{\bar{x}} = 34.3$; $\sigma_{\bar{x}} = .632$ **7.23** $\mu_{\bar{x}} = \$49,600$; $\sigma_{\bar{x}} = \$233.333$

7.25 $n = 400$ **7.27** a. $\mu_{\bar{x}} = 30.8$ b. $\sigma_{\bar{x}} = 3.747$ d. $\sigma_{\bar{x}} = 3.745$

7.35 $\mu_{\bar{x}} = 68$ miles per hour; $\sigma_{\bar{x}} = .671$ miles per hour; the normal distribution

7.37 $\mu_{\bar{x}} = 54$ years; $\sigma_{\bar{x}} = .722$ years; the normal distribution

7.39 $\mu_{\bar{x}} = \$70$; $\sigma_{\bar{x}} = \$2.635$; the normal distribution

7.41 $\mu_{\bar{x}} = \$45,360$; $\sigma_{\bar{x}} = \$659.640$; the normal distribution

7.43 86.64%   **7.45** a. $z = 2.44$ b. $z = -7.25$ c. $z = -3.65$ d. $z = 5.82$

7.47 a. .1530   b. .7611   **7.49** a. .0880   b. .9924   c. .0099

7.51 a. .0000 approximately   b. .9505

7.53 a. .0418   b. .2373   c. .9624   d. .0418

7.55 a. .9582   b. 1.0   c. .9936   d. .0344

7.57 .0124   **7.59** $p = .12$; $\hat{p} = .15$

7.61 7125 subjects in the population; 312 subjects in the sample   **7.63** sampling error $= -.06$

7.69 a. $\mu_{\hat{p}} = .18$; $\sigma_{\hat{p}} = .019$ b. $\mu_{\hat{p}} = .18$; $\sigma_{\hat{p}} = .014$

7.71 a. $\sigma_{\hat{p}} = .051$ b. $\sigma_{\hat{p}} = .070$ c. $\sigma_{\hat{p}} = .043$ d. $\sigma_{\hat{p}} = .062$

7.75 a. $p = .667$ b. 6   d. $-.067, -.067, .133, .133, -.067, -.067$

7.77 $\mu_{\hat{p}} = .31$; $\sigma_{\hat{p}} = .021$; the normal distribution

7.79 $\mu_{\hat{p}} = .65$; $\sigma_{\hat{p}} = .067$; the normal distribution   7.81 .9544

7.83 a. $z = -.82$ b. $z = 1.63$ c. $z = -1.43$ d. $z = 1.02$

7.85 a. .8949   b. .1711   7.87 a. .1426   b. .9429   7.89 .2033

7.91 $\mu_{\bar{x}} = 750$; $\sigma_{\bar{x}} = 10$; the normal distribution

7.93 $\mu_{\bar{x}} = \$98,401$; $\sigma_{\bar{x}} = \$1250$; the normal distribution

7.95 a. .9332   b. .1525   c. .8664   d. .0228

7.97 a. .8997   b. .5530   c. .7698   d. .2119

7.99 a. .0125   b. .8861   c. .7850   d. .0064

7.101 .0124   7.103 $\mu_{\hat{p}} = .70$; $\sigma_{\hat{p}} = .023$; the normal distribution

7.105 a. i. .1587   ii. .1935   b. .8384   c. .1151

7.107 a. .9544   b. .0456   c. .0062   d. .0030

Self-Review Test

1. b   2. b   3. a   4. a   5. b   6. b

7. c   8. a   9. b   10. a   11. c   12. a

14. a. $\mu_{\bar{x}} = 4.1$; $\sigma_{\bar{x}} = .142$; the normal distribution
    b. $\mu_{\bar{x}} = 4.1$; $\sigma_{\bar{x}} = .090$; the normal distribution
    c. $\mu_{\bar{x}} = 4.1$; $\sigma_{\bar{x}} = .045$; the normal distribution

15. a. $\mu_{\bar{x}} = \$12,570$; $\sigma_{\bar{x}} = \$1529.470$; shape is unknown
    b. $\mu_{\bar{x}} = \$12,570$; $\sigma_{\bar{x}} = \$1140$; approximately normal distribution
    c. $\mu_{\bar{x}} = \$12,570$; $\sigma_{\bar{x}} = \$764.735$; approximately normal distribution

16. a. i. .1203   ii. .1335   iii. .7486   b. .9736   c. .0013

17. a. i. .6967   ii. .0455   iii. .6480   b. .9198   c. .0143

18. a. $\mu_{\hat{p}} = .78$; $\sigma_{\hat{p}} = .059$; the normal distribution
    b. $\mu_{\hat{p}} = .78$; $\sigma_{\hat{p}} = .029$; the normal distribution
    c. $\mu_{\hat{p}} = .78$; $\sigma_{\hat{p}} = .014$; the normal distribution

19. a. i. .0764   ii. .8949   iii. .8289   iv. .1624   b. .9826   c. .0287   d. .0764

CHAPTER 8

8.11 a. 22.5   b. ±.83   c. 21.40 to 23.60   d. 1.10

8.13 a. 71.33 to 78.27   b. 70.68 to 78.92   c. 69.38 to 80.22   d. yes

8.15 a. 76.19 to 81.61   b. 77.09 to 80.71   c. 77.27 to 80.53   d. yes

8.17 a. 14.53 to 17.47   b. 14.76 to 17.24   c. 14.96 to 17.04

8.19 a. 53.93 to 56.71   b. 56.16 to 58.64   c. 54.95 to 57.55
    d. Confidence intervals of parts a and c cover μ but the confidence interval of part b does not cover μ.

8.21 a. 38.20   b. ±.81   c. 37.30 to 39.10   d. .90

8.23 a. \$79,595 to \$81,849   b. \$64,093.82 to \$66,422.18

8.25 a. 203; margin of error $= \pm 12.88$ b. \$192.15 to \$213.85   8.27 a. 9.93 to 10.07

8.29 35.98 to 36.06. The machine needs to be adjusted.   8.31 a. \$79.77 to \$90.23

8.39 **a.** 1.771   **b.** -2.080 **c.** -3.646 **d.** 2.787

8.41 **a.** $\alpha = .025$; right tail   **b.** $\alpha = .001$; left tail   **c.** $\alpha = .01$; left tail   **d.** $\alpha = .005$; right tail

8.43 **a.** 2.921   **b.** 2.069   **c.** 1.729

8.45 **a.** 12.25   **b.** 9.71 to 14.79   **c.** 2.54

8.47 **a.** 60.92 to 70.08   **b.** 61.73 to 69.27   **c.** 61.95 to 69.05

8.49 46.07 to 47.93   **8.51** $104.34 to $135.66   **8.53** **a.** 24.71 to 28.09   **8.55** 62.24 to 101.56

8.61 **a.** yes, sample size is large   **b.** no, sample size is not large   **c.** yes, sample size is large
 d. no, sample size is not large

8.63 **a.** .47   **b.** $\pm.033$ **c.** .427 to .513   **d.** .043

8.65 **a.** .583 to .677   **b.** .542 to .638   **c.** .624 to .716
 d. Confidence intervals of parts a and c cover .65 but the confidence interval of part b does not.

8.67 **a.** .687 to .753   **b.** .681 to .759   **c.** .668 to .772   **d.** yes

8.69 **a.** .549 to .791   **b.** .620 to .720   **c.** .639 to .701   **d.** yes

8.71 **a.** .266 to .454   **b.** .236 to .484   **c.** .294 to .426

8.73 **a.** .34; margin of error $= \pm.019$ **b.** .315 to .365

8.75 **a.** .26; margin of error $= \pm.019$ **b.** .241 to .279

8.77 .728 to .952   **8.79** .391 to .809   **8.83** **a.** $n = 118$ **b.** $n = 46$

8.85 **a.** $n = 299$ **b.** $n = 126$ **c.** $n = 61$ **8.87** $n = 16$ **8.89** $n = 221$

8.91 **a.** $n = 1046$ **b.** $n = 1359$ **8.93** **a.** $n = 1849$ **b.** $n = 601$ **c.** $n = 6807$

8.95 $n = 4161$ **8.97** $n = 1153$ **8.99** **a.** $6.75; margin of error $= \pm$.16   **b.** $6.59 to $6.91

8.101 23.989 to 24.041. The machine needs to be adjusted.

8.103 **a.** $161.97; margin of error is ± 15.95 **b.** $143.01 to $180.93

8.107 25.15 to 32.65   **8.109** $314.41 to $365.59   **8.111** 6.93 to 16.69   **8.113** 1.92 to 2.54

8.115 **a.** .18; margin of error $= \pm.106$ **b.** .040 to .320   **8.117** .117 to .683

8.119 $n = 60$ **8.121** $n = 1068$

Self-Review Test

1. **a.** population parameter; sample statistic
 b. sample statistic; population parameter
 c. sample statistic; population parameter

2. b   3. a   4. a   5. d   6. b

7. **a.** $2.35; margin of error is $\pm.1712$ **b.** $2.12 to $2.58   8. $255,509.69 to $339,218.31

9. **a.** .15; margin of error is $\pm.031$ **b.** .119 to .181   10. $n = 41$

11. $n = 1849$ 12. $n = 782$

CHAPTER 9

9.5 **a.** a left-tailed test   **b.** a right-tailed test   **c.** a two-tailed test

9.7 **a.** Type II error   **b.** Type I error

9.9 **a.** $H_0: \mu = 143{,}000$; $H_1: \mu > 143{,}000$; a right-tailed test
 b. $H_0: \mu = 15$; $H_1: \mu \neq 15$; a two-tailed test
 c. $H_0: \mu = 45$; $H_1: \mu < 45$; a left-tailed test
 d. $H_0: \mu = 35$; $H_1: \mu > 35$; a right-tailed test
 e. $H_0: \mu = 50$; $H_1: \mu \neq 50$; a two-tailed test

9.15   a.   Rejection region is to the left of -1.96 and to the right of 1.96. Nonrejection region is between -1.96 and 1.96.
   b.   Rejection region is to the left of -2.33. Nonrejection region is to the right of -2.33.
   c.   Rejection region is to the right of 2.05. Nonrejection region is to the left of 2.05.

9.17   statistically significant   **9.19**   a.   .025   b.   .05   c.   .01

9.21   a.   observed value of z is 6.67;   critical value of z is 1.65
   b.   observed value of z is 6.67;   critical values of z are 1.96 and -1.96

9.23   a.   reject H_0 if $z < -1.65$ b.   reject H_0 if $z < -1.96$ or $z > 1.96$ c.   reject H_0 if $z > 1.65$

9.25   a.   critical values: $z = -2.58$ and 2.58;   test statistic: $z = -1.33$;   do not reject H_0
   b.   critical values: $z = -2.58$ and 2.58;   test statistic: $z = 3.20$;   reject H_0

9.27   a. H_0: $\mu = 25$; H_1: $\mu \neq 25$;   critical values: $z = -2.58$ and 2.58;   test statistic: $z = 9.00$;   reject H_0
   b. H_0: $\mu = 12$; H_1: $\mu < 12$;   critical value: $z = -1.65$;   test statistic: $z = -1.49$;   do not reject H_0
   c. H_0: $\mu = 40$; H_1: $\mu > 40$;   critical value: $z = 1.28$;   test statistic: $z = 8.57$;   reject H_0

9.29 H_0: $\mu = 2116$; H_1: $\mu > 2116$;   critical value: $z = 1.65$;   test statistic: $z = 32.71$;   reject H_0

9.31 H_0: $\mu = 120{,}000$; H_1: $\mu > 120{,}000$;   critical value: $z = 1.96$;   test statistic: $z = 6.69$;   reject H_0

9.33   a. H_0: $\mu = 185$; H_1: $\mu > 185$;   critical value: $z = 2.33$;   test statistic: $z = 26.87$;   reject H_0
   b. P(Type I error) $= .01$

9.35   a. H_0: $\mu = 9252$; H_1: $\mu \neq 9252$;   critical values: $z = -2.24$ and 2.24;   test statistic: $z = 1.78$;   do not reject H_0
   b.   do not reject H_0

9.37   a. H_0: $\mu = 45$; H_1: $\mu < 45$;   critical value: $z = -1.96$;   test statistic: $z = -1.87$;   do not reject H_0
   b.   critical value: $z = -1.65$;   test statistic: $z = -1.87$;   reject H_0

9.39   a. H_0: $\mu = 36$; H_1: $\mu \neq 36$;   critical values: $z = -2.58$ and 2.58;   test statistic: $z = 1.90$;   do not reject H_0

9.45   a. $\beta = .6368$ b. $\beta = .9382$

9.47 $\beta = .8944$;   power of the test $= 1 - .8944 = .1056$

9.49   a. $\beta = .9429$ b. $\beta = .9082$ c. $\beta = .8315$ **9.51** $\beta = .7190$

9.53   a. $\beta = .9115$ b.   power of the test $= .0885$ **9.55**   a. $\beta = .8980$ b.   power of the test $= .1020$

9.57   a. p-value $= .0046$ b. p-value $= .0017$ c. p-value $= .0162$

9.59   a. p-value $= .0204$ b.   yes, reject H_0 c.   no, do not reject H_0

9.61 p-value $= .0910$ **9.63** p-value $= .0202$

9.65   a. p-value $= .0574$ b.   do not reject H_0 for $\alpha = .02$;   do not reject H_0 for $\alpha = .05$

9.67   a. p-value $= .0158$ b.   do not reject H_0 for $\alpha = .01$ c.   reject H_0 for $\alpha = .025$

9.71   a.   reject H_0 if $t < -2.977$ or $t > 2.977$ b.   reject H_0 if $t < -2.797$ c.   reject H_0 if $t > 2.080$

9.73   a.   critical value: $t = -1.753$;   observed value: $t = -1.800$
   b.   critical values: $t = -2.131$ and 2.131;   observed value: $t = -1.800$

9.75   a.   reject H_0 if $t < -1.721$ b.   reject H_0 if either $t > 2.080$ or $t < -2.080$
   c.   reject H_0 if $t > 1.721$

9.77   a.   critical value: $t = 2.131$;   test statistic: $t = 4.000$;   reject H_0
   b.   critical value: $t = 2.131$;   test statistic: $t = 1.086$;   do not reject H_0

9.79   a. H_0: $\mu = 60$; H_1: $\mu \neq 60$;   critical values: $t = -2.160$ and 2.160;   test statistic: $t = -1.663$;   do not reject H_0
   b. H_0: $\mu = 35$; H_1: $\mu < 35$;   critical value: $t = -2.807$;   test statistic: $t = -5.443$;   reject H_0
   c. H_0: $\mu = 47$; H_1: $\mu > 47$;   critical value: $t = 3.646$;   test statistic: $t = 2.828$;   do not reject H_0

9.81 H_0: $\mu \leq 30$; H_1: $\mu > 30$;   critical value: $t = 2.602$;   test statistic: $t = 2.400$;   do not reject H_0

9.83 H_0: $\mu = 850$; H_1: $\mu < 850$;   critical value: $t = -1.711$;   test statistic: $t = -1.630$;   do not reject H_0

9.85   a. H_0: $\mu = \$43{,}370$; H_1: $\mu < \$43{,}370$;   critical value: $t = -2.528$;   test statistic: $t = -2.598$;   reject H_0
   b. $\alpha = .01$

9.87   a.   yes, the claim is true
   b. H_0: $\mu \geq 1200$; H_1: $\mu < 1200$;   critical value: $t = -1.711$;   test statistic: $t = -4.118$;   reject H_0

9.89 $H_0: \mu = 18$; $H_1: \mu \neq 18$;   critical values: $t = -2.262$ and 2.262;   test statistic: $t = 2.692$;   reject H_0

9.93   **a.** not large enough   **b.** large enough   **c.** not large enough   **d.** large enough

9.95   **a.** reject H_0 if $z < -1.65$ or $z > 1.65$ **b.** reject H_0 if $z < -2.33$
       c. reject H_0 if $z > 1.65$

9.97   **a.** critical value: $z = 1.65$;   observed value: $z = 3.41$
       b. critical values: $z = -1.96$ and 1.96;   observed value: $z = 3.41$

9.99   **a.** reject H_0 if $z < -1.65$ **b.** reject H_0 if $z < -1.96$ or $z > 1.96$ **c.** reject H_0 if $z > 1.65$

9.101   **a.** critical values: $z = -2.58$ and 2.58;   test statistic: $z = -1.60$;   do not reject H_0
       b. test statistic: $z = 3.21$;   reject H_0

9.103   **a.** critical values: $z = -1.65$ and 1.65;   test statistic: $z = .60$;   do not reject H_0
       b. critical value: $z = -1.65$;   test statistic: $z = -4.12$;   reject H_0
       c. critical value: $z = 2.33$;   test statistic: $z = 1.23$;   do not reject H_0

9.105 $H_0: p = .45$; $H_1: p < .45$;   critical value: $z = -1.96$;   test statistic: $z = -2.26$;   reject H_0

9.107 $H_0: p = .12$; $H_1: p > .12$;   critical value: $z = 2.33$;   test statistic: $z = 4.32$;   reject H_0

9.109   **a.** $H_0: p = .66$; $H_1: p \neq .66$;   critical values: $z = -2.33$ and 2.33;   test statistic: $z = 2.36$;   reject H_0
       b. $P(\text{Type I error}) = .02$

9.111   **a.** $H_0: p \geq .35$; $H_1: p < .35$;   critical value: $z = -1.96$;   test statistic: $z = -1.26$;   do not reject H_0
       b. do not reject H_0

9.113   **a.** $H_0: p \leq .05$; $H_1: p > .05$;   critical value: $z = 1.96$;   test statistic: $z = 2.27$;   reject H_0
       b. critical value: $z = 2.33$;   test statistic: $z = 2.27$;   do not reject H_0

9.117   **a.** critical value: $z = 1.96$;   test statistic: $z = 2.10$;   reject H_0
       b. $P(\text{Type I error}) = .025$
       c. p-value $= .0179$;   do not reject H_0 at $\alpha = .01$;   reject H_0 at $\alpha = .05$
       d. $\beta = .9131$;   power of the test $= .0869$

9.119   **a.** critical values: $z = -2.33$ and 2.33;   test statistic: $z = 2.23$;   do not reject H_0
       b. $P(\text{Type I error}) = .02$
       c. p-value $= .0258$; H_0 is not rejected at $\alpha = .025$ or $\alpha = .005$
       d. $\beta = .8708$;   power of the test $= .1292$

9.121   **a.** $H_0: \mu \geq 2500$; $H_1: \mu < 2500$;   critical value: $z = -1.96$;   test statistic: $z = -1.77$;   do not reject H_0
       b. $P(\text{Type I error}) = .025$ **c.** no

9.123   **a.** $H_0: \mu = 4271$; $H_1: \mu \neq 4271$;   critical values: $z = -2.24$ and 2.24;   test statistic: $z = 2.05$;   do not reject H_0
       b. critical values: $z = -2.81$ and 2.81;   test statistic: $z = 2.05$;   do not reject H_0

9.125 $H_0: \mu \leq 15$; $H_1: \mu > 15$;   critical value: $z = 2.33$;   test statistic: $z = 2.25$;   do not reject H_0;
       journalist's conclusion is not fair

9.127   **a.** $\beta = .6141$ **b.** $\beta = .0853$

9.129 $H_0: \mu = 354.32$; $H_1: \mu \neq 354.32$;   test statistic: $z = 2.27$; p-value $= .0232$

9.131 $H_0: \mu = 1035$; $H_1: \mu \neq 1035$;   critical values: $t = -2.779$ and 2.779;   test statistic: $t = 8.753$;   reject H_0

9.133   **a.** $H_0: \mu = 236,400$; $H_1: \mu > 236,400$;   critical value: $t = 1.729$;   test statistic: $t = 3.247$;   reject H_0
       b. yes;   do not reject H_0

9.135 $H_0: \mu \leq 2$; $H_1: \mu > 2$;   critical value: $t = 2.718$;   test statistic: $t = 1.679$;   do not reject H_0

9.137   **a.** $H_0: p = .56$; $H_1: p > .56$;   critical value: $z = 2.33$;   test statistic: $z = 4.85$;   reject H_0
       b. $P(\text{Type I error}) = .01$

9.139 $H_0: p = .4$; $H_1: p \neq .4$;   critical values: $z = -2.58$ and 2.58;   test statistic: $z = -2.16$;   do not reject H_0

9.141   **a.** $H_0: p = .80$; $H_1: p < .80$;   critical value: $z = -2.33$;   test statistic: $z = -.79$;   do not reject H_0
       b. do not reject H_0

9.143 $H_0: p = .40$; $H_1: p \neq .40$;   test statistic: $z = -2.16$; p-value $= .0308$;   reject H_0 at $\alpha = .05$;   do not reject H_0 at $\alpha = .02$

9.145 $\beta = .9370$

Self-Review Test

1. a   2. b   3. a   4. b   5. a   6. a   7. a   8. b

9. c   10. a   11. c   12. b   13. d   14. c   15. a   16. b

17. a. $H_0: \mu = 1365$; $H_1: \mu \neq 1365$;   critical values: $z = -2.58$ and 2.58;   test statistic: $z = 4.46$;   reject H_0
     b. $P(\text{Type I error}) = .01$ c. do not reject H_0

18. a. $H_0: \mu \geq 15$; $H_1: \mu < 15$;   critical value: $t = -2.131$;   test statistic: $t = -2.778$;   reject H_0
     b. $P(\text{Type I error}) = .025$ c. critical value: $t = -3.733$;   do not reject H_0

19. a. $H_0: p = .73$; $H_1: p > .73$;   critical value: $z = 2.33$;   test statistic: $z = 1.80$;   do not reject H_0
     b. $P(\text{Type I error}) = .01$ c. do not reject H_0

20. a. $H_0: \mu = 60$; $H_1: \mu \neq 60$;   test statistic: $z = 2.36$; p-value $= .0182$
     b. Do not reject H_0 at $\alpha = .01$;   reject H_0 at $\alpha = .05$

21. a. $\beta = .9852$;   power of the test $= .0148$ b. $\beta = .9066$;   power of the test $= .0934$

22. a. p-value $= .0359$ b. Do not reject H_0 at $\alpha = .01$;   reject H_0 at $\alpha = .05$

23. a. $\beta = .9292$;   power of the test $= .0708$ b. $\beta = .3483$;   power of the test $= .6517$

CHAPTER 10

10.3   a. .76   b. .29 to 1.23

10.5 $H_0: \mu_1 - \mu_2 = 0$; $H_1: \mu_1 - \mu_2 \neq 0$;   critical values: $z = -1.96$ and 1.96;   test statistic: $z = 4.19$;   reject H_0

10.7 $H_0: \mu_1 - \mu_2 = 0$; $H_1: \mu_1 - \mu_2 > 0$;   critical value: $z = 2.33$;   test statistic: $z = 4.19$;   reject H_0

10.9   a. -4.75 b. -6.68 to -2.82
     c. $H_0: \mu_1 - \mu_2 = 0$; $H_1: \mu_1 - \mu_2 \neq 0$;   critical values: $z = -2.58$ and 2.58;   test statistic: $z = -6.34$; do reject H_0

10.11   a. .11   b. $-.04$ to .26
     c. $H_0: \mu_1 - \mu_2 = 0$; $H_1: \mu_1 - \mu_2 > 0$;   critical value: $z = 2.33$;   test statistic: $z = 1.24$;   do not reject H_0

10.13   a. -3.64 to -2.16 days
     b. $H_0: \mu_1 - \mu_2 = 0$; $H_1: \mu_1 - \mu_2 < 0$;   critical value: $z = -1.96$;   test statistic: $z = -9.15$;   reject H_0
     c. $P(\text{Type I error}) = .025$

10.15   a. \$1.87 to \$2.15
     b. $H_0: \mu_1 - \mu_2 = 0$; $H_1: \mu_1 - \mu_2 > 0$;   critical value: $z = 1.65$;   test statistic: $z = 28.27$;   reject H_0
     c. Do not reject H_0

10.17   a. $-.51$ to .01 minutes
     b. $H_0: \mu_1 - \mu_2 = 0$; $H_1: \mu_1 - \mu_2 < 0$;   critical value: $z = -1.96$;   test statistic: $z = -2.06$;   reject H_0
     c. p-value $= .0197$;   do not reject H_0 at $\alpha = .01$;   reject H_0 at $\alpha = .05$

10.21   a. 5.25   b. 1.21 to 9.29

10.23 $H_0: \mu_1 - \mu_2 = 0$; $H_1: \mu_1 - \mu_2 \neq 0$;   critical values: $t = -2.701$ and 2.701;   test statistic: $t = 3.514$;   reject H_0

10.25 $H_0: \mu_1 - \mu_2 = 0$; $H_1: \mu_1 - \mu_2 > 0$;   critical value: $t = 1.683$;   test statistic: $t = 3.514$;   reject H_0

10.27   a. -5.32 b. -8.33 to -2.31
     c. $H_0: \mu_1 - \mu_2 = 0$; $H_1: \mu_1 - \mu_2 < 0$;   critical value: $t = -2.080$;   test statistic: $t = -4.996$;   reject H_0

10.29   a. 1.50 to 16.50 items
     b. $H_0: \mu_1 - \mu_2 = 0$; $H_1: \mu_1 - \mu_2 > 0$;   critical value: $t = 2.120$;   test statistic: $t = 2.543$;   reject H_0

10.31   a. $-\$1041.25$ to -38.75
     b. $H_0: \mu_1 - \mu_2 = 0$; $H_1: \mu_1 - \mu_2 \neq 0$;   critical values: $t = -2.009$ and 2.009;   test statistic: $t = -1.806$; do not reject H_0

10.33   a. -14.52 to 4.52 minutes
     b. $H_0: \mu_1 - \mu_2 = 0$; $H_1: \mu_1 - \mu_2 < 0$;   critical value: $t = -2.412$;   test statistic: $t = -1.413$;   do not reject H_0

10.35   **a.** $4909.87 to $7610.13

      b. $H_0: \mu_1 - \mu_2 = 0$; $H_1: \mu_1 - \mu_2 > 0$;   critical value: $t = 2.400$;   test statistic: $t = 7.766$;   reject H_0

10.37   **a.**   10.30 to 40.50   **b.**   11.26 to 15.14   **c.**   28.53 to 40.67

10.39   **a.**   critical values: $t = -1.860$ and 1.860;   test statistic: $t = 8.040$;   reject H_0

      b.   critical value: $t = 1.721$;   test statistic: $t = 10.847$;   reject H_0

      c.   critical value: $t = -2.583$;   test statistic: $t = -7.988$;   reject H_0

10.41   **a.** -2.93 to .07

      b. $H_0: \mu_d = 0$; $H_1: \mu_d < 0$;   critical value: $t = -3.143$;   test statistic: $t = -2.338$;   do not reject H_0

10.43   **a.** -16.54 to -3.21

      b. $H_0: \mu_d = 0$; $H_1: \mu_d < 0$;   critical value: $t = -1.895$;   test statistic: $t = -2.809$;   reject H_0

10.45   **a.** -4.07 to .63

      b. $H_0: \mu_d = 0$; $H_1: \mu_d < 0$;   critical value: $t = -2.571$;   test statistic: $t = -2.951$;   reject H_0

10.49 $-.18$ to .06

10.51 $H_0: p_1 - p_2 = 0$; $H_1: p_1 - p_2 \neq 0$;   critical values: $z = -2.58$ and 2.58;   test statistic: $z = -1.32$;   do not reject H_0

10.53 $H_0: p_1 - p_2 = 0$; $H_1: p_1 - p_2 < 0$;   critical value: $z = -2.33$;   test statistic: $z = -1.32$;   do not reject H_0

10.55   **a.**   .04   **b.** $-.02$ to .10   **c.**   reject H_0 if $z > 1.96$ **d.** $z = 1.35$ **e.**   do not reject H_0

10.57   **a.** $-.20$ to $-.12$

      b. $H_0: p_1 - p_2 = 0$; $H_1: p_1 - p_2 < 0$;   critical value: $z = -2.33$;   test statistic: $z = -10.67$;   reject H_0

10.59   **a.**   .03 to .11

      b. $H_0: p_1 - p_2 = 0$; $H_1: p_1 - p_2 > 0$;   critical value: $z = 1.65$;   test statistic: $z = 4.29$;   reject H_0

10.61   **a.**   .03   **b.** $-.02$ to .08

      c. $H_0: p_1 - p_2 = 0$; $H_1: p_1 - p_2 \neq 0$;   critical values: $z = -1.96$ and 1.96;   test statistic: $z = 1.10$;

        do not reject H_0

10.63   **a.** $-.07$ to .01

      b. $H_0: p_1 - p_2 = 0$; $H_1: p_1 - p_2 < 0$;   critical value: $z = -1.96$;   test statistic: $z = -1.57$;   do not reject H_0

      c. p-value $= .0582$

10.65   **a.** -19.71 to 22.51

      b. $H_0: \mu_1 - \mu_2 = 0$; $H_1: \mu_1 - \mu_2 \neq 0$;   critical values: $z = -1.96$ and 1.96;   test statistic: $z = .17$;

        do not reject H_0

10.67   **a.**   .60 to 2.40

      b. $H_0: \mu_1 - \mu_2 = 0$; $H_1: \mu_1 - \mu_2 > 0$;   critical value: $z = 2.33$;   test statistic: $z = 4.30$;   reject H_0

10.69   **a.** $H_0: \mu_1 - \mu_2 = 0$; $H_1: \mu_1 - \mu_2 \neq 0$;   critical values: $t = -2.009$ and 2.009;   test statistic: $t = -1.058$;

        do not reject H_0

      b. $-.42$ to .18

10.71   **a.**   .37 to 13.63

      b. $H_0: \mu_1 - \mu_2 = 0$; $H_1: \mu_1 - \mu_2 > 0$;   critical value: $t = 2.416$;   test statistic: $t = 1.774$;   do not reject H_0

10.73   **a.** $-.64$ to .14

      b. $H_0: \mu_d = 0$; $H_1: \mu_d < 0$;   critical value: $t = -1.895$;   test statistic: $t = -2.236$;   reject H_0

10.75   **a.** $-.21$ to $-.08$

      b. $H_0: p_1 - p_2 = 0$; $H_1: p_1 - p_2 < 0$;   critical value: $z = -2.33$;   test statistic: $z = -5.96$;   reject H_0

10.77   **a.** $-.17$ to $-.07$

      b. $H_0: p_1 - p_2 = 0$; $H_1: p_1 - p_2 < 0$;   critical value: $z = -1.65$;   test statistic: $z = -4.44$;   reject H_0

Self-Review Test

1.   a

3.   **a.**   1.62 to 2.78

     b. $H_0: \mu_1 - \mu_2 = 0$; $H_1: \mu_1 - \mu_2 > 0$;   critical value: $z = 1.65$;   test statistic: $z = 9.86$;   reject H_0

4.   **a.** -2.72 to -1.88 hours

    b. $H_0: \mu_1 - \mu_2 = 0$; $H_1: \mu_1 - \mu_2 < 0$;   critical value: $t = -2.416$;   test statistic: $t = -10.994$;   reject H_0

5.   **a.** -3.43 to 1.43 items

    b. $H_0: \mu_d = 0$; $H_1: \mu_d \neq 0$;   critical values: $t = -2.447$ and 2.447;   test statistic: $t = -1.527$;   do not reject H_0

6.   **a.** $-.04$ to $.10$

    b. $H_0: p_1 - p_2 = 0$; $H_1: p_1 - p_2 \neq 0$;   critical values: $z = -2.58$ and 2.58;   test statistic: $z = .90$;   do not reject H_0

CHAPTER 11

11.3 $\chi^2 = 43.773$ **11.5** $\chi^2 = 36.191$ **11.7**   **a.** $\chi^2 = 5.009$ **b.** $\chi^2 = 3.565$

11.13   critical value: $\chi^2 = 11.070$;   test statistic: $\chi^2 = 5.200$;   do not reject H_0

11.15   critical value: $\chi^2 = 7.378$;   test statistic: $\chi^2 = .865$;   do not reject H_0

11.17   critical value: $\chi^2 = 13.277$;   test statistic: $\chi^2 = 4.100$;   do not reject H_0

11.19   critical value: $\chi^2 = 5.991$;   test statistic: $\chi^2 = 128.448$;   reject H_0

11.21   critical value: $\chi^2 = 9.210$;   test statistic: $\chi^2 = 2.556$;   do not reject H_0

11.25   **a.** H_0: columns and rows are independent; H_1: columns and rows are dependent

    c.   critical value: $\chi^2 = 13.277$ **d.**   test statistic: $\chi^2 = 11.373$ **e.**   do not reject H_0

11.27   critical value: $\chi^2 = 6.635$;   test statistic: $\chi^2 = 4.854$;   do not reject H_0

11.29   critical value: $\chi^2 = 2.706$;   test statistic: $\chi^2 = .545$;   do not reject H_0

11.31   critical value: $\chi^2 = 3.841$;   test statistic: $\chi^2 = .124$;   do not reject H_0

11.33   critical value: $\chi^2 = 7.378$;   test statistic: $\chi^2 = 1.748$;   do not reject H_0

11.35   **a.** 19.0693 to 109.5388 **b.** 20.7363 to 91.0252 **c.** 23.1684 to 73.5425

11.37   **a.** $H_0: \sigma^2 = 11$; $H_1: \sigma^2 \neq 11$

    b.   reject H_0 if $\chi^2 < 8.907$ or $\chi^2 > 32.852$

    c.   test statistic: $\chi^2 = 24.182$

    d.   do not reject H_0

11.39   **a.** $H_0: \sigma^2 \geq 2.5$; $H_1: \sigma^2 < 2.5$

    b.   reject H_0 if $\chi^2 < 12.401$

    c.   test statistic: $\chi^2 = 17.280$

    d.   do not reject H_0

11.41   **a.** $.0069$ to $.0389$

    b. $H_0: \sigma^2 = .01$; $H_1: \sigma^2 > .01$;   critical value: $\chi^2 = 32.852$;   test statistic: $\chi^2 = 26.600$;   do not reject H_0

11.43   **a.** $.3504$ to 1.1412; $.592$ to 1.068

    b. $H_0: \sigma^2 = .30$; $H_1: \sigma^2 \neq .30$;   critical values: $\chi^2 = 9.260$ and 44.181;   test statistic: $\chi^2 = 44.467$;   reject H_0

11.45   critical value: $\chi^2 = 11.345$;   test statistic: $\chi^2 = 5.526$;   do not reject H_0

11.47   critical value: $\chi^2 = 14.449$;   test statistic: $\chi^2 = 92.533$;   reject H_0

11.49   critical value: $\chi^2 = 11.070$;   test statistic: $\chi^2 = 3.336$;   do not reject H_0

11.51   critical value: $\chi^2 = 5.024$;   test statistic: $\chi^2 = .699$;   do not reject H_0

11.53   critical value: $\chi^2 = 5.991$;   test statistic: $\chi^2 = 9.608$;   reject H_0

11.55   critical value: $\chi^2 = 16.812$;   test statistic: $\chi^2 = 10.181$;   do not reject H_0

11.57   **a.** $H_0: \sigma^2 = 4.0$; $H_1: \sigma^2 > 4.0$;   critical value: $\chi^2 = 42.980$;   test statistic: $\chi^2 = 47.400$;   reject H_0

    b. 4.162 to 19.179

11.59   **a.** $H_0: \sigma^2 \leq .025$; $H_1: \sigma^2 > .025$;   critical value: $\chi^2 = 33.924$;   test statistic: $\chi^2 = 27.280$;   do not reject H_0

    b. $.0185$ to $.0621$

11.61   **a.** $s^2 = 10.9524$

    b. 3.9087 to 75.3606; 1.977 to 8.681

    c. $H_0: \sigma^2 = 10$; $H_1: \sigma^2 \neq 10$;   critical values: $\chi^2 = .676$ and 18.548;   test statistic: $\chi^2 = 6.571$;   do not reject H_0

Self-Review Test

1. b   2. a   3. c   4. a   5. b   6. b   7. c   8. b   9. a

10. critical value: $\chi^2 = 15.086$;   test statistic: $\chi^2 = 11.449$;   do not reject H_0

11. critical value: $\chi^2 = 26.217$;   test statistic: $\chi^2 = 40.637$;   reject H_0

12. critical value: $\chi^2 = 9.488$;   test statistic: $\chi^2 = 82.844$;   reject H_0

13. a. .0316 to .1457;   .178 to .382
    b. H_0: $\sigma^2 = .03$; H_1: $\sigma^2 > .03$;   critical value: $\chi^2 = 42.980$;   test statistic: $\chi^2 = 48.000$;   reject H_0

CHAPTER 12

12.3   a. 3.89   b. 3.61   c. 4.67   12.5   a. 2.52   b. 2.35   c. 3.91

12.7   a. 7.79   b. 5.41   12.9   a. 4.85   b. 3.22

12.13   a. H_0: $\mu_1 = \mu_2 = \mu_3 = \mu_4$; H_1: all means are not equal
     b. reject H_0 if $F > 3.78$
     c. SSB = 67.8663;   SSW = 255.0952;   SST = 322.9615
     d. 3 and 22   e. MSB = 22.6221;   MSW = 11.5952   f. $F = 3.78$ g. $F = 1.95$ i. do not reject H_0

12.15   a. $\bar{x}_1 = 13.571$; $\bar{x}_2 = 10.714$; $s_1 = 2.8200$; $s_2 = 2.8702$
     b. H_0: $\mu_1 = \mu_2$; H_1: $\mu_1 \neq \mu_2$;   critical values: $t = -2.179$ and 2.179;   test statistic: $t = 1.878$;   do not reject H_0
     c. critical value: $F = 4.75$;   test statistic: $F = 3.53$;   do not reject H_0
     d. conclusions are the same

12.17   b. critical value: $F = 3.29$;   test statistic: $F = 4.07$;   reject H_0

12.19   critical value: $F = 5.95$;   test statistic: $F = 1.63$;   do not reject H_0

12.21   critical value: $F = 4.02$;   test statistic: $F = 1.29$;   do not reject H_0

12.23   critical value: $F = 3.72$;   test statistic: $F = 6.55$;   reject H_0

12.25   a. critical value: $F = 5.74$;   test statistic: $F = 5.52$;   do not reject H_0 b. .01

12.27   a. critical value: $F = 5.42$;   test statistic: $F = .43$;   do not reject H_0

12.29   a. critical value: $F = 4.62$;   test statistic: $F = 6.56$;   reject H_0 b. do not reject H_0

Self-Review Test

1. a   2. b   3. c   4. a   5. a   6. a   7. b   8. a

10. a. critical value: $F = 6.11$;   test statistic: $F = 2.31$;   do not reject H_0 b. Type II error

CHAPTER 13

13.15   a. y-intercept $= 100$;   slope $= 5$;   positive relationship
     b. y-intercept $= 400$;   slope $= -4$;   negative relationship

13.17 $\mu_{y|x} = -5.582 + .289x$ 13.19 $\hat{y} = -83.714 + 10.571x$

13.21   a. \$40.00   b. the same amount   c. exact relationship

13.23   a. \$26.50 million   b. different amounts   . c. nonexact relationship

13.25   a. $\hat{y} = 2.3182 + 2.1869x$ c. \$56.99

13.27   b. $\hat{y} = 12.1683 + .5902x$ e. 19.25 units   f. 26.92 units

13.29   a. $\mu_{y|x} = 29.4824 + 1.4803x$ b. population regression line because information is on all teams; values of A and B d. 49.32%

13.31 $n - 2$ 13.35 $\sigma_\epsilon = 7.0756$; $\rho^2 = .04$ 13.37 $s_e = 4.7117$; $r^2 = .99$

13.39   a. $s_e = 11.2659$ b. $r^2 = .88$ 13.41   a. $s_e = 32.6362$ b. $r^2 = .97$

13.43   a. $\sigma_\epsilon = 5.4433$ b. $\rho^2 = .41$

13.45   a.   6.01 to 6.63
       b. $H_0: B = 0$; $H_1: B > 0$;   critical value: $t = 2.145$;   test statistic: $t = 59.792$;   reject H_0
       c. $H_0: B = 0$; $H_1: B \neq 0$;   critical values: $t = -2.977$ and 2.977;   test statistic: $t = 59.792$;   reject H_0
       d. $H_0: B = 4.50$; $H_1: B \neq 4.50$;   critical values: $t = -2.624$ and 2.624;   test statistic: $t = 17.219$;   reject H_0

13.47   a.   2.35 to 2.65
       b. $H_0: B = 0$; $H_1: B > 0$;   critical value: $z = 2.05$;   test statistic: $z = 39.12$;   reject H_0
       c. $H_0: B = 0$; $H_1: B \neq 0$;   critical values: $z = -2.58$ and 2.58;   test statistic: $z = 39.12$;   reject H_0
       d. $H_0: B = 1.75$; $H_1: B > 1.75$;   critical value: $z = 2.33$;   test statistic: $z = 11.74$;   reject H_0

13.49   a. -15.5304 to -7.1708
       b. $H_0: B = 0$; $H_1: B < 0$;   critical value: $t = -1.943$;   test statistic: $t = -6.645$;   reject H_0

13.51   a. $\hat{y} = 10.4986 + .8689x$ b.   .5559 to 1.1819
       c. $H_0: B = 0$; $H_1: B > 0$;   critical value: $t = 2.365$;   test statistic: $t = 8.323$;   reject H_0

13.53   a.   14.076 to 35.048
       b. $H_0: B = 0$; $H_1: B \neq 0$;   critical values: $t = -2.776$ and 2.776;   test statistic: $t = 10.784$;   reject H_0

13.59   a

13.63   a.   negative   b.   positive   c.   positive   d.   negative   e.   zero   f.   negative

13.65 $\rho = .21$ 13.67 $r = -.996$ 13.69   a.   positively   b.   .95

13.71 $r = .84$;   it has the same sign as b   13.73 $\rho = .64$

13.75   a. $SS_{xx} = 2104.8571$; $SS_{yy} = 1488.8571$; $SS_{xy} = 1726.8571$
       b. $\hat{y} = 43.4975 + .8204x$ d. $r = .98$; $r^2 = .95$ e. $s_e = 3.7985$ f.   129.64   g.   .5418 to 1.0990
       h. $H_0: B = 0$; $H_1: B \neq 0$;   critical values: $t = -2.571$ and 2.571;   test statistic: $t = 9.908$;   reject H_0

13.77   a. $SS_{xx} = 3.3647$; $SS_{yy} = 120.0000$; $SS_{xy} = 18.6500$
       b. $\hat{y} = 7.7113 + 5.5430x$ d. $r = .93$; $r^2 = .86$ e. $s_e = 1.8234$ f.   2.1973 to 8.8883
       g. $H_0: B = 0$; $H_1: B \neq 0$;   critical values: $t = -2.571$ and 2.571;   test statistic: $t = 5.576$;   reject H_0

13.79   a. $SS_{xx} = 31.7083$; $SS_{yy} = 591.3333$; $SS_{xy} = -33.1667$
       b. $\hat{y} = 72.6546 - 1.0460x$ d. $r = -.24$; $r^2 = .06$ e. $s_e = 11.7966$ f. -5.5123 to 3.4203
       g. $H_0: B = 0$; $H_1: B < 0$;   critical value: $t = -2.776$;   test statistic: $t = -.499$;   do not reject H_0

13.81   a.   13.871 to 16.629;   11.765 to 18.735   b.   62.359 to 67.721;   56.362 to 73.718

13.83   18.108 to 20.267;   15.838 to 22.537   13.85   111.512 to 123.155;   100.949 to 133.718

13.87   22.748 to 27.596;   18.574 to 31.770

13.89   b. $\hat{y} = 18.4254 + 2.2266x$ e.   47.37%   f. $r = .75$; $r^2 = .56$ g. $s_e = 7.5219$
       h. -2.2885 to 6.7417
       i. $H_0: B = 0$; $H_1: B \neq 0$;   critical values: $t = -4.604$ and 4.604;   test statistic: $t = 2.270$;   do not reject H_0

13.91   a.   positive relationship   b. $\hat{y} = -1.4340 + .8916x$ d. $r = .97$; $r^2 = .95$ e. $s_e = .9285$
       f.   .5708 to 1.2124   g. $H_0: B = 0$; $H_1: B > 0$;   critical value: $t = 2.571$;   test statistic: $t = 9.356$;   reject H_0

13.93   a. $\hat{y} = 2.0680 + .0714x$ b.   \$3.85 million   c. $r = .96$; $r^2 = .92$ d. $s_e = .1537$ e.   .0511 to .0917
       f. $H_0: B = 0$; $H_1: B > 0$;   critical value: $t = 3.143$;   test statistic: $t = 8.602$;   reject H_0

13.95   a. $\hat{y} = -331.8426 + 5.7746x$ c. $r = .97$; $r^2 = .94$ d.   216.74 pounds   e. $s_e = 13.0268$
       f.   3.6627 to 7.8865
       g. $H_0: B = 0$; $H_1: B \neq 0$;   critical values: $t = -3.707$ and 3.707;   test statistic: $t = 10.136$;   reject H_0

13.97   b. $SS_{xx} = 143.0000$; $SS_{yy} = 448.6667$; $SS_{xy} = 236.0000$ d. $\hat{y} = 131.5901 + 1.6503x$ f. $r = .93$
       g.   158.00 pounds

13.99   4.283 to 7.115;   1.696 to 9.702   13.101   19.901 to 25.478;   16.546 to 28.832

Self-Review Test

1.   d   2.   a   3.   b   4.   a   5.   b

6. b   7. true   8. true   9. a   10. b

15. a. The number of computers sold depends on experience.   b. positive   d. $\hat{y} = 16.6223 + 1.5413x$
    f. $r = .71$; $r^2 = .51$ g. 33.58 computers   h. $s_e = 6.6833$ i. -1.1968 to 4.2794
    j. $H_0: B = 0$; $H_1: B > 0$;   critical value: $t = 3.365$;   test statistic: $t = 2.270$;   do not reject H_0
    k. 22.1586 to 35.7468   l. 10.4754 to 47.4300

CHAPTER 14

14.7 a. $\hat{y} = 51.6 - .0599x_1 + .0850x_2 - .0048x_3$
      b. $a = 51.61$; $b_1 = -.05993$; $b_2 = .08497$; $b_3 = -.004773$
      c. $s_e = .8995$; $R^2 = 99.7\%$; $\overline{R}^2 = 99.6\%$; SST = 2918.00; SSR = 2909.91;
         SSE = 8.09; MSR = 969.97; MSE = .81
      d. For a: $s_a = 11.25$; $t = 4.59$; p-value = .000
         For b_1: $s_{b_1} = .01526$; $t = -3.93$; p-value = .003
         For b_2: $s_{b_2} = .02875$; $t = 2.96$; p-value = .014
         For b_3: $s_{b_3} = .008707$; $t = -.55$; p-value = .596
      e. 2901.82; 7.85; .24   f. 54.2648   g. 47.9411   h. $-.09393$ to $-.02593$ i. $-.00614$ to $.17608$
      j. $-.028839$ to $.019293$ k. 26.545 to 76.675
      l. $H_0: B_1 = 0$; $H_1: B_1 < 0$;   critical value: $t = -2.764$;   test statistic: $t = -3.927$;   reject H_0
      m. $H_0: B_2 = 0$; $H_1: B_2 > 0$;   critical value: $t = 2.764$;   test statistic: $t = 2.955$;   reject H_0
      n. $H_0: B_3 = 0$; $H_1: B_3 < 0$;   critical value: $t = -2.228$;   test statistic: $t = -.548$;   do not reject H_0
      o. $H_0: A = 0$; $H_1: A \neq 0$;   critical values: $t = -3.169$ and 3.169;   test statistic: $t = 4.588$;   reject H_0
      p. H_0: all B_i's are zero; H_1: all B_i's are not zero;   critical value: $F = 4.83$;   test statistic: $F = 1198.86$;   reject H_0

14.9 a. $\hat{y} = 15.1 + .167x_1 - .132x_2$ c. $s_e = 1.488$; $R^2 = 97.1\%$; $\overline{R}^2 = 96.4\%$
      d. 22.44   e. 24.44   f. .05295 to .28089
      g. $H_0: B_2 = 0$; $H_1: B_2 < 0$;   critical value: $t = -2.896$;   test statistic: $t = -1.92$;   do not reject H_0
      h. $H_0: B_1 = B_2 = 0$; H_1: Both B_i's are not zero;   critical value: $F = 8.65$;   test statistic: $F = 133.67$;   reject H_0
      i. $r_{1y} = .979$; $r_{2y} = -.940$; $r_{12} = -.912$

14.11 a. $\hat{y} = -9.02 + .176x_1 + .768x_2$ c. $s_e = 4.033$; $R^2 = 64.3\%$; $\overline{R}^2 = 55.4\%$
      d. SST = 364.91;   64.3% explained;   35.7% not explained   e. $26.65 thousand
      f. $29.61 thousand   g. .2981 to 1.2381
      h. $H_0: B_1 = 0$; $H_1: B_1 \neq 0$;   critical values: $t = -3.355$ and 3.355;   test statistic: $t = 2.08$;   do not reject H_0
      i. $H_0: B_1 = B_2 = 0$; H_1: both B_i's are not zero;   critical value: $F = 8.65$;   test statistic: $F = 7.22$;   do not reject H_0
      j. $r_{1y} = -.101$; $r_{2y} = .671$; $r_{12} = -.649$

14.13 one

14.15 a. $\hat{y} = 12.1 + .250x_1 - .0359x_2 - .79D$
      c. $s_e = 1.852$; $R^2 = 96.5\%$; $\overline{R}^2 = 95.3\%$; SST = 882.31; SSR = 851.45;
         SSE = 30.86; MSR = 283.82; MSE = 3.43
      d. $s_{b_3} = 1.100$;   observed $t = -.72$; p-value = .488   e. 31.39   f. 27.14   g. -3.283 to 1.693
      h. $H_0: B_3 = 0$; $H_1: B_3 < 0$;   critical value: $t = -2.821$;   test statistic: $t = -.72$;   do not reject H_0
      i. H_0: all B_i's are zero; H_1: all B_i's are not zero;   critical value: $F = 5.08$;   test statistic: $F = 82.77$;   reject H_0

14.17 a. $\hat{y} = 22.3 + .190x_1 - .0533x_2 - .190x_3 - 1.93D$
      c. $s_e = .6183$; $R^2 = 99.6\%$; $\overline{R}^2 = 99.2\%$
      d. 17.32375   e. 15.92   f. -4.2674 to $.3976$
      g. $H_0: B_4 = 0$; $H_1: B_4 \neq 0$;   critical values: $t = -2.571$ and 2.571;   test statistic: $t = -3.34$;   reject H_0
      h. H_0: all B_i's are zero; H_1: all B_i's are not zero;   critical value: $F = 11.39$;   test statistic: $F = 278.87$;   reject H_0

14.19 a. $\hat{y} = -36.6 + 4.19x_1 + 1.10x_2 - 2.79D$
      b. The reference group is *males* here. (Note that if you take *females* as the reference group, your answers will be different.)
      d. Males: $\hat{y} = -36.6 + 4.19x_1 + 1.10x_2$
         Females: $\hat{y} = -39.349 + 4.19x_1 + 1.10x_2$
      e. $s_e = 3.019$; $R^2 = 96.0\%$; $\overline{R}^2 = 94.4\%$ f. SST = 1804.67;   96% explained;   4% not explained
      g. $46.52 thousand   h. $43.73 thousand   i. $38.65 thousand   j. $41.44 thousand   k. -8.67 to 3.09

l. $H_0: B_3 = 0$; $H_1: B_3 < 0$; critical value: $t = -2.306$; test statistic: $t = -1.59$; do not reject H_0
m. H_0: all B_i's are zero; H_1: all B_i's are not zero; critical value: $F = 7.59$; test statistic: $F = 63.33$; reject H_0

14.21 a. positive b. $\hat{y} = 34.7 + .814x_1 - 5.05D_1 - 3.91D_2 + 5.84D_3$; yes
d. $s_e = 2.435$; $R^2 = 90.1\%$; $\bar{R}^2 = 86.1\%$
e. SST = 596.93; 90.1% explained; 9.9% not explained
g. 42.026; 28.182; 51.122   h. .325 to 1.303
i. -9.299 to $-.807$; -9.614 to 1.794; 1.155 to 10.535
j. 28.619 to 40.697
k. $H_0: B_1 = 0$; $H_1: B_1 > 0$; critical value: $t = 2.764$; test statistic: $t = 3.71$; reject H_0
l. $H_0: B_2 = 0$; $H_1: B_2 \neq 0$; critical values: $t = -2.764$ and 2.764; test statistic: $t = -3.77$; reject H_0
m. $H_0: B_3 = 0$; $H_1: B_3 < 0$; critical value: $t = -2.228$; test statistic: $t = -2.17$; do not reject H_0
n. $H_0: B_4 = 0$; $H_1: B_4 > 0$; critical value: $t = 2.764$; test statistic: $t = 3.95$; reject H_0
o. H_0: all B_i's are zero; H_1: all B_i's are not zero; critical value: $F = 3.48$; test statistic: $F = 22.67$; reject H_0

14.23 a. $\hat{y} = -9672 + .132x_1 - 1.13x_2$ c. $s_e = 199.2$; $R^2 = 97.9\%$; $\bar{R}^2 = 97.6\%$
d. SST = 27708840; 97.9% explained; 2.1% not explained   e. .074 to .190
f. $H_0: B_1 = 0$; $H_1: B_1 > 0$; critical value: $t = 2.131$; test statistic: $t = 6.69$; reject H_0
h. H_0: both B_i's are zero; H_1: both B_i's are not zero; critical value: $F = 6.36$; test statistic: $F = 341.76$; reject H_0

Self-Review Test

1. c   2. a   3. c
8. a. $\hat{y} = 4.09 + .332x_1 - .0077x_2 + 3.97D_1 - .679D_2$
c. $s_e = 1.727$; $R^2 = 97.5\%$; $\bar{R}^2 = 96.7\%$; SST = 1437.88; SSR = 1402.10;
SSE = 35.78; MSR = 350.53; MSE = 2.98
d. For a: $s_a = 5.460$; $t = .75$; p-value = .468
For b_1: $s_{b_1} = .03115$; $t = 10.64$; p-value = .000
For b_2: $s_{b_2} = .01240$; $t = -.62$; p-value = .544
For b_3: $s_{b_3} = 1.501$; $t = 2.64$; p-value = .012
For b_4: $s_{b_4} = .9362$; $t = -.73$; p-value = .482
e. 1362.72; 18.065; 19.764; 1.57   f. 29.523   g. 36.318   h. .237 to .427; $-.0456$ to .0302
i. .699 to 7.241; -2.719 to 1.361   j. -10.55 to 18.73
k. $H_0: B_1 = 0$; $H_1: B_1 > 0$; critical value: $t = 1.782$; test statistic: 10.64; reject H_0
l. $H_0: B_2 = 0$; $H_1: B_2 < 0$; critical value: $t = -2.681$; test statistic: $t = -.62$; do not reject H_0
m. $H_0: B_3 = 0$; $H_1: B_3 \neq 0$; critical values: $t = -3.055$ and 3.055; test statistic: $t = 2.64$; do not reject H_0
n. $H_0: B_4 = 0$; $H_1: B_4 < 0$; critical value: $t = -2.179$; test statistic: $t = -.73$; do not reject H_0
o. $H_0: A = 0$; $H_1: A \neq 0$; critical values: $t = -2.179$ and 2.179; test statistic: $t = .75$; do not reject H_0
p. H_0: all B_i's are zero; H_1: all B_i's are not zero; critical value: $F = 5.41$; test statistic: $F = 117.57$; reject H_0
9. a. positive   b. $\hat{y} = 10.6 + 5.79x_1 - 1.64D_1 + .15D_2$ d. $s_e = 2.974$; $R^2 = 49.4\%$; $\bar{R}^2 = 36.7\%$
e. SST = 209.750; 49.4% explained; 50.6% not explained
g. \$28.65 thousand; \$27.39 thousand; \$28.70 thousand
h. 1.88 to 9.70   i. -7.44 to 4.16; -5.35 to 5.65   j. -3.30 to 24.5
k. $H_0: B_1 = 0$; $H_1: B_1 > 0$; critical value: $t = 2.681$; test statistic: $t = 3.23$; reject H_0
l. $H_0: B_2 = 0$; $H_1: B_2 < 0$; critical value: $t = -2.179$; test statistic: $t = -.86$; do not reject H_0
m. $H_0: B_3 = 0$; $H_1: B_3 \neq 0$; critical values: $t = -2.681$ and 2.681; test statistic: $t = .08$; do not reject H_0
n. $H_0: A = 0$; $H_1: A \neq 0$; critical values: $t = -3.055$ and 3.055; test statistic: $t = 2.05$; do not reject H_0
o. H_0: all B_i's are zero; H_1: all B_i's are not zero; critical value: $F = 4.47$; test statistic: $F = 3.90$; do not reject H_0

CHAPTER 15

15.5 \$119 million   **15.7** 1.01   **15.9** \$1 million   **15.11** \$619.92 million
15.15 a. $\hat{y} = 46.4364 + 1.83636t$ b. 70.309

15.17　a.　$\hat{y} = 149.736 + 27.2796t$　　b.　$668.048 billion

15.19　a.　$\hat{y} = 50.4081 + 5.8406t$　　b.　$178.901 billion

15.21　a.　$\hat{y} = 90.7835 + 1.8477t$　　c.　135.128 million

15.31　a.　$\hat{y} = 1439.52 + 263.576t$　　b.　$8028.92 billion

15.33　a.　$\hat{y} = 189.609 + 1.8833t$　　b.　$234.81 billion

15.35　a.　$\hat{y} = 169.318 + 40.571t$　　c.　$1143.02 billion

Self-Review Test

2.　b　　　3.　c　　　4.　a　　　5.　c　　　6.　a　　　7.　a

14.　a.　$\hat{y} = 875.98 + 184.04t$　　c.　$5292.94 billion

INDEX

$$Z_{90\%} = 1.645$$
$$Z_{95\%} = 1.96 \qquad N \geq 30$$
$$Z_{99\%} = 2.58$$

TABLE VII STANDARD NORMAL DISTRIBUTION TABLE†

The entries in the table give the areas under the standard normal curve from 0 to z.

$P(z) = $ area or probability of z

z	.00	.01	.02	.03	.04	.05	.06	.07	.08	.09
0.0	.0000	.0040	.0080	.0120	.0160	.0199	.0239	.0279	.0319	.0359
0.1	.0398	.0438	.0478	.0517	.0557	.0596	.0636	.0675	.0714	.0753
0.2	.0793	.0832	.0871	.0910	.0948	.0987	.1026	.1064	.1103	.1141
0.3	.1179	.1217	.1255	.1293	.1331	.1368	.1406	.1443	.1480	.1517
0.4	.1554	.1591	.1628	.1664	.1700	.1736	.1772	.1808	.1844	.1879
0.5	.1915	.1950	.1985	.2019	.2054	.2088	.2123	.2157	.2190	.2224
0.6	.2257	.2291	.2324	.2357	.2389	.2422	.2454	.2486	.2517	.2549
0.7	.2580	.2611	.2642	.2673	.2704	.2734	.2764	.2794	.2823	.2852
0.8	.2881	.2910	.2939	.2967	.2995	.3023	.3051	.3078	.3106	.3133
0.9	.3159	.3186	.3212	.3238	.3264	.3289	.3315	.3340	.3365	.3389
1.0	.3413	.3438	.3461	.3485	.3508	.3531	.3554	.3577	.3599	.3621
1.1	.3643	.3665	.3686	.3708	.3729	.3749	.3770	.3790	.3810	.3830
1.2	.3849	.3869	.3888	.3907	.3925	.3944	.3962	.3980	.3997	.4015
1.3	.4032	.4049	.4066	.4082	.4099	.4115	.4131	.4147	.4162	.4177
1.4	.4192	.4207	.4222	.4236	.4251	.4265	.4279	.4292	.4306	.4319
1.5	.4332	.4345	.4357	.4370	.4382	.4394	.4406	.4418	.4429	.4441
1.6	.4452	.4463	.4474	.4484	.4495	.4505	.4515	.4525	.4535	.4545
1.7	.4554	.4564	.4573	.4582	.4591	.4599	.4608	.4616	.4625	.4633
1.8	.4641	.4649	.4656	.4664	.4671	.4678	.4686	.4693	.4699	.4706
1.9	.4713	.4719	.4726	.4732	.4738	.4744	.4750	.4756	.4761	.4767
2.0	.4772	.4778	.4783	.4788	.4793	.4798	.4803	.4808	.4812	.4817
2.1	.4821	.4826	.4830	.4834	.4838	.4842	.4846	.4850	.4854	.4857
2.2	.4861	.4864	.4868	.4871	.4875	.4878	.4881	.4884	.4887	.4890
2.3	.4893	.4896	.4898	.4901	.4904	.4906	.4909	.4911	.4913	.4916
2.4	.4918	.4920	.4922	.4925	.4927	.4929	.4931	.4932	.4934	.4936
2.5	.4938	.4940	.4941	.4943	.4945	.4946	.4948	.4949	.4951	.4952
2.6	.4953	.4955	.4956	.4957	.4959	.4960	.4961	.4962	.4963	.4964
2.7	.4965	.4966	.4967	.4968	.4969	.4970	.4971	.4972	.4973	.4974
2.8	.4974	.4975	.4976	.4977	.4977	.4978	.4979	.4979	.4980	.4981
2.9	.4981	.4982	.4982	.4983	.4984	.4984	.4985	.4985	.4986	.4986
3.0	.4987	.4987	.4987	.4988	.4988	.4989	.4989	.4989	.4990	.4990

†This table is the same as Table VII of Appendix B on page 844.